Practical Microstrip and Printed Antenna Design

For a listing of recent titles in the
Artech House Antennas and Electromagnetics Analysis Library,
turn to the back of this book.

Practical Microstrip and Printed Antenna Design

Anil Pandey

ARTECH HOUSE
BOSTON | LONDON
artechhouse.com

Library of Congress Cataloging-in-Publication Data
A catalog record for this book is available from the U.S. Library of Congress.

British Library Cataloguing in Publication Data
A catalog record for this book is available from the British Library.

ISBN-13: 978-1-63081-668-1

Cover design by John Gomes

685 Canton Street
Norwood, MA 02062

10 9 8 7 6 5 4 3 2 1

To my father and mother, B. P. Pandey and B .D. Pandey
and to my family, my wife Shraddha Pandey, my daughter Anushka Pandey,
and my son Atharva Pandey

Contents

Preface

The purpose of this book is to provide a practical antenna design guide to cover various real-world applications for those attempting to plunge into this area. There are many excellent textbooks on antenna theory, mathematical formulations, and design. This book focuses more on practical antenna design than theoretical analysis. It tries to make a good balance between the antenna fundamentals and the design of printed antennas. This book is all about practical-world antenna design beyond too many formulations and theories. The book chapters include the most useful recent work available from research in the printed antenna and microstrip antenna fields.

This book is the result of several years of my research work in antenna design and development for various applications, research investigations, and experience with practical antennas. The book is for new antenna engineers as well as for experienced designers, particularly those focusing on real-world practical antenna designs, as well as academics and researchers looking to keep up with this dynamically changing field. In industry, research labs, and academia, one may find it hard to get a general understanding of the field before engaging in any specialized techniques. Designers of planar microstrip antennas and printed antennas who develop antennas for wireless applications, the aerospace, automobile, and healthcare industries, fifth generation (5G), radars, phased array systems, and many other areas will find this book useful. Many of the antenna projects and topics chosen for discussion are those that I have found to be useful during my career.

Senior antenna designers and engineering managers can use this book as a guide to train and teach new hires. New hires and students can learn antenna designs by using the book and design projects discussed in this book and will be able to quickly gain expert-level proficiencies. Telecom network equipment engineers, who desire a stronger sense of antenna principles, as well as electronic engineering students studying electromagnetics, will find this book to be a good learning resource. I believe that this concise and readable book, with scope and depth, theoretical background, design projects, and recent progress, will be a welcome addition to the collection of books on this subject.

Several books have been published about printed and microstrip antenna design, which has many theoretical aspects. This book narrows the focus to those designs that have extensive utility and simplicity. Results from full-wave electromagnetic numerical methods such as finite difference time domain (FDTD), finite element method (FEM), or method of moments (MoM) are presented with input dimensions and parameters that were used to generate antenna designs. These designs are straightforward for the reader to reproduce and alter designs to aid their understanding.

Each chapter covers sufficient details to enable antenna designers, researchers, students, and scientists from all areas to follow and understand. The book begins with the elementary printed and microstrip antenna topics needed by engineers and students with no background in antenna theory to study and understand the basic design principles and features of antennas. Several antenna design topics and antenna projects are presented in this book for the first time. These include new designs in the areas of 5G, wearable antennas, automobile antennas, system-level antenna simulation, phased array beamforming designs, metamaterial antennas, and fractal antennas.

Applications of antennas are also changing with an increasing emphasis on the commercial use of antenna technology for wireless communication systems, automotive radar, remote sensing, 5G, Internet of Things (IoT), radars, medical devices, radio frequency identification (RFID), direct broadcast satellite televisions, and many other systems related to information infrastructures. These developments suggest that there will be no shortage of challenging problems in antenna engineering and that there is a clear need for engineers with creativity and understanding of practical design aspects of antenna engineering. I have tried to present the design approach and the logic behind these designs so that the reader can see and understand the process of applying fundamental concepts to arrive at useful results. In Chapters 1 and 2, the fundamentals of printed and microstrip antenna are stressed.

I have tried to organize the contents in a flow starting from single-element printed antennas and finishing with system-level designs to lead the readers in a logical way. The book covers the topic of antennas from three vantage points: printed antenna fundamentals, antenna design technique, and design of real-world application-based antennas. Moreover, this book contains some recent developments including my own research results. Electronics Design Automation (EDA) software based on electromagnetic numerical techniques are the essential tools of today's microwave engineer. Most of the antenna projects described in this book are designed using full-wave electromagnetic simulation techniques.

Chapter 1 discusses the various types of printed antennas and important antenna specifications that are needed to start an antenna design. Chapter 2 examines the most widely used printed antenna, microstrip antennas, to solidify concepts developed in this chapter and provides specific antenna design concepts for complex and array antenna designs in later chapters. Printed antennas are heavily used in wireless communication and mobile phone antennas. Chapter 3 gives an overview of antenna designs used in smartphones. Mobile phone designers are struggling every day with various compliance-related problems that see critical products fail to pass compliance certification and cause a big loss of profit. Chapter 4 introduces the concept of various compliances in mobile phone antenna design and methodology to measure these compliances. The frequency and polarization reconfiguration antenna design that is used in modern wireless communication systems is discussed in Chapter 5.

Printed antennas are not only used as a single element but are also very popular in arrays. Chapter 6 outlines the design of the printed array antenna for several applications. Microwave remote sensing antennas, synthetic aperture radar (SAR) antennas, and antenna pattern synthesis are described comprehensively using various antenna design methods in Chapter 7, while a high-power SAR antenna system is discussed in Chapter 8. For airborne applications, multifrequency antennas are required to collect data at different frequency bands without landing the aircraft for mounting of a different antenna at different frequency bands. In Chapter 9, the design and analysis of a multifrequency antenna using a shared aperture is presented.

Phased array antennas are common in modern communication systems and radar because of the benefit of far-field beam-shaping and -steering for specific, agile operational conditions. All design aspects of a phase array antenna with a real-world spaceborne application are discussed in Chapter 10. The arrival of the broadband system in the wireless communication area has required the design of antennas that must operate effectively over a wide range of frequencies. The design of broadband antennas is discussed based on electromagnetic bandgap (EBG), photonic bandgap (PBG), frequency selective service (FSS) fractal antenna, and many other latest technologies in Chapter 11.

The wireless communication industry is creating new products with advanced feature to support wireless services anywhere and anytime. Chapter 12 provides an idea of current research and development (R&D) trends in the design, analysis, and synthesis of broadband, multiband, and reconfigurable antennas for modern wireless communications. Automotive communication has witnessed a tremendous growth in recent years due to modern communication. Automobiles are equipped with many antenna systems such as navigation, Global System for Mobile communication (GSM), Bluetooth, satellite radio, Remote Keyless Entry Service (RKES), Tire Pressure Monitoring System (TPMS), and radars. Chapter 13 is focused on automobile antenna design.

The antenna is one of the most complex and important aspects of the RF system. Chapter 14 presents antenna design consideration in a complete transceiver system for the active phased array at C-band using multitechnology-based chips. The book is completed by Chapter 15 in which a system-level, three-dimensional, beam-steerable, high-gain phased array system for a 5G multiple-input and multiple-output (MIMO) network is designed.

All of the chapters are self-contained and include several antenna design projects, making the book suitable for self-study. A list of key references is included at the end of all chapters. In this book, the presentation principle is to explain printed antenna design concepts in simple and concise terms along with practical antenna design. Although I have avoided the textbook approach in which many equations and theories are presented with little or no explanations of actual context, a considerable amount of the material in this book is related to the design of printed antennas for both practical and educational use.

Acknowledgments

This book would not have been written without the encouragement of and interaction with many people. I thank my friends in the industry and academia for their support on projects and applications.

Much of the information contained in this book is the result of my research and investigation on printed antenna design carried out at the Keysight Technologies and Space Application Center (SAC). I would therefore like to acknowledge the support of my colleagues at the Keysight Technologies and SAC who have made various contributions.

I thank all my colleagues in Keysight Technologies for their support and collegiality thought the years. I especially thank Mohit Khanna, who worked with me on many antenna design projects and provided valuable guidance. Jan VanHese, Jack Sifri, HeeSoo Lee, Anurag Bhargava, and Marc Petersen gave much technical support on electromagnetic simulations and antenna design. In addition, I would like to acknowledge the support provided by the Boris Troyanovsky and Deepty Chauhan at Keysight Technologies.

I have benefited from many experts through their works and guidance. I would like to express my sincere thanks to them, especially to my colleagues at the Indian Space Research Organization (ISRO): S. B. Chakrabarty, V. K Singh, Sanjeev Kulshrestha, B. K. Pandey, S. B. Sharma, and Anil Solanki.

My family has also played a key role throughout the years, supporting me always when I needed it the most. Writing a book is tougher than I thought and more rewarding than I could have ever imagined. It requires significant time and effort to make complicated things easy to grasp and understood, and none of this would have been possible without my wife Shraddha Pandey and children Anushka and Atharva and I acknowledge them for patiently tolerating the extra work connected with writing a book.

I am particularly thankful to the reviewers who carefully reviewed the manuscript and gave valuable suggestions. I also offer sincere thanks to the personnel at Artech House for their help and patience throughout the preparation of the manuscript. Finally, I would like to thank my readers for their interest and passion in the subject.

CHAPTER 1

Printed Antennas

1.1 Introduction

Antennas are a fundamental component of any wireless communications systems. An antenna is a kind of transducer that converts a transmission line electrical signal into an electromagnetic wave in free space and vice versa. The technology related to antennas in the modern wireless system has been continuously improved with the corresponding increase in the number of and requirements for communications. Compared to other antenna options, printed antennas have some excellent features, such as low profile, low weight, and are easy to manufacture, along with good performance characteristics. The unique properties of printed antennas, such as mechanical conformability, durability, compactness, and cheap manufacturing costs make them suitable for many applications. Printed antennas have a range of applications in the commercial, medical, military, and aerospace sectors, and are often mounted on the exterior of spacecraft and aircraft as well as incorporated into wireless devices. In addition, the straightforward integration of printed antennas with electronics systems has opened new fields of application, such as 5G phased array and smart sensing systems.

Printed antennas are used in most classical microwave applications, including radars, telecommunications, wireless communication, satellites communication, mobile communications, aeronautical applications, the space industry, Global Positioning System (GPS), systems for detection and identification, and medical applications. In commercial sectors and industries, printed antenna uses include e-textiles, interactive point-of-sale displays, electronic toll collection, smart cards, and product packaging and equipment identification and tracking. These antennas are used at frequencies ranging from ultrahigh frequency (UHF) to millimeter waves. Printed antennas generally behave like resonators. This naturally leads to a narrow bandwidth, typically a of few percents. However, there are various techniques and design methodologies to make printed antennas behave differently and work in broadband applications.

1.2 Types of Printed Antennas

Focusing on applications, different types and technologies of printed antennas can be considered. Printed antennas originated from the use of planar microwave technologies such as microstrip, slot lines, coplanar waveguide (CPW), and customized printing. The different types of printed antennas are classified based on the shape of the radiating element, feeds, or ports that allow its excitation and radiation mechanism. Depending on their characteristics, we can find different classifications [1–25].

1.2.1 Microstrip Antenna

Among printed antennas, microstrip antennas are the most popular [1]. A microstrip antenna consists of a radiating patch with an electrically large ground plane separated by a dielectric material. For the radiating patch, different shapes are used. Rectangular [6] and circular shapes are the most commonly used. Patch antennas are comprised of a thin metallic patch placed above a conducting ground plane [7]. The patches are photo-etched on the dielectric substrate. The relative dielectric constant of a substrate is usually in the range of $2.2 < \varepsilon_r < 12$ [3]. Substrates with lower ε_r provide better efficiency, larger bandwidth, loosely bound fields for radiation into space. Higher values ($\varepsilon_r > 10$) may be used in special circumstances like for compact antenna designs. Due to its simple geometry, the half-wave rectangular patch is the most commonly used microstrip antenna [8,9]. A typical microstrip patch antenna structure is shown in Figure 1.1. Microstrip antenna analysis and design are discussed in detail in Chapter 2.

1.2.2 Slot Antenna

Printed slot antennas are formed by a dielectric substrate with a metallic plane [10,11]. The slot element is made by eliminating the pertinent metal on the metallic side; the most common shapes in slot antennas are rectangular and ring. Slot antennas have a bidirectional radiation pattern with a wider bandwidth than microstrip antennas. Their basic problem is the difficulty of integrating them into the electronics. To improve the possibility of integration, CPW feed schemes are normally used. To ensure unidirectional radiation patterns, a metallic ground plane is used as the reflector. Tapered slot antennas are used when bigger bandwidth and end-fire

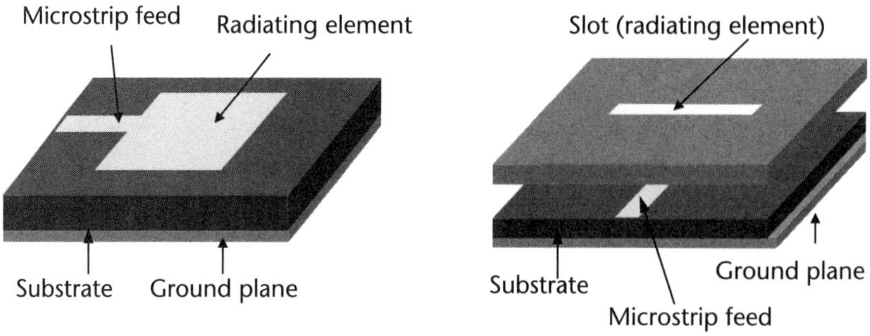

Figure 1.1 Microstrip patch antenna and slot antenna configuration.

radiation is required. Recently, slot antennas integrated into a substrate integrated waveguide (SIW) structures have attracted interest. The main advantages of slotted SIW antennas are their greater compactness and ease of integration when compared with microstrip antennas or traditional slot antennas, due to the guide wave behavior of the SIW structure.

1.2.3 Inverted-F Antenna

Inverted-F antennas (IFAs) are a very common antenna type in most smartphones and wireless devices. Visually, the antennas sometimes resemble an F shape [12–14]. The inverted-F antenna is basically a folded monopole with an inductive stub added to offset the capacitance introduced by locating to antenna element closer to the ground. The IFA can also be designed to deliver multiband operation by stacking two or more F type structures or the trace can be meandered to reduce its size.

Selective metallization is extensively used in mobile phones and is often used to combine multiple antennas onto a single plastic substrate. Other antenna types such as loops, inverted Ls, and monopoles can also be used side-by-side with the IFA to support the multiple radio technologies contained within today's mobile devices. IFAs are usually custom designed to fit within the available PCB space or product geometry. The design costs are small once amortized over the product's lifecycle and if a PCB trace-based design is selected, the IFA can be a very effective and low-cost solution.

1.2.4 Planar Inverted-F Antenna

The planar inverted-F antenna (PIFA) is a well-known variation of the classical microstrip antennas. Although it has a similar name as the IFA, the planar inverted F-antenna is, in fact, a different antenna type. Where the IFA is related to the monopole, the PIFA is related to the patch antenna [15–17]. As the name suggests, the PIFA structure is positioned planar to the PCB. This provides the antenna with an increased physical volume and hence a wider bandwidth than the IFA. In addition, as the PIFA can use air as its dielectric rather than a lossy PCB material, and it tends to be slightly more efficient. Like the off-board IFA, PIFAs can be constructed from formed metal (usually tin) or use selective metallization techniques on a plastic former. PIFA consists of a radiating patch and an electrically large ground plane. Both metallic components are connected by a metallic pin (via) or in some cases by

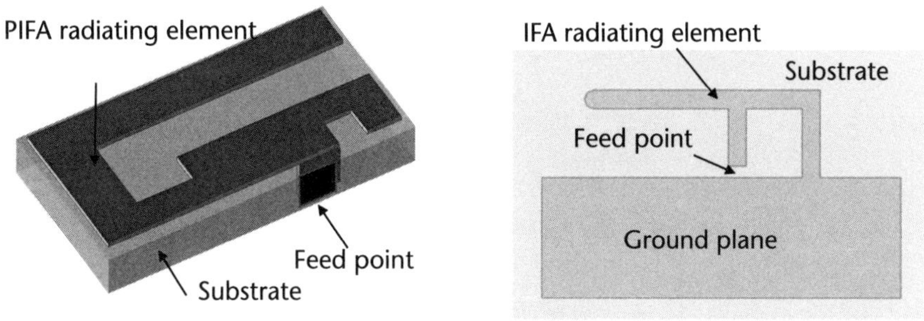

Figure 1.2 Planar inverted-F antenna and inverted-F antenna.

a metallic plane. This scheme reduces the final size of the radiating patch and avoids the use of a dielectric substrate. PIFAs have been widely used in mobile communications systems as mobile phones and laptops due to their compact design with multiband behavior capability. An electrically large ground plane is needed. The PIFA also has less radiation through the ground plane, which if used in a mobile phone or wireless headset can help reduce the amount of wasted RF energy sent toward the user's head and hence help reduce the specific absorption rate of the device.

1.2.5 Printed Dipole Antenna

Printed dipole antennas are widely used for applications based on personal wireless systems like Wi-Fi, Bluetooth, WiMAX, and WSNs based on these protocols. Such antennas consist of two radiating elements (arms of the antenna) fed in opposite phases. The antenna arms can be located on the same side of the substrate or on different sides. Depending on the location of the arms, different kinds of feeding schemes can be applied. The most used feeding schemes are coplanar waveguide lines, which are quarter-length line baluns and direct coaxial connector feed. Different shapes can be used as dipole arms [18,19].

1.2.6 Monopole Antenna

Printed monopole antennas consist of only one resonant element (one arm) directly fed with a transmission line. With an appropriate design procedure, some monopole antennas can provide a wider bandwidth than dipole antennas. This behavior makes monopole printed antennas suitable for compact and wideband applications [20]. Monopole antennas are widely used in mobile and small wireless communication devices. Design of a monopole antenna for Wi-Fi application in a mobile device is discussed in detail in Chapter 3.

1.2.7 Printed Inductor Antenna

Inductor antennas have gained high popularity due to their generalized use in radio frequency identification (RFID) applications. The working frequency of the printed inductor antenna in these cases is quite low when compared with typical

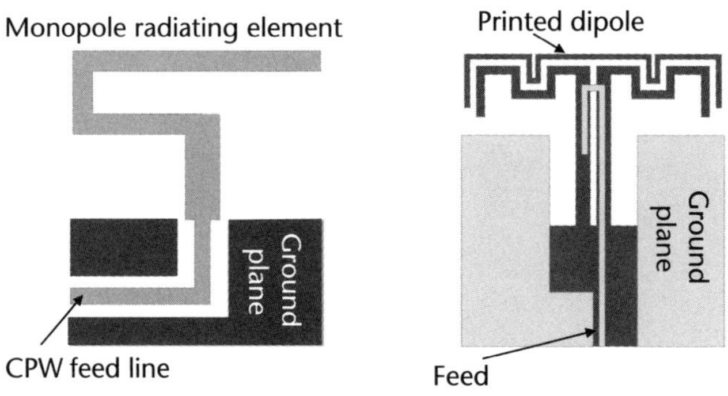

Figure 1.3 Monopole antenna and printed dipole antenna designs.

applications of printed antennas for wireless communications systems. In this sense, these can be considered electrically small antennas and they can be used in passive configurations. The antenna works in the near field region and the coupling is mainly inductive. The specific frequency depends on the application [21]. These antennas are also suitable for low-frequency RFID 13.56- and 900-MHz application. In these instances, the antenna of the tag is the main radiating element.

1.2.8 Printed Quasi-Yagi Antenna

Printed quasi-Yagi antennas are used when a directive behavior is required, by adding parasitic elements in a Yagi-Uda fashion. Printed Yagi-Uda antennas display the same structure and behavior as an original nonplanar Yagi-Uda antenna with the advantages of the printed technology. Planar quasi-Yagi antennas have several benefits compared to wire Yagi-Uda antenna, such as a simple structure, lightweight, low profile, low fabrication cost, broad bandwidth, unidirectional radiation with high gain, and appropriateness for integration with microwave circuits. For wireless sensor networks (WSNs), planar Yagi antennas are normally used to increase the communication distance [22]. The classic Yagi-Uda antenna consists of a driven element, a reflector, and one or more directors. The reflector and directors act as parasitic elements. Quasi-Yagi antennas typically have a narrow bandwidth but an end-fire radiation pattern and high gain make them suitable for many wireless applications.

1.2.9 Log-Periodic and Fractal Antenna

Log-periodic antennas are used to improve the bandwidth of printed antennas. The log-periodic dipole array (LPDA) is the most common form of log-periodic antenna. It consists of a few half-wave dipoles driven radiating elements of gradually increasing length, each consisting of a pair of metal rods. This kind of design approach has, theoretically, a frequency-independent behavior. The main disadvantage of these antennas is the size [23,24].

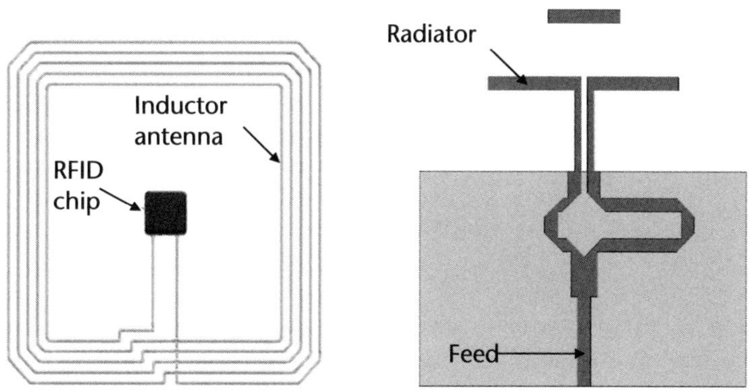

Figure 1.4 Printed inductor and quasi-Yagi-Uda antenna design.

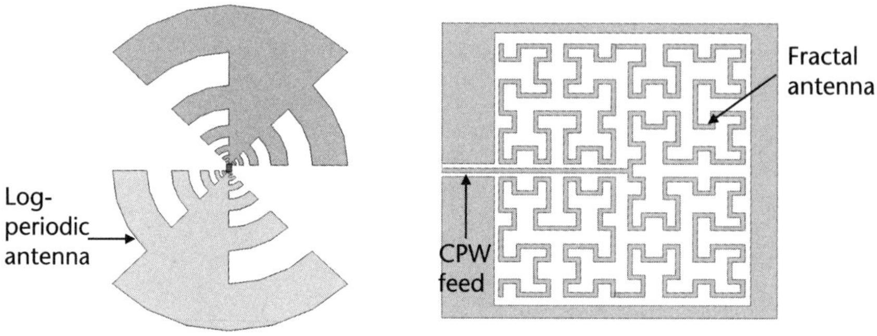

Figure 1.5 Log-periodic and fractal antenna design examples.

The fractal term was first coined by Benoit Mandelbrot in 1983 to classify the structure whose dimensions were not whole numbers. One of the properties of fractals geometry is that it can have an infinite length while fitting in a finite volume. There is an infinite number of possible geometries that are available to try as a design of the fractal antenna. The antenna size can be reduced significantly using fractal concepts, such as the Koch dipole, Koch loop, Koch monopole, and Minkowski loop. It can be also used for multiband applications and increase the bandwidth due to the self-similarity in the geometry, such as the Sierpinski dipole, fractal tree dipole, and Cantor slot patch [25–28]. In these antennas, self-similar shapes provide significant advancements than the fractal nature [28]. Fractal antenna designs are discussed in detail in Chapter 11.

1.2.10 Customized Printed Antenna

In modern wireless antenna applications, there is an increase in antenna complexity and miniaturization combined with a need to integrate many radio environments into one component. The rising adoption of customized printed antennas in the aerospace, defense, and mobile communication industry is projecting growth of this market by leaps and bounds. This is because a customized printed antenna with many resonant frequencies can be designed according to the shape and available space of the device. Customized printed antennas form an integral component of devices thereby enhancing the data transmission capabilities of these devices. Often multiple antenna structures are integrated into a single product. To achieve such a high level of integration the antenna structures are manufactured, for example, from flexible boards or via laser direct structuring (LDS), which allows the production of complex monopole or dipole antennas with a three-dimensionally curved shape.

1.2.11 Comparison of Planar Antennas

Several types of planar antennas are compared in Table 1.1. Patch and slots are resonant-type antennas that have a compact and low profile that can be readily integrated into large, active arrays. Inverted-F, PIFA, and customized antennas are appropriate choices for mobile communication as they require broadband antennas to cover all standards. Other more exotic planar antennas, such as quasi-Yagi

Table 1.1 Planar Antenna Comparison

Type	*Sketch*	*Radiation Pattern*	*Directivity*	*Bandwidth*
Microstrip Patch		Broadside	Medium	Narrow
Slot		Broadside/ bidirectional	Low/medium	Medium
Dipole	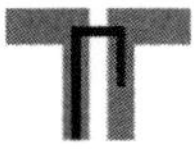	Broadside	Low	Medium
LPDA	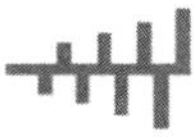	End-fire	Medium	Wide
Bow tie		Broadside	Medium	Wide
Circular loop		Broadside	Medium	Narrow
Spiral		Broadside	Medium	Wide
TSA		End Fire	Medium/high	Wide
Quasi-Yagi		End Fire	High	Wide
PIFA		Broadside	Medium	Medium
Monopole		Broadside	Low	Medium
Fractal		Broadside	High	Wide
Leaky wave		Scannable	High	Medium

antennas and microstrip leaky waves, demonstrate higher directivity and the ability for frequency scanning. Their compact size, broad bandwidth, and good patterns provide a viable alternative to antennas based on other technologies. As planar antenna technology continues to mature, the functionality of these antennas is expected to increase and drive penetration of commercial wireless communications to even higher frequency applications. The design of many of these antennas is discussed in detail thoroughout this book.

1.3 Important Specifications of Antenna Design

There are several antenna parameters that are useful to describe for antenna design engineers [1–5]. Before beginning antenna design work, it is essential to know typical antenna design specifications. It is easy to confuse some of the antenna specifications and how to use them in antenna design. For example, the gain of an antenna and amplified gain are two very different definitions.

The primary specifications or parameters of an antenna consists of frequency, gain, bandwidth, impedance, return loss, half-power beamwidth (HPBW), sidelobes, polarization, cross-polarization, and voltage standing-wave ratio (VSWR).

1.3.1 Operating Frequency

The antenna resonant point or design frequency f_c is the specifications of the antenna at which it resonates. Antenna operating frequency (or frequency bandwidth) is the range of frequencies within which the performance of the antenna, with respect to some characteristic, which conforms to a specified standard. The antenna center frequency between a high frequency point f_H and a low frequency point f_L is calculated as shown in (1.1):

$$f_c = \frac{f_H - f_L}{2} \tag{1.1}$$

1.3.2 Impedance

The impedance of an antenna relates the voltage to the current at the input terminal and it has real resistive and imaginary reactance parts. The inductive and capacitive elements of antenna impedance changes with frequency. An antenna will have two types of resistance associated with it: radiation and ohmic resistance, as shown in Figure 1.6(b). Radiation resistance converts electrical power into radiation. Ohmic resistance is the loss of the antenna's structure that converts electrical power into heat. For effective radiation and efficiency, the radiation resistance should be much higher than the ohmic resistance. For efficient transfer of energy, the impedance of the system, the antenna impedance, and the transmission line impedance must be the same.

1.3.3 Return Loss and VSWR

Antenna return loss measures how effectively power is delivered from a transmission line to an antenna. If the power incident on the antenna under test is P_{in} and the power reflected to the source is P_{ref}, the degree of mismatch between the incident and reflected power in the travelling waves is given by the ratio P_{in}/P_{ref}. Return loss is expressed either in terms of the VSWR present on the input to the antenna because of the reflection or as the relative level of the return signal in decibels. The

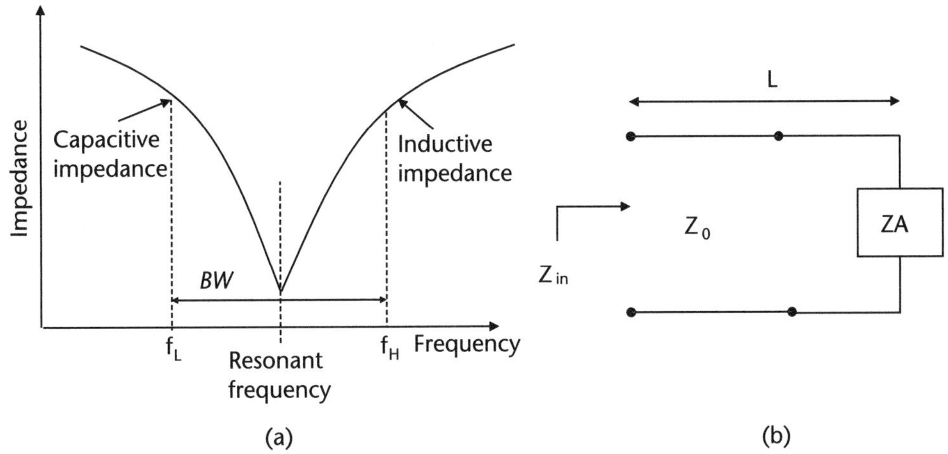

Figure 1.6 (a) Antenna impedance and bandwidth diagram, and (b) antenna resistance model.

negative of the magnitude of the reflection coefficient in decibels is return loss. Since power is proportional to the square of the voltage, return loss is given by (1.2):

$$RL(dB) = 10\log_{10}\frac{P_{in}}{P_{ref}} \tag{1.2}$$

Return loss and VSWR are related. VSWR is a measurement of reflected power from the source. It is calculated by measuring the forward voltage wave toward the load versus the voltage wave that is reflected from the load. A perfectly matched system has a VSWR of 1:1. The higher the first number, the worse the match and the more inefficient the system. In the case of antenna VSWR, this is usually 2:1; that is, 88.9% of the energy sent to the antenna by the transmitter is radiated into free space and 11.1% is either reflected into the source or lost as heat during the transmission process. Antenna bandwidth is also measured in terms of return loss, such as -10-dB return loss bandwidth.

1.3.4 Radiation Pattern

The variation of the power radiated by an antenna as a function of the direction away from the antenna at different locations in space is defined as an antenna radiation pattern. This power variation as a function of the arrival angle is measured in the antenna's far-zone. If the radiation pattern is the same in all directions, then it is called an isotropic antenna pattern. In practice, antennas with isotropic radiation patterns do not exist but an isotropic radiation pattern is used as a reference to compare gain and pattern of real antennas. For simplicity the radiation pattern can be drawn in a two-dimensional (2-D) plane for different azimuth and elevation angle referred to as azimuth plane pattern and elevation plane pattern. Radiation pattern can be a plot in polar and rectangular form. It is more informative to plot the radiation patterns in Cartesian (rectangular) coordinates, especially when the antenna radiation pattern consists of different sidelobes and where these sidelobes levels play an important role. Many antenna parameters like gain, directivity, the direction of the main lobe, half-power beamwidth, sidelobes, and null locations are calculated from the antenna radiation pattern, as shown in Figure 1.7.

1.3.5 Directivity and Gain

The ability of an antenna to focus radiated energy in a direction is measured with directivity. Directivity measures how much an antenna's peak radiated power density is in a direction than for a reference radiator (normally isotropic radiator or dipole) with the same source power. Gain is given with respect to a standard antenna. The two most common reference antennas to measure the directivity are the isotropic antenna and the resonant half-wave dipole antenna. The isotropic antenna radiates equally in all the directions. Real isotropic antennas do not exist, but they provide simple and uniform theoretical antenna patterns with which real antennas pattern is compared. The resonant half-wave dipole is a useful standard for comparing gain of antennas. Gain is measured in decibels or dBi (relative to an isotropic source) or dBd (with regard to a dipole antenna). A dipole has a gain of 2.15 dB more than isotropic or dBi = dBd + 2.15 dB.

It is important to note whether the gain is being expressed in dBd or dBi and convert appropriately when comparing gains of multiple antenna systems. Actual antenna performance is measured by the amount of energy that the antenna radiates into free space. Antennas are not 100% efficient and have internal losses. The gain of an antenna includes these losses. Antenna gain can be calculated using (1.3):

$$Gain = 4\pi \frac{Radiated\,intensity}{Total\,power} = 4\pi \frac{U(\theta,\varnothing)}{P_{in}} \tag{1.3}$$

Gain can be a confusing specification. Most engineers are familiar with the term about amplifiers, where the gain is a measure of how much an amplifier increases the input signal. But there is a significant difference between an amplifier's

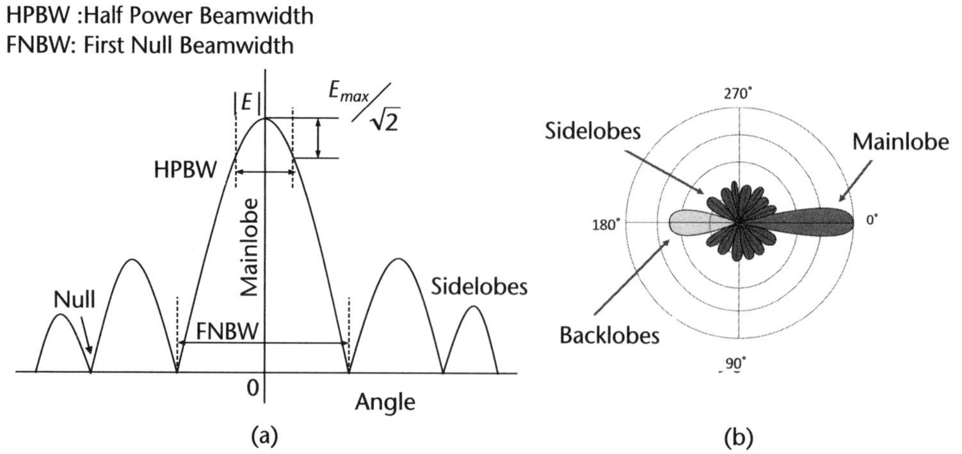

Figure 1.7 Antenna radiation pattern diagrams and parameters: (a) rectangular plot of radiation pattern, and (b) polar plot of radiation pattern.

gain and an antenna's gain. The amplifier puts energy into the system, making it an active device.

1.3.6 Antenna Efficiency

Efficiency is a measurement of how much energy put into the antenna gets radiated into free space rather than lost as heat on the antenna's structure or reflected into the source. The antenna's impedance and VSWR at the center frequency play a big role in this measurement. In a high-efficiency antenna, most of the input power delivered at the antenna terminal is radiated away while a low-efficiency antenna has most of the power lost, reflected back, or absorbed within the antenna.

1.3.7 Bandwidth

The bandwidth of an antenna is defined as the range of frequencies over which the antenna can properly radiate or receive energy. Bandwidth is typically quoted in terms of VSWR or decibels. For instance, an antenna may be described as operating at 50 to 100 MHz with a VSWR<1.5 or –10 dB. This statement implies that the reflection coefficient is less than 0.2 across the quoted frequency range. Hence, of the power delivered to the antenna, only 4% of the power has reflected the transmitter. Bandwidth can be defined in terms of radiation patterns or VSWR/reflected power. Bandwidth is often expressed in terms of percent bandwidth because the percent bandwidth is constant relative to frequency. Typically, our antennas have a nominal 10% bandwidth. This may increase or decrease depending on the antenna and performance required.

1.3.8 Half-Power Beamwidth

The beamwidth is the aperture angle from where most of the energy is radiated from the antenna. The angle over which the radiated energy falls to half (50%) its peak level is as the HPBW or the -3-dB point. The unit of HPBW is radians or degrees.

1.3.9 Sidelobes

Sidelobes are the lobes in the radiation pattern that are not the main beam. In good antenna design, the power density in the sidelobes is generally much less than that in the main beam. Sidelobe levels are measured in decibels relative to the peak of the main beam, for example -13 dB. In array antennas there are several techniques to control sidelobe levels depending on the application.

1.3.10 Polarization

All electromagnetic waves and radiation are polarized. The orientation of the electrical field or E-plane describes the polarization of an antenna. The polarization can be linear, circular, or elliptical. Linear polarization is usually horizontal or vertical. Dual-pole antennas can produce both vertical and horizontal polarization via separate ports. Circular polarization is produced when the E-plane of the antenna radiated wave rotates along the propagation axis. Depending on the direction of spin, the polarization is right or left. Antennas are never perfectly polarized. The amount of energy in the plane perpendicular to the E-plane or opposite hand of circular polarization is measured with a cross-polarization. It is typically quoted relative to the peak gain of the antenna.

1.4 Analysis and Simulation Software for Antenna Design

The transmission line model, cavity model, and full-wave model (which include primarily integral equations/moment method) are the most popular models for the analysis of printed antennas. The transmission line model gives good physical insight and it is the simplest of all models, but it is less accurate. The cavity model is more accurate and gives a good result but is complex in nature. These techniques make a simplifying assumption about the geometry of a problem to apply closed-form solutions. However, most electromagnetic coupling problems are simply unpredictable to be modeled using this approach. As the frequency of operation is increased, it has been realized, and now it is widely known, that the quasi-static analyses of microstrip circuit elements are not accurate enough, and a more rigorous full-wave analysis is required. Full-wave techniques analyses the entire geometry provided without making a priori assumptions about which field interaction are most significant. The full wave models are very accurate, versatile, and can treat single elements, stacked elements, finite and infinite arrays, arbitrarily shaped elements, and coupling. Almost all commercial printed antennas are designed in the industry using a full-wave analysis technique. Full wave electromagnetic analysis means that the solution is obtained by solving Maxwell's equations for the electric and magnetic fields in the time domain or the frequency domain.

1.4.1 Transmission-Line Circuit Model

A microstrip patch, operating at its fundamental mode, is essentially a $\lambda/2$ long microstrip transmission line and can be represented by an equivalent circuit network [29,30]. For a rectangular patch, its radiation is basically generated from its two edges with two equivalent slots along the resonating dimension, as shown in Figure 1.8. The microstrip radiator can be characterized by these two slots separated by a transmission line, where each slot is represented by a parallel circuit of conductance (G) and susceptance (B). The complete patch antenna can be represented by the equivalent network shown in Figure 1.8.

This transmission line model is simple, computationally fast, and intuitively appealing, but it suffers from limited accuracy. For examples, this model does not consider mutual coupling between the two radiating slots and it lacks the radiation from the nonradiating edges of the patch [31].

1.4.2 Cavity Model

Microstrip radiator can be considered as an open cavity bounded by the patch element and ground plane. The open edges can be represented by radiating magnetic walls. Just like a completely enclosed metallic cavity, such a cavity will support

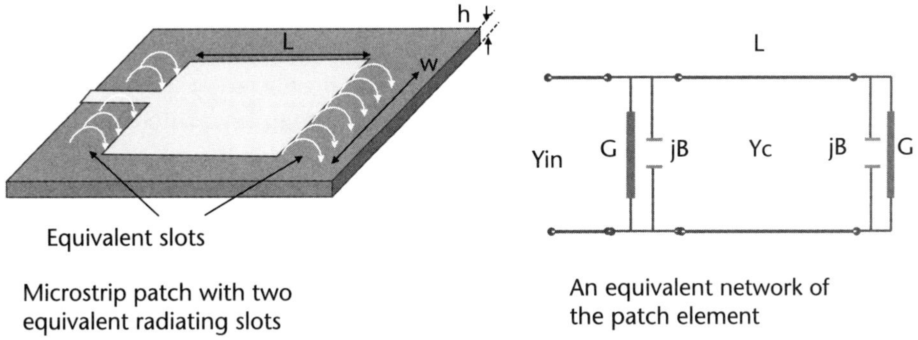

Figure 1.8 Transmission line model (TLM) of microstrip patch antenna.

multiple discrete. As shown in Figure 1.8, its total electric field in the cavity can be expressed as the sum of the fields associated with each sinusoidal mode [32]:

$$E_z(x,y) = \sum_m \sum_n C_{mn} \cos\left(\frac{m\pi}{w}\right) x \cos\left(\frac{n\pi}{L}\right) y \tag{1.4}$$

where C_{mn} is a constant that depends on the feed location, w, and L dimensions, and the dielectric constant. As substrate is considered thin, the fields are assumed to be z-directed only with no variation in the z-direction. The most important dominant mode is the TM_{10} mode, that is generated if the dimension L is approximately $\lambda g/2$. The equivalent edge magnetic currents can be determined by knowing the fields at the edges of the patch and it can be integrated to find the far-field radiation patterns. The cavity model technique permits to know the mode structure underneath the patch to understand physical radiation mechanisms, such as its resonating and cross-polarization behaviors. However, when the substrate becomes thick, because of certain assumptions the field has no z-variation, and cavity model solution is not very accurate.

1.4.3 Full-Wave Analysis Technique

There are various antenna full wave analysis and simulation tools based on different electromagnetic numerical techniques. Software for antenna design can be selected based on antenna type and size. Choosing the appropriate technique for solving an antenna problem is important, because by choosing the wrong solver one can either get incorrect results or the simulation will take excessively long to compute. Several important electromagnetic simulation technologies have developed over the years. Out of these simulations technique, the method of moments (MoM), finite element (FEM), and finite difference time domain (FDTD) solutions are used almost in all commercial electronic design automation (EDA). In principle, while these full-wave analysis technologies could be used to solve the same problems there are often good practical reasons why one simulator is better suited to solving a problem type.

While selecting a full-wave solver, the first selection criteria are based on geometry, whether antenna geometry is planar in nature or it is three-dimensional. For planar structures, MoM provides the most efficient and fastest simulation method, and for true 3-D antenna structures, such as parabolic dish, horn, and waveguide antennas, FEM or FDTD are more appropriate. Large problems are better suited to FDTD simulation, such as an antenna placed on a vehicle or aircraft, the analysis of antenna performance in the presence of human body models, and several large array antenna simulations.

1.4.3.1 MoM

Numerical techniques based on the method of weighted residuals are called moment methods. This is mostly a frequency domain method where a single frequency is considered at a time. The current density of the antenna is the main variable used to derive other antenna parameters. It is implemented by replacing the antenna with an equivalent surface current density. The surface current density is then discretized into a set of current-density elements called basis functions or expansion with variable amplitudes. Then Green's function for the problem is used to represent the electric and/or magnetic fields everywhere in terms of current-density elements on the surface of the antenna. The boundary conditions for both fields are then enforced on the surface of the antenna using testing functions or weighted functions as pioneered by Harrington [33] for solving electromagnetic problems. MoM-based mesh on a microstrip line and current distribution on the line is shown in Figure 1.9.

All weighted residual techniques begin by establishing a set of trial solution functions with one or more variable parameters. The residual is a measure of the difference between the trial solution and the true solution. The variable parameters are obtained in a manner that guarantees the best fit of the trial functions based on a minimization of the residuals. The equation solved by moment- method technique is generally in form of electric field integral equation (EFIE) or the magnetic field integral equation (MFIE). Both equations can be derived from Maxwell's equations by considering the problem of a field scattered by a perfect conductor (or a lossless dielectric). These equations are of the form that the MoM technique can be applied to configurations of conductors only, homogeneous dielectrics only, or very specific conductor-dielectric geometries depending on the form of field integral equation used. They are not effective when applied to an arbitrary configuration with complex or inhomogeneous dielectrics and the interior of conductive enclosures or thin plates.

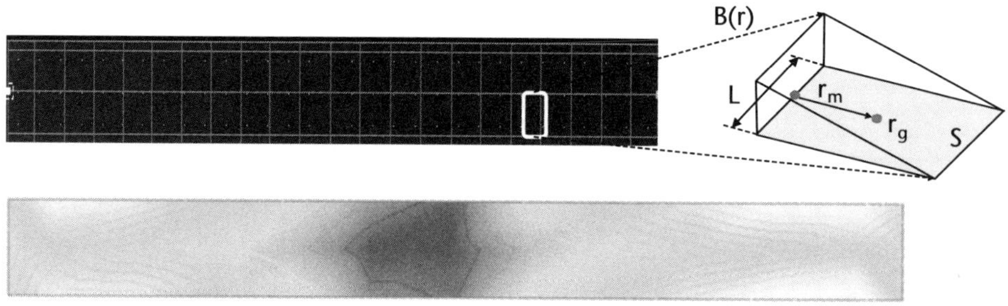

Figure 1.9 MoM mesh on a microstrip line and simulated current distribution over the line.

1.4.3.2 FEM

FEM is mainly a frequency domain method that is sometimes preferred because of its ability to model inhomogeneous media. FEM is a mathematical method of analyzing complex structures by subdividing them into many small elements (the elements can be small where geometry details exist and much larger elsewhere) and analyzing these small structures. The elements can be of many shapes such as triangles and squares for 2-D for tetrahedral and cuboids for 3-D. The field in each element is approximated using mathematical expressions (solutions of the weak form of Helmholtz equations) as described by Silvester et al. [34]. Each edge of the element is termed as a node and magnetic and electric field at each node is evaluated, and then the complete problem is solved numerically. FEM does not model unbounded radiation problems as effectively as moment method techniques.

FEM simulation uses the finite element method to calculate the electromagnetic field solution from which *S*-parameters data is computed. In the finite element method, full-problem space is divided into many smaller regions, as shown in Figure 1.10. Electromagnetic fields are calculated in each subregion (element) with a local function. In FEM-based computer-aided design (CAD) tools, the geometric model is automatically divided into many tetrahedra cells, where a single tetrahedron is formed by four equilateral triangles. An example of how a structure is discretized using finite elements is shown in Figure 1.10. The model contains information about the material constants, device geometry, excitation, and boundary constraints.

1.4.3.3 FDTD

FDTD is one of the most widely used modeling techniques in many engineering applications [35–38] introduced by Yee in 1966. The FDTD method has the advantage of being relatively simple in its formulation and its definition of the geometry (grid-layout) because no complicated and time-consuming mesh generation is needed. Just like FEM, FDTD techniques also require the complete volume to divide into the mesh. In general, the mesh must be uniform, but it can be also controlled based on geometry and applications. The FDTD technique is a time domain simulation method. FDTD methods are good numerical modeling techniques for complex

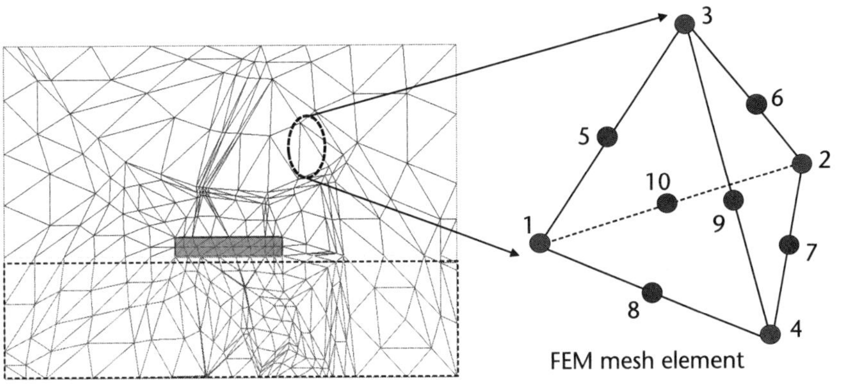

Figure 1.10 FEM mesh for a microstrip line simulation model and a tetrahedral mesh element.

inhomogeneous configurations. Many FDTD formulations and approaches model unbounded problems better than the finite element modeling technique. The FDTD method is a popular numerical technique for computer-aided printed antennas and arrays designs. Complex microstrip antenna characteristics, including the effects of parasitic elements and aperture-coupled feeds, can be solved using this technique. Figure 1.11shows FDTD mesh on a microstrip line and a Yee cell.

Mutual coupling effects between elements is a tricky problem in array antenna design, and an FDTD solver is used to model mutual coupling problems. One main advantage of the FDTD approach is that it gives broadband output from a single execution. However, the main reason for using the FDTD approach for large problem size is the excellent scaling performance of the method as the problem size grows.

1.4.4 Comparison between MOM, FEM, and FDTD

Computational electromagnetics (CEM) based on various techniques solves Maxwell's equations that satisfy a given set of boundary conditions and initial conditions. In doing so, these methods enable complex electrodynamic systems to be modeled. Among the methods used for CEM, are integral-equation (IE) solvers, differential-equation (DE) solvers, asymptotic techniques, and some other numerical methods. Most commercial electromagnetic simulation software is based on the MoM, FEM, and/or FDTD full-wave analysis techniques.

The MoM and multilevel fast multipole method (MLFMM) are based on IE-solving techniques while FEM and FDTD use differential-equation-solving techniques. Asymptotic techniques are used for large antennas based on the optical concept and include physical optics (PO), geometric optics (GO), and the uniform theory of diffraction (UTD). The FEM numerical technique, is a very popular simulation tool for high-frequency structure and antenna designs. Some CAD tools have multiple EM solvers like FDTD-, FEM-, or MoM-based simulators. A comparison among these solvers is shown in Table 1.2.

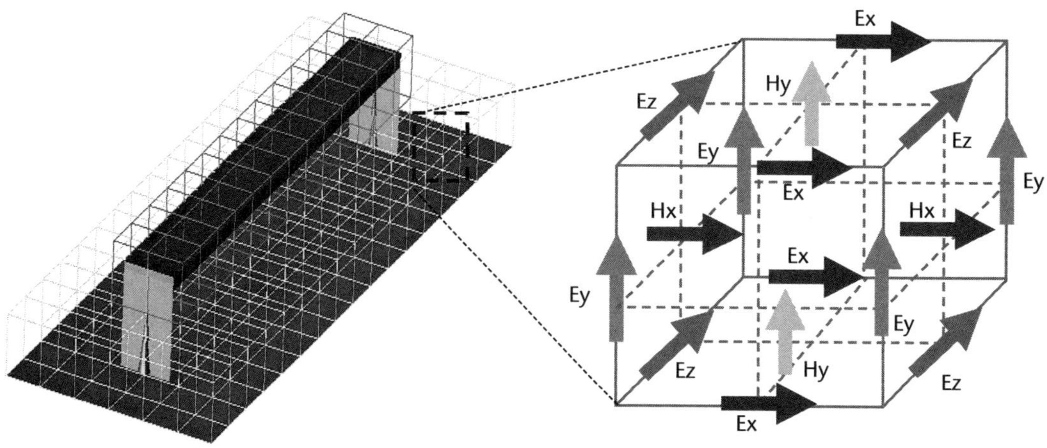

Figure 1.11 FDTD mesh for a microstrip line simulation model and a Yee cell.

Table 1.2 Comparison between MOM, FEM, and FDTD

Parameters	*MOM*	*FEM*	*FDTD*
Method	Frequency-domain	Frequency-domain	Time-domain
Principal	Need frequency-dependent Green's function	Variational principal (minimizing energy function)	A discrete solution of Maxwell's equations
Main advantage	Fast simulation	Very flexible with suitable mesh generation	Can solve electrically large structure
Equations	Integral equation	Differential equations	Differential equations
Suitable for	Wire antenna, planar antenna	Arbitrary shapes, single or few frequencies	Electrically large structures, Broadband
Not suited for	Electrically large structures, various material, broadband type	Electrically large structures, broadband investigation	Multiport structure Hight Q structures

1.5 RF and Microwave Frequency Band and Applications

RF and microwave frequency ranges have been divided into many frequency bands, each represented as shown in Figure 1.12.

Different sources define different frequency ranges as microwaves and it includes UHF and EHF (mmWave) bands also. The frequencies from 1 GHz to 100 GHz is generally defined as microwave frequency band. There are many organizations that assign these bands. The most common being the IEEE Bands followed by NATO Radio Bands and ITU Bands. These frequency bands are used in different applications as listed in Table 1.3.

Antennas, which are used to transmit and receive data, play a crucial role in most of these communication systems. Among various microwave antennas, printed antennas are the most popular and used technologies. The increasing compactness of modern microwave communication and radar equipment commonly demands that the associated antennas have flat profiles. Planar and printed monopole antennas are good candidates for use in wireless technology because of their wide impedance bandwidth and nearly omnidirectional azimuthal radiation pattern.

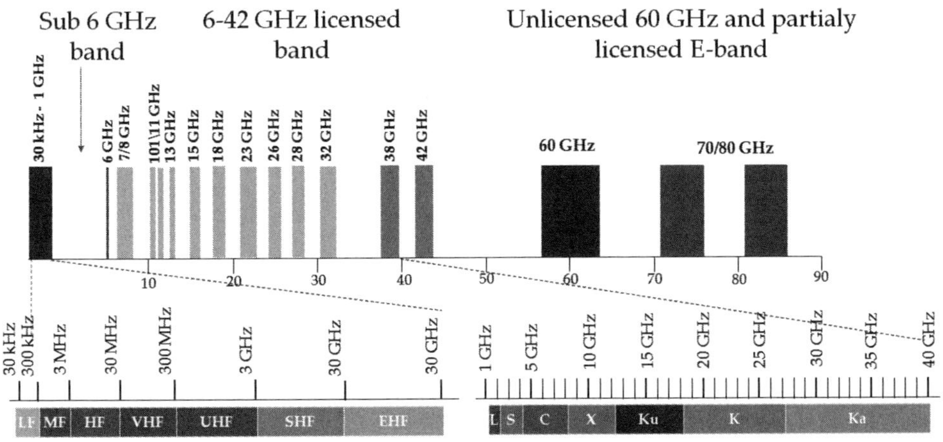

Figure 1.12 RF and Microwave frequency spectrum.

Table 1.3 RF and Microwave Frequency Spectrum and Applications

Band	*Frequency Range*	*Application*
Low frequency(LF)	30 to 300 kHz	Navigation, RFID, time standard, submarines communication
Medium frequency (MF)	300 kHz to 3 MHz	Marine/aircraft navigation, AM broadcast
High frequency (HF)	3 MHz to 30 MHz	Mobile radio, shortwave broadcasting
Very high frequency (VHF)	30 MHz to 300 MHz	Land mobile, FM/TV broadcast, amateur radio, aircraft communications
Ultrahigh frequency (UHF)	300 MHz to 3 GHz	RFID, mobile phone, wireless LAN, PAN
L	1 to 2 GHz	Mobile satellite service (MSS), GNSS/GPS, remote sensing, military telemetry, mobile phones (GSM), air route surveillance radar (ARSR), airborne warning & control (AWACS)
S	2 to 4 GHz	MSS, remote sensing, WiFi, Bluetooth, mobile communication, XM radio, airport surveillance radar (ASR), radio astronomy, wireless LAN, ZigBee, missile guidance, biomedical
C	4 to 8 GHz	satellite communication, WiMax, WLAN, DSRC, weather radar systems, WiFi and ISM band applications, radar altimeter
X	8 to 12 GHz	Radar, FSS military, fire control radars, police speed radar
Ku	12 to 18 GHz	Broadcast satellite services, VSAT, airborne Doppler navigator (ADN)
K	18 to 27 GHz	Radar, satellite communications, astronomical observations
Ka	27 to 40 GHz	Satellite communications, mapping radars
Q	30 to 50 GHz	terrestrial microwave communications, radio astronomy, automotive radar
U	40 to 60 GHz	5G cellular, intersatellite communication
V	50 to 75 GHz	New WLAN, 802.11 ad/WiGig
E	60 to 90 GHz	5G cellular
W	75 to 110 GHz	Automotive radar
F	90 to 140 GHz	Experimental, radar, radio astronomy
D	110 to 170 GHz	Experimental, radar, directed-energy weapon, millimeter wave scanner

There are many new techniques for designing printed antennas and arrays and antenna design for many of these applications will be covered in subsequent chapters.

In this introductory chapter, printed antennas and various types of printed antenna and their applications are discussed. Important antenna specifications that are basic requirements for any antenna designer are explained. For commercial antenna design and analysis, the full-wave analysis technique based on FEM, FDTD, or MoM is used. These three main simulation techniques are covered in this chapter. The next chapter focuses on microstrip antenna design, which is a frequently used printed antenna type.

1.6 Overview of the Book

Printed antennas have a variety of beneficial properties including compactness, conformability, low weight, and cheap manufacturing costs. This book offers a practical design guide to state-of-the-art printed antenna technology used for all types of commercial and industrial systems. The book is divided into six sections covering fundamental printed radiating elements, printed antennas for wireless and mobile systems, array antennas, phased array antennas, and advanced concepts and applications of printed antenna technology. Figure 1.13 shows how these sections are arranged within the book.

After introductory Chapter 1, microstrip antenna technology is discussed. Various types of printed antenna design, analysis methods, and applications are then

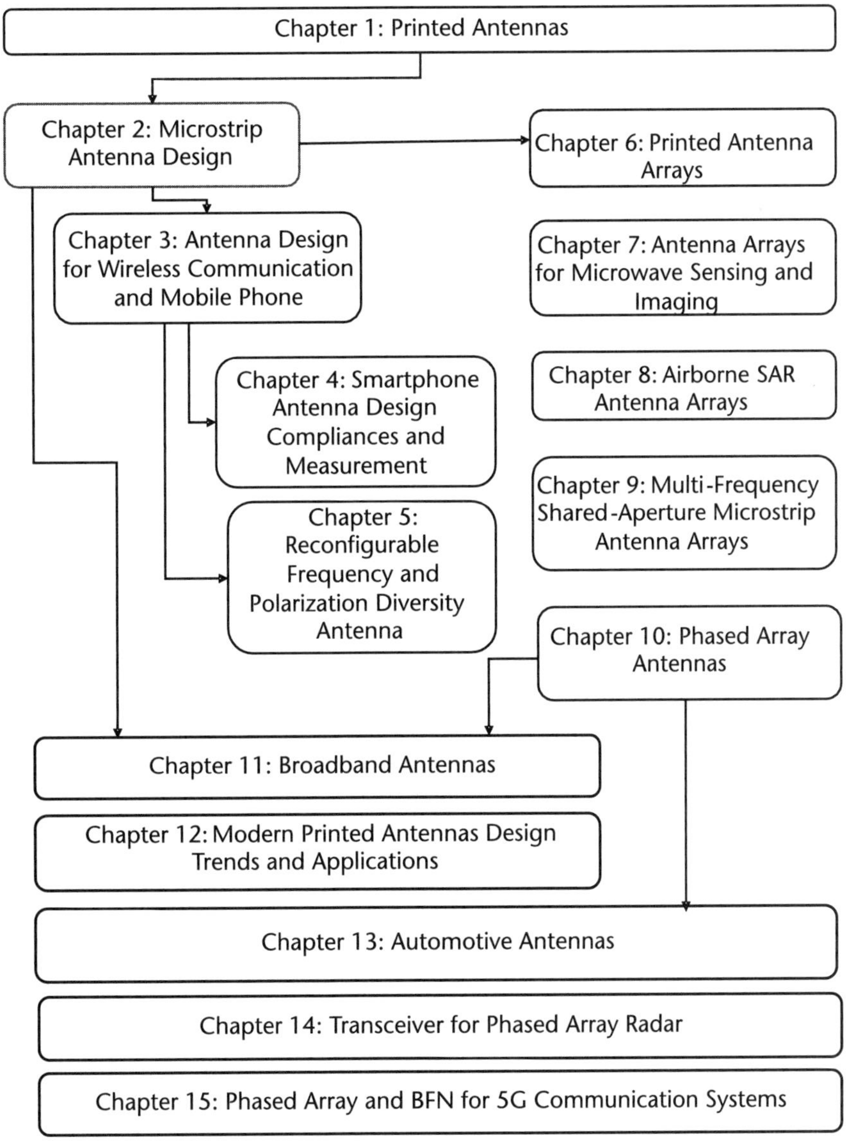

Figure 1.13 Book chapters and content arrangement.

described in detail. Three chapters are then devoted to the design and analysis of printed antennas for wireless and mobile communication systems. The following four chapters describe array antenna technology and its applications. Phased array antennas and advanced modern technologies used in printed and microstrip antennas are covered in the final four chapters. The book is then drawn together with a chapter on system-level design of printed antennas. The following is an overview of each chapter.

Chapter 1: Printed Antennas. This chapter has provided the context for printed antennas and the design process of antennas. Printed antennas have been described together with an explanation of current terminology. Likely design parameters were given together with a review of the design process in which analysis tools should serve.

Chapter 2: Microstrip Antenna Design. To design microstrip antennas, an appreciation of antenna working principle and design process is necessary. The chapter begin with a discussion of the microstrip antennas working principle, design process flow, radiating elements, substrate selection, and feeding mechanisms. Important aspects of microstrip antenna design and knowledge-based design approaches are outlined. Then the discussion converges on practical design of microstrip antenna using the electromagnetic numerical technique.

Chapter 3: Antenna Design for Wireless Communication and Mobile Phone. Printed antennas are heavily used in wireless communication and mobile phone antennas. This chapter gives an overview of antennas used in smartphone design and WLAN antennas as well as an introduction to equivalent circuit modeling of circuits and antenna. Design of GSM, Bluetooth, and MIMO antenna for mobile applications is discussed. Finally, antenna simulation and its effect on the actual environment, along with the human body and phone components, are discussed.

Chapter 4: Smartphone Antenna Design Compliances and Measurement. Smartphone designers are struggling every day with various compliance-related problems that fail critical products to pass compliance certification and causes a big loss of profit. This chapter introduces the concept of various compliance in smartphone antenna design and methodology to measure these compliances.

Chapter 5: Reconfigurable Frequency and Polarization Diversity Antenna. In wireless communication systems antennas are designed and deployed efficiently to effectively utilize the available frequency resources. In this chapter, the frequency and polarization reconfiguration of an antenna is discussed.

Chapter 6: Printed Antenna Arrays. Printed antennas are used not only as a single element but are also very popular in arrays. This chapter outlines the design of printed array antennas for several applications. Array feeding topology, broadband antenna arrays, linear and planar array, and conformal antenna arrays are discussed in detailed.

Chapter 7: Antenna Arrays for Microwave Sensing and Imaging. Large-scale airborne and spaceborne phased arrays are some of the most important enabling techniques for microwave remote sensing. Synthetic aperture radars (SARs) are widely used for remote sensing and mapping within the aerospace and defense industries. In this chapter, microwave remote sensing antennas, SAR antennas, and antenna pattern synthesis are described comprehensively using various antenna design methods.

Chapter 8: Airborne SAR Antenna Arrays. Airborne radars typically use a variation of a cosec square shaped beam, which can get the required elevation coverage when the received power is independent of the radar range for a constant height target. This chapter covers antenna array synthesis, cosec-square patterns, and high-power SAR antennas for airborne remote sensing applications.

Chapter 9: Multifrequency Shared-Aperture Microstrip Antenna Arrays. For airborne applications, multifrequency antennas are required to collect data at different frequencies without landing the aircraft for mounting of the different antenna at different frequency bands. In the chapter, a design and analysis of a multifrequency antenna using a shared aperture are presented.

Chapter 10: Phased Array Antennas. Phased array antennas are common in communications and radar and offer the benefit of far-field beam shaping and steering for specific, agile operational conditions. All design aspects of the phase array antenna with a real-world spaceborne application is discussed in this chapter.

Chapter 11: Broadband Antennas. The advent of the broadband system in the area of wireless communication has necessitated the design of antennas that must operate effectively over a wide range of frequencies. In this chapter, design of broadband antennas is discussed based on EBG, PBG, FSS fractal antenna, and many other latest technologies.

Chapter 12: Modern Antennas Design Trends and Applications. The wireless communication industry is developing new products with advanced features to support wireless services anywhere, any time. The rapid development of communication technologies such as wearable computers, cell phone technology, and personal area networks (PANs) for remote retrieval and monitoring of the surroundings requires antennas suitable to operate with multiband characteristics in wireless communication devices. This chapter provides an idea of current R&D trends in design, analysis, and synthesis of broadband, multiband, and reconfigurable antennas for modern wireless communications.

Chapter 13: Automotive Antennas. Automotive communication has witnessed a tremendous growth in recent years due to modern communication. In addition to the traditional AM/FM, present-day automobiles are equipped with many antenna systems, such as navigation, Global System for Mobile (GSM) communication, Bluetooth, satellite radio, remote keyless entry service (RKES), and tire-pressure monitoring system (TPMS). Apart from these services, higher-end cars are equipped with digital audio broadcasting (DAB), satellite in-vehicle TV, satellite digital audio radio service (SDARS), electronic toll collection (ETC), short-range and long-range radar system for adaptive cruise control (ACC), parking assistance, and collision detection and avoidance. Antenna designs for automotive applications are discussed in this chapter.

Chapter 14: Transceiver for Phased Array Radar. The antenna is one of the most complicated aspects of RF design. This chapter presents antenna design consideration in a complete transceiver system for the active phased array at the C-band using multitechnology-based chips.

Chapter 15: Phased Array and BFN for 5G Communication Sytems. The book is completed by a chapter in which a system level three-dimensional beam steerable high gain phased array system for 5G multiple-input and multiple-output (MIMO) network is designed. Phased-arrays operating at millimeter-wave or very high frequencies are already an active area of research by the emerging 5G cellular market.

The beamforming network (BFN) is the main component of multiple beams antennas. Beamforming networks are complex networks used to precisely control the phase and amplitude of radio frequency (RF) energy passing through them, which is conveyed to the radiating elements of an antenna in the 5G system. This chapter covers the design aspect of a phased array and beamforming network for a 5G communication system.

References

[1] Waterhouse, R. B., *Microstrip Patch Antennas: A Designer's Guide*, Norwell, MA: Kluwer Academic Publishers, 2003.

[2] Sangeetha, R. G., *Microstrip Patch Antenna Array with Omnidirectional Pattern*, Lambert Academic Publishing, 2015.

[3] Al-Naiemy, Y., *Design, Fabrication, and Testing of Microstrip Antennas*, Lambert Academic Publishing, 2013.

[4] Bhartia, P., I. Bahl, R. Garg, and A. Ittipiboon, *Microstrip Antenna Design Handbook*, Norwood, MA: Artech House, 2001.

[5] Kumar, G., and K. P. Ray, *Broadband Microstrip Antennas,* Norwood, MA: Artech House, 2002.

[6] Pan, B. C., and T. J. Cui, "Broadband Decoupling Network for Dual-Band Microstrip Patch Antennas," *IEEE Transactions on Antennas and Propagation*, Vol. 65, No. 10, pp. 5595–5598, October 2017.

[7] Mendes, C., and C. Peixeiro, "A Dual-Mode Single-Band Wearable Microstrip Antenna for Body Area Networks," *IEEE Antennas and Wireless Propagation Letters*, Vol. 16, pp. 3055–3058, 2017.

[8] Pandey, B K, A. K. Pandey, S. B. Chakrabarty, S. Kulshrestha, A. L. Solanki, and S. B. Sharma, "Dual Frequency Single Aperture Microstrip Patch Antenna Element for SAR Applications," *IETE Technical Review*, Vol. 23, No. 6, Noveber–December 2006, pp. 357–366.

[9] Pandey, A., "Design of a Cosecant Square-Shaped Beam Pattern SAR Antenna Array Fed with Square Coaxial Feeder Network," *2013 European Radar Conference (EuRAD),* Nuremberg, Germany, November 13, 2013, pp. 1699–1702.

[10] Antoniades, M. A., A. Dadgarpour, A. R. Razali, A. Abbosh, and T. A. Denidni, "Planar Antennas for Compact Multiband Transceivers Using a Microstrip Feedline and Multiple Open-Ended Ground Slots," *IET Microwaves, Antennas & Propagation*, Vol. 9, No. 5, 2015, pp. 486–494.

[11] Das, S., P. Chowdhury, A. Biswas, P. P. Sarkar, and S. K. Chowdhury, "Analysis of a Miniaturized Multiresonant Wideband Slotted Microstrip Antenna with Modified Ground Plane," *IEEE Antennas and Wireless Propagation Letters*, Vol. 14, 2015, pp. 60–63.

[12] Liu, Y., Y. M. Zhou, G. F. Liu, and S. X. Gong, "Heptaband Inverted-F Antenna for Metal-Rimmed Mobile Phone Applications," *IEEE Antennas and Wireless Propagation Letters*, Vol. 15, 2016, pp. 996–999.

[13] Chaabane, G., V. Madrangeas, M. Chatras, E. Arnaud, L. Huitema, and P. Blondy, "High-Linearity 3-Bit Frequency-Tunable Planar Inverted-F Antenna for RF Applications," *IEEE Antennas and Wireless Propagation Letters*, Vol. 16, 2017, pp. 983–986.

[14] Deng, J., J. Li, L. Zhao, and L. Guo, "A Dual-Band Inverted-F MIMO Antenna with Enhanced Isolation for WLAN Applications," *IEEE Antennas and Wireless Propagation Letters*, Vol. 16, 2017, p. 2270.

[15] Asadallah, F. A., J. Costantine, and Y. Tawk, "A Multiband Compact Reconfigurable PIFA Based on Nested Slots," *IEEE Antennas and Wireless Propagation Letters*, Vol. 17, No. 2, February 2018, pp. 331–334.

[16] Yan, S., V. Volskiy, and G. A. E. Vandenbosch, "Compact Dual-Band Textile PIFA for 433-MHz/2.4-GHz ISM Bands," *IEEE Antennas and Wireless Propagation Letters*, Vol. 16, 2017, pp. 2436–2439.

[17] Behdad, N., and K. Sarabandi, "A Compact Antenna For Ultrawide-Band Applications," *IEEE Transactions on Antennas and Propagation*, Vol. 53, No. 7, July 2005, pp. 2185–2192.

[18] Sim, C. Y. D., C. C. Chen, X. Y. Zhang, Y. L. Lee, and C. Y. Chiang, "Very Small-Size Uniplanar Printed Monopole Antenna for Dual-Band WLAN Laptop Computer Applications," *IEEE Transactions on Antennas and Propagation*, Vol. 65, No. 6, June 2017, pp. 2916–2922.

[19] Saghlatoon, H., T. Björninen, L. Sydänheimo, M. M. Tentzeris, and L. Ukkonen, "Inkjet-Printed Wideband Planar Monopole Antenna on Cardboard for RF Energy-Harvesting Applications," *IEEE Antennas and Wireless Propagation Letters*, Vol. 14, 2015, pp. 325–328.

[20] Ahmad, A., F. A. Tahir, and F. Khan, "A Planar GPS/GLONASS/LTE/WWAN Antenna for Ultra-Slim Smartphones," *2017 11th European Conference on Antennas and Propagation (EUCAP)*, Paris, 2017.

[21] Ghaffar, F. A., S. Yang, H. M. Cheema, and A. Shamim, "A 24 GHz CMOS Oscillator Transmitter with an Inkjet Printed On-Chip Antenna," *2016 IEEE MTT-S International Microwave Symposium (IMS)*, San Francisco, CA, 2016, pp. 1–3.

[22] Tang, M. C., T. Shi, and R. W. Ziolkowski, "Flexible Efficient Quasi-Yagi Printed Uniplanar Antenna," *IEEE Transactions on Antennas and Propagation*, VOL. 63, NO. 12, December 2015, pp. 5343–5350.

[23] Booker, H. G., "Slot Aerials and Their Relation to Complementary Wire Aerials (Babinet's Principle)," *Journal of the Institution of Electrical Engineers–Part IIIA: Radiolocation,* Vol. 93, No. 4, 1946, pp. 620–626.

[24] Dyson, J., "The Unidirectional Equiangular Spiral Antenna," *IRE Transactions on Antennas and Propagation*, Vol. 7, No. 4, October 1959, pp. 329–334.

[25] Wei, X., J. Liu, and Y. Long, "Printed Log-Periodic Monopole Array Antenna with a Simple Feeding Structure," *IEEE Antennas and Wireless Propagation Letters*, Vol. 17, No. 1, January 2018, pp. 58–61.

[26] El Hamdouni, A., et al., "A Novel Design of a CPW-Fed Printed Fractal Antenna for UWB Applications," *2015 Third International Workshop on RFID And Adaptive Wireless Sensor Networks (RAWSN)*, Agadir, Morocco, 2015, pp. 38–41.

[27] Sabban, A., "New Fractal Compact Printed Antennas," *2017 IEEE International Symposium on Antennas and Propagation & USNC/URSI National Radio Science Meeting*, San Diego, CA, 2017, pp. 2197–2198.

[28] Gianvittorio, J. P., J. Romeu, S. Blanc,h and Y. Rahmat-Samii, "Self-similar Prefractal Frequency Selective Surfaces for Multiband and Dual-Polarized Applications," *IEEE Transactions on Antennas and Propagation,* Vol. 51, No. 11, November 2003, pp. 3088–3096.

[29] Gianvittorio, J. P., and Y. Rahmat-Samii, "Fractal Element Antennas: A Compilation of Configurations with Novel Characteristics," *IEEE Antennas and Propagation Society International Symposium, Transmitting Waves of Progress to the Next Millennium, 2000 Digest.* Held in conjunction with USNC/URSI National Radio Science Meeting, Salt Lake City, UT, 2000, Vol. 3, pp. 1688–1691.

[30] Derneryd, A. G., "A Theoretical Investigation of the Rectangular Microstrip Antenna," *IEEE Transactions on Antennas and Propagation*, Vol. AP-26, July 1978, pp. 532–535.

[31] Pues, H., and A. Van de Capelle, "Accurate Transmission-Line Model for the Rectangular Microstrip Antenna," *Proceedings of the IEE*, Vol. 131, Part H, December 1984, pp. 334–340.

[32] Richards, W. F., Y. T. Lo, and D. Harrison, "An Improved Theory for Microstrip Antennas and Applications," *Transactions on Antennas and Propagation*, Vol. AP-29, January 1981, pp. 38–46.

[33] Harrington R. F., *Field Computation by Moment Methods*, New York, the Macmillan Co., 1968.

[34] Silvester P. P., and R. L. Ferrari, *Finite Element for Electrical Engineers*, Third Edition, Oxford, U. K., Cambridge University Press, 1991.

[35] Zhang X., J. Fang, and K. Mei, "Calculations of The Dispersive Characteristics of Microstrips by the FDTD Methods," *IEEE Transactions on Microwave Theory and Techniques*, Vol. 26, February 1988, pp. 263–267.

[36] Shibata T., T. Havashi, and T. Kimura, "Analysis of Microstrip Circuits Using Three Dimensional Full-Wave Electromagnetic Fields Analysis in the Time-Domain," *IEEE Transactions on Microwave Theory and Techniques*, Vol. 36, June 1988, p. 1064.

[37] Feix, N., M. Lalande, and B. Jecko, "Harmonical Characterization of a Microstrip Bend via the Finite Difference Time Domain Method *IEEE Transactions on Microwave Theory and Techniques*, Vol. 40, No. 5, May 1992, pp. 955–961.

[38] Taflove A., "The Finite-Difference Time-Domain Method for Electromagnetic Scattering and Interaction Problems," *IEEE Transactions on Electromagnetic Compatibility*, Vol. EMC-22, August 1980, p. 191.

[39] J. R., James, and Hall P. S. (eds.), *Handbook of Microstrip Antennas*, Stevenage, UK: IET, 1989.

Selected Bibliography

Adin, I., G. Bistue, R. Mittra, C. Quemada, and K. F. Warnick, *Numerical Analysis for Electromagnetic Integral Equations,* Norwood, MA: Artech House, 2008.

Bahl, I. J., and P. Bhartia, *Microstrip Antennas,* Dedham, MA: Artech House, 1980.

Balanis, C. A., *Antenna Theory Analysis and Design,* Fourth Edition, Hoboken, NJ: John Wiley & Sons, 2016.

Dennis, S., *Electromagnetic Simulation Using the FDTD Method*, Piscataway NJ: Wiley-IEEE Press, July 2000.

Harrington, R. F., *Field Computation by Moment Methods*, New York: The MacMillan Co., 1968.

Hubing, T. H., "Calculating the Currents on Wires Attached to Opposite Sides of a Thin Plate," Applied Computation Electromagnetics Society 1990, pp. 9–13.

Kara, M., "Design Consideration for Rectangular Microstrip Antenna Elements with Various Substrate Thickness," *Microwave and Optical Technology Letters*, Vol. 19, No. 2, October 1998, pp. 111–120.

Kin-Lu, W., *Compact and Broadband Microstrip Antennas*, New York: John Wiley & Sons, 2002, pp. 28–30.

Kunz, K. S., and R. J. Luebbers, *The Finite Difference Time Domain Method for Electromagnetics*, Boca Raton, FL: CRC Press, 1993.

Pozar, D. M., and D. H. Schaubert, *Microwave Antennas*: *The Analysis and Design of Microstrip Antennas and Arrays*, New York: IEEE Press, 1995.

Sheen, D., S. Ali, M. Abouzahra, and J. Kong, "Application of Three-Dimensional Finite Element to the Analysis of Planar Microstrip Circuits," *IEEE Transactions on Microwave Theory and Techniques*, Vol. 38, July 1990, pp. 849–857.

Stutzman, W. L., and G. A. Thiele, *Antenna Theory and Design*, Second Edition, New York: John Wiley & Sons, 1998.

Sullivan, D. M, *Electromagnetic Simulation Using the FDTD Method,* New York: IEEE Press, 2000.

Taflove, A., and M. Brodwin, "Numerical Solution of Steady State Electromagnetic Scattering Problems Using the Time Dependent Maxwell's Equations," *IEEE Transactions on Microwave Theory and Techniques*, Vol.23, No.1, August 1975, pp. 623–630.

Yee, K., *Numerical Solution of Initial Boundary Value Problems Involving Maxwell's Equations in Isotropic Media, IEEE Transactions on Antennas and Propagation*, Vol. 14, No. 3, 1966, pp. 302–307.

Yu, W., W. Li, Elsherbeni, A., and Y. Rahmat-Samii, *Advanced Computational Electromagnetic Methods & Applications,* Norwood, MA: Artech House, 2015.

CHAPTER 2

Microstrip Antenna Design

2.1 Introduction

Applications of printed and microstrip patch antennas are increasing in popularity for use in wireless applications due to their low-profile structure. Microstrip antennas are extremely compatible with embedded antennas in handheld wireless devices such as cellular phones. The communication and telemetry antennas on missiles should be thin and conformal and are often microstrip patch antennas. Satellite communication and remote sensors are other areas where microstrip antennas have been used successfully. Because of their extremely thin profile, printed antennas have found main applications in military aircraft, missiles, rockets, and satellites. There are so many systems that use this type of antenna, such as remote-controlled television, cellular phones, satellite communications, spacecraft, radars, wireless phones, and wireless computer networks. Use of antennas in the aircraft and spacecraft has also increased the demand for a low-profile antenna that can provide reliable communication. The microstrip patch is combined with electronic components and integrated circuits. Another major advantage of the microstrip patch is that it can be conformal or flush-mounted planar with other surfaces, such as an airplane radome or mobile body, with only a minimum of space required for the feed line.

2.2 Microstrip Antenna

Microstrip antennas are planar resonant cavities that leak from their edges and radiate. A microstrip patch antenna contains a metallic radiating patch on one side of a dielectric substrate that has a ground plane on the other side. The patch is made of conducting material such as gold or copper. The radiating patch and the feed lines are photoetched on the dielectric substrate. The patch shape can be square, rectangular, circular, triangular, elliptical, or some other common shape.

Low dielectric constant substrates materials are preferred for maximum radiation. Rectangular and circular microstrip patch configurations are the most commonly used configuration. Other shapes can also be designed but these are complex to analyze and require significant numerical computations. By changing the shape of the structure, flexibility in resonant frequency, pattern, polarization, and impedance can be achieved. A microstrip antenna is characterized by its length, width, input impedance, gain, and radiation pattern. The length is a very critical parameter because it governs the resonant frequency of the antenna. Length of the patch is nearly a half wavelength in the dielectric. For compact microstrip patch antennas, substrates with higher dielectric constants are used but these are less efficient and result in a narrower bandwidth. Hence a trade-off must be realized between the antenna dimensions and antenna performance. There is no fixed rule to find the width of a patch. Normally, the width of a patch is comparable to the operating wavelength and the thickness of the substrate is kept very small compared to wavelength.

Some of the benefits of microstrip patch antennas include [1] small profile, light weight, inexpensive fabrication, and planar configuration. Many feeding configurations are possible for feeding the microstrip patch structure, such as aperture feeds, probe feeds, microstrip line feeds, inset feed, and proximity feeds, where each method has advantages depending on the application. Despite these advantages, microstrip antennas present major challenges to the designer due to an inherently narrow bandwidth, the spurious radiation, low efficiency due to dielectric losses and conductor losses, lower power handling capability, and tolerance problems [2]. Considerable research has been done to overcome these limitations, notably in increasing the bandwidth.

2.2.1 Microstrip Antenna Working Principle

A microstrip patch radiates from fringing fields around its edges, as shown in Figure 2.1. When the metallic patch is excited by feed, a charge distribution is being established between the ground plane and the underneath of the patch. There is a positive charge below the patch and the ground plane is charged to negative and because of that, attractive forces are being set up between the patch and ground plane. A general analysis of patch considers the patch as a resonant cavity with metal (electric) walls of the patch and the ground plane and magnetic or impedance walls around the edges. Impedance match happens when a patch resonates as a resonant cavity. The antenna can achieve peak efficiency with perfect impedance matching. The field lines are uniform along the width while vary sinusoidally along the length.

The current is maximum at the center of the patch, while the electric field is maximum at the two radiating edges. Unlike the patch, a normal transmission line radiates less power because the fringing fields are matched by nearby counteracting

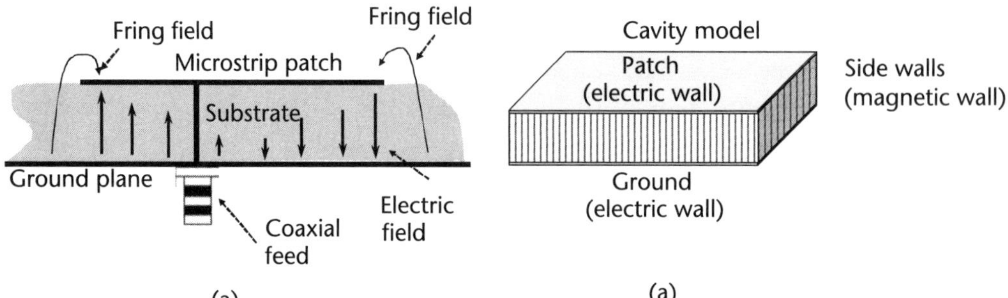

Figure 2.1 Microstrip patch (a) radiation mechanism and (b) cavity model representation.

fields. Power radiates from open circuits and from discontinuities such as transitions and corners, but the amount depends on the radiation conductance load to the line relative to the patches. The edges of a patch act as slots whose excitations depend on the internal fields of the cavity.

A thick dielectric substrate having a low dielectric constant is desirable for antenna design because this provides better efficiency, larger bandwidth, and better radiation, although such a configuration leads to a larger antenna size. To design a compact microstrip patch antenna, higher dielectric constants must be used, which are less efficient and result in a narrower bandwidth. Therefore, a compromise must be reached between antenna dimensions and antenna performance.

2.2.1.1 Cavity Model of Rectangular Microstrip Patch

In the cavity model, the microstrip patch is represented by assuming that the material of the substrate is truncated and does not extend beyond the edges of the patch as shown in Figure 2.1. The four side walls of the substrate act as the magnetic walls and these represent four narrow apertures (slots) through which radiation takes place. The inside area of the dielectric substrate is presented as a cavity limited by electric walls (conductors) on the top and bottom. The radiated field is calculated using equivalence principle. The field inside the cavity is assumed equal to zero, and its influence on the field in the infinite region outside is represented by the equivalent surface currents on the surface of the cavity [2]. Therefore, the tangential H components at the slots equal to zero:

$$J_s = \hat{n} \times H = 0 \tag{2.1}$$

Only the equivalent magnetic current density has substantial contribution to radiated field.

$$M_s = -\hat{n} \times E \tag{2.2}$$

The radiation from an (x-y) slot of constant M_s current can be found using the electric vector potential F. Since M_s has only a y component, $F = F_y$ also has only y component:

$$F_y = -\frac{\varepsilon E_0 W h}{2\pi r} . e^{-jk_0 r} . \frac{\sin X}{X} . \frac{\sin Y}{Y} \tag{2.3}$$

where

$$X = \frac{k_0 h}{2} \sin\theta \cos\varnothing$$

$$Y = \frac{k_0 W}{2} \sin\theta \sin\varnothing$$

E-plane pattern (xz plane, $\varnothing = 0°$, $0° \leq \theta \leq 180°$)

$$f_E(\theta) = \frac{\sin\left(\frac{k_0 h}{2}\sin\theta\right)}{\frac{k_0 h}{2}\sin\theta}.\cos\left(\frac{k_0 L_{eff}}{2}\cos\theta\right) \quad (2.4)$$

H-plane pattern (xy plane, $\theta = 90°$, $-90° \leq \varnothing \leq 90°$).

$$f_H(\theta) = \cos\varnothing.\frac{\sin\left(\frac{k_0 h}{2}\cos\varnothing\right)}{\frac{k_0 h}{2}\cos\varnothing}.\frac{\sin\left(\frac{k_{0W}}{2}\sin\varnothing\right)}{\frac{k_{0W}}{2}\sin\varnothing} \quad (2.5)$$

2.2.2 Microstrip Antenna Design Process

Microstrip antenna design steps are shown in Figure 2.2. The most vital step is to finalize design specifications. Design specifications are driven based on the application. Normally antenna specifications are given by the system engineer after doing the system-level analysis and simulation of a complete system such as a 5G network and radar system. A wireless mobile communication antenna has a completely different specification than that of a satellite or radar antenna system. Important antenna design specifications were covered in Chapter 1, Section 1.3. Antenna specifications such operating frequency, bandwidth, gain, radiation pattern, and polarization are driven by system requirements. The first step in designing a printed antenna is to choose an appropriate substrate. After collecting the design requirement, the next step is to select the radiating element, the size of the antenna, the feeding mechanism, and the electromagnetic simulation technique needed to get antenna performance. Proper feed selection for single-radiation element and feed network design for array antennas are another crucial design step.

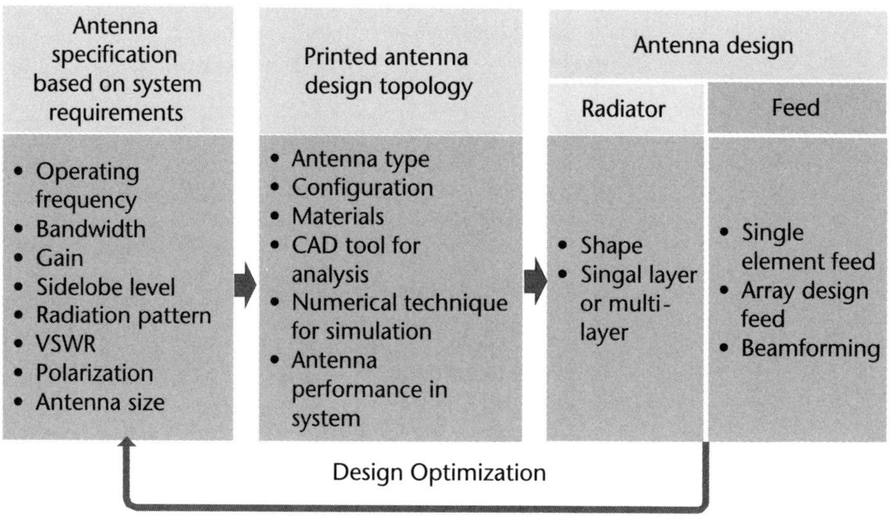

Figure 2.2 Microstrip patch antenna design flow.

2.2.3 Radiating Element Design of Microstrip Antenna

The dimensions of a single microstrip patch antenna as shown in Figure 2.3 are first calculated using approximation formulas. There are several formulas with different levels of approximation accuracy developed by various researchers [3–8] for thin substrate and effective length L less than a half-wavelength of the chosen operating frequency; however, Kara [9] developed compact formulas for thick substrates. The formulas by Kara were used to calculate the dimensions of the patch as illustrated next based on the transmission-line model of Derneryd [4].

The patch conductor can be of any shape. For a rectangular patch, the length L of the patch is usually $0.3\lambda_0 < L < 0.5\lambda_0$, where λ_0 is the free-space wavelength. The patch is very thin, so that $t << \lambda_0$ (where t is the patch thickness). The height h of the dielectric substrate is usually $0.003\,\lambda_0 \leq h \leq 0.05\lambda_0$. The dielectric constant of the substrate (ε_r) is typically in the range of $2.2 \leq \varepsilon_r \leq 12$. For efficient radiation, low dielectric constant material is preferred over high dielectric constant material. The patch width, W, is usually the first to be chosen. For thin substrates, the empirical approximation by Bahl and Bhartia [10] as shown in (2.6) is used:

$$W = \frac{\lambda_0}{2\sqrt{0.5(\varepsilon_r + 1)}} \tag{2.6}$$

where λ_0 and ε_r are the wavelength in free space and dielectric constant of the substrate, respectively. The effective permittivity in the substrate is evaluated using the formula by Schneider [11] by means of (2.7):

$$\varepsilon_{eff} = \frac{\varepsilon_r + 1}{2} + \frac{\varepsilon_r - 1}{2}\left(\frac{1}{\sqrt{1 + 12h/w}}\right) \tag{2.7}$$

where h is the thickness of the substrate and the terms are as explained above. Hall and Wood [12] give an approximation for obtaining the length L of the microstrip

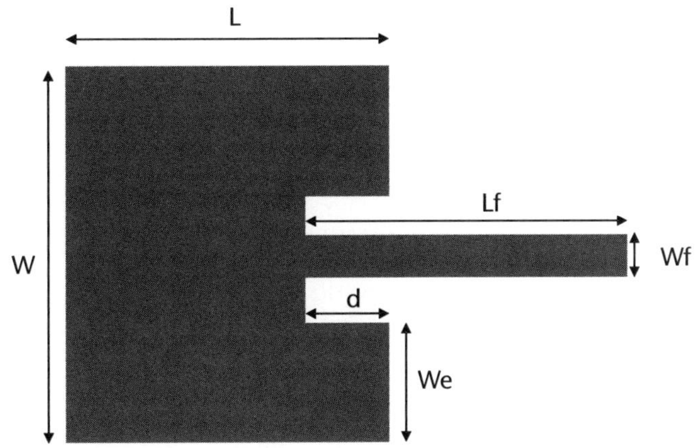

Figure 2.3 Inset fed rectangular microstrip antenna design.

patch antenna element with the substrate thickness less than $0.0815\lambda_d$ in this kind of media as

$$L = \frac{c_0}{2f_r\sqrt{\varepsilon_{eff}}} \tag{2.8}$$

with $c_0 = 2.998 \times 10^8$ meter/second being the velocity of the electromagnetic wave in free space. Because of the fringing field at the edges of the patch [13], a fringe factor, ΔL, must be subtracted from (2.8) to obtain the effective length. This fringe factor is given by (2.9):

$$\Delta L = 0.412h\frac{(\varepsilon_{eff} + 0.300)\left(\frac{W}{h} + 0.264\right)}{\varepsilon_{eff} - 0.258 \left(\frac{W}{h} + 0.813\right)} \tag{2.9}$$

Hence by means of (2.8) and the line extension, the effective length, L_{eff}, of a rectangular antenna element is [13–14]:

$$L_{eff} = L - 2\Delta L \tag{2.10}$$

2.2.4 Criteria for the Substrate Selection

The substrate in microstrip antennas is primarily needed for the mechanical support of the antenna. To provide this support, the substrate should consist of a dielectric material, which may affect the electrical performance of the antenna, circuits, and transmission line impedance. Some important parameters that should be considered while selecting the substrate material in the design of antennas are surface wave excitation, dispersion of the dielectric constant, loss tangent of the substrate, copper loss, effects of temperature, mechanical requirements (conformability, machinability, solderability, weight, elasticity, etc.), and cost. The criterion for the selection of substrate necessitates the following considerations:

- Substrates with higher thickness and a high dielectric constant result in smaller bandwidths and lower efficiencies due to the possibility of surface-wave excitation.
- Higher dielectric constant materials with longer lengths of feed line result in increased losses and chances of spurious radiation.
- The substrate thickness should be made as thin as possible to eliminate multiple surface waves. A height of $0.01\lambda_0$ to $0.05\lambda_0$ is considered ideal in this regard.
- It is preferable to use low dielectric constant material and a substrate of optimum thickness for containing the spurious feed radiation to a minimum.

A microstrip patch element launches surface wave modes in addition to the fields radiated into space. Surface waves give rise to end-fire radiation and it can be excited at the dielectric-to-air interface. Moreover, a surface wave can lead to unwanted coupling between array elements. The power thrown into the surface waves is a power that will ultimately be lost, and hence the excitation of surface waves lowers the overall radiation efficiency of the antenna. The phase velocity of surface waves is strongly dependent on the dielectric constant ε_r and thickness h of the substrate. The dependence of the dielectric constant ε_r and loss tangent on the frequency is referred to as frequency dispersion. Generally, the cost of printed antennas depends on the substrate material and connectors only. FR4 is one of the commercially available low-cost materials for printed antennas above the 1-GHz range. Some commonly used substrate materials in printed and microstrip antenna design are listed in Table 2.1 [2].

2.2.5 Losses in Microstrip Antennas

Losses in microstrip antennas are due to conductor loss, radiation loss, and dielectric loss. These losses affect microstrip antenna radiation efficiency. Conductor loss is calculated using (2.11):

$$\begin{aligned} &\alpha_c = 8.686\log\left(R_s/2WZ_0\right) \text{ dB/length} \\ &R_s = \sqrt{\pi f \mu \rho} \text{ skin resistance} \end{aligned} \tag{2.11}$$

Table 2.1 Commonly used Substrates for Printed and Microstrip Antenna Design

Material	*Dielectric Constant*	*Loss Tangent*
Foam	1.08	0.0010
Unreinforced PTFE, Cuflon	2.1	0.0004
RT/duroid 5880	2.2	0.0009
Teflon-ceramic,	2.3	0.0010
RT/duroid 5870	2.33	0.0012
polyolefin-glass,	2.42	0.0010
Teflon-quartz,	2.47	0.0006
Teflon glass	2.55	0.0015
Polystyrene quartz	2.6	0.0005
RT/duroid 6202	2.94	0.0015
Fused quartz	3.78	0.0001
FR4	4.4	0.0200
Rogers TMM6	6.0	0.0023
Polyolefin-ceramic	7.5	0.0010
Alumina	9.8	0.0001
Teflon-ceramic	10.3	0.0020
RT/Duroid 6010.2LM	10.7	0.0023
Silicon	11.9	0.0400
GaAs	12.9	0.0600

Dielectric loss is calculated by using Equation (2.12):

$$\alpha_d = 27.3\left(\frac{\varepsilon_r}{\sqrt{\varepsilon_{eff}}}\right)\left(\frac{\varepsilon_{eff}-1}{\varepsilon_e-1}\right)\left(\frac{\tan\delta}{\lambda}\right) \tag{2.12}$$

where tanδ is thedielectric loss coefficient.

Low dielectric loss tangent substrate materials have less dielectric loss, which is good for antenna applications.

2.3 Feed Techniques for Single-Element

Microstrip patch antenna elements can be fed by a variety of methods. These methods can be classified into two categories: direct or indirect contact. In the direct contact technique, the power is fed directly to the radiating patch using a connecting element such as a microstrip line or coaxial connector. In the indirect contact scheme, electromagnetic field coupling is done to transfer power between the microstrip line and the radiating patch. The four most popular feed techniques used are the microstrip line, coaxial probe, proximity coupling, and aperture coupling.

2.3.1 Microstrip Line Feed

In the microstrip line feed technique, a conducting metallic strip directly connects the path element with an RF power source. This kind of feed plan has the advantage that the feed can be etched on the same substrate along with the patch element to provide a planar structure. The feed line is smaller in width as compared to the patch. As shown in Figure 2.4, there are many different configurations of microstrip line feeds.

In edge- and offset edge-fed schemes, the feed line starts from the edge of the patch, while in inset Feed, the microstrip line starts from a location inside the patch. The objective of the inset cut in the patch is to match the impedance of the feed line to the patch without the need for any additional matching element. This

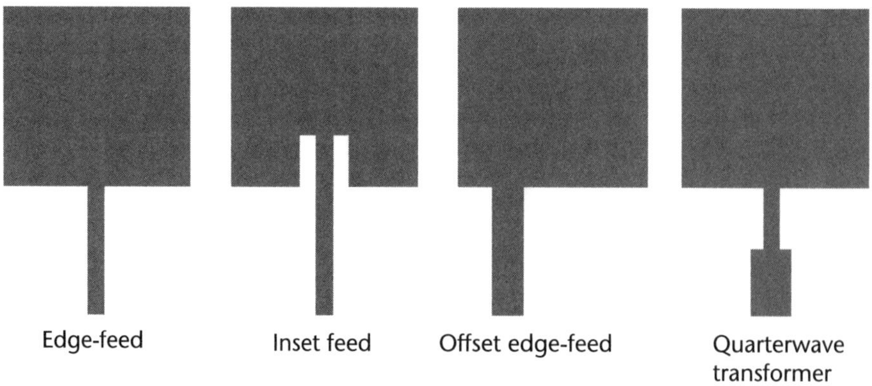

Figure 2.4 Different feeding mechanisms in a microstrip patch antenna.

is achieved by properly controlling the inset position and impedance of the line. Inset-fed is an easy feeding scheme as it provides ease of fabrication and simplicity in modeling as well as impedance matching. One disadvantage of the microstrip line feed is undesired cross-polarized radiation.

2.3.2 Coaxial Feed

The coaxial feed or probe feed for feeding microstrip patch antennas is a very common technique. The inner conductor of the coaxial feed extends through the dielectric and is soldered to the radiating patch, while the outer conductor is connected to the ground plane. The coaxial feed can be placed at any desired location inside the patch to match with its input impedance. This feed method has low spurious radiation. However, it provides narrow bandwidth and is difficult to model since a hole must be drilled in the substrate and the connector protrudes outside the ground plane, thus not making it completely planar for thick substrates ($h > 0.02\lambda_0$) and because of this, it is difficult to use this feed mechanism in array applications. Another disadvantage is for thicker substrates, which provide broad bandwidth; the increased probe length makes the input impedance more inductive, leading to matching problems. Practically, 50Ω SMA connectors are coaxial fed. The feeding probe point location can be located at the point (X_f, Y_f) in the x-y coordinates, as shown in Figure 2.5. The location points are given by (2.13):

$$X_f = \frac{L}{2} \text{ and } R\left(Y_f\right) = R(y=0)\cos^2\left(\frac{\pi y_0}{L}\right) \tag{2.13}$$

where $R(y = 0)$ is the edge impedance of the patch.

2.3.3 Aperture-Coupled Feed

In the aperture-coupled feed technique, the radiating patch and the microstrip feed line are placed on two different layers and separated by the ground plane, as shown in Figure 2.6. Coupling between the patch and the feed line is made through an

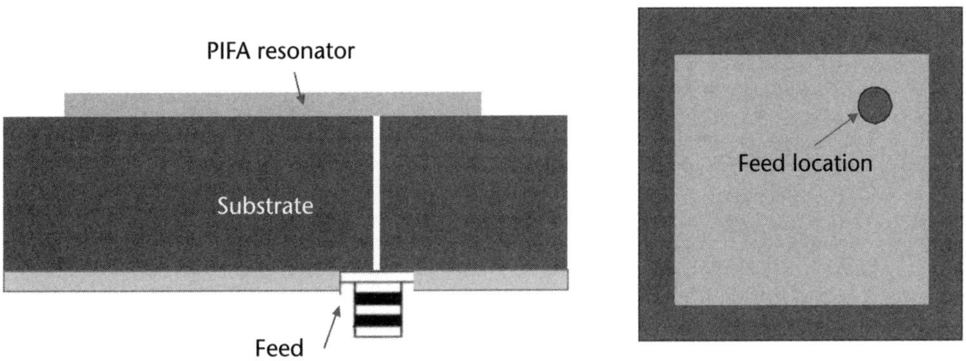

Figure 2.5 Coaxial-fed microstrip patch antenna (side and top views).

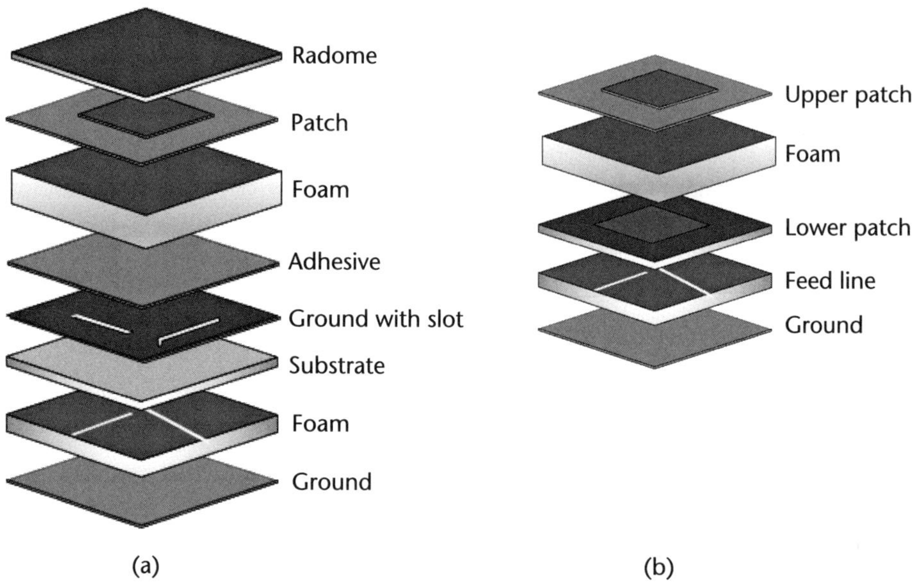

Figure 2.6 Microstrip antenna feed mechanism: (a) aperture-coupled antenna, and (b) proximity-coupled antenna.

aperture or a slot in the ground plane. The coupling slot is usually centered under the patch, leading to lower cross-polarization due to the symmetry of the configuration. In the dual-polarized antenna, there are two coupling slots, one for vertical polarization and other for horizontal polarization.

The amount of coupling from the feed line to the patch is determined by the size, shape, and location of the aperture. Since patch and the feed line are seprated by the ground plane, spurious radiation is minimized. A high dielectric material is used at the bottom for the feed line and a thick, low dielectric constant material is used for the radiator at the top substrate to optimize radiation from the patch. This feeding scheme provides narrow bandwidth.

2.3.4 Proximity- or Electromagnetically Coupled Feed

The proximity-coupled feed technique is also called an electromagnetically couple feed. As shown in Figure 2.6, the main advantage of this feed technique is that it eliminates spurious feed radiation and provides very high bandwidth (as high as 14%) due to an overall increase in the thickness of the microstrip patch antenna. This scheme also provides choices between two different dielectric material; one for the patch and another for the feed line to optimize the individual performances. Matching can be realized by controlling the length of the feed line and the width-to-line ratio of the patch. The major disadvantage of this feed structure is that it is difficult to fabricate because the two dielectric layers need a proper alignment. A comparison of all major microstrip antenna feeding structures is shown in Table 2.2.

Table 2.2 Feeding Techniques Comparison

Characteristics	*Microstrip Line Feed*	*Coaxial Feed*	*Aperture-Coupled Feed*	*Proximity-Coupled Feed*
Spurious feed radiation	More	Less	Less	Minimum
Reliability	Better	Poor due to soldering	Good	Good
Ease of fabrication	Easy	Soldering and drilling needed	Alignment required	Alignment required
Bandwidth	2%–5%	2%–5%	2%–5%	14%
Polarization purity	Good	Poor	Excellent	Poor

2.4 Design and EM Simulation of Microstrip Antennas

Design, analysis, and electromagnetic simulation of a single microstrip patch antenna analysis, design, and electromagnetics operating at 2.4 GHz is discussed in this section. The designed antenna is used for industrial, scientific, medical (ISM) applications and WLAN. Consideration is given to a practical design technique, including substrate selection and antenna measurements. The microstrip antenna is characterized by its length L, width W, and substrate thickness h.

2.4.1 Design Process

A single-element microstrip antenna design flow is shown in Figure 2.7. Design starts with antenna specification. Antenna specification is derived from the application requirement. Based on specification and cost, the substrate material is selected. Antenna radiating parameters and feed design parameters are calculated based on empirical equations (2.6 to 2.15). Antenna performance is optimized in electromagnetic simulation based EDA tools. It is important to carefully select a design and simulation EDA tool and electromagnetic simulation software for efficient antenna design as mentioned in Chapter 1.

After the antenna design parameter optimization, antenna design PCB is generated for fabrication. Finally, all antenna parameters are measured before actual use of the antenna. To design a rectangular patch antenna, its length L, width W, copper layer thickness t, and thickness of a substrate h information is needed.

2.4.2 Feed-Point Location of a Single Patch Antenna

Edge feed is used to feed the antenna. The edge impedance Z_A of the patch antenna is calculated by using (2.14):

$$Z_A = 90\left(\frac{\varepsilon_r^2}{\varepsilon_r - 1}\right)\left(\frac{L}{W}\right)^2 \tag{2.14}$$

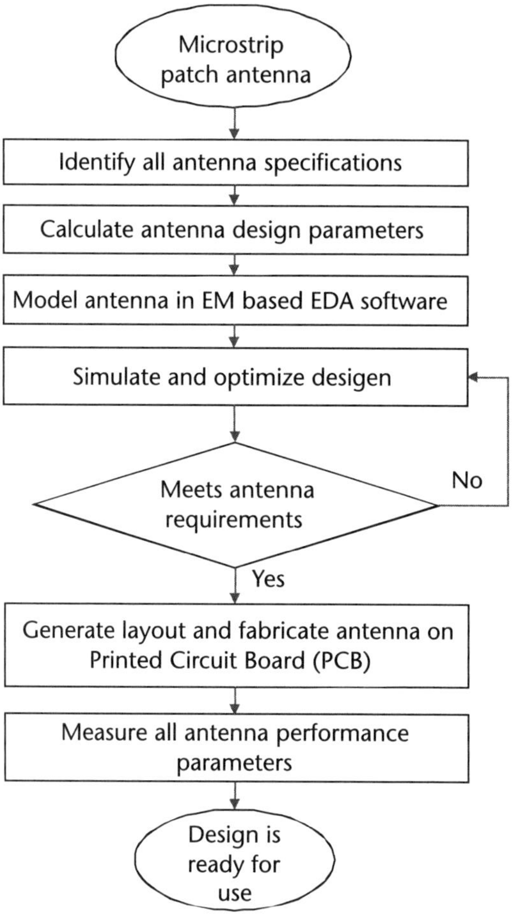

Figure 2.7 Microstrip antenna design and fabrication process.

Figure 2.8 shows a patch antenna with a quarter-wavelength matching section. Input impedance Z_{in} is matched with a 50-ohm microstrip feed line impedance Z_0. If the edge impedance of the patch antenna is Z_A, then the input impedance seen from the beginning of the quarter-wavelength line is

$$Z_{in} = Z_0 = \frac{Z_L^2}{Z_A} \tag{2.15}$$

Input impedance Z_{in} can be changed by the selection of Z_L, so that $Z_{in} = Z_0$ and the antenna is matched by impedance. The impedance Z_L value varies with the width of the quarter-wavelength strip.

2.4.3 Microstrip Patch Antenna Design Parameters

Patch antenna design parameters are shown in Table 2.3. Rogers 5870 with ε_r equal to 2.33 is used as the dielectric material. The thickness t of a metallic layer must be a substantially smaller than the wavelength of the free space ($t << \lambda_0$) for a given

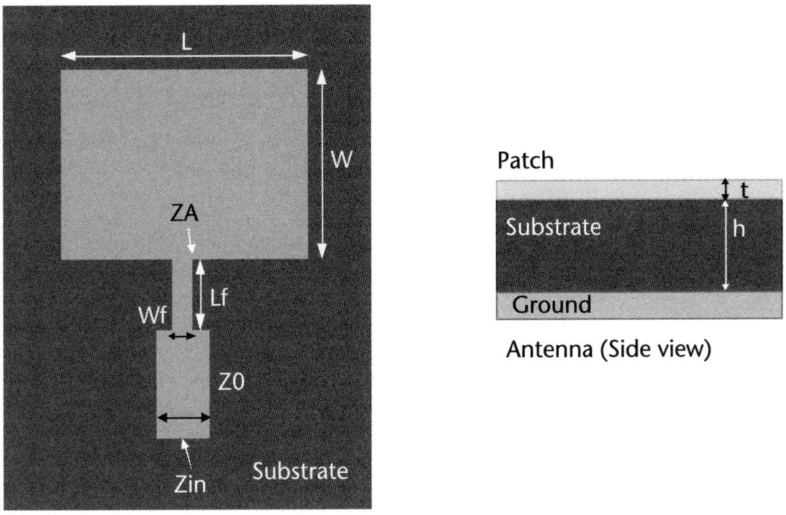

Figure 2.8 Rectangular microstrip patch antenna design parameters.

Table 2.3 Typical Design Specifications of a Microstrip Antenna

Specifications	*Parameters*	*Value*
Antenna radiating element	Patch length L	40 mm
	Patch length W	47 mm
Antenna feedline	Edge impedance (Z_A)	360
	Feedline impedance	50 ohm
	Quarter-wave transformer L	22 mm
	Quarter wave Transformer W	0.3 mm
Substrate	Dielectric constant (ε_r)	2.33
	Thickness of the substrate (h)	0.5 mm
	Loss tangent ($\tan\delta$)	0.008
	Copper metal thickness (t)	0.035 mm
Antenna specifications	Operating frequency (f_0)	2.4 GHz
	Polarization	Linear
	Feed type	Edge feed
Design requirement	Return loss	< −15 dB
	Bandwidth	10 MHz
	Gain	6 dBi
	Half-power beamwidth	71 Deg

frequency, in our case 2.4 GHz. Calculation of width W and length L for a single patch antenna is performed using (2.6) and (2.8), respectively. ΔL is the fringe factor calculated by (2.9). For an operating frequency of 2.4 GHz, the wavelength λ_0 equals 0.125m.

2.4.4 Patch Antenna Design and Analysis Result at 2.4 GHz

In a microstrip patch antenna simulation process, the first step is the creation of the layout geometry together with the definition and assignment of material. The edge-fed antenna is modeled and simulated in Advanced Design System (ADS) software.

A quarter-wave transformer is used to match patch edge impedance to 50Ω feedline impedance as shown in Figure 2.9. The next step is to define port excitation. Finally, antenna performance is optimized using the full-wave electromagnetic solver FEM. The patch antenna simulated return loss plot is shown in Figure 2.9. The obtained return loss of patch is -17 dB at 2.4 GHz. Like VSWR, return loss is indicated by how well the matching between the transmitter and antenna has taken place. The ideal value of the return loss is around –10 dB, which corresponds to a VSWR of less than 2. These results show that a microstrip patch antenna mainly radiates in the vertical direction.

The peak gain is 6.1 dBi at 2.4 GHz and with the main lobe in the direction of theta = 0 degrees and phi = 0 degrees, as shown in Figure 2.10. The half-power beamwidth of the antenna is 71 degrees.

2.4.5 Coaxial-Fed Patch Antenna for Coal Mines

In coal mines, the size of microcell zones can be efficiently increased by using bidirectional antennas, which preferentially radiate the antenna beam along the longitude. The special channel property and safety production in coal mines bring up new challenges in antenna design, such as low profile, easy fabrication, low cost, low wind resistance, antiexplosion, and good bidirectional radiation pattern. The performance of the system is degraded by severe signal attenuation inside the coal mine where the wall conductivity or dielectric constant is relatively high. Broadband VHF/UHF transmission is normally used as it is difficult to predict the polarization of the reflected component for such applications. Coaxial-fed microstrip antennas are a very suitable choice as they meet most of coal mine antenna requirements. Figure 2.11 shows a microstrip patch antenna fed with coaxial feed designed in the UHF band for such an application. This antenna is developed for end use in coal mines where the antenna is flush-mounted on coal strata. The antenna is used in determining the angle of arrival of the reflective component when RF is launched in coal strata with the antenna embedded in it. Thus, it will be able to measure the angle of arrival of any reflective component due to the presence of a waterbed at a

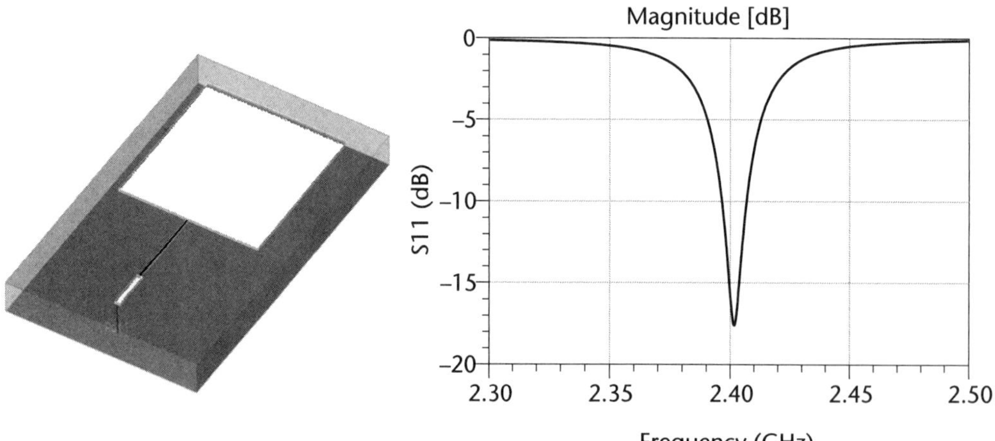

Figure 2.9 Three-dimensional model of microstrip antenna and simulated return loss plot using FEM.

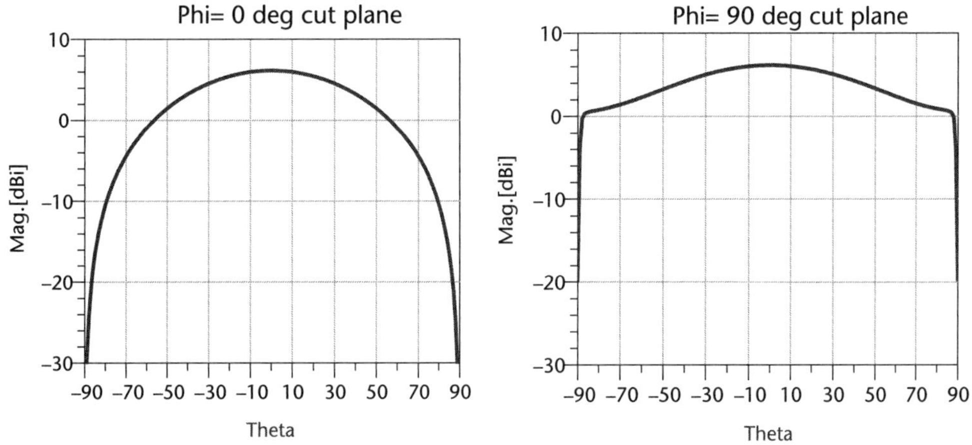

Figure 2.10 Antenna radiation patterns at phi = 0 and phi = 90-degree cuts.

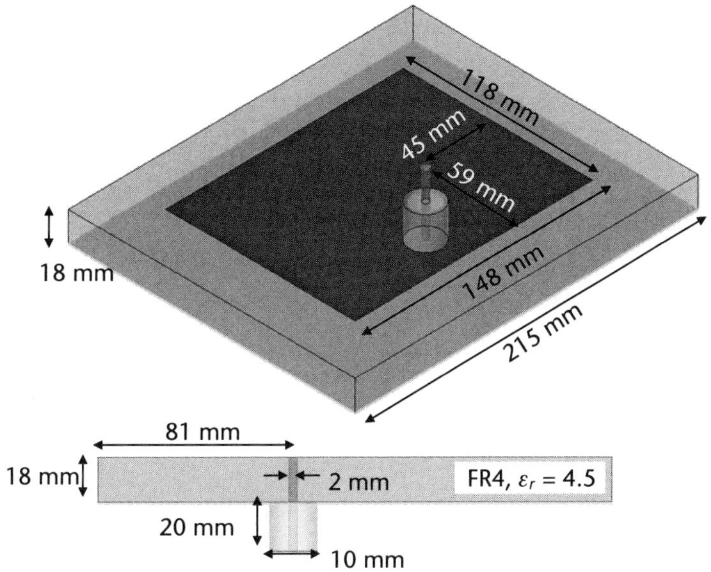

Figure 2.11 Coaxial-fed microstrip patch antenna design for coal mine.

distance. The antenna is fed with a 50Ω coaxial line at a feeding location of 59 mm by 45 mm on the patch, where patch impedance is also 50Ω. This type of feeding scheme can be placed at any desired location inside the patch to match with the desired input impedance and also has low spurious radiation.

The antenna is fabricated on an 18-mm thick FR4 substrate of 4.5 dielectric constant. Patch size is 148 mm × 118 mm. The effect of the key geometrical parameters on antenna radiation and impedance matching performance has been studied by the FEM numerical technique. The antenna operating frequency is 440 MHz, as shown in Figure 2.12. It is a 10-MHz narrowband (2.2% -10 dB bandwidth) antenna. As discussed in the previous section, coaxial feed antennas are normally

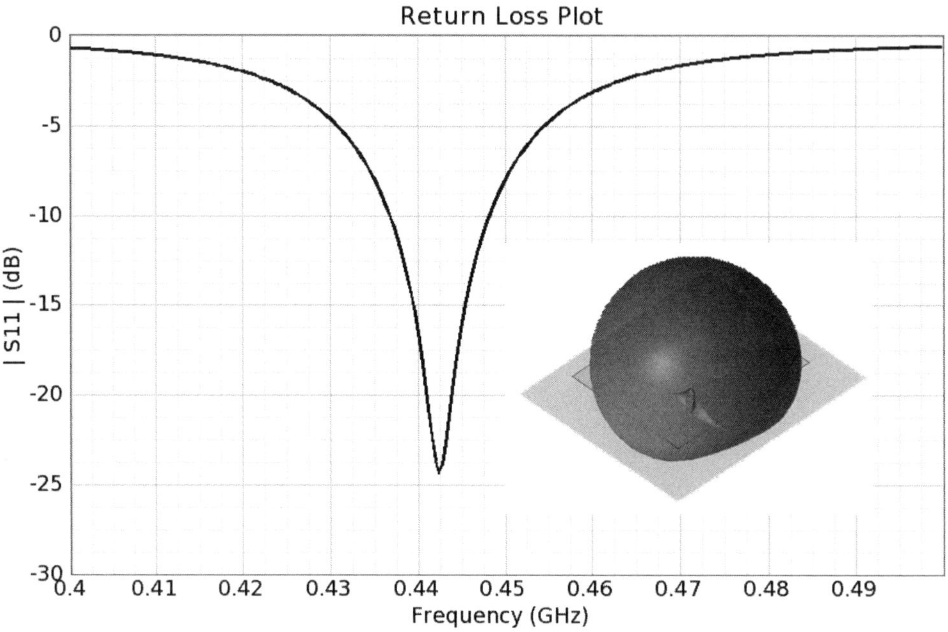

Figure 2.12 Antenna return loss plot of a coaxial-fed microstrip patch antenna.

narrowband antennas. Antenna gain is 7.1 dBi at 440 MHz and antenna radiation efficiency is nearly 60%.

2.5 Circularly Polarized Microstrip Patch Antennas

In a circularly polarized antenna, the electric-field vector at a given point in space traces a circle as a function of time. The conditions to accomplish this are if the electric field vector possesses all the following:

- The field must have two orthogonal linear components;
- The two components must have the same magnitude;
- The two components should have a time-phase difference of odd multiples of 90°.

Circular polarization (CP) in patch antennas can be achieved using either a single feed or a multifeed technique. CP designs of microstrip antennas are mainly on circular, square, and triangular microstrip patches. CP antennas are used in GPS, GNSS, and many other wireless communication applications.

2.5.1 Single-Fed Circularly Polarized Patch Antenna

Typical configurations for singly fed CP microstrip antennas are shown in Figure 2.13. Because a patch with single-point feed generally radiates linear polarization, to radiate CP it is necessary for two orthogonal patch modes with equal amplitude and inphase quadrature to be induced. This can be achieved by slightly perturbing a

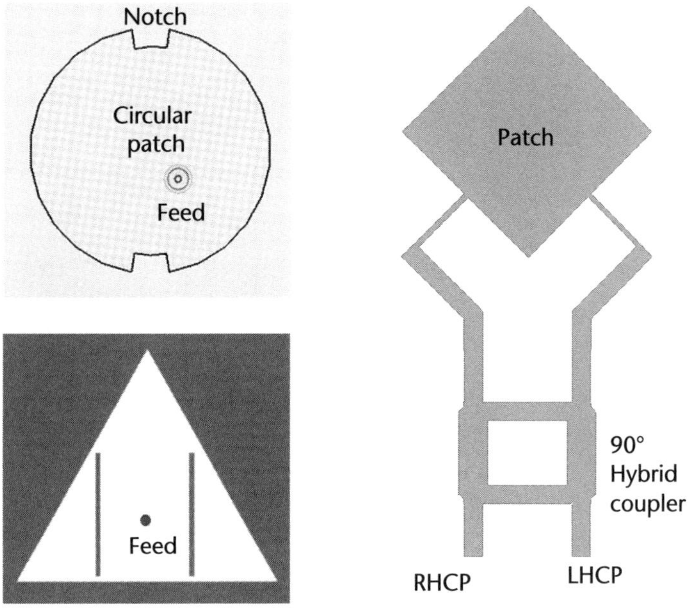

(a) (b)

Figure 2.13 Configuration of circularly polarized patch antenna: (a) single-fed, and (b) dual-orthogonal fed.

patch at appropriate locations with respect to the feed. Perturbation configurations for generating CP operate on the principle of detuning degenerate modes of a symmetrical patch by perturbation segments, as shown in Figure 2.13.

The fields of a singly fed patch can be resolved into two orthogonal degenerate modes 1 and 2. Proper perturbation segments will detune the frequency response of mode 2 such that, at the operating frequency, the axial ratio rapidly degrades while the input match remains acceptable. CP operation of single-feed triangular microstrip antennas using a nearly equilateral triangular patch [15], an equilateral triangular patch with an inserted slit [16], or a tip-truncated equilateral triangular patch [17] has also been designed by many antenna engineers. Due to their physically smaller patch size as compared to the circular or square microstrip antennas, these circularly polarized triangular microstrip antennas are suitable for applications in systems where limited space is available.

There are several kinds of the patch elements such as square, circular, triangular, and ring shapes that have been used to realize CP radiation using a single feed [18]. The antenna shown in Figure 2.14 is a millimeter-wave CP antenna fed with a single coaxial line. The antenna finds application in 5G microcell, where a high-speed data rate is required. As frequency increases, complexities in system design also increase. Circular polarization is achieved by using two notches cut in the patch. The antenna is fabricated on Teflon ($\varepsilon_r = 2.0$) and has a substrate thickness 0.254 mm.

An antenna resonating at 29 GHz and 10 dB bandwidth of the antenna is 1.2 GHz (from 27.9 to 29.1 GHz). The designed antenna has an axial ratio (AR) value of 0.31 dB at center frequency. Single-feed microstrip antennas usually have very narrow 3-dB axial ratio bandwidth. This antenna has 0.8% 3 dB AR bandwidth.

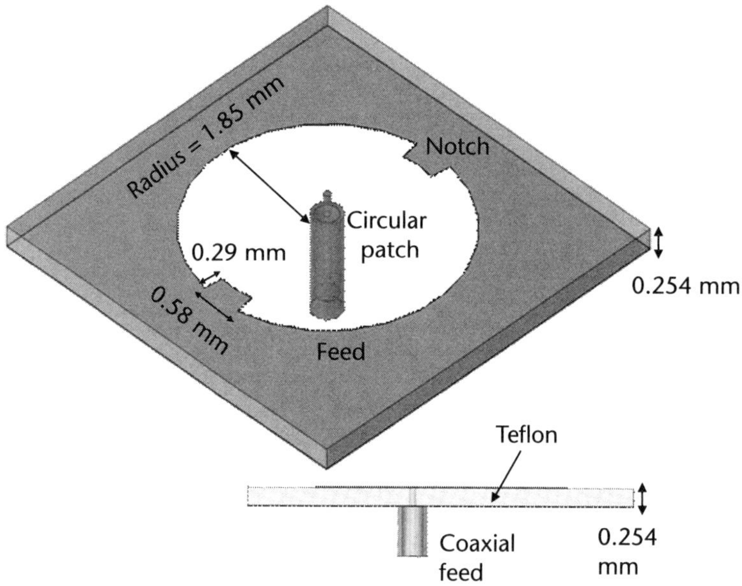

Figure 2.14 Coaxial-fed circularly polarized millimeter-wave microstrip antenna.

2.5.2 Dual-Orthogonal Fed Circularly Polarized Patch

The fundamental configurations of a dual-orthogonal fed circularly polarized patch using an external power divider is shown in the Figure 2.15. Use of a dual-feed technique in GPS and GNSS applications, is the most direct way to generate circular polarized radiation from a square patch. The two orthogonal modes required for the generation of circular polarization can be simultaneously excited using dual-orthogonal feeds. In designing, the patch is first matched to the feed lines by either properly selecting the feed locations or using impedance transformers. The two feeds are then connected to the output ports of a power divider circuit, which provides the required amplitude and phase excitations. Square patch circularly polarized antenna dimension is shown in Figure 2.15.

The dual-orthogonal feed excites two orthogonal modes with equal amplitude but inphase quadrature. The quadrature hybrid splits the input RF signal into two outputs with equal magnitude but 90° out of phase. The quadrature hybrid provides a broad axial ratio bandwidth. These power splitters can be easily constructed from various planar transmission lines. Microstrip feed lines excite the patch antenna in TM_{01} and TM_{10} modes so that it radiates both vertically polarized wave and a horizontally polarized wave simultaneously. The antenna is designed on 0.5-mm thick RT Duroid 5870 substrate with a dielectric constant of 2.33 at 2.4 GHz. The microstrip patch antenna consists of a patch and a 3-dB hybrid coupler. Figure 2.16 shows the simulated return loss and the phase difference between the two output ports, which is about to 90°.

Figure 2.17 shows the return loss, axial ratio, current distribution, antenna 3-D radiation pattern, and phi-0-degree cut pattern at 2.4 GHz. An axial ratio below 3 dB corresponds to the axial ratio bandwidth that it is extended from 2.1 to 2.7 GHz. The simulated 10-dB impedance bandwidth is from 2.2 to 2.8 GHz (about 25%). Antenna gain is 6.8 dBi.

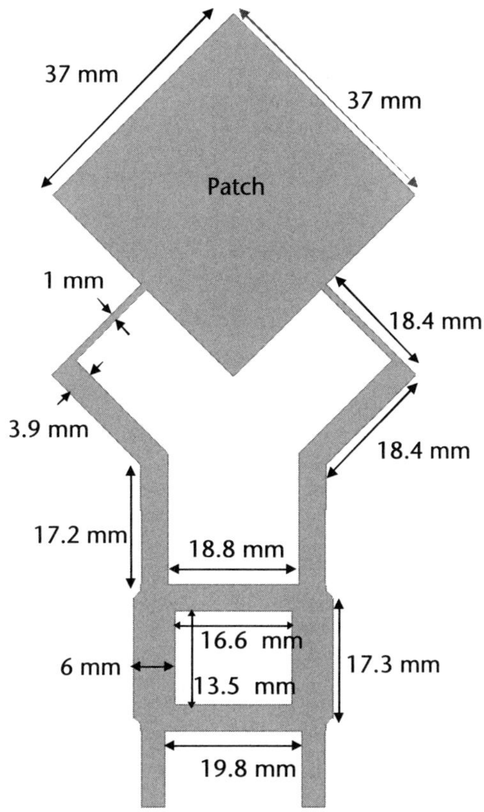

Figure 2.15 Design dimension of dual-orthogonal fed circularly polarized microstrip patch antenna.

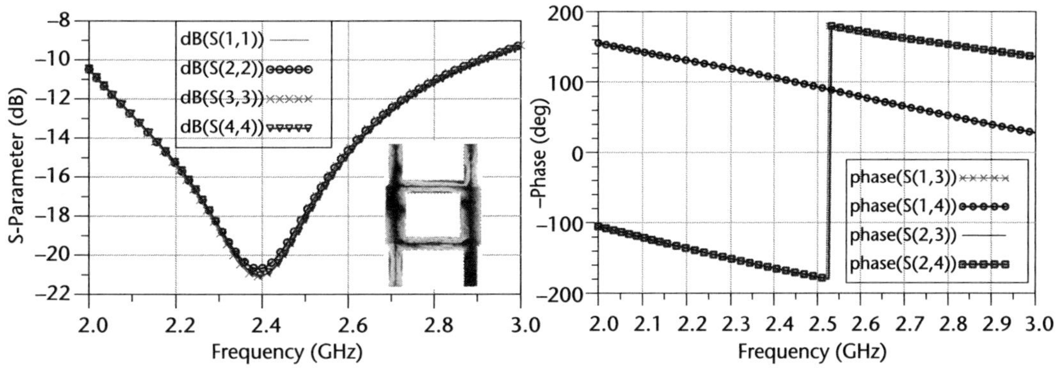

Figure 2.16 Return loss and phase plot of quadrature hybrid coupler.

2.6 Antenna Measurements

After antenna fabrication, all antenna parameters are measured to know the antenna performance. Typical parameters of antennas are resonance frequency point, bandwidth, radiation pattern, gain, directivity, beamwidth, polarization, and impedance. During antenna measurements we ensure that the antenna meets the required specifications and we can characterize the antenna parameters. The measurement

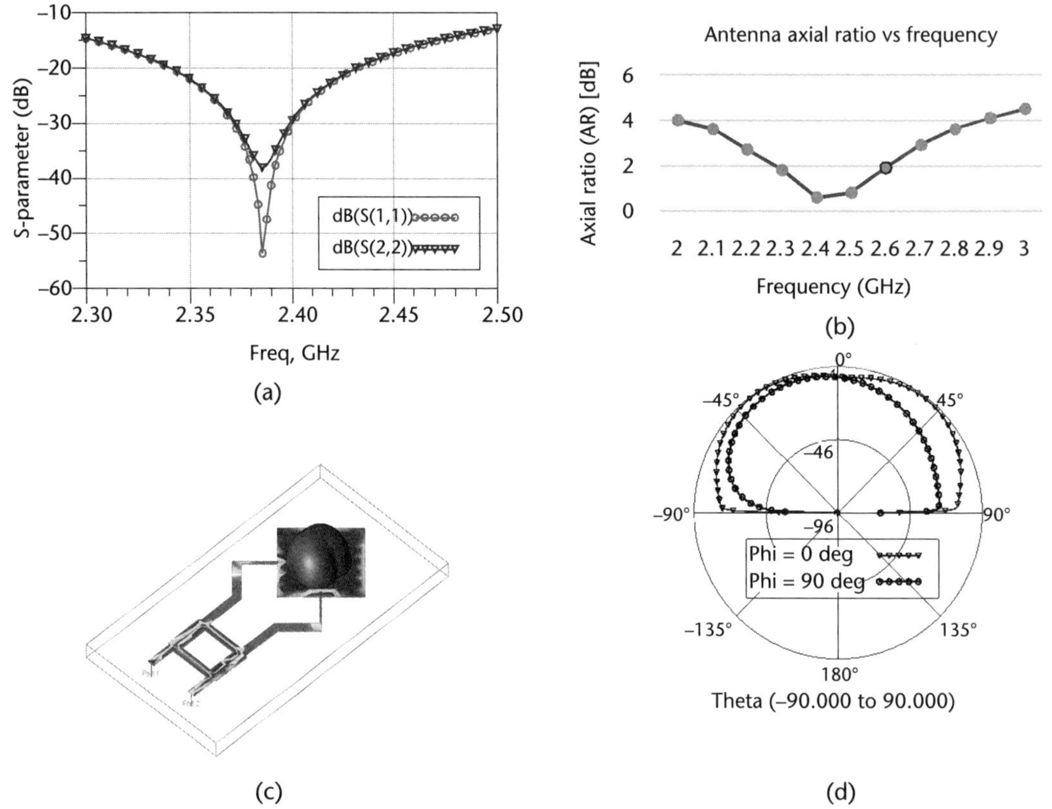

Figure 2.17 Antenna performance results: (a) return loss plot, (b) axial ratio plot, (c) current distribution over the patch, (d) radiation pattern at phi = 0 and phi = 90 degrees.

equipment includes scalar and vector network analyzers, antenna ranges, antenna positioners, pattern recorders, signal generators, and antenna gain standards. Currently, sophisticated computer systems are used to provide automated control of pattern measurements as well as fast calculations related to antenna directivity, two-dimensional to three-dimensional pattern conversion, and near-to-far field transformations (in compact antenna ranges). *S*-parameters are measured using a vector network analyzer (VNA) while antenna radiation characteristics such as far-filed patterns, gain, and efficiency are measured using different antenna pattern measurement techniques. To measure the far-field pattern, it should be illuminated with a uniform plane wave.

2.6.1 S-Parameter and Input Impedance

Antenna impedance is a fundamental parameter of all antennas operating at RF. Therefore, for the best accuracy the impedance should be measured in an environment that will most closely resemble the environment where it is intended to operate.

Impedance measurements are carried out using a VNA. This tool is used to measure the input impedance as a function of frequency. It can plot return loss (S11), and the VSWR, both of which are frequency-dependent functions of the antenna impedance. The *S*-parameter measurement setup is shown in Figure 2.18.

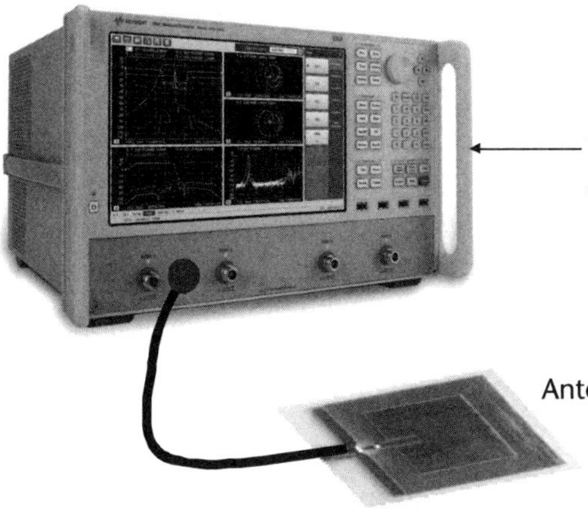

Figure 2.18 Antenna S-parameter and input impedance measurement setup using a vector network analyzer.

2.6.2 Radiation Pattern Measurements

A radiation pattern is the antenna radiated field as a function of the direction in space. The radiated field is measured at various angles at a constant distance from the antenna. Usually the antenna radiation pattern is measured in a far-field antenna range. The antenna under test is placed in the far-field distance from the transmitting antenna. Because of the size required to create a far-field range for large antennas, near-field techniques are employed. Near-field techniques allow one to measure the fields on a surface close to the antenna (usually 3–10 wavelength).

Near-field distances are transferred to far-field by using the Fourier transform. The far-field distance or Fraunhofer distance, *R*, is given in (2.16):

$$R = 2D^2/\lambda \tag{2.16}$$

where *D* is the maximum antenna dimension and λ is the antenna wavelength. The radiation pattern graphs can be drawn using Cartesian (rectangular) coordinates or as a polar plot. The antenna main beam is measured between the points that the maximum relative field intensity E decays to 0.707E. Half of the radiated power is concentrated in the antenna main beam. This antenna beam is called the 3-dB beam width. Radiation to the undesired direction is concentrated in the antenna sidelobes. The antenna radiation pattern is usually measured in free-space ranges as shown in Figure 2.19.

2.7 Summary

In this chapter, the microstrip patch antenna, which is the most popular printed antenna, was discussed in detail. A practical coaxial-fed rectangular patch antenna at the resonant frequency of 440 MHz has been designed for coal mine application

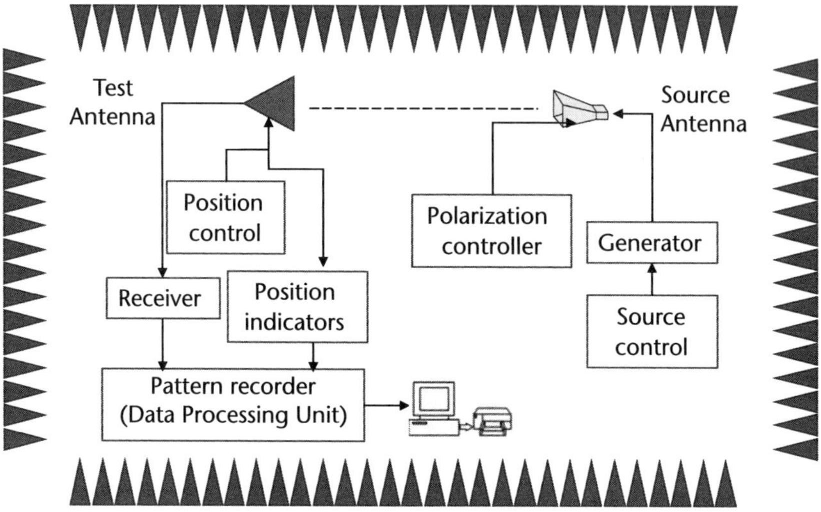

Figure 2.19 Antenna radiation pattern measurement setup in an anechoic chamber.

and simulated using an electromagnetic solver. Important antenna specifications, design parameters, and design flow were discussed through this antenna. A single microstrip antenna is the building block for complex array antennas so it very important to understand the design and working principle of the single radiator and its feeding mechanism. The radiation patterns and other important parameters of gain, efficiency, and return loss have been shown. Basic antenna parameters measurement can be done using VNA and far-field measurement ranges. In the next chapters, complex array antenna design will be presented for different applications.

References

[1] Pozar, D. M., and D. Schaubert (eds.), *Microstrip Antennas: The Analysis and Design of Microstrip Antennas and Arrays*, Piscataway, NJ: IEEE Press, 1995.

[2] Garg, R., P. Bhartia, I. Bahl, and A. Ittipiboon, *Microstrip Antenna Design Handbook*, Norwood, MA: Artech House, 2001.

[3] Kara, M., "Formulas for the Computation of the Physical Properties of Rectangular Microstrip Antenna Elements with Various Substrates Thicknesses," *Microwave and Optical Technology Letters*, Vol. 12, No. 4, 1996, pp. 234–239.

[4] Derneryd A. G., "A Theoretical Investigation of the Rectangular Microstrip Patch Antenna Element," *IEEE Transactions on Antennas and Propagat*ion, Vol. AP-26, July 1978, pp. 532–535.

[5] Lee, K. F., K. M. Luk, and H. W. Lai, *Microstrip Patch Antennas*, Second Edition, London: World Scientific Publishing Company, July 2017.

[6] Kara, M., "Closed-Form Expression for the Resonant Frequency of Rectangular Microstrip Antenna Elements with Thick Substrate," *Microwave Optical Technology Letters,* Vol. 12, No. 3, June 1996, pp. 131–136.

[7] James, J. R., and Hall P. S. (eds.), *Handbook of Microstrip Antennas*, Vols. I and II, London: Peter Peregrinus (IEE), 1989.

[8] Sainati, R. A., *CAD of Microstrip Antennas for Wireless Applications*, Norwood, MA: Artech House Publishers, 1996.

[9] Kara, M., "Empirical Formulas for the Computation of the Physical Properties of Rectangular Microstrip Antenna Elements with Thick Substrates," *Microwave Optical Technology Letters*, Vol. 14, No. 2, February 1997, pp. 115–121.

[10] Bahl, I. J., and Bhatia P., *Microstrip Antennas*, Dedham, MA: Artech House, 1980.

[11] Schneider, M., "Microstrip Lines for Microwave Integrated Circuits," *Bell System Technical Journal,* Vol. 48, 1969, pp. 1421–1444.

[12] James J. R., P. S. Hall, and C. Wood, *Microstrip Antenna Theory and Design*, Stevenage, UK: Peter Peregrinus (IEE), 1981.

[13] Hammerstad, E. O., "Equations for Microstrip Circuits Design," *5th European Microwave Conference*, Hamburg, Germany, September 1975, pp. 268–272.

[14] Bahl I. J., "Build Microstrip Antennas with Paper-Thin Dimensions," *Microwaves,* October 1979, p. 50.

[15] Suzuki Y., N. Miyano, and T. Chiba, "Circularly Polarised Radiation from Singly Fed Equilateral-Triangular Microstrip Antenna," *IEE Proceedings H-Microwaves, Antennas and Propagation*, Vol. 134, April 1987, pp. 194–198.

[16] Lu, J. H., C. L. Tang, and K. L. Wong, "Circular Polarization Design of a Single-Feed Equilateral-Triangular Microstrip Antenna," *Electronics Letters,* Vol. 34, 1998, pp. 319–321.

[17] Tang, C. L., J. H. Lu, and K. L. Wong, "Circularly Polarized Equilateral Triangular Micro Strip Antenna with Truncated Tip," *Electronics Letters,* Vol. 34, 1998, pp. 1277–1278.

[18] Wong, K. L., *Compact and Broadband Microstrip Antennas*, New York: John Wiley & Sons, 2002.

Selected Bibliography

Ahn, C. H., Y. J. Ren, and K. Chang, "A Dual-Polarized Cylindrical Conformal Array Antenna Suitable for Unmanned Aerial Vehicles," *International Journal of RF and Microwave Computer Aided Engineering,* Vol. 21, No. 1, 2011, pp. 91–98.

Balanis, C., *Antenna Theory: Analysis and Design*, Hoboken, NJ: Wiley, 2012.

Bokhari, S. A., J.- F. Zurcher, J. R. Mosig, and F. E. Gardiol, "A Small Microstrip Patch Antenna with a Convenient Tuning Option", *IEEE Transactions on Antennas and Propagation*, Vol. 44, No. 11, November 1996.

Deshmukh A. A., and K. P. Ray, "Compact Broadband Modified Triangular Microstrip Antennas," *IET Microwaves, Antennas & Propagation*, Vol. 9, No. 11, 2015, pp. 1205–1212.

Ávila-Navarro, E., J. A. Carrasco, and C. Reig, "Printed Dipole Antennas for Personal Communications Systems," *IETE Technical Review,* Vol. 27, No. 4, 2010, pp. 286-292.

Hayes, G. J., Y. Liu, J. Genzer, G. Lazzi, and M. D. Dickey, "Self-Folding Origami Microstrip Antennas," *IEEE Transactions on Antennas and Propagation*, Vol. 62, No. 10, October 2014, pp. 5416–5419.

Iwasaki, H., "A Circularly Polarized Small Size Microstrip Antennas with Cross Slot," *IEEE Transactions on Antennas and Propagation,* Vol. 44, No. 10, 1996, pp. 1399–1401.

Javor, R. D., X. D. Wu, and K. Chang, "Design and Performance of a Microstrip Reflect Array Antenna." *IEEE Transactions on Antennas and Propagation*, Vol. 43, No. 09, September 1995, pp. 932–939.

Lee, M. S., and Y.H. Kim, "Design and Performance of a 24-GHz Switch-Antenna Array FMCW Radar System for Automotive Applications", *IEEE Transactions on Vehicular Technology*, Vol. 59, No. 5, 2010, pp. 2290–2297.

Mendes C., and C. Peixeiro, "A Dual-Mode Single-Band Wearable Microstrip Antenna for Body Area Networks," *IEEE Antennas and Wireless Propagation Letters*, Vol. 16, 2017, pp. 3055–3058.

Pandey, A., "Efficient Synthetic Aperture Radar Antenna Array Design Using Multiple Electromagnetic Solvers," *Antenna System & Technology*, February 2003.

Song, Q., and X. Zhang, "A Study on Wideband Gap Coupled Microstrip Antenna Array," *IEEE Transactions on Antennas and Propagation*, Vol. 43, No. 03, Mar. 1995, pp. 313–317.

Sun, W., Y. Li, Z. Zhang, and Z. Feng, "Broadband and Low-Profile Microstrip Antenna Using Strip-Slot Hybrid Structure," *IEEE Antennas and Wireless Propagation Letters*, Vol. 16, 2017.

Xu, J., et al., "A Wideband F-Shaped Microstrip Antenna," *IEEE Antennas and Wireless Propagation Letters,* Vol. 16, 2017, pp. 829–832.

Wang Y., and Z. Du, "Dual-Polarized Dual-Band Microstrip Antenna with Similar-Shaped Radiation Pattern," *IEEE Transactions on Antennas and Propagation,* Vol. 63, No. 12, December 2015, pp. 5923–5928.

CHAPTER 3

Antenna Design for Wireless Communication and Mobile Phones

3.1 Introduction

The antenna of a mobile phone, which is a terminal equipment for transmitting and receiving electromagnetic waves, plays a vital role in mobile communication. A dramatic growth has been seen the mobile industry; it has evolved from analog to digital second generation (2G) GSM communication, then to high data rate third generation (3G) wideband code division multiple access (WCDMA), further to packet-optimized 3.5G High Speed Packet Access (HSPA) and fourth generation (4G) (Long Term Evolution (LTE) and LTE Advanced) systems and now moving towards fifth generation (5G) wireless communication system. As the smartphone era unfolds, the operating frequency band of the mobile phone is becoming wider and the space for a smartphone antenna is limited. It is very significant to design a compact wideband antenna for mobile phone application with a limited space. In recent years, the printed antenna has been widely adapted in wireless devices because it can be printed directly on the PCB along with other circuitry at a low cost. Many printed broadband antennas for the mobile phone have been proposed in the open literature [1–6]. The printed inverted-F monopole antenna designed for the multifrequency bands with compact size is one of the most used antenna choices in a cell phone.

The trend in mobile devices has been to provide more memory, faster access, improved processors, additional connectivity with Wi-Fi, GPS, 3G, and 4G world access, all with longer battery life in thinner, sleeker packages. Compound this with the desire of mobile operators to expand their available band allocations and mobile communication standards, and what results is a difficult and complex antenna design. The main design challenges of modern mobile phone antenna are the built-in structure, multisystem in multiband, including all wireless communication standards 2G, 3G, 4G, and other noncellular RF bands, and meeting all compliance standards and requirements such as specific absorption rate (SAR), over-the-air (OTA), and hearing aid compatibility (HAC). This chapter gives an overview of the design of some important antennas for mobile.

3.2 Mobile Communication Standards

In mobile communication, the common 2G, 3G, and 4G bands are GSM850 (824–894 MHz), GSM900 (890–960 MHz), GSM1800 (1,710–1,880 MHz), GSM1900 (1,850–1,990 MHz), UMTS2100 (1,920–2,170 MHz), LTE2300 (2,305–2,400 MHz), and LTE2500 (2,500–2,690 MHz) as shown in Table 3.1. The realization of the compact multiband antenna through excellent design and proper impedance matching is the main design objective of cell phone antennas. In recent years, to meet the requirements of current mobile phone built-in antennas [7–11], monopole antenna, PIFA antenna, and many other types of antennas have been designed. A navigation antenna is specially provided in most phones, which mainly covers the GPS band.

LTE provides the essential and far reaching vital solutions for wireless communication systems. It provides high-speed internet along with broader bandwidth and high throughput. During its initial stages in Europe, LTE was assigned 1.8-GHz and 2.6-GHz bands. Currently, due to the increase in the necessity of additional bandwidth, the 790–865-MHz band has been added to the LTE operating system.

The main challenges for the mobile device antenna are small size, built-in, multiband, and coexistence of a multiradio system and a multiple-input-multiple-output (MIMO) system. There have been two major requirements in the antenna design. One is that antennas for mobile terminal require small size, built-in, and multiband operation. Another is that mobile antennas should pass various compliances before production, depending on the functionality of the system, its service areas, and the quality and quantity of data to be transmitted. For a compact mobile terminal, environmental conditions are also a serious issue to be considered in the antenna design. Proximity effects due to materials near the antenna element, such as circuit components, acoustic components, camera, and printed circuit board grounding quality, may degrade antenna performance. In modern antenna design, the proximity effect is treated with the integration concept, in which nearby materials are included in an antenna system as an integral part of the radiator. For MIMO and diversity applications in Wi-Fi and 4G systems, the multi-antenna system requires low mutual coupling and low pattern correlation between antennas to realize good diversity performance. The presence of mutual coupling at the receive

Table 3.1 Mobile Communication Standards and Frequencies

Gen	*Standard*	*Band*	*Uplink (MHz)*	*Downlink (MHz)*	*Full Name*
2G	GSM	GSM850	824–849	869–894	Global System for Mobile communication
		GSM900	880–914	925–959	
		GSM1800	1,710–1,784	1,805–1,879	
		GSM1900	1,850–1,910	1,930–1,990	
3G	UMTS	UMTS Band I	1,920–1,980	2,110–2,170	Universal Mobile Teli-communication System
		UMTS Band II	1,850–1,910	1,930–1,990	
		UMTS Band V	824–849	869–894	
4G	LTE	LTE2300	2,305–2,400	2,350–2,360	Long Term Evaluation- Advanced
		LTE2500	2,500–2,690		
5G	Research is ongoing for 28-GHz and 60-GHz frequency bands.				

array antenna leads to additional correlation between spatial channels and results in a loss of MIMO system capacity [10].

Antenna modeling and simulation in handheld devices, using electromagnetic simulation tools, have improved significantly by allowing the simulation of the antenna behavior in complex environments surrounding the antenna such as the antenna along with the human head and hand and the presence of nearby components such as cameras, batteries, displays, and speakers. Some of important electromagnetic parameters to characterize antennas for mobile handheld devices are impedance mismatch, radiation efficiency, diversity gain, MIMO capacity, total radiated power (TRP), total isotropic sensitivity (TIS), SAR, average fading sensitivity (AFS), and data bit throughput (TPUT).

3.3 Mobile Phone Antennas

Smartphone antenna design is reasonably complicated because of different communication standards and regulatory requirements. In addition, there are multiple antennas on a smartphone as shown in Figure 3.1. Most of these antennas are PIFAs or a custom printed antenna. A PCB ground plane that supports the circuit board and touch screen also acts as a ground plane for all the antennas. The size of the ground plane affects antenna bandwidth, so, for better performance, the whole available size of the chassis should be used as a ground plane. PIFA bandwidth reduces with a small-size ground plane. Some important antennas used in a

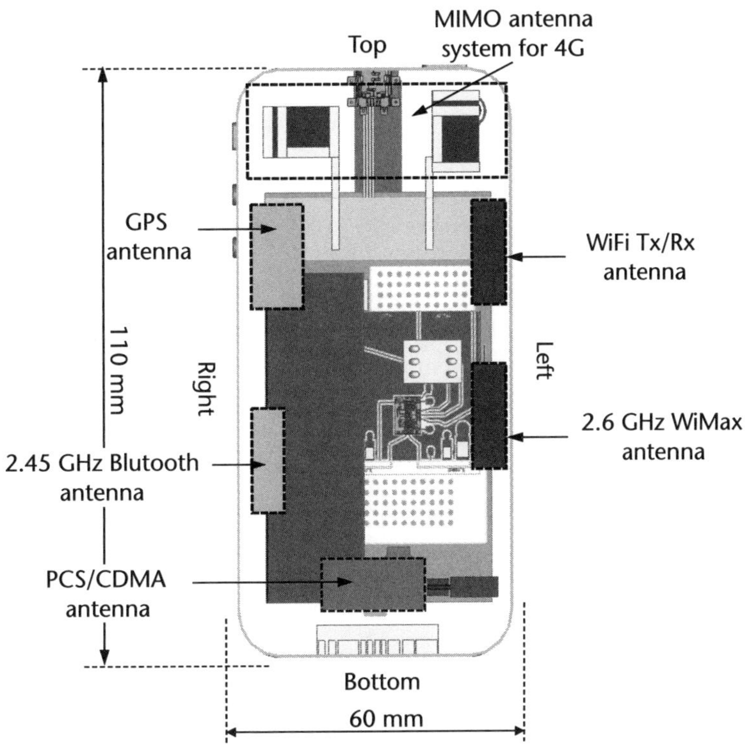

Figure 3.1 Various wireless communication antennas in a smartphone.

smartphone for various wireless standards such as GSM, LTE, WiMAX, WLAN, and Bluetooth are 2G/3G mobile antenna, 4G mobile antenna, diversity antenna, GPS antenna, Wi-Fi antenna, near-field communication (NFC) antenna, Bluetooth antenna, and 5G communication antenna.

3.4 Multiband Antenna Design for Mobile Phones

Mobile communication requires small, low-profile, and low-cost antennas. Before 1998, most of the mobile phones used the external antenna, which is simple and easy to reuse. The most popular design was the dual-band nonuniform helical antenna proposed by Egorov and Ying [12]. The Hagenuk phone, produced by a Danish company in 1996, was the first mainstream GSM phone with an internal antenna. Internal antennas are used in almost all modern smartphones. The internal antenna can increase the mechanical robustness of a mobile terminal; also, the internal antenna housing can be used as an acoustic cavity to improve the audio performance. The main types of internal antennas for mobile phone are the PIFA [13], the folded monopole antenna [14], a meander line antenna, and the loop antenna [15]. The PIFA is usually located at the top of the phone behind the printed wire board (PWB). However, the folded monopole antennas are usually used at the bottom of the handset to have less head loss and risk of high SAR. PIFA bandwidth depends on the antenna size and ground plane size. The bandwidth of the monopole is much larger than a PIFA. The monopole can be capacitive, coupled, or matched to the terminal housing to excite the chassis mode to gain some extra bandwidth when the chassis length is larger than a quarter-wavelength.

3.4.1 Feeding Structures

There are two types of feeding structures: the inductive feed and the capacitive feed. An inductive feed is a wire/pogo pin with a direct electrical connection between the PCB and the antenna. Antenna designers can fine-tune the antenna impedance matching by controlling the width and length of the feed line. A capacitive feed couples the energy to the main radiating element [16] and can also be self-resonant for matching purposes at a high-frequency band. It is either placed adjacent to the main antenna or directly underneath the antenna. Manufacturing capacitive feeds directly underneath the antenna is difficult because the feeding is highly dependent on the separation between the main antenna and the capacitive feed. This separation acts as a matching network with variable serial-capacitance loading. The antenna designs presented in this chapter used these feeding structures.

To understand the design process, design steps, optimization technique, and complexity associated with mobile antennas, four different kinds of antennas based on monopole, inverted-F, and PIFA are discussed in this chapter. These antenna designs are also applicable for any other mass wireless communication system. Practically, smartphone antenna shape and size are governed by the antenna body and available space, but the design principle and challenges are same presented in this chapter.

3.5 Dual-Band Monopole Antenna Design for WLAN

A compact planar monopole antenna with a dual band is discussed in this section. The antenna occupies an overall area of 18 × 12 mm^2. The antenna combines two folded strips and is fed by a CPW transmission line, which radiates at 2.4 and 5.2 GHz. In this antenna, by optimally selecting the antenna dimensions, dual-band resonant modes with a much wider impedance matching at the higher band can be produced. With the fast development of wireless communication systems, there is a high demand to increase the data rate of wireless communication systems [17]. Wireless local area network systems such as Bluetooth, Wi-Fi, and ZigBee always operate in the 2.4-GHz ISM band. The growing demand for WLAN technology has led to an interest in integrating the 2.4-GHz (IEEE802.11b/g) and 5.2-GHz (IEEE 802.11a) frequency bands into a single device that requires a dual-band antenna. The WLAN antenna not only has a wider frequency band, smaller size, and is easier to install, but it also has a high radiation efficiency [18]. Due to the space limitation, the WLAN antenna should be compact and should support multiband operations [19].

At present, extensive studies of dual-band microstrip antennas applied in WLAN have been performed, and various promising antennas that work in a dual-band have been put forward, such as PIFAs [20–22], planar monopole antennas [23], dipole antennas [24], dual-band monopole antennas with different lengths [25] or sleeves [26], bent folded monopoles [27], and slot antennas [28, 29]. These antennas are simple in structure and low in production cost, which is suitable for the use of WLAN devices. However, the designs of the antenna may increase cost or are complicated to be fabricated or the overall dimensions of the antenna are large in size. In the design of multiband antenna, one critical technique is to reduce the radiation caused by the currents distributed on the ground plane to improve the radiation patterns and impedance bandwidth [30]. Multiband antennas have also been greatly researched. In that, CPW-fed antennas show better antenna performance.

To cover all the three bands of WLAN (i.e., lower band of 2.4 GHz (802.11b/g standard) and upper bands of 5.15–5.35 GHz and 5.725–5.825 GHz (802.11a standard)), two monopoles are used: one working for the lower band and the other working for the higher band [31–33]. However, there is a limit in most of these antennas to obtain the wideband characteristic especially at the 5-GHz band for WLAN applications operating simultaneously at 5.2 and 5.8 GHz. Similarly, several dual-band slot antennas were proposed for wholly covering 2.4/5.2-GHz WLAN bands [34–42]. Although the reported antennas cover all 2.4/5-GHz WLAN bands, they are not very compact in size, so there is a demand for designing compact multiband antennas. Table 3.2 shows the comparison of the antenna size and the operating bands of the proposed antennas, with antennas reported in [30–42]. Antenna size variation is between 500 m^2 and 5,000 m^2.

3.5.1 Antenna Design

The geometry of the designed antenna is shown in Figure 3.2. The radiator and ground plane of the antenna are etched on a piece of PCB with an overall size of 18 × 12 × 1.6 mm^3. The antenna has two monopole arms resonating at 2.4 GHz

Table 3.2 Comparison of Designed Antenna with Other Compact Antennas

	Published Designs				
S. No.	*Reference*	*Design*	*Antenna Size (mm)*	*Total Area (mm²)*	*Frequency Bands Covered*
1	[30]		25 × 25	625	2.4/5.2/5.8 GHz
2	[31]		40 × 20	800	2.45/5.2/5.8 GHz
3	[32]		33 × 28	924	2.4/5.2/5.8 GHz
4	[33]		22 × 20	440	2.4/5.2 GHz
5	[34]		75 × 75	5,625	2.4/5.2/5.8 GHz
6	[35]		50 × 75	3,750	2.4/5.2 GHz
7	[36]		60 × 45	2,700	2.4/5.2/5.8 GHz
8	[37]		43 × 41	1,763	2.4/5.2/5.8 GHz
9	[38]		50 × 35	1,750	2.4/5.8 GHz
10	[39]		40 × 40	1,600	2.4/5.2/5.8 GHz
11	[40]		35 × 24	840	2.4/3.4/5.2/5.8 GHz
12	[41]		40 × 15	600	2.4/5.2 GHz
13	[42]		32 × 16	512	2.4/5.2/5.8 GHz
14	Designed antenna		18 × 12	216	2.4/5.2/5.8 GHz

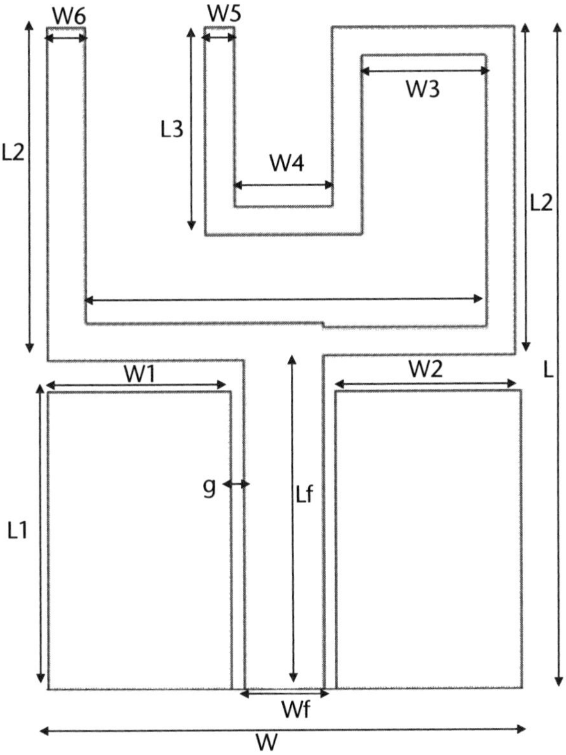

PARAMETERS	VALUE
L	12.0 mm
W	17.5 mm
Wf	2.00 mm
Lf	8.70 mm
g	0.30 mm
L1	7.90 mm
L2	8.67 mm
L3	5.49 mm
W1	4.70 mm
W2	4.70 mm
W3	3.13 mm
W4	2.45 mm
W5	0.75 mm
W6	1.00 mm

Figure 3.2 Geometry of the designed dual-band monopole antenna.

and 5.2 GHz. It consists of a coplanar ground plan, a radiation structure, and a feed line. A CPW transmission line of 50Ω, which consists of a signal strip width of 2 mm and a gap distance of 0.3 mm between the single strip and the coplanar ground plane, is used for feeding the antenna. In addition, a parametric study of the antenna might be referential for antenna engineers.

The design dimensions of the antenna are shown in Figure 3.2. It is printed on the FR4 antenna substrate with a relative dielectric constant of 4.4. The size of the substrate is W and L, and the thickness is 1.6 mm. The ground plane is printed on the same side of the radiating element, whose length and width are represented by L_1, W_1, and W_2, respectively.

The antenna is fed by a CPW line with a characteristic impedance of 50Ω, and the size of the microstrip line is $W_f \times L_f$. The radiating element of the antenna is composed of two rectangular strips. Each of the radiating element is folded and is fed with a CPW located at a distance offset from the middle of the loop. The lengths of the strips should be integer multiples of half the effective wavelength λ_{eff} at the desired frequencies to form resonant antennas. The electrical length of the whole monopole strip is equivalent to the quarter-wavelength of the lower-frequency band of 2.4 GHz, while the L-shape branch, which is a part of the folded loop, is mainly used to generate the higher-frequency band of 5.2 GHz, and the corresponding resonant quarter-wavelength is obtained by adjusting the feeding network location. Further, the antenna element has been purposely folded, not only to reduce the antenna physical dimensions but also to reduce the coupling effect between antenna elements. At the lower frequency, the quality factor Q for such a resonator is high due to the limited size of the radiation apertures at two open ends, and the high Q leads to limited operation bandwidth.

The size of the antenna is mainly determined by the size of the monopole governing the lower band of the WLAN application. The lower-frequency band resonator has a longer arm length compared to a higher-frequency band resonator. The impedance bandwidth and resonant frequencies obtained can be varied by variation in the dimensions of folded strips. The width of the feed line has been fixed at 2 mm to provide 50Ω impedance matching.

3.5.2 Mathematical Modeling

The antenna can be characterized by an equivalent circuit of several lumped elements. This approach is based on Foster canonical forms that describe the antenna frequency characteristics [43, 44]. To represent a distributed circuit by a lumped element model, its transfer function should be first determined and then the equivalent circuit network should be found giving the same response for a required excitation as considered a distributed structure. This lumped element model provides a compact description of the distributed circuit. To establish the network model of distributed circuit, the transfer function is computed using different approaches. One such approach to perform a synthesis of compact lumped element models for linear lossy reciprocal multiport is based on the Foster canonical forms. The Foster reactance theorem states that the reactance of a passive, lossless, two-terminal (one-port) network always strictly monotonically increases with frequency. The reactance of inductors and capacitors independently increase with frequency and, from that basis, a passive lossless network can be constructed using lumped inductor and capacitors. Ronald Martin Foster presented the proof of the theorem in 1924. Generally, antennas are linear, passive elements that their input impedances can be represented by Foster canonical forms.

Normally, antennas are linear, passive elements whose input impedances can be represented by Foster canonical forms, as shown in Figure 3.3. Dipole and monopole antennas are electric antennas; this Foster canonical form is suitable for modeling electric antennas, which behave as an open circuit at the DC input signal. The antenna input impedance $Z_{in}(\omega)$ is modeled by means of the equivalent network depicted in Figure 3.3:

$$Z_{in}(\omega) \cong j\omega L_0 + \frac{1}{j\omega C_0} + \sum_{n=1}^{N_{max}} \frac{R_n}{1 + jQ_n\left(\frac{\omega}{\omega_n} - \frac{\omega_n}{\omega}\right)} \tag{3.1}$$

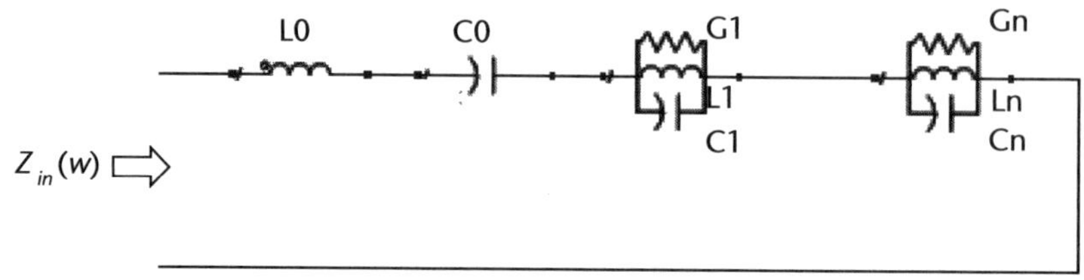

Figure 3.3 Foster canonical form of the equivalent circuit for the antenna.

where $Q_n = \omega_n R_n C_n$ and $\omega_n = (L_n C_n)^{-0.5}$. N_{max} is the number of modes needed to properly describe the frequency behavior of the antenna input impedance. Ω is the operating radial frequency, and Ω_n is the radial frequency of the nth resonant mode. C_0 is the quasi-static input capacitance, and L_0 is an inductance that considers the higher-order modes and feeding effects, while C_n, L_n, R_n, and Q_n are the capacitance, inductance, resistance, and quality factor, respectively, describing the lumped resonance processes that take place in the antenna structure [45].

The antenna equivalent circuit model with actual components is shown in Figure 3.4(a). The values of resistors, capacitors, and inductors are extracted from the Advanced Design System (ADS) schematic circuit simulation results, where these values are adjusted manually to obtain the proper response. C_0 and L_0 represent the capacitance and inductance of the monopole antenna when the antenna operates at a lower frequency and are properly chosen to resonate at the first resonance frequency of the antenna. Figure 3.4(b) shows the reflection coefficient curves of the considered antenna as obtained from an equivalent lumped element circuit model.

3.5.3 Antenna Design Parametric Study

During the antenna design phase, the antenna designers perform a series of parametric studies to achieve the desired antenna performance, particularly tuning the resonant frequencies and impedance levels. In this design, the critical antenna parameters are the resonating length of strips and the width of the strip. A complete

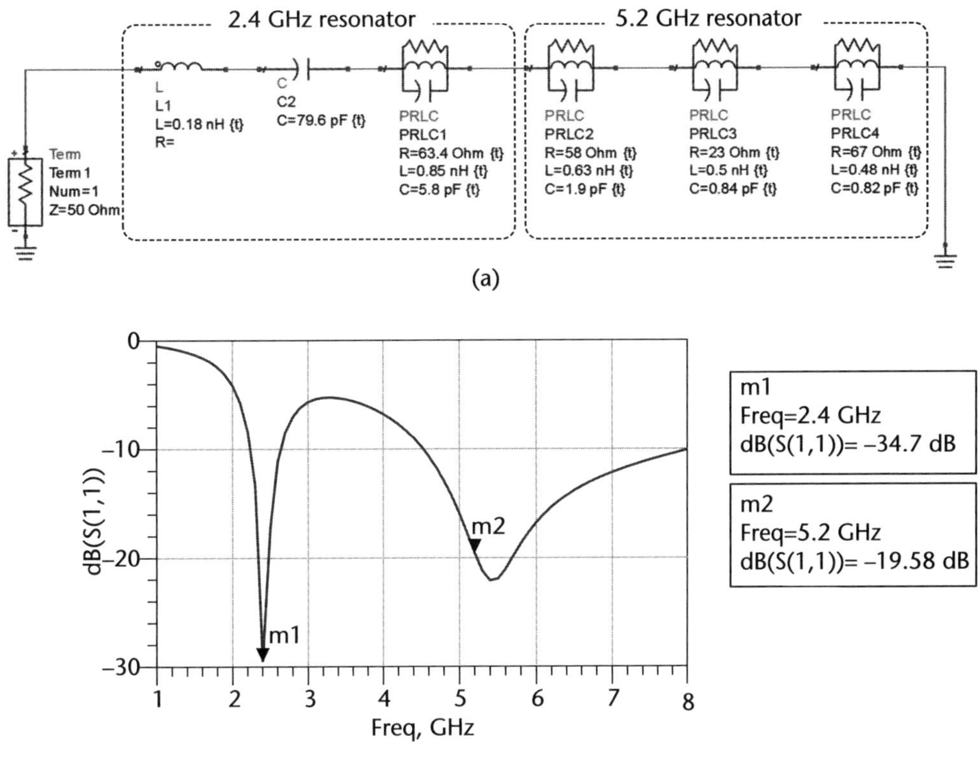

(b)

Figure 3.4 (a) Equivalent lumped-element circuit model of antenna and (b) reflection coefficient response for the antenna.

parametric analysis is performed for different parameter values by varying one parameter and keeping others constant.

3.5.3.1 Influence of the Variation of the Straight Strip Length L2

The effect of L_2 is investigated, keeping the other parameter constant. The simulation is carried out for L_2 values of 12 mm, 15 mm, 19 mm, and 22 mm. Figure 3.5 shows that the 5.2-GHz resonant frequency depends on the parameter L_2, and the second resonance frequency of 2.4 GHz is almost independent of the variation in L_2. The variation in the value of L_2 from 12 mm to 22 mm leads to shifting in first resonant frequency from 6.5 GHz to 4.5 GHz. The longer L_2 shifted the resonating frequency below 5.2 GHz, while the shorter L_2 shifted the frequency above 5.2 GHz.

3.5.3.2 Influence of the Variation of the Straight Strip Length L3 + W4 and CPW Feed Line Gap g

Figure 3.6 shows the antenna return loss (S_{11}) with a different value of length L_3 + W_4. The performance and antenna resonance frequency of low band change if the length (L_3+ W_4) is increased or decreased. However, the high band does not change much around 5.2 GHz. The 2.4-GHz frequency is mainly determined by this length (L_3 + W_4) of the strip.

The frequency shifts at the lower-frequency band as the length of this strip increases. A variation in the value of L_3 from 22 mm to 32 mm in step size of 3 mm leads to shifting in first resonant frequency from 3.2 GHz to 2.2 GHz. The antenna returns loss (S_{11}) with a different value of the gap between CPW feed line and coplanar grounds is shown in Figure 3.6. The variation of g is not shifting resonance frequency significantly, but the performance and antenna resonance frequency of the high band change if the g is increased or decreased. Optimum antenna perfor-

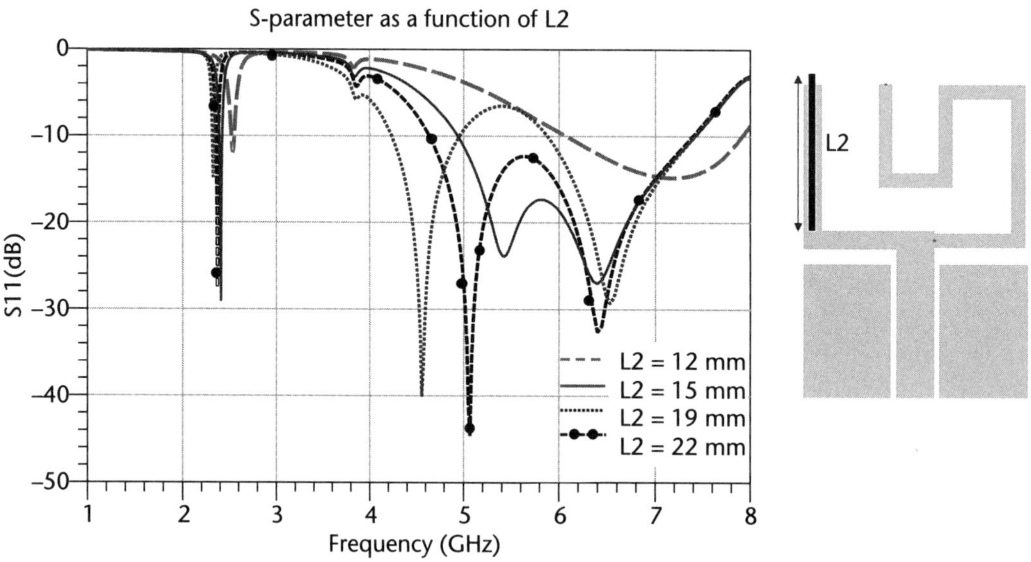

Figure 3.5 Return loss plot as a function of the length L2 variation.

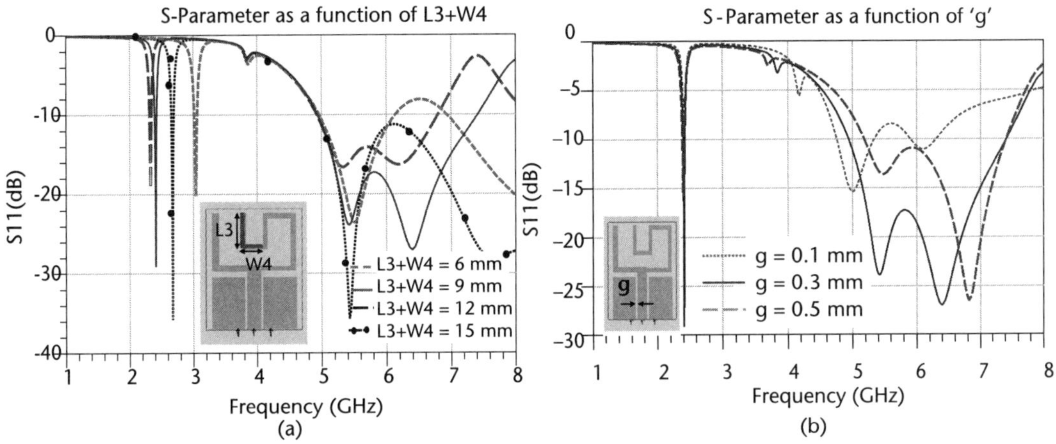

Figure 3.6 S-parameter result as a function of (a) length L3 + W4 and (b) gap g.

mance is achieved with $g = 0.3$ mm. At a higher frequency, there is more shift in resonant frequency as well as a change in the return loss value.

3.5.4 Simulated Result of the Optimized Design

After getting the optimized design parameter with parametric analysis, the antenna is simulated using an electromagnetic (EM) solver. The antenna has a bandwidth of 40 MHz at 2.4 GHz and 2,700 MHz at 5.2 GHz, and the maximum gain for a signal operating an element of the designed antenna is –0.16 dBi at 2.4 GHz and 2.1 dBi at 5.2 GHz.

3.5.4.1 S-Parameter Result of the Design

The simulated return loss of an optimized antenna shows that two operating bands are generated in 2.4 GHz and 5.2 GHz. The antenna elements resonated between 2.37–2.42 GHz and 4.9–7 GHz with a minimum $|S_{11}| < -10$ dB. The optimized antenna gives a bandwidth of about 40 MHz over a lower band of 2.4 GHz and about 2.7 GHz over an upper band of 5.2/5.8 GHz, which are very compatible with the 802.11n WLAN standard. VSWR is less than 2 obtained in the entire impedance bandwidth range. The value of VSWR at resonant 2.4 GHz is 1.09 and 1.05 at a resonance frequency of 5.2 GHz, which is well below the desired value of 2. A simulated input impedance curve as the function of frequency is shown in Figure 3.7. It is observed that at a resonant frequency of 2.4 GHz, the antenna provides an input impedance of value $(53.45 + j\,22.12)\Omega$, which is close to impedance provided by a feed line; similarly, at 5.2 GHz, this value is $(52.16 + j\,30)$. The simulated VSWR and input impedance variations indicate good impedance matching between the feed line and the antenna structure.

3.5.4.2 Electrical Field Plot

The electric field distributions with respect to the two frequencies of 2.4 GHz and 5.2 GHz are shown in Figure 3.8(a, b), respectively, At the dark region, the current

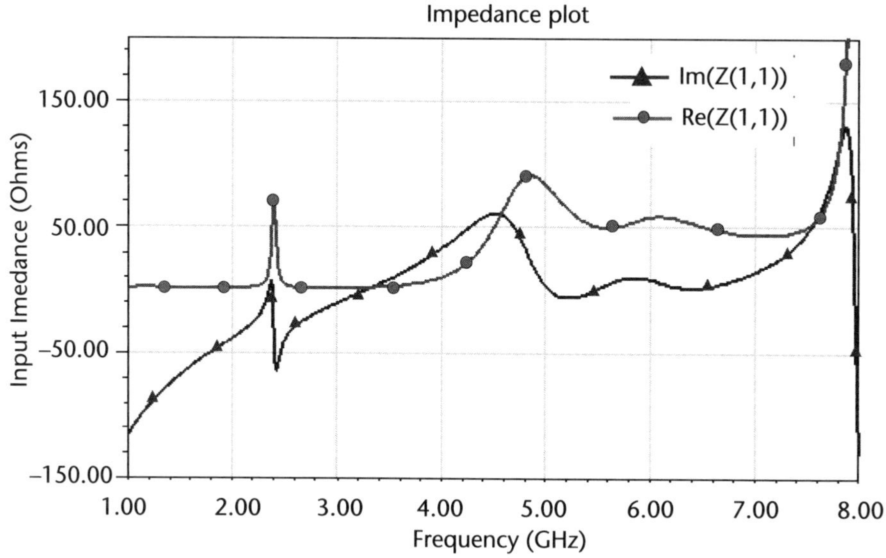

Figure 3.7 Simulated input impedance plot of a dual-band monopole antenna.

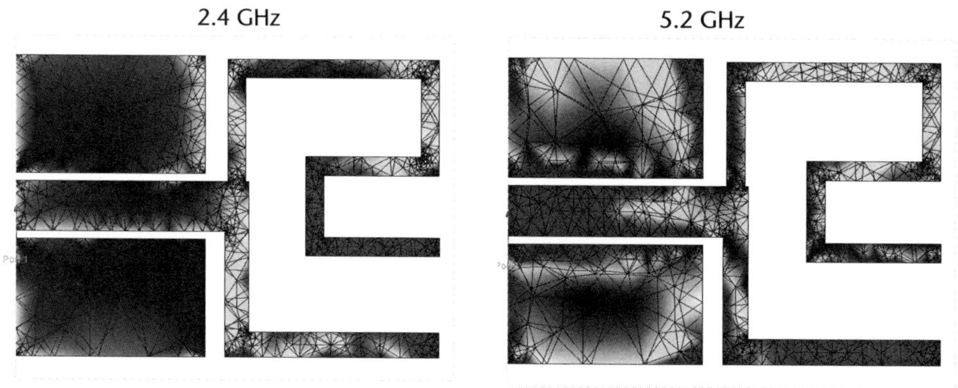

Figure 3.8 E-field distribution on antenna at (a) 2.4 GHz and (b) 5.2 GHz.

has an increasingly larger value. Conversely, moving toward the lighter region, the current has increasingly smaller value.

It is observed that, at the resonance frequency, the monopole works as an antenna generating a strong surface current, mainly on the edge of the strip. In addition, a significant current distribution is also observed along the two straight strip edges. The CPW feed line and the ground plane coupling effect may usually affect the impedance matching condition. It means that the effect of the feed line gap *g* provides impedance matching for the designed antenna. Surface currents appeared on the inner side of CPW feed line and the ground plane outer side edges. The currents on the two side edges of the ground plane have the same phase with those along the CPW feed line, thus enhancing the copolarization radiation in the far-field zone.

3.5.4.3 Antenna Radiation Pattern

The simulation result of an antenna radiation pattern is shown in Figure 3.9. In addition, simulated radiation patterns in two principal planes (*xz*-plane and *yz*-plane) at resonant frequencies of 2.4 and 5.2 GHz are presented in Figure 3.10. The radiation pattern is nearly omnidirectional for both bands that are shown in Figure 3.10 of the 2-D cut of the radiation pattern. It demonstrates that the antenna gain variation within 2.4 GHz and 5.2 GHz is minimal. The antenna radiation pattern shows a broadside pattern over the entire bandwidth with a maximum gain of about 2.1 dBi. The gain of the antenna is from –2.1 to 2.1 dBi. The peak antenna gain of 2.21 dBi is experienced at a resonant frequency of 5.2 GHz and 5.8 GHz.

3.6 Inverted-F Bluetooth Antenna

Bluetooth is a wireless technology standard for exchanging data over short distances from fixed and mobile devices, using short-wavelength radio transmissions

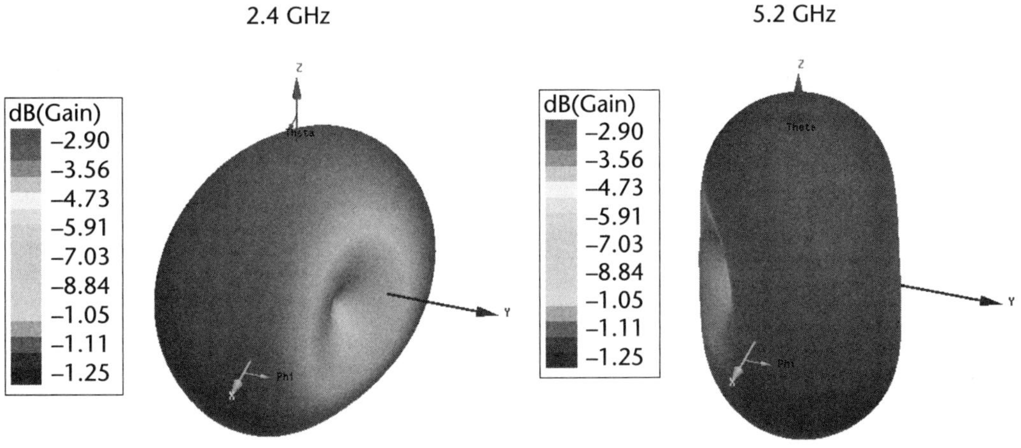

Figure 3.9 Antenna radiation pattern at 2.4 GHz and 5.2 GHz.

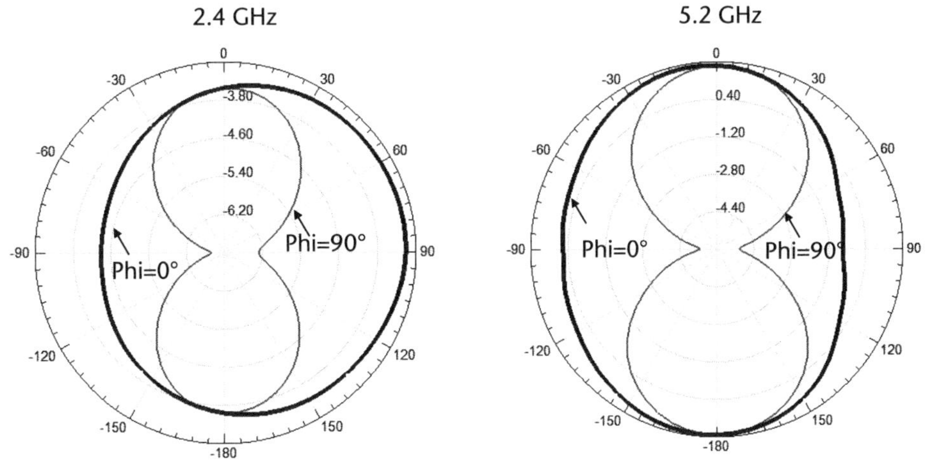

Figure 3.10 Azimuth and elevation radiation patterns at 2.4 GHz and 5.2 GHz.

in the ISM band (IEEE 802.11b and IEEE 802.11b/g standard). It creates a personal area network (PAN)with high levels of security that can connect several devices. Bluetooth technology was created by telecom vendor Ericsson in 1994 and it was originally conceived as a wireless alternative to RS-232 data cables. In a mobile phone, Bluetooth technology permits a hands-free headset for incoming voice calls, automatic synchronization of a cellphone with other devices, and the ability to print and fax. The Bluetooth antenna is designed for low-range operation because many applications only require a few meters of range.

Many compact antennas for Bluetooth have been reported in the literature [46, 47]. Some of these antennas are narrowband, while some have relatively larger dimensions. There are very compact and multiband antennas designed to be integrated into mobile terminals. These antennas have printed geometries and multiband operation is achieved by incorporating parasitic elements, slots, capacitive loading, and shorting pins. The Bluetooth antenna shown in Figure 3.11 is an IFA, employs a shunt stub for impedance matching. Four parameter variables (L, H, S, and D) control the behavior of the structure. The feed is placed from the upper arm of the IFA to the ground plane. The upper arm of the IFA has a length that is roughly a quarter-wavelength. The antenna is mounted on the plastic body of a mobile device that also acts as a substrate with dielectric constant of 2.8, and the plastic body thickness is 1.2 mm.

The resonating length of the printed antenna is 24 mm at 2.45 GHz. The antenna resonating length depends on the thickness of the substrate and the dielectric constant. Copper clearance is required around the radiating element, which is fed from a point along it. The feed position is used to control the input impedance into the antenna. The ground plane required on one side of the antenna is mobile PCB ground plane. The upper arm is shorted to the ground plane to the left of the feed. The microstrip feed line impedance is 50Ω. A PIFA can be also conformed as per a mobile housing structure as shown in Figure 3.11(b). To get the antenna performance, EM simulation is carried out using an FDTD solver. In general, the FDTD grid size should be dense where most of the electric fields are confined (edges of antenna strip). The generated electric effect is measured on other parts of the mobile phone, which is very important to meet compliance.

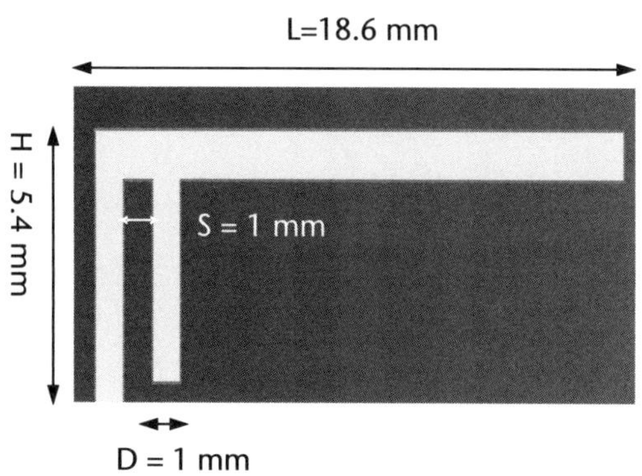

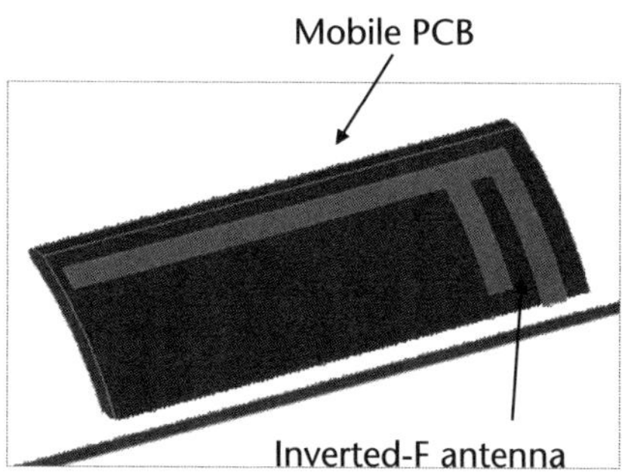

Figure 3.11 Inverted-F Bluetooth (a) antenna planar structure and (b) antenna wrapped on mobile body.

The simulated return loss S_{11} is displayed in Figure 3.12. The antenna resonating frequency is 2.45 GHz with a bandwidth of more than 100 MHz and efficiency greater than 50%. Antenna-resonating frequency depends upon the antenna-resonating length L. The variation of S-parameter data with L (1-mm variation) is shown in Figure 3.12.

The antenna is resonating at 2.4 GHz for $L = 18.6$ mm. The polarization of this antenna is vertical, and the radiation pattern is roughly donut-shaped, with the axis of the donut in the vertical direction as shown in Figure 3.13. Antenna gain is 0.27 dBi. The 2-D far-field cut plot (phi = 0°). After optimizing a stand-alone Bluetooth antenna design, it is important to simulate the antenna in the presence of other components and reoptimize the design to get the desired antenna performance.

3.7 LTE Band PIFA

As discussed in the previous section, the most popular small built-in antenna in mobile phones is the PIFA-type antenna. The PIFA radiator is mounted just above the ground layer; this ground layer acts as a metallic plate between the antenna radiation plane and the human brain when being used; thus, the impact of radiation on the human brain is reduced significantly. Typically, these antennas have good specific absorption rate (SAR) properties. The PIFA has the advantages of small size and omnidirectional pattern, and antenna production is unanimously good. Because of the shorting pin at the end, the PIFA is resonant at a quarter-wavelength; in that way, it reduces the required space needed on the phone.

To further reduce the length of a PIFA, it is common to use capacitive loading in PIFA. In this technique, a capacitance is added to the antenna between the feed point and the open edge, as shown in Figure 3.14. The impedance of the PIFA can be controlled via the distance of the feed to the short pin. Impedance decreases if the feed point is moved close to the shorting pin. The impedance can

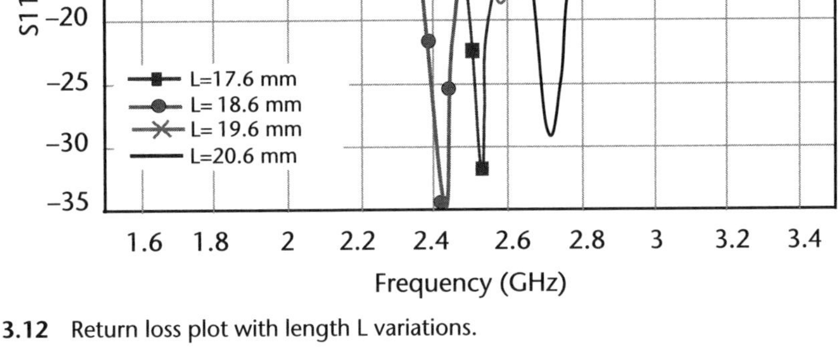

Figure 3.12 Return loss plot with length L variations.

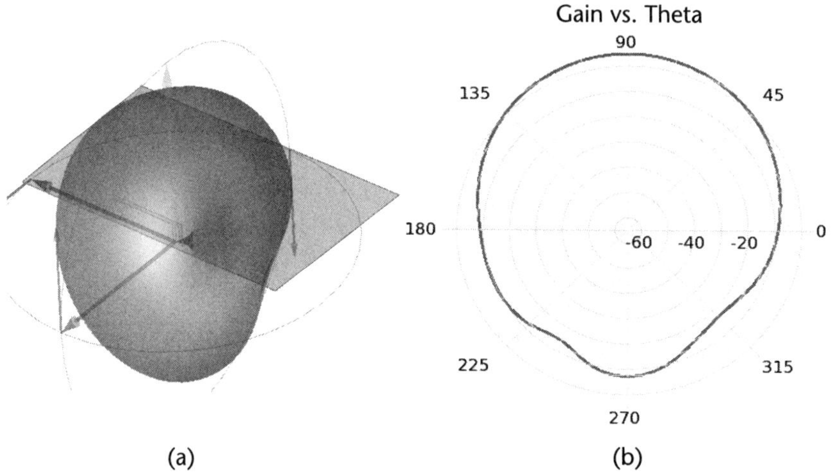

Figure 3.13 Antenna radiation pattern: (a) 3-D plot and (b) phi = 00 cut pattern plot.

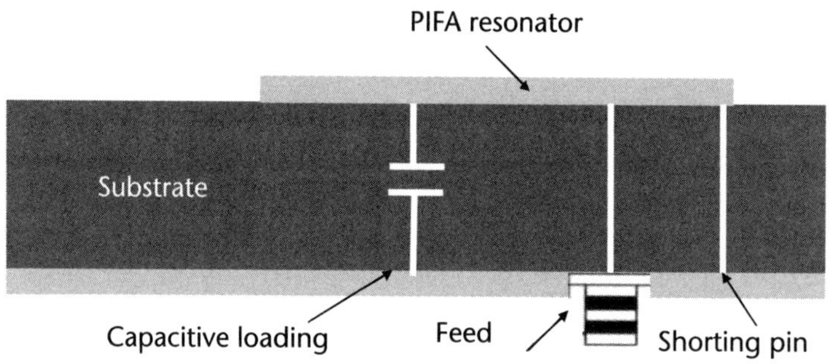

Figure 3.14 PIFA configuration.

be increased by moving it farther from the short edge. A PIFA may be in different shape. Minimal antenna size and tight integration with other device components are major challenges in LTE antennas. Due to the volume limitations, a traditional PIFA antenna solution does not meet the requirements of the LTE antenna RF and brings greater challenges to antenna designers. LTE antenna design, shown in Figure 3.15, is a customized PIFA to meet LTE band requirements. The 4G/LTE provides a comprehensive and secure all-IP based mobile broadband solution to all kinds of mobile communication devices. WiMAX and first-release LTE have been the dominant technologies in this market. These technologies have been designed for channel bandwidths of 5 to 20 MHz (optionally up to 40 MHz) and offer peak data rates of ~100 MB/sec for high mobility devices and 1 GB/s for low-mobility devices. The resonant wavelength of a PIFA can be calculated using (3.2):

$$L1 + L2 - W = \lambda_0 \tag{3.2}$$

where $\lambda_0 = {}^{c_0}\!\big/_{(f_0\sqrt{\epsilon_r})}$.

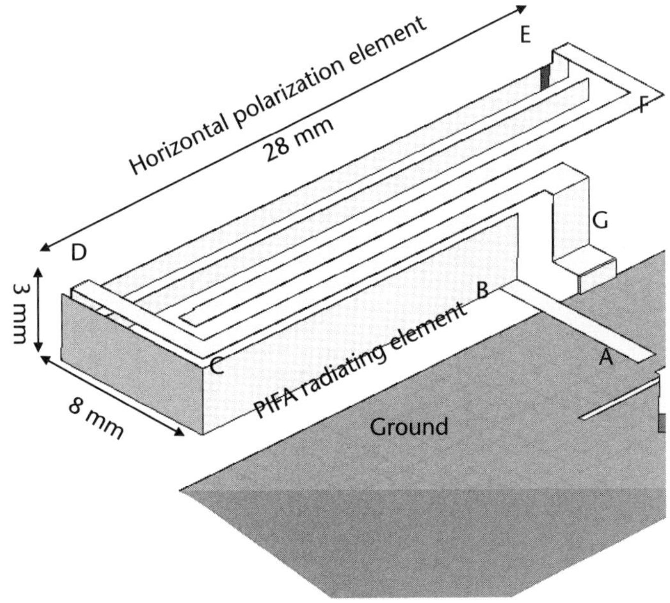

Figure 3.15 Dual-band LTE PIFA design dimensions.

A PIN diode at location E is used to control the resonating frequency of the antenna. This antenna resonates at the 800-MHz and 1,400-MHz band. The antenna is designed on FR4 with $\varepsilon_r = 4.4$ and a thickness of 0.8 mm. Isolation between both resonating bands is very high because only one band radiates at a time.

Figure 3.16 shows the simulated return loss of the LTE antenna. The antenna is tuned over 800 MHz and 1,400 MHz. The obtained simulated bands are 760–850 MHz and 1,380–1,470 MHz with a reflection coefficient better than –10 dB, covering the 800-MHz and 1,400-MHz bands. The 3-D simulated radiation patterns are at frequencies 800 MHz and 1,400 MHz, shown in Figure 3.17. Radiation patterns are nearly omnidirectional, suitable for mobile phone communication. As current nulls increase at higher frequencies, the radiation patterns show some distortions. Antenna-simulated gain varies from 3.7 dBi to 2.13 dBi.

3.8 MIMO Antenna Design for LTE Mobile Terminals

To increase the data transmission rate, MIMO technology has become an important feature in LTE communication systems. MIMO technology can linearly increase channel capacity with an increase in the number of antennas, without needing additional bandwidth or power [48]. A multiband internal antenna, including a diversity antenna in the DCS/PCS/UMTS bands and a MIMO antenna in the LTE17/LTE4 bands, is shown in Figure 3.18.

The designed antenna consists of a monopole to achieve the multiband operation. It also has a compact dimension by using 3-D, front-to-front arrangement. This antenna can cover two bands while fitting within tight volumes, providing high diversity performance and achieving reasonable mean effective gain (MEG) across both the bands of operation. The MIMO antenna, as shown in Figure 3.18,

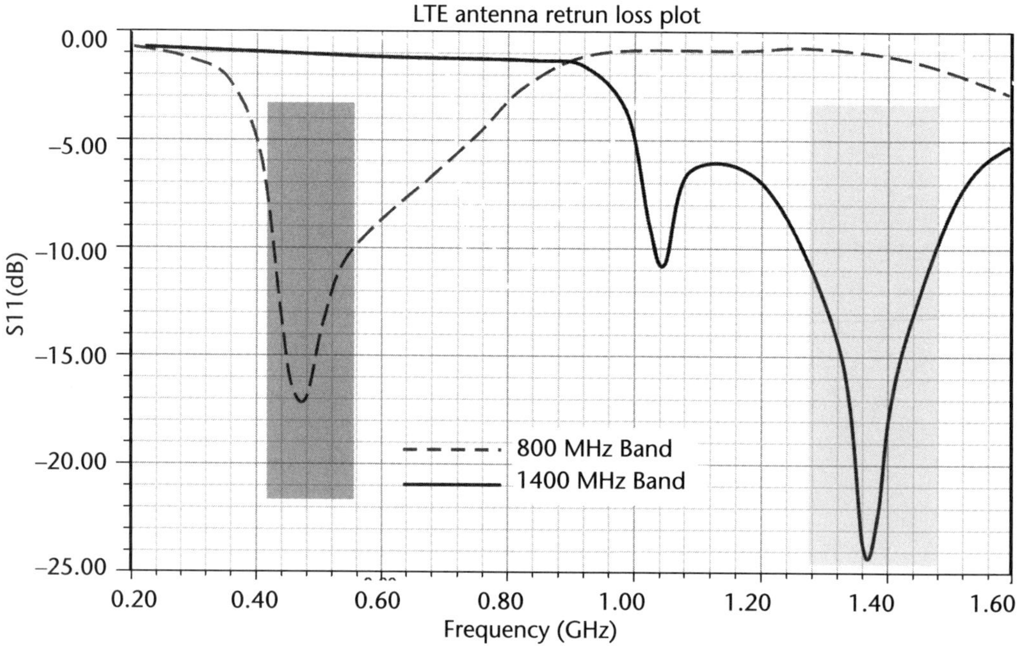

Figure 3.16 Antenna return loss plot of dual-band LTE PIFA.

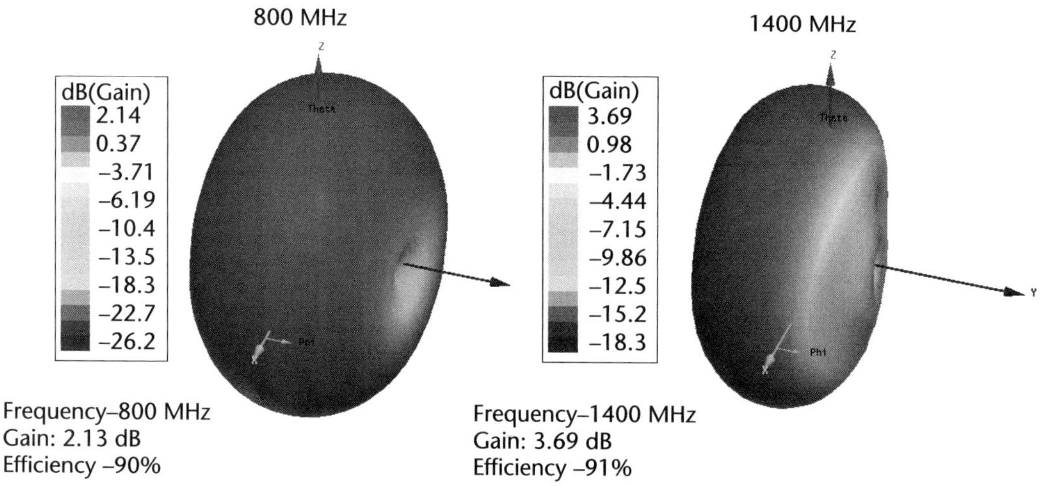

Figure 3.17 Radiation pattern plot at 800 MHz and 1,400 MHz.

has a metallic strip of a total length of 98 mm. The 2-mm-wide radiating antenna element is placed on an FR-4 substrate block of dimensions 16 × 10 × 5 mm^3 and a relative dielectric constant of 4.4. The radiating bottom strip has a 2-mm feed. This multiband monopole antenna covers 890-MHz and 1,800-MHz bands. As the fractional bandwidth of an antenna is approximately the inverse of its Q [49] and bending a metal antenna to permit it to efficiently occupy a small volume can reduce its Q, a folded 3-D monopole enjoys simultaneous benefits: enhanced bandwidth and smaller required volume. Smartphone ground plane impact antenna

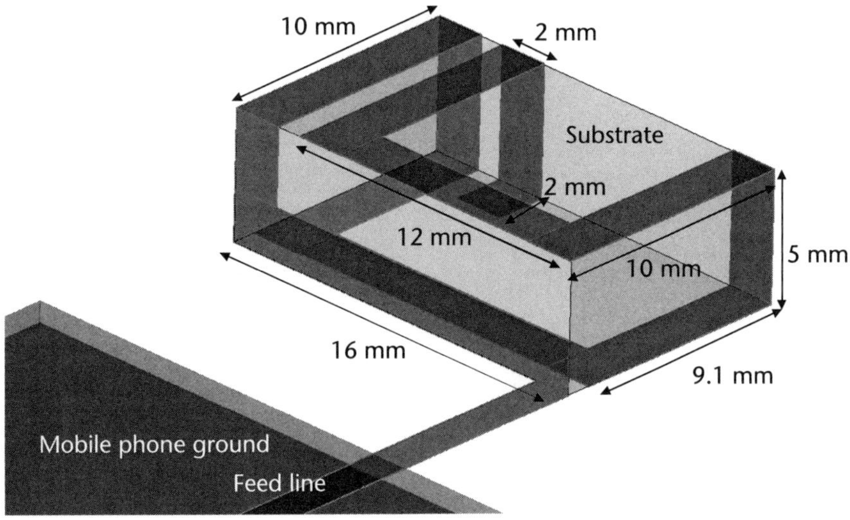

Figure 3.18 The configuration and design dimensions of the MIMO antenna.

performance must be included in the simulation and testing. The ground plane size of the antenna is 60 mm × 110 mm. For this antenna, the conductive strips are primary radiation elements and constitute two monopole radiators, identified here as monopole 1 and monopole 2. Monopole 1 is the strip length just after the antenna feed and before the strip on the substrate while the monopole 2 is the total strip length. The initial geometric parameters are optimized by using the FDTD numerical technique. The total electrical length of the monopole is set to a quarter-wavelength at the fundamental mode of 890 MHz. Figure 3.19 shows simulated reflection coefficients of the antenna. These curves indicate that the resonant frequencies of the antenna are in GSM900 and GSM1800 frequency bands.

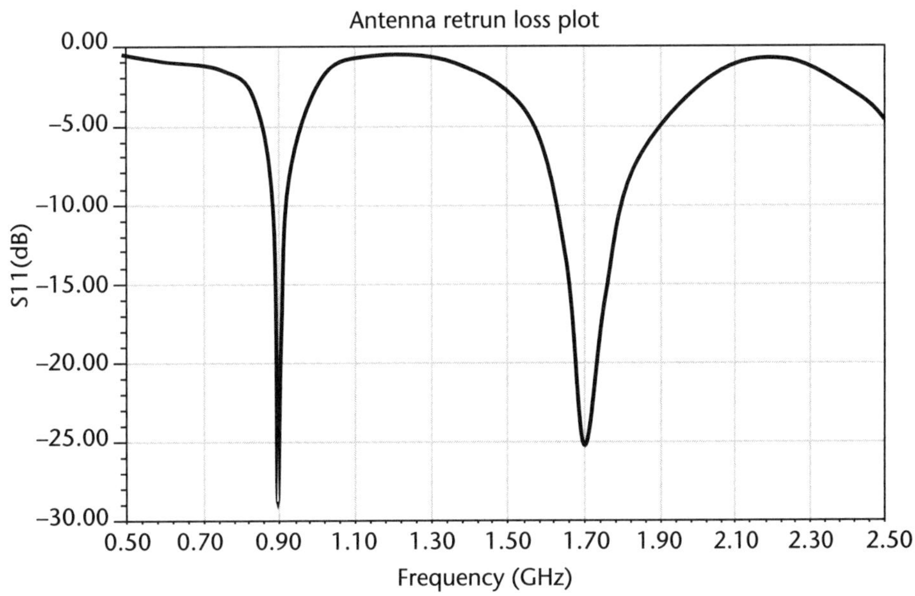

Figure 3.19 Return loss plot of a single-element LTE MIMO antenna.

To enhance the understanding of the multiband operation, the current distributions on the monopole are shown in Figure 3.20. The radiation 3-D gain patterns were simulated using FDTD. Figure 3.20 illustrates results for a 80-MHz operating frequency. After optimization of the antenna element, it is simulated along with other components of the mobile phone.

The mobile device has a maximum form factor of 115 × 65 × 12 mm^3. Major electromechanical components include the other antennas radiation elements, a frame for supporting the antenna, a PCB, a plastic, mobile screen, and shielding metals. The design is slightly further fine-tuned to get optimum performance in the presence of other components of the mobile device.

3.9 Mobile Antenna Performance with Components and Standard Anthropomorphic Model Head

Antenna performance changes in the presence of other phone components and the housing structure in a smartphone. After the completion of standalone antenna design and optimization, the next step is to integrate the antenna model in a mobile phone along with other components and perform EM simulation. Integrated simulation enables the evaluation of the coupling effects of neighboring objects such as the camera, shields, traces, battery, and flash capacitors, as well as the influence of dielectric materials such as the housing and display screen. Antenna return loss, gain, and radiation pattern are affected by other electronics components of smartphone. Figure 3.21 shows electric field variation in the phone and antenna radiation pattern with other components of the mobile phone and human head model.

The smartphone performance, having a different antenna type, is analyzed in the presence of a human body, with emphasis on the head and hand to know the impact of the standard anthropomorphic model (SAM) on antenna performance. In accordance with IEEE standards, such as 1528, the SAM is used as the head model. The presence of the SAM is simulated by a head and shoulder only, since the handset antenna near-field radiation is mainly interacted with the presence of head, shoulder, and hand-hold rather than the other parts of the user's body. The frequency-dependent dielectric properties of the SAM tissue and liquid can be

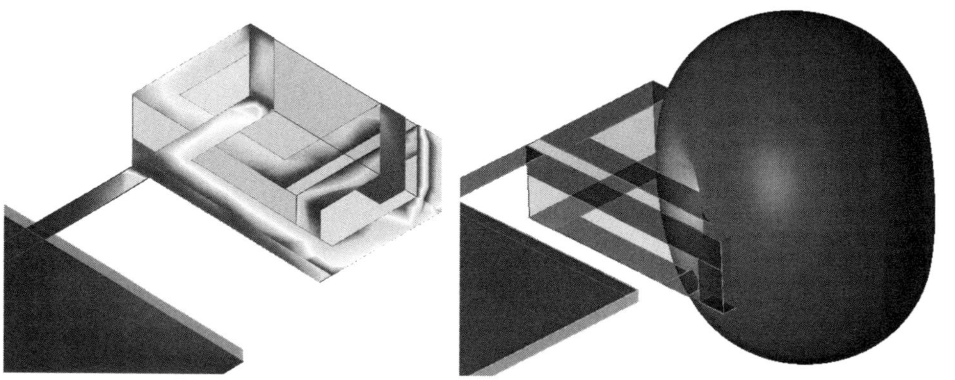

Figure 3.20 Electric current distribution on antenna and radiation pattern plot at 890 MHz.

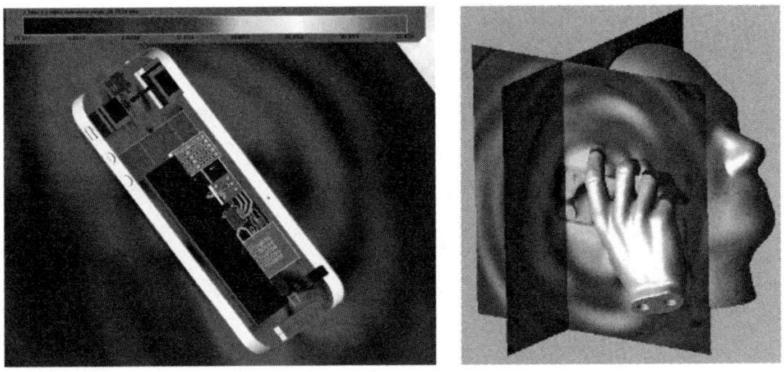

Figure 3.21 E-Field plot with phone components and SAM head model.

modeled as a dispersive material in the simulation tool. Such simulation can give important insight into how the SAM phantom or the homogeneous body models affect the performance of the mobile phone. The antenna return loss plot in Figure 3.22 indicates that deterioration is the return loss in the presence of a phone and the SAM. As can be seen, the antenna resonant frequency shifts at a higher side at 3 GHz when it is simulated with the mobile model. Reoptimization and tuning are needed to bring down Bluetooth antenna resonance at 2.45 GHz. LTE MIMO antenna behavior is different, in this case, resonance frequency shifts at a lower-frequency band as indicated in Figure 3.22(b). Antenna gain, radiation pattern, and electric field distribution are also affected by the hand and head of the human body. These simulations are a must for mobile phone antenna designers to get antenna performance in a real working environment.

In conjunction with radiation efficiency and return loss, the total radiated power (TRP) and the total isotropic sensitivity (TIS) are relevant for the complete phone. The TIS is a measure of the handset receiving performance and TIS measurements verify how the receiver antenna design performs. The objective is to measure the sensitivity of the mobile phone due to noise impairment and to find the lowest power level on which the phone can still maintain a reliable connection. A TRP measurement verifies the transmitter or transmitter antenna design. The TRP is measured by measuring the antenna radiated power at many points around the mobile phone and summing the values to the total power radiated.

The TIS and TRP together determine the effectiveness of the mobile phone and the maximum range at which the phone can operate from the base station with some given level of performance. Figure 3.23 shows the TIS values for the Bluetooth and LTE MIMO antenna in different scenarios. The lowest TIS value is measured for a mobile phone with the SAM with a tilted hand. In practice, the measurement process is often iterative: first, design all antennas separately and then simulate the antenna in the presence of other radiating elements and reoptimize the design and location of the antenna.

3.9.1 Cross-Talk of Bluetooth Antenna with MIMO Antenna

In mobile phones, several antennas are placed in a small volume. The main problem of colocating many antennas in a confined volume is mutual coupling or cross-talk between all antennas, inevitably leading to an increased design complexity.

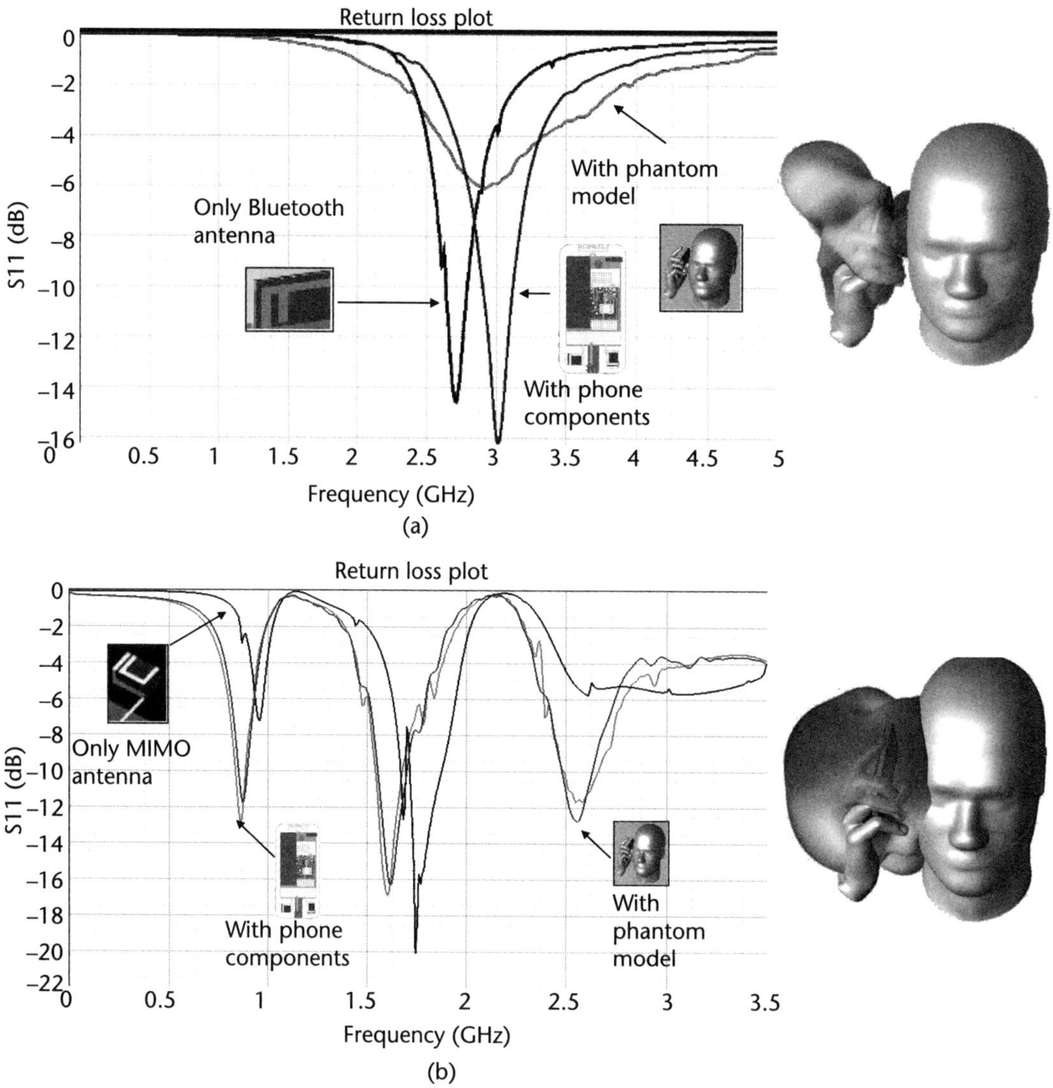

Figure 3.22 Simulated result of (a) Bluetooth and (b) LTE MIMO antenna along with phone components and the SAM head.

Designing and placing multiple antennas in a mobile phone are often extremely challenging; taking the mutual effects of other antenna elements into account can be difficult and time-consuming. Figure 3.24 shows mutual coupling better than –20 dB between Bluetooth and MIMO antenna. The mutual coupling can be reduced by the appropriate placing of the antenna, shielding, and feeding techniques.

In this chapter, basic mobile communication standards and antennas used in smartphones were discussed. Antenna design challenges and the types of antennas used in a smartphone were presented. Four different types of antennas used in modern smartphone devices are presented. Two antenna systems, one for Bluetooth and other for MIMO application, were modeled and simulated using a phone model and a SAM head and hand model. Important antenna specifications, design parameters, and design flow were discussed through these two antennas. Designing these stand-alone antennas is not sufficient to use these antennas commercially

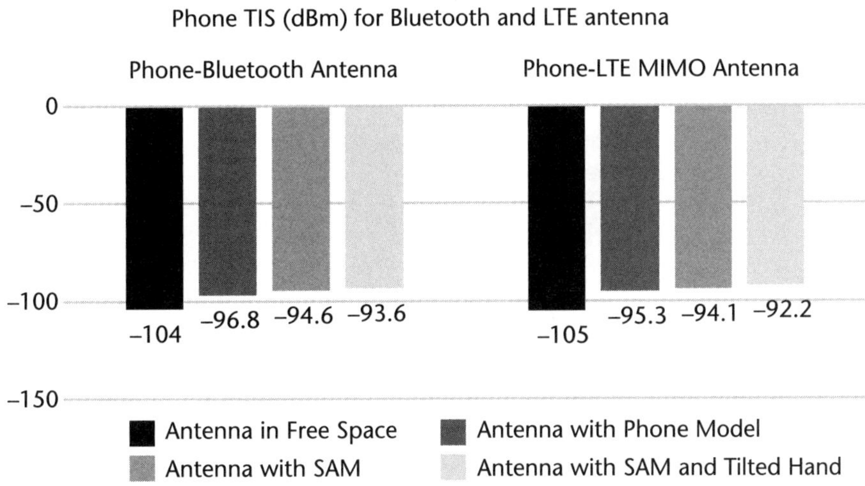

Figure 3.23 Phone TIS for Bluetooth and MIMO antenna.

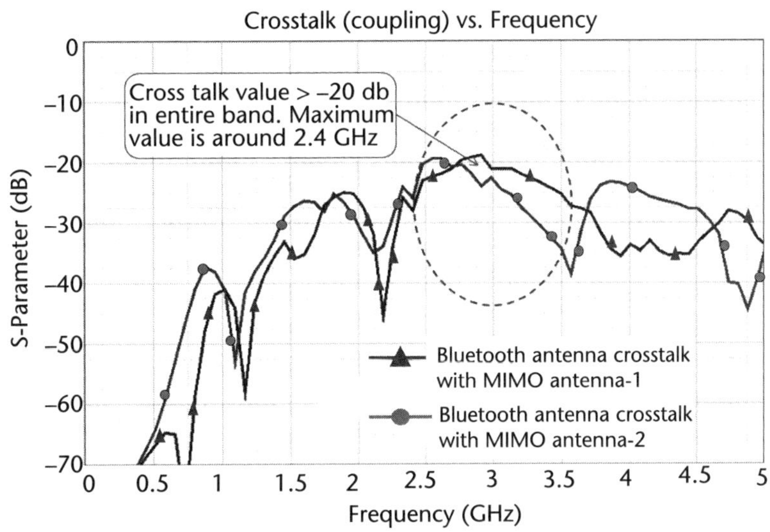

Figure 3.24 Cross-talk (coupling) between Bluetooth and MIMO antenna.

in mobile devices without meeting various mobile compliances. In Chapter 4, all these compliance and qualification processes are discussed, along with practically designed antennas.

References

[1] Ngan, T. L., et al., "The Enhanced Workflow and Efficiency of the Wireless Local Area Network (WLAN)-Based Direct Digital Radiography (DDR) Portable Radiography," *J. Digit. Imaging*, Vol. 28, 2015, pp. 302–308.

[2] Wang, Z., et al., "Embroidered Multiband Body-Worn Antenna for GSM/PCS/WLAN Communications," *IEEE Transactions on Antennas and Propagation*, Vol. 62, 2014, pp. 3321–3329.

[3] Ryu, H. K., E. Kim, and J. -M. Woo, "Miniaturization of Printed Inverted-F Antenna Using Chip Coupler for Bluetooth Applications," *Electronics Letters*, Vol. 46, No. 13, 2010.

[4] Jiang, B. T., and J. F. Mao, "Design of a PIFA-IFA-Monopole in Dual-SIM Mobile Phone for GSM/DCS/Bluetooth Operations," *ICMMT 2008 Proceedings*, 2008.

[5] Lee, E., P. Hall, and P. Gardner, "Novel Compact Wideband or Multi-Band Planar Monopole Antenna," *Proceedings of the IEEE Antennas and Propagation Society International Symposium*, Salt Lake City, UT, July 16–21, 2000, pp. 624–627.

[6] Khodabakhshi, H., and A. Cheldavi, "Irradiation of a Six-Layered Spherical Model of Human Head in the Near Field of a Half-Wave Dipole Antenna," *IEEE Transactions on Microwave Theory and Techniques*, Vol. 58, 2010, pp. 680–690.

[7] Chen, H. D., and H. T. Chen, "A CPW-Fed Dual-Frequency Monopole Antenna," *IEEE Transactions on Antennas and Propagation,* Vol. 52, No. 4, 2004, pp. 978–982.

[8] Ali, M., et al., "Dual-Frequency Strip-Sleeve Monopole for Laptop Computers," *IEEE Transactions on Antennas and Propagation*, Vol. 47, No. 2, 1999, pp. 317–323.

[9] Lu, J., and B. Huang, "Planar Multi-Band Monopole Antenna with L-Shaped Parasitic Strip for WiMAX Application," *Electronics Letters*, Vol. 46, No. 10, 2010, pp. 671–672.

[10] Fletcher, P. N., M. Dean, and A. R. Nix, "Mutual Coupling in Multi-Element Array Antennas and Its Influence on MIMO Channel Capacity," *Electronics Letters*, Vol. 39, No. 4, February 20, 2003, pp. 342–344.

[11] Teng, X. Y., et al., "A Compact CPW-Fed Omni-Directional Monopole Antenna for WLAN and RFID Applications," *Progress in Electromagnetics Research Letters*, Vol. 32, 2012, pp. 91–99.

[12] Egorov, I., and Z. Ying, "A Non-Uniform Helical Antenna for Dual-Band Cellular Phones," *IEEE Antennas and Propagation Society International Symposium*, Salt Lake City, UT, 2000, pp. 652–655.

[13] Nie, Y., and L. Song, "A Compact Triband Fractal PIFA Antenna for Mobile Handset Applications," *2013 6th International Congress on Image and Signal Processing (CISP),* Hangzhou, 2013, pp. 1468–1472.

[14] Teng, P. L., and K. L. Wong, "Planar Monopole Folded into a Compact Structure for Very-Low-Profile Multi-Band Mobile Phone Antenna," *Microw. Opt. Technol. Lett.*, Vol. 33, No. 3, 2002.

[15] Hayasida, S., H. Morishita, and K. Fujimoto, "A Wideband Folded Loop Antenna for Handset," *IEICE Trans.*, Vol. J86B, No. 9, 2003, pp. 1806–1815.

[16] Chen, H., Y. L. Ban, and L. J. Ying, "Compact PIFA Using Capacitive Coupled-Fed for LTE/GSM/UMTS WWAN Operation in the Mobile Application," *IEEE Proceedings of the Cross Strait Quad-Regional Radio Science and Wireless Technology Conference (CSQRWC '11)*, Harbin, China, July 2011.

[17] Ngan, T. L., et al., "The Enhanced Workflow and Efficiency of the Wireless Local Area Network (WLAN)-Based Direct Digital Radiography (DDR) Portable Radiography," *J. Digit. Imaging*, Vol. 28, 2015, pp. 302–308.

[18] van Rooyen, M., J. W. Odendaal, and J. Joubert, "High-Gain Directional Antenna for WLAN and WiMAX Applications," *IEEE Antennas and Wireless Propagation Letters*, Vol. 16, 2017, pp. 286–289.

[19] Mehdipour, A., T. Denidni, and A. R. Sebak, "Multi-Band Miniaturized Antenna Loaded by ZOR and CSRR Metamaterial Structures with Monopolar Radiation Pattern," *IEEE Transactions on Antennas and Propagation*, Vol. 62, 2014, pp. 555–562.

[20] Islam, M. S., et al., "Converting a Wireless Biotelemetry System to an Implantable System Through Antenna Redesign," *IEEE Transactions on Microwave Theory and Techniques*, Vol. 62, 2014, pp. 1890–1897.

[21] Choudhury, S., A. Mohan, and D. Guha, "Wideband Quasi-Omnidirectional Planar Inverted F-Antenna for Compact Wireless Systems," *IEEE Antennas and Wireless Propagation Letters*, Vol. 17, No. 7, July 2018, pp. 1305–1308.

[22] Makris, I., et al., "Modified Designs for UWB Planar Monopole Antennas," *2007 Loughborough Antennas and Propagation Conference*, Loughborough, 2007, pp. 249–252.

[23] Ghanbari, L., et al., "Planar Low VSWR Monopole Antenna for UWB and LTE Communication," *2016 16th Mediterranean Microwave Symposium (MMS)*, Abu Dhabi, 2016, pp. 1–4.

[24] Feng, B., et al., "A Dual-Wideband and High Gain Magneto-Electric Dipole Antenna and Its 3D MIMO System with Metasurface for 5G/WiMAX/WLAN/X-Band Applications," *IEEE Access*, Vol. 6, 2018, pp. 33387–33398.

[25] Xiaomu, H., S. Yan, and G. A. E. Vandenbosch, "Wearable Button Antenna for Dual-Band WLAN Applications with Combined On and Off-Body Radiation Patterns," *IEEE Transactions on Antennas and Propagation*, Vol. 65, No. 3, March 2017, pp. 1384–1387.

[26] Rahim, T., F. A. Abbasi, and J. Xu, "Dual Band Sleeve Dipole Antenna for WLAN Applications," *2016 10th European Conference on Antennas and Propagation (EuCAP)*, Davos, 2016, pp. 1–3.

[27] Lin, Y. -D., and P. -L. Chi, "Tapered Bent Folded Monopole for Dual-Band Wireless Local Area Network (WLAN) Systems," *IEEE Antennas and Wireless Propagation Letters*, Vol. 4, 2005, pp. 355–357.

[28] Li, S., Y. Mao, and W. Yu, "A Novel Dual Notched UWB Antenna for Wireless Body Area Network," *2016 IEEE/ACES International Conference on Wireless Information Technology and Systems (ICWITS) and Applied Computational Electromagnetics (ACES)*, Honolulu, HI, 2016.

[29] Wen, R., "Compact Planar Triple-Band Monopole Antennas Based on a Single-Loop Resonator," *Electron. Lett.*, Vol. 49, No. 15, July 2013, pp. 916–918.

[30] Chen, Z. N., T. S. P. See, and X. Qing, "Small Printed Ultrawideband Antenna with Reduced Ground Plane Effect," *IEEE Transactions on Antennas and Propagation*, Vol. 55, No. 2, February 2007, pp. 383–388.

[31] Chang, T. N., and J. J. Jiang, "Meandered T-Shaped Monopole Antenna," *IEEE Transactions on Antennas and Propagation*, Vol. 57, No. 12, December 2009, pp. 3976–3978.

[32] Chu, Q. X., and L. H. Ye, "Design of Compact Dual-Wideband Antenna with Assembled Monopoles," *IEEE Transactions on Antennas and Propagation*, Vol. 58, No. 12, December 2010, pp. 4063–4066.

[33] Yeh, S. H., and K. L. Wong, "Dual-Band F-Shaped Monopole Antenna for 2.4/5.2 GHz WLAN Application," *IEEE Antenna Propagation Society International Symposium*, Vol. 4, June 2002, pp. 72–75.

[34] Hsiao, H. M., et al., "Novel Dual-Broadband Rectangular-Slot Antenna for 2.4/5-GHz Wireless Communication," *Microw. Opt. Technol. Lett.*, Vol. 46, 2005, pp. 197–201.

[35] Kuo, Y. -L., and K. -L. Wong, "Printed Double-T Monopole Antenna for 2.4/5.2 GHz Dual-Band WLAN Operations," *IEEE Transactions on Antennas and Propagation*, Vol. 51, No. 9, September 2003, pp. 2187–2192.

[36] Tsai, L. C., "A Triple-Band Bow-Tie-Shaped CPW-Fed Slot Antenna for WLAN Applications," *Progress in Electromagnetics Research*, Vol. 47, 2014, pp. 167–171.

[37] Liu, W. C., "Broadband Dual-Frequency Cross-Shaped Slot CPW-Fed Monopole Antenna for WLAN Operation," *Microw. Opt. Technol. Lett.*, Vol. 46, 2005, pp. 353–355.

[38] Panda, J. R., and R. S. Kshetrimayum, "A Printed 2.4 GHz/5.8 GHz Dual-Band Monopole Antenna with a Protruding Stub in the Ground Plane for WLAN and RFID Applications," *Progress in Electromagnetics Research C*, Vol. 117, 2011, pp. 425–434.

[39] Ren, W., "Compact Dual-Band Slot Antenna for 2.4/5 GHz WLAN Applications," *Progress in Electromagnetics Research B*, Vol. 8, 2008, pp. 319–327.

[40] Song, Y., et al., "Multiband CPW-Fed Triangle-Shaped Monopole Antenna for Wireless Applications," *Progress in Electromagnetics Research*, Vol. 70, 2007, pp. 329–336.

[41] Chang, M., and W. Weng, "A Dual-Band Printed Dipole Slot Antenna for 2.4/5.2 GHz WLAN Applications," *2011 IEEE International Symposium on Antennas and Propagation (APSURSI)*, Spokane, WA, 2011, pp. 274–277.

[42] Zhao, G., et al., "Compact Ring Monopole Antenna with Double Meander Lines for 2.4/5 GHz Dual-Band Operation," *Progress in Electromagnetics Research*, Vol. 72, 2007, pp. 187–194.

[43] Van Valkenburg, M. E., *Introduction to Modern Network Synthesis*, New York: Wiley, 1960.

[44] Wang, S. B. T., A. M. Niknejad, and R. W. Brodersen, "Circuit Modeling Methodology for UWB Omnidirectional Small Antennas," *IEEE Journal on Selected Areas in Communications*, Vol. 24, No. 4, April 2006, pp. 871–877.

[45] Caratelli, D., et al., "Circuit Model and Near-Field Behavior of a Novel Patch Antenna for WLAN Applications," *Microwave and Optical Technology Letters*, Vol. 49, No. 1, January 2007, pp. 97–100.

[46] Jung, M., Y. Kim, and B. Lee, "Dual Frequency Meandered PIFA for Bluetooth and WLAN Applications," *Proceedings of the IEEE International Antennas and Propagation Symposium,* Vol. 2, June 2003, pp. 958–961.

[47] Naik, P. R., B. A. Singh, and S. Thakur, "Dual Band Compact Printed Monopole Antenna for Bluetooth (2.54GHz) and WLAN (5.2GHz) Applications," *2016 International Conference on Global Trends in Signal Processing, Information Computing and Communication,* Jalgaon, 2016, pp. 320–323.

[48] Wallace, J., et al., "Experimental Characterization of the MIMO Wireless Channel: Data Acquisition and Analysis," *IEEE Transactions on Wireless Communications*, Vol. 2, No. 2, March 2003, pp. 335–343.

[49] Vainikainen, P., et al., "Resonator-Based Analysis of the Combination of Mobile Handset Antenna and Chassis," *IEEE Transactions on Antennas and Propagation*, Vol. 50, No. 10, October 2002, pp. 1433–1444.

CHAPTER 4

Smartphone Antenna Design Compliances and Measurement

4.1 Introduction

The desire for internet connection anytime and anywhere has spurred cellular carriers to shift their services from voice-only networks to more data-centric services. Novel antenna technologies such as MIMO have increased data rates while improving coverage. Smartphone designers are struggling every day with various compliance-related problems that fail critical products to pass compliance certification and cause a big loss of profit. When selecting a radio technology, product designers should consider not only the performance requirements, but also the regulatory constraints. Compliance testing is testing to determine whether a product or system meets some specified standard that has been developed for efficiency or interoperability [1–10]. Standards organizations such as the American National Standards Institute (ANSI), the U.S. Federal Communications Commission (FCC), and the International Electrotechnical Commission (IEC) have certain criteria that a product must meet before compliance is recognized. A product cannot be sold unless it is known to meet those requirements specified in the standards. The first step towards regulatory compliance is to thoroughly investigate the applicable electromagnetic compatibility (EMC) specifications for the target geographic market.

Nowadays new advances in EDA technologies are providing accurate predictive simulations in a reasonable amount of time for compliance measurements [11]. With careful engineering utilizing cutting-edge, full-wave field-solvers based on MoM, FDTD, and FEM numerical techniques along with a hardware boost using heterogeneous massive CPU/GPU parallel processing (CUDA) technology, the designer can measure all compliance-related parameters such as radiations, SAR, HAC, and EMC. These simulators work as a virtual compliance test laboratory, providing the engineers with an early-stage compliance testing to find the optimum trade-off between performance and cost.

A smartphone has more advanced computing capability, connectivity and user applications than a feature phone. The smartphone combines the functions of a basic mobile phone, GPS navigation, media players, digital cameras, pocket video cameras, high-resolution touchscreens, and web browsers. Most of the operators already switched to LTE network coverage and plans and because of that many smartphone manufacturers are switching to LTE-based phones. LTE is a radio platform technology that allows operators to achieve even higher peak throughputs than HSPA+ in higher spectrum bandwidth. The prime objective of LTE is to provide high data speed. A typical smartphone along with important electronics component is represented in Figure 4.1. Smartphone antennas design challenges, simulation flow, diversity technique, and various compliance measured in mobile devices are discussed in this chapter with practical mobile phone designs. Even all these challenges are going to increase several times when the 5G phone becomes operational in the near future.

4.2 Design Challenges in Smartphone Systems

Just designing various chips and antennas on the smartphone alone is not enough. These components, such as antennas, influence overall circuit performance. Increased digital to RF interference is required to isolate digital and RF circuits from each other and mitigate digital to RF noise interference. Even with shielding, problems can arise if the RF circuit is not properly terminated to the ground layer of the PCB. Antenna resonances and couplings are influenced by the limited size of the mobile ground plane because of the surface current flows on the ground plane. For all compliance measurement and simulation, a single EM simulation technology may not solve all problems. Both planar MoM and 3-D full-wave field-solver FDTD and FEM, along with a hardware boost using heterogeneous massive central processing unit and graphics processing unit (CPU/GPU) parallel processing (CUDA) technology, make it possible to use a simulated system in less time in the

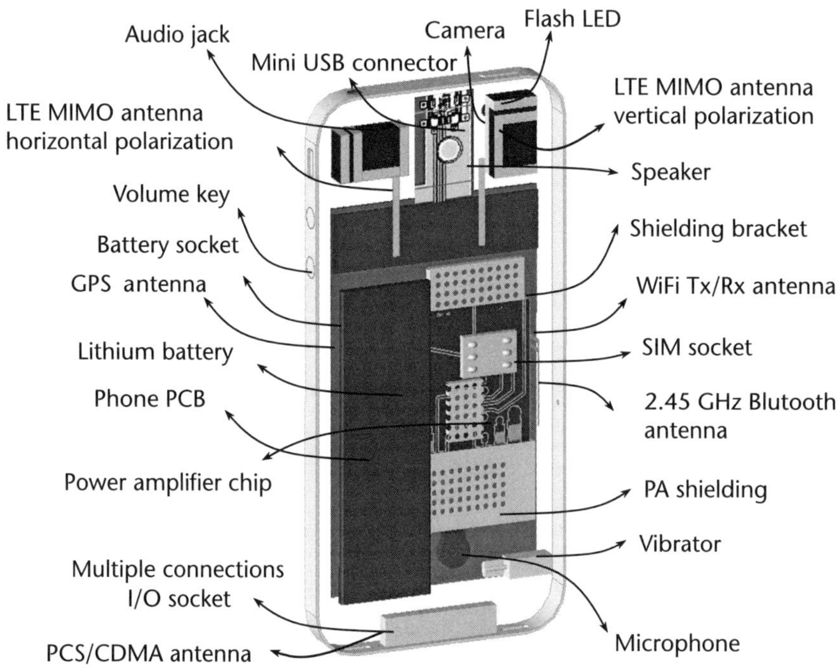

Figure 4.1 Various wireless antennas and components in smartphone.

presence of all mobile compoents. Challenging requirements such as CPU performance, wireless connectivity, memory capacity, bandwidth, display, and multimedia performance are driving new technologies in smartphone design and making a system complex. Radiation from antennas affects the performance of other electronic components in the phone as well as human tissues.

4.3 Smartphone Antenna Simulation Workflow

Antenna simulation workflow in a mobile phone is shown in Figure 4.2. The design and simulation of WLAN, GSM, Bluetooth, and MIMO antennas are discussed in Chapter 3. Mobile antennas, along with other active components of antennas such as power amplifiers, low noise amplifiers (LNAs), shields, and phone body, are simulated in a real environment. In a smartphone simulation model, the near-field environments are created using the housing and enclosed circuitry of a production-style smartphone. The major components that provide the near-field EM interaction with antennas include a liquid crystal display (LCD), a battery, a dielectric frame, phone RF board and chips, shields, and an exterior plastic shell. All individual designed components should work properly along with other devices in the phone; this is ensured with system-level simulation. Poor antenna matching can manifest itself in a high level of spurious emissions. Compliance simulation and measurement are a basic requirement to qualify a mobile device for commercial usage.

4.4 Smart Antenna System Using the MIMO and Diversity Technique

In a communication system, fading is variation of the attenuation of a signal with various variables as shown in Figure 4.3. In the case of long-distance

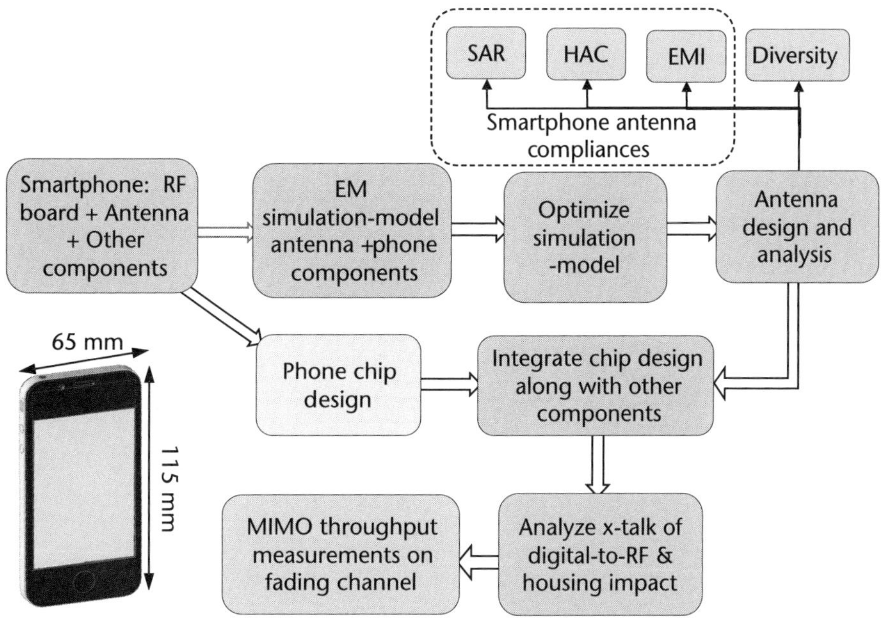

Figure 4.2 Smartphone antenna design and simulation workflow.

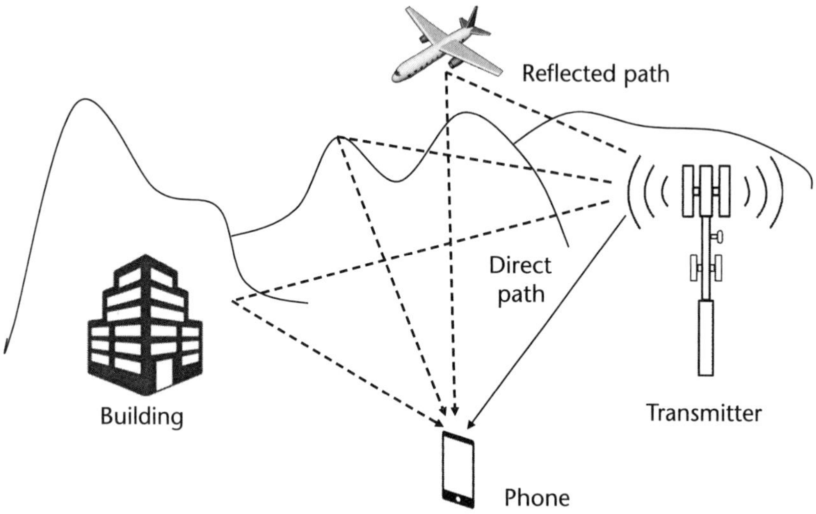

Figure 4.3 Communication signal fading environment.

communication, signal fading is a common feature for a single-input single-output (SISO) antenna system. At times, the received signal drops below the mean signal level, which ultimately leads to communication being lost. To receive the required signal strength, the power level needs to be increased, which further invites interference with other channels. To increase the power level, some active devices are used, which also introduces some nonlinear behavior into the channels. The MIMO antenna system (sometimes called smart antenna technology) using spatial diversity reception or transmission increases the probability of reception, spectral efficiency, and reliability as compared to single-antenna communication systems. Deploying multiple antennas at both transmitter and receiver ends achieves a higher data rate without increasing the total transmission power or bandwidth in a multipath wireless channel [12–16].

This antenna system is used in a wireless communication system, especially when the signal propagates through some corrupted environments. For mobile communication, the diversity scheme has already been implemented to mitigate the fading effects of multipath scenario. In a multipath rich wireless channel, deploying multiple antennas at both the transmitter and receiver achieves high data rate without increasing the total transmission power or bandwidth. In a MIMO system, the most important parameter is the increase in channel capacity because of the use of multiple transmitting and receiving antennas [17]. This antenna system is widely applicable to increase the range and reliability of Wi-Fi LAN, Bluetooth, personal digital assistants (PDAs), digital communication system (DCS), WLAN, and UMTs. It also plays a significant role in the implementation of next generation wireless equipment.

4.4.1 MIMO Antenna Configuration

Communication systems are broadly categorized into four fundamental ways in which multiple transmit and receive antennas can be used to communicate through the radio channel. Figure 4.4 shows different, general redundancy configurations:

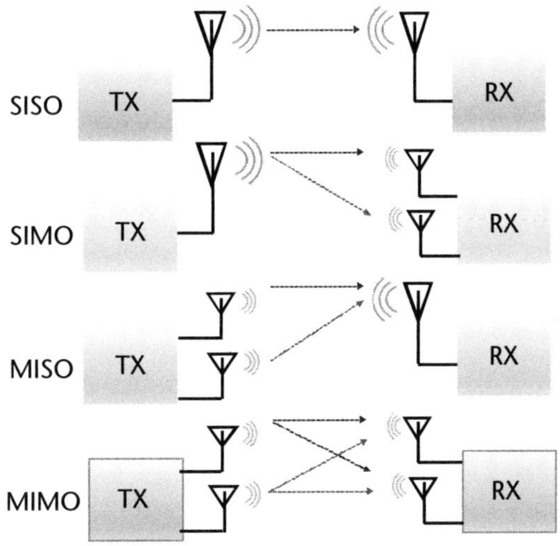

Figure 4.4 Redundancy configurations using multiple antennas.

SISO, single-input multiple-output (SIMO), multiple-input single-output (MISO), and MIMO.

SISO systems do not provide any type of robustness or capacity improvement and represent a basic wireless communication link. SIMO systems are configurations that provide receive-side diversity and provide additional robustness, but no capacity improvement. MISO provides transmitter diversity by coupling the channel at different points in space such that the links will not have the same fading characteristics and the spatial sum of the signals will be dominated by the stronger of the two signals. This configuration is better known as transmit diversity. In MIMO systems, multiple antennas are used both at transmitter and receiver ends to make use of reflected signals to provide gains in channel robustness and throughput. This provides for both redundancy and channel capacity improvement [18–22]. See Figure 4.4.

4.4.2 Antenna Diversity

Apart from the antenna configurations, there are two types of MIMO with respect to how data is transmitted across the given channel. The existence of multiple antennas in a system means the existence of different propagation paths. To improve the reliability of the system, the same data is sent across the different propagation (spatial) paths. This is called spatial diversity or simply diversity. To improve the data rate of the system, different portions of the data may be placed on different propagation paths and this is called spatial multiplexing. These two systems are listed here:

1. *Implementation using diversity techniques:* This system provides diversity gain to improve the reliability.
2. *Implementation using spatial-multiplexing techniques:* This system provides degrees of freedom or multiplexing gain to improve the data rate of the system.

In diversity techniques, the same information is sent across independent fading channels to combat fading. When multiple replicas of the same data are sent across independently fading channels, the amount of fade suffered by each copy of the data will be different. This guarantees that at least one of the data copies will suffer less fading compared to rest of the copies. Thus, the chance of properly receiving the transmitted data increases, which results in improving the reliability of the entire system as well as reducing the cochannel interference significantly. This technique is referred as spatial diversity in the communication system. To combat the fading effect, an array of antennas is used.

Diversity techniques are an effective means that enhances the immunity of the communication system to the multipath fading. A diversity antenna system provides many receiving branches or ports from which the diversified signals are derived and fed to a receiver. The receiver then combines the incoming signals from all the branches to produce a combined signal with improved quality in terms of signal strength or signal-to-noise ratio (SNR). The performance of a diversity antenna system primarily depends on the branch correlation and signal level difference between branches. There are four different diversity techniques as listed here. Spatial and polarization diversity techniques are realized through antenna systems.

- *Time diversity:* Signals transmitted at different times will get uncorrelated due to the time dependence of the multipath environment.
- *Frequency diversity:* Signals transmitted at different frequencies will get uncorrelated due to the frequency dependence of the multipath environment.
- *Spatial diversity:* Signals received at different locations can get uncorrelated due to the small-scale fading.
- *Polarization diversity:* Signals received with different polarizations can get uncorrelated due to different fading behavior for horizontal/vertical polarization.

4.4.3 MIMO Antenna Design Challenges

Despite many advantages, the MIMO antenna system suffers from a serious problem caused by strong mutual coupling between antenna elements, which causes signal interference between them and a higher value of mutual coupling lowers antenna efficiency, so that, in general, 0.5λ or more distance between antennas is needed to avoid the degradation of antenna performance. As is common in cellular phone handsets, higher levels of mutual coupling, which can easily occur in a handset, these systems can significantly degrade antenna efficiencies and produce an undesirable signal correlation between multiple antenna outputs. High signal correlation is associated with the loss of spectral efficiency and can degrade system throughput and battery efficiency. To achieve the desired low signal correlation, these communication systems require high diversity [13] and reduced coupling, which is also identified as a high port-to-port isolation.

4.4.4 Antenna Diversity Simulation and Measurement

The issue of signal fading in a multipath environment is a major problem. Using antenna diversity techniques can overcome the problem. There are two different approaches to obtain diversity gain: one is based on antenna power measurement at the receiver side, which is termed as the power-level approach, and the other is based on the analysis from both the embedded element efficiency and correlation coefficient, which is termed as the correlation approach. The second approach is described in this section to get diversity gain.

Antenna diversity always requires two far-zone patterns and two antenna systems as illustrated in Figure 4.5. For each far zone, only one antenna is fed, and the second antenna is left open. These two antennas are completely characterized by their far-zone patterns. Antenna far-zone patterns are valid for both receiving and transmitting. FDTD simulation technique is used to obtain the radiation field function and the S-parameters for calculating the correlation coefficient and embedded element efficiency. The antenna design presented in Chapter 3, Section 3.8, is used as an antenna element for a mobile terminal, which contains two PIFAs on a mobile PCB. It is discussed that the diversity antenna system can provide a good diversity performance by properly placing the two antennas through theoretical analysis and EM simulation.

4.4.5 Antenna Diversity Modeling

Using EDA tools, a complex wireless environment can never model in a deterministic way. It is even difficult to know beforehand in which environment a mobile phone antenna will be used. The complete environment (transmitter + obstacles) is described as a probability density function (PDF) that provides direction information with the highest change to receive a signal. Indoor and outdoor environments are modeled as uniform PDF and Gaussian PDF or user-defined PDF, respectively. PDF is used to model communication channels and the environment as shown in Figure 4.6. This function "weights" different directions in the far-zone sphere such that certain far-zone direction is considered more than others in the mean effective

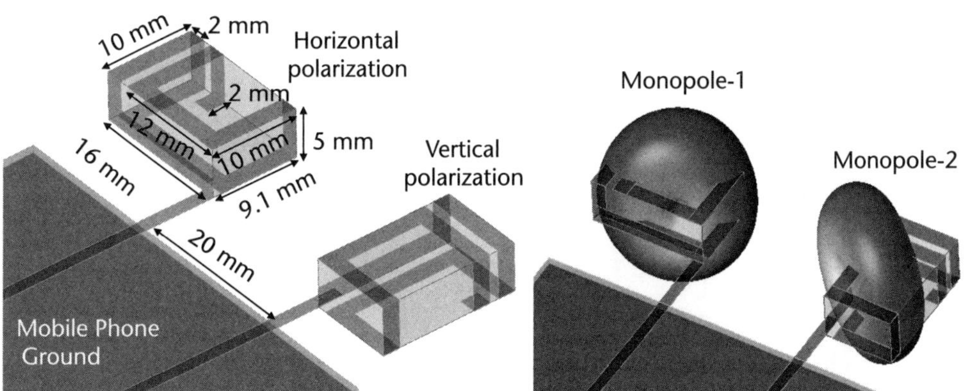

Figure 4.5 Horizontal and vertical polarization antenna elements and radiation pattern for antenna diversity system.

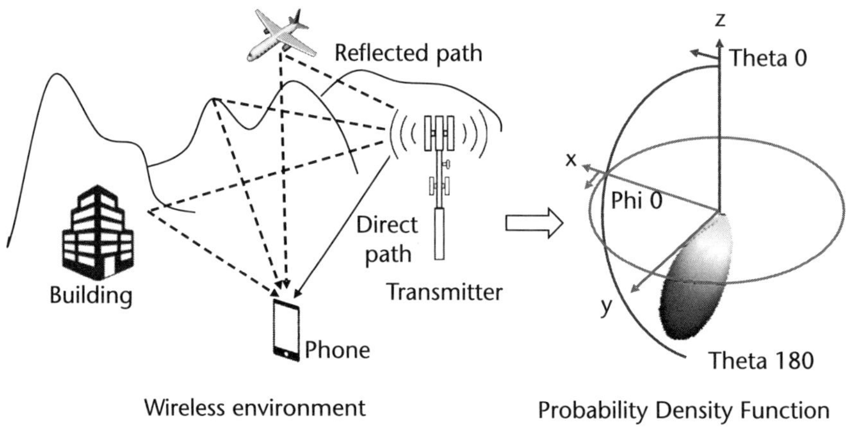

Figure 4.6 Stochastic representation of the environment.

gain and antenna diversity calculations. It has a maximum value of 1 and a minimum value of 0.

4.4.5.1 Diversity Modeling Calculations

The following are the diversity modeling calculations:

- *Complex correlation:* Given the stochastic model of the environment and antenna far-zone patterns, it gives information on how similar signals are received by the two antennas on average.
- *Complex correlation:* Real part of correlation – j^* complex part of the correlation.
- *Envelope correlation close to 1:* Very similar signals. There is no gain by using two antennas. If one signal is very weak, the other one will most likely also be very weak.
- *Envelope correlation close to 0:* Very unsimilar signals. There is significant gain by using two antennas. If one signal is very weak, the other can still be very strong with a high likelihood.
- *Rule of thumb:* Aim at an envelope correlation < 0.5.

Correlation from the 3-D pattern is given by

$$R_{ij} = \int_0^{\pi}\int_0^{2\pi} [XPD.E_{\theta i}(\theta,\varphi).E^*_{\theta j}(\theta,\varphi) + E_{\varphi i}(\theta,\varphi).E^*_{\varphi j}(\theta,\varphi)].P(\theta,\varphi)\sin\theta d\theta d\varphi \tag{4.1}$$

On average over complete 3D sphere

Is there a preferred polarization?

Polarizations of antenna far-zone

Stochastic model for the environment

Correlation from the 2-D pattern is given by

$$R_{ij} = \int_0^{2\pi} [XPD.E_{\theta i}\left(\frac{\pi}{2},\varphi\right).E^*_{\theta j}\left(\frac{\pi}{2},\varphi\right) + E_{\varphi i}\left(\frac{\pi}{2},\varphi\right).E^*_{\varphi j}\left(\frac{\pi}{2},\varphi\right)].P(\varphi)d\varphi \tag{4.2}$$

XPD is cross-polarization discrimination of the incident multipath field. It can be expressed by the ratio of time-averaged vertical power to time-averaged horizontal power in the fading environment. Users can input a linear value or a decibel value.

- Small XPD = strong phi-polarization;
- Large XPD = strong theta-polarization.

The two antennas are completely characterized by their far-zone patterns. Two antenna patterns are always needed for diversity calculation because largest performance improvement has been observed with two antennas as shown in Figure 4.7. Only minor additional performance improvement is noticed when using more than two antennas. Antenna systems with N antennas are characterized by an $N \times N$ correlation matrix, which is built by considering the 2 2 antenna system. Complex correlations calculated from 3-D and 2-D patterns are shown in Table 4.1.

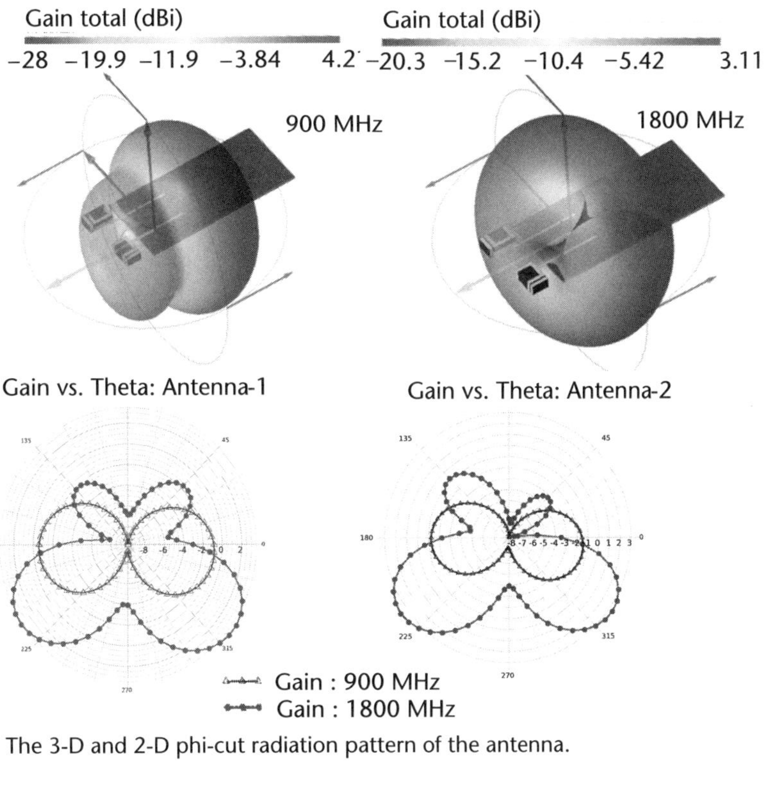

Figure 4.7 The 3-D and 2-D phi-cut radiation pattern of the antenna.

Table 4.1 Complex Correlation from 3-D and 2-D Antenna Patterns

	Envelope Correlation	*Complex Correlation*	*Complex Correlation Magnitude*
From 3-D pattern	0.1395	0.3703–0.0491*j*	0.3735
From 2-D pattern	0.1305	0.3567–0.0578*j*	0.3613

4.5 Mobile Phone Compliances and Measurement

Mobile communications devices, such as mobile phones, wireless modems, and base stations (including mobile phone towers), produce EM energy (EME) to transmit a signal. For mobile phones, HAC, SAR, EMC, and EMI compliances have been framed for serious reasons based on the context and failing to adhere to the compliance may be a violation based on the laws in the specific industry or country.

One important issue to be considered in designing antennas for a mobile phone is the reduction of SAR values, which could be caused by RF radiation from multiple radio and antennas in the mobile phone. The SAR should be as low as possible, especially when in contact with the human brain. The handset with multiple transmitters and antennas will increase the complexity of the near-field problem. In addition, HAC is required for mobile phones in some countries, which will also increase the complexity of antenna design.

4.5.1 EMC and EMI Compliance

All electronics circuits have the possibility of radiating of picking up unwanted electrical interference, which can compromise the operation of one or other of the circuit. EMC is the concept of enabling different electronics devices to operate without mutual interference, electromagnetic interference (EMI) when they are operated close to each other. The first stage in the mobile phone antenna design process is to design an antenna to meet the EMC compliance test requirements and any EMC testing that is undertaken. For good EMI performance, it is necessary to ensure that adequate filtering and shielding are in place to prevent unwanted radiation taking place or to allow in unwanted signals. Much of this can be undertaken by placing adequate filtering and metallic shielding. It is also necessary to ensure that the phone is adequately screened. Within the phone, any sources of signals or areas that are susceptible to radiation should be isolated from another component.

4.5.2 HAC Compliance

A hearing aid is an electroacoustic device that is placed in or behind the user's ear and is designed to amplify and modulate sounds for the user. When mobile phones are used near some hearing devices (hearing aids and cochlear implants), the user may detect a humming, buzzing, or whining noise. Some hearing devices are more sensitive than others to this interference noise. Similarly, cell phones also vary in the amount of interference that they generate. The FCC has approved standards and passed regulations for digital wireless device use with hearing aids to ensure that everyone benefits from advances in technology. HAC-compliant devices carry a rating of M3, M4, T3, or T4. Cell phones that comply with the FCC's HAC rule must receive a minimum rating of M3 for RF emissions and T3 for telecoil coupling.

4.5.3 SAR Compliance

The SAR is a measure of the amount of RF energy absorbed by the human body when using a mobile phone. The SAR value is measured in terms of watts per kilogram or milliwatts per gram. The RF exposure limits are expressed in the terms of SAR, which is a measure of the electric and magnetic field strength and power density for transmitters operating at frequencies from 300 kHz to 100 GHz. The FCC and other federal governmental agencies around the world require that any wireless device is evaluated to meet the RF exposure limits set forth in the governmental SAR regulations.

SAR testing uses standardized models of the human body and head that are filled with liquids with different dielectric constants and represents the RF absorption characteristics of different human tissues. A mobile phone is tested while operating at its highest power level in all the operating frequency bands and in various specific positions against the dummy head and body. This setup represents the way that different users typically hold a cell phone, including to each side of the head to determine compliance. A robotic probe takes a series of measurements of the electric field at specific pinpointed locations in a very precise, grid-like pattern within the dummy head and torso, and this data is processed to calculate the SAR value and the compliance of the cell phone.

4.6 HAC

HAC testing includes measuring the electric and magnetic fields produced by a phone to ensure that it does not interfere with the sound quality of a hearing aid. HAC testing was originally introduced in 1970 with the purpose of making it possible to couple a wired telephone with a hearing aid. The wireless telephone industry has developed M and T ratings to assist hearing aid users in finding wireless devices that may be compatible with their hearing aids. Wireless devices with HAC compliance that are rated have the rating displayed on their box together with other relevant approval markings.

Early in 1996, the FCC called together a summit between the hearing industry, the wireless industry, and consumers to resolve the compatibility issue between hearing aids and cellular phones. Because of those efforts, in 1996, ANSI ASC 63 formed task group C63.19 to develop a measurement standard for HAC with wireless communications devices; the goal of this task group was to develop a set of tests that would evaluate the compatibility of hearing aids with cellular phones. The challenges presented to the task group were formidable. The availability of HAC handsets is observed by the FCC. The target is a market coverage of at least 50% M-rated handsets and 2 handsets with T-rating per provider. Since 2005, the FCC has required that cellular telephone providers and manufacturers offer a growing number of handsets, which are compatible with hearing aids according to the standard ANSI C63.19. This standard specifies the audio magnetic field for T-coil coupling as well as the allowable RF interference level for E-field and H-field as a function of hearing aid RF susceptibility. In the T-coil mode, a hearing aid receives its signal not by a microphone but through an inductive pickup coil. The intended magnetic signal for the t-coil is that produced by the audio signal driving the phone receiver element or some other inductive circuit element. In the ideal case, a hearing aid delivers the same audio level whether in audio or t-coil mode.

4.6.1 M and T Ratings

These are the M and T ratings:

- *M-rating:* An M rating measures compatibility with acoustic hearing aids, and the higher the number (e.g., M3, M4, and so forth), the more compatible the handset. Usually, handsets rated M3 or T3 by the ANSI are considered compliant with FCC requirements. Wireless devices rated M3 or M4 meet FCC requirements. M4 is the better or higher of the two ratings.
- *T-rating:* A T rating measures compatibility with induction hearing aids. The higher the T number, the more likely the user will be able to use the phone with hearing aid on the telecoil setting. Wireless devices rated T3 or T4 meet FCC requirements and are likely to be more usable with a hearing device's telecoils (T switch or telephone switch) than unrated wireless devices. T4 is the better or higher of the two ratings.

4.6.2 HAC Measurement

The standardized method of measuring the HAC [23] defines the tests to be done in free-space (without the user) on a square HAC plane (50 mm × 50 mm) at a 15-mm distance from the surface of the phone. The center of the square plane is aligned with the earpiece of the mobile device. In this work, the earpiece is determined to be located at a 10-mm distance from the edge of the chassis of the terminal as shown in Figure 4.8. The HAC plane is divided into nine equal-sized subgrids. The maximum of the electric and magnetic fields is measured for each subgrid. As per the HAC standard, three adjacent subgrids (except the center subgrid) can be excluded for both electric and magnetic fields so that one of the excluded subgrids must be the same for both electric and magnetic fields. The electric and magnetic field strengths should be below the given limits to fulfill the HAC standard in the remaining six subgrids. HAC sensors gather data on a 5 cm × 5 cm arbitrarily oriented rectangle in free space. These data are used to determine if a wireless device will generate electrical and magnetic fields large enough to interfere with a hearing aid.

The maximum field limits for the HAC standard M3 (articulation weighting factor decibels for GSM systems) are shown in Table 4.2. As can be seen, the standard field values are much stricter above 960 MHz. Thus, it is assumed that it is more difficult to fulfill the standard at the upper GSM frequencies [24].

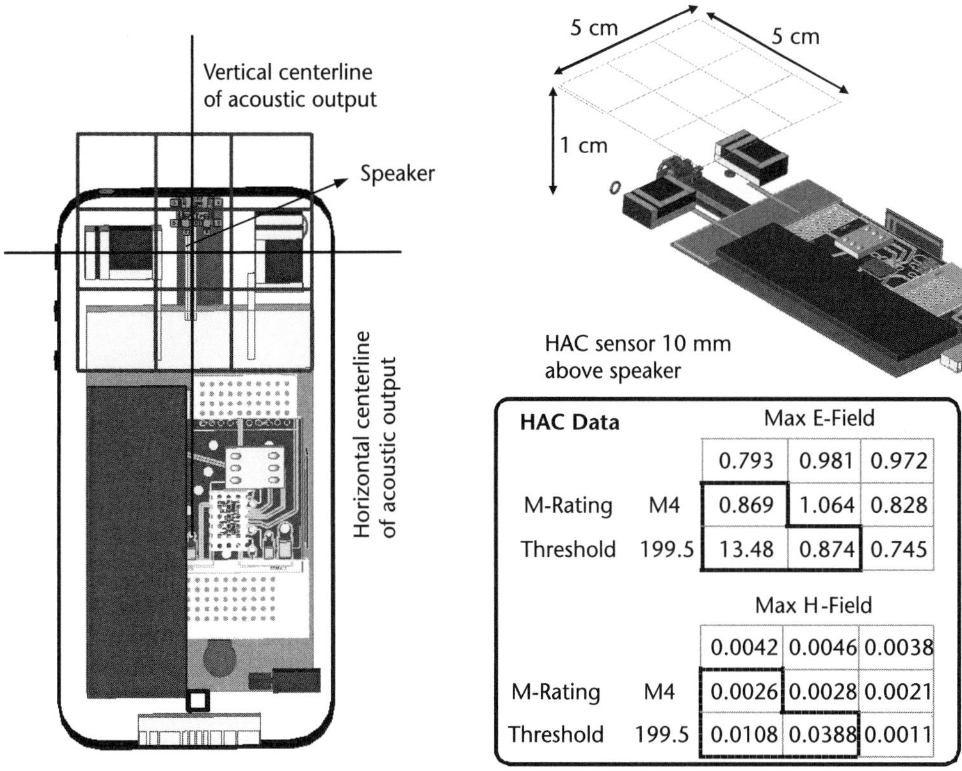

Figure 4.8 HAC sensor 10 mm above a speaker's 3 × 3 measurement grid. Simulated electric and magnetic fields of the phone on the HAC plane (peak values) with the allowed exclusions according to the HAC standard.

Table 4.2 Near-Field Limits of the HAC Standard

HAC Category M3 Limits (Articulation Weighting Factor (AWF) = –5 dB)	*Below 960 MHz*	*Above 960 MHz*
Maximum electric field	266.1 V/m	84.1 V/m
Maximum magnetic field	0.80 A/m	0.25 A/m

4.7 SAR

Mobile phones and other wireless devices emit RF. The radiation affects human health, and this has become one of the common areas of interest for both technical and medical researchers. Mobile phone users are exposed to significant amount of EM energy radiated from the transmitting antennas and directly or mutually coupled headsets, and it is seen that absorption of EM waves in the human head and other body parts causes adverse biological effects [25–28]. The phone radiation level is measured by the SAR, and it varies from device to device. To properly quantify the response of the biological structure, in terms of the incident and induced field, of the energy absorbed and maintained inside the body, the SAR was introduced and defined.

The SAR is a measure of the rate of RF energy absorption by the body from the source being measured and it is defined as the power absorbed per mass of tissue and has units of watts per kilogram. The SAR is normally averaged either over a small sample volume or over the whole body (typically 10g or 1g of tissue). The SAR can be defined as the power absorbed in a mass of tissue as given in

$$SAR = \frac{\sigma |E_x|^2}{2\rho} + \frac{\sigma |E_y|^2}{2\rho} + \frac{\sigma |E_z|^2}{2\rho} \tag{4.3}$$

where σ = electrical conductivity, $|E_{xyz}|$ = magnitude of the electric field (V/m), and ρ = material density (kg/m^3).

Many factors can influence the SAR in a biological structure irradiated by electromagnetic energy. Among these, tissue geometry, dielectric properties, and the orientation of the body relative to the source play a vital role. The internal field and, through it, the SAR distribution are nonuniform because the RF energy induced in the body is scattered and absorbed at various interfaces. The peak spatial averages of the SAR values associated with the use of wireless handsets are mainly functions of geometry and depend on the square of the magnitude of the RF current density distribution inside the device and to the geometric position of the device relative to the head. European regulators require a 10-g cube localized peak SAR < 2 W/kg and U.S. FCC regulations require a 1-g cube localized peak SAR < 1 W/kg [24–27].

4.7.1 SAR Testing

SAR testing uses standardized models of the human head and body that are filled with liquids that simulate the RF absorption characteristics of different human tissues as displayed in Figure 4.9. Each cell phone is tested while operating at its highest power level in all the operating frequency bands and in various specific positions against the dummy head and body to determine compliance. This is the way that different users typically hold a cell phone, including to each side of the head.

In the SAR compliance test, the phone is precisely placed in various common positions next to the head and body. A robotic probe takes a series of measurements of the electric field at specific pinpointed locations in a very accurate, grid-like pattern within the dummy head and torso as shown in Figure 4.10. Only the highest SAR values for each frequency band are included in the final authorization to demonstrate compliance with the FCC's RF guidelines.

SAR is measured in three ways:

- Whole body average, which is the total amount of energy dissipated in the human body, expressed as an average of the total mass of the human body;
- A 10-g cube localized peak, which is the highest level of energy dissipation in any 10-g cube of tissue in the human body;

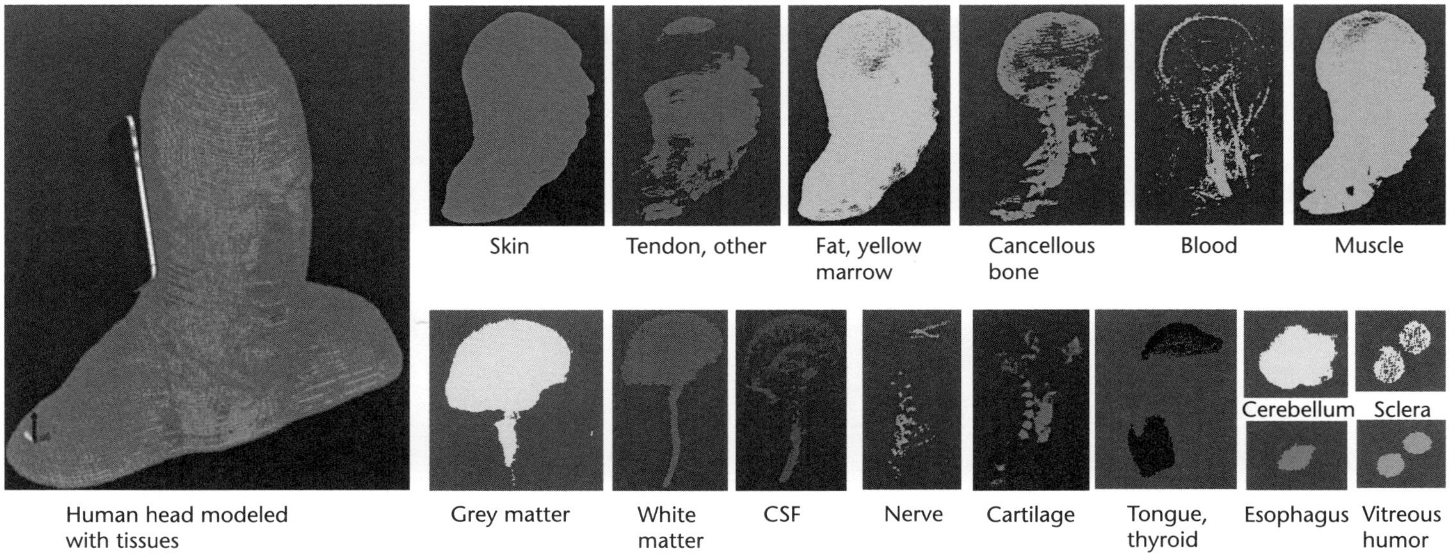

Figure 4.9 Smartphone with a human head modeled with tissues.

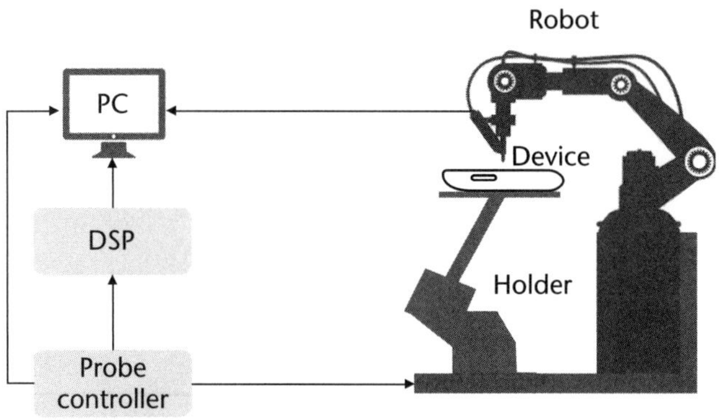

Figure 4.10 Mobile phone SAR measurement setup.

- A 1-g cube localized peak, which is the highest level of energy dissipation in any 1-g cube of tissue in the human body.

The SAR analysis and simulation flow, along with the SAM head and hand model and different antenna systems placed in a cell phone, is shown in Figure 4.10. There are various companies providing standardized models for SAR testing. For simulation, one of the most commonly used numerical models is FDTD based in the time domain because the FDTD solver can handle large simulation space and tissue material properly. The FEM or MoM can also be utilized to calculate SAR by solving Maxwell's equations in a heterogeneous electrical model of human body parts.

4.7.2 SAR Measurement

The field images of SAR values may be saved for any plane containing lossy tissue. The SAR may be adjusted for any desired input power and the peak SAR value and location are computed. The whole-body average SAR, that is, the SAR of each cell divided by the number of cells, is computed. The maximum calculated SAR value for a 10-g cube of tissue in the human body and a 1-g cube is given in Table 4.3 for the mobile phone and antenna system. Simulated peak 10-g and 1-g SAR values are compared with the corresponding measured values for two typical commercial mobile phones working at the GSM 900 band. A shown in Figure 4.11, the mobile phone is placed touching the head model.

Variations of peak 10-g and 1-g SARs with the distance measured along the Y-axis (ZX-plane) in the mid-coronal plane at 930 MHz were obtained the using

Table 4.3 Maximum Calculated SAR Value for 10g and 1g

Frequency	*Maximum SAR Value (10g)*	*Maximum SAR Value (1g)*
900 MHz	4.702e-6 W/kg	7.181e-6 W/kg
1,800 MHz	1.668e-4 W/kg	2.422e-4 W/kg
2,400 MHz	1.332e-3 W/kg	2.054e-3 W/kg

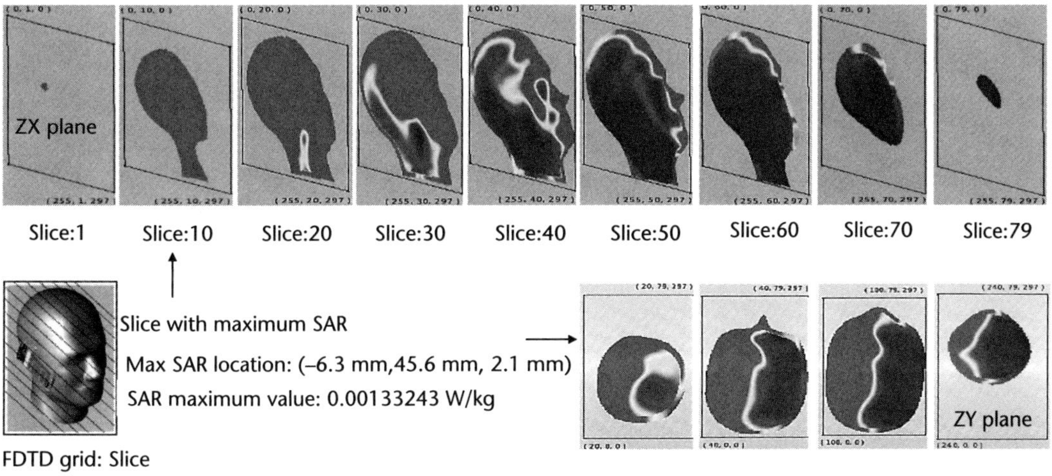

Figure 4.11 Variations of peak 10-g and 1-g SARs with distance.

FDTD solver. The stem plots of peak gram averaged SAR versus distance contain many hotspots. The density of the hotspots decreases with the increase of the SAR value and vice versa. The maximum SAR value is found corresponding to each maximum peak 10-g and 1-g SAR as shown in Figure 4.11.

4.8 EMI and EMC in Mobile Phones

The EMI and EMC are often used when referring to regulatory testing of electronic devices. As the density of wireless devices continues to increase, the concern about EM radiation and its effects from sources producing EMI also increases. The EMI and EMC can be defined as:

- *EMI:* EM emissions from a system or device that interferes with the normal operation of another device or system.
- *EMC:* The ability of equipment or system to function satisfactorily in its EM environment without introducing an intolerable EM disturbance to anything in that environment.

The EMC problem is between a source (an emission culprit) and a victim (susceptible device), which are coupled through a means, either conducted through wires or radiation through space or as shown in Figure 4.12. These electrical influences can be generated by either radiated or conductive EMI sources. Radiated sources include anything electromechanical or electrical, including power lines, antennas, traces on a PCB, motors, and even the chips on the cell phone board. Conductive EMI is a form of electrical noise pollution on the power supply traces of cell phone and this can be caused by induced voltage spikes from other devices within a system.

As shown in Figure 4.12, the radiated emission is RF energy that reaches susceptible device via radiation while conducted emission is RF energy that reaches susceptible device via conductive connections. EMI is a worldwide concern and

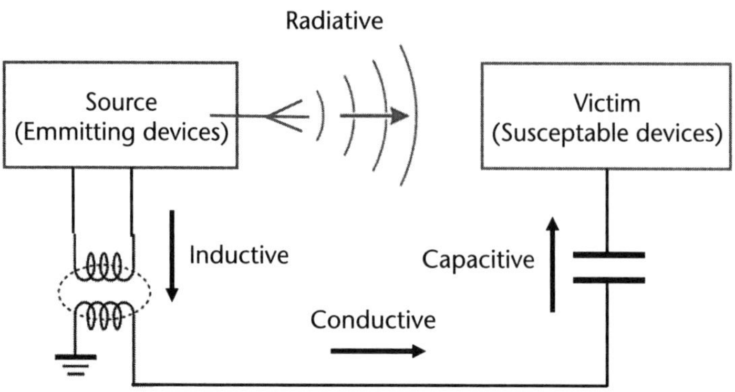

Figure 4.12 EMI coupling modes.

many governmental agencies and industry bodies have supplied specifications with which all electrical, electronic equipment and electromechanical systems must comply. These specifications are shaped to ensure that proper EMC techniques are followed by cell phone and other electronic device manufacturers during the design and fabrication of their products. Without following these specifications, even for simple appliances, the designer may be faced with unfamiliar standards governing radio interference and user safety.

EMC is becoming a critical issue within electronic components. It is needed to maintain the required level of EMC in mobile phones. With the increase of system frequencies, wireless applications and the demand for lower supply voltages, the cell phone becomes more vulnerable to the negative effects of EMI. When electronic equipment performs its designed functions without causing or suffering unacceptable degradation due to EM interference to or from other equipment, it is well designed and EM compatible. The minimization and control of EMC are out of necessity. Mobile phone manufacturers and compliance regulatory agencies are responsible to take care of EMC issue during design. Figure 4.13 shows all types of EMC tests.

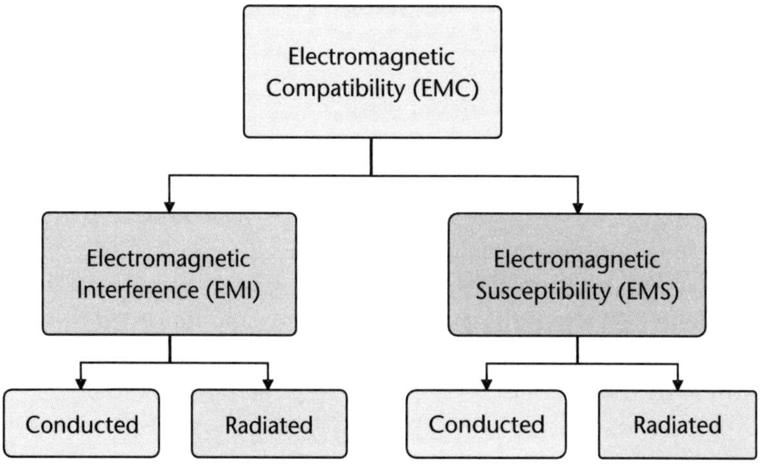

Figure 4.13 Types of EMC tests.

- *Radiated emissions test:* In radiated emission test, EM disturbances originating from an electronic device through radiation that may cause interference or malfunction to another electronic device in the same environment is measured. This is also known as EMI. Radiated emissions are propagated through the air directly from the source device to victim device. In EMC testing, radiated emissions measurements are carried out using a spectrum analyzer and or an EMI receiver and a suitable measuring antenna.
- *Conducted emissions test:* In a conducted emission test, EMI or disturbances that originate from source device and propagate along interconnected cables such as wired ports such as power conductors are measured. In this testing, conducted emissions measurements are made on an EMI receiver via an ISN (impedance stabilization network) located within the test chamber.
- *Radiated susceptibility test:* In this test, the ability of an electronic device to function normally in an EM radiation environment without experiencing interference or malfunction due to the radiation emissions emanating from another electronic device is measured. In radiated susceptibility testing, RF signal generators, amplifiers, and antennas are used to produce an EM field at varying frequencies.
- *Conducted susceptibility test:* In this testing, an EM field is generated by an RF signal generator and amplifier and then this EM field is coupled to the victim device thorough a power cable. Normally, conducted immunity testing is applicable to AC and DC ports and signal cables longer than 3m in length.

4.8.1 EMC Simulation Model

Normally, EMC testing is performed towards the end of the product development cycle that is one of the major risks to uncover an unexpected problem, forcing a design modification and a delay in the product release. EMC measurements carried out using certified test laboratories can be both time consuming and expensive that why many designers prefer to perform EMC simulation before the actual product of phones. As the design evolves, EMC aspects can be analyzed by comparing different component locations on the board or optimizing trace routing paths to minimize coupling.

EMC simulation using the FEM for mobile phone devices is performed to see the electric field level at different electronic components inside the phone [28–31] due to antennas being placed in the mobile phone. Figure 4.14 shows the E-field and H-field distribution on the smartphone board, when the Bluetooth antenna is operational at 2.4 GHz. EMC problems must be analyzed over a broad frequency range including all device components, connectors, antennas, cables, and housing. At some components, as shown in Figure 4.15, the electric field level is high, and if this value is beyond a maximum permissible limit to operate that component efficiently, then it is important to reduce the radiation level by applying different techniques such as placing a metallic shield or radiation absorbing tapes. This is important to also meet EMC compliance. Designing a shield enclosure is often a

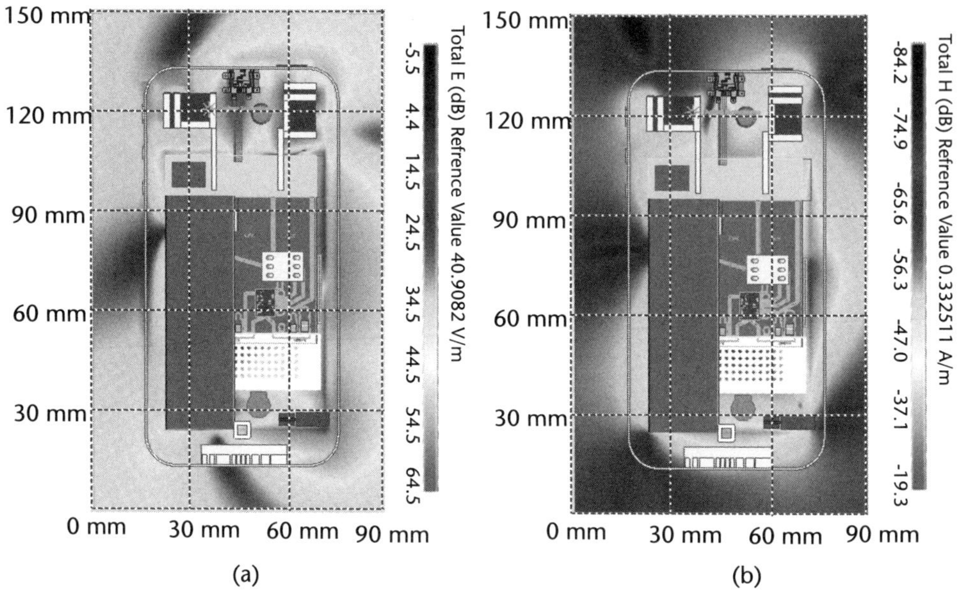

Figure 4.14 Bluetooth antenna radiation effect: (a) E-field and (b) H-field plot and values at different parts of a smartphone.

Components	Distance from antenna port	E-field value (dB)	H-field value (dB)
Speaker	49.3 mm	58.4	-38.2
Microphone	43.4 mm	43.5	-44.0
Camera	46.8 mm	31.2	-47.1
LCD	24.2 mm	34.7	-55.7
PA Chip	19.3 mm	28.9	-54.1
USB	57.0 mm	53.4	-47.5
VCO	23.1 mm	42.3	-46.6

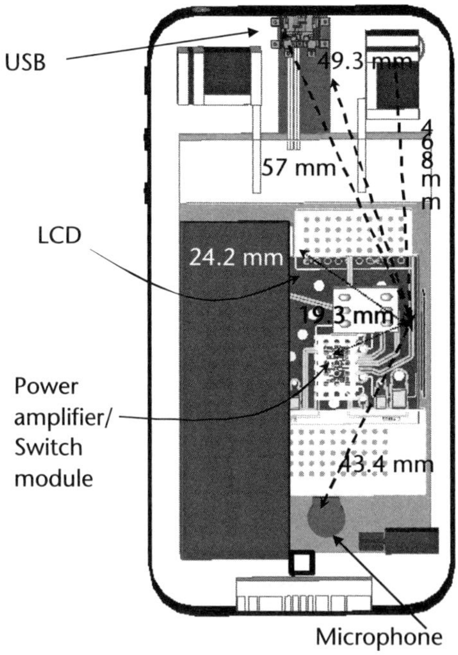

Figure 4.15 Electric field and magnetic field values at different electronic components inside a phone.

trade-off between suppressing radiation and still providing sufficient airflow for heat transportation and cooling of the electronic device.

The electric field value reduces significantly after putting a metallic shield enclosure over a power amplifier (PA) and switch module. The simulated shielding effectiveness for a PA from 200 MHz to 10 GHz is shown in Figure 4.16. A perfectly conducting enclosure that surrounds the component prevents external radiation within that volume from electrically coupling to anything outside that volume. This type of enclosure is also called a Faraday cage.

4.8.2 EMI Simulation Model

To overcome the EMI issue, the designer must either remove the source of the interference or protect the circuit from being affected (EMC) as discussed in the previous section. EMC at the device or system level may have to comply with certain emission standards, so that the device does not affect other equipment or appliances. To meet these, the designer should think about EMI suppression starting from the board level [32]. Smartphone electronics has a strong integration of features such as high-speed digital links (such as double data rate (DDR), high-definition multimedia interface (HDMI), and USB3.1) and sensitive wireless functions (such as Wi-Fi 802.11 and Bluetooth). These digital interfaces are potential EMI aggressors. If an antenna is placed quite close to any potential EMI aggressors, Wi-Fi sensitivity can be affected and RF throughput can be significantly degraded.

Radiated EM coupling analysis is performed to investigate the EMI coupling on a component using near field EM data and then overlay this data with FCC and CISPR (International Special Committee on Radio Interference) standards for compliance measurement. In the United States, digital device regulations were established under FCC Part 15 Subpart J in 1979, internationally by the IEC standard CISPR 22 in 1985, and in the European Union under the EMC Directive in 1989 [33, 34]. The FCC has different sets of regulations for different types of digital devices. Devices that are marketed for use in commercial, industrial, or business environments are classified as Class A digital devices. Devices that are marketed for use in residential environments, notwithstanding their use in commercial, industrial, or

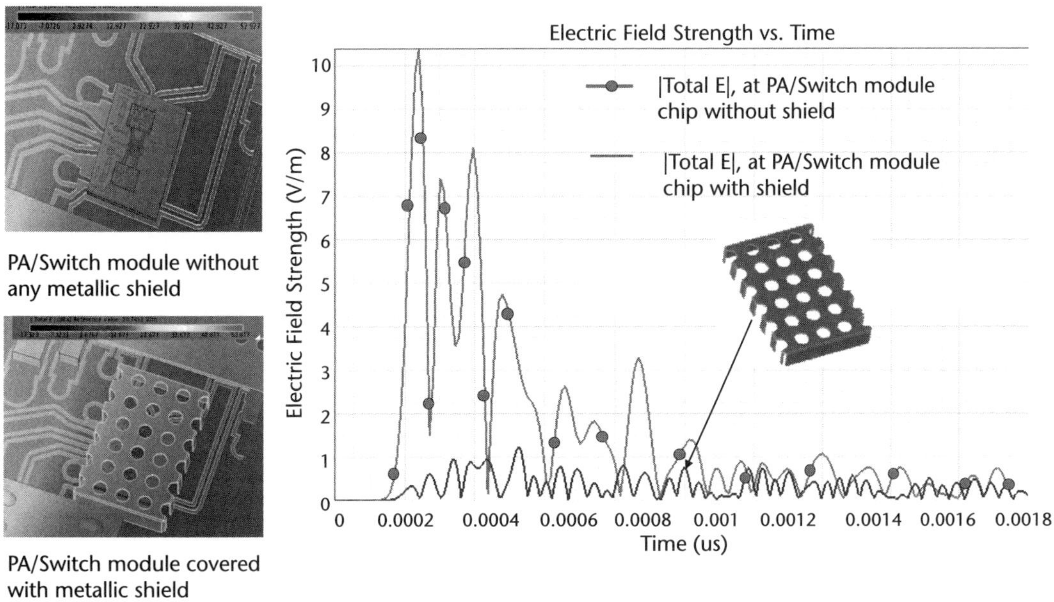

Figure 4.16 Electric field at power amplifier module with and without a shield.

business environments are classified as Class B digital devices. Radiation emission measurements at 1m, 3m and 10m are shown in Figure 4.17.

The method for measuring radiated emissions varies depending on the type of device being measured. Class A digital devices must be measured at 10m from the product and Class B devices are to be measured at 3m from the product. CISPR 22 regulations have been also adopted by several countries. CISPR 22 regulations require that radiated emissions measurements for Class A devices be measured at 30m and Class B devices be measured at 10m. Radiated emission from the Bluetooth antenna is measured at 3m and 10m as shown in Figure 4.18 using both antenna near-field and far-field approaches. Simulated data is compared with FCC

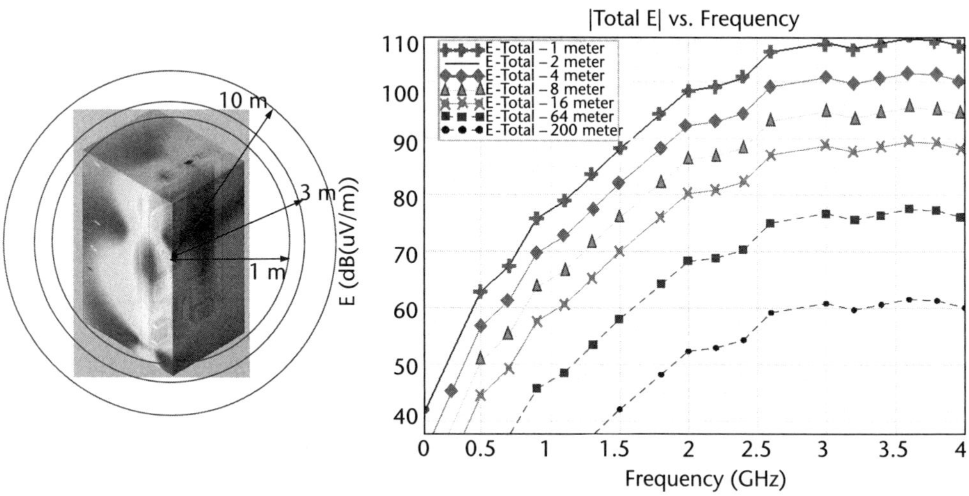

Figure 4.17 Radiation emission measurements at 1m, 3m, and 10m.

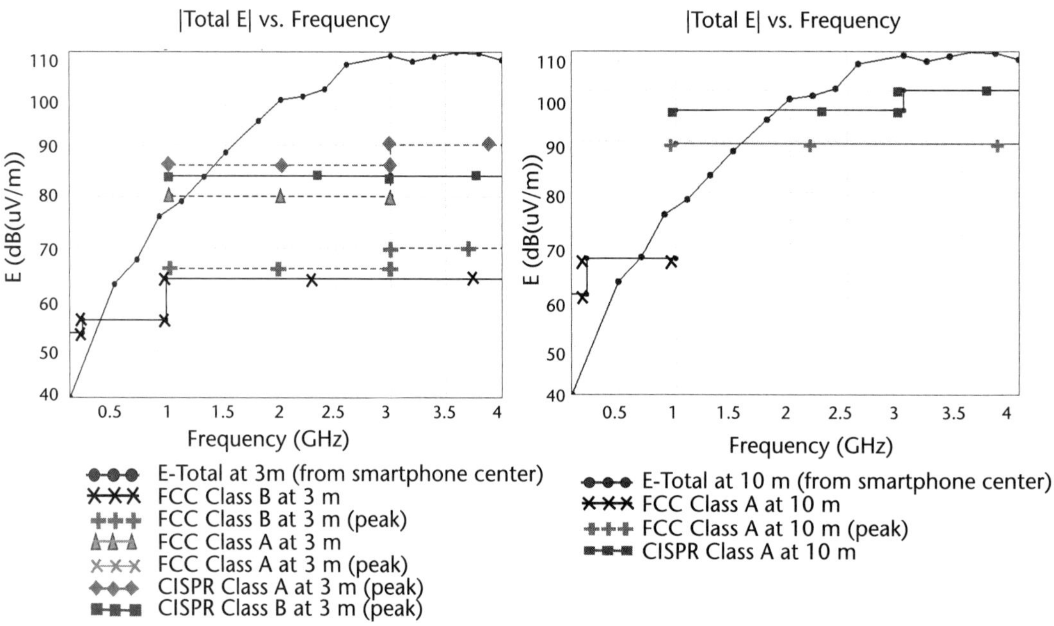

Figure 4.18 EMI limits: FCC and CISPR class A & B at 3m and 10m.

and CISPRR values. If these values are not as per compliance, then the EMI reduction technique is applied before the production of mobile devices.

4.9 Summary

In this chapter, compliance parameters and its measurement for mobile phone antennas were discussed. A product cannot be sold unless it is known to meet those requirements specified in the standards. Compliance is appreciated when a product passes a series of tests without occurring some specified mode of failure. For mobile phones, various compliances such as HAC, SAR, and EMC/EMI were discussed in detail. Antenna diversity and MIMO systems were discussed to mitigate the signal fading effect.

References

[1] Electromagnetic Compatibility (EMC)—Part 4-3: Testing and Measurement Techniques—Radiated, Radio-Frequency, Electromagnetic Field Immunity Test, *IEC Standard 61000–4-3*, Ed. 3.0, 2006.

[2] Kitchen, R., *RF and Microwave Radiation Safety Handbook*, 2nd ed., New York: Elsevier, 2001.

[3] Maeda, A., and K. Ohizumi, "A Study into the Effect of Smartphone Development on Antenna Design," *2013 IEEE Symposium on Product Compliance Engineering (ISPCE)*, Austin, TX, 2013, pp. 1–6.

[4] Durney, C. H., H. Massoudi, and M. F. Iskander, *Radiofrequency Radiation Dosimetry Handbook*, Brooks Air Force Base, Aerospace Medical Division (AFSC), 1986.

[5] Lee, A. K., et al., "Numerical Implementation of Representative Mobile Phone Models for Epidemiological Studies," *Journal of Electromagnetic Engineering and Science*, Vol. 16, No. 2, 2016, pp. 87–99.

[6] Balzano, Q., O. Garay, and T. Manning, "Electromagnetic Energy Exposure of the Users of Portable Cellular Telephones," *IEEE Transactions on Vehicular Technology*, Vol. 44, May 1995, pp. 390–403.

[7] Cicchetti, R., and A. Faraone, "Estimation of the Peak Power Density in the Vicinity of Cellular and Radio Base Station Antennas," *IEEE Transactions on Electromagnetic Compatibility*, Vol. 46, No. 2, May 2004, pp. 275–290.

[8] Tesche, F. M., M. V. Ianoz, and T. Karlsson, *EMC Analysis Methods and Computational Models*, New York: Wiley, 1997.

[9] Gosselin, M. C., et al., "Dependence of the Occupational Exposure to Mobile Phone Base Stations on the Properties of the Antenna and the Human Body," *IEEE Transactions on Electromagnetic Compatibility*, Vol. 51, No. 2, May 2009, pp. 227–235.

[10] Methods for the Assessment of Electric, Magnetic and Electromagnetic Fields Associated with Human Exposure, *IEC, Int. Electrotechnical Commission Stand.* 62209-1, 2005.

[11] Meyer, F. J. C., K. D. Palmer, and U. Jakobus, "Investigation into the Accuracy, Efficiency and Applicability of the Method of Moments as Numerical Dosimetry Tool for the Head and Hand of a Mobile," *Applied Computational Electromagnetics Society*, University of Mississippi, 2001.

[12] Björnson, E., E. G. Larsson, and M. Debbah, "Massive MIMO for Maximal Spectral Efficiency: How Many Users and Pilots Should Be Allocated?" *IEEE Transactions on Wireless Communications*, Vol. 15, No. 2, February 2016, pp. 1293–1308.

[13] Zheng, L., and D. N. C. Tse, "Diversity and Multiplexing: A Fundamental Tradeoff in Multiple-Antenna Channels," *IEEE Transactions on Information Theory*, Vol. 49, No. 5, May 2003, pp. 1073–1096.

[14] Yang, P., et al., "Transmit Antenna Selection for Multiple-Input Multiple-Output Spatial Modulation Systems," *IEEE Transactions on Communications*, Vol. 64, No. 5, May 2016, pp. 2035–2048.

[15] Larsson, E. G., et al., "Massive MIMO for Next Generation Wireless Systems," *IEEE Communications Magazine*, Vol. 52, No. 2, February 2014, pp. 186–195.

[16] Paulraj, A. J., et al., "An Overview of MIMO Communications: A Key to Gigabit Wireless," *Proc. IEEE*, Vol. 92, No. 2, February 2004, pp. 198–218.

[17] Kong, L. K., et al., "Near-Capacity Cooperative Space-Time Coding Employing Irregular Design and Successive Relaying," *IEEE Transactions on Communications*, Vol. 58, No. 2, August 2010, pp. 2232–2241.

[18] Vaughan, R. G., and J. Bach Andersen, "Antenna Diversity in Mobile Communications," *IEEE Transactions on Vehicular Technology*, Vol. VT-36, November 1987.

[19] Schwartz, M., W. R. Bennett, and S. Stein, *Communication System and Techniques*, New York: McGraw-Hill, 1965.

[20] Kildal, P. S., and K. Rosengren, "Correlation and Capacity of MIMO Systems and Mutual Coupling, Radiation Efficiency and Diversity Gain of Their Antennas: Simulations and Measurements in Reverberation Chamber," *IEEE Communications Magazine*, Vol. 42, No. 12, December 2004, pp. 102–112.

[21] Rappaport, T. S., *Wireless Communications: Principles and Practice*, Boston, MA: Pearson Education, 2002.

[22] Blanch, S., J. Romeu, and I. Corbella, "Exact Representation of Antenna System Diversity Performance from Input Parameter Description," *Electronic Letters*, Vol. 39, No. 9, May 1, 2003, pp. 705–707.

[23] American National Standard for Method of Measurements of Compatibility Between Wireless Communication Devices and Hearing Aids, ANSI C63.19-2007, *Amer. Nat. Standards Inst.*, New York, 2007.

[24] Holopainen, J., et al., "Near-Field Control of Handset Antennas Based on Inverted-Top Wavetraps: Focus on Hearing-Aid Compatibility," *IEEE Antennas and Wireless Propagation Letters*, Vol. 8, 2009, pp. 592–595.

[25] Khan, O. M., et al., "Specific Absorption Ratio Evaluation of Penta Band Antenna for LTE/GSM Mobile Handset Applications," *2017 International Conference on Electrical and Computing Technologies and Applications (ICECTA)*, Ras Al Khaimah, 2017, pp. 1–3.

[26] Wang, M., et al., "Evaluation and Optimization of the Specific Absorption Rate for Multiantenna Systems," *IEEE Transactions on Electromagnetic Compatibility*, Vol. 53, No. 3, August 2011, pp. 628–637.

[27] Zhao, K., et al., "SAR Study of Different MIMO Antenna Designs for LTE Application in Smart Mobile Handsets," *IEEE Transactions on Antennas and Propagation*, Vol. 61, No. 6, June 2013, pp. 3270–3279.

[28] Carcamo, H. K., C. Lanyon Olivares, and C. Ahumada Vera, "Numerical Finite Difference Method of Electric Field Generated to Determine the SAR Generated in a Microstrip Antenna Type Applied to a WBAN," *IEEE Latin America Transactions*, Vol. 14, No. 6, June 2016, pp. 2921–2926.

[29] Hadjem, A., et al., "Analysis of Power Absorbed by Children's Head as a Result of New Usages of Mobile Phone," *IEEE Transactions on Electromagnetic Compatibility*, Vol. 52, No. 4, November 2010, pp. 812–819.

[30] Sim, D. -U., and S. -O. Park, "A Triple-Band Internal Antenna: Design and Performance in Presence of the Handset Case, Battery, and Human Head," *IEEE Transactions on Electromagnetic Compatibility*, Vol. 47, No. 3, August 2005, pp. 658–666.

[31] Arnaut, L. R., and C. S. Obiekezie, "Stochastic Analysis of Wideband Near-Field Emissions From Dipole Antennas and Integrated Circuits," *IEEE Transactions on Electromagnetic Compatibility*, Vol. 56, No. 1, February 2014, pp. 93–101.

[32] Triverio, P., et al., "Geometrically Parameterized Circuit Models of Printed Circuit Board Traces Inclusive of Antenna Coupling," *IEEE Transactions on Electromagnetic Compatibility*, Vol. 52, No. 2, May 2010, pp. 471–478.

[33] Braun, S., "A Novel Time-Domain EMI Measurement System for Measurement and Evaluation of Discontinuous Disturbance According to CISPR 14 and CISPR 16," *2011 IEEE International Symposium on Electromagnetic Compatibility*, Long Beach, CA, 2011, pp. 480–483.

[34] Heirman, D. N., "ANSI C63.4 and CISPR 22-Harmony at Last? EMC Test Standards," *IEEE 1997, EMC, Austin Style. IEEE 1997 International Symposium on Electromagnetic Compatibility. Symposium Record (Cat. No.97CH36113)*, Austin, TX, 1997, pp. 112–117.

Selected Bibliography

Hossain, M. I., M. R. I. Faruque, and M. T. Islam, "Original Research: Analysis on the Effect of the Distances an Inclination Angles Between Human Head and Mobile Phone on SAR," *Progress in Biophysics and Molecular Biology*, Vol. 119, November 1, 2015, pp. 103–110.

Kivekas, O., et al., "Bandwidth, SAR, and Efficiency of Internal Mobile Phone Antennas," *IEEE Transactions on Electromagnetic Compatibility*, Vol. 46, No. 1, February 2004, pp. 71–86.

Lee, A. K., et al., "SAR Comparison of SAM Phantom and Anatomical Head Models for a Typical Bar-Type Phone Model," *IEEE Transactions on Electromagnetic Compatibility*, Vol. 57, No. 5, October 1, 2015, pp. 1281–1284.

Okuyucu, S., et al., "A Numerical Parametric Study of SAR from a Dual-Band PIFA Based Antenna for Mobile Communication," *2017 10th International Conference on Electrical and Electronics Engineering (ELECO)*, Bursa, 2017, pp. 1004–1008.

Park, H. H., H. B. Park, and H. S. Lee, "A Simple Method of Estimating the Radiated Emission from a Cable Attached to a Mobile Device," *IEEE Transactions on Electromagnetic Compatibility*, Vol. 55, No. 2, April 2013, pp. 257–264.

Prado, G. V., and M. A. Y. Medina, "Design and Implementation of a Non-Ionizing Radiation Measuring System Evaluated with an Unmanned Aerial Vehicle," *2015 Asia-Pacific Conference on Computer Aided System Engineering*, Quito, 2015, pp. 52–57.

Rao, Q., S. Ali, and D. Wang, "A Multi-Band Antenna for Potential 700 MHz Handheld Device Applications," *IEEE Antennas and Propagation Society International Symposium*, Charleston, SC, 2009, pp. 1–4.

Triverio, P., et al., "Geometrically Parameterized Circuit Models of Printed Circuit Board Traces Inclusive of Antenna Coupling," *IEEE Transactions on Electromagnetic Compatibility*, Vol. 52, No. 2, May 2010, pp. 471–478.

Viswanath, P., D. N. C. Tse, and R. Laroia, "Opportunistic Beamforming Using Dumb Antennas," *IEEE Transactions on Information Theory*, Vol. 48, No. 6, June 2002, pp. 1277–1294.

CHAPTER 5

Reconfigurable Frequency and Polarization Diversity Antennas

5.1 Introduction

The antenna is the most critical part of wireless communication systems and must be designed and deployed efficiently so that it can effectively utilize the available frequency resources ensuring uninterrupted concurrent multiple services to the users. In this perspective, we have three antenna methodologies: individual antennas, multiband antennas, and frequency-reconfigurable antennas. The application of these three antenna approaches on a wireless platform has its own merits and demerits [1–3]. Increased antenna complexity, large space required to accommodate multiple wireless services at a time, and coupling due to little spacing between individual antennas are major design challenges. Due to comparatively more space requirements, both single and multiband antennas are not suitable for wireless handheld devices or MIMO applications. Multiband antennas possess comparatively wide bands causing difficulty in achieving the desired antenna miniaturization. They also exhibit poor isolation between different operating bands, for which the chances of jamming or unavailability of a wireless service are increased.

Antenna reconfigurability is the capacity to change an individual radiator's fundamental operating characteristics through electrical, mechanical, or other means. Thus, as per this definition, the traditional phasing of signals between elements in an array to achieve beam steering and beamforming does not make the antenna reconfigurable because the basic operating characteristics of an antenna remain unchanged in this case. Ideally, reconfigurable antennas should be able to alter their impedance bandwidths, operating frequencies, polarizations, and radiation patterns independently to accommodate changing operating requirements. Antenna reconfigurability in many circumstances provides numerous advantages. For example, the ability to tune the antenna's operating frequency could be utilized to change operating bands, filter out interfering signals, or tune the antenna to account for a new environment. If the radiation pattern of the antenna could be changed, it could be redirected toward the access point and use less power for transmission, resulting in significant savings in battery power. The ability of reconfigurability to antenna arrays provides additional degrees of freedom that may result in wider instantaneous frequency bandwidths, more extensive scan volumes, and radiation patterns with control on sidelobe distributions.

5.2 Reconfigurable Antennas

The reconfiguration of an antenna is achieved by altering the radiated fields of the antenna's effective aperture [4–10]. It is based on a focused rearrangement of the antenna currents or a reconfiguration of the antenna's radiating edges. This redistribution of properties results in a change in the functionalities of the antenna. These functional changes allow users to propose reconfigurable antennas for various wireless communication platforms. There are four reconfiguration properties that a reconfigurable antenna can achieve. An antenna can exhibit a reconfigurable frequency of operation, a reconfigurable radiation pattern, a reconfigurable polarization behavior, or a combination of any of these properties [4]. Based on the reconfigurable antenna property, the reconfiguration technique is selected. An antenna designer selects a technique that satisfies the imposed constraints and at the same time completes the antenna design task efficiently.

There are various techniques to implement reconfigurable antennas, as indicated in Figure 5.1. These reconfiguration techniques are divided into four major categories: electrical, optical, mechanical, and material changes. Antennas based on electronic switching components to redirect their surface currents are called electrically reconfigurable. Antenna reconfiguration is achieved using electrical switches such as varactors, pin diodes, microelectromechanical systems or recently digital tunable capacitors [11]. An antenna that depends on on photoconductive switching elements is called an optically reconfigurable antenna. By altering the structure of the antenna, physically reconfigurable antennas can be achieved. Finally, reconfigurable antennas can be realized through the use of smart materials such as ferrites and liquid crystals. Phase change materials are also extensively used for a reconfigulable antenna [12, 13]. These have advantages over ferrite and liquid crystals as they have less losses, high reliability, and excellent linearity. Also, their biasing can

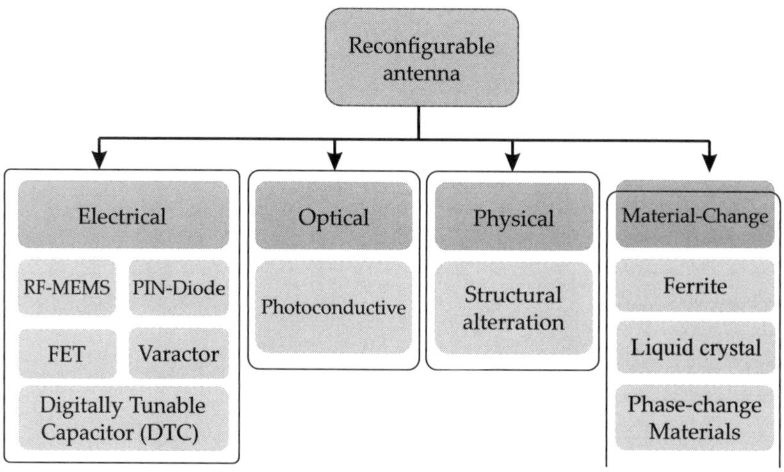

Figure 5.1 Various techniques to design reconfigurable antennas.

happen remotely so their biasing network does not affect the performance of the antenna, which is a big advantage compared to other methodologies.

In this chapter, electrical reconfiguration techniques based on the use of switches to connect and disconnect antenna parts are discussed. PIN diodes or varactors are a faster and more compact alternative to RF microelectromechanical systems (MEMS) [14–16]. RF-MEMS technology is commonly used in satellite and defense applications antenna systems and now poised to improve smartphone antenna performance in the near future by way of higher antenna efficiency [17]. MEMS-based applications need the charge pumps for high-voltage applications, which is one issue with this technology. MEMS require relatively high and variable bias voltages to be actuated [18], but recent RF-MEMS include the charge pumps on the MEMS chip and can work even with voltages as low as 3V. Reconfigurable antennas using PIN diodes [19] have a more dynamic reconfiguration ability.

5.2.1 Electrically Reconfigurable Antenna

The electrically reconfigulable antenna can be classified into four groups based on antenna reconfigurable property (e.g., frequency, radiation pattern, polarization, or combination of these):

1. Frequency-reconfigurable antennas;
2. Polarization reconfigurable antennas;
3. Radiation pattern reconfigurable antennas;
4. Hybrid reconfigurable antennas.

5.2.2 Frequency-Reconfigurable Antenna

Frequency-reconfigurable antennas have great potential to address most of the challenges coming in single and multiband antennas. They need comparatively less space to operate on multiple wireless standards with high out-of-band rejection. They are also more effective for space diversity and MIMO implementation. Cognitive radio will demand multiband reconfigurable front-end RF systems on mobile devices such as mobile phones, PDAs, and netbook computers. In this chapter, a dual-polarized frequency-reconfigurable antenna design and configuration are described for the wireless portable communication device. The antenna mathematical modeling and operating principle of the antenna is discussed. The effective length of the antenna, and hence its operating frequency, can be changed by adding or removing part of the antenna length through electronic, mechanical, or other means. There are different kinds of switching technologies. PIN diode and FET have used in frequency tunable monopole and dipole antennas for various frequency bands.

5.2.3 Polarization Reconfigurable Antenna

Antenna polarization reconfiguration can help to provide immunity to interfering signals in varying environments as well as provide an additional degree of freedom to improve link quality as a form of switched antenna diversity. The current flow direction on the antenna translates directly into the polarization of the electric field in the far field of the antenna. To achieve polarization reconfigurability, the antenna structure and/or feed configuration should alter in ways that change the path of current flows on the antenna without significant changes in impedance or frequency characteristics. Polarization reconfigurations can take place between different kinds of linear polarizations, between right and left-handed circular polarizations, or between circular, linear, and elliptical polarizations. The mechanisms to achieve polarization reconfigurability are mainly the same as those described for frequency reconfigurability earlier, although their implementations are necessarily different. By inserting a switch (PIN diode or FET) into the antenna geometry and by controlling the current direction, different polarizations are achieved. Polarization reconfigurability is also accomplished by including switches in the feed excitation slots rather than the surface of the patch.

5.2.4 Pattern Reconfigurable Antenna

Reconfigurable antennas with its pattern selectivity can enhance system security, improving communication reliability and reducing energy loss to better the comprehensive performance of the whole system. One of the most effective and common methods to change radiation patterns independently from frequency behavior is the use of electrically tuned or switched parasitic elements. Parasitic element lengths are changed with FET or PIN diode switches or varactors, which, in turn, alter the magnitudes and phases of the currents on the parasitic elements relative to the driven element. Tilts in the main beam in one plane can then be switched or swept as the lengths of the parasitic elements are changed. These methods possess several attractive qualities: isolation of the driven element(s) from the tuned element(s) and a range of available topologies and functionalities. Fundamentally, tuning of antenna radiation patterns in this manner relies on the mutual coupling between closely spaced driven and parasitic elements, resulting in effective array behavior from a single feed point. Therefore, changes in radiation patterns are achieved through changes in the coupling between the elements, which, in turn, change the effective source currents on both the driven and parasitic elements.

5.2.5 Hybrid Reconfigurable Antenna

The ability of the antenna to reconfigure the selective operating frequency, bandwidth, radiation pattern and polarization characteristics are termed as a hybrid reconfigurable antenna. The fundamental theory of operation behind any hybrid reconfigurable antenna is the same as of an ordinary reconfigurable antenna, and design and control in this case are obviously more complex. The configuration of a hybrid reconfigurable antenna with frequency and polarization is shown in Figure 5.2.

5.2.6 Antenna Switching Techniques

The fundamental role of a switch or relay is to make or break an electric circuit. In static and quasi-static terminology, a switch operates simply as either a conduction path or a break in the conduction path. In RF antenna systems, switch function typically entails controlling and directing the flow of RF energy along a desired RF path. Usually, this path may include any of the RF subsystems leading to the

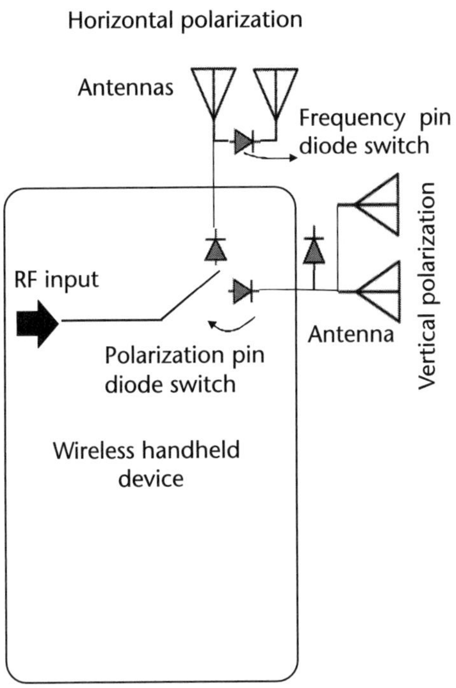

Figure 5.2 Design concept of a reconfigurable frequency and polarization antenna.

antenna feed distribution network. In reconfigurable antennas, switches are utilized to control the direction and flow of the RF current. There are several important characteristics that must be evaluated for all RF switch applications, particularly reconfigurable antenna designs. Like electrical switches, RF and microwave switches come in different configurations providing the flexibility to create complex matrices and automated test systems for many different applications. The following is a list of typical RF switch configurations and their usage:

1. Single-pole-double-throw (SPDT or 1:2) switches route signals from one input to two output paths. Using only a single-polar-double-throw (SPDT) switch, the polarization switching can be achieved.
2. Single-pole-multiple-throw (SPnT) switches allow a single input to multiple (three or more) output paths.
3. Double-pole-double-throw (DPDT) switches can be used to switch between two inputs and two outputs as a drop-out switch, for signal reversal, or to bypass a test component.

5.3 Reconfigurable Frequency and Polarization Antenna Design

Many multiband antennas have been designed to cover a wide bandwidth. However, most of them could not cover 800 MHz and 1,400 MHz with dual-polarization capability, because the antennas required significant space, as the electrical length of $\lambda_g/4$ at 800 MHz [20]. To make an antenna compact, a typical solution is to use the coupled-feed technique [21, 22], which generates a dual resonance. In this technique, a coupled gap between the feeding line and the radiator element is essential; therefore, the ground clearance of the antenna should be increased. A couple of studies focused on a compact size [23, 24], with the designed antennas having sizes under 500 mm^2 with an efficiency around 40%. Nevertheless, they are not suitable for mobile devices because the obtained efficiency of the antennas is less than 50% in the lower band. Moreover, the antenna designed in [23] did not have a –6-dB reflection coefficient in the desired band. In [25], a printed monopole antenna with a C-shaped ground plane was introduced to improve the operating bandwidth. This antenna had a compact size of 350 mm^2, but it does not cover dual resonance and dual-polarization need. To reduce the antenna size and still cover two polarizations, novel techniques are required, such as the frequency and polarization reconfigurable technique. Until now, many reconfigurable antennas for mobile device applications have been studied. However, most of them could not cover 800 MHz and 1,400 MHz with dual-polarization capability [26–28]. The conclusion of the above discussion is that there is usually a critical trade-off between the antenna operating frequency band, antenna efficiency, antenna gain, and reduction in the antenna size.

In this design, a frequency-reconfigurable antenna with a compact size that occupies an area of 950 mm^2 is presented. Here, it is noteworthy that very few of the reported mobile antennas have a small ground clearance within 15 mm. Also, the efficiency is proportional to the size of the antenna. A small-sized antenna has a lower efficiency. Nevertheless, the efficiency of the designed antenna is more than 88% in all frequency bands. Consequently, this design has much better out-of-band rejection, radiation efficiency, reflection efficiency, and positive gain. The planar geometry of the antenna makes it suitable to find applications in thin-profile, handheld wireless devices. Two orthogonal printed PIFAs are designed [29]. As shown in Figure 5.3, two PIN diodes are used to switch between horizontal and vertical

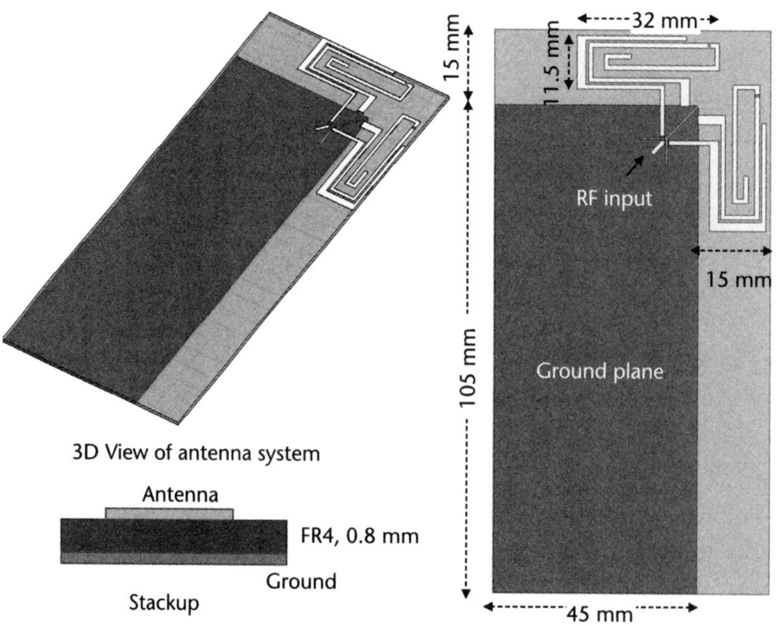

Figure 5.3 Designed antenna geometry and substrate stack-up.

polarization to select the desired polarization. Another two PIN diodes used in antenna branches are used to tune the antenna at 800 MHz and 1.4 GHz.

Figure 5.4 shows the geometry of the designed L-shaped PIFA. Planar antenna is printed over a commercially available FR-4 dielectric substrate with a permittivity of 4.4 and a thickness of 0.8 mm. The antenna has a compact design and the size of the designed structure is 32 × 14 × 2 (unit) mm^2. It should be noted that this antenna differs in some important ways from a conventional PIFA. In a conventional PIFA, the short is directly connected to the feeding element; in this antenna, instead of putting the short at the feeding strip, it is placed at the end of an electromagnetically coupled strip that is parallel to the feed strip. The designed antenna mainly consists of radiating elements and four PIN diodes (part number: BAP64-03). Reconfigurability is achieved by varying the length of the parasitic shorted strip using a PIN diode with simple bias circuitry. On the back side of the main board, a ground plane of 105 × 60 mm^2 is printed to serve as system ground of the mobile phone. Both elements of the designed antenna are fabricated near the substrate boundary to achieve a compact structure. Antenna design dimensions are shown in Figure 5.4. Two PIN

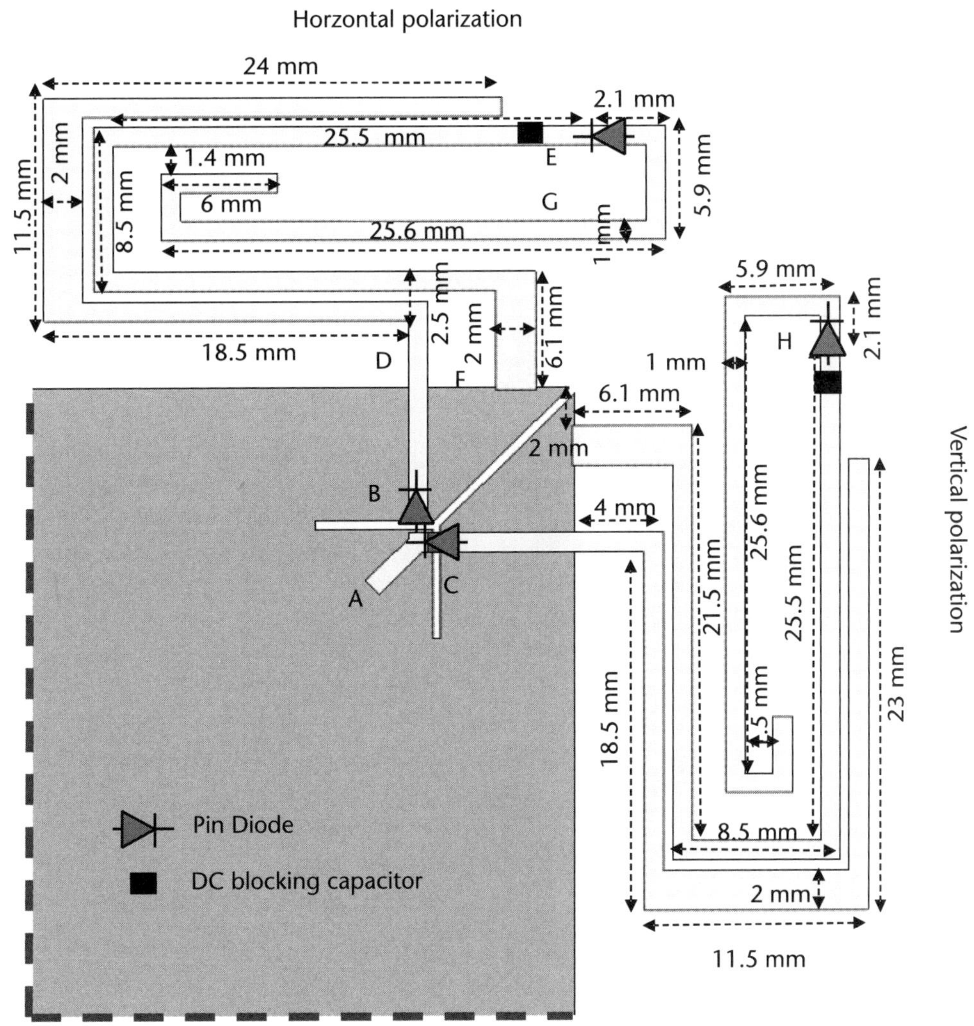

Figure 5.4 Antenna (top view) design dimensions.

diodes are placed at E and H to control resonance frequency, and another two PIN diodes are placed at B and C to control antenna polarization, The PIN diodes operate at +5V, and their consumption currents are 3 mA. The insertion loss is less than 0.5 dB in the operating band. Two RF chokes are used to improve the electrical isolation between the antenna and the bias line.

A DC block capacitor is embedded into the coupling strip to block DC in the antenna. The structure is composed of two independent elements. The driven feeding element is a monopole fed by a 50Ω microstrip line. The overall length (58 mm) of this monopole is equal to about a quarter-wavelength at 1.4 GHz. An additional resonance is created using an electromagnetically coupled strip (93 mm) using a shorted parasitic element. This shorted parasitic element is connected to the ground plane through a planar inverted via. The feeding strip is connected to a 50Ω microstrip line arranged on the front side of the main board. The shorted strip is connected to the system ground plane through a via-hole in the main board. There is a coupling gap of 0.5 mm between the feeding strip and the shorted strip. The shorted strip acts as a second radiator and generates its own resonance modes. The feeding strip has a length of about 93 mm (close to a quarter-wavelength at 800 MHz), thus generating the lowest resonant mode at about 800 MHz with reflection coefficient greater than −10 dB. The antenna operating mode and PIN diodes functions are listed in Table 5.1.

5.3.1 Antenna Mathematical Modeling

A PIFA and its supporting PCB are shown diagrammatically in Figure 5.5. This configuration can be decomposed into radiating and balanced modes as previously shown in [30]. The input impedance is given by:

$$Z_1 = \frac{(1+\alpha)^2 Z_R(Z_L + Z_B) + \alpha^2 Z_L Z_B}{(1+\alpha)^2 (Z_L + Z_R) + Z_B} \tag{5.1}$$

where Z_L is the load impedance applied to the unfed side of the strip, Z_R is the radiating mode impedance, and Z_B is the impedance of the short-circuit transmission line formed by the strip, derived from the balanced mode. The current sharing factor, α, is given by

$$\alpha = \frac{I_{R2}}{I_{R1}} \tag{5.2}$$

Table 5.1 Antenna Operating Modes and PIN Diode Function

Antenna Mode	*Polarization*	*Resonant Frequency*	*PIN Diodes* B	C	E	H
1	Horizontal	1.4 GHz	ON	OFF	OFF	OFF
2	Horizontal	800 MHz	ON	OFF	ON	ON
3	Vertical	1.4 GHz	OFF	ON	OFF	OFF
4	Vertical	800 MHz	OFF	ON	ON	ON

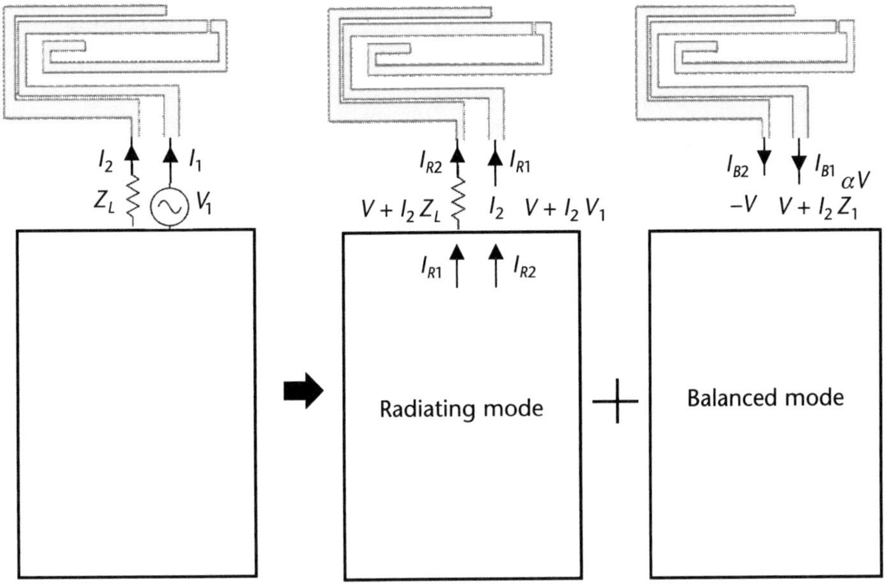

Figure 5.5 Antenna radiating and balanced mode operations.

I_{R1} and I_{R2} are the radiating mode currents at the feed and load, respectively. For the case of $Z_L = 0$, this equation simplifies to

$$Z_1 = \frac{(1+\alpha)^2 Z_R Z_B}{(1+\alpha)^2 (Z_R) + Z_B} \tag{5.3}$$

This is like the well-known expression for a folded dipole—the radiating mode is impedance transformed by a factor $(1 + \alpha)^2$ and adds in parallel with the balanced mode.

The equivalent circuit is illustrated in Figure 5.6. It should be noted that the impedance transformation in this mode is high, since I_{R2} is greater than I_{R1} due to the position of the slot. This tends to produce an impedance that is too high. It is also worthy of note that the bandwidth can be enhanced if Z_B and Z_R are resonant at the same frequency (when it becomes a double-tuning circuit). Also of interest is the open circuit case. Setting $Z_L = \infty$ gives

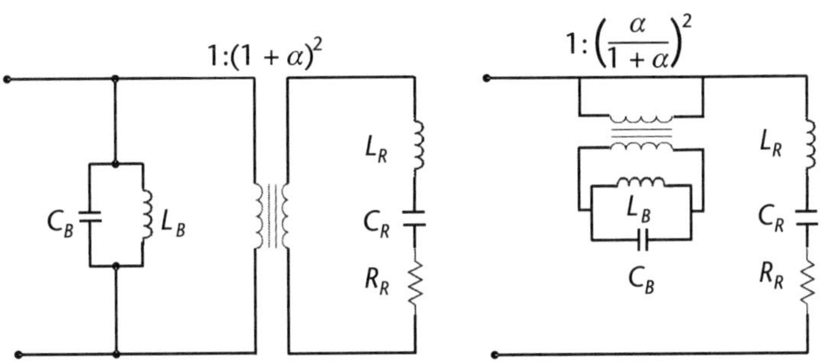

Figure 5.6 Equivalent circuit with a short-circuit load and open-circuit load.

$$Z_1 = Z_R + \left(\frac{\alpha}{1+\alpha}\right)^2 Z_B \tag{5.4}$$

The balanced mode impedance is transformed by a factor $(\alpha/(1+\alpha))^2$ and adds in series with the radiating mode. The balanced mode causes the antenna to resonate twice at lower and higher frequencies than determined by the radiating mode (controlled by L_R and C_R). The low-frequency and high-frequency resonances occur when the radiating mode capacitance and inductance are cancelled by the inductance and capacitance of the transformed balanced mode, respectively. There is no impedance transformation with an open circuit load, which tends to produce an impedance that is too low. With a reactive load, the antenna can be tuned over a wide range of impedances: from below the first resonance up to the second resonance of the open circuit mode. $Z_B = -j\Omega L_B/(1 - \Omega L_B C_B)$ and $Z_R = R_R + j\Omega L_R + (1/j\Omega C_R)$. The equivalent antenna circuit is validated using circuit simulation as shown in Figure 5.7 for 1.4 GHz.

Values used in the subsequent detailed design are $\propto = 1$, $R_R = 8.8$, $L_R = 4.6$ nH, $C_R = 1.9$ pF, $L_B = 2.3$ nH, and $C_B = 0.46$ pF. Using these values, the S-parameter for open-circuit and short-circuit loads is shown in Figure 5.7. With inductive loading, any frequency between the first resonance of the open circuit mode and the resonance of the short-circuit mode can be achieved. Capacitive tuning allows frequencies between the resonance of the short-circuit mode and the second resonance of the open-circuit mode to be covered. From the equivalent circuits with open-circuit and short-circuit loads, the open-circuit mode exhibits resonance at a lower resistance than the short-circuit mode.

For switching application, BAP64-03 PIN diode is sued in this antenna design. Equivalent circuit model of the diode in ON and OFF modes are shown in Figure 5.8. The PIN diode BAP64-03 is commonly used as the frequency-reconfigurable

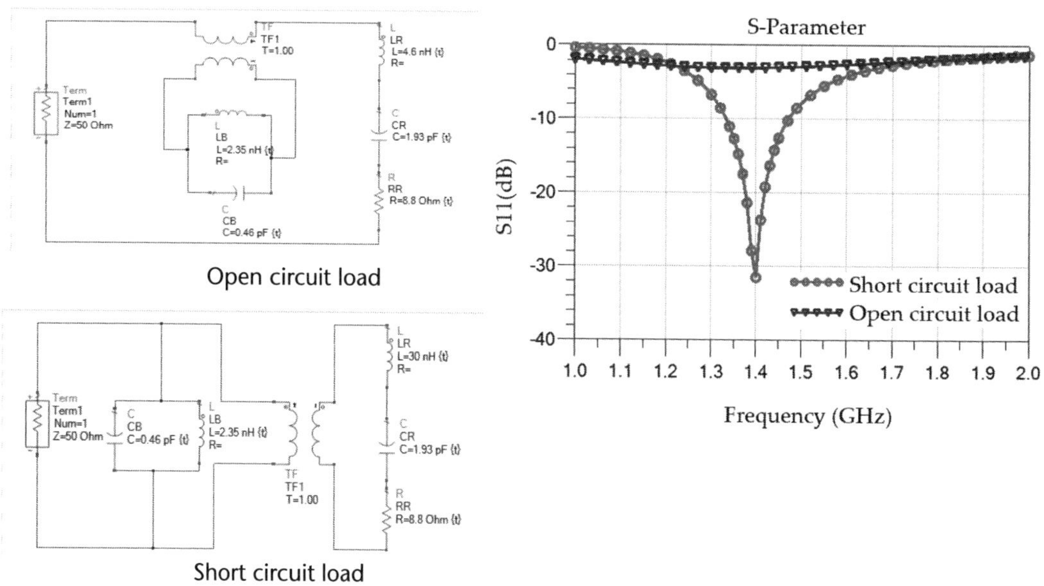

Figure 5.7 Simulation model in ADS schematic with short-circuit and open-circuit loads and simulated S-parameters.

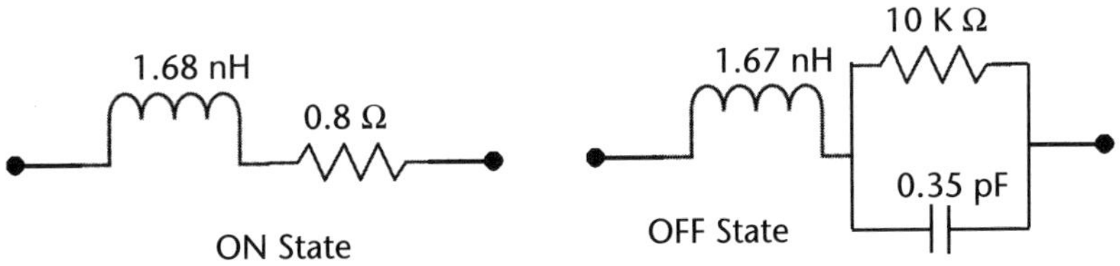

Figure 5.8 Equivalent circuit of BAP64-03 PIN diode ON state and OFF state.

antenna design. The circuit of the PIN diode switch in ON state is equivalent to a lumped element with 0.8Ω resistor series with the 1.68-nH inductor. The equivalent circuit of the PIN diode switch in the OFF state may be considered as a 1.67-nH series with a parallel circuit of 10 KΩ and 0.35 pF. A 100-pF DC block was connected to DC blocking in the antenna. When the diode is in the OFF state, the current will flow along the smaller strip. That means the resonant frequency will be at a higher frequency. When the diode is in the ON state, the major current will flow along the larger strip and then the major resonant frequency will be lower frequency. By changing the path of the current flow, the different resonant frequency can be achieved.

5.3.2 Antenna Operating Principle and Results

The horizontal antenna first resonant frequency is at 1,400 MHz, which can cover the LTE1400 bands. This resonant frequency is formed by section A-B-D-E as shown in Figure 5.4, by turning PIN diode placed at E OFF. The second resonant frequency at 800 MHz is formed by section A-D-E-F-G. To understand the operating principle of the antenna, Figure 5.9 shows the simulated reflection coefficient for different configurations of the designed antenna. As is clear from S_{11} results, the first resonant mode is centered at about 1,400 MHz. By adding an electromagnetically coupled shorted strip to the feeding strip (the designed antenna in the ON state), the antenna resonant mode shifts at about 800 MHz.

The PIN diodes are placed inside the inner strip. If the PIN diode switch is OFF, the geometrical of electrical path is small-tuned to 1,400 MHz. By changing the voltage on the PIN diode switch, it will be ON, and the electrical path of the inner strip becomes large-tuned to 800 MHz. To control the PIN diode, switch ON and OFF, a Vcc DC bias of +5V is applied through the inner strip. To have an impedance match to the desired band and reduce the RF signal to the DC source, a DC block capacitance of 100 pF is connected. The designed antenna is optimized using the FEM for the best return loss performance both for horizontal and vertical polarizations. By combining two working states of all four PIN diodes, the antenna is successfully tuned over 800 MHz and 1,400 MHz for horizontal and vertical polarization.

For horizontal polarization, the obtained simulated bands are 760–850 MHz and 1,380–1,470 MHz with a reflection coefficient better than –10 dB, covering 800 MHz and 1,400-MHz bands. For vertical polarization, the achieved bands are 760–810 MHz and 1,310–1,470 MHz, with a reflection coefficient better than –10 dB. It is always a good practice to allow the bandwidth of the antenna to operate below the –10-dB return-loss threshold. In the designed design, most of the bands are covered with a –10-dB reflection coefficient.

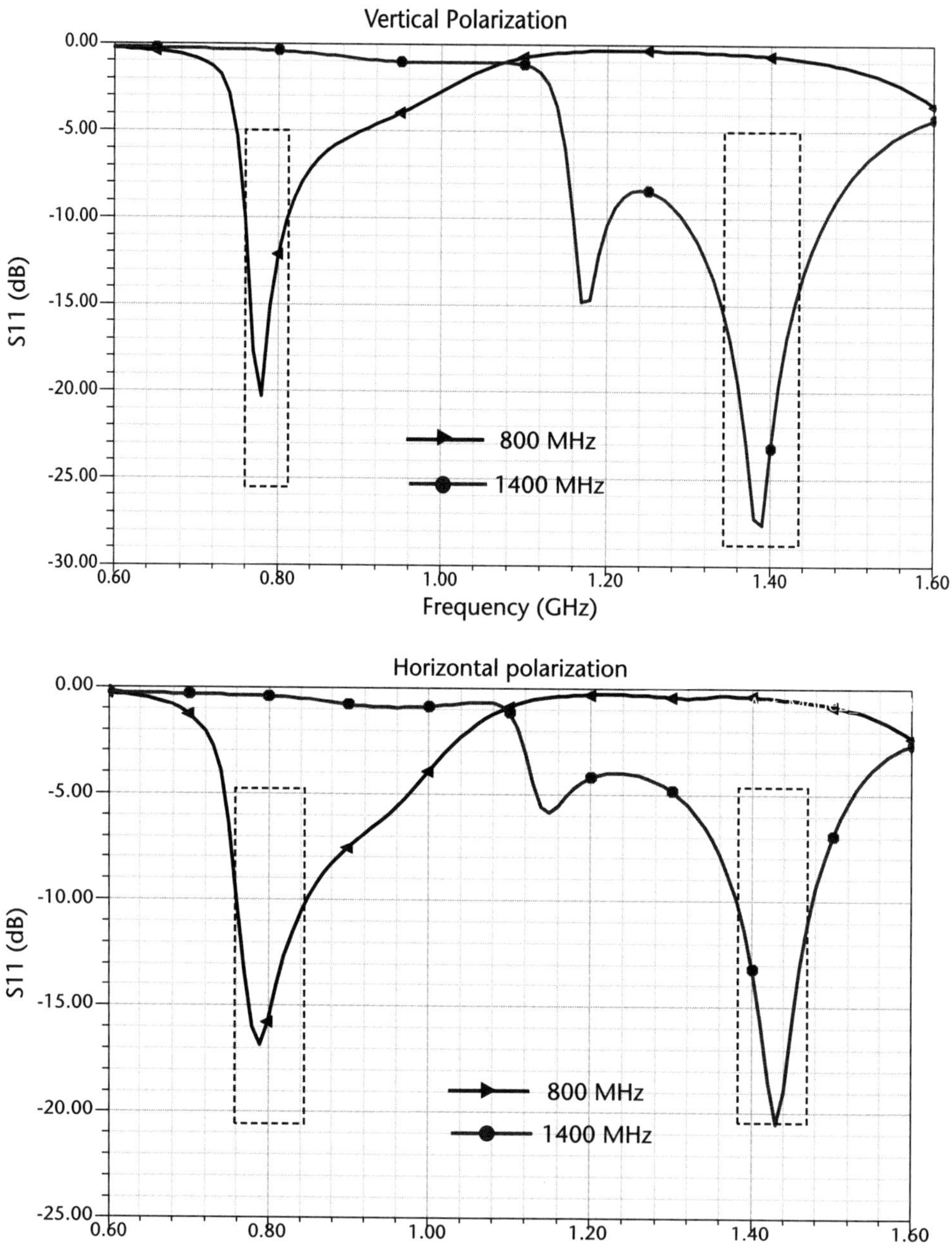

Figure 5.9 Antenna return loss plots for vertical and horizontal polarization.

Antenna 3-D simulated radiation patterns are at frequencies 800 MHz and 1,400 MHz for horizontal and vertical polarizations and are shown in Figure 5.10. Radiation patterns are nearly omnidirectionally suitable for mobile phone communication. As current nulls increase at higher frequencies, the radiation patterns show some distortions.

Table 5.2 listed simulated gain, radiation efficiency, and antenna bandwidth data. The antenna simulated gain varies from 3.7 dBi to 2.13 dBi, and the simulated

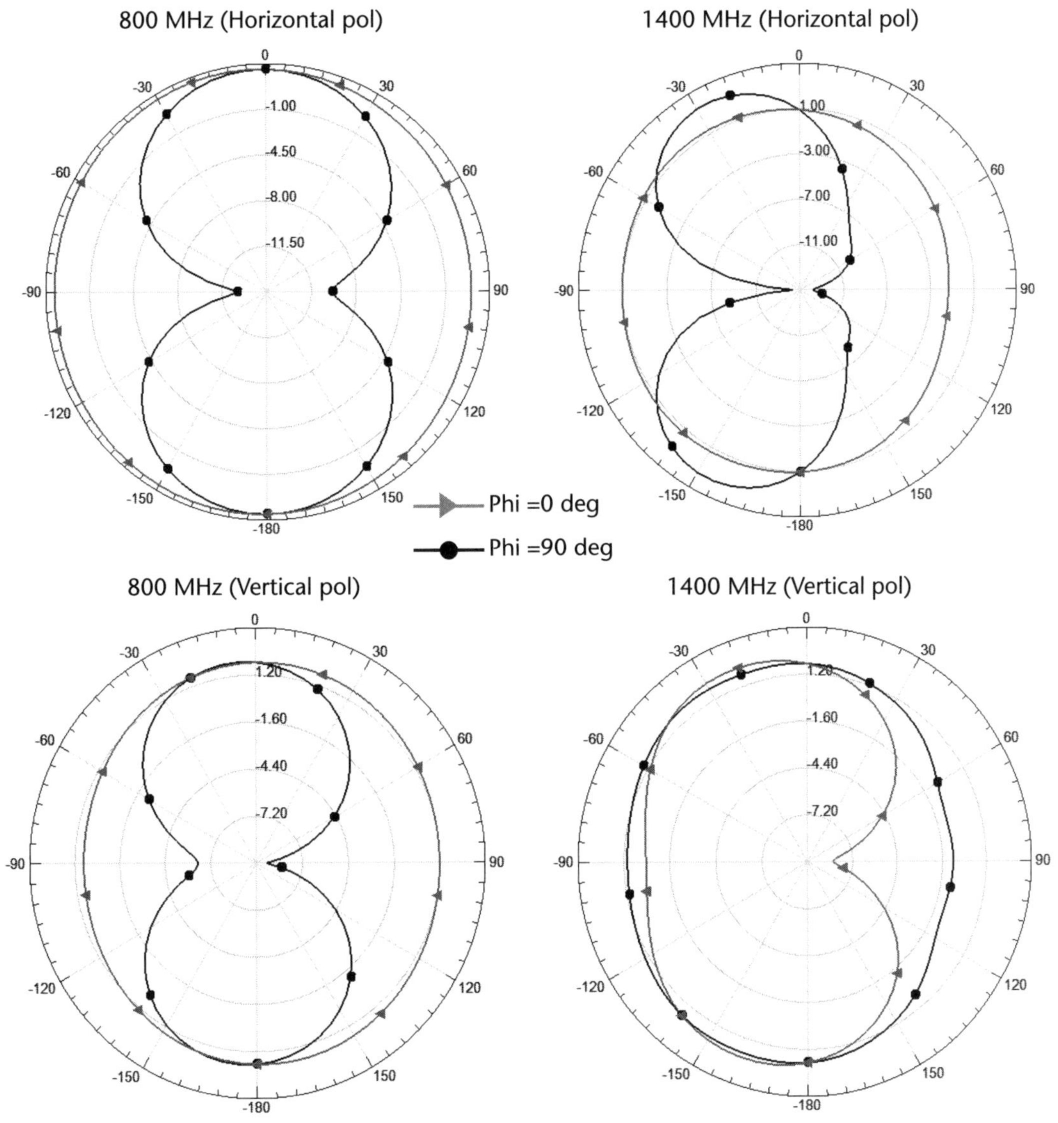

Figure 5.10 Antenna radiation patterns for horizontal and vertical polarization.

Table 5.2 Simulated Antenna Results

			Antenna Results		
Antenna Mode	*Polarization*	*Resonant Frequency*	*Gain (dB)*	*Efficiency (%)*	*Bandwidth (–10 dB)*
1	Horizontal	1.4 GHz	3.69	91	90 MHz
2	Horizontal	800 MHz	2.13	90	90 MHz
3	Vertical	1.4 GHz	2.58	92	50 MHz
4	Vertical	800 MHz	2.14	88	150 MHz

efficiency is about 88% to 92%. To further confirm the working principle of antenna and to clearly show which parts of the antennas are responsible for the radiation in the different bands, the study of the E-field distribution has been made. The surface current distribution at resonant frequencies is shown in Figure 5.11.

From the surface, the current distribution in the OFF state, and it can be observed that the resonant modes at about 1,400 MHz are mainly contributed by the feeding strip. However, due to strong coupling at higher frequencies, relatively small surface currents are also observed on the shorted strip at this band. Similarly, in the ON-state case, strong surface currents are observed on the shorted strip at about 800 MHz. At 1,400 MHz, the current is mainly located on the feeding monopole element, and at 800 MHz, strong currents appear on the coupled monopole. Antenna polarization is characterized by electric field direction.

5.4 Summary

In this chapter, various components, categorization, and the design process of printed reconfigurable antennas were presented. The various wireless applications where reconfigurable antennas are applied were discussed. A planar frequency-reconfigurable and polarization-reconfigurable antenna for two-band operation is designed to understand the working principle and design flow of such antennas. With the help of four PIN diodes, multifrequency band operation can be achieved for horizontal and vertical polarization. The modeling, control, and optimization of reconfigurable antennas were also detailed, and mechanisms were presented. This

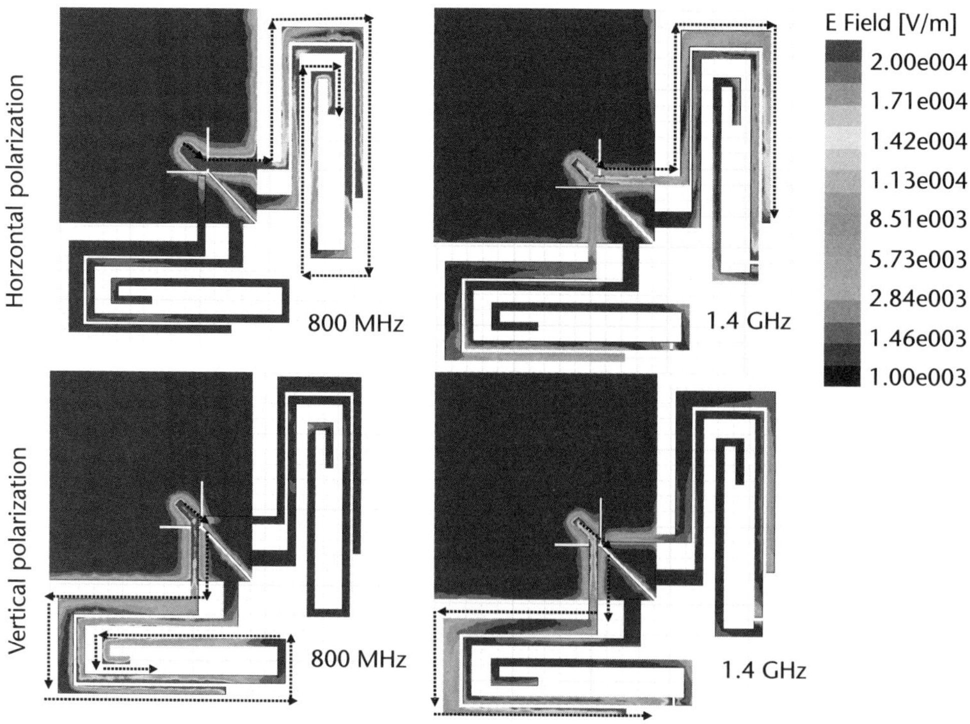

Figure 5.11 E-Field distribution in all four operating modes.

chapter emphasized the development of reconfigurable antennas to service evolving and futuristic practical wireless communication applications and proposed optimal solutions for more efficient and holistic designs.

References

[1] Park, Y. K., and Y. Sung, "A Reconfigurable Antenna for Quad-Band Mobile Handset Applications," *IEEE Transactions on Antennas and Propagation*, Vol. 60, No. 6, June 2012, pp. 3003–3006.

[2] Angus, C. K. M., et al., "Reconfigurable Multiband Antenna Designs for Wireless Communication Devices," *IEEE Transactions on Antennas and Propagation*, Vol. 55, No. 7, 2007, pp. 1919–1928.

[3] Songnan, Y., et al., "Frequency Reconfigurable Antennas for Multiradio Wireless Platforms," *IEEE Microwave Magazine*, Vol. 10, 2009, pp. 66–83.

[4] Christodoulou, C. G., et al., "Reconfigurable Antennas for Wireless and Space Applications," *Proc. IEEE*, Vol. 100, No. 7, July 2012, pp. 2250–2261.

[5] Balanis, C. A., *Modern Antenna Handbook*, New York: John Wiley & Sons, 2011.

[6] Bernhard, J. T., *Reconfigurable Antennas*, San Rafael, CA: Morgan and Claypool, 2007.

[7] Pringle, L. N., et al., "GTRI Reconfigurable Aperture Design," *IEEE Antennas and Propagation Society International Symposium (IEEE Cat. No.02CH37313)*, San Antonio, TX, 2002, pp. 473–476.

[8] Chen, R., and J. Row, "Single-Fed Microstrip Patch Antenna with Switchable Polarization," *IEEE Transactions on Antennas and Propagation*, Vol. 56, No. 4, April 2008, pp. 922–926.

[9] Erdemli, Y. E., R. A. Gilbert and J. L. Volakis, "A Reconfigurable Slot Aperture Design over a Broad-Band Substrate/Feed Structure," *IEEE Transactions on Antennas and Propagation*, Vol. 52, No. 11, November 2004, pp. 2860–2870.

[10] Behdad, N., and K. Sarabandi, "Dual-Band Reconfigurable Antenna with a Very Wide Tunability Range," *IEEE Transactions on Antennas and Propagation*, Vol. 54, No. 2, February 2006, pp. 409–416.

[11] Peregrine Semiconductor, "Digitally Tunable Capacitors," *Microwave Journal*, June 14, 2013.

[12] Ha, S. D., et al., "Quick Switch: Strongly Correlated Electronic Phase Transition Systems for Cutting-Edge Microwave Devices," *IEEE Microwave Magazine*, Vol. 15, No. 6, September-October 2014, pp. 32–44.

[13] Teeslink, T. S., et al., "Reconfigurable Bowtie Antenna Using Metal-Insulator Transition in Vanadium Dioxide," *IEEE Antennas and Wireless Propagation Letters*, Vol. 14, 2015, pp. 1381–1384.

[14] Brown, E. R., "RF-MEMS Switches for Reconfigurable Integrated Circuits," *IEEE Transactions on Microwave Theory and Techniques*, Vol. 46, No. 11, November 1998, pp. 1868–1880.

[15] Zhang, S., et al., "A Pattern Reconfigurable Microstrip Parasitic Array," *IEEE Transactions on Antennas and Propagation*, Vol. 52, No. 10, October 2004, pp. 2773–2776.

[16] Anagnostou, D. E., et al., "Reconfigurable UWB Antenna with RF-MEMS for On-Demand WLAN Rejection," *IEEE Transactions on Antennas and Propagation*, Vol. 62, No. 2, February 2014, pp. 602–608.

[17] Rebeiz, G. M., *RF MEMS: Theory, Design, and Technology*, New York: Wiley-Interscience, 2003.

[18] Nabki, F., et al., "Low-Stress CMOS-Compatible Silicon Carbide Surface-Micromachining Technology—Part II: Beam Resonators for MEMS Above IC," *Journal of Microelectromechanical Systems*, Vol. 20, No. 3, June 2011, pp. 730–744.

[19] Perruisseau-Carrier, J., P. Pardo-Carrera, and P. Miskovsky, "Modeling, Design and Characterization of a Very Wideband Slot Antenna with Reconfigurable Band Rejection," *IEEE Transactions on Antennas and Propagation*, Vol. 58, No. 7, July 2010, pp. 2218–2226.

[20] Lin, K. C., C. H. Lin, and Y. C. Lin, "Simple Printed Multiband Antenna with Novel Parasitic–Element Design for Multistandard Mobile Handset Applications," *IEEE Transactions on Antennas and Propagation*, Vol. 61, No. 1, pp. 488–491.

[21] Lee, C. T., and K. L. Wong, "Planar Monopole with a Coupling Feed and an Inductive Shorting Strip for LTE/GSM/UMTS Operation in the Mobile Phone," *IEEE Transactions on Antennas and Propagation*, Vol. 58, No. 7, July 2010, pp. 2479–2483.

[22] Lu, J. H., and Y. S. Wang, "Internal Uniplanar Antenna for LTE/GSM/UMTS Operation in a Tablet Computer," *IEEE Transactions on Antennas and Propagation*, Vol. 61, No. 5, May 2013, pp. 2841–2846.

[23] Wong, K. L., and T. W. Weng, "Small-Size Triple-Wideband LTE/WWAN Tablet Device Antenna," *IEEE Antennas Wireless Propag. Lett.*, Vol. 12, November 2013, pp. 1516–1519.

[24] Wong, K. L., and M. T. Chen, "Small-Size LTE/WWAN Printed Loop Antenna with an Inductively Coupled Branch Strip for Bandwidth Enhancement in the Tablet Computer," *IEEE Transactions on Antennas and Propagation*, Vol. 61, No. 12, December 2013, pp. 6144–6151.

[25] Lu, J. H., and J. L. Guo, "Small-Size Octaband Monopole Antenna in an LTE/WWAN Mobile Phone," *IEEE Antennas Wireless Propag. Lett.*, Vol. 13, March 2014, pp. 548–551.

[26] Sung, Y., "A Frequency-Reconfigurable Antenna for Internal Mobile Handset Applications," *Microw. Opt. Technol. Lett.*, Vol. 56, No. 6, June 2014, pp. 1366–1371.

[27] Lee, D. A., and Y. Sung, "Reconfigurable PIFA/Loop Antenna for Mobile Handset Applications," *Microw. Opt. Technol. Lett.*, Vol. 56, No. 9, September 2014, pp. 2034–2037.

[28] Lee, J. H., and Y. Sung, "Reconfigurable Hexa-Band Planar Inverted-F Antenna Using a Pin Diode for Mobile Handset," *Microw. Opt. Technol. Lett.*, Vol. 55, No. 8, August 2013, pp. 1926–1928.

[29] Chuang, H. -R., and L. -C. Kuo, "3-D FDTD Design Analysis of a 2.4-GHz Polarization-Diversity Printed Dipole Antenna with Integrated Balun and Polarization-Switching Circuit for WLAN and Wireless Communication Applications," *IEEE Transactions on Microwave Theory and Techniques*, Vol. 51, No. 2, February 2003, pp. 374–381.

[30] Boyle, K. R., and L. P. Ligthart, "Radiating and Balanced Mode Analysis of PIFA Antennas," *IEEE Transactions on Antennas and Propagation*, Vol. 54, No. 1, January 2006, pp. 231–237.

Selected Bibliography

Boukarkar, A., et al., "Compact Mechanically Frequency and Pattern Reconfigurable Patch Antenna," *IET Microwaves, Antennas & Propagation*, Vol. 12, No. 11, September 12, 2018, pp. 1864–1869.

Christodoulou, C., J. Constantine, and Y. Tawk, *Antenna Design for Cognitive Radio*, Norwood, MA: Artech House, 2016.

Jiang, Z., and F. Yang, "Reconfigurable Sensing Antennas Integrated with Thermal Switches for Wireless Temperature Monitoring," *IEEE Antennas Wireless Propag. Lett.*, Vol. 12, 2013, pp. 914–917.

Jin, G. P., D. L. Zhang, and R. L. Li, "Optically Controlled Reconfigurable Antenna for Cognitive Radio Applications," *IET Electron. Lett.*, Vol. 47, No. 17, 2011, pp. 948–950.

Xu, H., et al., “A Compact and Low-Profile Loop Antenna with Six Resonant Modes for LTE Smartphone,” *IEEE Transactions on Antennas and Propagation*, Vol. 64, No. 9, September 2016, pp. 3743–3751.

Xu, Z., et al., “Reconfigurable MIMO Antenna for Integrated-Metal-Rimmed Smartphone Applications,” *IEEE Access*, Vol. 5, 2017, pp. 21223–21228.

CHAPTER 6

Printed Antenna Arrays

6.1 Introduction

Printed antennas are used as a single element and are also very popular in arrays [1–16]. Microstrip antennas are well suited for arrays due to their low profile and low weight with conformability and low manufacturing cost. In various radar and communications systems, planar microstrip array antennas are greatly desired. Antenna arrays are used to increase the directivity and gain, scan the beam of an antenna, and enhance various other functions, which would be difficult with a single-element antenna. The main limitation of microstrips is that they radiate efficiently only over a narrow band of frequencies and they cannot operate at the high-power levels of the coaxial line, waveguide, or even stripline. This can be reduced with the help of various array configurations, dielectric materials, feeding methods, and ground planes.

Phased array antenna is a combination of multiple antenna systems, in which, antenna radiation pattern can be reinforced in a direction and suppressed in undesired directions [17–24]. The direction of phased array radiation can be electronically rotated, avoiding the need for any mechanical rotation mechanism. Phased arrays are used in a broad range of applications since the advent of this advanced technology because of these unique capabilities. Elements of an array may be spatially distributed to form a linear, circular, planar, or volume array. Linear arrays have antenna elements arranged in a line, while planar arrays have elements distributed on a plane. Planar array configurations are widely used in both communication and radar systems where a narrow pencil beam is required in both the azimuth and elevation planes. Individual array elements can be positioned along a rectangular grid to form a planar array for better control of the beam shape and position in space. Microstrip patch arrays are also versatile as they can be used to synthesize a required antenna pattern that cannot be achieved with a single element. The elements can be fed by a single line or by multiple lines in a feed network topology as per specifications and requirements. In this chapter, the printed array antenna working principle and designs are presented for several modern applications.

6.2 Microstrip Antenna Arrays

In an antenna array, the signals of all individual elements are coordinated to achieve better antenna gain, directivity, and SNR over those of a single antenna. The design of finite arrays demands to group several radiating elements in a symmetrical pattern so that the desired radiation is obtained in a specific direction. This can be realized only when fields due to individual patches get combined in phase in the desired direction and cancel each other in all other directions. In other words, each patch output is combined constructively to obtain the fields radiated by the array. It is important to note that radiation pattern of an individual element in the array is the same when it operates in standalone mode. The original pattern of the individual patch gets multiplied by the antenna array factor that takes into consideration the amplitudes and phases of the feed current. In the array, due to closer proximity between patches, there is an interaction between the elements. This leads to coupling among the radiating patches as each patch element induces currents to the other adjacent patches. Interelement spacing and location between the arrays affect the radiation pattern as well as antenna parameters.

6.3 Array Antenna Radiating Elements

The most basic property of the array is that the relative displacements of the antenna elements with respect to each other introduce relative phase shifts in the radiation vectors, which can then add constructively in some directions or destructively in others [25]. Many theoretical studies on antenna arrays have been published [25–28]. A linear array antenna is the primary configuration and can be applied to planar arrays. Each element is an electric or magnetic current source, which gives rise to a radiated field, the solution to Maxwell's equations [27]. In the array design below are controlling parameters that are used to shape the overall pattern of the antenna:

- The geometrical configuration of the overall array (linear, circular, planar);
- Interspacing between array antenna elements;
- Phase and amplitude of excitation to each element;
- Radiation pattern of each element.

The far-zone field of a uniform array is equal to the product of the field of a single element and the array factor. This is called pattern multiplication.

$$E_{Total} = [E_{Single\ Element}]x\ [Array\ Factor] \tag{6.1}$$

The array factor [29] is given by:

$$AF = \sum_{n=1}^{N} e^{j(n-1)}\varphi \tag{6.2}$$

where $\phi = kd\cos\theta + \beta$. The pattern for an N-element array with equal spacing and a constant phase shift between elements is given by some algebraic manipulation of (6.2) as:

$$E_{Total} = [E_{Single\ Element}]x\frac{1}{N}\frac{\sin[N(kd\sin(\theta)-\beta)/2]}{\sin[(kd\sin(\theta)-\beta)/2]} \tag{6.3}$$

The spacing should be determined such that mutual coupling among the antenna elements is minimized (often <–20 dB) and any grating lobe is invisible during beam scanning. To avoid grating lobes, the array spacing must satisfy:

$$d < \frac{\lambda}{1+|\sin\theta_{max}|} \tag{6.4}$$

where d is the array spacing and $\theta_{\max}$is the maximum scan angle [30]. In array design, the interelement spacing lies between λ_g and $\lambda_g/2$. Arrays having element spacing between λ_g to $2\lambda_g$ will radiate not only a broadside beam but also two grating lobes at angles (measured from the broadside) given by:

$$\theta_0 = \sin^{-1}\left(\frac{n\lambda_0}{d}\right) \tag{6.5}$$

where $n = \pm 1$ corresponds to the appearance of the first set of grating lobes, $n = \pm 2$ corresponds to the appearance of the second set of grating lobes, and so forth. d is the element spacing.

6.3.1 Linear Microstrip Arrays

Linear arrays form a building block for planar arrays and are used for fan-shaped coverage. A linear array of a microstrip patch is shown in Figure 6.1 with equal space d between elements. All elements are exited with equal amplitude and phase to get a high gain antenna beam direction in broadside. Each patch element can be excited either individually using a separate port or with a transmission-line feed network.

The length of the microstrip feed line between patches determines the phase distribution across the antenna array elements. The separation between two array

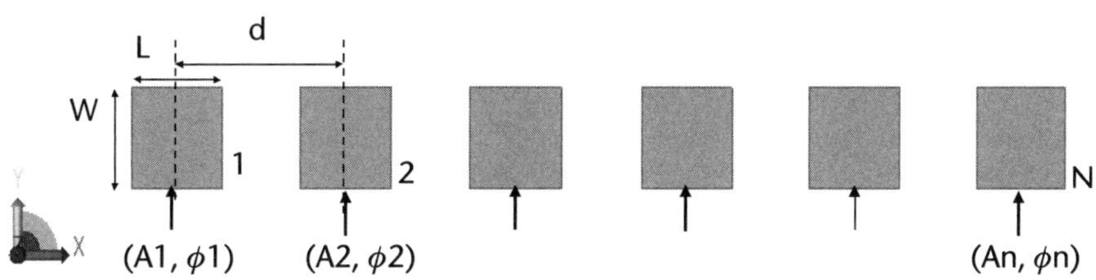

Figure 6.1 Linear microstrip antenna array configuration.

elements (i.e., an interelement spacing) is a vital parameter in array design. The closely spaced array elements suffer from strong mutual coupling effects. This affects the input and radiating characteristics of the antenna array and degrades the performance. The variation in the spacing changes the sidelobe level and the beamwidth of the main lobe. Many designs of linear array antennas are discussed in this chapter.

6.3.2 Planar Microstrip Antenna Arrays

In the planar array, all the elements are arranged in one plane as shown in Figure 6.2. In addition to placing elements along a line to form a linear array as discussed in previously, individual radiators can be placed along a rectangular grid to form a rectangular or planar array. Planar arrays provide additional variables that can be used to control and shape the pattern of the array in two directions. A planar array provides a large aperture and is used to form a pencil beam by varying the relative phase of each element. Planar arrays are also used as a reflector behind the active plane. The array factor, depending on the array geometry and excitation, is the pattern of an array of all the elements in the plane.

6.4 Feed Techniques for Array Antenna

Feed network designing is the most important part in array design. The feed network for an array application is a major design concern in terms of complexity and size. There are several kinds of feeding networks of an array antenna. Array feed network topology depends on antenna type, available space, pattern shape, sidelobe level, losses, and geometry. The efficiency of microstrip antenna arrays may

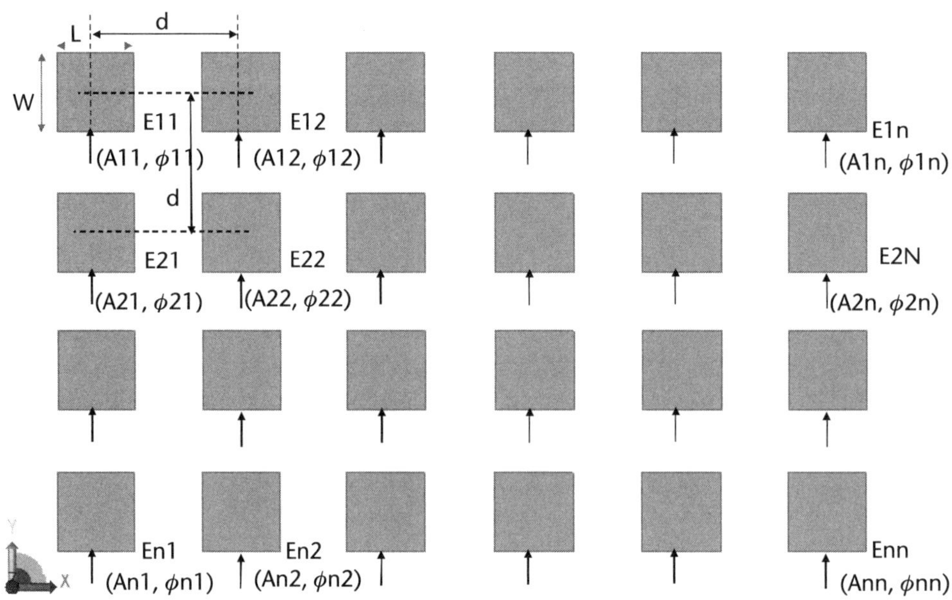

Figure 6.2 Microstrip planar antenna with equal spacing between elements.

be improved by reducing losses in the feed network. Losses in the microstrip feed network are due to dielectric loss, conductor loss, and radiation loss.

6.4.1 Corporate Feed Network

For a passive array antenna, the most common feeding network type is the corporate feed network as shown in Figure 6.3 [31–33]. This feeding technique integrates the RF feed lines with the radiating elements on the same substrate layer. As there are many microstrip lines in a corporate feed, it is lossy, particularly at high frequencies. Losses in the feed line degrade the microstrip antenna arrays' efficiency. The efficiency of microstrip antenna arrays can be improved by using a waveguide feed network. However, a waveguide-based feed network results in a significant increase in the antenna weight and dimensions. A corporate feed is also called parallel feed network. The corporate feed is composed of a power divider with transmission lines to the individual radiators. Although a corporate feed method is physically large and expensive, it is also the most broadband. The overall bandwidth of array is limited by the bandwidth of the radiators as well as feedlines. Corporate feeds are a suitable choice for phased arrays where phase shifters are inserted in the transmission lines to steer the beam direction electrically.

6.4.1.1 High Port Isolations with Wilkinson Power Divider

A Wilkinson power divider is used for high port isolation while maintaining a matched condition on all ports, especially in the case of a dual-polarized antenna. In the printed antenna array, a simple way to implement any type of uniform or nonuniform excitation is by using microstrip lines, achieving different feed amplitudes through Wilkinson or the T-junction power splitters. The T-junction is lossless and simple and takes up little area in the PCB, but its ports are not matched and there is no isolation between output ports. The Wilkinson divider divides the power in a similar way to the T-junction, but, unlike T-junctions, all ports are matched. In this power divider, reflected power is dissipated in the isolation resistor. The isolation

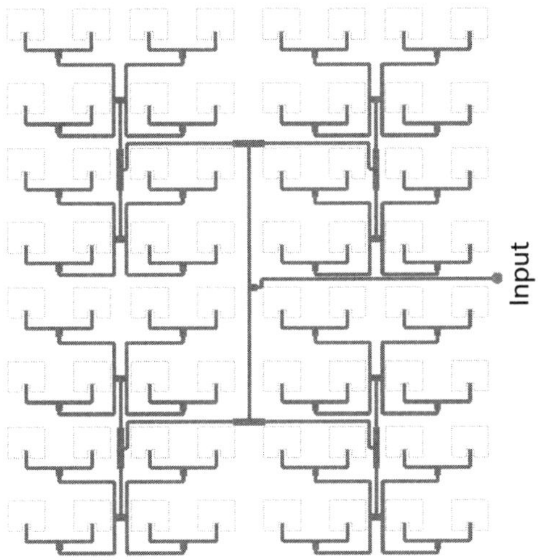

Figure 6.3 Microstrip array antenna fed with a corporate feed network.

resistor used for the Wilkinson divider introduces minimal loss, especially when microstrip transmission lines are used along with low-loss PCB substrate material.

Ernest Wilkinson invented this power divider in 1960 [34]. The S-parameter matrix for two-port equal power division is given by

$$[S] = \frac{-j}{\sqrt{2}} \begin{bmatrix} 0 & 1 & 1 \\ 1 & 0 & 0 \\ 1 & 0 & 0 \end{bmatrix} \tag{6.6}$$

The S-parameter indicates that the network is reciprocal ($S_{ij} = S_{ji}$), all ports are matched ($S_{ij} = 0$), and its lossless network is reciprocal ($[S]^H[S] = [I]$). The Wilkinson power divider is the most common topology [35–38] for the 3-dB power divider, where two parallel quarter-wavelength transmission lines are used for impedance transformation. For 50Ω input and output, the central $\lambda_g/4$ transmission line impedance is 70.7Ω. Usually, at high frequencies, the transmission lines must be separated for some distances to reduce parasitic coupling. Figure 6.4 shows a 3-dB Wilkinson power divider with a surface mount chip resistor designed for a center frequency of 7 GHz. RO 3003 of a 0.38-mm-thick substrate is used for design. This divider has a 50Ω microstrip line with a 0.96-mm width for easy fabrication. A 70.7Ω microstrip line width is 0.54 mm. The simulated S-parameter data as shown in Figure 6.4 illustrates high port isolation.

The 1:4 divider has been fabricated and measured at 7 GHz as shown in Figure 6.5. In a corporate feed network, a 1:4 Wilkinson power divider plays an important role to achieve good isolation between the output ports.

6.4.2 Inline Series Feed Network

Series-fed microstrip patch antenna arrays are used in many applications in wireless communication and radar array systems. In a series-fed array topology, the input signal is coupled serially to the antenna elements as it progresses. A series-fed network is compact as compared to corporate feed, which is one of the main

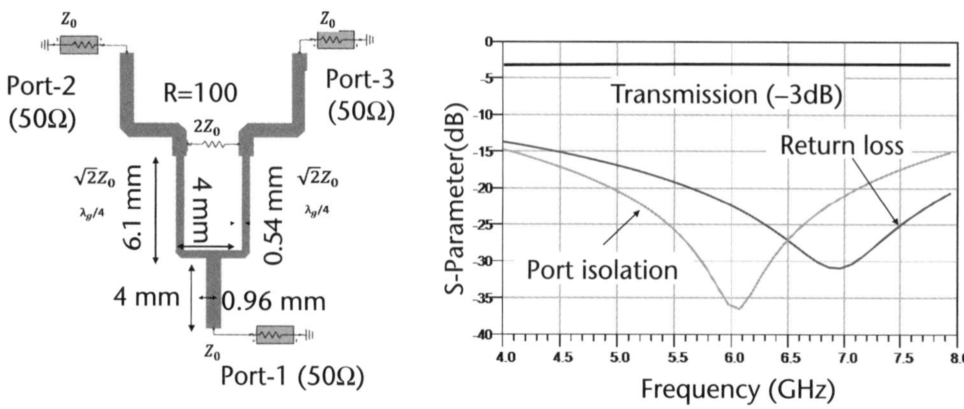

Figure 6.4 Wilkinson power divider design and S-parameter data.

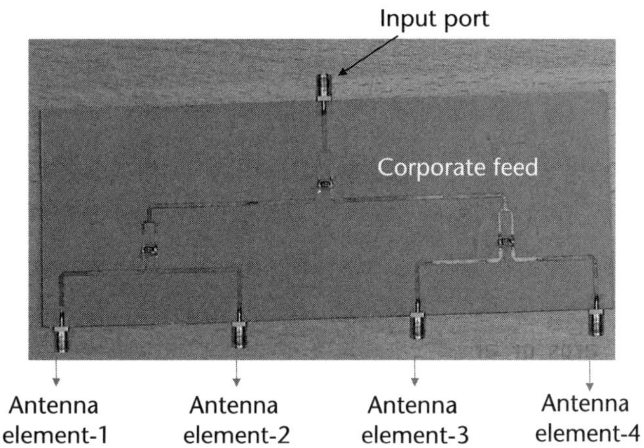

Figure 6.5 Fabricated 1:4 Wilkinson power divider.

advantages that make them more attractive than their parallel-fed counterparts. The small size of series-fed arrays results in less radiation and insertion losses, which is why, in many cases, the series-fed arrays are more efficient than corporate fed arrays. One major disadvantage of series feed configurations is narrow bandwidth and an inherent phase difference caused by the differences in the lengths of feed lines.

Figure 6.6 shows an end-fed series feed network. Radiators are coupled to a single transmission feed line and energy decays along the line length. Another configuration is a middle series-fed array as shown in Figure 6.7, in which input RF power is fed at the center of the feed network. Efficient design and fabrication of planar array antennas with low sidelobe levels are a challenging task. For reducing the sidelobe level of a uniform array, one technique utilizes amplitude tapering in which the excitation amplitudes of the array elements generally decrease with distance from the center of the array. To obtain the amplitude tapering in a linear array, a series-fed network with variable length and impendences is used.

6.4.3 Multilayer Feed Network

For broadband applications, the feeder network in the array antenna should preferably be made using a corporate network. However, for large linear arrays, the

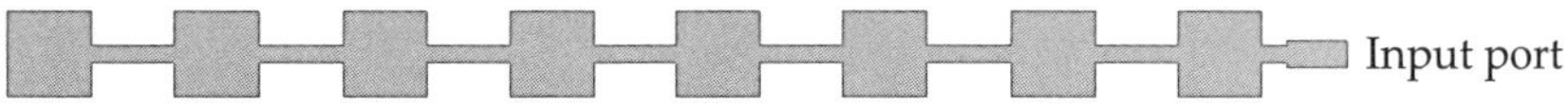

Figure 6.6 Inline end series-fed microstrip patch antenna.

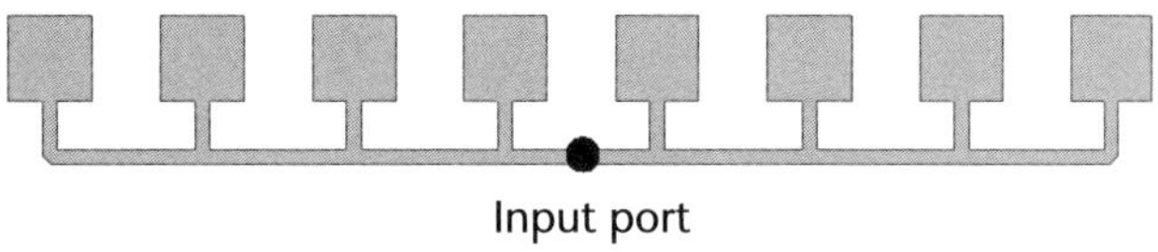

Figure 6.7 Middle series-fed microstrip patch antenna.

corporate feed network in one layer occupies a large area because the number of stages is N where 2^N is the number of output ports. For larger arrays, the number of stages increases accordingly, and thus the realization of corporate feed networks is not feasible for larger arrays. One possible solution to use the corporate feeder network for large linear arrays is to use the different stages of the feeder network in different layers, thus breaking the feed network into many parts and placing them on separate layers that can be joined by vias. An alternative solution for this is to use a series feed network, which may be realized using one stage only, but the serious limitation with this configuration of a feed network is that, when the frequency is changed, there is a progressive phase difference and the beam is scanned, resulting in frequency scanning. To solve this problem, the series and the corporate combination may be used so that the linear phase error is symmetrical about a mid-point so that the overall phase error is of quadratic nature. The quadratic phase error results in the broadening of beamwidth and the null feeling. Thus, the realization of the corporate feed network in a multilayer configuration facilitates achieving a broadband dual-polarization operation in the limited interelement space in the planar array antennas.

Printed antennas integrated with the feed network and active circuitry in a multilayer architecture as shown in Figure 6.8 are the latest technology trend. This approach allows the integration of antenna elements with complex processing circuitry maintaining the fabrication and functional modularity at different levels while overcoming some fundamental constraints. Different circuit levels are isolated from each other by placing metal planes in between, leading to parallel-plate-type circuit environments. Multilayer feed design configuration is selected according to the requirement and specifications. It depends upon available space, the power-handling capability of individual feed layers, complexity, and some critical RF parameters. The use of slot as a means of power transfer between layers makes the use of multilayer feeding networks attractive. The transfer of power between two adjacent layers in a multilayer structure can be done using either a pin, which makes an electrical contact, or via coupling apertures in the ground plane separating the layers. Aperture coupling leads to inherently symmetric feed network

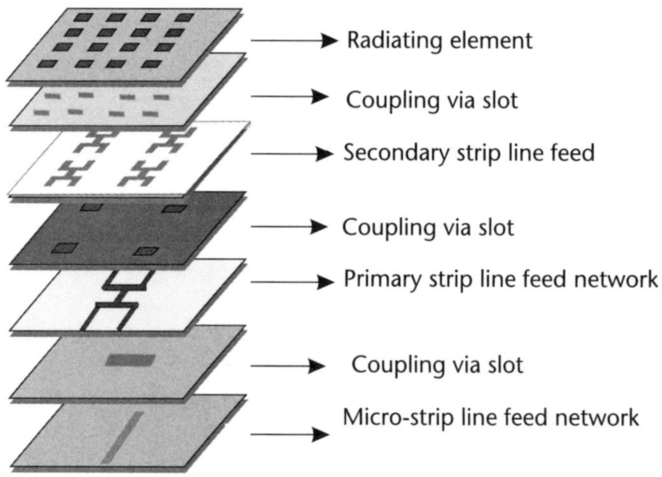

Figure 6.8 A multilayer feed architecture of the printed antenna.

layouts, which further simplify the design. Multilayer feed networks are very effective in highly integrated SiP (system-in-package) RF modules in which antenna elements are connected to the output of different components placed on the top and bottom of the chip. MMIC and LTCC-based modules along with integrated antennas use this feed network topology.

6.4.4 Feed Network for Shaped Beam Pattern Arrays

Some systems, such as airborne SAR, require a shaped antenna pattern along with dual-polarization and low cross-polarization capability. For such a requirement, the feeding network must incorporate matching circuits enabling matched wideband operation of antenna elements together with different types of power dividers and combiners providing proper, usually nonuniform, power distribution to antenna elements. A feed network called the beamforming network (BFN) is used to provide dual-linear polarization with low cross-polarization capability as well as good isolation between the individual ports. Such feed networks can be designed using a multilayer structure where different parts of BFN are placed on different layers. Sometimes special feed networks are designed to obtain shaped antenna radiation patterns like the $cosec^2$ radiation pattern. Each port amplitude and phase is calculated using null perturbation technique that generates the $cosec^2$ shape pattern. The feed network for a shaped pattern array is covered in detail in Chapter 8.

6.4.5 Feed Network for Active Antenna Arrays

Active phased arrays are usually composed of a feed network and many phase shifters. In the active phased-array antenna, transmit power is produced by many TX/RX (transceive) modules. In these antennas, each element or group of elements are connected to a separate transceiver module. Active antennas are primarily used in remote-sensing satellite, adaptive antenna, 5G phased array antennas, and radars. Feed networks are used to distribute the output signal of the transmitter to the radiation elements, and phase shifters control the phase of the signals at each radiating element to form a beam in the desired direction. Feed network design is a very critical task for such a system because of impedance matching with active components and effectively controlling amplitude and phase distribution across the network. Feed network design for active phased array radar and 5G FD-MIMO is covered in Chapter 15. In some cases, a combination of various feed networks also called the hybrid feed topology is used.

6.5 Printed Antenna Array Design for Rainfall Radar

In this section, a microstrip antenna array design is explained with a rainfall radar antenna called the Precipitation Occurrence Sensor System (POSS) design, as shown in Figure 6.9. The design has been done for the C-band application at 5.8 GHz. An array consisting of 64 elements of 8×8 is fed with the corporate feed network. RT Duroid 5870 substrate of dielectric constant 2.33 and height of substrate equal to 0.5 mm are used. The antenna gain of 25.1 dBi is achieved. The bandwidth of antenna is 150 MHz for a return loss of –10 dB. The array antenna is excited by

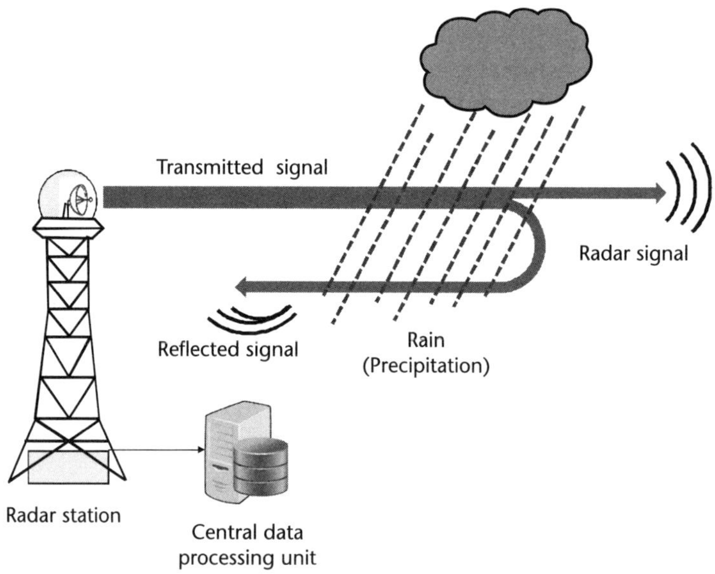

Figure 6.9 Working principle of a Doppler weather rainfall radar.

uniform amplitude and phase feeding method. In rainfall measurement, the antenna plays a critical role in terms of accuracy and performance. Such a system requires a low-cost, small-size, minimal-weight antenna capable of producing good performance at the desired frequency.

A microstrip antenna is very suitable for this application; it is very economical, is easy for fabrication, installation, and requires very little space. However, the gain of the antenna is very low; hence, to improve the directivity, antenna arrays are used. The complete design flow of an array antenna is shown in Figure 6.10. The design starts with the optimization of the single radiating element. This is a very crucial step for optimum gain and bandwidth performance of an array antenna. The array antenna feed is designed based on the available space, the power-handling requirement, and the shape of the beam and the bandwidth of the antenna as discussed in the previous section. Antenna topology decides the polarization and shape of the antenna; it may be a rectangular array, a circular array, or a customized shape. First, a linear array is optimized because it is easy to make a prototype

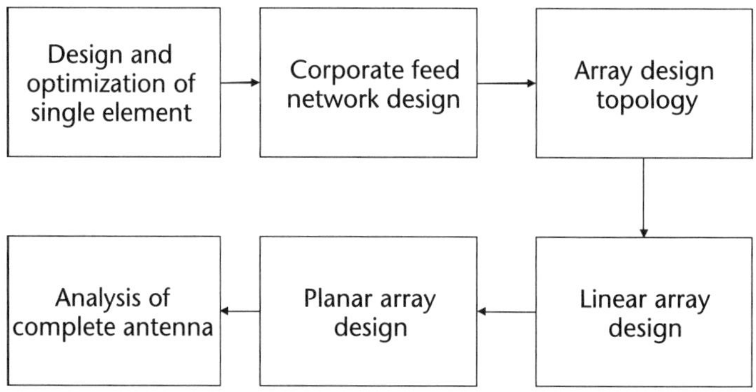

Figure 6.10 Printed array antenna design flow diagram.

of the linear array from the cost and fabrication point of view. Finally, the complete planar array is designed and simulated using EM solvers. To meet the desired bandwidth, gain, half-power beamwidth, low cross-polarization, low sidelobe levels, and antenna main lobe direction, various innovating techniques are applied. In the next section, these techniques are discussed in various presented array designs in this book.

6.5.1 Antenna Design Specifications

Rainfall antenna design specifications are driven by the gain and antenna beamwidth requirement. For a typical gain, this antenna is 35 dBi and the half-power beamwidth requirement is 1°. All important antenna design specifications are listed in Table 6.1.

6.5.2 Design and Optimization of a Single Element

The microstrip patch antenna array has been designed at 5.8 GHz on an RT Duroid 5870 substrate with the relative permittivity (ε_r) of 2.33. The antenna element is fed with 50Ω inset feed with equal power division and phase. The patch is designed using the conventional formulas mentioned in Chapter 2. The edge impedance of the patch is very high and generally varies from 100Ω to 200Ω. Impedance decreases gradually from the edge to the center of the patch and at one where impedance is 50Ω; a feed line of 50Ω with a width of 1.45 mm is used to feed the patch as shown in Figure 6.11. The size of the substrate is chosen in such a way that there remains at least $\lambda/4$ distance from the edge of the patch to the edge of the substrate.

The antenna is simulated using the full-wave analysis technique MoM. The single antenna gain and bandwidth are 6.7 dBi and 140 MHz, respectively. Microstrip antennas are a narrowband antenna. In the next section, some techniques are discussed to increase an antenna bandwidth. In any array, the single microstrip element is building a block of design; hence, it is very necessary to optimize the

Table 6.1 Antenna Design Specifications

Antenna Size	*Value*
Operating frequency	5.8 GHz
Polarization	Linear
3-dB beamwidth	1°
Gain	35 dB
First sidelobe level	–26 dB
Cross-polarization	–27 dB
VSWR	1.5
Antenna weight	80 kg
Antenna size	1.2m × 1.2m
Shape	Rectangular

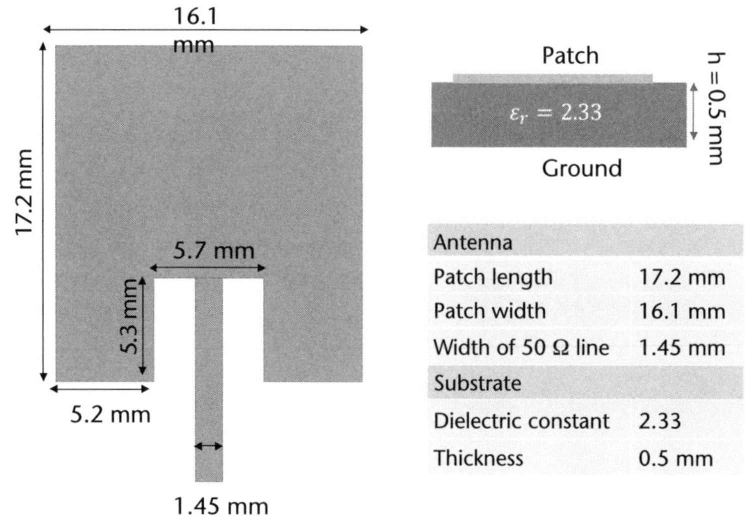

Antenna	
Patch length	17.2 mm
Patch width	16.1 mm
Width of 50 Ω line	1.45 mm
Substrate	
Dielectric constant	2.33
Thickness	0.5 mm

Figure 6.11 Single microstrip antenna element dimensions.

design for the best return loss, bandwidth, and gain before proceeding to the array design.

6.5.3 Design of the Corporate Feed Network

The array is fed with a corporate feed, in which equal power is available for each element in the array. The spacing between two elements is important, which varies from 0.5λ to λ. Spacing less than a half-frequency space wavelength is generally not used because of feed-line overcrowding. As we increase the spacing between the elements, it has been observed that the sidelobes increase significantly and the beamwidth reduces. When the spacing between the elements has reduced, the sidelobes reduce and the beamwidth increases. Thus, there is a trade-off between the sidelobe level and beamwidth of the array. If the spacing between the elements exceeds λ, then grating lobes are observed. In this design, interelement spacing of 0.75λ is selected so that we get narrow beam and sufficient space for the feed lines.

In a corporate feed network, the microstrip transmission line, bend, and junction with a quarter-wave transformer are the three main components, as shown in Figure 6.12. The width of the transmission line is calculated based on the input impedance of the antenna and the feeding location on the patch. For a bending design, there is no best way to bend a microstrip or stripline transmission line. Bend discontinuity changes the line characteristic impedance; without compensation, the bend adds shunt capacitance. However, the small capacitance usually does not change the circuit's performance very much. Another problem associated with bends is that it can cause far more damage to the intended performance of a highly tuned circuit: the effective length of the transmission line becomes shorter than the center-line length.

When a 90° bend is made of a microstrip transmission line, it adds a small amount of capacitance. Mitering the bend chops off some capacitance, restoring the line back to its original characteristic impedance. Figure 6.13 shows the important parameters of a mitered bend. The optimum mitered bend equations for a

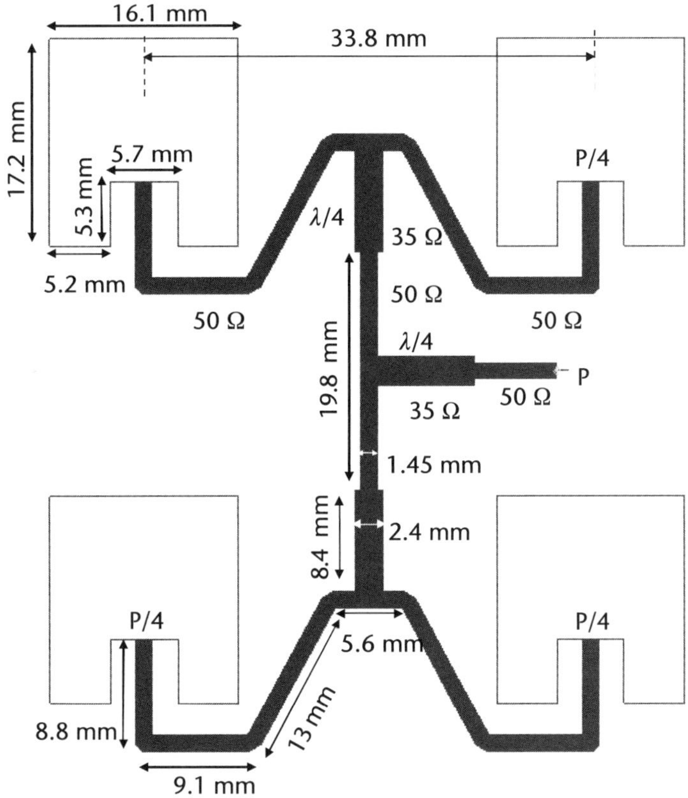

Figure 6.12 A 2 × 2 planar array fed with a corporate feed network.

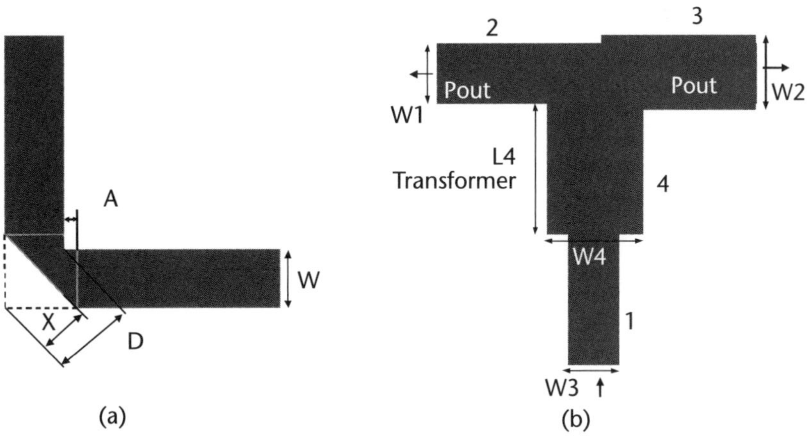

Figure 6.13 Feed network elements: (a) bend and (b) junction with quarter-wave transformer.

microstrip with a width W is given in (6.7) to (6.9). In this antenna array design, the bend has $X = D/2$ and $A = 0$. This bend gives good compensation for higher frequencies.

$$D = W\sqrt{2} \tag{6.7}$$

$$X = W\sqrt{2}\left(0.52 + 0.65e^{\left(-1.35\frac{W}{h}\right)}\right) \tag{6.8}$$

$$A = X\sqrt{2} - W \tag{6.9}$$

A junction is formed where feedlines from two patch elements meet. A junction between two dissimilar width sections introduces a large discontinuity and changes the impedance value at the junction. A standard T-junction power divider is shown in Figure 6.13. The input power P_{in} is delivered to the intersection on a microstrip of width W_1 and impedance Z_1. The line then branches into two arms with power P_{out}, width W_2 and W_3, and impedance Z_2 and Z_3. The design equations for this divider are:

$$Z_{junction} = \frac{Z_2 Z_3}{Z_{2+} Z_3} \tag{6.10}$$

A mismatch in impedance between two points on a transmission line can be compensated with a quarter-wave transformer. The quarter-wave transformer is a very suitable matching technique. In this technique, the junction impedance is matched to the characteristic feedline impedance through a quarter-wavelength of transmission. The input impedance considering the matching section of line is given by:

$$Z_4 = \sqrt{Z_{junction} Z_1} \tag{6.11}$$

To match the output port impedance line with the input port, a quarter-wave transformer with a line width W_4 and an impedance Z_4 is used. In this designed feed network, T-junction is a 3-dB power divider with $Z_1 = Z_2 = Z_3 = 50\Omega$. The junction impedance is 25Ω and the quarter-wave transformer impedance is 35.35Ω.

6.5.4 Planar Antenna Design

A linear 1 × 2 array structure is designed and then it is converted into a 4-element array as shown in Figure 6.14. The antenna center frequency is 5.8 GHz and is fed with a 50Ω impedance line. To match an antenna impedance with an input port, three quarter-wave transformers are used. The interelement spacing between antenna elements depends on the desired beamwidth and can be determined using:

$$\theta_m = 2\left[\frac{\pi}{2} - \cos^{-1}\left(\frac{\lambda}{Nd}\right)\right] \tag{6.12}$$

where θ_m is the first null beamwidth, d is the separation between array elements, and λ is the free-space wavelength. All elements are excited in phase and the beam

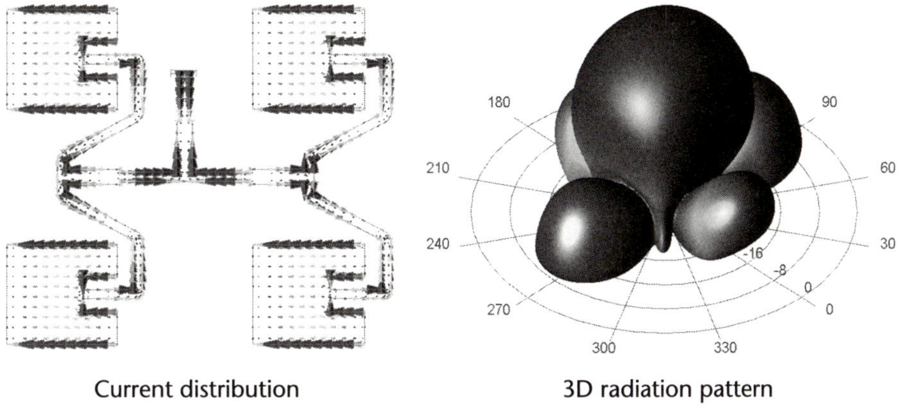

Figure 6.14 The 2 × 2 microstrip antenna array and simulated 3-D radiation pattern plot.

direction is towards boresight. The current distribution and 3-D antenna radiation pattern are shown in Figure 6.14.

The antenna half-power beamwidth also reduces with array size (number of elements in an array antenna) as shown in Figure 6.15. The antenna beamwidth reduces with an increased number of elements. An array of 8 × 8 and 16 × 16 element

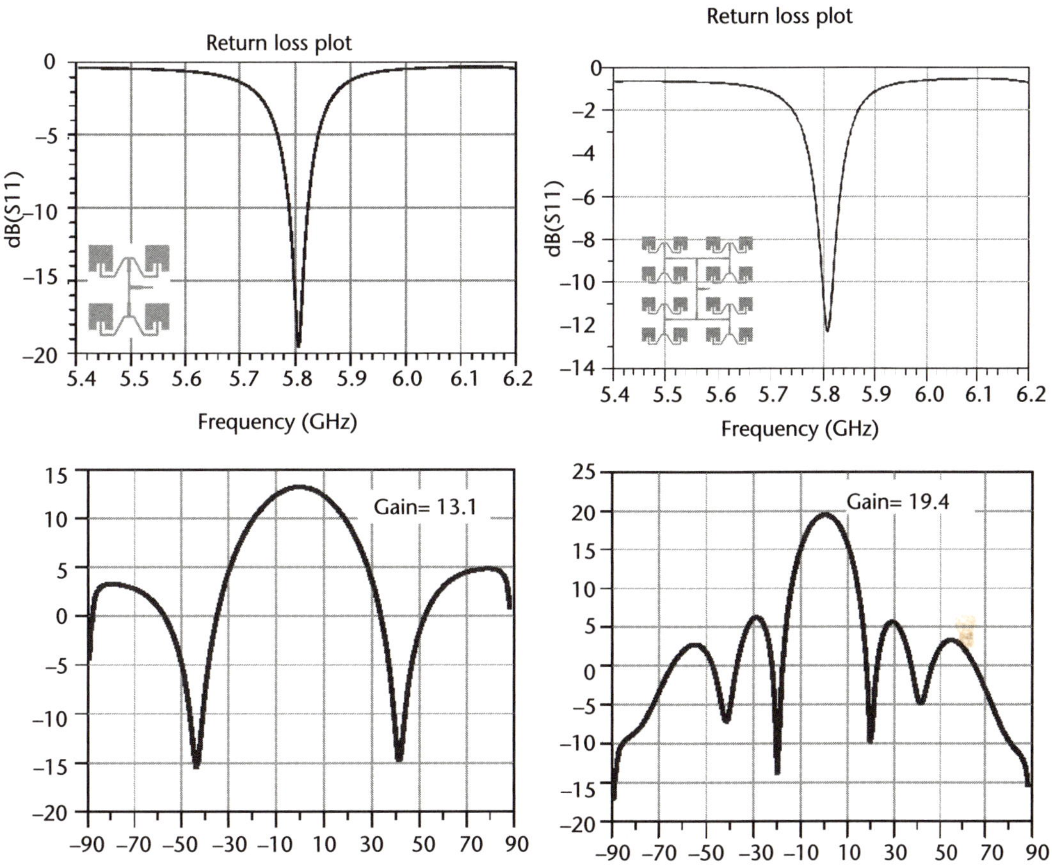

Figure 6.15 Return loss and gain pattern for 2 × 2 and 4 × 4 arrays at 5.8 GHz.

is simulated before making complete 32 × 32 element array. The gain of the 8 × 8 planar array is 26 dBi while the 16 × 16 antenna gain is 30 dBi at 5.8 GHz. The return loss plot and the radiation pattern cut of 2 × 2 and 4 × 4 antennas are shown in Figure 6.15.

6.5.5 Complete Antenna Design and Simulation

To get gain of 35 dBi and a beamwidth of 1°, a planar array of 32 × 32 is designed and simulated. The size of the complete antenna array is 1.2m × 1.2m. This antenna is placed in the weather radar radome design. All antenna performance parameters for different array sizes are summarized in Table 6.2. With an increased array size, the antenna simulation time and system memory requirement increase. Also, antenna bandwidth decreases with increased array size because of feed network losses.

The complete antenna is simulated using MoM, and this is considered a large problem to solve using the EM problem. Simulated results are shown in Figure 6.16. In the corporate feed configuration, the antenna array with *N* elements are fed by a 1:*N* power divider network with identical path lengths from the feed point to each element. The advantages of this topology include design simplicity, flexible choice of element spacing, and broader bandwidth, and they are amenable to integration with other devices such as amplifiers and phase shifters. The disadvantage of this type of feed is that it requires more space for the feed network. For large arrays, the length of feed lines running to all elements is prohibitively long, which results in a high insertion loss. The insertion loss is even more pronounced at millimeter-wave frequencies, thereby adversely degrading the gain of the array. At higher frequencies, the feed lines placed on the same plane as the patches will also radiate and interfere with the radiation from the patches.

Return loss better than −10 dB is achieved in the band. Figure 6.17 shows an antenna pattern for 256 elements. Antenna results and EM simulation time are summarized in Table 6.2. The gain of the antenna increases in an array and the beamwidth reduces. In an array antenna, the gain increase by 3 dB by doubling the number of element in the array. The 16 × 16 array has 6 dB more gain as compared to 8 × 8 array antenna because a total number of elements increased 4 times.

Table 6.2 Summary of Antenna Design Result

Array Type	*Array Size*	*Total Elements*	*Gain (dBi)*	*Bandwidth (MHz)*	*HPBW (Degree)*	*Simulation Time (Relative to Single Element)*
Single	1 × 1	1	6.7	40	78	1.0×
Linear	1 × 2	2	9.7	38	39	2.5×
Planar	2 × 2	4	13.1	35	39	5.5×
Planar	4 × 4	16	19.4	32	19	19.5×
Planar	8 × 8	64	25.1	27	10	95.0×
Planar	16 × 16	256	30.2	25	4.5	1,700×
Planar	16 × 32	512	32.8	24	2.1	7,000×
Planar	32 × 32	1,024	35.8	22	0.9	35,000×

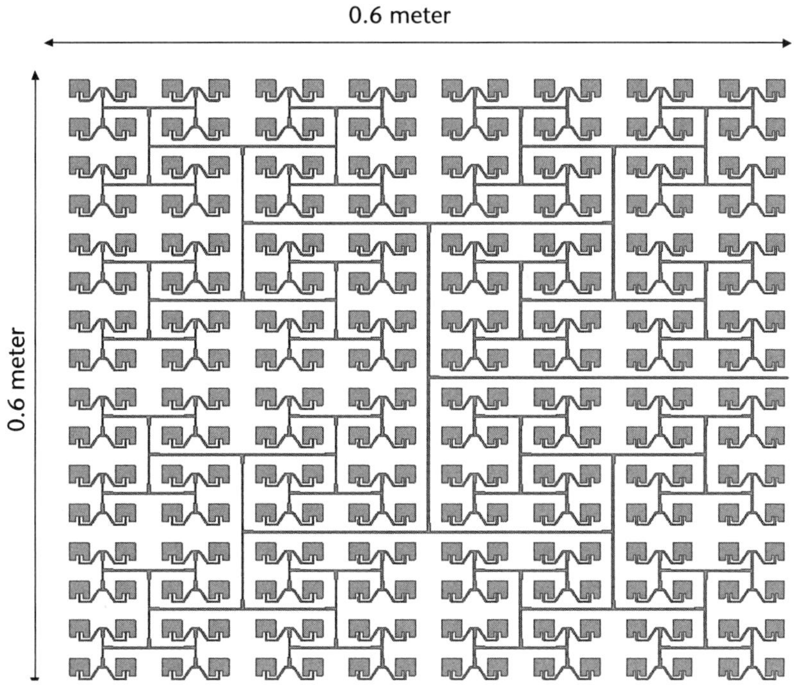

Figure 6.16 A 16 × 16 element antenna array fed with a corporate feed network.

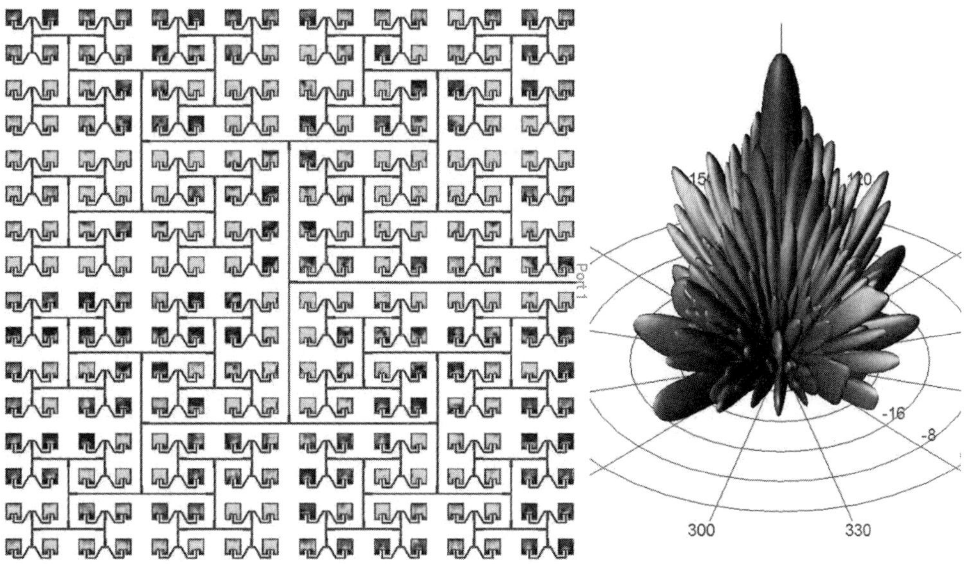

Figure 6.17 Current distribution and 3-D radiation pattern plot of a 256-element antenna array fed with a corporate feed.

6.5.6 Antenna Fabrication

The planar array antenna of 8 × 8 elements fed with a corporate feed network is printed using photolithography processes as shown in Figure 6.18. Antenna array copolarized and cross-polarized radiation patterns (E and H planes) are taken in an anechoic chamber and are shown in Figure 6.19. Cross-polarization is better than –30 dB. Gain is measured relative to the standard horn antenna and is 25.1 dB. Minor asymmetry in the pattern is observed because of some misalignment during integration and a chance of improvement is possible. The radiation pattern is also measured using the planar near-field pattern measurement technique. Simulated data matches very well with measured data.

6.5.7 Circular Aperture Array Antenna

In many applications, the circular aperture array antenna as shown in Figure 6.20 is required based on the available system space. The symmetry method is preferred for a regular subaperture because the symmetry network is beneficial in cancelling out spurious radiation.

A circular-shaped array can be divided into regular subapertures. The number of elements is chosen to be $2N$ (N = 1, 2, 3 ...), which can be designed by symmetrical methods. A 4 × 4 element subarray is first designed and then, by duplicating the subarray, a larger 2^N-element array is designed. The feed network needs to be redesigned for the corner subarrays as indicated in Figure 6.20. Some antenna elements are removed to make it a circular aperture and this process feed-network

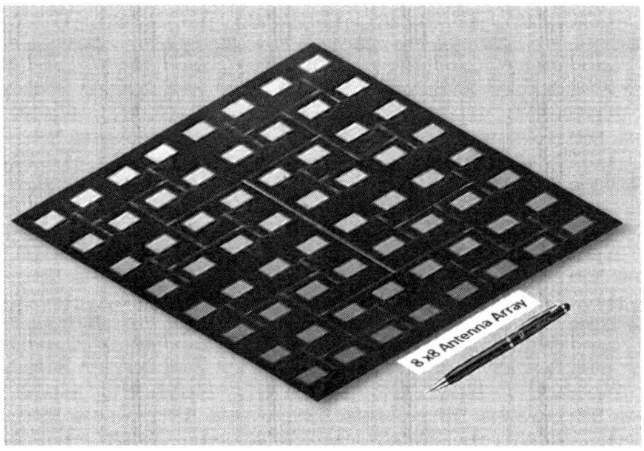

Figure 6.18 Fabricated 8 × 8 antenna array.

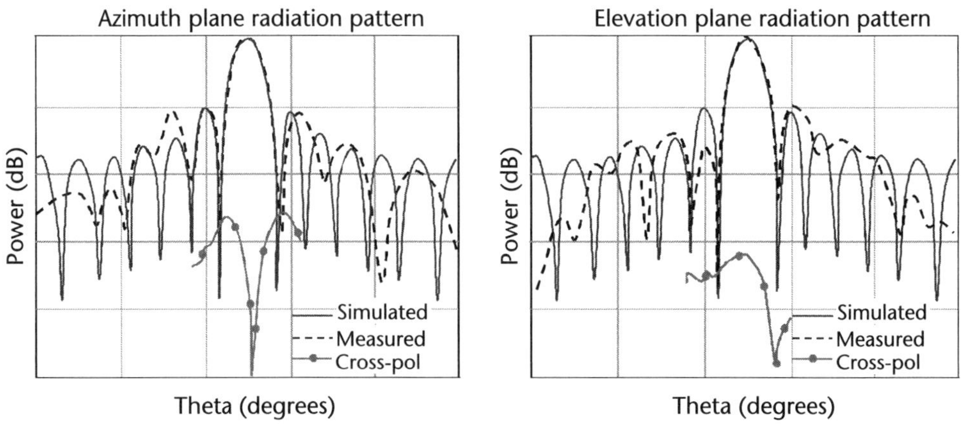

Figure 6.19 Measured azimuth and elevation plane radiation pattern of an 8 × 8 antenna array.

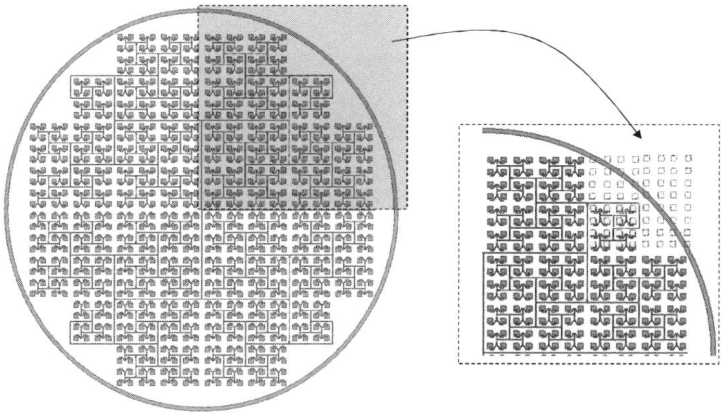

Figure 6.20 Circular aperture microstrip antenna array.

impedance for a truncated subarray is calculated based on the current division in this array segment.

6.6 Bandwidth Enhancement Technique of Microstrip Array Antenna

One of the major disadvantages of microstrip patch antenna is their narrow bandwidth. Conventional microstrip patch antennas have a small bandwidth of 1% to 3% around its resonant frequency. Many systems that could profit from the use of these radiators employ other types of elements due to the requirement for more bandwidth. Researchers have investigated techniques to improve the impedance bandwidth of this class of antennas [39, 40]. The antennas can be made of wideband aperture coupled microstrip array antennas [41], a U-slot with a designed π-shaped stub [42], a double rectangular patch with bridges [43], stacked layered structures [44], resonator antennas with capacitive coupled parasitic patch element [45], and circular edge truncation [46]. The bandwidth of microstrip radiator is directly proportional to substrate thickness h. However, if h is greater than about $0.05\lambda_0$, the probe inductance (for coaxial feed) becomes large enough for an impedance matching network. A foam substrate gives a high bandwidth; by using a thick foam substrate, a bandwidth of about 10% can be achieved. The bandwidth of a rectangular patch is also proportional to the patch width W (but we need to keep $W < 2L$).

The purpose of the design of wideband antenna elements is to arrange them into an array configuration to form a wideband array system. However, the elements in the array are fed by a feed network. Therefore, a wideband feed network is also required to match with the wideband elements and provide overall wideband array system. As discussed earlier, a corporate feed network has been proven to provide a wide bandwidth with good return loss and low transmission loss [32, 33] as compared to the series feed network. By using special feeding techniques like aperture coupling and stacked patches, bandwidths of 100% can be achieved. Edge-coupled patches and multilayer structures are two of the most widely used techniques to enhance the bandwidth of microstrip antennas. The methods developed for effective wideband patch antenna design are based on one or more of the following principles:

- Slots or parasitic elements: Additional resonances are introduced so that, in conjunction with the main resonance, an overall broader band response is obtained.
- Multilayer or thick substrates of low permittivity are used.
- Techniques to reduce the mismatch problem are associated with thick substrates.

6.6.1 Edge-Coupled Patches

Edge-coupled patches increase the impedance bandwidth of a microstrip patch through the introduction of additional resonant patches. By doing so, a few closely spaced resonances can be created. Only one of the elements is driven directly. The other patches are coupled through proximity effects as shown in Figure 6.21. With the edge-coupled approach, impedance bandwidths of up to 25% can be achieved. The advantages of the edge-coupled approach include the fact that the structure is coplanar in nature and that it can be fabricated on a single-layer substrate. However, this approach also has a few drawbacks. Because the different patches radiate with different amplitudes and phases at different frequencies, the radiation patterns change significantly over the operating frequencies and reduce operating bandwidth of the antenna. The enlarged size of the structure is also not good for array designs. The realization of the feed network in an array configuration is another challenge for this antenna kind of structure because of limited available space for feed lines.

6.6.2 Multilayer Structure

Multilayer or stacked microstrip patch is a very popular and effective technique to increase the bandwidth of a microstrip patch antenna, especially in array design. In this structure, the presence of additional dielectric substrates above the main patch causes a shift of the operating frequency. Figure 6.21(b) shows the geometry of such an antenna element where the bottom patch is driven by a feed element and the top patch, which is located on a different substrate layer, is proximity-coupled to the

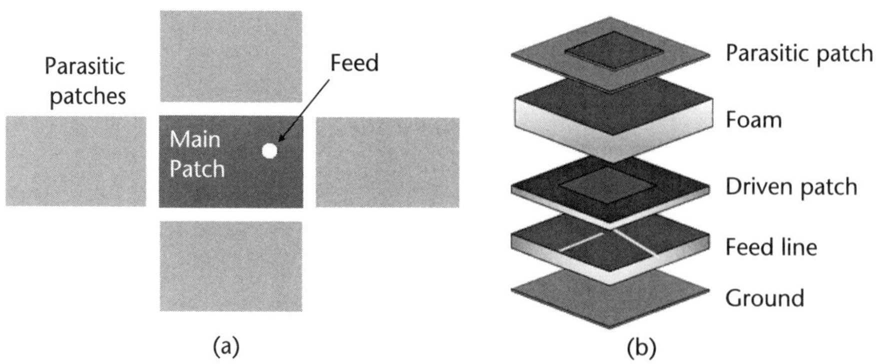

Figure 6.21 (a) Edge-coupled patches and (b) stacked patches

bottom one. In practice, the patches placed on different layers have the slightly different sizes to generate two distinct resonances.

The stacked-patch configuration has many advantages over the edge-coupled configuration. Since it does not increase the surface area of the element, it can be used in array configurations without the danger of creating grating lobes. Its radiation patterns and phase center also remain relatively constant over the operating frequency band. One drawback of stacked patches is that it requires more than one substrate layer to support the patches. This technique is used in many of designs presented in this book to get the desired bandwidth as per applications of array antennas. The rainfall antenna design presented previously is a narrowband antenna that is not suitable for many applications such as airborne SAR where a wideband antenna is required. The next designed antenna is a multilayer wideband microstrip array antenna.

6.7 Multilayer Antenna Design with Wideband Performance

The spacecraft radar antenna design should meet the requirement of high gain and wide bandwidth, keeping the weight as low as possible in the limited available space. A multilayer, lightweight direct-fed microstrip planar array antenna at 4.5 GHz is designed using foam material to meet a wide bandwidth requirement. The foam material is sandwiched between the patch and the ground plane to improve the radiation efficiency and bandwidth performance in the limited space over the single-layer conventional microstrip patch antenna. The antenna configuration is shown in Figure 6.22, where patch and feed lines are printed on the same substrate and the foam layer is placed between the patch and the ground plane to improve radiation efficiency and bandwidth performance by reducing the effective dielectric constant of the combined sandwich layer.

Terminating the patch with a high impedance line can control spurious radiation from the feed line network and cross-polar performance is improved by employing the proper phase cancellation technique. Antenna designing materials are same as used in the previous section only a foam of 2.5 mm is inserted between two 0.5-mm-thick RT Duroid 5870 dielectrics. Both patch geometry and feedline are printed on the top substrate layer, for which the bottom substrate supports the ground plane. Dielectric constant and height of individual substrate layers change the effective dielectric constant for antenna design. A low dielectric constant, which

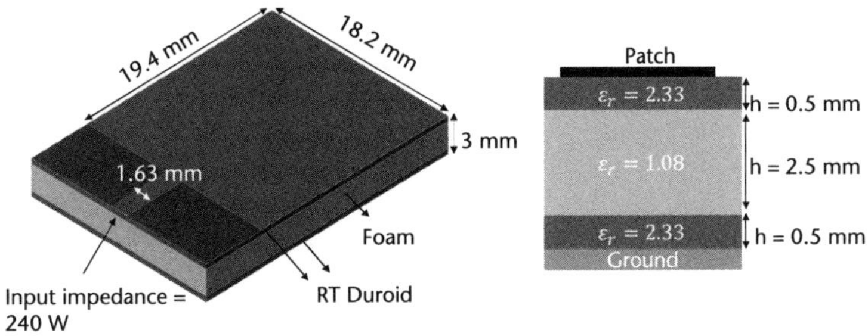

Figure 6.22 Multilayer microstrip patch antenna configuration and design dimensions.

enhances the radiation efficiency for the patch, can give rise to spurious radiation from the feed line network where high dielectric constant is generally preferred to confine the EM field. Foam material would be the best candidate for the patch, but it cannot be directly printed on foam material, which is the reason that two substrate layers are used. The single patch is optimized using the MoM EM simulation technique.

A patch antenna needs to be reoptimized because the insertion of foam and another layer of RT Duroid substrate changes the effective dielectric constant. The return loss plot of a single patch antenna with and without foam material is compared in Figure 6.23. Without foam antenna, the bandwidth is 0.6% (35 MHz) and with foam bandwidth it increased to 4.8% (280 MHz), which is an 8 times increase. This patch element is used in an array configuration to design a broadband antenna array.

6.8 Millimeter-Wave Antenna Array for Automotive Radar

The collision avoidance radar is one of the most important components of the advanced vehicle control and safety system in the future Intelligent Transport System (ITS) for the millimeter-wave ultrawideband (UWB) applications such as short-range radar (SRR) for vehicles. The radar system operating at a millimeter-wave band has the advantages of better resolution and higher data rates if compared with applications in the lower-frequency spectrum. The operation frequency of an automotive radar system is typically in the 59–61-GHz and 76–77-GHz bands for frequency-modulated continuous wave (FMCW) or pulse-Doppler operation. The higher bandwidth allows for W-band UWB systems with improved distance resolution. Microstrip antenna arrays are the most common choice in many antenna types because of their lower profile and price for automotive applications as shown in Figure 6.24.

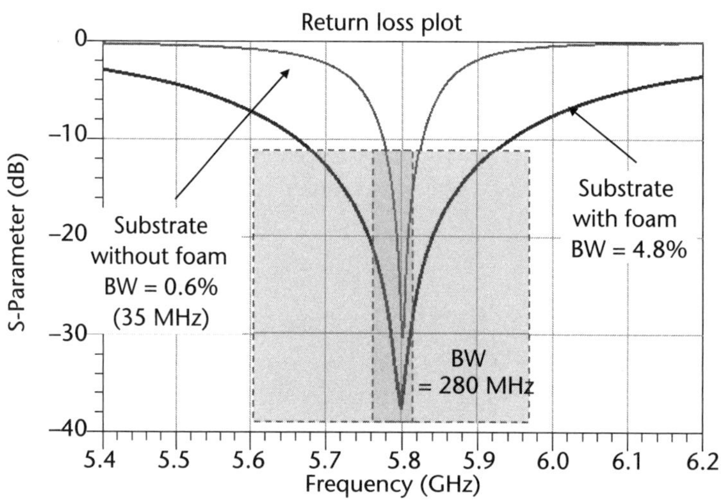

Figure 6.23 Return loss plot of antenna with and without a foam layer.

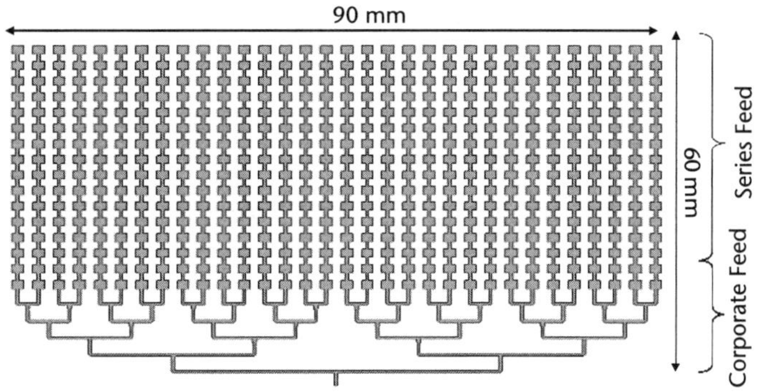

Figure 6.24 Designed millimeter-wave 32 × 16 element antenna array.

6.8.1 Equivalent Model of a Series-Fed Antenna Array

The series-fed array antenna consists of several parallel arranged and serially fed patch rows, which are supplied by a corporate feed network as shown in Figure 6.24. The linear array is composed of two components, the feedlines and the patches. This arrangement reduces the feed line lengths and parasitic radiation. Sometimes the aperture distribution is tapered to meet the desired side lobe suppression. The taper is realized by varying the patch width along the rows and using suitable power dividers in the corporate feed network. Derneryd's transmission line model of a series-fed patch array antenna is shown in Figure 6.25 [47]. In this model, the microstrip patch element is represented by two radiating slots, which are separated by a transmission line. The radiating slot is modeled as a conductance G and a susceptance B in parallel, representing the radiation and the fringing fields, although this model is not accurate when the number of patch elements in the array

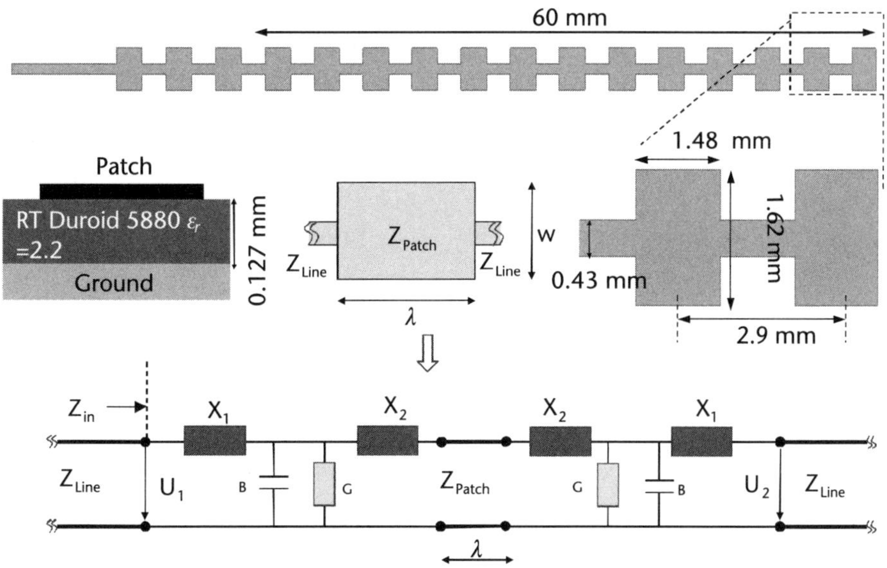

Figure 6.25 Series-fed linear array design and electrical equivalent circuit model.

is large. To improve this model, the step-in width between the feeding line and the patch is represented by two additional reactances X_1 and X_2. Using the MoM simulated scattering parameters, the parameters X_1, X_2, B, G, and λ of the mode can be extracted.

To achieve the broadside radiation pattern, the elements must be in phase, which means that the phase difference between two elements must be 360°. The array would be resonant at the center frequency, so the equivalent circuit of the array can be simplified to n radiation conductance in parallel or n radiation resistances in series (n is the number of patches).

6.8.2 Series-Fed Antenna Array Design

In the millimeter-wave frequency band, the series-fed method is better than the parallel-fed method because of the simple feeding network and low insertion loss. For that reason, the series-fed method has been more widely adopted in millimeter-wave radar. This feeding concept enables beam steering in one direction in an easy way and a minimal spacing. In the millimeter-wave band, the loss of microstrip structure becomes considerable with frequency increasing [48]. To minimize the loss at high frequency, a series-fed method is often chosen to minimize and simplify the feed network for the antenna array. Series feed network has some advantage compared to corporate feed such as compact size, low loss, and lower sidelobe level, but it also has certain disadvantages such as narrow bandwidth and antenna beam tilt with frequency.

The linear array of 16 elements with –20-dB sidelobe is designed at the desired center frequency 60 GHz according to the resonant principle [49]. A planar array of a low sidelobe level is realized with the combination of 16 linear subarrays with certain power distribution. The excitation distribution is designed by the T-junction feed network. The characteristic impedance of feedline is calculated for 50Ω line. The array is printed on the substrate Rogers RT Duroid 5880 (0.127-mm thickness and dielectric constant = 2.2). Linear array antenna dimensions are shown in Figure 6.25. A single 60-GHz radiating patch size is 1.48 mm × 1.62 mm. The series feed line width is 0.43 mm for all antenna elements. The linear array antenna performance results are shown in Figure 6.26. The antenna gain is 18.6 dBi at 60 GHz and return loss is better than –10 dB over the band. The bandwidth of antenna is 30 MHz (0.05%). The antenna half-power beamwidth in the ∅ = 0° plane is 40°, while in the ∅ = 90° plane the half-power beamwidth is 76°. The antenna simulation summary for different array sizes is given in Table 6.3.

The planar antenna array is constructed by using 64 linear arrays fed with a corporate microstrip feed network as shown in Figure 6.27. The antenna resonance frequency is at 60.1 GHz. The half-power beamwidth is the same in the elevation direction as of the linear array because there is no change in the number of radiating elements (16 elements) in this direction.

6.9 Conformal Printed Antennas

In many applications such as a missile, satellite, spacecraft, and aircraft, a directive antenna mounted on a curved body is required. In the case of mobile

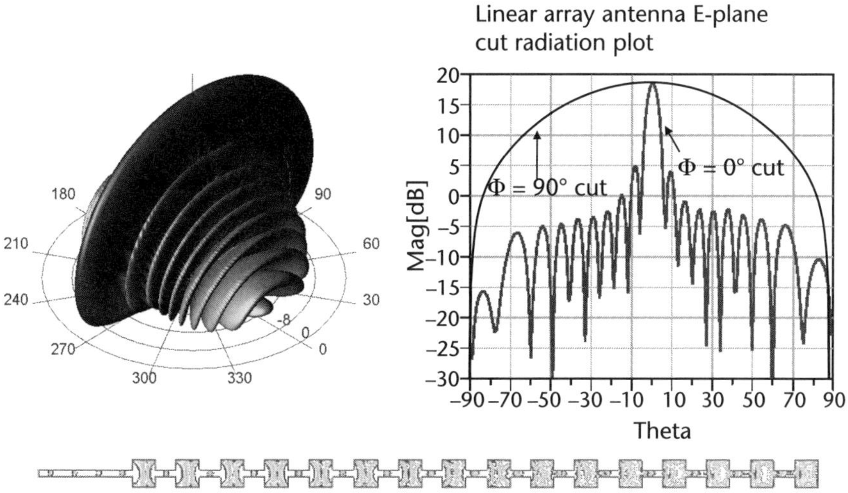

Figure 6.26 E-field and radiation pattern plot of a linear antenna array.

Table 6.3 Summary of the Antenna Array

Array Type	*Array Size (Elements)*	*Total Elements*	*Gain (dBi)*	*HPBW (Degrees)*
Linear	1 × 16	16	9.7	76 and 4
Planar	2 × 16	32	13.1	38 and 4
Planar	4 × 16	64	19.4	18 and 4
Planar	8 × 16	128	25.1	8.5 and 4
Planar	16 × 16	256	30.2	4 and 4
Planar	32 × 16	512	32.8	2 and 4

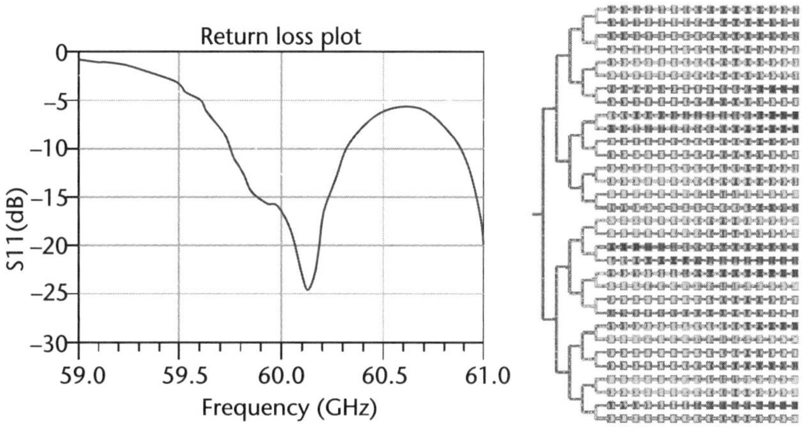

Figure 6.27 Antenna return loss plot and E-field plot on 32 × 16 element array.

communications with small vehicles, a large planar surface to dispose of microstrip array antennas is generally not available, and it is necessary to use conformal arrays. Conforming the antenna according to the application surface saves space and is often essential for structural reasons. Specifically, these radiators can become integrated with the structures on which they are mounted and, consequently, do not cause extra drag and are less visible to the human eye. In addition, the fast-progressing development of high-frequency applications for telecommunications and navigation in automotive, aeronautics, and space increases the need for antennas, which can be integrated seamlessly into predefined structures. For both applications, conformal antenna arrays are excellently qualified. Compared to planar antennas, their steering range is significantly larger and the microstrip technology offers the possibility of manufacturing antennas as a compact building block fitting into nearly arbitrary structures. The conformal antenna can be placed in the belly of aircraft that eliminate separate radome design for the antenna. Besides this, shaping and resolution can be improved by using the conformal antenna as more elements can be accommodated. Another important feature of conformal antennas is their radiation characteristics. These antennas usually provide broader beams than their planar counterparts. This feature is an advantage when omnidirectional patterns are required, which normally are achieved by applying an array. Conformal antennas can be any geometrical shape, although the main structures investigated so far are spherical, cylindrical, and conical.

Microstrip antenna technology is the most suitable for conformal applications because of its ability to conform to nonplanar structures. Antenna patches are placed above the device's conducting plane with a dielectric substrate separating the patch from the conducting plane. The properties of conformal array depend strongly on whether it is small compared to the radius of curvature of the mounting body, in which case it behaves nearly like a planar array or whether it is comparable or large to the radius. In many conformal antennas, the resonance frequency is the same for various curvatures. However, as curvature increases, the pattern broadens.

In conformal antennas, the effect of curvature on the performance of conformal microstrip antennas on cylindrical bodies for the TM_{10} mode is very important. Curvature affects the fringing field and the fringing field affects the antenna parameters. Conformal antennas find a lot of applications in the aerospace defense system because of space limitations. The development of efficient design techniques for conformal microstrip radiators, assisted by computational EM tools, is desirable to establish clear procedures that bring about reductions in computational time, along with high accuracy results. Nowadays, the commercial availability of high-performance 3-D EM tools allows computer-aided analysis and optimization that replace the design process based on an iterative experimental modification of the initial prototype.

6.9.1 Patch Antenna Printed on Circular Cylindrical Structure

Antennas on a uniform curved surface are the simplest conformal antennas. The main objective of such an antenna is to improve the azimuth coverage, obtaining even in some cases omnidirectional coverage. The circular cylindrical antenna is frequently used in conformal antenna applications since it is one of the most straightforward nonplanar geometry. Design and analysis of radiating the microstrip patch antenna mounted on a cylindrical surface are discussed in this section. A conformal patch antenna design procedure involves parameters related to cylindrical microstrip elements to realize the desired resonant frequency, input impedance, and radiation patterns. The analysis and synthesis of the conformal array are slightly more complicated than a linear array with uniform spacing because the array elements point in different directions.

6.9.1.1 Cylindrical Microstrip Antenna Analysis for Large Radii

The circular cylindrical microstrip patch wrapped around a grounded dielectric cylinder is shown in Figure 6.28. It is fed by an inset feedline. The antenna is optimized for 5.7 GHz. The parameters of the antenna are: relative permittivity of the substrate $\varepsilon_r = 2.2$, the diameter of the ground plane cylinder = 40 mm, and length and width of the patch = 18.27 mm. A microstrip line is used to feed the patch antenna. Antenna design specifications are given in Table 6.4.

RT Duroid of thickness 0.5 mm and relative permittivity of 2.2 are used as a substrate material for antenna design. Two options are designed, one along the curvature of the cylinder and other along the axial direction of the cylinder as shown in Figure 6.28. The antenna design was first simulated for the planer and subsequent simulation was performed for the cylindrical surface having radius of 20 mm. As for bending increases, the performance of the antenna for conformal surface degrades.

Table 6.4 Antenna Design Specifications

Design Parameters	*Value*
Design frequency	5.7 GHz
Patch antenna dimensions	18.27 × 18.27 mm
Substrate thickness	0.5 mm
Substrate permittivity	2.2
Substrate loss tangent	0.009
Radius of cylinder	20 mm

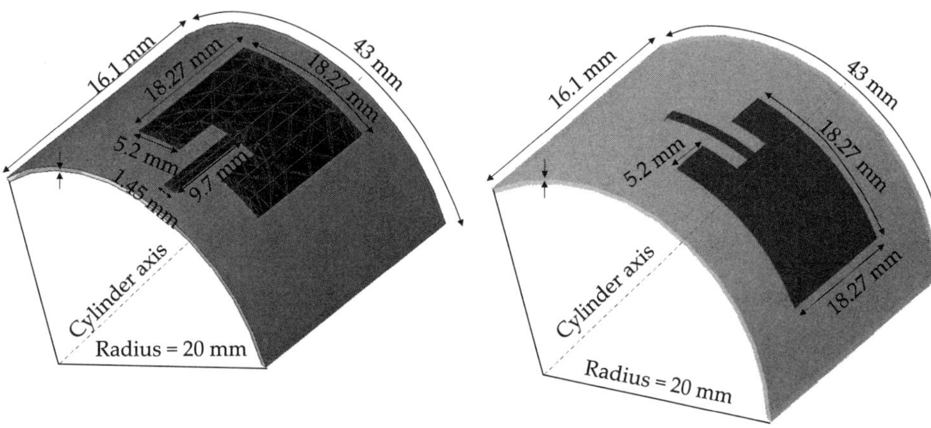

Figure 6.28 Conformal microstrip patch antenna along the axial direction and along the curvature of the cylinder.

The comparison of the return loss (S_{11}) graph shows that an increase in array size results in significant changes in frequency in conformance to the planar antenna due to the effect of curvature and mutual coupling because of interelement spacing. The effect of the curvature of the conformal antenna on resonant frequency been presented by Krowne [50] as:

$$[f_r]_{mn} = \frac{1}{2\sqrt{\mu\epsilon}}\sqrt{\left(\frac{m}{2\theta a}\right)^2 + \left(\frac{n}{2b}\right)^2} \tag{6.13}$$

where a is a radius of the cylinder, $2b$ is a length of the patch antenna, 2θ is the angle bounded the width of the patch, ϵ represents electric permittivity, and μ is the magnetic permeability as shown in Figure 6.29.

In most of the cases, the analysis of conformal antennas can be based on approximate techniques, and when the antenna has very large radii of curvature, it may be often analyzed as if it were planar. To analyze the characteristics of cylindrical patch antennas, the cavity model approach is suitable only for very thin substrate cases. Full-wave analysis is the most accurate approach for conformal antenna analysis. The knowledge of the mutual coupling among the radiating elements and the radiation patterns of individual elements are essential ingredients for conformal antenna analysis. The EM field on the surface of the antenna is calculated to find other performance parameters of the antenna, such as radiation pattern.

The simulation results are collected for different radii of the cylinder. The return loss does not have a substantial variation in the frequency band, and the overall variation is only 3 to 4 dB. The antenna gain is varying 0.8 dBi when the patch is placed along the curvature of the cylinder by increasing the cylinder radius from 15 mm to 40 mm. The gain variation is 1 dBi when the patch is placed along the cylinder axis. The simulated gain of the single element over different radii of the cylindrical surface is shown in Table 6.5.

It is observed that gain of the single radiator is more when the feed is along the axis of the cylinder. Also, there is an increase in gain with an increase in the radius of the cylinder. The antenna is simulated using FEM simulation techniques. Return loss comparison of the conformal antenna with planar antenna configuration is shown in Figure 6.30. There is slight resonance frequency shift at a higher frequency when the antenna is placed along the cylinder axis while the resonance

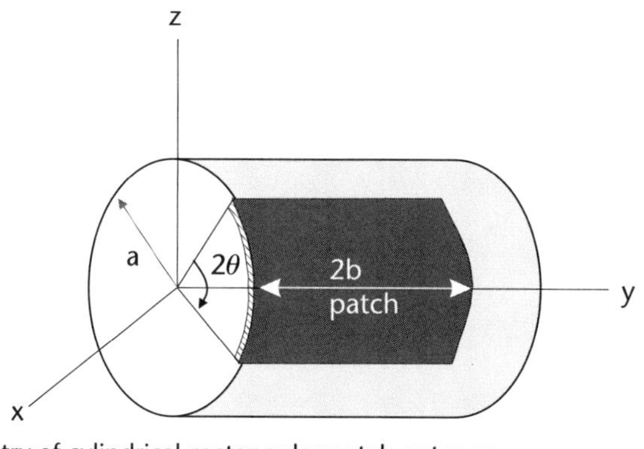

Figure 6.29 Geometry of cylindrical-rectangular patch antenna.

Table 6.5 Gain Comparison of the Antenna with Feedline Position over the Cylinder

Radius of Cylinder (mm)	*Gain of Patch Fed Along the Curvature of Cylinder (dBi)*	*Gain of Patch Fed Along the Axis of Cylinder (dBi)*
15	6.2	6.4
20	6.6	6.8
30	6.8	7.4
40	7.0	7.4

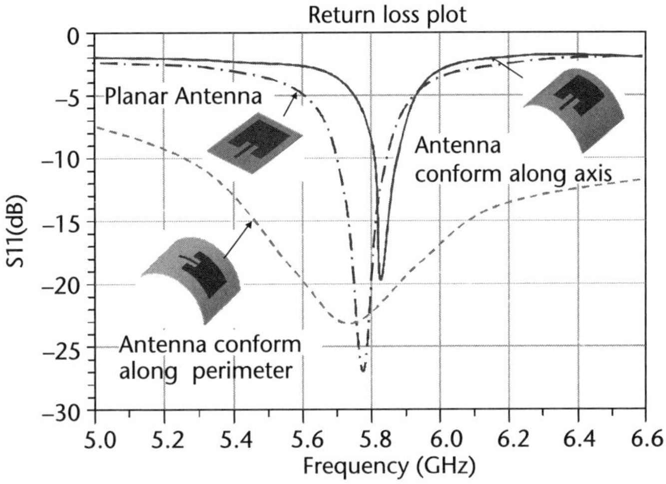

Figure 6.30 Simulated antenna return loss plot planar microstrip and conformal antennas.

frequency shifted at the lower side when the antenna was placed along the perimeter of the cylinder.

The comparison of the E-field radiation patterns can be seen in Figure 6.31. In the $\phi = 0^\circ$ plane, shown in Figure 6.31(a), the radiation from the doubly curved geometries is broader since there is a curved surface, whereas, for the singly curved geometries, it is straight. Typically, singly curved geometries are theoretically analyzed with the infinite ground plane, so it is an infinite long cylinder in the case of cylindrical antennas. In the elevation plane, the antennas are considered as an infinite planar antenna, not allowing back radiation and presenting very narrow beamwidth. In the $\theta = 90^\circ$ plane, this conformal antenna has the same curvature of $0.3\lambda_0$. As illustrated in Figure 6.31(b), that leads to approximately the same beamwidth in the hemisphere of maximum radiation. The beam in the elevation plane is narrow, while the beam is relatively broad in the plane around the cylinder axis in the hemisphere where the patch is located.

6.9.1.2 Cylindrical Microstrip Antenna Analysis for Small Radii

A conformal patch antenna has a very different performance when the radius of the cylinder is less, and the cylinder circumference is comparable to the antenna size as shown in Figure 6.32. For small radii, the size of the antenna does not match

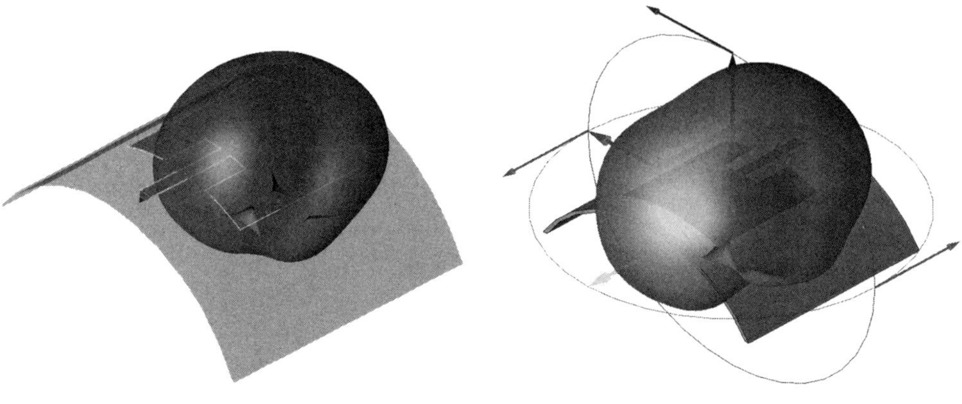

(a) (b)

Figure 6.31 Conformal microstrip patch antenna radiation pattern: (a) antenna along the axial direction of cylinder and (b) antenna along the curvature of the cylinder.

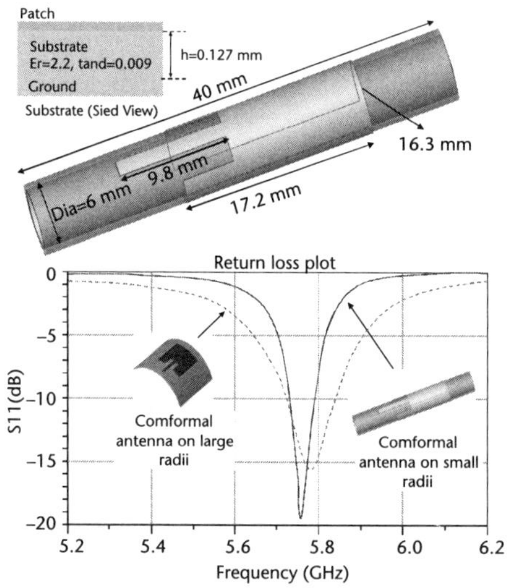

Figure 6.32 (a) Patch antenna mounted on small radii cylinder and (b) antenna return loss plot.

exactly with that of the planar since the effects of the curvature are more significant. When the radius of the curved structure is large, the antenna can be analyzed as the planar one, although for a structure with smaller radii, more rigorous analysis methods should be used. Cylindrical conformal structures, with radii greater than one half-wavelength, are used as prospective candidates for mobile communications systems, cellular base stations, and Telemetry and Telecommand (TTC) communication that is essential to maintain space missions due to their full field of view advantage [51]. The simulations for small radii antenna are performed based on the full-wave analysis.

Nearly omnidirectional patterns using a single radiating element are achieved with wraparound patch antennas in the azimuth direction, while patch has dipole-like radiation patterns in the plane that include the length of the cylinder, resulting in almost null radiation in that direction as shown in Figure 6.33.

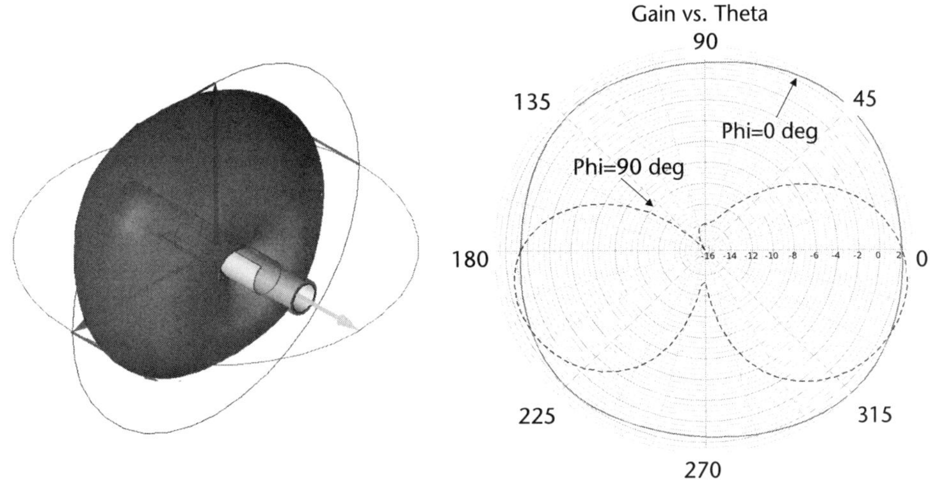

Figure 6.33 Small radii conformal microstrip antenna 3-D and 2-D radiation patterns.

6.9.2 Antenna on Missile Surface

The high-speed airborne craft, rocket, and satellite require the antenna to neither influence their air motive force performance nor destroy their mechanical structure and intensity. The moving missile is required to realize the data transmission between the missile and the launcher for the high-precision attack. It leads to how to effectively design the missile antennas when considering the least disturbance to the flight dynamics and to the air frame structure. It should have a low profile and can be mounted on the surface of a missile. Obviously, the cylindrically conformal antenna is the preferred choice. Such kinds of antennas have always been the focus of the research, and several forms of practical antennas were reported, such as the printed dipole [52], microstrip patch [53, 54], printed slot, and substrate integrated waveguide (SIW) antennas [55]. Antenna performance such as gain can be controlled by varying the shape of the conformal antennas. It is possible to reach nearly omnidirectional coverage, while, with planar antennas, the almost full coverage is limited to approximately ±60°.

Conformal antennas also have certain drawbacks. The major disadvantages are the increased design complexity and costs in the manufacturing of conformal antennas on an arbitrary shape. In addition, their size is restricted when a package of electronics is desired because they cannot be included in strongly curved structures. The technology of conformal antennas is not as fully established as it is for the planar ones, because many of their analysis tools are still in development. As discussed previously, for very large radii, conformal antennas can be considered as planar ones for analysis simplification, so their performance would be very similar to planar antennas. However, for small radii, planar design techniques cannot be applied and the designed rules are not established. Furthermore, circular polarization is more difficult to achieve in curved structures with the inconvenience that the maximum values of XPD are lower than for their planar counterparts. In a missile system, a main important characteristic of the antenna is its radiation pattern. The anti-interference and anti-multipath effect ability of antennas can be improved by utilizing the directionality.

For the high-speed whirling cylindrical carrier as the guided missile, its request for the antenna is to produce the isotropic radiation pattern in the plane that is perpendicular to the axis of the missile approximately, but in the plane that includes the axis of the missile, the radiation pattern should be offered as the hemisphere beam. For this purpose, different kinds of cylindrical conformal antennas should be adopted for different diameters of cylinders separately. The missile tracking system works at 2.4 GHz by using position data from the onboard GPS receiver and/or Inertial Navigation System (INS). Apart from the GPS antenna, the extra high gain antenna in the missile can serve many purposes. It can be used when the main guidance is not yet fully developed or when GPS is automatically cut off at any cruise speed over 1.5 mach. It can also be used for sending various missile parameter and monitoring data to the ground control station or sending the video signal from a camera attached to the head of the missile to the ground for visual guidance purposes. A microstrip conformal microstrip patch antenna array is designed at 2.4 GHz using series-fed 8 × 8 elements as shown in Figure 6.34. The antenna array is mounted on the body of 4.3-m-long missile system. The diameter of missile body is 100 mm. The antenna length is 815 mm and a total number of elements is 64 in an 8 × 8 array. The antenna simulation data and radiation pattern are shown in Figure 6.35.

6.9.3 Conformal Microstrip Antenna Array Fed with Waveguide

In high-power applications, the antenna input signal is distributed to subarrays of microstrip antennas. Among all microwave transmission lines, waveguides feature extremely low losses up to very high frequencies and can handle high power, provided that their internal surfaces are finished smoothly. Figure 6.36 shows a 220-element, waveguide-fed, microstrip patch array. To reduce the feed loss and cater to

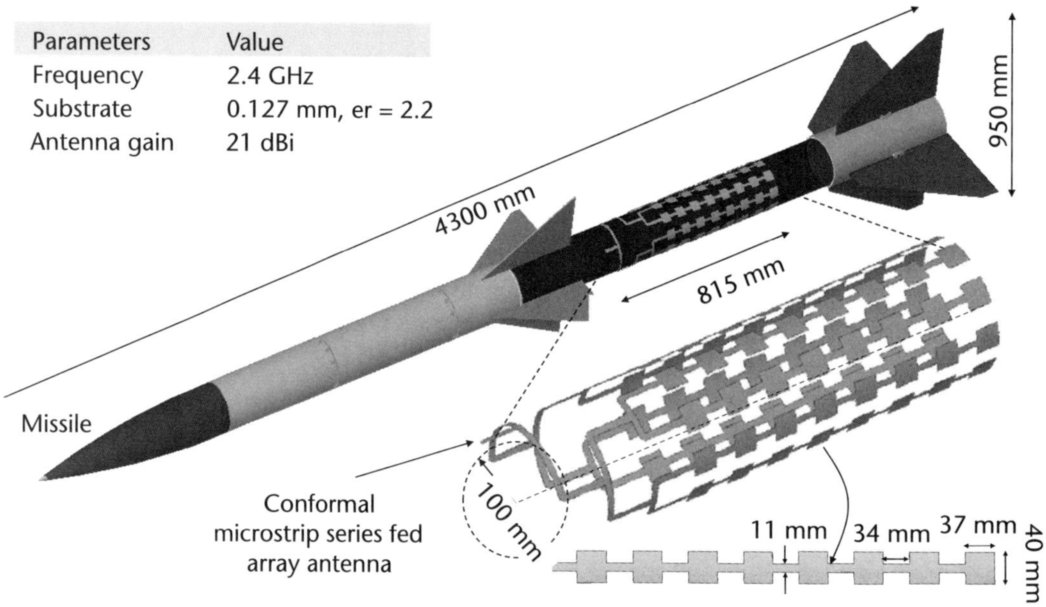

Figure 6.34 Conformal series-fed microstrip antenna array mounted on the missile.

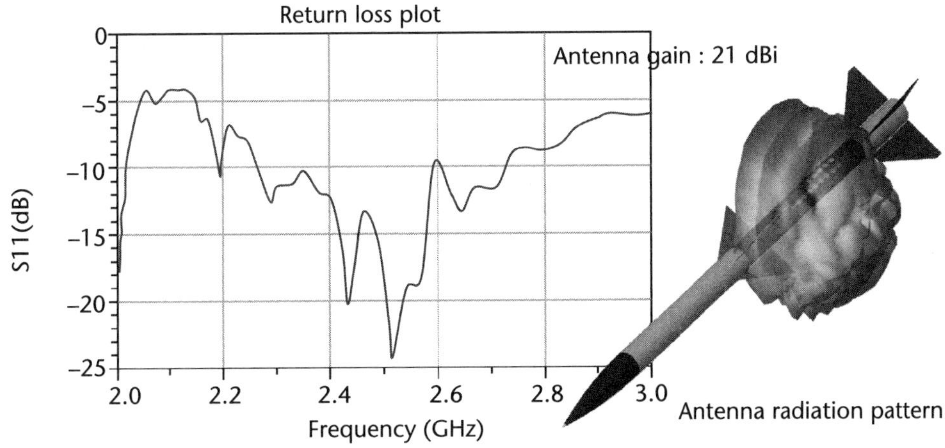

Figure 6.35 Return loss and radiation pattern plot of a conformal antenna mounted on the missile surface.

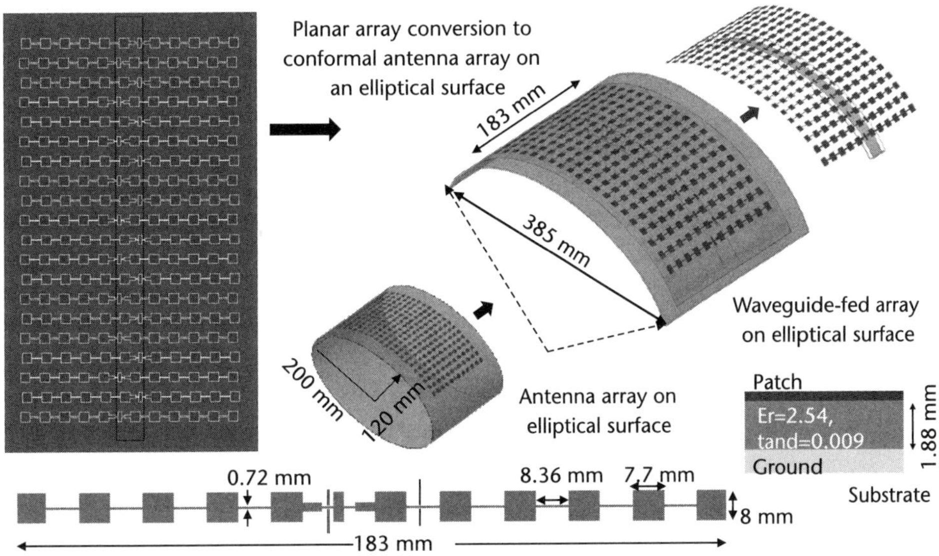

Figure 6.36 Large waveguide-fed conformal antenna array on an elliptical surface.

the high-power requirement, a slotted waveguide is used instead of the main feed line to feed the microstrip patch array. First, a rectangular microstrip patch antenna array is designed and optimized, and then this geometry is conformed over an elliptical curved surface and the design is reoptimized. As resonant frequency does not significantly change with a curved surface, the same dimensions of the patch of the planar surface are modeled over an elliptical surface. The width of the feed lines and the patch dimensions are calculated in the same way as the microstrip planar antenna array.

To simplify the manufacturing procedure, the slots of the waveguide are etched on the ground plane of the dielectric substrate, and the etched ground plane serves as the upper wall of the waveguide. A coaxial probe at the center of the bottom wall of the waveguide is used to excite the slotted waveguide. The two ends of the waveguide are shorted. After optimizing the rectangular array, this is converted in conformal antenna array as shown in Figure 6.36.

The main advantage of this center-fed array structure is that it avoids beam squint with frequency. This array of a waveguide-fed microstrip antenna can be used in many modern communication systems, such as a mobile base station antenna. The resonant frequency is 11.5 GHz and the gain of the array is more than 30 dBi.

6.10 Summary

In this chapter, several applications of microstrip antenna arrays were presented and discussed. Microstrip antenna array concepts with an application such as rainfall and weather radar antenna, automotive antennas were covered in this chapter. One of the major benefits of microstrip antennas is the simplicity of array construction. Millimeter-wave microstrip antenna arrays with high efficiency are designed for automotive radar. A feed network is the most important design aspect in microstrip antenna. The efficiency of microstrip antenna arrays may be improved significantly by reducing losses in the feed network. Microstrip antennas are narrowband antennas, but there are several techniques to increase the bandwidth of the array antenna. The power radiated from a discontinuity may be evaluated from the computed S-parameters using the full-wave analysis technique. The gain limitation in microstrip antenna arrays can be overcome by employing active microstrip arrays. In some applications, the reflect arrays are employed. In reflect arrays, the received power is transmitted by the same antenna array. The optimization of the antenna structure and feed network dimensions allows maximizing the power rate dissipated on the load. The feed network design is the most important and critical design step in the array antenna. Depending on the application, the proper feed topology or a combination of multiple feed schemes can be selected. Several array antenna feeding techniques, including the most popular corporate and series feed, were discussed. Rainfall radar antenna array was discussed step by step to explain the complete design process and associated challenges in the planar array antenna. To overcome the bandwidth limitation of convention microstrip antennas, the broadbanding technique is applied. One such scheme is the multilayer stacking of the patch element. In the next chapter, phased array antennas are discussed.

References

[1] Pozar, D. M., "Microstrip Antennas," *IEEE Proc.*, Vol. 80, January 1992, pp. 79–91.

[2] Schaubert, D. H., "Microstrip Antennas," *Electromagnetism*, Vol. 12, 1992, pp. 381–401.

[3] Pozar, D. M., and D. H. Schaubert, *Microstrip Antennas: The Analysis and Design of Microstrip Antennas and Arrays*, New York: IEEE Press, 1995.

[4] Liu, Y., et al., "RCS Reduction of a Patch Array Antenna Based on Microstrip Resonators," *IEEE Antennas and Wireless Propagation Letters*, Vol. 14, 2015, pp. 4–7.

[5] Alhalabi, R. A., and G. M. Rebeiz, "High-Efficiency Angled-Dipole Antennas for Millimeter-Wave Phased Array Applications," *IEEE Transactions on Antennas and Propagation*, Vol. 56, No. 10, October 2008, pp. 3136–3142.

[6] Khan, O., et al., "Hybrid Thin Film Antenna for Automotive Radar at 79 GHz," *IEEE Transactions on Antennas and Propagation*, Vol. 65, No. 10, October 2017, pp. 5076–5085.

[7] Sun, D., and L. You, "A Broadband Impedance Matching Method for Proximity-Coupled Microstrip Antenna," *IEEE Transactions on Antennas and Propagation*, Vol. 58, No. 4, April 2010, pp. 1392–1397.

[8] Zhang, L., W. Zhang, and Y. P. Zhang, "Microstrip Grid and Comb Array Antennas," *IEEE Transactions on Antennas and Propagation*, Vol. 59, No. 11, November 2011, pp. 4077–4084.

[9] Li, Y., et al., "A Fixed-Frequency Beam-Scanning Microstrip Leaky Wave Antenna Array," *IEEE Antennas and Wireless Propagation Letters*, Vol. 6, December 2007, pp. 616–618.

[10] Hamberger, G. F., et al., "A Planar Dual-Polarized Microstrip 1-D-Beamforming Antenna Array for the 24-GHz Band," *IEEE Transactions on Antennas and Propagation*, Vol. 65, No. 1, January 2017, pp. 142–149.

[11] Bilgic, M. M., and K. Yegin, "Wideband Offset Slot-Coupled Patch Antenna Array for X/Ku-Band Multimode Radars," *IEEE Antennas and Wireless Propagation Letters*, Vol. 13, 2014, pp. 157–160.

[12] Shen, T. M., et al., "Antenna Design of 60-GHz Micro-Radar System-in-Package for Noncontact Vital Sign Detection," *IEEE Antennas and Wireless Propagation Letters*, Vol. 11, 2012, pp. 1702–1705.

[13] Nilsson, E., and L. Baath, "Radar Interferometric Measurements with a Planar Patch Antenna Array," *IEEE Sensors Journal*, Vol. 7, No. 7, July 2007, pp. 1025–1031.

[14] Di Bari, R., et al., "Dual-Polarized Printed S-Band Radar Array Antenna for Spacecraft Applications," *IEEE Antennas and Wireless Propagation Letters,* Vol. 10, 2011, pp. 987–990.

[15] Tanyer-Tigrek, F. M., I. E. Lager, and L. P. Ligthart, "A CPW-Fed Printed Loop Antenna for Ultra-Wideband Applications, and Its Linear-Array Performance," *IEEE Antennas and Propagation Magazine*, Vol. 52, No. 4, August 2010, pp. 31–40.

[16] Menzel, W., and A. Moebius, "Antenna Concepts for Millimeter-Wave Automotive Radar Sensors," *Proceedings of the IEEE*, Vol. 100, No. 7, July 2012, pp. 2372–2379.

[17] Chou, H. T., M. Y. Lee, and C. T. Yu, "Subsystem of Phased Array Antennas with Adaptive Beam Steering in the Near-Field RFID Applications," *IEEE Antennas and Wireless Propagation Letters*, Vol. 14, 2015, pp. 1746–1749.

[18] Haider, N., A. G. Yarovoy, and A. G. Roederer, "L/S Band Frequency Reconfigurable Multiscale Phased Array Antenna with Wide Angle Scanning," *IEEE Transactions on Antennas and Propagation,* Vol. 65, No. 9, September 2017, pp. 4519–4528.

[19] Lambard, T., et al., "Ka-Band Phased Array Antenna for High-Data-Rate SATCOM," *IEEE Antennas and Wireless Propagation Letters,* Vol. 11, 2012, pp. 256–259.

[20] Mohammed, J. R., "Phased Array Antenna with Ultra-Low Sidelobes," *Electronics Letters,* Vol. 49, No. 17, August 15, 2013, pp. 1055–1056.

[21] Chazelas, J., et al., "Potential Technological Breakthroughs for Phased Array Antennas," *IEEE Aerospace and Electronic Systems Magazine*, Vol. 24, No. 11, November 2009, pp. 22–27.

[22] Tanha, M., et al., "Overlapped Phased Array Antenna for Avalanche Radar," *IEEE Transactions on Antennas and Propagation*, Vol. 65, No. 8, August 2017, pp. 4017–4026.

[23] Babakhani, B., S. K. Sharma, and N. R. Labadie, "A Frequency Agile Microstrip Patch Phased Array Antenna with Polarization Reconfiguration," *IEEE Transactions on Antennas and Propagation,* Vol. 64, No. 10, October 2016, pp. 4316–4327.

[24] Croq, F., and A. Papiernik, "Large Bandwidth Aperture Coupled Microstrip Antenna," *Electronics Letters*, Vol. 26, August 1990, pp. 1293–1294.

[25] James, J. R., and P. S. Hall, *Handbook of Microstrip Antennas*, Stevenage, UK: IEE Electromagnetic Waves Series 28, 1989.

[26] Josefsson, L., and P. Persson, "Conformal Array Antenna Theory and Design," *IEEE Antennas and Propagation Society*, Sponsor, 2006, pp. 15–40.

[27] Orfanidis, S. J., D. Ramaccia, and A. Toscano, "Electromagnetic Waves and Antennas," Dept. of Electrical and Computer Engineering, Rutgers University, Piscataway, NJ, 2002.

[28] Balanis, C. A., *Antenna Theory: Analysis and Design*, 3rd Edition, New York: John Wiley, 2005.

[29] Stutzman, W. L., and G. A. Thiele, *Antenna Theory and Design*, New York: Wiley, 1998.

[30] Lin, Y. -D., and S. -N. Tsai, "Coplanar Waveguide-Fed Uniplanar Bow-Tie Antenna," *IEEE Transactions on Antennas and Propagation*, Vol. AP-45, 2000, pp. 305–306.

[31] Topak, E., et al., "A Novel Millimeter-Wave Dual-Fed Phased Array for Beam Steering," *IEEE Transactions on Microwave Theory and Techniques,* Vol. 61, No. 8, August 2013, pp. 3140–3147.

[32] Kim, D., et al., "4 x 4-Element Corporate-Feed Waveguide Slot Array Antenna with Cavities for the 120 GHz-Band," *IEEE Transactions on Antennas and Propagation,* Vol. 61, No. 12, December 2013, pp. 5968–5975.

[33] Vosoogh, A., and P. S. Kildal, "Corporate-Fed Planar 60-GHz Slot Array Made of Three Unconnected Metal Layers Using AMC Pin Surface for the Gap Waveguide," *IEEE Antennas and Wireless Propagation Letters,* Vol. 15, 2016, pp. 1935–1938.

[34] Wilkinson, E. J., "An N-Way Hybrid Power Divider," *IRE Trans. Microwave Theory Tech.*, Vol. 8, January 1960, pp. 116–118.

[35] Lim, J. -S., et al., "A 4:1 Unequal Wilkinson Power Divider," *IEEE Microwave and Wireless Components Letters*, Vol. 11, No. 3, March 2001, pp. 124–126.

[36] De, D., et al., "Design and Analysis of Various Wilkinson Power Divider Networks for L Band Applications," *2016 3rd International Conference on Signal Processing and Integrated Networks (SPIN)*, Noida, 2016, pp. 67–72.

[37] Tourche, K., et al., "Design and Realization of a Two-Way Microstrip Wilkinson Power Splitter," *2017 Seminar on Detection Systems Architectures and Technologies (DAT)*, Algiers, 2017, pp. 1–5.

[38] Croq, F., and A. Papiernik, "Stacked Slot-Coupled Printed Antenna," *IEEE Microwave and Guided Wave Letters*, Vol. 1, October 1991, pp. 288–290.

[39] Ning, C. Z., and Y. W. C. Michael, "Broadband Planar Antennas: Design and Applications," *Institute for Infocomm Research*, Singapore, 2006.

[40] Shavit, R., "Dielectric Cover Effect on Rectangular Microstrip Antenna Array," *IEEE Transactions on Antenna and Propagation*, Vol. 42, 1994, pp. 1180–1184.

[41] Mohanna, S., and N. Ghassemi, "Wideband Aperture Coupled Microstrip Array Antennas for Radar Applications," *18th International Conference on Radar and Wireless Communications,* June 2010, pp. 1–2.

[42] Lau, K. L., K. M. Luk, and K. F. Lee, "Wideband U-Slot Microstrip Patch Antenna Array," *IEE Proc.-Microw. Antennas Propag.*, Vol. 148, No. 1, February 2001, pp. 41–44.

[43] Jung, C. W., and F. De Flaviis, "A Dual-Band Antenna for WLAN Applications by Double Rectangular Patch with 4-Bridges," *IEEE Antennas and Propagation Society International Symposium*, Vol. 4, June 2004, pp. 4280–4283.

[44] Egashira, S., E. Nishiyama, and A. Sakitani, "Stacked Microstrip Antenna with Wide Band and High Gain," *IEEE Antennas and Propagation Society International Symposium*, Vol. 3, May 1990, pp. 1132–1135.

[45] Pues, H. F., and A. R. Van De Capelle, "Impedance Matching of Microstrip Resonator Antenna," *Proceedings of the North American Radio Science Meeting*, Quebec, 1980, p. 189.

[46] Elsherbeni, A., et al., *Wideband Slot and Printed Antennas*, New York: Wiley, 2005.

[47] Derneryd, A., "Linearly Polarized Microstrip Antennas," *IEEE Transactions on Antennas and Propagation*, Vol. 24, No. 6, November 1976, pp. 846–851.

[48] Bhartia, P., K. V. S. Rao, and R. S. Tomar, *Millimeter-Wave Microstrip and Printed Circuit Antennas*, Norwood, MA: Artech House, 1991, pp. 52–55.

[49] Freese, J., et al., "Synthesis of Microstrip Series-Fed Patch Arrays for 77GHz-Sensor Applications," *Microwave Conference*, 2000.

[50] Krowne, C.M., "Cylindrical-Rectangular Microstrip Antenna," *IEEE Transactions on Antennas and Propagation*, Vol. AP-31, 1983, pp. 194–199.
[51] Geng, J. P., et al., "The Development of Curved Microstrip Antenna with Defected Ground Structure," *Progress in Electromagnetics Research*, Vol. 98, 2009, pp. 53–73.
[52] Wang, P., et al., "A Wideband Conformal Endfire Antenna Array Mounted on a Large Conducting Cylinder," *IEEE Transactions on Antennas and Propagation*, Vol. 61, No. 9, September 2013, pp. 4857–4861.
[53] Kulshrestha, S., et al., "Rectangular-Cylindrical Conformal Printed Antenna for SAR," *International Conference on Antenna Technology ICAT 2005*, Space Applications Center, Ahmedabad, ICAT-2005, February 2005, pp. 395–401.
[54] Sriharsha, C., et al., "Synthesis of Shaped Radiation Pattern for Conformal Antenna Array on Cylindrical Surface," *International Conference on Antenna Technology ICAT 2005*, Space Applications Center, Ahmedabad, February 2005, pp. 419–422.
[55] Bayraktar, O., and O. Aydin Civi, "Circumferential Traveling Wave Slot Array on Cylindrical Substrate Integrated Waveguide (CSIW)," *IEEE Transactions on Antennas and Propagation*, Vol. 62, No. 7, July 2014, pp. 3557–3566.

Selected Bibliography

Croq, F., and D. M. Pozar, "Millimeter Wave Design of Wide-Band Aperture Coupled Stacked Microstrip Antennas," *IEEE Transactions on Antennas and Propagation*, Vol. 39, December 1991, pp. 1770–1776.
Geng, J. P., et al., "The Development of Curved Microstrip Antenna with Defected Ground Structure," *Progress in Electromagnetics Research*, Vol. 98, 2009, pp. 53–73.
Josefsson, L., and P. Persson, "Conformal Array Antenna Theory and Design," *IEEE Antennas and Propagation Society*, Sponsor, 2006, pp. 15–40.
Mueed, A., J. D. Xu, and G. Mehdi, "Cavity Backed Embedded Antenna on Cylindrical Surface for Conformal Applications," *Proc. 14th IEEE INMIC*, December 2011, pp. 351–356.
Quan, X. L., et al., "Analysis and Design of a 45° Slant-Polarized Omnidirectional Antenna," *IEEE Transactions on Antennas and Propagation*, Vol. 62, No. 1, January 2014, pp. 86–93.
Wang, P., et al., "A Wideband Conformal Endfire Antenna Array Mounted on a Large Conducting Cylinder," *IEEE Transactions on Antennas and Propagation*, Vol. 61, No. 9, September 2013, pp. 4857–4861.

CHAPTER 7

Antenna Arrays for Microwave Sensing and Imaging

7.1 Introduction

Large-scale airborne and spaceborne phased arrays are one of the most important enabling techniques for remote microwave sensing and imaging (MSI). SAR is widely used for MSI within the aerospace and defense applications. It provides day-and-night and weather-independent images for civil and military applications. The antenna plays an important role in the SAR systems, and many kinds of antennas have been applied in SAR. For example, the microstrip patch antenna array was used first in SeaSat-A, which was launched in 1978. The waveguide slots antenna array was used in SIR-C and Kosmos1870 subsequently. X-band SAR payloads with electronic beamforming system for all-weather spaceborne remote sensing or imaging are increasingly being deployed by the small satellite industry, especially with the availability of commercial off-the-shelf components [1]. Antenna researchers have been striving to develop suitable X-band wideband arrays with the dual-polarization property. One of the already-built X-band spaceborne SAR systems is TerraSAR-X [2], but the X-band dual-polarized slotted waveguide subarray employed in that system is bulky, expensive, and difficult to manufacture, thus making the subarray inappropriate for a low-cost small satellite. The use of SAR for remote-sensing applications is particularly suited for tropical countries. The microwave signal can penetrate rain, clouds, haze, fogs and precipitation with very little attenuation, thus allowing operation in unfavorable weather conditions that preclude the use of the visible/infrared system. SAR is an active sensor that provides its own source of illumination. As there is no dependency on sunlight, it can therefore operate day or night. These systems can illuminate with a variable look angle and can select wide area coverage. It has been shown to be very useful over a wide range of applications, including sea and ice monitoring, mining, oil pollution monitoring, oceanography, snow monitoring, and classification of Earth terrain. The potential of SAR in a diverse range of applications led to the development of several airborne and spaceborne SAR systems.

The spacecraft antenna design should meet the requirement of high-gain, wide bandwidth keeping weight as low as possible in the limited available space. The necessary features of SAR devices include wide bandwidth, optimum-shaped radiation patterns, and high power. A SAR antenna system must be light-weight and low profile, so planar antennas are generally used in these systems. The bandwidth limitations of microstrip antennas can be overcome with multilayer-stacked, electromagnetically coupled printed antennas. Another limitation is associated with the power-handling capability of the microstrip patch antennas, and this cannot be overcome in the SAR system where a pulse peak power of several kilowatts is typically used. For these types of applications, a hybrid antenna where a feeder implemented in the waveguide or square coaxial line (SCL) can be used. High input power levels within the feeder networks are converted to lower-power levels by using cascaded power dividers. These distributed lower-power levels are then fed to a microstrip patch antenna. Square coaxial feed technology has volume and weight advantages over a waveguide feeder.

Medical imaging is an important tool for visualizing the interior of the body for clinical analysis and identifying any abnormality, usually in a noninvasive way. Using nonionizing EM waves at microwave frequencies for sensing and imaging for medical diagnosis is emerging as a low-cost and low health risk alternative to the imaging technologies. MSI offers a noninvasive way for the diagnosis of functional and pathological tissue conditions [3–5]. Sensing applications of microwave involve single point applications wherein microwave signals transmitted by an antenna are received by the same or a different antenna after reflection or scattering by tissues. In such a case, no 2-D or 3-D images are formed. Examples of such cases can be heartbeat detection or changes in the microwave signal caused because of the pathological accumulation of water in the brain called cerebral edema [6]. In the imaging applications, a map or an image either in 2-D or 3-D is formed that shows different tissue electrical properties or the location of strong scatterers that is usually a tumor inside the body.

7.2 Microwave Remote Sensing and Imaging

Because of their long wavelengths microwave signals have special properties that are important for remote sensing; compared to the visible and infrared, microwaves have special properties that are important for remote sensing. Longer-wavelength microwave signal can penetrate through cloud cover, dust, haze, and all but the heaviest rainfall as the longer wavelengths are not susceptible to atmospheric scattering, which affects shorter optical wavelengths. This property permits the detection of microwave energy under almost all weather and environmental conditions so that data can be collected at any time.

As shown in Figure 7.1, the two classes of sensors are passive and active. In passive sensors energy leading to radiation received comes from an external source (e.g., the Sun), and in active sensors energy is generated from within the sensor system and beamed outward, and the fraction returned is measured. Sensors can be nonimaging (measures the radiation received from all points in the sensed target, integrates this, and reports the result as an electrical signal strength) or imaging (the electrons released are used to excite or ionize a substance such as silver in film or a display). The most common form of imaging active microwave sensors is radar. The sensor transmits a microwave signal towards the target and detects

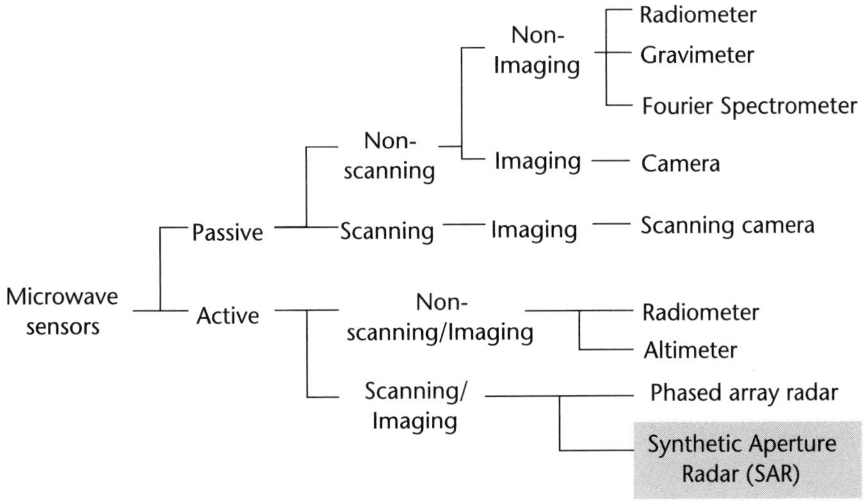

Figure 7.1 Type of microwave remote sensors.

the backscattered portion of the signal. The strength of the backscattered signal is measured to discriminate between different targets and the time delay between the transmitted and reflected signals determine the distance (or range) to the target. The SAR is an active imaging sensor.

7.2.1 Microwave Remote-Sensing Frequency Bands

As detailed in Figure 7.2, radar remote sensing uses the microwave portion of the EM spectrum, from a frequency of 0.3 GHz to 300 GHz. Microwave radars sense EM radiations in the microwave region of the EM spectrum. Most spaceborne radars operate at wavelengths between 0.5 cm and 75 cm. Shorter wavelengths such as X-band imagery are reflected from the top of the canopy, while longer wavelengths such as L-band imagery normally go down to the ground and are reflected from there.

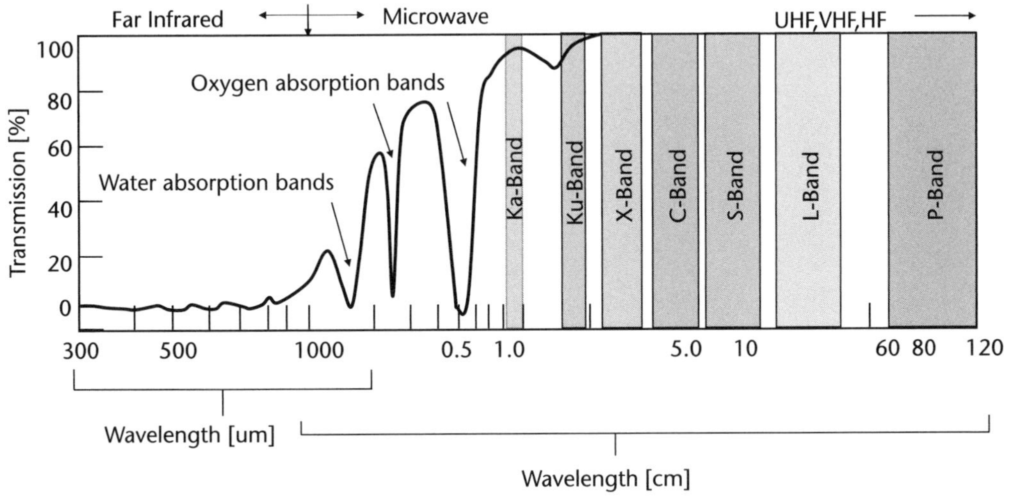

Figure 7.2 Frequency bands used in remote-sensing radars.

A remote-sensing radar system has three primary functions. It transmits microwave signals towards an area. It receives the part of the transmitted energy backscattered from the objects in that area and then it observes the strength (detection) and the time delay (ranging) of the return signals. Radar resolution is the ability of the radar to distinguish two objects in a frequency band. Spaceborne and airborne radar can collect data while flying over a large distance and then can process the data as if it came from a physically long antenna. Frequency bands commonly used in various kinds of radar applications are listed in Table 7.1

7.2.2 Antenna Polarization in Remote-Sensing Applications

Transmitted and received radar signals propagate in a certain plane called as the polarization. The propagation planes are typically horizontal (H) or vertical (V). Vertically polarized waves interact with the vertical stalks of the plant canopy, while horizontally polarized waves penetrate through plant canopy. The different combinations of the polarized wave provide different image characteristics:

- *Single polarization:* Both transmitted and received signals have the same polarization in a radar system.
- *Dual polarization:* The radar system has one polarization to transmit the signal and another polarization or both the polarizations to receive the signal.
- *Quad polarization:* Horizontal and vertical polarizations are used for alternate pulses to transmit the signal and with both simultaneously to receive the signal.

7.3 SAR

SAR is a radar system that utilizes the aircraft or satellite flight path of the platform to simulate an extremely large antenna or aperture electronically, and that generates high-resolution remote-sensing imagery. It uses polarized radiation and therefore can exploit polarization signatures of the imaged scatters for obtaining more information about the image. It has been widely used for Earth remote sensing for more than 40 years. It provides high resolution all the time and weather-independent

Table 7.1 Commonly Used Frequency Bands in Radar Remote Sensing

Band	*Frequency*	*Wavelength*	*Applications*
P	0.3–1 GHz	30–100 cm	Used for research and experimental applications; significant penetration capabilities
L	1–2 GHz	15–30 cm	Penetrates vegetation to support observations applications over vegetated surfaces for monitoring ice and glacier dynamics
S	2–4 GHz	7.5–15 cm	Used for medium-range metrological applications (e.g., rainfall measurement)
C	4–8 GHz	3.75–7.5 cm	Penetration capability of vegetation or solid is limited and restricted to the top layers; useful for sea-ice surveillance
X	8–12 GHz	2.4–3.75 cm	Widely used for military reconnaissance, mapping, and surveillance

images. These systems are used in many applications ranging from geoscience and climate change research, environmental and Earth system monitoring, 2-D and 3-D mapping, change detection, and security-related applications up to planetary exploration. SAR systems are based on a pulsed radar installed on a platform with a forward movement, and it has a side-looking imaging geometry. The radar system transmits EM pulses with high power and receives the echoes of the backscattered signal in a sequential way as shown in Figure 7.3.

The pulse repetition frequency typically ranges from a few hundred to a few thousand hertz for airborne and spaceborne systems, respectively. The swath width varies typically from a few kilometers to 30 km in the airborne case and from 30 to 500 km in the spaceborne case. The transmitted pulse interacts with the objects on the Earth's surface, and only a portion of this signal is backscattered to the receiving antenna. In a monostatic radar, the receiving antenna is the same as the transmit antenna and in a bistatic or multistatic radar, transmitting and receiving antennas are different. The microwave energy scattered back to the spacecraft is measured. The phase and amplitude of the backscattered signal depend on the physical properties (i.e., roughness, shape) and electrical properties (i.e., permittivity) of the imaged object. Depending on the operating frequency band, substantial penetration can occur to model objects as a volume (e.g., vegetation, ice and snow, dry soil). The simplest radar system provides a 2-D reflectivity map of the imaged area. High backscattered signals are identified as bright spots in the radar images and flat, smooth surfaces as dark areas. The direction of the flight is denoted as the azimuth and the line of sight is denoted as the slant range direction.

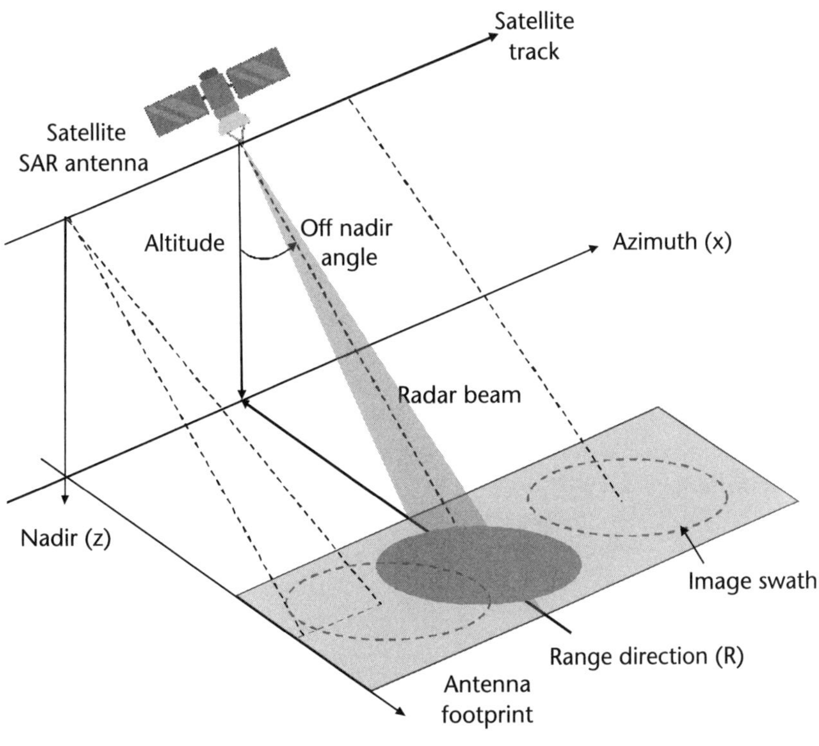

Figure 7.3 Imaging concept for a spaceborne SAR system.

The SAR makes use of the radar principle to form an image by utilizing the time delay of the backscattered signal. For high-resolution imaging of the Earth's surface, a very long antenna is required that is not feasible for a spacecraft to carry. To overcome this limitation, SAR capitalizes on the motion of the spacecraft to emulate a large antenna. The SAR processor stores all the radar returned signals as amplitudes and phases for the time T from one position to another position. With this stored data, it is possible to reconstruct the signal, which would have been obtained by an antenna of length vT, where v is the platform speed. With the changes in the line-of-sight direction along the radar platform trajectory, a synthetic aperture is produced by signal processing, which has the effect of lengthening the antenna. Making T large enough makes the synthetic aperture large, and hence a higher resolution can be achieved.

7.3.1 Antenna Types Used in SAR Systems

The antenna must be selected carefully to assure a good resolution for the SAR. Printed microstrip antennas are the most suitable choice for a SAR antenna. A variety of printed antennas, such as microstrip patch, microstrip printed dipole, and slot antennas, can be used as the radiating element. Printed antennas fulfill SAR requirements on the radiation pattern, polarization purity, return loss, and power-handling capability. A high antenna gain with a sharp pattern specification is preferable. The minimum width of the main lobe of the radiation pattern in azimuth (i.e., in the direction of flight) is determined by the required resolution. Generally, a SAR antenna would have a narrow beam in the azimuth direction and a wide-shaped beam in the elevation direction. The resolution in the azimuth direction is obtained using the Doppler shift caused by the movement of the aircraft. The radiation pattern in elevation (i.e., in the vertical plane perpendicular to the flight plane) must ensure a reasonably even illumination of the area to be mapped. Table 7.2 shows the antenna systems used in various SAR projects.

7.4 Dual-Polarized SAR Antennas Design

The design of a dual-polarized, lightweight, multilayer microstrip planar array antenna at 9.6 GHz for a remote-sensing application is discussed in this section. An 8 $\times$ 32 element dual-polarized printed planar antenna is designed using RT5870 and foam material. The foam material is sandwiched between the patch and the ground plane to improve the radiation efficiency and bandwidth performance in the limited space over the single-layer conventional microstrip patch antenna. Generally, impedance bandwidth of patch antennas varies from 2% to 3% for single-layer direct coupling or probe coupling [7–9] and can be further enhanced by coupling the power from the stripline through an aperture (slot) in the ground plane and/or printing patch effectively on the foam material whose dielectric constant is nearly equal to that of air [10–12]. In such a case, the feed network is isolated from the radiating element by a ground plane, which prevents spurious radiation. The resonant or nonresonant slot couples energy from the stripline to the patch. If the slot is resonant, it provides another resonance in addition to the patch resonance, resulting in an antenna with 10% to 15% impedance bandwidth. However, the resonant

Table 7.2 SAR Antennas Used in Various Projects

Name of the Project	*Frequency Band*	*Antenna Type*	*Gain*	*Beamwidth (Degrees)*	
				Az	*El*
NASA AIRSAR	P-band	Microstrip	14 dBi	19	38
	L-band	Microstrip	18 dBi	8	44
	C-band	Microstrip	24 dBi	2.5	50
DLR E-SAR	P-band	Microstrip	12 dBi	30	60
	L-band	Microstrip	17 dBi	18	35
	S-band	Microstrip	NA	20	35
	C-band	Microstrip	17 dBi	19	33
	X-band	Horn	17.5 dBi	17	30
CCRS C/X SAR	C-band	Horn	26 dBi	3.3	25
	X-band	Horn	28.5 dBi	1.4	26
NASDA X/L-band SAR	L-band	Microstrip	18 dBi	2.3	40
	X-band	Waveguide-slot	26.5 dBi	2.3	40
DCRS EMISAR	L-band	Microstrip	17.1 dBi	10	42
	C-band	Microstrip	26 dBi	2.4	31
Lynx	Ku-band	Dish	NA	3.2	7
PHARUS C-band SAR	C-band	Microstrip	NA	2.3	24

slot has a strong back lobe, which substantially reduces the gain of the antenna. The nonresonant slot has a better front-to-back ratio, which results in an improvement in efficiency with a narrow bandwidth. Hence, multilayer microstrip antennas are difficult to optimize when maximum efficiency is required in a limited allowable available space with the weight as low as possible.

The SAR antenna is designed as smaller units of subarrays. The typical design specification of a dual-polarized SAR antenna is given in Table 7.3. The single antenna radiating element configuration is shown in Figure 7.4, where patch and feed lines are printed on the same substrate and foam layer is sandwiched between

Table 7.3 Design Specifications of Dual-Polarized SAR Linear Antenna Array

Parameters	*Specifications*
Operating frequency	9.6 GHz
Bandwidth	200 MHz, 3.7%
Polarization	Perpendicular (VV)
Gain	14.6 dBi
Maximum antenna width	0.6m
Power	1W (continuous)
Impedance	50
Connection	Standard SMA coaxial 50Ω
Conformity	Planar surface mount
Elevation incidence	60°
Elevation beamwidth	40°–50°

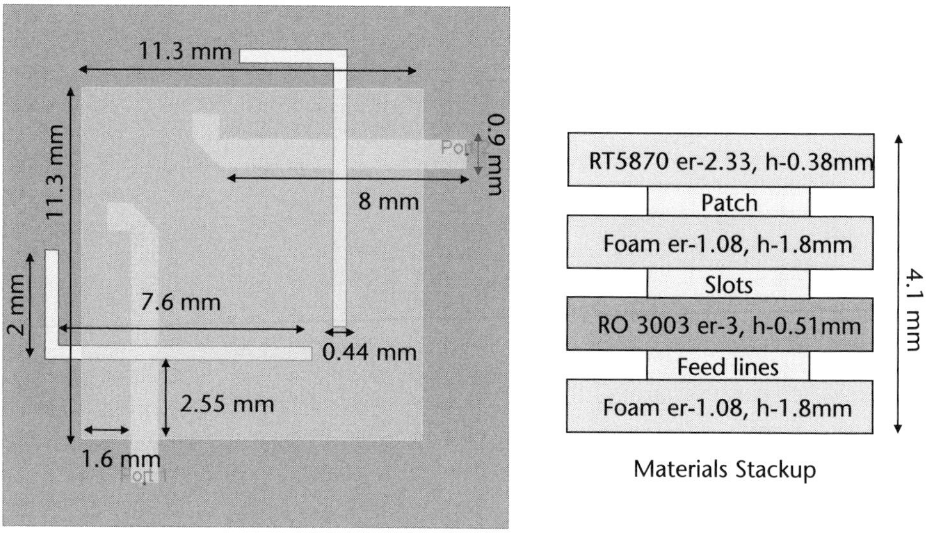

Figure 7.4 Dual-polarized multilayer microstrip antenna radiator.

the patch and the ground plane to improve radiation efficiency and bandwidth performance by reducing the effective dielectric constant of the combined sandwich layer. Terminating the patch with a high-impedance line can control spurious radiation from the feed-line network and cross-polar performance is improved by employing a proper phase cancellation technique. The design of the antenna begins with the proper selection of material, its dielectric constant, and the height of the substrate, which plays a crucial role when weight and gain should be optimized simultaneously.

A low dielectric constant, which enhances the radiation efficiency for the patch, can give rise to spurious radiation from the feed-line network where a generally high dielectric constant is usually preferred to confine the EM field. Foam material would be the best candidate for the patch, but it cannot be directly printed on foam material. The antenna is printed on RT Duroid 5870 with a dielectric constant of 2.33 and a height of 0.38 mm, while the feed line is printed on RO 3003 for mechanical rigidity and the electrical requirement of the feed line.

7.4.1 Design of Microstrip Patch Radiator

A strip-slot coupled patch has been used as a radiating element for wide bandwidth and low cross-polarization performance. The requirement of dual polarization can be fulfilled by selecting a square patch fed at two points to restore the symmetry of the structure. The MoM is used for predicting the impedance, cross-polarization, front-to-back ratio, and the bandwidth requirement to improve the efficiency of the antenna.

The propagation constant for a wave in the microstrip substrate must be accurately known to predict the resonant frequency, resonant resistance, and other antenna properties. The most sensitive parameter in microstrip antenna performance is the estimation of the dielectric constant of the substrate material, and that the manufacturer's tolerance on relative dielectric constant is sometimes inadequate.

The change in operating frequency of a thin substrate microstrip antenna due solely to a small tolerance related change of substrate dielectric constant may be expressed as:

$$\frac{\delta f}{f_0} = -\frac{1}{2}\frac{\delta \varepsilon_r}{\varepsilon_r} \tag{7.1}$$

where f_0 is the resonant frequency of a microstrip antenna assuming that a magnetic wall boundary condition, ε_r is the relative dielectric constant, and δ represents the small change. The relative frequency change for small dimensional changes may be expressed in terms of linear dimensions or in terms of temperature change as follows:

$$\frac{\delta f}{f_0} = -\frac{\delta l}{l} = -\alpha_T \delta T \tag{7.2}$$

where δ_T is thermal expansion coefficient. Polytetrafluoroethylene (PTFE) substrates reinforced with either glass woven web or glass random fiber are also very commonly used substrates because of their desirable electrical and mechanical.

The radiation of waves from the patch can be predominant when the substrate is thick and has a low permittivity. Foam materials provide very low permittivity values, very close to that of air. The surface waves become significant when the substrate is thick and has a large permittivity. It should be noted that the substrate requirement for the circuit and those needed for the antenna are contradictory. The design of a dual-polarized aperture coupled patch antenna is shown in Figure 7.5. Three different dielectric substrates are used in this design. The material properties are listed in Table 7.4.

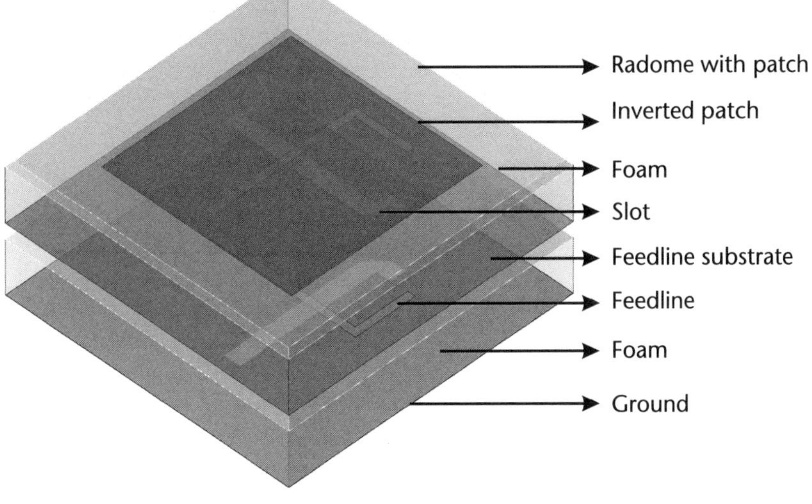

Figure 7.5 Multilayer patch antenna configuration and different layers.

Table 7.4 Material Properties of a Multilayer Antenna

Material Used	ε_r	*h (mm)*	*Loss Tangent*
RT Duroid 5870	2.33	0.38	0.001
Foam	1.08	1.8	0.008
Rogers RO 3003	3.0	0.51	0.001

Rogers RO 3003 is selected as the material for the feed line having a dielectric constant of 3, a height of the substrate of 0.51 mm, and a loss tangent of 0.001. RT Duroid 5870 material is used as radome of the antenna for environmental protection. The antenna is printed on the back side of this substrate.

7.4.2 Antenna Performance with Different Types of Slots

Antenna performance is simulated to measure coupling of the EM energy and slot-type effect on impedance variation with three different types of slots as shown in Figure 7.6. The front-to-back ratio and cross-polarization are calculated over the frequency band, and it has been found that the dog-bone slot enhances the coupled energy, but, in the case of dual polarization, the electric field vector has a higher component of cross-polarization and is not suitable for dual polarization. A thin rectangular aperture gives much stronger coupling. The coupling can be increased by using wider or longer rectangular apertures. The transverse electric field disappears at the end of the aperture for a simple rectangular slot. Coupling increases as the electrical length of the slot increases, which increases the resonant input impedance. The same concept also applies for the aperture width [13–19].

Aperture coupling using a slot is an indirect method of feeding the patch. The feed line creates an electric field in the aperture, which induces surface currents on the patch. Fringing fields are created by the patch edges perpendicular to the feed line that radiate into free space. The coupling level is mainly determined by the length of the coupling slot, as well as the back-radiation level. The slot should therefore be made no larger than what is required for impedance matching; it should be $0.25_{\lambda 0}$ and the width of the slot should be $0.1L$, where L is the length of the patch radiator. The width of the feed line affects the coupling to the slot, besides controlling the characteristic impedance of the feed line. Thinner feed lines couple

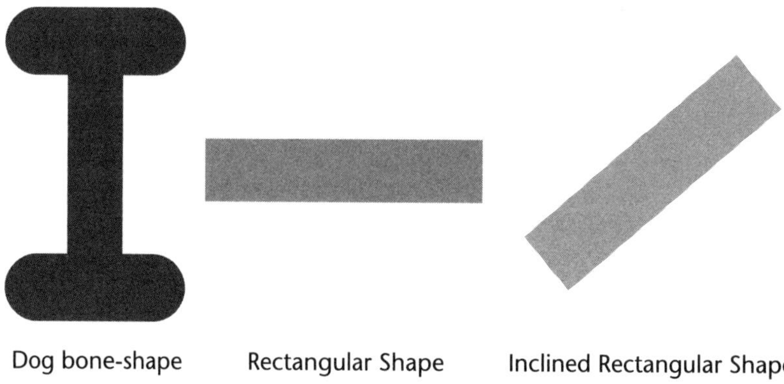

Figure 7.6 Different shapes of the slot.

more strongly to the slot to a certain degree. For maximum coupling, the feed line should be positioned at right angles to the center of the slot [20]. Slanting the feed line from the slot will reduce the coupling, as will positioning the feed line towards the edge of the slot. The stub is typically slightly less than $\lambda_g/4$ in length; shortening the stub will move the impedance locus in the capacitive direction on the Smith chart. A summary of cross-polarization and the front-to-back ratio results for the dog bone-shaped slot, the rectangular slot, and the inclined rectangular shape slot is given in Table 7.5.

A single-element fabricated microstrip antenna is shown in Figure 7.7. Different layers of the microstrip antenna are integrated using adhesive. The measured bandwidth of the antenna is 750 MHz (7.8%). The antenna is resonating at 9.6 GHz. For maximum coupling, the patch should be cantered over the slot. Moving the patch relative to the slot in the E-plane (resonant) direction decreases the coupling level while moving the patch relative to the slot in the H-plane direction has little effect. The excess reactance of the slot coupled antenna is tuned with a tuning stub.

7.5 Array Antenna Design for X-Band SAR

Once a suitable radiating element is designed, the next step is to find an array configuration that gives satisfactory beam characteristics (direction and width). To analyze the complete radiation pattern of the SAR arrays, it is necessary to know the electrical characteristics of the single element as discussed in earlier. The design of a linear subarray antenna begins with optimization of the following parameters:

- Position of the radiating elements;
- Spacing between the elements;

Table 7.5 Results Summary of Different Slots

	Rectangular Slot		*Dog-Bone Slot*		*Tilted Slot*	
Frequency	*Cross-Polarization*	*Front-to-Back Ratio*	*Cross-Polarization*	*Front-to-Back Ratio*	*Cross-Polarization*	*Front-to-Back Ratio*
9.2	−24.25	−16.63	−24.52	−19.25	−23.40	−18.9
9.3	−23.63	−17.64	−24.0	−19.40	−23.00	−18.7
9.4	−22.47	−18.04	−24.0	−19.40	−23.05	−18.8
9.5	−22.49	−18.80	−23.7	−20.50	−22.55	−18.6
9.6	−22.40	−18.84	−23.75	−21.41	−22.52	−18.6
9.7	−22.59	−18.34	−23.7	−21.00	−22.50	−18.5
9.8	−22.59	−17.54	−24.5	−21.00	−22.49	−18.0
9.9	−22.43	−16.59	−24.7	−20.05	−22.44	−17.7
10.0	−22.40	−15.80	−25.01	−19.10	−22.40	−17.3
10.1	−22.40	−15.00	−25.2	−19.00	−22.41	−16.2
10.2	−22.0	−15.00	−25.35	−18.00	−22.41	−15.9

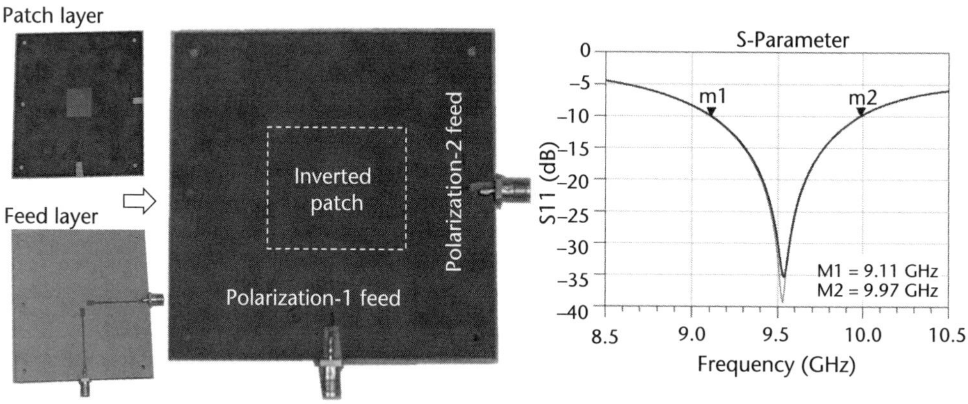

Figure 7.7 Fabricated antenna and return loss plot.

- Amplitude and phase distribution for each radiating element;
- Feed-line optimization.

A dual-polarized linear array of 1 × 8 element is shown in Figure 7.8. Interelement spacing is 0.75λ. The antenna feed network design is a most crucial step in the realization of the array antenna. For a linear array of isotropic elements as shown in Figure 7.9, the relationship of the main beam direction (θ), the distance between the elements (d), the wavelength (λ), and the phase difference ($\Delta\varphi$) are given by

$$\frac{\varnothing}{360^\circ} = \frac{d \sin\theta}{\lambda} \tag{7.3}$$

The direction in which the radiation from an array of linearly placed uniform antennas undergoes constructive interference is determined by the amount of extra

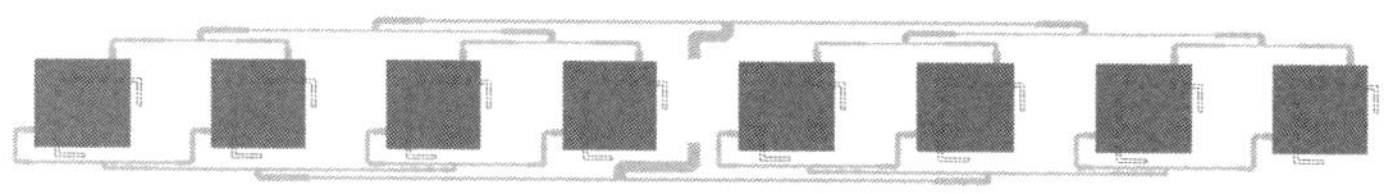

Figure 7.8 Layout design of a linear array SAR antenna.

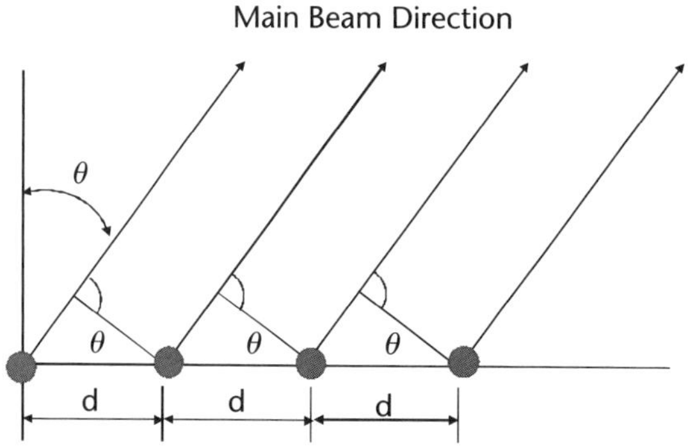

Figure 7.9 Antenna element arrangement in the linear array and beam direction.

path length that the wave needs to travel to be in phase with the wave from the adjacent antenna.

7.5.1 The Design of the Feed Network

The design of the feed network begins with the computation of the impedance that the patch offers at the point of feeding. The impedance of the patch at feeding point is 50Ω at the center frequency. This impedance value is utilized as terminating impedance for the feed network. A quarter-wave transformer is used to transform 50Ω impedance to 100Ω as shown in Figure 7.10.

Amplitude distribution can be changed by changing the impedance value of the line, hence changing the width of the line. Also, phase distribution can be changed by varying the length of the line. This high current taper ratio demands a power ratio of −9.0 dB and impedance ratio of 1:200Ω. The width of the microstrip line for 200Ω at X-band is difficult to fabricate without transformation of this impedance at a lower value impedance. Impedance transform is utilized to achieve a high current taper ratio. A MATLAB code can be developed to compute the possible impedance line width for different impedance lines, which can be printed by changing the characteristic impedance of the line of a quarter-wave transformer. A linear array of 8 elements is simulated using the MoM solver. As shown in Figure 7.11,

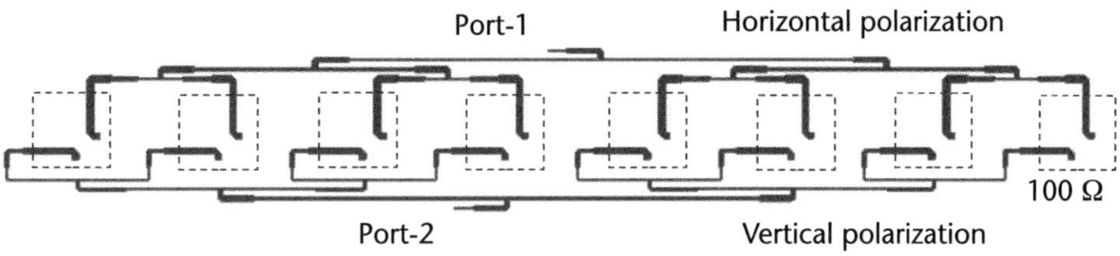

Figure 7.10 A feed network design for a dual-polarized linear array antenna.

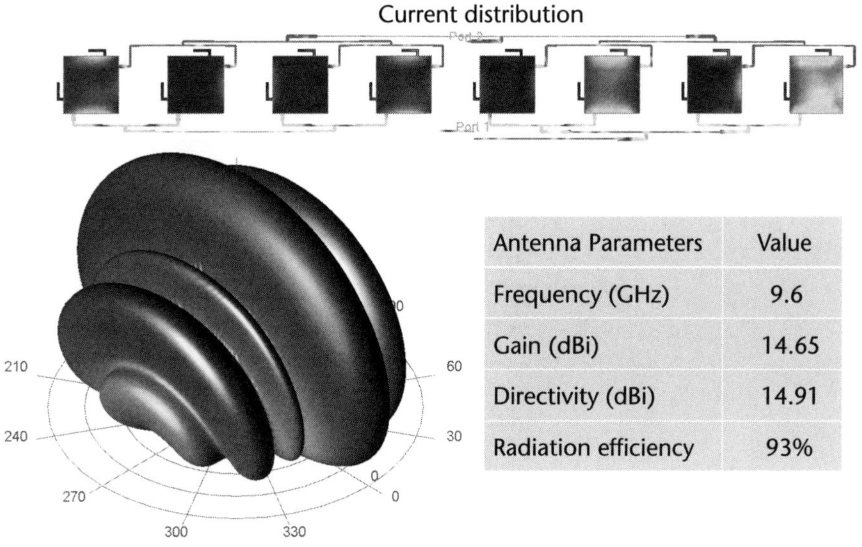

Antenna Parameters	Value
Frequency (GHz)	9.6
Gain (dBi)	14.65
Directivity (dBi)	14.91
Radiation efficiency	93%

Figure 7.11 Current distribution and antenna radiation pattern of the linear array.

the antenna has a fan beam pattern in the $\varphi = 0°$ plane while the beam pattern is narrow in the $\varphi = 90°$ cut plane. The antenna gain is 14.65 dBi, while the efficiency is 93%.

7.5.2 Antenna Performance Results

The 8-element linear array is fabricated and integrated into a multilayered configuration. Antenna return loss is measured to determine the operating bandwidth and resonance frequency. Antenna bandwidth is 700 MHz (9.2–9.9 GHz), with 9.6 GHz as the center frequency. The radiation pattern measurement of copolarization and cross-polarization for both vertical and horizontal polarization is performed at a far-field test range. A summary of the results is given in the Table 7.6 for both horizontal and vertical polarizations. As shown in Figure 7.12, ripples are high for one polarization, and the reason for this high ripple is the coupling between the coupled arm and the microstrip line that is very near to the slot, which changes the impedance of the patch antenna at the feed point.

The specification of the –20-dB sidelobe level in the azimuth plane with very little spacing between the elements makes the design of the feed network complex. This spacing is kept on the higher edge to avoid grating lobes in the visible region. While designing the feed network for the subarray, if the location of the microstrip

Table 7.6 Summary of the Linear Array Results

	Azimuth			*Elevation*		
Frequency (GHz)	*Cross-Polarization*	*3-dB Bandwidth*	*SLL (Sidelobe Level)*	*Cross-Polarization*	*3-dB Bandwidth*	*Ripple*
9.2 (H)	–28.0 dB	70°	–20.0 dB	–23.0 dB	40°	0.2 dB
9.6	–28.4 dB	65°	–19.5 dB	–22.0 dB	40°	1.0 dB
9.9	–25.0 dB	63°	–16.2 dB	–20.0 dB	37°	1.0 dB
9.2 (V)	–20.0 dB	65°	–25.0 dB	–19.0 dB	36°	2.5 dB
9.6	–25.0 dB	57°	–26.0 dB	–25.0 dB	36°	1.5 dB
9.9	–25.0 dB	56°	–22.0 dB	–20.0 dB	—	High

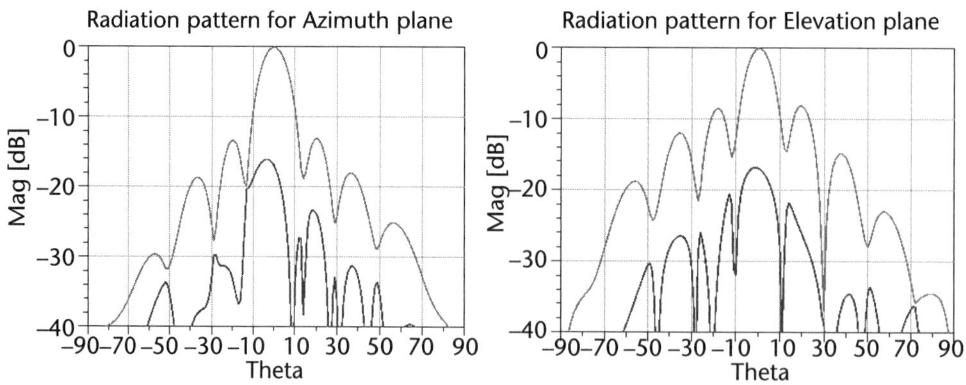

Figure 7.12 Antenna radiation pattern cuts in the azimuth and elevation planes.

line is very near to the slot for one polarization, then it might create the frequency shift. The impedance loci of the antenna do not affect only the imaginary part of the impedance but also the real part when the open stub length of the line is increased or decreased. This phenomenon is more prominent when the microstrip line is very near the slot.

7.5.3 Coupling in Feed Lines

The coupling between the feed lines of the same polarization deteriorates the excitation coefficients and hence the radiation performance of the patch antenna in an array environment. Simulation is carried out to check the coupling between the feed lines, and it was found that more spacing between the lines can reduce the problem, but an engineering compromise must be made for distance between the feed lines so that the whole feedline network for both the polarizations is accommodated in the allowed space and coupling can be reduced to a significant amount. Coupling between the feed lines is more significant in the elevation plane where shaping is stringent. In the azimuth plane, coupling can be reduced with more spacing between the lines.

7.6 Summary

In this chapter, array antennas for microwave remote-sensing application were discussed. Planar and printed antennas are very common in airborne and spaceborne SARs. The design and development of a dual-polarized antenna at X-band for SAR application were presented to prove the concept of dual polarization. Aperture coupled multilayer structures are the best choice when low cross-polarization is required over a large bandwidth. The design technique presented here will find a direct application for a dual-polarized antenna at other frequencies of interest. Also, the performance of the antenna can be improved further by taking mutual coupling effect into account within the patches, and this effect can be utilized to detune the impedance seen by the patch. In airborne, spaceborne, weather monitoring, and other high-performance applications where aerodynamic profile, cost, weight, mechanical structure, and ease of installation are important constraints, printed array antennas are preferred.

References

[1] Kramer, H. J., and A. P. Cracknell, "An Overview of Small Satellites in Remote Sensing," *Int. J. Remote Sens.*, Vol. 29, No. 15, 2008, pp. 4285–4337.

[2] Grafmuller, B., A. Herschlein, and C. Fischer, "The TerraSAR-X Antenna System," *Proc. IEEE Int. Radar Conf.*, Arlington, VA, May 2010, pp. 222–225.

[3] Semenov, S., "Microwave Tomography: Review of the Progress Towards Clinical Applications," *Philos. Trans. A Math. Phys. Eng. Sci.*, August 2009, pp. 3021–3042.

[4] Fear, E. C., et al., "Confocal Microwave Imaging for Breast Cancer Detection: Localization of Tumours in Three Dimensions," *IEEE Transactions on Biomedical Engineering*, Vol. 49, No. 8, August 2002, pp. 812–822.

[5] Semenov, S. Y., and D. R. Corfield, "Microwave Tomography for Brain Imaging: Feasibility Assessment for Stroke Detection," *Int. J. Antennas Propag.*, Vol. 8, 2008.

[6] Lin, J. C., and M. J. Clarke, "Microwave Imaging of Cerebral Edema," *Proceedings of the IEEE*, Vol. 70, No. 5, May 1982, pp. 523–524.

[7] Pozar, D. M., "Microstrip Antennas," *IEEE Proc.*, Vol. 80, January 1992, pp. 79–91.

[8] Frank, R., *Understanding Smart Sensors*, Norwood, MA: Artech House, 2013.

[9] Pozar, D. M., and D. H. Schaubert, *Microstrip Antennas: The Analysis and Design of Microstrip Antennas and Arrays*, New York: IEEE Press, 1995.

[10] Gauthier, G. P., A. Courtay, and G. M. Rebeiz, "Microstrip Antennas on Synthesized Low Dielectric-Constant Substrates," *IEEE Transactions on Antennas and Propagation,* Vol. 45, August 1997, pp. 1310–1314.

[11] Croq, F., and A. Papiernik, "Large Bandwidth Aperture Coupled Microstrip Antenna," *Electronics Letters*, Vol. 26, August 1990, pp. 1293–1294.

[12] Croq, F., and A. Papiernik, "Stacked Slot-Coupled Printed Antenna," *IEEE Microwave and Guided Wave Letters,* Vol. 1, October 1991, pp. 288–290.

[13] Sullivan, P. L., and D. H. Schaubert, "Analysis of an Aperture Coupled Microstrip Antenna," *IEEE-AP*, Vol. AP-34, No. 8, August 1986.

[14] Targonski, S. D., R. B. Waterhouse, and D. M. Pozar, "Wideband Aperture Coupled Stacked Patch Antenna Using Thick Substrates," *Electronics Letters*, Vol. 32, No. 21, October 1996, pp. 1941–1942.

[15] Edimo, M., A. Sharaiha, and C. Terret, "Optimised Feeding of Dual Polarised Broadband Aperture Coupled Printed Antenna," *Electronics Letters*, Vol. 28, No. 19, September 1992, pp. 1785–1787.

[16] Ravindra, V., et al., "A Dual-Polarization X-Band Traveling-Wave Antenna Panel for Small-Satellite Synthetic Aperture Radar," *IEEE Transactions on Antennas and Propagation*, Vol. 65, No. 5, 2017, pp. 2144–2156.

[17] Pandey, B. K., et al., "Dual Frequency Single Aperture Microstrip Patch Antenna Element for SAR Applications," *IETE Technical Review*, Vol. 23, No. 6, November-December 2006, pp. 357–366.

[18] Mao, C. -X., S. Gao, and T. Rommel, "Low-Profile Aperture-Shared X/Ka-Band Dual-Polarized Antenna for DBF-SAR Applications," *2017 International Workshop on Antenna Technology: Small Antennas, Innovative Structures, and Applications (iWAT)*, Athens, 2017, pp. 104–107.

[19] Adamiuk, G., et al., "SAR Architectures Based on DBF for C-and X-Band Applications," *Proceedings of 10th European Conference on Synthetic Aperture Radar (EUSAR 2014)*, June 2014, pp. 3–5.

[20] Heinstadtm J., "Efficient Numerical Strategies for the Computation of Microstrip-Patches and Microstrip-Arrays," Doctoral Thesis, D17, 1995.

Selected Bibliography

Akbar, P. R., et al., "X-Band Parallel-Plate Slot Array Antenna for SAR Sensor Onboard 100 kg Small Satellite," *2015 IEEE International Symposium on Antennas and Propagation & USNC/URSI National Radio Science Meeting*, Vancouver, BC, 2015, pp. 208–209.

Di Bari, R., et al., "Dual-Polarized Printed S-Band Radar Array Antenna for Spacecraft Applications," *IEEE Antennas and Wireless Propagation Letters*, Vol. 10, 2011, pp. 987–990.

Cerutti-Maori, D., I. Sikaneta and C. H. Gierull, "Optimum SAR/GMTI Processing and Its Application to the Radar Satellite RADARSAT-2 for Traffic Monitoring," *IEEE Transactions on Geoscience and Remote Sensing*, Vol. 50, No. 10, October 2012, pp. 3868–3881.

Kahrizi, M., T. K. Sarkar and Z. A. Maricevic, "Analysis of a Wide Radiating Slot in the Ground Plane of a Microstrip Line," *IEEE-MTT*, Vol. 41, No. 1, January 1993, pp. 29–37.

Kothapudi, V. K., and V. Kumar, "A 6-Port Two-Dimensional 3x3 Series-Fed Planar Array Antenna for Dual-Polarized X-Band Airborne Synthetic Aperture Radar Applications," *IEEE Access*, Vol. 6, 2018, pp. 12001–12007.

Mittermayer, J., et al., "The TerraSAR-X Staring Spotlight Mode Concept," *IEEE Transactions on Geoscience and Remote Sensing*, Vol. 52, No. 6, June 2014, pp. 3695–3706.

Mao, C., et al., "Low-Cost X/Ku/Ka-Band Dual-Polarized Array with Shared Aperture," *IEEE Transactions on Antennas and Propagation*, Vol. 65, No. 7, July 2017, pp. 3520–3527.

Pokuls, R., J. Uher, and D. M. Pozar, "Microstrip Antennas for SAR Applications," *IEEE Transactions on Antennas and Propagation*, Vol. 46, No. 9, September 1998, pp. 1289–1296.

Shafai, L. L., et al., "Dual-Band Dual-Polarized Perforated Microstrip Antennas for SAR Applications," *IEEE Transactions on Antennas and Propagation*, Vol. 48, No. 1, January 2000, pp. 58–66.

Yang, J., C. Liu, and Y. Wang, "Imaging and Parameter Estimation of Fast-Moving Targets with Single-Antenna SAR," *IEEE Geoscience and Remote Sensing Letters*, Vol. 11, No. 2, February 2014, pp. 529–533.

Ruiz Rodon, J., et al., "Geosynchronous SAR Focusing with Atmospheric Phase Screen Retrieval and Compensation," *IEEE Transactions on Geoscience and Remote Sensing*, Vol. 51, No. 8, August 2013, pp. 4397–4404.

Santosa, C. E., and J. T. S. Sumantyo, "Broadband Circularly Polarized Microstrip Antenna for Airborne X Band CP-SAR," *2017 IEEE International Symposium on Antennas and Propagation & USNC/URSI National Radio Science Meeting*, San Diego, CA, 2017, pp. 2307–2308.

CHAPTER 8

Airborne SAR Antenna Arrays

8.1 Introduction

A SAR system places a stringent requirement on its antenna radiation properties. Typically, a SAR antenna would require a high degree of polarization purity because the backscatter is a function of the wave polarization. A high degree of cross-polarization would reduce the accuracy of the ground data interpretation. The radiation pattern of a typical SAR antenna should be shaped in the main beam. An airborne radar has a range difference between the near target and the far target, within the main beam. The difference between them should be compensated in the gain of the antenna. Furthermore, the sidelobes and back lobes of the antenna should be suppressed to avoid reflection. The reflection from aircraft will result in radiation pattern distortion and gain loss, while the reflection from the opposite track will result in data corruption. The reflection from under the wing will cause multipath propagation, whereby the reflection from the ground reaches the antenna via a different path.

Airborne radars typically use a variation of a cosec square-shaped beam, which can get the required elevation coverage where the received power is independent of the radar range for a constant height target [1]. The desirable features of the antenna for airborne SAR applications include a shaped radiation pattern and a wide bandwidth capability, a good cross-polarization isolation, and a high-power capability. Shaped radiation across the track improves the target dynamic range and compensates the requirement of the sensitivity time control (STC) correction at the receiver end. Wide bandwidth performance results in finer resolution. The planar array antenna is preferred as compared to the reflector type antenna for less air drag.

Due to the project complexity and cost factor, the antenna designer first simulates the design using commercial electronic design automation (EDA) tools to get performance data before real production. The antenna design cycle time plays a big part in determining a product launch schedule. Simulation tool capability and the way that tools are used play important roles in the design cycle time and cost. Simulation technologies based on different numerical techniques can change the way that complex integrated array antennas are designed. One solution enables the use of separate solvers for the radiator and 3-D feeder networks separately.

8.2 SAR Antenna Pattern Synthesis

Antenna arrays are designed to produce desired patterns, and they are optimized under a set of specified constraints. The constraints can be sidelobe level, first null-to-null beamwidth, VSWR, and impedance. It is always possible to compute the amplitude and phase excitations for a specified radiation pattern. The process of choosing the parameters of an antenna array to produce desired radiating characteristics is known as pattern synthesis. In the array pattern synthesis, the desired output is usually the excitation of the radiating elements, which can produce an acceptable pattern shape. The synthesis of the antenna array with the desired radiation pattern is a nonlinear optimization problem.

Antenna arrays provide an efficient means to detect and process signals arriving from different directions. An array of radiating elements can have its beam pattern modified with an amplitude and phase distribution called the weights of the array. After preprocessing the antenna outputs, signals are weighted and summed to give the antenna array beam pattern. The antenna array pattern synthesis problem consists of finding weights that satisfy a set of specifications on the beam pattern. Antennas with a constant height pattern or a cosecant squared pattern are specially designed for air-surveillance radar sets as discussed previously. These permit an adapted distribution of the radiation in the space and cause a more ideal beam scanning. Such an antenna pattern can get the required elevation coverage where the received power is independent of the radar range for a constant height target. It is a mechanism to achieve a more uniform signal strength at the input of the receiver as a target moves with a constant height within the beam. The antenna pattern synthesis problem is needed in many array antenna applications. One of the key advantages of array antennas is that the array excitation can be accurately controlled to produce extremely low sidelobe patterns or very accurate approximations of chosen radiation patterns. Antenna synthesizing methods fit into three main classes of synthesis:

1. Synthesis of various antenna radiation pattern shapes;
2. Synthesis of low sidelobe;
3. Synthesis that optimizes an array parameter, such as gain and SNR.

In general, the antenna synthesis problem can be stated as the inverse of the antenna analysis problem to find out a set of design parameters that gives the desired antenna output such as the far-field pattern, sidelobe, and gain. The theory of pattern synthesis is concerned with determining the relationship between the far-field pattern of an array and the amplitude and phase distribution of the individual elements and their spacing.

To find a solution, many analytical techniques such as the Woodward-Lawson [2], Orchard-Elliot [3], Taylor method, Fourier transform method [4], Schelkunoff method [5], and Dolph-Chebyshev methods are used [6]. In general, the goal in antenna array synthesis is determining the complex excitations, amplitudes and phases of the elements for producing desired radiation pattern. The Woodward synthesis technique and power pattern synthesis methods are very appropriate for the synthesis of shaped-beam patterns. Many optimization methods such as particle swarm optimization, genetic algorithm [7, 8], and simulate annealing [9] have been useful for the synthesis of antenna array.

8.2.1 Cosecant Square Shaped Pattern

Cosecant squared pattern antennas are frequently used in the base station of wireless communication systems or surveillance radar systems to achieve an equivalent power coverage in the desired distance. Cosecant-squared pattern is one such pattern which is generally employed for long-range systems requiring higher gain near the horizon with low gain at higher elevation angles. During detection of an aircraft flying in space, it will be observed at a closer range at higher elevation angles, so use of such pattern significantly limits the power available to aircraft at higher elevation angles thereby providing a uniform signal strength to the aircraft throughout its journey. The specific gain variations [1] in the cosecant squared region compensate the propagation loss in different distances and maintain the received power unchanged.

Figure 8.1 is showing a microwave remote sensing aircraft flying at a constant height H. As it can be observed that image swath has one low gain and one high gain point. Thus, due to this variation in the range of swath angle, the antenna power distribution is different on these two points. To receive uniform power by the receiver, the radiation shape needs to be modified to cosecant-squared shape.

8.2.2 Woodward-Lawson Pattern Synthesis Technique

Woodward-Lawson pattern synthesis technique is used to compute the excitation coefficients. A pattern produced with a standard design is used in the iterative sampling method. In an iterative sampling method, correction is applied to the pattern obtained from a standard design to make the approximation pattern more nearly satisfy the specifications on the closeness fit. A correction pattern is derived by

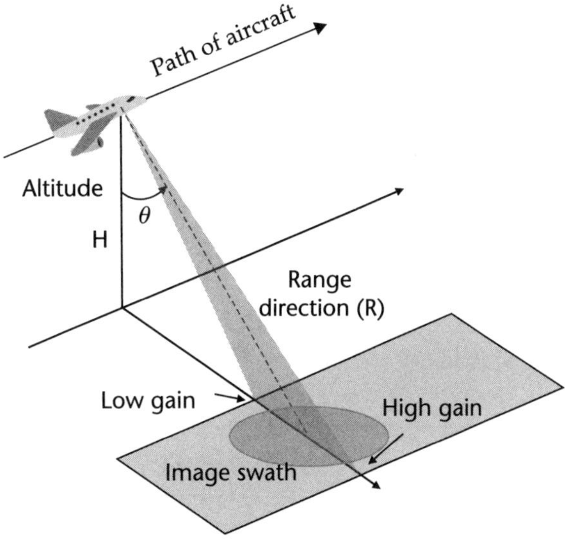

Figure 8.1 Airborne remote sensing radar system.

comparing the desired pattern with the pattern produced from synthesis method. This is the first iteration. If the resulting pattern is not satisfactory, additional corrections may be applied to the first iteration pattern. If the second iteration is also not yet satisfactory, further iterations are applied. Lastly, this iteration process converges to give the desired pattern over the region where the corrections are applied.

Iterative procedure begins with the principal plane pattern of line source of length l represented in terms of the aperture distribution $A(t)$

$$F(u) = \int_{-l}^{l} A(t)e^{iut}\,dt \tag{8.1}$$

where $u = (\pi l/\lambda)\sin(\theta)$, t is length normalized to wavelength and $A(t)$ is represented by the sum of uniform traveling waves with a linear phase change of $2n\pi$ radians over the length of the aperture,

$$A(t) = \frac{1}{2}\sum_{-N}^{N} b_n e^{-jn\pi t} \tag{8.2}$$

where N is the largest whole number of wavelengths contained within the length of the array, so the pattern factor is given by

$$F(u) = \sum_{-N}^{N} b_n \frac{\sin(u - n\pi)}{(u - n\pi)} \tag{8.3}$$

It can be made to coincide exactly with any given function $F(u)$ at $2N + 1$ equally spaced sampling points $u = 0, \pm\pi, \pm 2\pi, \ldots, \pm N\pi$ over the visible region by choosing each coefficient b_n to have the sample value:

$$b_n = F(n\pi) \tag{8.4}$$

The pattern of uniform linear array is given by:

$$F(\theta) = b_n \frac{\sin\left[\frac{N}{2}kd(\cos\theta - \cos\theta_m)\right]}{N\sin\frac{kd}{2}(\cos\theta - \cos\theta_m)} \tag{8.5}$$

For $2N + 1$ element, the array factor becomes:

$$AF(\theta) = \sum_{-N}^{N} b_n \frac{\sin\left[\frac{N}{2}kd(\cos\theta - \cos\theta_m)\right]}{N\sin\frac{kd}{2}(\cos\theta - \cos\theta_m)} \tag{8.6}$$

The N array amplitude weights $a(m)$ is calculated using the inverse discrete Fourier transform (DFT) formulas ((8.4) to (8.6)). Then, replace the weights by their windowed versions using any symmetric length-N window [10]. The final expressions for the windowed weights are, for odd $N = 2M + 1$:

$$a(m) = w(m)\frac{1}{N}\sum_{i=0}^{N} A(\psi_i)e^{-jm\psi_i}, m = 0, \pm 1, \pm 2, \ldots, \pm M \tag{8.7}$$

and for even $N = 2M$,

$$a(\pm m) = w(\pm m)\frac{1}{N}\sum_{i=0}^{N} A(\psi_i)e^{\pm j(m-\frac{1}{2})\psi_i}, m = 1, 2, \ldots, M \tag{8.8}$$

The power of Woodward Lawson pattern synthesis method lies in the ability to specify any shape for the array factor through its frequency samples. Woodward pattern synthesis technique is utilized to compute the complex excitation coefficients for $N = 8$ elements in a linear with $d = 0.75\lambda$. The corresponding wavenumber ψ will be $\psi = kd\cos(\theta)$. The design of the $cosec^2(\theta)$ shape pattern requires that the array factor itself have a secant dependence. As the secant pattern is defined only up to an angle θ_{max}, theoretical array factor in the normalized form can be defined:

$$f(x) = \begin{cases} \dfrac{\cos\theta_{max}}{\cos\theta} & \Big| \, if\, 0 \le \theta \le \theta_{max} \\ 1 & \Big| \, if\, \theta_{max} < \theta \le 90^\circ \end{cases} \tag{8.9}$$

A MATLAB code is written to solve these equations to get amplitude ration and phase value for each element [10].

```
%matlab code%
N=8;
K=(N-1)/2;
d=0.75;
thetamax=42;
psimax = 2*pi*d * cos(thetamax*pi/180);
Ai = ones(1,K+1);
psi = 2*pi*(0:K)/N;
j = find(psi);
Ai(j) = psimax*(psi(j)>=psimax)./psi(j) + (psi(j)<psimax);
Ai = [Ai, Ai(K:-1:1)];
a = woodward(Ai, 0) / N;
aw = a .* (0.54 - 0.46*cos(2*pi*(0:N-1)/(N-1)));
```

The required amplitude and phase distribution for the elevation plane pattern is shown in Table 8.1.

Table 8.1 Excitation Coefficients Calculated Using Woodward Lawson Pattern Synthesis Technique

Element No.	*Amplitude Ratio*	*Phase*
1	0.59	–53.7°
2	0.38	–24.8°
3	0.46	36.0°
4	2.3	52.6°
5	3.77	16.0°
6	2.70	–19.7°
7	0.91	–7.5°
8	1.00	0.0°

Elevation plane pattern requires a $\text{cosec}^2(\theta)$ shape pattern to avoid Sensitivity Time Control correction at the receiver end so that large beamwidth can be provided with higher gain. This kind of antenna will find application for the Airborne Synthetic Aperture Radar (ASAR) and other remote sensing antennas. To have the shaping pattern not only amplitude of the excitation coefficient will change, but also the phase distribution will be different to all the radiating elements to fill the nulls occurring in the visible region. Amplitude will also be tapered to suppress the sidelobe to –20 dB and hence increase the efficiency of the antenna. Hence, complex excitation coefficients make the design challenging.

8.2.3 Theoretical Predicted Patterns

After computing the aperture distribution, the radiation pattern can be obtained to compare with the required pattern as shown in Figure 8.2.

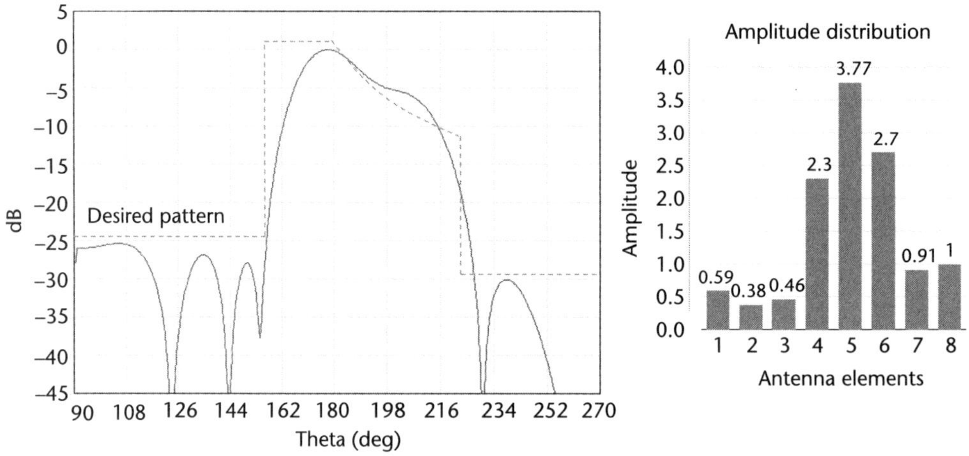

Figure 8.2 Theoretical prediction of elevation plane cosec2 antenna pattern.

8.3 Design of Airborne SAR Antenna

In the airborne application, mainly the antenna size and weight are restricted. To obtain a high-resolution image, it is required either to use an impractically long antenna or to employ wavelengths so short that the radar must contend with severe attenuation in the atmosphere. In the SAR system, the forward motion of the actual antenna is used to synthesize a very long antenna. SAR permits the possibility of using longer wavelengths and still achieving good resolution with antenna structures of a reasonable size. The antenna array design, synthesis, and simulation discussed in this section are an airborne high-power wideband cosecant squared pattern SAR array antenna with a low sidelobe level. In this antenna, a shaped radiation pattern within 35° is achieved with 8 elements. The antenna pattern ripple level is better than 0.5 dB. The feeder network for dual polarization is designed with corporate feeding for eight elements within the limited available interelement spacing. Two linear arrays (8 × 2) are combined at the microstrip feed network level and then it is directly fed from one of the outputs of the 1 × 8 square coaxial line. The input to the square coaxial line is given by the waveguide.

Normally, for remote-sensing applications, the frequency range from 1 to 30 GHz is used. In the 1–10-GHz frequency range, the transmissivity through the air is almost 100%. Therefore, a SAR operating in this frequency band is always able to image the Earth's surface independent of the precipitation or cloud cover. A SAR antenna array is designed to operate at C-band (5.4 GHz). To achieve a bandwidth of 10% at the array level, a multilayer, broadband, single-element antenna is designed. A multilayer stacked radiating array of 8 × 8 elements at C-band with a center frequency of 5.4 GHz for a peak power of 5 kW and a pulse repetition frequency (PRF) of 1.3% is fed with a 1:8 square coaxial feeder network. Antenna beam shaping in the elevation plane is accomplished by obtaining the complex excitation distribution employing the null perturbation technique. Many sets of coefficients giving same power patterns were obtained and the one that gave the minimum current taper ratio is selected for the realization of the antenna. This SAR system is configured with a swath of 50 km operating at an altitude of 4 km.

8.3.1 SAR Antenna Design Flow and Methodology

The high-power SAR antenna is a complex system and a single EM solver is not sufficient to design, simulate, and synthesize complete antenna. With a proper understanding, the sequence of design processes and complexity of a SAR antenna array, the simulation times and resources can be reduced significantly. The hybrid EM simulation approach is used to reduce design cycle by 50% to 60%. This high-power SAR array antenna design has been divided into two segments, the design of a multilayer planar radiating array antenna and a high-power 3-D BFN. A radiating microstrip antenna array is designed with MoM, as this is a very fast solver for the planar kind of structures. A high-power SCL feed network of the antenna is designed using the FEM-based 3-D EM solver. The complete integrated antenna simulation is carried out using the FDTD solver due to complex nature and large problem size.

8.3.2 SAR Antenna Design

The selection of radiating element for planar array is decided as per the requirement of large bandwidth, low cross-polarization with high gain. The multilayer-stacked electromagnetically coupled printed antenna overcomes the bandwidth limitation of the conventional single-layer microstrip antenna. Although the aperture coupled antenna offers the advantage of optimizing the feeder network and patch independently, the impedance matching is poor due to improper grounding when coaxial feeding below the ground is required for the array. The multilayer printed antenna as the basic radiating element is efficient for a large bandwidth. In the design discussed in this section, microstrip patch is effectively printed on the foam material whose dielectric constant is nearly equal to air to enhance bandwidth and to minimize the surface-wave propagation. Because the patch cannot be printed on the foam material, it is printed on the opposite face of the 10-mil dielectric sheet. This dielectric is also used as radome to protect the antenna against the environment.

Electromagnetically coupled feeding is used to avoid the soldering connection to patches and making it more reliable. RT Duroid material 5880 with a thickness 0.79 of mm is used for the feeder network to meet the high-power requirement. The upper patch dimensions are taken to be larger than the lower patch and are fabricated on RT5880 with h = 10 mil. The Rohacell foam, which is 5 mm thick, is placed between the upper and lower radiating patches for bandwidth enhancement. The design of a single element is optimized for a return loss better than –17 dB and cross-polarization better than –25 dB. The optimized antenna dimension and material used are shown in Figure 8.3. The measured results show VSWR (2:1) bandwidth of 800 MHz (17%, –10-dB return loss) with cross-polarization better than –22 dB.

The antenna configuration is shown in Figure 8.3. It has four layers, with an upper patch dimension of 22.6 mm by 19.6 mm and a lower patch dimension of 18.3 mm by 18.3 mm. Broadband behavior is obtained with a staggered tuning of upper and lower patch elements. Staggered tuning is a technique used in the design of broadband antennas whereby each layer is tuned to a slightly different frequency. In comparison to synchronous tuning (where each layer is tuned identically), it produces a wider bandwidth. A linear array design is created with an unequal microstrip distributed network for each of the eight antenna elements to generate a shaped antenna pattern after the single-element design is optimized. All the layers' array antennas are bonded together using nonconducting adhesive

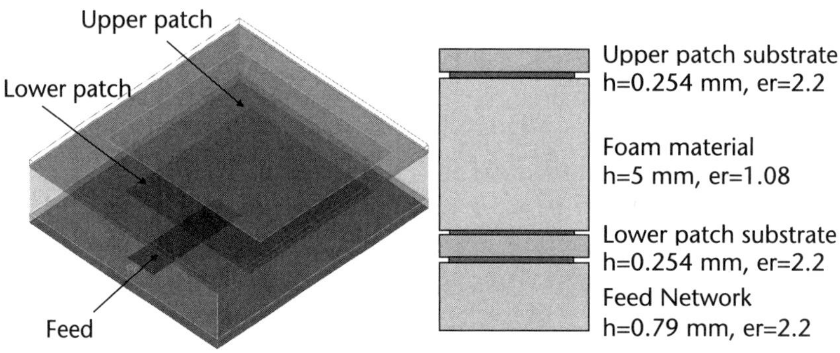

Figure 8.3 Single radiating element and substrate stack-up.

cyanate ester. In antenna simulation, it is better to take the effect of the adhesive layer for better accuracy. Antenna dimensions and simulated S-parameter data are shown in Figure 8.4.

8.3.3 Shaped Pattern BFN

In RF transceiver systems, beamformers are used between the RF signal source and the antenna radiating elements to shape the resulting radiated EM field to illuminate a target or region of space with a precisely controlled beam. To generate a cosec square antenna beam pattern, a BFN is required. The BFN controls the amplitude weight and phase of the feed arrays, which are fed to different elements of the array antenna. The BFN described in this section has a unique architecture by which excitation amplitude and phase can be controlled to shape the antenna beam.

The complex excitation coefficients for the required shape are computed using the null perturbation technique with an element spacing of 0.8λ where roots of the polynomial are moved both in the radial and angular directions in the shaped region and only angular displacement for the nonshaped region [3]. Distribution coefficients are given in Table 8.2. As shown in Figure 8.5, the microstrip feed network is first simulated and optimized by the ADS circuit simulator and optimizer by modeling an asymmetrical coupled line to consider the effect of the coupling between the lines. Because of the better phase control of the order of 5-degree, this results in fewer ripples in the shaped patterns. The design is optimized in multiple smaller parts representing T-Networks. Then all parts are combined to get overall effect of BFN.

Layout design is created for the optimized distributed network as shown in Figure 8.6. This network is simulated using EM solver MoM and a minor adjustment in length was carried out for the required phase matching. Amplitude and phase distribution at each output port are realized as per the computed value shown in Table 8.2 to get cosecant square shaped pattern. Because microstrip lines are placed close to each other in this BFN, coupling between lines affects the feed network performance.

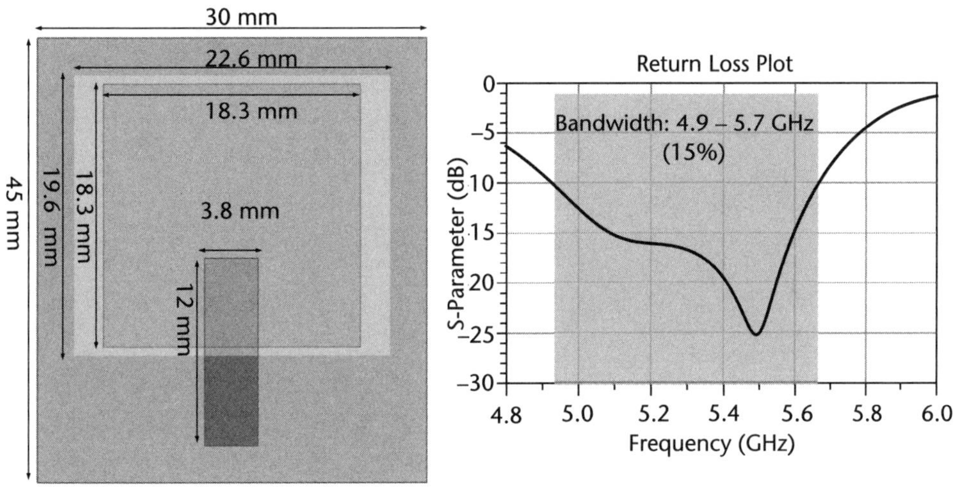

Figure 8.4 Radiating element design dimensions and return loss plot.

Table 8.2 Amplitude and Phase Distribution Coefficient of 1:8 BFN

Element Number	*Amplitude Distribution*	*1-W Power Distribution at Each Element*	*Power at Each Element (dBm)*	*Phase (Degrees)*
1	1	0.003W	–14.7	0
2	0.58	0.011W	–19.4	100.3
3	1.98	0.134W	–8.7	–107.5
4	3.37	0.380W	–4.2	–59.4
5	2.72	0.256W	–5.9	–35.9
6	2.02	0.136W	–8.6	–33.6
7	1.09	0.040W	–13.9	–21
8	0.6	0.010W	–19.3	–53.17

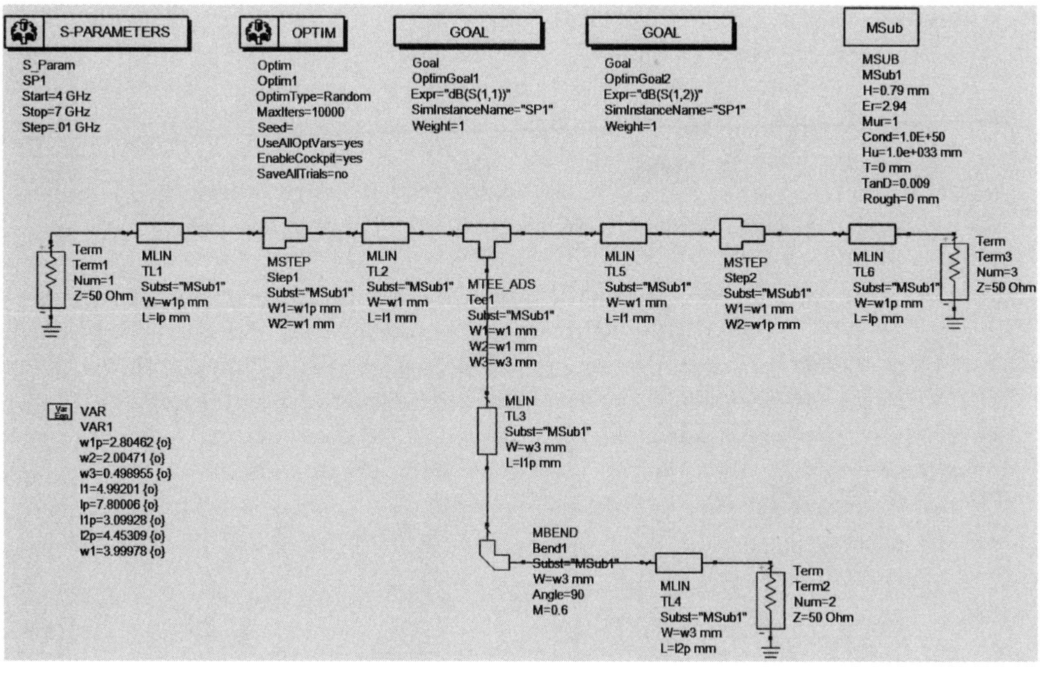

Figure 8.5 ADS circuit simulation schematic of Tee-network of BFN.

8.3.4 Antenna Array Design Realization and Simulation

First, a linear antenna array (N = 8, d = 0.8λ) was realized in printed microstrip technology, which works in C-band (center frequency f_0 = 5.4 GHz) with a frequency bandwidth of 17%. Then, using 8 optimized linear arrays, a planar array of an 8 × 8 element is designed with an interarray spacing of 0.8λ. A linear array antenna shown in Figure 8.7 has been simulated using MoM. The antenna pattern is a cosecant square in the elevation plane while it is a fan beam in the azimuth plane.

A planar array as shown in Figure 8.8 is formed by placing eight linear arrays with 0.8λ interarray spacing. The antenna performance is same as that of the

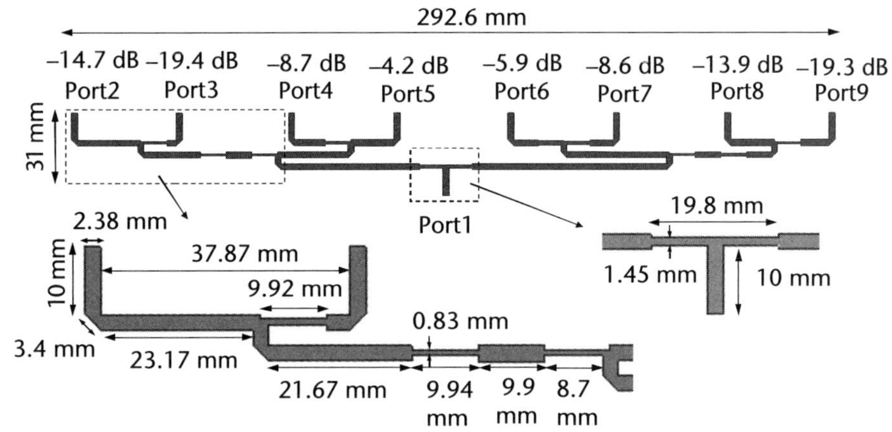

Figure 8.6 A 1:8 BFN using a microstrip line.

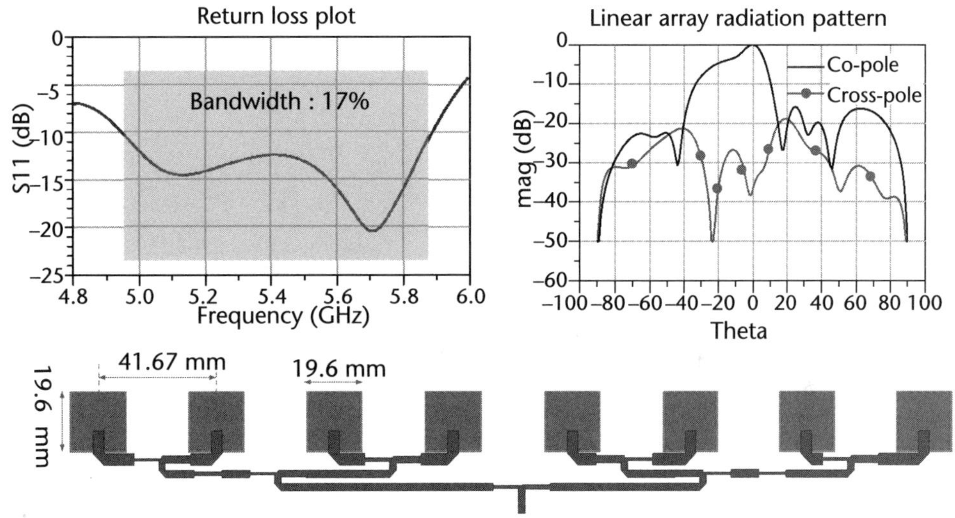

Figure 8.7 A linear antenna array with eight radiating elements.

linear-array antenna performance. All eight ports are simulated using the FEM solver. The antenna radiation pattern in azimuth and elevation plane is shown in Figure 8.9. In azimuth plane combines radiation pattern is pencil beam. More number of element in array provide sharper pattern in azimuth direction.

The antenna is fabricated layer by layer. All these layers are aligned and bonded together using a nonconducting adhesive cyanate ester. The measured return loss performance, along with the simulation of the antenna, shows the 2:1 VSWR bandwidth of the antenna is better than 350 MHz. The performance of the measured patterns and its comparison with simulated patterns are shown in Figure 8.10. A very good matching observed in the measured patterns over simulated patterns confirms the validity of the technique and shows that shaping in elevation is achieved with a 0.5-dB ripple over the frequency. This array antenna tile cannot handle the high-power (8 kW) requirement of a SAR system; for that, an SCL-based BFN design is described next.

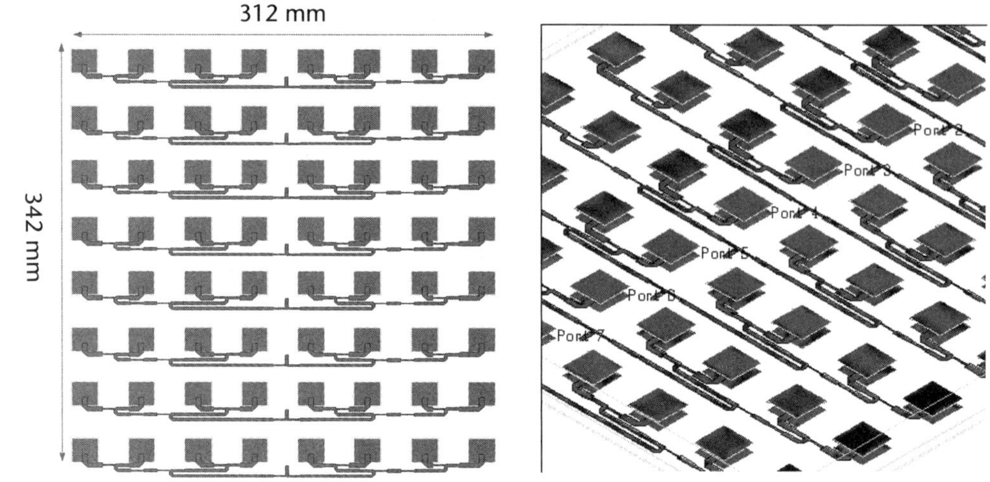

Figure 8.8 The planar antenna array (8 × 8 elements) and current distribution plot.

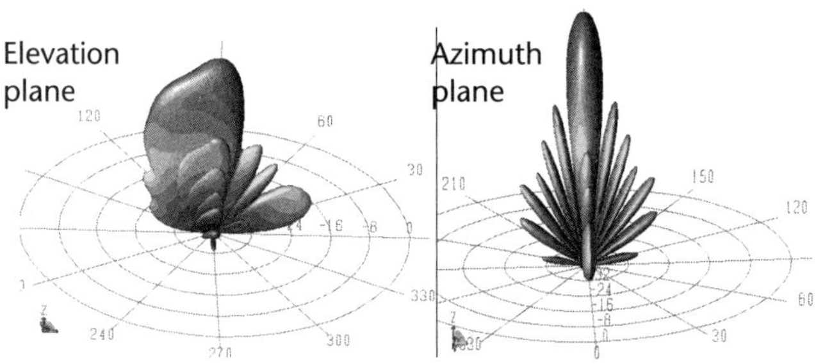

Figure 8.9 Elevation and azimuth plane radiation pattern of planar antenna array.

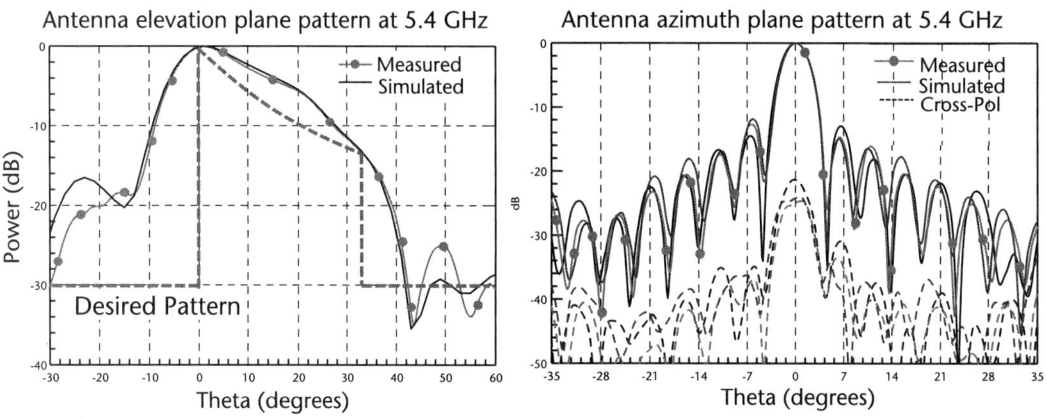

Figure 8.10 Comparison of measured and simulated azimuth and elevation plane patterns.

8.4 High-Power Feed Network Design

Microwave remote-sensing applications require antennas with shaped beams like cosecant-squared ($cosec^2$) sector shape, and elliptical beam. As discussed in previous sections, this requires the development of complex BFN to generate the necessary amplitude and phase distributions in the radiating elements of the antenna array to yield the desired shaped beam [11]. Usually, planar antennas are well suited for SAR applications. The most commonly used radiating element used in the planar array is the microstrip patch antenna, which is inherently low-profile and lightweight. However, there is a serious limitation with microstrip patch antennas in terms of the problem associated with the power-handling capability. Therefore, microstrip patch antennas cannot be directly used for SAR systems where a pulse peak power of several kilowatts is used. For such applications, a hybrid antenna where the waveguide feeding networks can be used to handle the high input power and proper transitions can be made to the microstrip line or directly to the microstrip patch, which can handle the split lower power. BFN represents the heart of any array antenna system. It consists of many power dividers to adjust the phase and amplitude of the signal at each feeding element. The key requirements for the BFN are wideband operation, high precision power division (typically 0.5 dB in amplitude and 5° in phase), high-power capability, and reliable technology for airborne and spaceborne applications.

Although a BFN made of a hollow waveguide is best suited from the point of view of power handling, but from the point of view of weight and compactness, the BFNs are generally not made solely of waveguides. In this case, rectangular (or, more specifically, square) coaxial lines are used as a transmission medium for the BFN in the lower-frequency range such as L-band and C-band up to X-band. While circular coaxial lines have been used extensively in the past, rectangular coaxial lines may be preferable in some applications if the presence of flat rather than circular surfaces offer mechanical advantages and it is desired to have an unambiguously defined plane of polarization. The loss characteristics and power-handling capability of rectangular coaxial line are comparable to that of circular coaxial line. However, manufacturing components with a rectangular cross-section are more economical and this type of coaxial line is predominantly used in the feed systems employing many components to form the BFN.

8.4.1 Rectangular Coaxial Line

The idea of a square coaxial radiator started in 1979 when Frank Hanna and Wilson Ishu developed the first prototype at the University of Illinois at Chicago, which, due to its symmetric aperture in theory, should have equal E and H plane patterns [12]. To obtain a low-loss coaxial transmission line, for a given operating frequency range, the cross-sectional area of the coaxial structure should be optimized [13]. Figure 8.11 shows an air-filled square coaxial transmission line.

The appropriate choice of a cross-section to be used in a specific design is a trade-off between having a low attenuation constant for the transmission line and predicting at what frequency the higher modes will start to propagate. When higher modes are present, these lead to a dispersive transmission line, so for most applications, it is desired to propagate only a TEM mode. The dominant mode in the rectangular coaxial line is TEM. For the TEM mode of propagation, the distributed

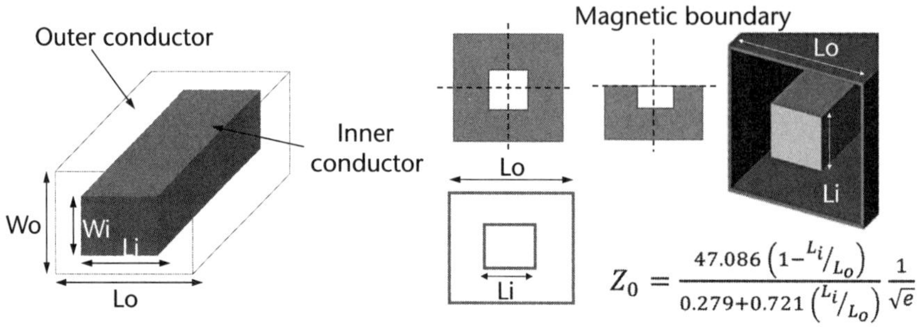

Figure 8.11 Rectangular and square coaxial line design configuration and EM simulation model.

circuit theory can be applied. The bandwidth of the rectangular coaxial line for the TEM mode propagation starts from DC to the cutoff frequencies of the first higher-order modes. Analytically, the higher-order modes in the rectangular coaxial lines can be derived using a formulation similar to that applied to the ridge waveguide [14–16].

The transverse resonance technique applied to the solution of ridge waveguide can also be used to solve the eigenvalue problems in the rectangular coaxial medium. Obviously, different boundary conditions will lead to different zeros of the system determinant and thus to different cutoff frequencies. The characteristic impedance of the rectangular coaxial line can be computed from the following equation;

$$Z_0 = 59.952 \ln\left(\frac{1+\frac{L_o}{W_o}}{\frac{L_i}{W_o}+\frac{W_i}{W_o}}\right) \tag{8.10}$$

The above equation has 10% accuracy for $\frac{W_i}{W_o} < 0.3$ and $\frac{L_o}{L_i} < 0.8$.

In the case of the SCL characteristic, impedance is calculated using the following equation, which has 1% accuracy:

$$Z_0 = \frac{47.086(1 - {}^{Li}/_{Lo})}{0.279 + 0.721({}^{Li}/_{Lo})}\frac{1}{\sqrt{e}} \tag{8.11}$$

8.4.2 Corona and Multipaction Breakdown in the High-Power Feed Network

The corona effect can be defined as a local discharge initiated by gas ionization, when the electric field has reached a threshold [17, 18]. If gases were only composed of neutral particles, the corona effect would never occur. Electrons are necessary elements to the beginning of the discharge mechanism. Electrons absorb part of the EM energy and they start a movement of oscillation. Under certain pressure conditions, they will collide with neutral particles. Because of these collisions, the peripheral electrons are freed. They also oscillate and collide with other neutral particles. If these new free electrons are not absorbed anywhere else, their density will increase: plasma is created, and the discharge phenomenon begins. Corona and multipaction calculations are important to determine the maximum power-handling capability of the square-coaxial feed network without any breakdown in operating frequency range.

Other breakdown phenomena exist such as the multipactor effect and the Paschen discharge. As far as a multipactor is concerned, initial electrons are also required. However, contrary to the corona effect, initial electrons will shell the metallic wall, thus starting the emission of secondary electrons. These ones are accelerated thanks to the electric field and will collide with the metallic wall on the other side. The physical quantities, which are of main importance for this effect, are the electric field magnitude, the distance between the walls, and the type of material used for the metallic wall. The mean free path of electrons helps to make the distinction between the corona effect and the multipactor, and this parameter depends on the pressure. The higher the pressure is, the more electrons have a chance to collide with molecules before reaching the wall. The two phenomena thus have their own pressure domain. The multipactor effect occurs when the ambient pressure is sufficiently low that the electron mean free path is longer than the electron separation distance. When the electron's mean free path is lower that the distance between the walls, the corona effect will occur.

Multipaction represents a possible failure mechanism of the electronic system in space in very low pressure because it can destroy RF components or it can significantly raise the system noise levels. A proper component design reduces the probability of a multipactor-free operation of an electronic system.

The air breakdown level is given by the air rigidity, which is >30 kV/mm.

- For a P power in a 50Ω impedance line, the root mean rectangular voltage is $V_{eff} = \sqrt{P.R}$V.
- For a g critical gap, the root mean rectangular field is $E_{eff} = V_{eff}/g$.
- For P = 2 kW and g = 1.8 mm, the field Eeff is 175 V/mm.

8.4.3 Design of the Feed Network

The design of SCL begins with a choice of impedance with which the patch is to be fed and it is taken as a 50Ω line and with the outer dimension $a = 9$ mm and the inner dimension $b = 3.5$ mm. The design of the feeder network requires the individual analysis of the following sections:

- 90° corner and Tee network;
- 1:2, 1:4, and 1:8 power dividers;
- SCL to microstrip transition.

The design starts with a 1:2 power divider followed by the design of 1:4 to 1:8. Power splitting to even more ratios can be accomplished in the multilayer corporate configuration to house the total feeder network in the limited available space. The coupling between different levels is done by transitions designed especially for this purpose. Using this SCL, different power dividers are designed for equal power divisions at C-band, as shown in Figure 8.12. A 1:2 power divider is designed for equal power division (–3 dB). The input and output port impedance is 50Ω. The outer and inner conductors have dimensions of 9 mm and 3.5 mm, respectively.

8.4.3.1 Rounded Bend

The bends in the power divider can be either rounded type or chamfered bend. The design and optimization of 90° bend are important to transform waveguide impedance smoothly at bend point otherwise it will introduce mismatch. In this design bend is optimize using FEM solver as shown in Figure 8.13. It is necessary to ensure that the bending and twisting are accomplished in the correct manner; otherwise, the electric and magnetic fields will be excessively distorted, and the signal will not propagate in the way required, causing loss and reflections.

8.4.3.2 A 1:4 Power Divider

The choice of the dimension for the inner and outer conductors is based on two considerations. First is the power-handling capability of the transmission line. The gap between the inner and outer conductors should be enough to handle the transmitted power. On the other side, the second consideration is the generation of higher-order modes. A large gap between the inner and outer conductors may result

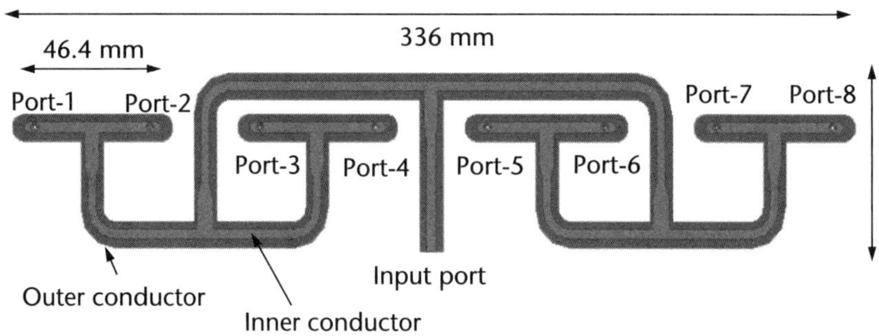

Figure 8.12 The top view of a 1:8 power divider based on SCL.

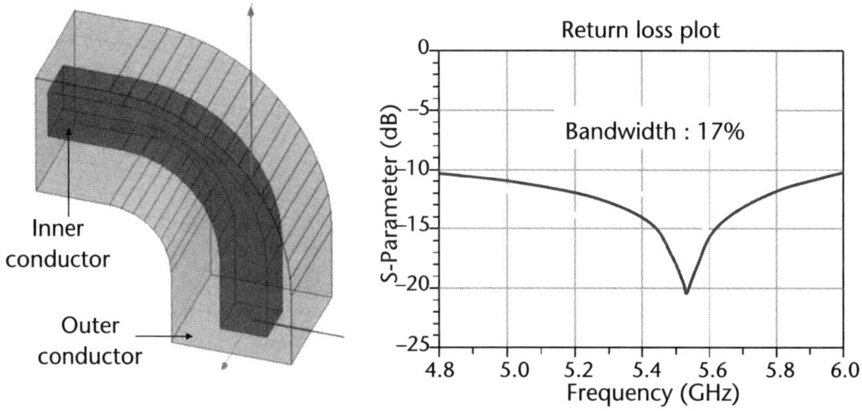

Figure 8.13 A 90° rounded bend and return loss plot.

in the generation of a higher-order mode. These modes should be avoided. Hence, the choice of dimension is a compromise between power handling and higher-order modes. A $\lambda/4$ transformer is used to match 50Ω at the input to 25Ω at the junction. The design and simulated results are shown in Figure 8.14. The simulated result shows a better than –20-dB return loss over 1 GHz.

8.4.3.3 A 1:8 Equal Power Divider

The 1:4 power divider is further progressed to 1:8 equal power division (–9 dB). The input and output sections are kept at 50Ω impedance. This power divider is designed and simulation is performed for a complete power divider with 8-port outputs. The $\lambda/4$ transformers are tuned to achieve the return loss better than –25 dB at 5.3 GHz. The Chebyshev multisection matching transformer is used for larger bandwidths. This technique provides a larger bandwidth than the binominal multisection transformer for a given number of transmission line sections but at the expense of an increased ripple over the passband of the matching network. The Chebyshev transformer is realized by equating the overall reflection coefficient to Chebyshev polynomials. Using these polynomials, matching networks are designed with a reflection coefficient at or below some prescribed level over a wide bandwidth.

The shrinkage of bandwidth is again due to transformers used at different levels. Square coaxial feeder network modules are optimized for a return loss better than –15 dB in the desired frequency band as shown in Figure 8.16. For these designs, an average of 20 to 25 design iterations is needed. A fast simulation and less memory by applying magnetic symmetry plane simply reduce design time by 50% to 60%. Optimized designed TEE, 90° bend, and transition network are integrated to form a 1:8 feeder network.

8.4.3.4 Transitions

A feeder network using an SCL is used to feed power to a radiating element such as a microstrip patch. A transition from an SCL to a microstrip line is required that will act as an interface between the two. Moreover, the power from a traveling-wave

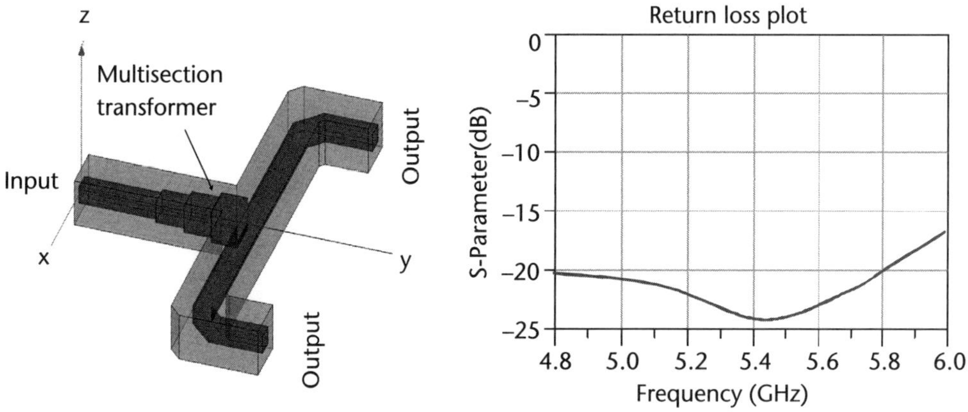

Figure 8.14 A 1:4 SCL power divider and return loss plot.

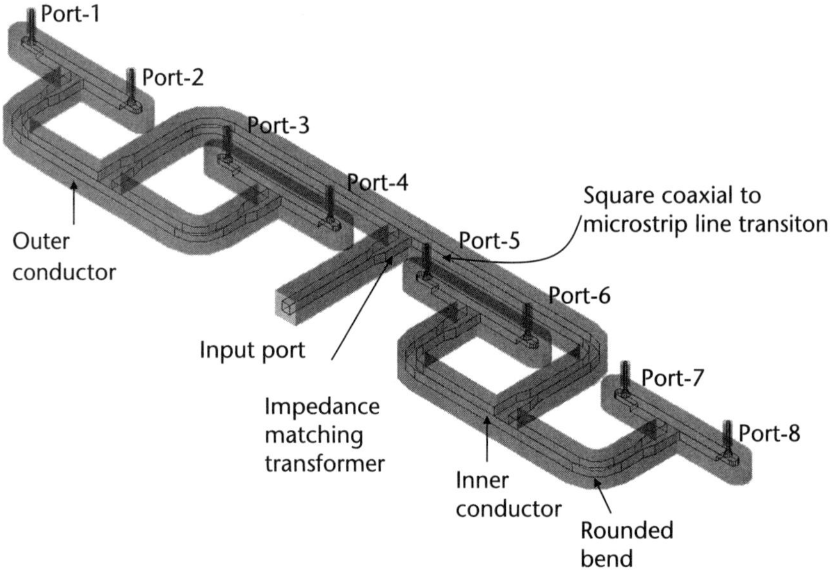

Figure 8.15 A 3-D view of a 1:8 power divider with microstrip line transition.

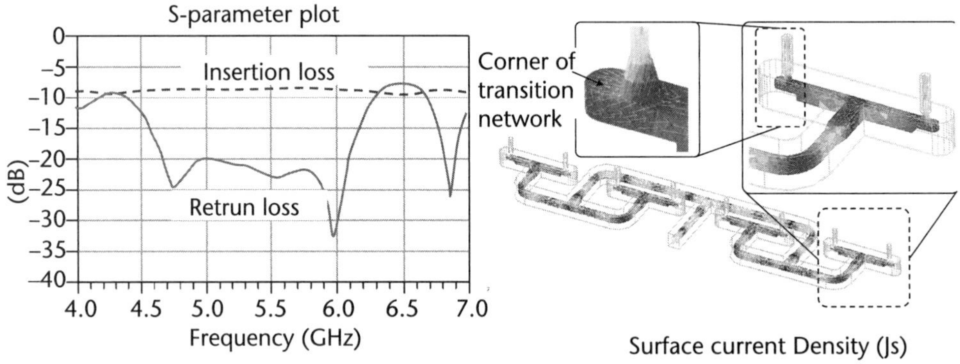

Figure 8.16 Return loss plot and surface current distribution of a 1:8 power divider.

tube (TWTA) or solid-state power amplifiers (SSPA) is to be coupled to the feed network. For high-power applications, the TWTA output is in waveguide. The transition from waveguide to SCL is required to couple power from the input to the SCL network. Sometimes it is also required to split the network into two levels so that the whole network can be accommodated in the available space. The key requirements of these transitions for practical applications would be low loss, broadband performance, tolerance to fabrication, and compact size. A transition to couple power between two levels of SCLs is also required. All these transitions are to be designed and analyzed separately.

8.4.3.5 Waveguide to SCL Transition

The coupling between waveguide and SCL is achieved by using an end launcher type of probe coupling. Numerous transition structures have been studied and now are available for microwave and millimeter-wave transceiver applications. Among them, simple probe-type transitions providing low cost, simplicity, and broadband performances have been typically utilized for microwave and millimeter-wave applications [19, 20]. The depth of the probe and its distance from the shorted wall of waveguide are critical for the performance. These two parameters are optimized to achieve the best performance. The termination of probe in the SCL is done on the inner conductor. The aperture size of the waveguide is 40 mm × 20 mm. The simulated results show that at 5.3 GHz the return loss is better than 30 dB. The operation of the waveguide-to-coaxial line transition can be easily understood if we assume that an input power is applied to the SCL. In this transition structure, the center conductor of the coaxial line is inserted at the $\lambda/4$ distance from the waveguide wall. This transition structure can be considered as an impedance transformer structure for the impedance matching between a coaxial line and waveguide.

8.4.3.6 SCL to Microstrip Line Transition

The transition from the SCL to the microstrip line is the most critical to design. To achieve return loss and power coupling from this transition, the shape of the SCL is modified at the location of the transition. Different perturbations with the shape lead to working the transition in different frequency ranges. SCL to microstrip line transition is shown in Figure 8.17.

The transitions with short-circuited square-coaxial waveguide of $\lambda_g/4$ guided wavelength are very popular because their principle of mode transformation is almost the same with that of ordinary transitions of a waveguide and a coaxial cable. The probe transition connects a microstrip line and a waveguide and a probe at one end of the microstrip line is inserted into the SCL whose one end is short-circuited by the back-short waveguide.

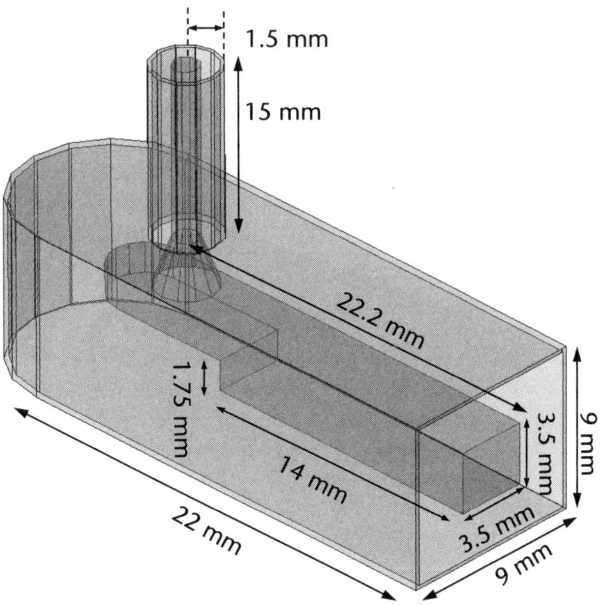

Figure 8.17 SCL to microstrip line transition design.

The performance is achieved by shaping both the outer and inner conductors. The inner and outer conductors are made circular at the end. The location of the end point of the inner conductor with respect to the end point of the outer conductor is an important parameter of optimization.

Moreover, a section of the inner conductor is cut at the end. The length and depth of the cut affect the coupling. Also, a cone-like tapered section is placed at the inner conductor, which guides the impedance transformation from a 50Ω SCL to a 50Ω circular coaxial line. The location of the center of the cone on the inner conductor affects the matching of two sections. This location can be optimized for return loss improvement. The other end of the 50Ω circular coaxial line is directly fed to the microstrip patch feed line. This transition is the most versatile of all the transitions. The simulated return loss of this transition works very well from 2 GHz to 7 GHz for a return loss better than –20 dB in this frequency band as shown in Figure 8.18.

8.4.4 High-Power Calculation of the SCL

In square-coaxial feed network design, a sufficient margin is taken in the design to avoid any damage because of the high-power operation. The electric field strength in the SCL is evaluated by modeling the complete network using the FEM. The electric field strength is computed at the junction where the minimum gap has been created because of the change in the inner conductor dimension for the required impedance transformation. The following shows the result of the calculation for a typical 8 kW of peak power and 72W of average power:

- Electric field strength in square coaxial line = 63.92 V/cm (from the FEM solver).
- Average voltage for 1-W average power = 17.75V.
- Average voltage for 72-W average power = 150.67V.
- Peak voltage for 8-kW peak power with 0.90 duty cycle = 1.588 × 103V.
- Peak electric field strength for rectangular coaxial dimensions = 577.532 V/mm.

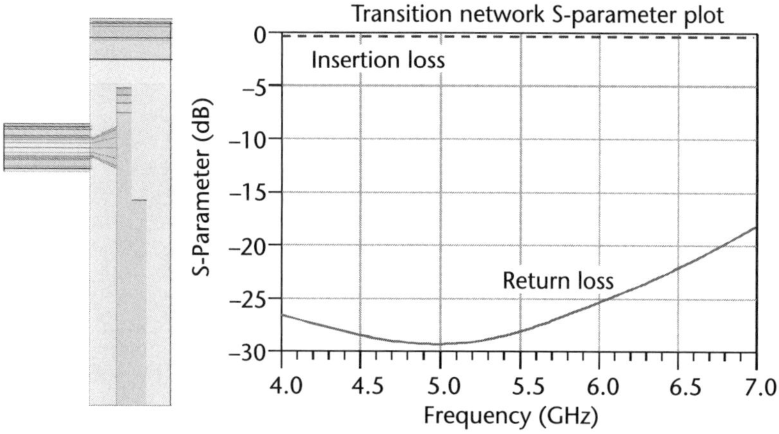

Figure 8.18 S-parameter plot of the SCL to microstrip line transition.

- Peak electric field strength for air breakdown = 3,000 V/mm.
- Peak electric field strength for dielectric breakdown = 2,000 V/mm.
- Breakdown margin for air = 14.31 dB.
- Breakdown margin for dielectric = 10.789 dB.

The gap gives around 10-dB margin and the cutoff frequency for dominant higher-order TE10 mode is 10 GHz.

8.5 Integrated Array Antenna Design

Complete SAR antenna design is designed by integrating planar microstrip radiating array with high-power square-coaxial feed network. All layers and components are shown in Figure 8.19. A complete antenna system is simulated using the FDTD technique. This is one panel of the antenna, for high gain antenna multiple panels are used on top of the SCL feed network.

SCLs are designed on aluminum by cutting grooves and placing the inner conduct metal with the help of PTFE spacers. The printed antenna is placed on top of this aluminum block and coaxial pins of each 8-output port is fed to the microstrip antenna array.

8.6 Summary

In this chapter, the airborne SAR antenna system was discussed. Antenna pattern synthesis is very important in these types of antennas. A multilayer printed antenna with EM coupling is designed and developed in C-band for wide bandwidth and low cross-polarization. Beam shaping in the elevation plane is realized by obtaining the complex excitation distribution employing the null perturbation technique.

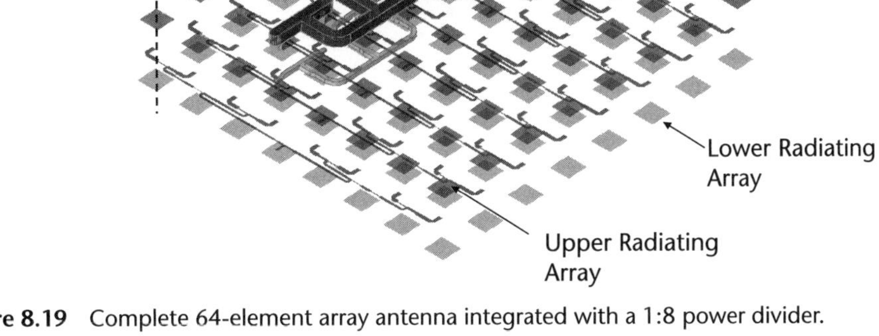

Figure 8.19 Complete 64-element array antenna integrated with a 1:8 power divider.

The design of a wideband aperture coupled microstrip linear array antenna that consisting of eight elements in the elevation plane and another eight elements in the azimuth plane has been described. The complete antenna design with MoM, FEM, and FDTD with efficient techniques reduces the design cycle by 60% compared to the traditional single solver approach.

References

[1] Rengarajan, S. R., and A. G. Derneryd, "Application of Compound Coupling Slots in the Design of Shaped Beam Antenna Pattern," *IEEE Transactions on Antennas and Propagation*, Vol. 41, No. 1, January 1983, pp. 5–9.

[2] Woodward, P. M., and J. D. Lawson, "The Theoretical Precision with Which an Arbitrary Radiation-Pattern May Be Obtained from a Source of Finite Size," *Journal of the Institution of Electrical Engineers - Part III: Radio and Communication Engineering*, Vol. 95, No. 37, September 1948, pp. 363–370.

[3] Orchard, H. J., R. S. Elliott, and G. J. Stern, "Optimizing the Synthesis of Shaped Beam Antenna Patterns," *Proc. IEE*, Vol. 132-H, No. 63, February 1985.

[4] Steyskal, H., "On the Problem of Antenna Power Pattern Synthesis for Linear Arrays, FDA Reports," *Research Institute of National Defence*, Stockholm, Sweden, Vol. 5, No. 3, May 1971, pp. 1–16.

[5] Schelkunoff, S. A., "A Mathematical Theory of Linear Arrays," *Bell Syst. Tech. J.*, Vol. 22, No. 1, January 1943, pp. 80–107.

[6] Dolph, C. L., "A Current Distribution for Broadside Arrays Which Optimizes the Relationship Between Beam Width and Side-Lobe Level," *Proc. IRE*, Vol. 34, June 1946, pp. 335–348.

[7] Marcano, D., and F. Duran, "Synthesis of Antenna Arrays Using Genetic Algorithms," *IEEE Antennas and Propagation Magazine*, Vol. 42, No. 3, 2000, pp. 328–332.

[8] Donelli, M., et al., "Linear Antenna Synthesis with Hybrid Genetic Algorithm," *Prog. Electromagn. Res.*, Vol. 49, 2004, pp. 1–22.

[9] Farhat, N. H., and B. Bai, "Phased-Array Antenna Pattern Synthesis by Simulated Annealing," *IEEE Proc.*, Vol. 75, No. 6, 1987, pp. 842–844.

[10] Orfanidis, S. J., *Electromagnetic Waves and Antennas*, Rutgers University, 1999.

[11] Pandey, A. K., "Design of a Cosecant Square-Shaped Beam Pattern SAR Antenna Array Fed with Square Coaxial Feeder Network," *2013 European Microwave Conference*, Nuremberg, 2013.

[12] Lee, C. Q., et al., "Design of a Square Waveguide Feed," *Proceedings of USNC/URSI Meeting*, Boulder, CO, November 1979, p. 134.

[13] Pandey, B. K., et al., "Square Coaxial Beam Forming Network for Multilayer Microstrip Antenna," *2007 Asia-Pacific Microwave Conference*, Bangkok, 2007, pp. 2477–2480.

[14] Yubo, T., Z. Bing, and W. Xin, "Numerical Method for Solving Characteristic Parameters of Square Coaxial Line," *2007 IEEE Antennas and Propagation Society International Symposium*, Honolulu, HI, 2007, pp. 73–76.

[15] Hurtado, M., H. E. Lorente, and C. H. Muravchik, "Analysis of a Square Microstrip Antenna with an Eccentric Slot," *IEEE Antennas and Propagation Society International Symposium*, Boston, MA, 2001, pp. 498–501.

[16] Oikawa, N., et al., "Coaxial-Line Connection of a Circularly-Polarized Patch Array Antenna on the Thick Resin and a 60 GHz-Band Quadrature Oscillator Chip," *2014 IEEE iWEM*, Sapporo, 2014, pp. 169–170.

[17] Woode, A., and J. Petit, "Diagnostic Investigations into the Multipactor Effect, Susceptibility Zone Measurements and Parameters Affecting a Discharge," *ESTEC Working Paper*, Vol. 1556, November 1989.

[18] Vaughan, J. R. M., "Multipactor," *IEEE Transactions on Electron Devices*, Vol. 35, No. 7, July 1988, pp. 1172–1180.

[19] Ko, D. H., et al., "V-Band Waveguide-to-Coplanar Waveguide Transition for 60 GHz Wireless LAN Application," *IEEE MTT-S Int EUMC*, 2004, pp. 641– 644.

[20] Kim, K. W., C. H. Na, and D. S. Woo, "New Dielectric-Covered Waveguide-to-Microstrip Transitions for Ka-Band Transceivers," *IEEE MTT-S Int. Microwave Symp Dig.*, Vol. 2, Philadelphia, PA, 2003, pp. 1115–1118.

Selected Bibliography

Alessandri, F., M. Mongiardo, and R. Sorrentino, "Computer Aided Design of Beam Forming Networks for Modern Satellite Antennas," *IEEE Transactions on Microwave Theory and Techniques*, Vol. MTT-40, No. 6, June 1992, pp. 1117–1127.

Angeletti, P., G. Toso, and G. Ruggerini, "Array Antennas with Jointly Optimized Elements Positions and Dimensions Part II: Planar Circular Arrays," *IEEE Transactions on Antennas and Propagation*, Vol. 62, 2014, pp. 1627–1639.

Bucci, O. M., et al., "Antenna Pattern Synthesis: A New General Approach," *Proc. IEEE*, Vol. 82, March 1994, pp. 358–371.

Buckley, M. J., "Synthesis of Shaped Beam Antenna Patterns Using Implicitly Constrained Current Elements," *IEEE Transactions on Antennas and Propagation*, Vol. 44, February 1999, pp. 192–198.

Deshpande, M., and N. Das, "Rectangular Microstrip Antenna for Circular Polarization," *IEEE Transactions on Antennas and Propagation*, Vol. 34, No. 5, May 1986, pp. 744–746.

Elliot, R. S., "Antenna Pattern Synthesis," *IEEE Antenna and Propagation Society Newsletter*, October 1985, pp. 5–9.

Guo, J., and Y. Lu, "Novel Aperiodic Phased Array with Reduced Number of Active Chains for Space-Borne SAR," *IEEE Fifth Asia-Pacific Conf. Synthetic Aperture Radar*, 2015, pp. 20–23.

Hu, Y., et al., "Characterization of the Input Impedance of the Inset-Fed Rectangular Microstrip Antenna," *IEEE Transactions on Antennas and Propagation*, Vol. 56, No. 10, October 2008, pp. 3314–3318.

Kumar, G., and K. P. Ray, *Broadband Microstrip Antenna*, Norwood, MA: Artech House, 2003.

Mailloux, R. J., *Phased Array Antenna Handbook*, Norwood, MA: Artech House, 1994.

Milovanovic, A., and B. Koprivica, "Analysis of Square Coaxial Lines by Using Equivalent Electrodes Method," *Proceedings of the Joint INDS'11 & ISTET'11*, Klagenfurt, 2011, pp. 1–6.

Sharma, P., and K. Gupta, "Analysis and Optimized Design of Single Feed Circularly Polarized Microstrip Antennas," *IEEE Transactions on Antennas and Propagation*, Vol. 31, No. 6, November 1983, pp. 949–955.

Sharma, S. B., et al., "Dual-Polarized Shaped-Beam Printed Antenna for Airborne SAR Applications," *Microwave Opt. Technology Letter,* Vol. 44, 2005, pp. 383–385.

Uher, J., J. Bornemann, and U. Rosenberg, *Waveguide Components for Antenna Feed Systems: Theory and CAD*, Norwood, MA: Artech House, 1993.

Zhang, X. -M., et al., "Cosecant-Square Pattern Synthesis with Particle Swarm Optimization for Nonuniformly Spaced Linear Array Antennas," *2008 8th International Symposium on Antennas, Propagation and EM Theory*, 2008, pp. 193–196.

Zurcher, J. F., and F. E. Gardiol, *Broadband Patch Antennas*, Norwood, MA: Artech House, 1995.

CHAPTER 9

Multifrequency Shared-Aperture Microstrip Antenna Arrays

9.1 Introduction

High-resolution airborne and spaceborne imaging of the Earth is often carried out using SAR techniques, as discussed in the previous chapter. Early synthetic aperture radar (SAR) systems were single-frequency and single-polarization systems, but time has seen a growing interest in multifrequency, dual-polarization SAR systems. The reason for this drift is the additional amount of geophysical information, possibly to extract from multifrequency, polarimetric SAR data [1]. Multifrequency systems such as tactical radar systems and SAR systems are common in both military and civil applications because their information capacity is significantly higher than single-frequency counterparts. A multiband antenna array can be realized by simply positioning several single-band antenna arrays side by side. However, this approach is not desirable because it significantly increases the aperture size and antenna weight. The multifrequency shared aperture antenna arrays can minimize the array size and weight by sharing the same aperture for each band. Therefore, this approach is very attractive for compact systems like space-borne SAR and vehicle-based tactical radars.

An antenna array that operates in multifrequency bands with dual-polarization capability in each band and whose elements are integrated together into a common physical space by sharing the single aperture falls into the category of multiband dual-polarized shared-aperture (MBDP-SA) array. In the multifrequency shared aperture array design, antenna designers face some design challenges such as the following:

- In multifrequency antenna array design, the requirement of avoiding grating lobes at the higher frequency dictate the interelement spacing. This condition unnecessarily increases the number of elements at the lower-frequency band [2].
- Moreover, only a moderate-frequency ratio between the operating frequencies can be achieved by this approach. For frequency ratios on the order of

4:1 or more, designing a multifrequency microstrip antenna is a complex matter.

- Antenna design becomes more challenging, especially if other spaceborne SAR critical issues such as size, system bandwidth, beam scanning, and polarization purity come into the picture.

In multifrequency SAR, the higher-band wave reflects primarily at the crown of the tree, while the lower-band wave can penetrate the leaves and reflect at the root of the forest as shown in Figure 9.1. Longer wavelength bands (L-bands) can penetrate forest canopy. These wavelengths are used to estimate forest biomass and detect the amount of wood in a forest. The shorter wavelengths (X-bands) are used to detect smaller features like leaves and twigs. The longer the microwave wavelength, the greater the penetration of canopy. The back-scattering matrix of a ground target depends on frequency, polarization, and incidence angle of the incident wave.

Theoretically, an ultrawideband (UWB) array such as a Vivaldi antenna can also be employed to minimize the aperture size, because it can cover multibands with only a single aperture [3]. However, these arrays are less suitable for the SAR applications because the element spacing is based on the smallest wavelength in the operating bandwidth to avoid the grating lobes at a higher frequency. So at a lower frequency, the elements are too densely distributed. More elements and corresponding transceiver (T/R) modules are therefore required as compared to the single frequency microstrip patch antenna. Hence, UWB arrays are less feasible due to the high cost, large weight, and difficulties in the heat control of a large amount of T/R modules. The MBDP-SA array is more suitable for the multiband SAR antenna than the UWB.

9.1.1 SAR Imaging

In SAR imaging, a very long aperture is synthesized through the integration of the magnitude and phase information of the returned echoes from an object over the entire time it is in view of the radar. Because comparatively long-wavelength radar energy penetrates through (even dense) water vapor, radar imaging is often possible in situations where optical sensing is severely limited or even impossible due to persistent cloud cover. The strength of the signal received by the antenna for a given ground resolution cell (represented by the radar backscatter coefficient σ) is

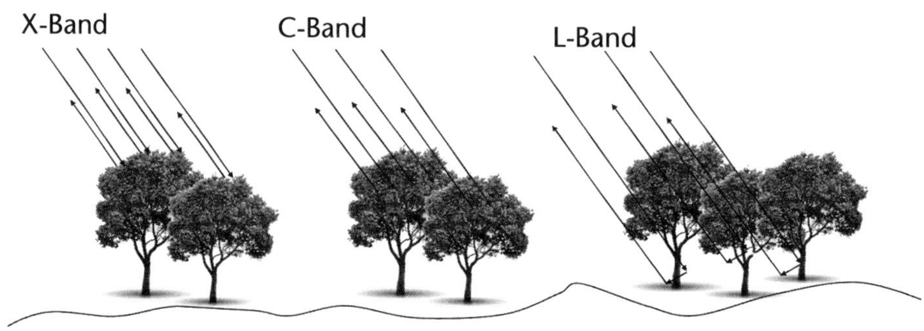

Figure 9.1 Short-wavelength X-band radar energy reflects from canopy surface while long-wavelength L-band energy penetrates through canopy and reflects from stems and terrain surface.

mainly a function of the wavelength of the characteristics of the imaged feature, the microwave energy, and the geometry of the image acquisition.

In the context mapping, the wavelength of the radar determines whether the SAR backscatter is dominated by surface scattering or volume scattering. In the case of a relatively short wavelength (for example, X-band), the microwave energy interacts with the surface of the Earth and the energy is scattered by small-scale components on the surface, such as the foliage and small branches or trees. In contrast, radar energy with relatively long wavelengths (like 24 cm for L-band) will penetrate deep and reflect from large-scale components including large objects and the terrain surface. The SAR signal logged from a surface is proportional to the surface roughness. High-frequency X-band is very sensitive to even a small-scale roughness. The same surface is much flatter by the L-band SAR due to the difference in wavelength. The penetration of the microwave signal in terrain is directly dependent on wavelength. This means that longer-wavelength SAR has more signal or sensitivity to the volume content of terrain (e.g., soil moisture). This interprets that while X-band can detect even slight details, L-band is suitable to observe larger objects, which would be unclear with X-band. Combining the two is possible to monitor both geometrical objects and small details.

Many SAR systems have the capability to send and receive energy with different polarizations. Because radar signal can be depolarized upon interaction with various surface objects, independently recording the reflection of like-polarized energy (e.g., horizontal send – horizontal receive (HH) or vertical send – vertical receive (VV)) and cross-polarized energy (e.g., horizontal send – vertical receive (HV) or vertical send – horizontal receive (VH)) can yield valuable information regarding the characteristics of imaged features and can be particularly useful in the analysis of vegetation type and structure. Commonly applied bands for radar applications and bands and polarizations for different applications are listed in Table 9.1.

Most of the current space-borne SAR systems operate from L-band to X-band due to the atmosphere attenuation. The operating frequency below L-band will suffer reflections from the ionosphere, while radiated energy higher than the X-band radiation is increasingly absorbed by water and oxygen molecules in the atmosphere and thus becomes sensitive to weather condition. Ka-band is also frequently used for SAR applications (climate surveillance or fine imaging), as it offers less

Table 9.1 Commonly Applied Bands for Radar Applications

			Applications			
Band	*Frequency Range (GHz)*	*Radar Frequency (GHz)*	*Geology and Water*	*Agriculture and Ocean*	*Forest*	*Soil*
L	1–2	1.215–1.4	L	—	L	—
S	2–4	2.3–2.5 and 2.7–3.7	—	—	—	—
C	4–8	5.25–5.925	C	C	C	C
X	8–12	8.5–10.68	—	X	X	—
Ku	12–18	13.4–14 and 15.7–17.7	—	—	—	—
K	18–26	24–24.25	—	—	—	—
Ka	26–40	33.4–36	—	—	—	—

attenuation to signal due to its location between the water and oxygen absorption peak at 23 GHz and 63 GHz, respectively.

9.2 Dual-Frequency Shared Aperture Antenna Array

The antenna plays a very important role in determining the overall system performance of airborne and spaceborne SAR systems. SAR system performance such as azimuth and elevation resolution, imaging ambiguity, and the width of the mapping area is directly determined by the performance of its antenna. The main driving force for frequency and polarization selection of a SAR system is the application targeted for. For most the applications, SAR systems generally operate at L-band, C-band, and X-band. There is an increasing demand for simultaneous multifrequency SAR systems to get images of targets of different signatures during one visit to a region of the Earth. Additionally, for airborne applications, SAR data at different frequencies like C-band and X-band can be recorded without landing the aircraft for the mounting of the different antenna at different frequency bands if a multifrequency SAR system is used onboard the aircraft. Antennas at different frequency bands will occupy a larger area, which may be prohibited by the available space in the radome of the aircraft. Thus, there is a strong need to develop antennas that fulfill the requirement of multifrequency operation for SAR applications [4].

In this chapter, the design and development of a dual-frequency shared aperture antenna array is discussed [5–15]. Rectangular slots are used as a radiator at X-band, and the inset fed patch has been used as radiating elements at C-band. A parametric study by varying various dimensions of the slot and simulated return loss curves are also presented in this chapter. The dual-band operation of an antenna has been applied in a lot of systems, such as wireless communications, radars, deep space communications, and radio astronomical telescopes. The requirements for antennas are varied in different missions.

9.3 Design of C-Band and X-Band Radiating Element

A dual-frequency antenna with a shared aperture places a severe constraint on space for the elements and the associated feed networks and makes a multilayer geometry a necessity. The C-band and X-band radiating elements and the pertinent feed networks are placed on separate layers. The design dimensions of the single C-band and X-band radiating elements are shown in Figure 9.2.

The bottom layer is a substrate material of Rohacell foam of 6.71-mm thickness with a dielectric constant of 1.08 backed by a ground plane. The foam material has low dielectric constants close to 1 because they are filled with small gas bubbles. This allows for the designing of antennas with high efficiency [16]. The ground plane serves as an effective shield between the radiating patches and the feed network improving polarization purity. Above this foam layer, there is a layer of the substrate material of 0.79-mm thickness of RT-Duroid5880 with a dielectric constant of 2.2 for the electromagnetically coupled antenna feed network for X-band antenna.

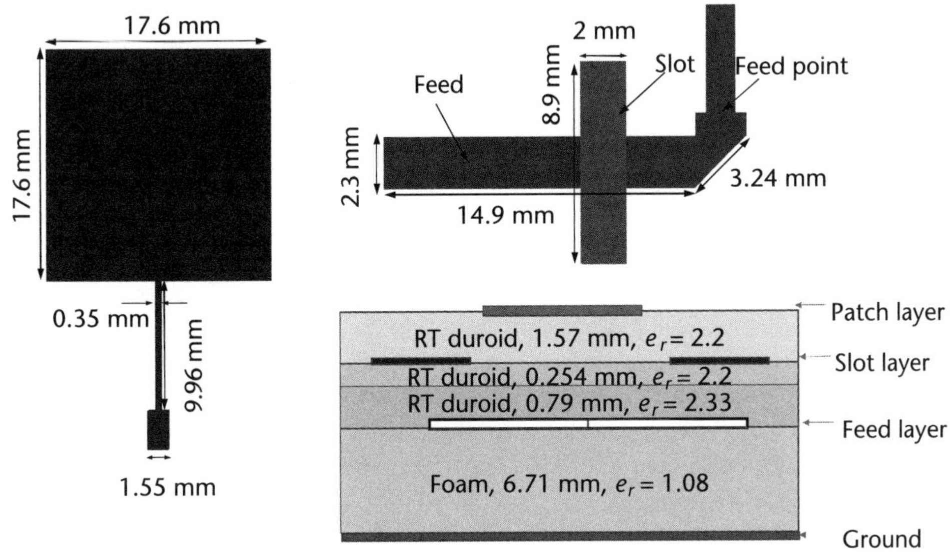

Figure 9.2 C-band and X-band antenna configuration, design dimensions, and substrate stackup.

Slots are cut at the top side of this substrate and feedlines are etched on the bottom side to excite the slots. Above the slot, there is another substrate layer of 1.57-mm thickness of RT-Duroid5880 for the C-band microstrip patch that is etched on the top side of this substrate. The patch antenna at C-band is edge-fed by a microstrip line. The slot at X-band is fed by the microstrip line beneath the slot layer. Feed for the slot is isolated from the patch by a ground layer. Thickness and permittivity of the substrate are determined as a trade-off between cross-polar radiation and bandwidth. The bottom ground layer is required to avoid radiation in the undesired direction as the slot has the characteristic of bidirectional radiation. Complete assembly is sharing a common aperture with the slot and patch radiating in the presence of each other without affecting their individual radiation pattern.

9.3.1 Design of Patch Antenna at C-Band

Microstrip edge-fed patch is selected as radiating element at C-band. In edge-fed patch configuration, the input impedance of the patch is very high at the edge because of which the width of the microstrip line feeding the patch becomes very small. In this design, the width of the 180Ω line is 0.35 mm, which is transformed into 100Ω impedance using a $\lambda/4$ transformer. The antenna design parameters have been optimized using the finite element method (FEM) solver. The final design dimensions of the patch, which yields the desired RF performance in the presence of the X-band slots cut in the ground plane of the patch, are shown in Figure 9.2. The simulated gain of the single patch is 7.1 dBi at 5.3 GHz.

9.3.2 Design of the Slot Antenna at X-Band

At X-band, the printed slot is employed as a radiating element. This slot is cut in the ground plane of the C-band patch and it is fed by a microstrip line below a dielectric thickness of 0.79 mm and a relative permittivity of 2.2. The microstrip line of 50Ω impedance and a width of 2.3 mm is used as a slot feedline. Below this microstrip feedline, another ground plane at the separation of 5.0 mm is used to avoid radiation from the slot in the lower half plane. The RF performance of the slot must be optimized in the presence of the C-band patch which acts as a parasitic element for slot radiator. For the dual-polarized antenna, rectangular shape slot is a preferred option as it yields better cross-polar performance.

Figure 9.3 shows the simulated return loss performance of a single slot radiator at X-band. Return loss bandwidth is 800 MHz at a center frequency of 9.6 GHz for the voltage standing wave ratio (VSWR) of 2:1. At 9.6 GHz, the simulated gain of a single slot is 4.5 dBi. After designing a single radiating element at X-band, its sensitivity analysis is carried out for various dimensional tolerances. The length and width dimensions of the rectangular slot as shown in Figure 9.2 are varied, which results in a frequency shift in the return loss performance. Good impedance matching is obtained on wide bandwidths around the working frequency.

9.4 Array Design

Integrated antenna array structure using both C-band and X-band linear arrays are shown in Figure 9.4. For eight linear arrays (each having 4 C-band and 16 X-band elements), the total array size is 17.6 cm × 36 cm.

Both C-band and X-band radiating elements are arranged in such a way that these do not block each other's radiation pattern. One C-band patch radiator is placed between four X-band slot radiators. This topology helps to avoid placing two antennas for C-band and X-band separately and gives real estate to the system in which they are used.

9.4.1 Design of the C-Band Linear Array

In the C-band linear array, all four elements placed in one row and are fed through a series feed network. The entire row is then fed in the center by a corporate feed network. The design dimensions of the series feed network are shown in Figure 9.5. Multiple $\lambda/4$ transformers are used to transform 50Ω input impedance to the impedance of the patch at the edge. Figure 9.5 shows the simulated result of the linear

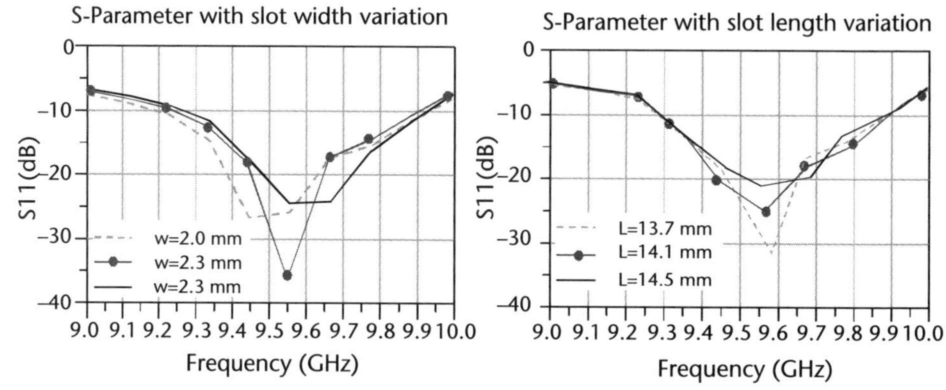

Figure 9.3 Return loss plot of the X-band antenna with the width and length variation of the slot.

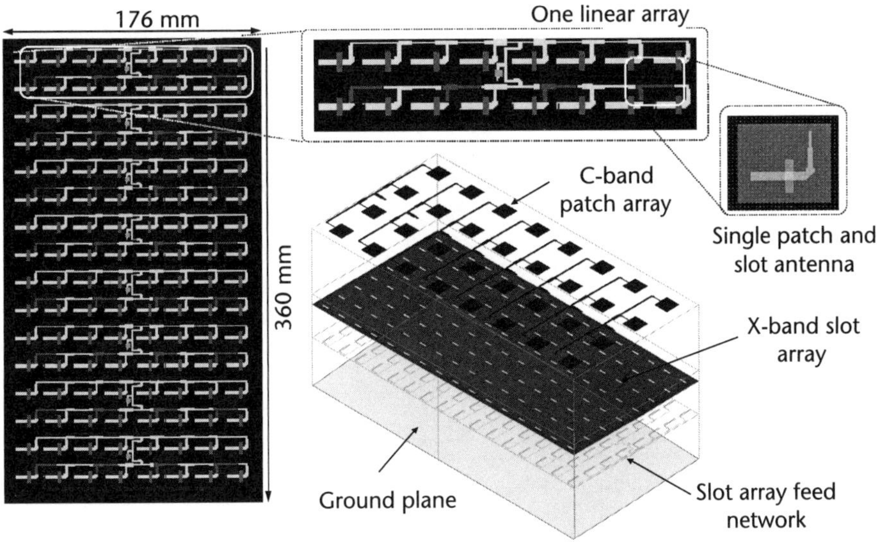

Figure 9.4 The C-band and X-band shared aperture antenna array configuration.

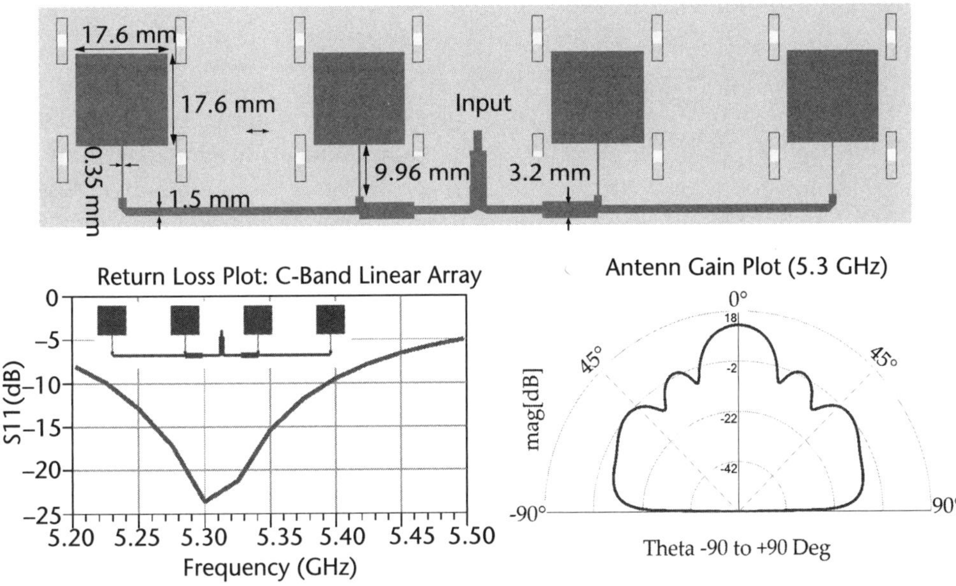

Figure 9.5 The C-band linear array antenna and simulated results.

array. The antenna is resonating at 5.3 GHz, and the operating bandwidth is 180 MHz (5.22–5.4 GHz) and gain of the linear array antenna is 13 dBi.

9.4.2 Design of the X-Band Linear Array

In the X-band linear array, eight elements placed in one row are fed through a series feed network. Two rows are combined in the center and fed with corporate feed network as shown in Figure 9.6. Such an antenna feeding mechanism is called a

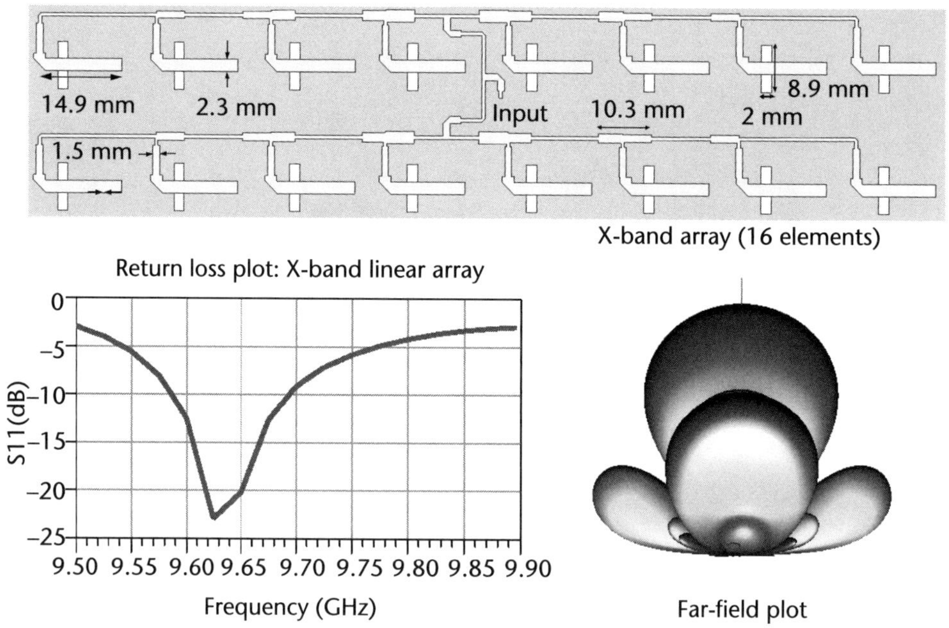

Figure 9.6 The X-band linear array antenna and simulated results.

hybrid feed network, which is a very efficient way to combine multiple antenna elements in a limited space in an array design. The simulated performance of X-band linear antenna, in the presence of the C-band linear array, is shown in Figure 9.6. The antenna is resonating at 9.63 GHz and the operating bandwidth is 120 MHz (9.58–9.7 GHz). The 16-element antenna gain is 15.6 dBi.

9.4.3 Integrated Array

A single linear array is duplicated in Y directions to form a planar array as shown in Figure 9.7. The spacing between patch elements is 0.8λ in both the X and Y directions. The interelement spacing between patches is optimized for the maximum gain of the array. The spacing between slots is subjected to a condition that the slot array is within the inner periphery of the patch array. Slots are spaced at 0.8λ in the X direction and 0.6λ in the Y direction. The total aperture of the patches is square in geometry, while that of slots is rectangular. In a planar antenna array, the total number of elements in C-band is 32 (4×8 elements), while total number of elements in X-band is 128 (8×16 elements).

A simulated return loss plot of a planar array is shown in Figure 9.7. The 10-dB bandwidth of a 4×8 planar array is 100 MHz at C-band and 300 MHz at X-band. Isolation between two antennas is better than 27 dB over the entire operating frequency band. The fabricated antenna is shown in Figure 9.8. The radiation pattern of the antenna is measured at the anechoic chamber. The measured gain of a patch array is 13.5 dBi at a center frequency of 5.3 GHz, and the gain of a slot array is 14.8 dBi at a center frequency of 9.6 GHz.

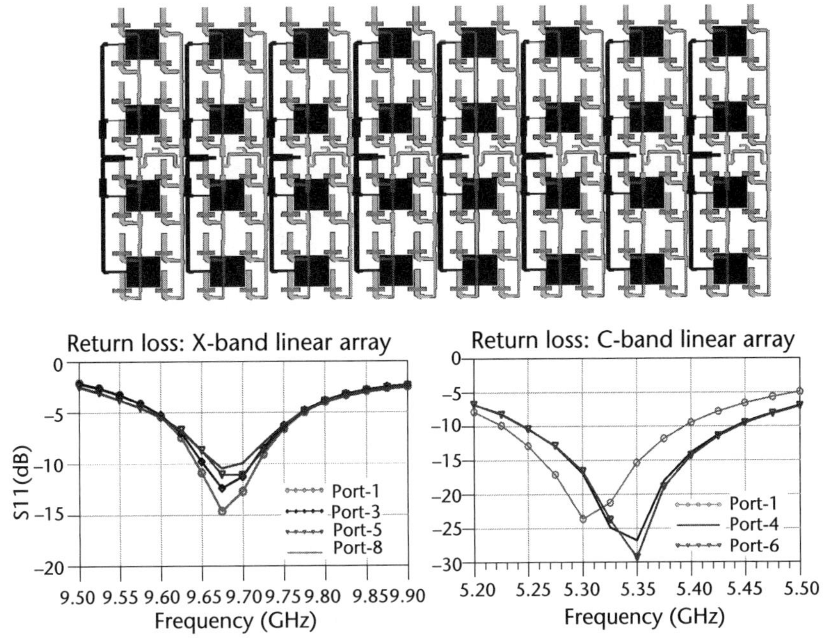

Figure 9.7 Dual-frequency shared-aperture planar antenna array and return loss plot.

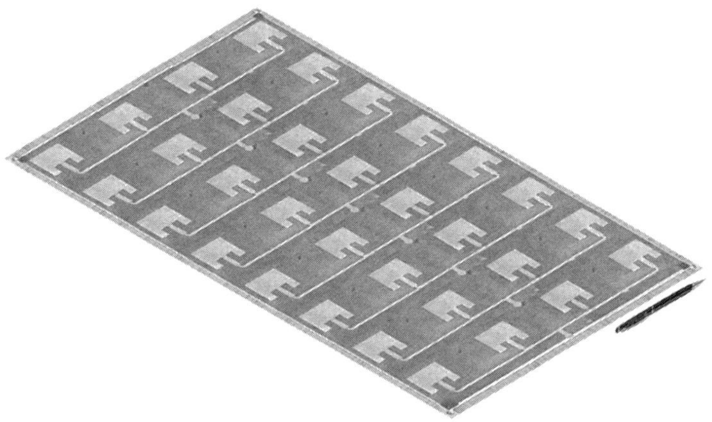

Figure 9.8 Fabricated planar antenna array.

9.5 Summary

In this chapter, the method of assembling two frequency band shared-aperture subarrays to form a large shared-aperture array was introduced. The planar array for the SAR application was constructed from microstrip patches at the C-band slots at X-band in a single aperture. An array prototype was fabricated and measured to validate the method. This array design method can be extended to the shared-aperture arrays with more than two frequency bands. The advantage of this configuration is that the isolation between the two frequency bands is better than 27 dB, which ensures proper functioning of the system at an individual frequency band.

References

[1] National Aeronautics and Space Administration, "Spaceborne SAR: Current Status and Future Directions," NASA Technical Memorandum 4679, April 1995.

[2] Shafai, L. L., et al., "Dual-Band Dual-Polarized Perforated Microstrip Antennas for SAR Applications," *IEEE Transactions on Antennas and Propagation*, Vol. 48, January 2000, pp. 58–66.

[3] Loui, H., J. P. Weem, and Z. Popovic, "A Dual-Band Dual-Polarized Nested Vivaldi Slot Array with Multilevel Ground Plane," *IEEE Transactions on Antennas and Propagation*, Vol. 51, No. 9, September 2003, pp. 2168–2175.

[4] Pokuls, R., J. Uher, and D. M. Pozar, "Dual Frequency and Dual Polarization Microstrip Antennas for SAR Applications," *IEEE Transactions on Antennas and Propagation*, Vol. 46, September 1998, pp. 1289–1296.

[5] Pandey, B. K., et al., "Dual Frequency Single Aperture Microstrip Patch Antenna Element for SAR Applications," *IETE Technical Review*, Vol. 23, No. 6, November-December 2006, pp. 357–366.

[6] Uher, R. P. J., and D. M. Pozar, "Dual-Frequency and Dual-Polarization Microstrip Antennas for SAR Applications," *IEEE Transactions on Antennas and Propagation*, Vol. 46, No. 9, September 1998, pp. 1289–1296.

[7] Pozar, D. M., and S. D. Targonski, "A Shared-Aperture Dual-Band Dual Polarized Microstrip Array," *IEEE Transactions on Antennas and Propagation*, Vol. 49, No. 2, February 2001, pp. 150–157.

[8] Zhong, S. -S., et al., "Tri-Band Dual Polarization Shared-Aperture Microstrip Array for SAR Applications," *IEEE Transactions on Antennas and Propagation*, Vol. 60, No. 9, September 2012.

[9] Wong, H., K. -L. Lau, and K. -M. Luk, "Design of Dual-Polarized L-Probe Patch Antenna Arrays with High Isolation," *IEEE Transactions on Antennas and Propagation*, Vol. 52, No. 1, January 2004, pp. 45–52.

[10] He, S., and J. Xie, "Analysis and Design of a Novel Dual-Band Array Antenna with a Low Profile for 2400/5800-MHz WLAN Systems," *IEEE Transactions on Antennas and Propagation*, Vol. 58, No. 2, February 2010, pp. 391–396.

[11] Qu, X., et al., "Design of an S/X Dual-Band Dual-Polarised Microstrip Antenna Array for SAR Applications," *IET Microw. Antennas Propag.*, Vol. 1, No. 2, 2007, pp. 513–517.

[12] Hsu, S. -H., Y. -J. Ren, and K. Chang, "A Dual-Polarized Planar-Array Antenna for S Band and X-Band Airborne Applications," *IEEE Transactions on Antennas and Propagation*, Vol. 51, No. 4, August 2009, pp. 70–78.

[13] Rahmat-Samii, Y., et al., "Advanced Precipitation Radar Antenna: Array-Fed Offset Membrane Cylindrical Reflector Antenna," *IEEE Transactions on Antennas and Propagation*, Vol. 53, No. 8, August 2005, pp. 2503–2515.

[14] Naishadham, K., et al., "A Shared-Aperture Dual-Band Planar Array with Self-Similar Printed Folded Dipoles," *IEEE Transactions on Antennas and Propagation*, Vol. 61, No. 2, February 2013, pp. 606–613.

[15] Pokuls, R., J. Uher, and D. M. Pozar, "Dual-Frequency and Dual- Polarization Microstrip Antennas for SAR Applications," *IEEE Transactions on Antennas and Propagation*, Vol. 46, No. 9, September 1998, pp. 1289–1296.

[16] Rostan, F., and W. Wiesbeck, "Aperture-Coupled Microstrip Patch Phased Arrays in C- and X-Band: A Contribution to Future Multi-Polarization Multi-Frequency SAR Systems," *Proceedings of International Symposium on Phased Array Systems and Technology*, Boston, MA, 1996, pp. 141–146.

Selected Bibliography

Chiou, T. W., H. C. Tung, and K. L. Wong, "A Dual-Polarization Wideband Circular Patch Antenna with Hybrid Feeds," *Microwave Opt. Techno. Lett.*, Vol. 26, July 5, 2000, pp. 37–39.

Krishna, D., et al., "Compact Dual-Polarized Square Microstrip Antenna with Triangular Slots for Wireless Communication," *IEE Electronics Letters*, Vol. 42, No. 16, August 2006, pp. 894–895.

Lindmark, B., et al., "Dual Polarized Array for Signal-Processing Applications in Wireless Communications," *IEEE Transactions on Antennas and Propagation*, Vol. AP-46, No. 6, June 1998, pp. 758–763.

Nystrom, I., and D. Karlsson, "Reduction of Back Radiation and Cross-Coupling in Dual Polarized Aperture Coupled Patch Antennas," *IEEE Antennas Propagat. Soc. Int. Symp. Dig.*, 1997, pp. 2222–2225.

Skora, J., A. Sanz, and R. Fabbra, "Dual Polarized Phased Array Antenna for Airborne L-Band SAR," *IEEE MTT-S International Conference*, Long Beach, CA, July 2005, pp. 192–196.

Vallecchi, A., and G. B. Gentili, "Design of Dual-Polarization Series-Fed Microstrip Arrays with Low Losses and High Polarization Purity," *IEEE Transactions on Antennas and Propagation*, Vol. 53, No. 5, May 2005, pp. 1791–1798.

CHAPTER 10

Phased Array Antennas

10.1 Introduction

The phased array antenna is a multi-element antenna system in which the radiation pattern can be reinforced in a direction and suppressed in undesired directions by controlling the phase of individual elements. The direction of phased array radiation can be electronically steered avoiding the need for any mechanical rotation. These specific capabilities have found a broad range of applications for phased arrays since the arrival of this technology. In a phased array antenna with a very large number of regularly spaced radiating elements along with phase shifters, the gain realized at the beam peak is equal to the number of antenna elements times the gain realized in the same direction when only one element is excited. Antenna beams are formed by shifting the phase of the signal emitted from each radiating element to provide constructive/destructive interference to steer the beams in the desired direction. Thus, in the phased array, the main beam of the array is steered by applying a progressive phase shift across the array aperture by exciting all the antenna elements simultaneously.

Phased array radars are distinguished from other radars by phased array antennas that can steer the beam electronically in space. This provides a greater flexibility, makes the system much more versatile by being able to radiate the signal in the volume of space, and optimizes the search and track functions. Alhough scanning antenna arrays had been developed for various electronic communications applications, it was the radar system for which the phased array was broadly used for the first time. Radars demand fast beam scanning, wideband performance, and the ability to adapt the beam patterns to avoid jammers that could be addressed by phased array antennas. Recent growth in communication systems and civilian radar-based sensors has drawn increasing interest in utilizing phased array technology for commercial applications.

Phased array antennas are becoming popular for various applications such as automotive driver assist systems, satellite communications, and advanced radar. These antennas are widely used in communication and radar systems. They offer the benefit of far-field beam shaping and steering as per the requirements of agile operational conditions. Phased array antennas have been used for military radar applications for many decades. Another important use of phased arrays is in automotive collision avoidance radar or ACC technology [1]. These arrays are also used in base stations to enhance the signal coverage in an area while minimizing interference to other areas [2]. In the healthcare sector, satellite biomedical applications are among the emerging applications that take advantage of phased arrays [3–10]. These arrays are used in microwave imaging to detect early-stage breast cancer. In this system, a wideband impulse is emitted by an antenna array placed at the breast surface. The receiver uses beamforming technique to focus the backscattered signal from the malignant tumor and compensate for the frequency-dependent propagation effect.

The performance of phased array antennas is limited by the characteristics of antenna arrays and the beam-scanning effects. The work on phased array technology was initiated more than four decades ago, but the development of the array technology was slow primarily because of the high cost of the components. The cost per active array element is still not low, but in the last 15 years, the demands for phased array radar systems have significantly increased. Many new technologies for better performance have resulted in the incorporation of features such as beam agility, multitarget tracking, antenna pattern control, electronic countermeasures, and multimoding in these systems. The complexity and cost issues involved in developing phased array antennas are being addressed through new functionalities in electronic design automation (EDA) software that supports antenna designers with the means to develop new antenna system architectures and component specifications, as well as implement the physical design of individual components and verify performance prior to prototyping.

10.2 Phased Array Types

In general, there are two types of phased arrays: passive and active. Passive arrays use a central transmitter and receiver but have phase shift capability at each radiating element or subarray. In active arrays, the high-power generation for transmit and low-noise amplification on receive are distributed as well as the phase control at each radiating element. Active arrays provide added system capability and reliability, but they did not receive much attention until the last 15 years because they were too expensive and complex. Active arrays are becoming the preferred approach for many radar systems and communication systems requiring rapid scanning, with the advent of relatively low-cost GaAs monolithic microwave integrated circuits (MMICs), automated assembly of microwave components, and low-cost, high-speed, high-throughput digital processor [11, 12].

10.2.1 Passive Phased Arrays

In passive phased arrays, the power from a single transmitter is distributed to the array elements through a passive beamformer network. Usually, in a passive array, there is no element amplitude control; only bilateral phase shifters are used at each radiating element to provide the required phase shift for scanning. These phase shifters are situated directly behind the antenna elements (i.e., they transfer the full transmit power and the raw receive signals). Therefore, the phase shifter losses increase antenna losses, which reduce the overall system efficiency and increase the system noise figure. The design challenge in a passive array is to minimize the losses in the feed network and the phase shifters to increase the system efficiency and sensitivity. This necessity often limits the type of RF feed network to the waveguide that increases the weight of an array system appreciably. If very low sidelobe levels are required, a separate receive feed network with the appropriate amplitude weighting is used in a passive phased array.

A block diagram of a typical passive phased array system is shown in Figure 10.1. A centralized high-power transmitter, which generally consists of microwave tubes such as traveling-wave tubes or crossed-field amplifier tube (CFA or amplitron), provides the power to the radiating antenna elements through a BFN. High-power ferrite or diode phase shifters are controlled at each radiating element to electronically steer the beam to the desired direction. In the receive mode, the outputs of the antenna elements and phase shifters are combined using a low-power BFN.

In the receive mode, LNAs are used to amplify the signal at the output of the beamformers. Passive array systems have some inherent performance limitations and inefficiencies. For instance, the transmit beamformer typically has significant losses, and the transmitter must generate a large amount of power to overcome these losses. Because of losses, a significant portion of the RF power generated by the transmitter is dissipated as heat before being radiated. A high-power centralized transmitter generally employs a microwave tube. These tubes operate at lower duty factors and have limited waveform flexibility. High receive feed network losses, particularly when low sidelobes are required, significantly degrade receive sensitivity. Also, the transmit noise from a centralized source is often a limitation in clutter-driven radar applications.

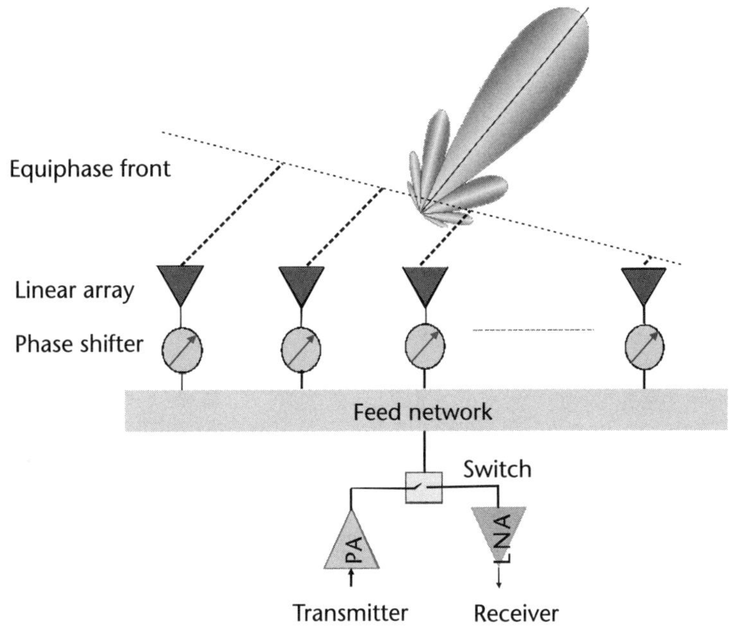

Figure 10.1 Passive phased array antenna system.

10.2.2 Active Phased Arrays

In an active phased array system, a transmit/receive module (TRM) is used at each antenna element to provide both amplitude and phase control. The central transceiver module used in a conventional passive array is replaced by the distributed TRM as shown in Figure 10.2. The main advantage of an active array is that the system sensitivity is increased because the system noise figure is set, and the RF power is generated at the aperture. Another advantage is that the TRMs provide complete flexibility in amplitude and phase control for both the transmit and receive signals. Also, in an active phased array, the feed networks need not be optimized for minimum losses, thereby allowing design flexibility and the ability to minimize size and weight.

The active phased array has gained importance in the past 5 years mainly because, with this technology, it is possible to realize affordable electronic scanning radar sensors. TRMs are main elements in any active phased array radar system. RF transmit power and received low-noise amplification in an active phased array are distributed at the aperture, which reduces the effect of losses in the transmit/receive (T/R) BFNs. Adequate transmit gain in the TRM allows a low RF input power to the antenna. In the receive mode, a high-gain LNA reduces the noise contribution from secondary stages of the TRMs and beamformers. By reducing the impact of transmit and receive feed network losses, large active phased array antennas provide increased radar sensitivity in the range of 8–12 dB compared to passive arrays. In addition, TRMs provide wider bandwidth and higher average transmit power as compared to passive phased arrays.

Because active phased array antennas contain large numbers of active components, the antenna performance degrades gracefully as components fail over time. The electrical block diagram of an active phased array antenna is shown in Figure 10.3. The system-level analysis to cross-check antenna performance along with other electronic components is very important in such a complex system.

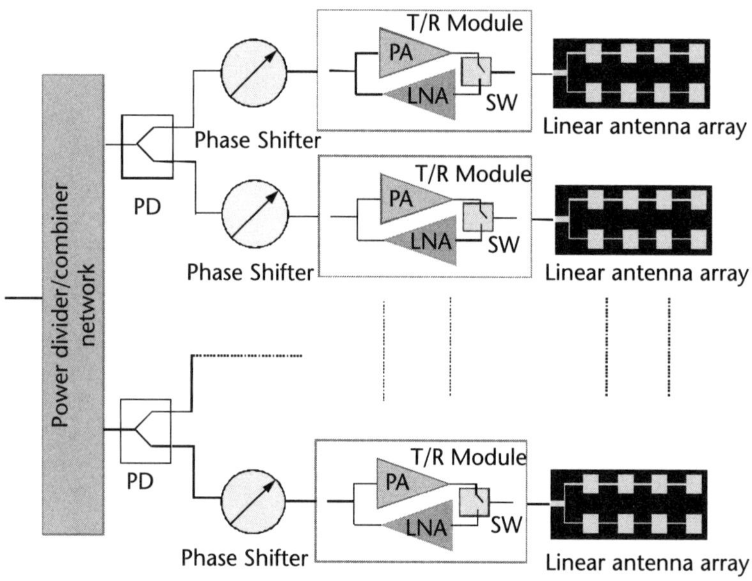

Figure 10.2 Active phased array antenna system.

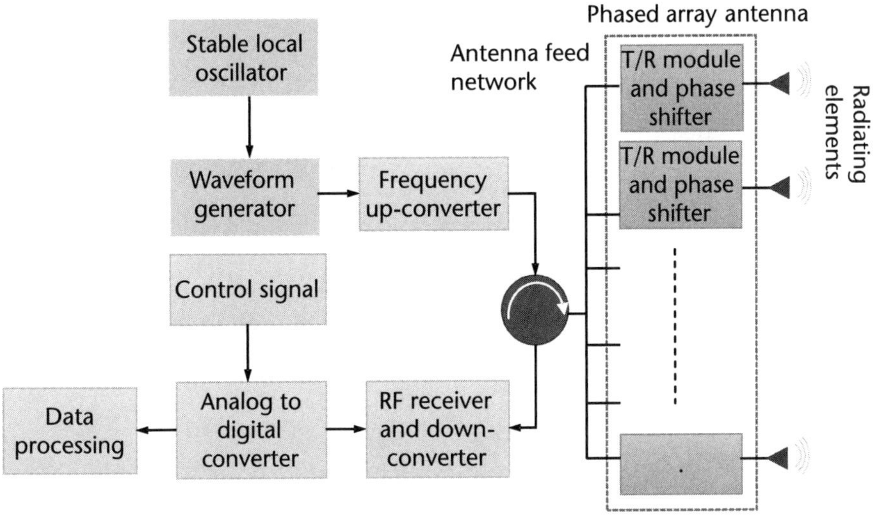

Figure 10.3 Electrical block diagram of an active phased array radar system.

The operating frequency, bandwidth, waveform parameters, and array performance requirements will vary depending on the applications served by the radar. The effective isotropic radiated power (EIRP) and antenna beamwidth determine the total number of antenna elements and the required TRM output power. The sidelobe levels determine the amplitude and phase characteristics of the TRMs and feed network. The phase noise, reliability, stability, and maintainability requirements all influence the array architecture as well as the characteristics of the TRMs and power supplies.

Table 10.1 lists the key parameters of an active phased array system. Nearly all high-performance radars under development today employ an active array antenna. Active phased array antennas are going to play big role in 5G communication system. For current and next-generation advanced spaceborne SAR antennas, high-speed beam steering, high gain, and precise beam shaping are mandatory. Compared to other antenna options such as multifed reflector antenna or passive phased arrays, the active phased array antenna provides the only practical solution to all these requirements, with successful applications in a series of spaceborne SAR systems. One notable disadvantage of space SAR active phased array antenna is the great power consumption. To compensate extreme high path loss, the transmitting RF power of the SAR antenna must be on a very high level. This power consumption usually exceeds the maximum power generated by the solar array onboard spaceborne SAR satellites.

10.2.3 Hybrid Phased Arrays

A hybrid phased array combines features of both passive and active phased arrays. A central transmitter feeds the array as in a conventional passive phased array, but an LNA is placed at each element in front of the phase shifters to improve the overall system noise figure. A limiter may be required in front of the LNA for protection,

Table 10.1 Typical Specification of an Active Phased Array

Specification	*Value*
Electronic specification	
Antenna type	Microstrip phased array antenna
Operation frequency	Tx: 14.2 GHz, Rx: 12.5 GHz
Bandwidth	500 MHz
Antenna gain	35 dBi
Beamwidth	Azimuth: 1.2°, elevation: 4°
Polarization mode	Linear (H, V)
G/T	9.6 dB/K (without radome)
Cross-pol isolation	≥25 dB
EIRP	>49/1 beam dBW
Phase shifter bit number	6
Operating mode	Satellite-switched time division multiple access (SS-TDMA) continuous-wave mode
VSWR	<2 (50Ω reference)
Efficiency	60%
Mechanical specification	
Antenna size	1,200 × 900 × 120 mm (length * width * height)
Aperture size	1,200 mm × 900 mm
Number of radiating elements	256
Weight	25 kg
Operation temperature	–40°C~65°C
Scanning	Electronics
Scanning range	Azimuth: 360° continuous Elevation: 20°– 60°
Max. steering step size	5°
Electrical specification	
Power consumption	<700W
Power supply	AC220V ± 10%, 50 Hz ± 2 Hz
Rated consumption	120W (without block upconverter (BUC))

as in an active phased array. A separate receive feed network is used for sidelobes. In a hybrid array, it is also possible to distribute medium power transmitters (e.g., in each column of the array) to provide increased system reliability. For very large phased arrays systems with wide instantaneous bandwidths, array architecture with true time delay (TTD) is required to prevent signal distortion and beam squinting. In these arrays, the length of the transmission line feeding each element provides the differential phase shift. However, to scan the beam, the effective line length of the feed lines to each element must be changed accordingly [13].

10.3 Phased Array Theory

The radiation pattern of linear and planar phased arrays is a function of each element's physical structure, interelement distance, excitation, and array lattice.

Typically, similar radiating elements are arranged in a linear, rectangular, circular, triangular, or customized lattice with periodic spacing between antenna elements. When similar radiators are used in a large rectangular array, the radiation pattern is the product of two factors: first is the element factor that is a function of the radiating element physical realization; and second is the array factor that is a function of the array geometry and the element excitations. The radiation pattern of phased array is the product of the element and array factor. Antenna arrays with the radiators arranged in concentric circles have certain advantages such as lower first sidelobes compared to a rectangular array lattice.

To understand the array factor, consider an array of identical radiators as shown in Figure 10.4. The $2N + 1$ elements, equally spaced by a distance d along the z-axis, are arranged in a line. All elements have same current distributions that differ only in magnitude and phase. In is the magnitude of the current on the nth element. For this linear array, the array factor $F_a(\theta, \varnothing)$ is proportional to the radiated field at a point in space described by $(\theta, \varnothing)$.

$$F_a(\theta, \varnothing) = \sum_{n=-N}^{n=N} \frac{I_n}{I_0} e^{jnkd\cos\theta} \tag{10.1}$$

where $k = 2\pi/\lambda$ and λ is the wavelength. For a linear array, $F_a(\theta, \varnothing)$ is rotationally symmetric about the z-axis (independent of $\varnothing$).

If the currents are equal and in phase, (10.1) reduces to (10.2):

$$F_a(\theta, \varnothing) = \sum_{n=-N}^{n=N} e^{jnkd\cos\theta} \tag{10.2}$$

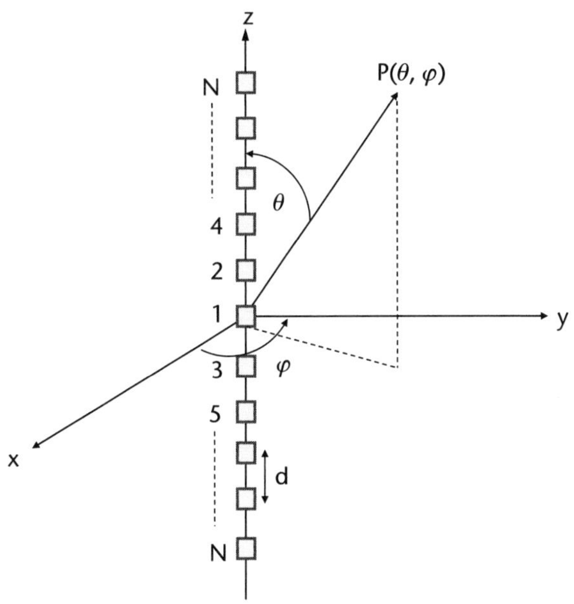

Figure 10.4 A linear array with the radiating elements equally spaced along the z-axis.

The array factor is equal to a sum of phasors of unity magnitude with progressive multiples of the basic angle φ:

$$\varphi = \frac{2\pi d}{\lambda}\cos\theta \tag{10.3}$$

If the array length L is very large compared to a wavelength ($L>>\lambda$) and the element spacing d is less than a wavelength ($d< \lambda$), then the array pattern gives a narrow beam broadside to the array axis ($\theta = \pi/2$) with several sidelobes.

In the linear array, the first sidelobe is 13.5 dB below the peak of the main beam, and the peak of each successive sidelobe is even lower as its distance from the main beam increases. Tapering the magnitude of the currents across the array elements can decrease the sidelobes. The central element is fed with the largest current and the magnitudes of the currents in the other elements are symmetrically tapered. As a result, the sidelobes are reduced, but the width of the main beam (beamwidth) is increased.

The directivity of a uniformly weighted linear array of N isotropic radiators spaced $\lambda/2$ apart is equal to N independent of the main beam scan angle. For nonuniformly weighted linear arrays, the directivity is given by:

$$D = \frac{2d}{\lambda}D_0 \tag{10.4}$$

where D_0 is given by

$$D_0 = \frac{\left[\sum_{n=-N}^{n=N} I_n\right]^2}{\sum_{n=-N}^{n=N} I_n^{\,2}} \tag{10.5}$$

In a linear array, elements with currents all have equal amplitudes and a uniform progressive phase as follows:

$$F_a(\theta,\Phi) = \sum_{n=-N}^{n=N} e^{jn(kd\cos\theta-\alpha_z\}} \tag{10.6}$$

The uniform progressive phase factor α_z changes the peak beam position from broadside to another angle in space θ_0 given by

$$\theta_0 = \cos^{-1}\left(\frac{\alpha_z}{2\pi}\frac{\lambda}{d}\right) \tag{10.7}$$

10.3.1 Grating Lobes

The active impedance of each radiating antenna element in a phased array varies with the scan angle, because of mutual coupling between the elements. Mutual coupling is fundamentally unavoidable in a closely spaced infinite array of elements. Mutual coupling reduces antenna efficiency and coupling increases with closer spacing. Another phenomenon associated with a phased array is that grating lobes can exist when the antenna elements are not closely spaced; for example, in a linear or planar array with more than $\lambda/2$ spacing, there is high chance of grating lobe occurrence as shown in Figure 10.5. In this case, the array will have equally strong radiation in other directions. These unintentional beams of radiation are known as grating lobes.

If the element spacing d is too large compared to a wavelength, the grating lobe will appear in the radiation pattern. This will happen at an angle θ [13], where

$$kd\cos\theta - \alpha_z = \pm 2\pi \tag{10.8}$$

$$\cos\theta = \cos\theta_0 \pm \frac{\lambda}{d} \tag{10.9}$$

To prevent a second main beam, the spacing must be chosen to satisfy the following condition:

$$\frac{d}{\lambda} < \frac{1}{1+|\cos\theta_0|} \tag{10.10}$$

The equation for the maximum interelement spacing is a function of the wavelength of operation and the maximum look angle θ.

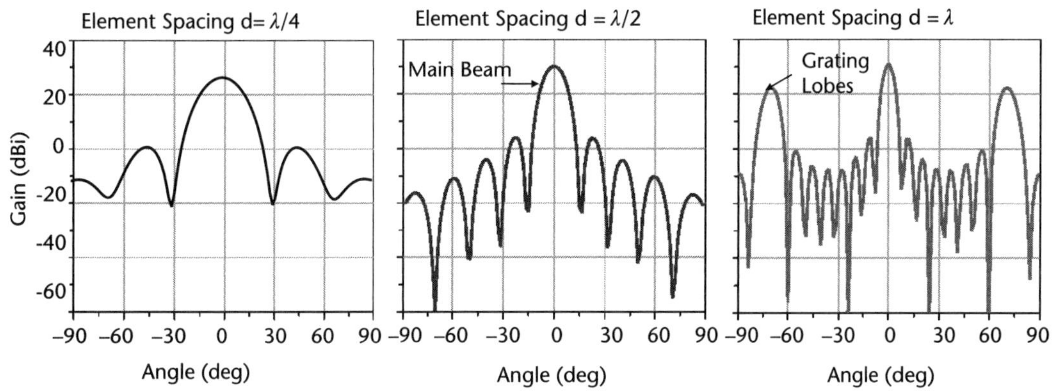

Figure 10.5 Linear antenna array radiation pattern with different interelement spacing.

$$d_{max} = \frac{\lambda}{1+\sin\theta} \tag{10.11}$$

Thus, for 45° scan angles, the maximum interelement spacing is 0.59λ. The elements must be spaced one half-wavelength apart to prevent a second main beam when scanning close to endfire ($\theta = 0$ or π). Figure 10.5 shows a grating lobe appearing in real space when the beam of a 16-element array with elements equally spaced at $d = \lambda$ of a wavelength is scanned off broadside.

10.4 Active Phased Array Antenna Design

In a phased array radar, the antenna is one of the most critical components from a design point of view due to the requirement of enhanced capabilities in terms of dual polarization, beam agility, high gain, and reliability. Because of the size and weight constraints, the printed multilayer planar array antenna is the most popular choice in a phased array system. The challenges in active array antenna design are to achieve high gain, dual linear polarization with low cross-polar radiation, and accommodation of dual feed networks within the limited space. In the case of a dual-polarized phased array, the most critical aspect of design is the realization of dual-polarized feed networks, which require an innovative design approach. Apart from the complexity of electrical design, modeling, and simulation using EDA tools, the fabrication of antenna with required tolerance, assembly, and alignment of various layers are of paramount importance [14–19]. A phased array antenna design operating in X-band is discussed in this section. This design covers antenna specifications for active phased array, different options of multilayer radiators, array antenna feed network configuration, design sensitivity analysis, and measurement philosophy.

Phased array antenna design starts with the antenna design specification provided by the system engineer. Based on the system-level requirement, the system engineer provides the design requirement specification for antennas such as the maximum available space to mount the antenna, gain, beamwidth, bandwidth, cross-polarization level, sidelobe level, and some other parameters as listed in Table 10.2.

10.4.1 Antenna Configuration

The physical aperture size of an active phased array antenna is determined based on the required antenna beamwidth in the principal planes. Form an antenna design fabrication and assembly point of view, an antenna is divided into multiple panels and then each panel consists of several linear arrays as shown in Figure 10.6. This antenna consists of 4 foldable panels of a size of 0.37m × 0.53m. To scan the beam in the elevation plane, 16 linear arrays each having 24 elements are designed with a distributed power scheme. Linear arrays are fed at the center by T/R (transceiver) modules for digital phase control. Several types of antennas such as the open-ended waveguide horn array, slotted array, and printed microstrip array can be used to meet antenna specification requirements. The horn antenna array is not suitable

Table 10.2 Design Specifications of X-Band Active Phased Array Antenna

	Specifications		
Parameters	*Linear Array*	*One Tile*	*Complete Antenna*
Size ($L \times W$)	0.53m 0.02m	0.53m × 0.35m	1.5m × 0.52m
Antenna beamwidth	2.7° × 73°	2.7° × 2.7°	0.6° × 1.6°
Gain	20.7 dB	35.3 dB	43.4 dB
Gain bandwidth	1.1 dB	1.1 dB	1.1 dB
Antenna bandwidth	50 MHz	47 MHz	45 MHz
Return loss	–10 dB	–10 dB	–10 dB
Cross-polarization	–21 dB	–20 dB	–20 dB
Frequency	9.5 GHz (X-band)		
Polarization	Dual linear polarization		

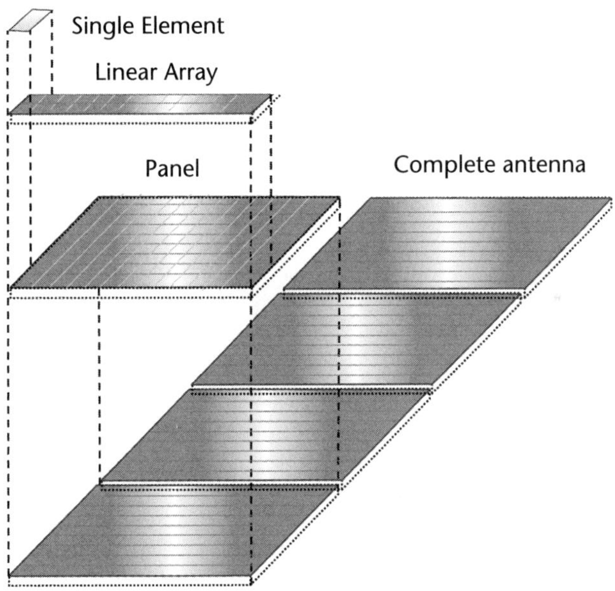

Figure 10.6 Designed X-band phased array antenna configuration and assembly.

due to its large size and weight. The slotted waveguide antenna array has narrow bandwidth characteristics and, because of that, it is also not feasible. A low-profile and low-weight printed multilayer planar array antenna configuration is the best option for this spaceborne SAR application due to the weight constraints in the satellite payload.

A single element of the phased array antenna should be broadband and high gain so that the overall antenna gain requirement can be met with fewer elements. In an array antenna, the antenna gain increases about 3dB when the number of elements doubles. For such a bandwidth requirement, the conventional direct fed microstrip patch cannot be used. Multilayer microstrip patch antennas are suitable to meet the functional requirements. Two options of multilayer microstrip patch antennas can be used to increase antenna bandwidth and gain:

1. Aperture-coupled microstrip antenna (ACMA);
2. Proximity EM coupled microstrip antenna (PECMA).

10.4.1.1 Aperture Coupled Microstrip Antenna

In the aperture coupled microstrip antenna configuration, the radiating element and the feed line are separated by a ground plane layer that consists of slots to couple power from the feed line to the radiator as shown in Figure 10.7(a). The feed line is isolated from the radiator to avoid spurious radiation from the feed line, which would otherwise distort the radiation pattern of the radiator. Different substrate material can be selected for radiator and feed line to optimize the performance in terms of gain, bandwidth, isolation, and front-to-back ratio of the radiator. The optimum performance is obtained by putting the radiator on a substrate of lower dielectric constant as near to that of air and the feed line on a substrate of a higher dielectric constant. The preferred configuration for a feed line is to have an asymmetric stripline configuration. In this arrangement, the feed line is sandwiched between two ground planes where the upper ground plane contains coupling slots and the lower ground plane acts as a ground plane for the feed line. The ground

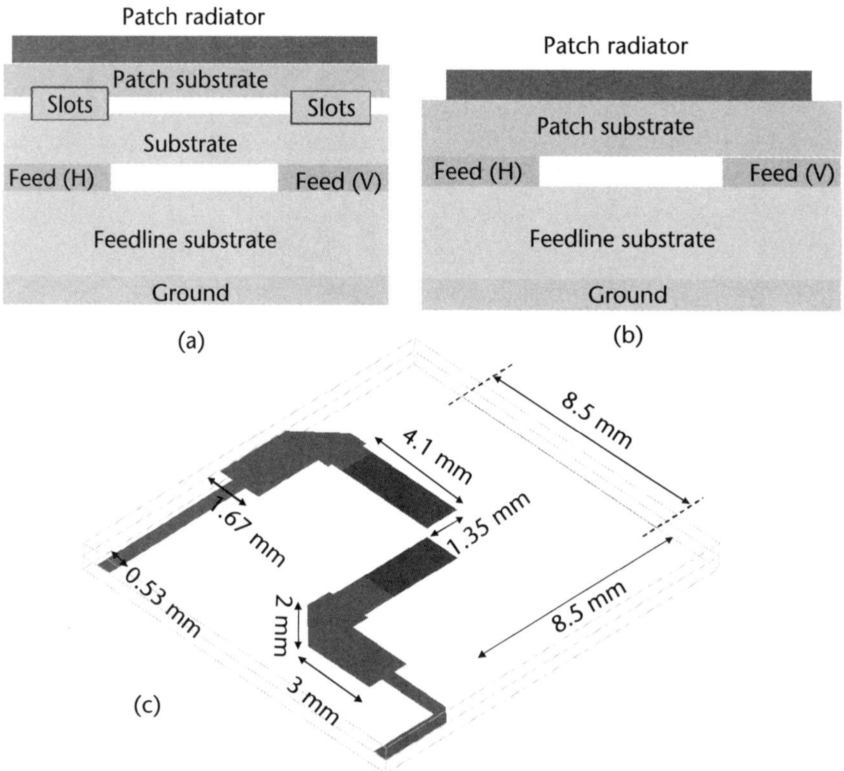

Figure 10.7 (a) Aperture-coupled and (b) electromagnetic microstrip antenna stack-ups; and (c) single-element electromagnetically coupled dual-polarized radiator.

plane slot and patch center are positioned above the microstrip feed line $\lambda_g/4$ from the open termination. On microstrip lines above a ground plane, a voltage null and current maximum occur $\lambda_g/4$ from an open termination.

A major problem in this configuration is the impedance matching of a 50Ω coaxial line with the microstrip feed line when coaxial connector pins protrude from below the ground plane as required. Using shorting pins connecting the lower and upper slot ground planes near the coaxial connector probe solves this matching problem, but this results in a very unreliable configuration due to the enormous number of shorting pins and problems associated with their soldering and assembly.

The equivalent circuit model of an aperture-coupled microstrip patch antenna is shown in Figure 10.8. The antenna ground plane slot acts as an impedance transformer and parallel LC circuit ($L_{aperture}$ and $C_{aperture}$) in series with the microstrip feed line [20]. The ground plane slot resonant behavior is represented with the LC circuit. The 1:N impedance transformer represents the effects of the impedance of the patch antenna being coupled through the ground plane slot. The patch antenna is modeled as two transmission lines terminated by parallel RC components (C_{fring} and R) due to patch edge fringing fields.

10.4.1.2 Proximity EM Coupled Stacked Patch Antenna

In a proximity EM coupled stacked patch configuration, the patch is excited through an EM coupled feed line. In this configuration, two dielectric substrates are used such that the feed line is in the interior of the substrates and the radiating patch on top of the other substrates shown in Figure 10.7(c). Decoupling is obtained by placing the patch and feed at different levels as shown in Figure 10.7(b). It offers better port isolation between the feed lines for vertical and horizontal polarizations. In the case of the broadband array requirement, two stacked radiating patched are separated by a thick layer of foam of low permittivity to enhance the radiation efficiency and bandwidth performance. Compared to aperture-coupled configuration, this configuration has been found to be optimum from the design and realization point of view and has been preferred for the development of the antenna. A single radiating antenna is optimized for a minimum gain of 7.7 dB. The optimization was carried out in terms of the patch dimension, substrate thickness, and dielectric

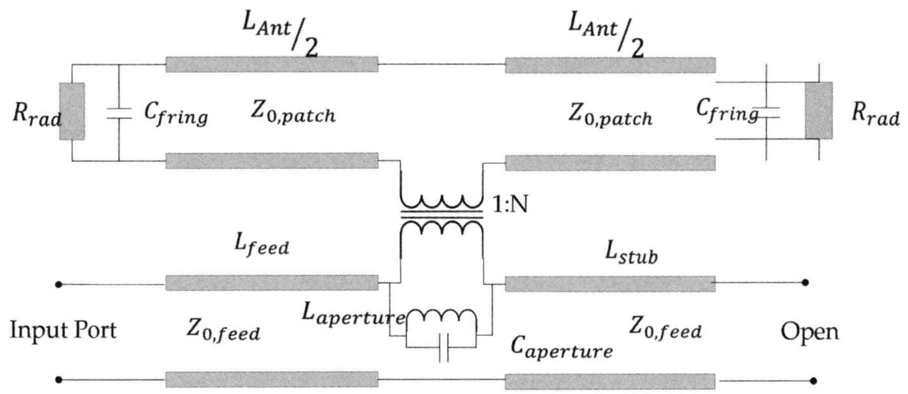

Figure 10.8 Equivalent circuit of aperture-coupled patch antenna.

constant and dielectric loss tangent. The optimized electrical performance of the various options of single radiating elements is given in Table 10.3.

Of these two single radiating element options, the stacked EM coupled patch is preferred because of the following advantages:

1. This configuration meets the antenna array bandwidth requirement.
2. The center pin of the 50Ω coaxial line can be matched with the printed feed line without any spurious radiation issue.
3. This configuration is slightly less sensitive to fabrication errors and misalignment errors as compared to the aperture-coupled configuration.

One disadvantage of this configuration is that the feed lines are not isolated from the radiating patches and the radiation from the feed lines can result in a degradation in cross-polarization. However, cross-polarization can be further improved in the planar array environment using cross-polarization suppression techniques. Beside these two configurations, it is possible to increase the absolute bandwidth of the microstrip antenna by simply using thicker substrates. However, thick substrate introduces several problems such as the excitation of surface waves, which distorts the normal radiation pattern and introduces additional loss, and the excitation of higher-order modes with Z-dependence, which introduces further distortions on the pattern and impedance characteristics. The RT Duroid 6002 substrate (thickness of 15 mils and dielectric constant of 2.94) is used both for the radiating element and the feed line substrate material.

10.4.2 Dual-Polarized Feed Network

The array feed network design begins with the optimization of a 1:2 feed followed by 1:6, 1:12, and finally 1:24 to feed all the 24 elements of the linear array as shown in Figure 10.9. The feed network design is the most complex part in an array dual-polarized array antenna design. Before the start of the design, the current distribution is calculated at the input of each element and, based on this, the microstrip line width is calculated for each value of impedance. Then these lines are combined to form a feed network and optimization is carried out to find out optimal dimensions of these lines. In this design, one major design constraint is that the feed line networks for both the polarizations should be accommodated within limited interelement spacing. The spacing between the lines is as close as 0.9 mm at some points. The coupling between the feed network lines should be considered during the design of the complete feed network, as these lines run very close to each other. Feed network design is optimized using ADS optimization and the MoM simulation technique is applied for the EM analysis of the feed network. A corporate feed

Table 10.3 Performance of a Single Radiating Element

Configuration	*Return Loss Bandwidth*	*Isolation (dB)*	*Gain (dB)*
Aperture coupled	80 MHz	–30.0	7.5
Proximity EMC patch	90 MHz	–24	7.7

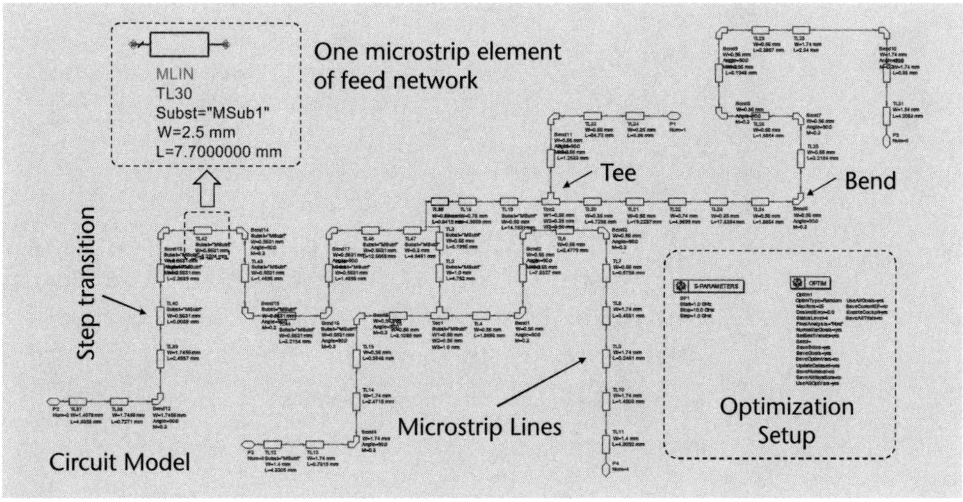

Figure 10.9 Feed network design and simulation in ADS.

network is a better choice from a bandwidth point of view, but it was not possible to realize such a feed due to the limited available space for the dual linear networks. The series feed networks can be easily accommodated within the given space, but they result in phase spreading over the bandwidth. Therefore, a combination of corporate and series feed is used in the design of 1:24 dual-polarized feed networks. The layout diagram of the dual polarized feed network is given in Figure 10.10.

It is shown that input power is divided using both corporate and series power dividers. Additional line lengths are added to achieve equal phase at all the 24 ports of the power divider at the center frequency. The feed network has been optimized for an equal amplitude at all 24 ports within ±0.3 dB at the center frequency and ±0.6 dB at the band edges.

In addition to this, the feed network of the linear array was designed using the cross-polarization suppression technique to reduce cross-polarization radiation in the planar array environment [21] compared with that of single radiator as shown in Figure 10.11. Option-3 is applied in this antenna array design. In this configuration, the antenna element mirroring is performed in both the horizontal and vertical planes. With this arrangement, the array cross-polarization is improved significantly, but sometime undesired sidelobes will appear in the horizontal and vertical planes. A complete linear antenna array is simulated using MoM-based electromagnetic simulation. A return loss plot for horizontal and vertical polarizations is shown in Figure 10.12. Linear array resonance is at 9.47 GHz and –10-dB bandwidth of antenna is 50 MHz.

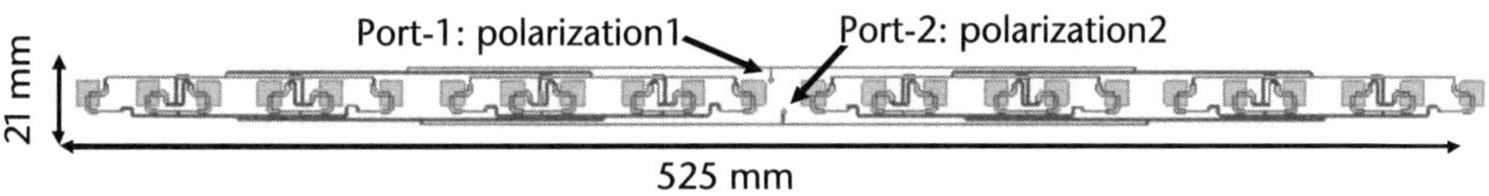

Figure 10.10 Dual-polarized linear array design with a hybrid feed network.

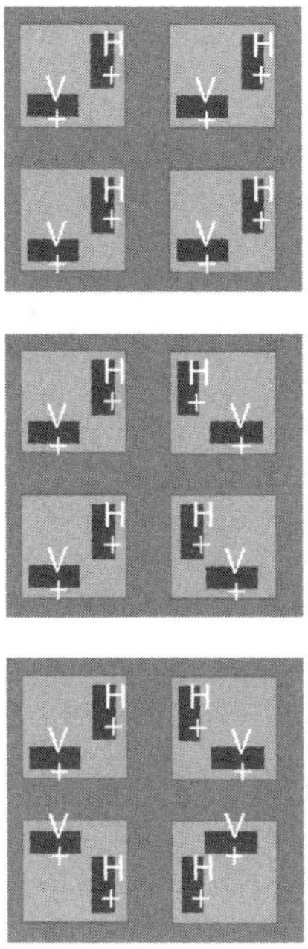

Figure 10.11 Cross-polarization suppression techniques.

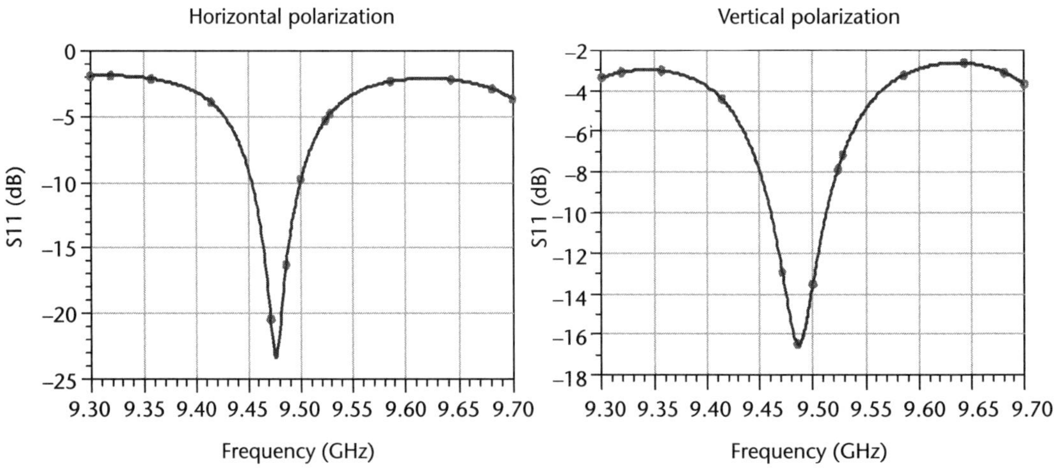

Figure 10.12 Return loss plot of horizontal and vertical polarization linear array.

This is a narrowband array meeting the system design requirement, but, in many cases, broadband array antenna is needed for spaceborne SAR application. Port isolation between horizontal and vertical polarization is better than 35 dB across the operating frequency band.

The radiation pattern plot of a linear array antenna at three different frequencies (9.45, 9.48, and 9.50 GHz) in the antenna band is shown in Figure 10.13. The peak gain of array is 21.1 dBi. The sidelobe variation is within 1.5 dB. The gain variation obtained over the frequency band was within 0.6 dB.

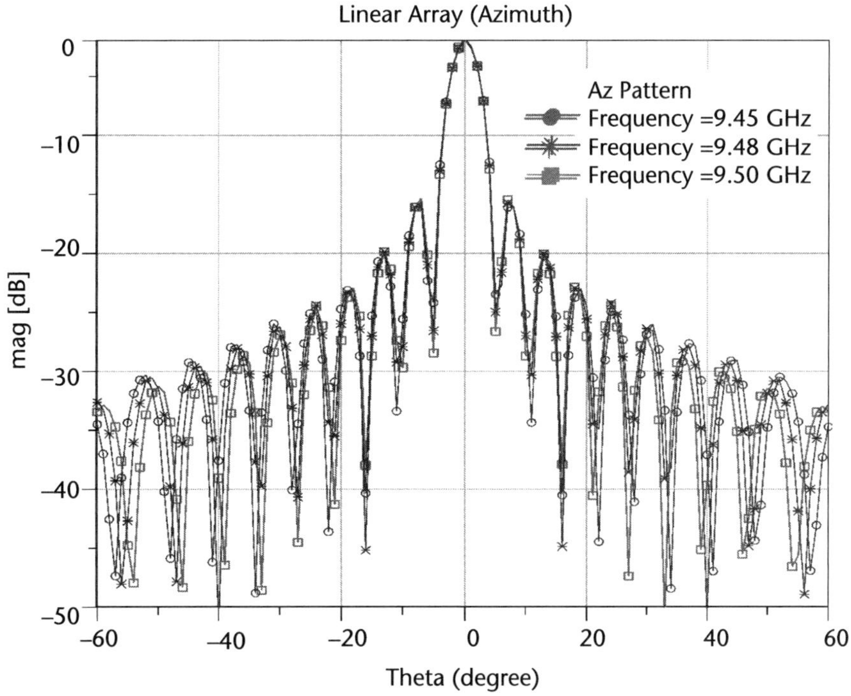

Figure 10.13 Azimuth pattern plot of linear array antenna.

10.5 Planar Array Antenna Design

A planar array is formed by placing 16 linear arrays at the required interelement spacing to get the beamwidth in the elevation plane as shown in Figure 10.14. The cross-polarization of the planar array gets further deteriorated as compared to a single linear array. The major challenge of the planar array design is to control the isolation between two linear arrays to improve the cross-polar radiation

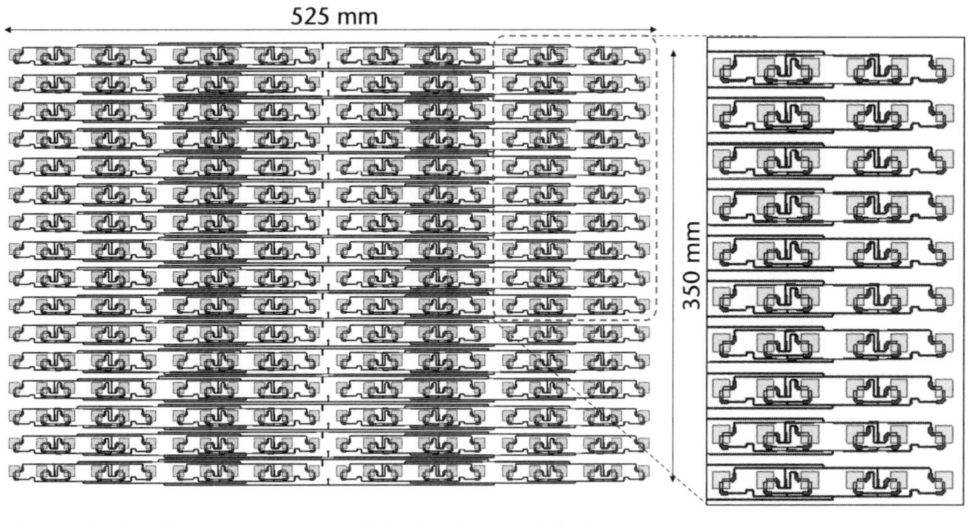

Figure 10.14 Planar antenna array (20 × 16 elements) design.

performance. Option-3 of cross-polarization suppression is also applied in the planar antenna array design.

Using this technique, the cross-polar radiation is further reduced by –9 dB than that of a single linear array. Planar antenna return loss and radiation pattern are shown in Figure 10.15. By combining four planar antennas, a complete dual-polarized active phased array antenna is constructed as shown in Figure 10.16. The peak gain at 9.5 GHz of this antenna is 43.4 dBi. A total 128 TRM (64 for horizontal polarization and 64 for vertical polarization) is required to drive the antenna and scan the antenna beam.

10.6 Measurement of Phased Array Antenna

The radiation pattern measurement of the antenna can be measured either using far-field measurement techniques or a planar near-test facility (PNF). The return loss and isolation between ports are measured using a vector network analyzer (VNA). To test antenna beam scanning, T/R with automated test setup consisting of digital control unit (DCU) and control computer are used. The TRMs are controlled to

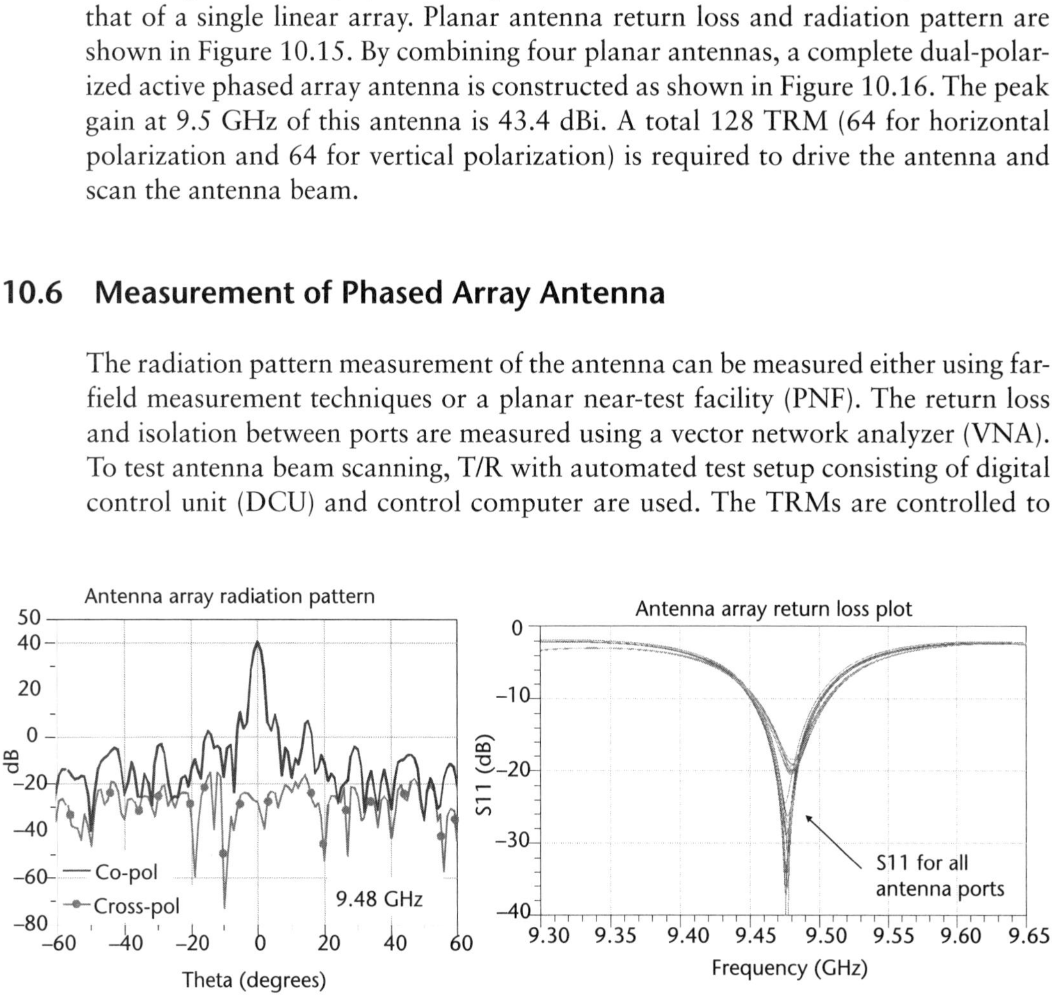

Figure 10.15 Return loss and radiation pattern plot of planar array antenna.

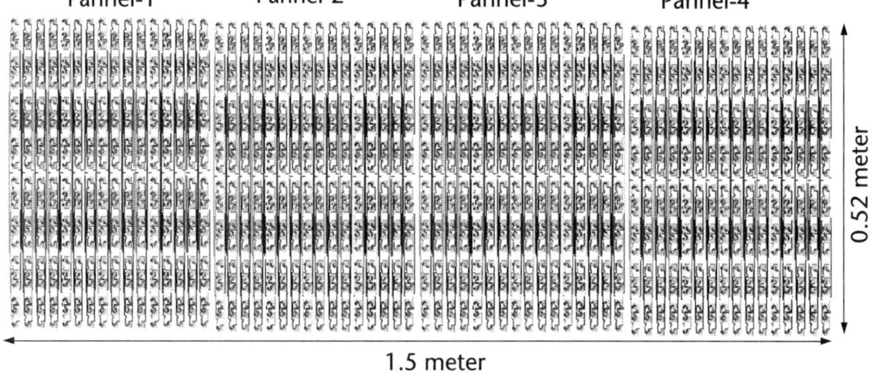

Figure 10.16 Layout of complete dual-polarized phased array antenna design.

get amplitude and phase. In this case, 16 TRMs are connected to 16 linear arrays for either vertical or horizontal polarization. To control the phase and amplitude between linear arrays for beam scanning, DCU controls the phase and amplitude of T/R. This DCU will feed 6-bit phase and amplitude information to R-modules. Extensive testing is required to characterize antenna for all possible beam locations. The performance of the antenna is checked over the complete operating bandwidth with a center frequency of 9.5 GHz. Continuous beam scanning can be achieved using amplitude and phase only control required for beam location. It has a DCU to cater to any amplitude and phase combination for each linear array fed at the center with maximum of 16 such arrays. A 6-bit phase shifter meets the minimum phase step criteria of 5.6° with a maximum of 360. Any phase beyond 360 repeats the cycle again.

10.6.1 Microwave Holography

Microwave holography is a method for the diagnosis and alignment of phased array antennas. The microwave holography measurement is conducted using near-field measurement (PNF) technique. This technique is also known as back-transformation. Microwave holography [22] is a fast inversion technique that relies on the measurement of the magnitude and phase of the wave scattered from the phased array antenna. In this technique, far-field data is back-projected using the inverse Fourier transformation to compute amplitude and phase at the aperture. A microwave sensor scans completely the array antenna aperture and collects electromagnetic field data at each location along the X-axis and Y-axis. The scanning data gives a 2-D power plot as shown in the Figure 10.17. This aperture distribution is compared with the given amplitude and phase distribution to locate the faulty array

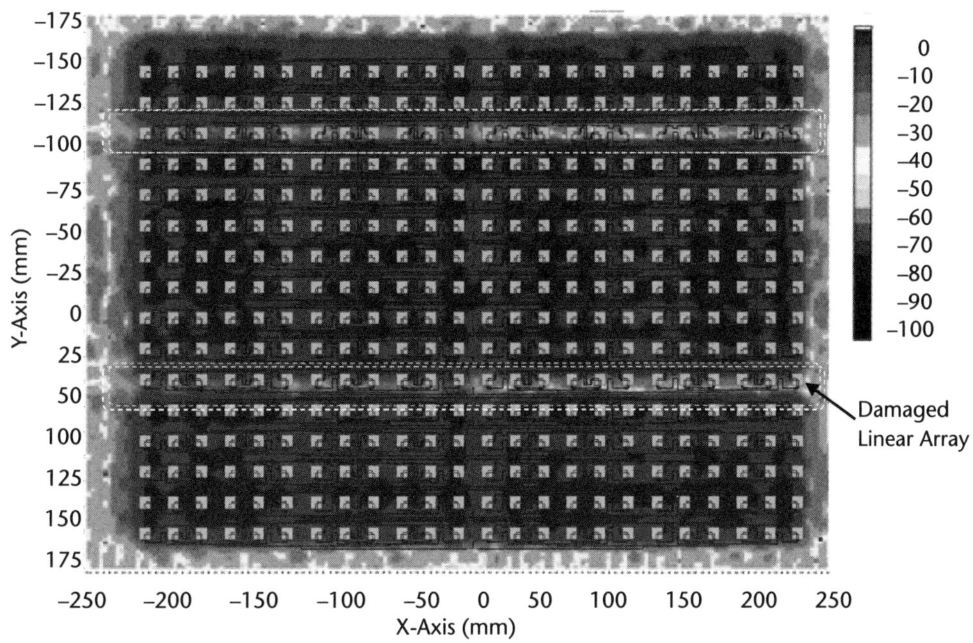

Figure 10.17 Microwave holography of a planar antenna array showing two faulty linear arrays in the design.

element. In this example, there are two faulty array elements as shown in Figure 10.17. The plot clearly shows a sudden drop of power in the form of a line that shows a faulty element of the array. Thus, this technique is very useful to locate faulty elements of the larger arrays. This analysis ascertains that back-projected data can be used for aperture distribution analysis.

10.7 Summary

Phased array radar technology has been traditionally used mainly in military or remote-sensing applications. Due to the proliferation of new consumer and commercial applications such as Internet of Things (IoT) and smart city and smart home applications, the demand on radar technology continues to grow, as it can enhance the product features and increase the product's competitiveness. Such features may include occupancy detection, gesture detection and motion, and speed detection for smart home applications. In this chapter, phased array antenna technology was discussed that is applicable for various applications. In future 5G systems, to handle the bandwidth needed for greater than 10-Gbps data rates, the millimeter-wave spectrum along with massive MIMO phased array is going to be a driving technology. There are many phased array antennas that do everything required, for example, the fire control radar and an active electronic scanning array (AESA). Phased array antennas are used in missile guidance systems to control the flight path. In this chapter, phased array antenna design was discussed for spaceborne remote sensing.

References

[1] Kolak F., and C. Eswarappa, "A Low Profile 77 GHz Three Beam Antenna for Automotive Radar," *2001 IEEE MTT-S International Microwave Symposium Digest*, Phoenix, AZ, 2001, pp. 1107–1110.

[2] Sawahashi, M., et al., "Coordinated Multipoint Transmission/Reception Techniques for LTE Advanced," *IEEE Wireless Communications Mag.*, Vol. 17, No. 3, June 2010, pp. 26–34.

[3] Irishina, N., N. Moscoso, and N. Dorn, "Microwave Imaging for Early Breast Cancer Detection Using a Shape-Based Strategy," *IEEE Transactions on Biomedical Engineering*, Vol. 56, No. 4, April 2009, pp. 1143–1153.

[4] Klemm, M., et al., "Radar-Based Breast Cancer Detection Using a Hemispherical Antenna Array Experimental Results," *IEEE Transactions on Medical Imaging,* Vol. 28, No. 7, 2009, pp. 969–981.

[5] Wo, J. Y., et al., "Effect of Very Small Tumor Size on Cancer-Specific Mortality in Node-Positive Breast Cancer," *J Clin. Oncol.*, Vol. 29, May 2011, pp. 2619–2627.

[6] Rosen, A., M. A. Stuchly, and A. V. Vorst, "Applications of RF/Microwaves in Medicine," *IEEE Transactions on Microwave Theory and Techniques*, Vol. 50, No. 3, March 2002, pp. 963–974.

[7] Fenn, A. J., *Adaptive Phased Array Thermotherapy for Cancer*, Norwood, MA: Artech House, 2008.

[8] Nikolova, N., "Microwave Imaging for Breast Cancer," *IEEE Microwave Magazine*, Vol. 12, No. 7, 2011, pp. 78–94.

[9] Balanis, C. A., *Antenna Theory Analysis and Design*, New York: Wiley-Interscience, 2005.

[10] Mailloux, R. J., *Phased Array Antenna Handbook*, Norwood, MA, Artech House, 2005.

[11] Livne, A., "System Design Considerations in Fiber-Optic Networks for Active Array Antennas." *Proc. 19th Europe, Microwave Conf.*, 1989.

[12] Lockerd, R. M., and G. E. Crain, "Airborne Active Element Array Radars Come of Age," *Microwave Journal,* January 1990.

[13] Hansen, R. C., *Significant Phased Array Papers*, Dedham, MA: Artech House, 1973.

[14] Jyoti, R., et al., "Multilayer Printed Active Phased Array Antenna for Radar Imaging Satellite," *2014 IEEE International Microwave and RF Conference (IMaRC)*, Bangalore, 2014, pp. 205–208.

[15] Tapan, M., et al., "Synthetic Aperture Radar Payload On-Board RISAT-1: Configuration, Technology and Performance," *Current Science*, Vol. 104, No. 4, February 25, 2013, pp. 446–461.

[16] Woelder, K., and J. Granholm, "Cross-Polarization and Sidelobe Suppression in Dual Linear Polarization Antenna Arrays," *IEEE Transactions on Antennas and Propagation*, Vol. 45, No. 12, December 1997, pp. 1727–1740.

[17] Serra, A. A., et al., "A Wide-Band Dual-Polarized Stacked Patch Antenna," *IEEE Antennas and Wireless Propagation Letters*, Vol. 6, 2007, pp. 141–144.

[18] Chang, F. S., et al., "Compact Vertical Patch Antenna for Dual-Band WLAN Operation," *IEE Electronics Letters*, Vol. 44, No. 10, 2008.

[19] Jackson D. R., and P. Manghnani, "Analysis and Design of a Linear Array of Electromagnetically Coupled Microstrip Patches," *IEEE Transactions on Antennas and Propagation*, Vol. 38, May 1990, pp. 754–759.

[20] Civerolo, M., and D. Arakaki, "Aperture Coupled Patch Antenna Design Methods," *2011 IEEE International Symposium on Antennas and Propagation (APSURSI)*, Spokane, WA, 2011, pp. 876–879.

[21] Woelder, K., and J. Granholm, "Cross-Polarization and Sidelobe Suppression in Dual Linear Polarization Antenna Arrays," *IEEE Transactions on Antennas and Propagation*, Vol. 45, No. 12, December 1997, pp. 1727–1740.

[22] Farhat, N. H., and W. R. Guard, "Holographic Imaging at 70 GHz," *Proc. of the IEEE*, Vol. 58, No., 12, 1970, pp. 1955–1956.

Selected Bibliography

Aboserwal, N. A., et al., "Source Current Polarization Impact on the Cross-Polarization Definition of Practical Antenna Elements: Theory and Applications," *IEEE Transactions on Antennas and Propagation*, Vol. 66, No. 9, September 2018, pp. 4391–4406.

Brookner, E., "Phased-Array and Radar Breakthroughs," *Proc. IEEE Radar Conf.*, 2006, pp. 37–42.

Cesario, J., "Electromagnetic Coupling Between a Metal Nanoparticle Grating and a Metallic Surface," *Optics Letters*, Vol. 30, No. 24, December 15, 2005, pp. 3404–3406.

Hong, L., et al., "Design and Implementation of an Active Array Antenna with Remote Controllable Radiation Patterns for Mobile Communications," *IEEE Transactions on Antennas and Propagation*, Vol. 62, No. 2, February 2014, pp. 913–921.

Ip, K. H. Y., T. M. Y. Kan, and G. V. Eleftheriades, "A Single-Layer CPW FED Active Patch Antenna," *IEEE Microwave and Guided Wave Letters*, Vol. 10, No. 2, 2000, pp. 64–66.

Kang, M., et al., "Efficient Synthesis of Antenna Pattern Using Improved PSO for Spaceborne SAR Performance and Imaging in Presence of Element Failure," *IEEE Sensors Journal*, Vol. 18, No. 16, August 15, 2018, pp. 6576–6587.

Kasemodel, J. A., C. Chen, and J. L. Volakis, "Wideband Planar Array with Integrated Feed and Matching Network for Wide-Angle Scanning," *IEEE Transactions on Antennas and Propagation*, Vol. 61, No. 9, September 2013, pp. 4528–4537.

Kumar, C., et al., "Dual Circularly Polarized Spherical Phased-Array Antenna for Spacecraft Application," *IEEE Transactions on Antennas and Propagation*, Vol. 61, No. 2, February 2013, pp. 598–605.

Landsberg, N., and E. Socher, "Design and Measurements of 100 GHz Reflectarray and Transmitarray Active Antenna Cells," *IEEE Transactions on Antennas and Propagation*, Vol. 65, No. 12, December 2017, pp. 6986–6997.

Pokuls, R., J. Uher, and D. M. Pozar, "Dual Frequency and Dual Polarization Microstrip Antennas for SAR Applications," *IEEE Transactions on Antennas and Propagation*, Vol. 46, September 1998, pp. 1289–1296.

Rowe, W. S. T., and R. B. Waterhouse, "Broadband Microstrip Patch Antennas for MMICs," *Electronics Letters*, Vol. 36, No. 7, March 30, 2000, pp. 597–599.

Talukder, P., et al., "A 24 GHz Active Antenna in Flip Chip Technology with Integrated Frontend," *IEEE MTT-S Int. Microw. Symp. Dig.*, June 2000, pp. 1776–1779.

Talisa, S. H., et al., "Benefits of Digital Phased Array Radars," *Proceedings of the IEEE*, Vol. 104, No. 3, March 2016, pp. 530–543.

Yan, J., et al., "A Dual-Polarized 2–18-GHz Vivaldi Array for Airborne Radar Measurements of Snow," *IEEE Transactions on Antennas and Propagation*, Vol. 64, No. 2, February 2016, pp. 781–785.

CHAPTER 11

Broadband Antennas

11.1 Introduction

The arrival of broadband systems in the wireless communication area has demanded the design of antennas that must operate effectively over a wide range of frequencies. The broadband antenna is an antenna with a wide bandwidth. The definition of a broadband antenna is somewhat arbitrary and depends on the type of the antenna. If the impedance and pattern of an antenna do not change significantly over about an octave ($f_U/f_L = 2$) or more, the antenna is classified as a broadband antenna. Broadband antennas are generally made of structures that do not change abruptly in their physical dimensions, but instead utilize materials with smooth boundaries. Smooth physical geometry tends to produce antenna patterns and input impedance that also change smoothly with frequency, which is very important for a broadband antenna operation.

Typical broadband antennas include the log-periodic antenna, traveling-wave antenna, helical antenna, biconical antenna, frequency-independent antenna, spiral antenna, and complementary antenna. However, most of these existing broadband antennas are nonplanar and may not be directly mounted on a metallic platform. Many new technologies have emerged in recent years and, for such emerging technologies, printed antennas and antennas in the package are preferred. The fundamental and basic principles of broadband design of microstrip and printed antennas are [1–5]:

1. Low Q-factor of the magnetic wall cavity under the patch that can be achieved with low dielectric constant or larger thickness of the substrate.
2. Multiple resonances: Various geometries and perturbations to introduce multiple resonances as well as input impedance matching such as reactive loading by shaped slot, notch, cuts, pin, or post.
3. Customized patch geometry: Very irregular and unconventional patch shape optimized using the Genetic Algorithm.
4. Suppression of surface waves in a thick substrate: Periodic patterns on the ground plane or on any substrate produce an EBG structure on one face of which the microstrip element or arrays are printed.
5. Frequency-dependent substrate or ground plane: Multiple layers of FSS can reflect at respective frequency bands.

In this chapter, the designs of printed broadband antennas are discussed based on EBG, PBG, FSS fractal antenna, and many other latest technologies.

11.2 Metamaterials

In 1968, Veselago [6] hypothesized the possibility of having simultaneous negative values for permittivity and permeability. Thereafter, he also delved into the physical realization of $-\mu$ and $-\varepsilon$. Materials that display these properties are known as a metamaterial. Metamaterials are material that has properties that are not found in nature. Metamaterials possessing these properties are also frequently named double negative (DNG) material and Negative Refractive Index (NRI).

In addition to the materials with simultaneously negative permeability and negative permittivity, the single negative metamaterials have also drawn a great interest as shown in Figure 11.1. Over the years many researchers worked on microwave antenna composites with metamaterials [7–9].

11.3 Antenna Designs Based on EBG

In broadband antenna applications, the antenna cavity is usually loaded with absorbers to eliminate the backward radiation, but in doing so the radiation efficiency of the antenna is reduced. To enhance the radiation efficiency of such antennas, EBG structures are used. EBG structures that exhibit metamaterial behavior are researched intensively in the application of microstrip antennas. The magnetic conductors improve the radiation efficiency of antennas, but do not exist. One method to create an artificial magnetic conductor (AMC) is to design periodic structures that have electrical characteristics such as a magnetic conductor in the desired frequency band; one such example is an EBG structure. When EBG structures interact with EM waves, different electrical properties are observed at different frequencies. Periodic structures pass certain frequency bands, reject some frequency bands, and behave like a magnetic conductor in a band of frequencies, known as the bandgap [10, 11]. The EBG terminology came from the photonic bandgap (PBG) phenomena in optics that are realized by periodical structures. EBG structures are equivalent to

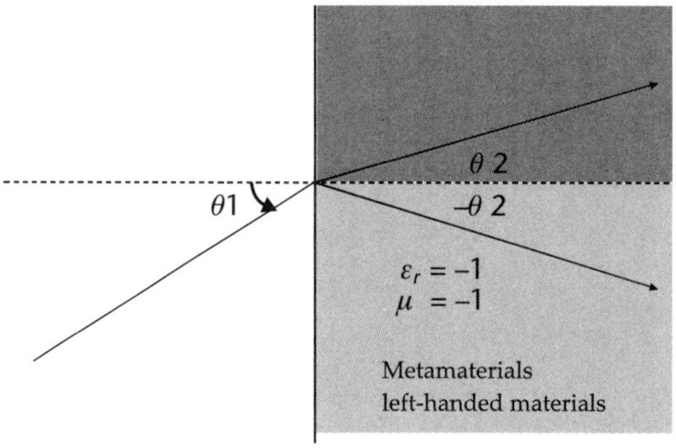

Figure 11.1 Negative refraction occurring at the interface between air and NRI metamaterials.

a tank circuit with capacitance represented by the gap between the patches and the inductance represented by the distance between the patches and the ground plane. EBG structures are usually compact and have low loss and can be easily integrated with antennas but operate over a narrow band of frequencies.

The EBG of a structure can be determined from the reflection phase, the dispersion diagram, and the transmission characteristics. The EM nature of a surface is predictable from the reflection phase characteristics, such that a 180° reflection phase implies a perfect electric conductor (PEC) surface and a 0° reflection phase resembles a perfect magnetic conductor (PMC) surface. EBG structures can be classified based on reflection phases for the same reflection magnitude. These structures can be designed to have a reflection phase of 0° for a specific frequency, which makes EBGs behave like a magnetic conductor, hence the name AMC. The bandgap of EBG structures is defined as the band of frequencies where the reflection phase from the EBG surface is between 90° and –90°.

EBG structures exhibit unique EM properties that have led to a wide range of applications in antenna and propagation fields [12–17]. There are two major advantages associated with EBG ground planes. First, these structures exhibit the possibility of creating a PMC because of in-phase reflection phase characteristics. This feature can cause efficient radiation for antennas laid adjacent to the EBG ground plane [12]. Second, such structures prevent the propagation of EM waves in a certain frequency band. Therefore, these can be used to block surface waves that usually corrupt antenna performance. Bandgap features are revealed in two ways: the in-phase reflection coefficient and the suppression of surface-wave propagation. The surface-wave suppression feature helps to improve an antenna's performance such as increasing the antenna gain and reducing back radiation.

Depending on the applications and geometries as shown in Figure 11.2, many other different names have been used for periodic structures. These applications include FSS, filtering, and EBGs. EBG structures are adopted to reduce surface waves and increase the gain of antennas and they were first introduced as FSS. In an antenna system that operates at multiple frequency bands, EBG structures (FSS layers) can be used to isolate bands.

The performance of an antenna is significantly affected by the surrounding medium and the presence of conducting ground planes. A conducting surface close

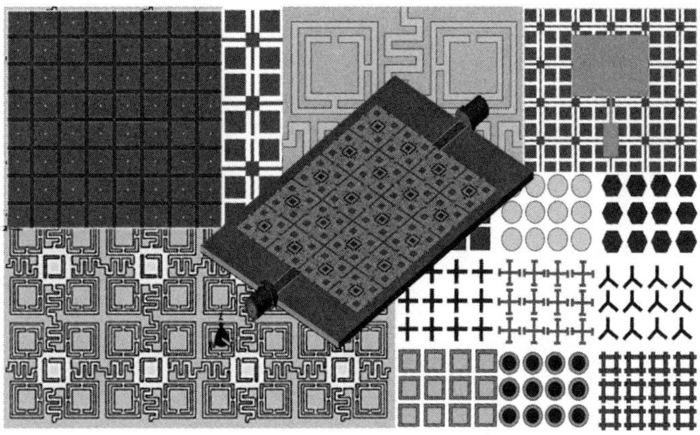

Figure 11.2 Various configurations of EBG structures [12–17].

to a radiating antenna typically degrades the performance of the antenna. From the conducting surface, a parallel electric field is reflected out of phase. When this current interacts with the forward propagating EM wave, they often add destructively and thus affect the radiation characteristics of the antenna. A tangential current close to a conducting plane is equivalent to two current elements, which are in opposite directions, as compared with the current in free space without the conducting plane. This property of the current causes the radiation from the current elements to cancel each other, thus affecting the radiation characteristics of the antenna. However, when an antenna is placed close to a parallel magnetic conductor, the electric field is reflected in phase and thus interacts constructively with the forward propagating electric field, which enhances the radiation in the forward direction instead of completely cancelling it. A tangential current element close to the magnetic conducting plane is equivalent to two electric current elements, in the same direction for the free space equivalent without the magnetic conducting plane. This behavior causes the radiation from the current elements to add constructively, thus enhancing the radiation characteristics of the antenna.

EBG structure classification is shown in Figure 11.3. Uniform EBG (U-EBG) structures are typically periodic structures consisting of metal patches that are separated by small gaps and vias that connect the patches to the ground plane. U-EBG structures are equivalent to a magnetic surface at the frequency of resonance and thus have very high surface impedance; this makes the EBG structures useful when mounting an antenna close to the metallic ground plane, provided that the antenna currents are parallel to the EBG structure. These structures can also be used between antennas in an array to decrease coupling between adjacent antenna elements, helping to eliminate blind scan angles. Utilizing EBG in a patch antenna with an air gap appears to perform five key functions:

1. Increase operation bandwidth;
2. Reduce sidelobe levels;
3. Increase front-to-back ratios;
4. Increase directivity and consequently gain improvement;

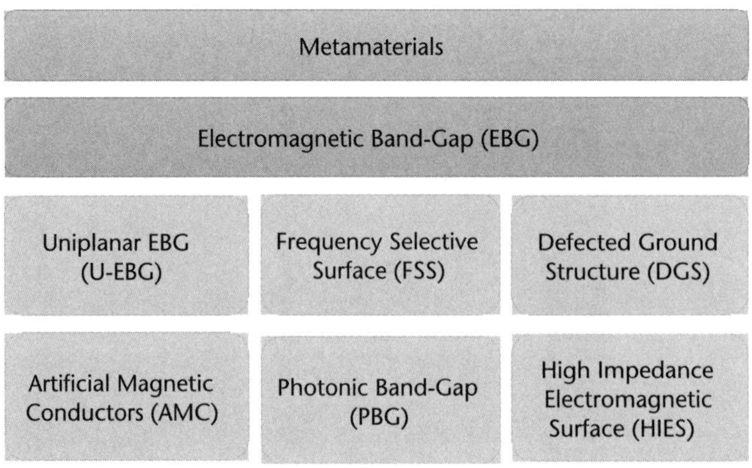

Figure 11.3 EBG structure types and classifications.

5. Harmonic control (suppression of resonance at the harmonic frequencies of the antenna).

Because of rapid progress in computational EM techniques various computational methods have developed for EBG analysis. These analyses are accurate, but very time-consuming, as Maxwell's equations need to be solved inside the unit cell by applying appropriate boundary conditions on unit cell walls. One of the advantages of using existing full-wave solvers is that the results allow the extraction of the properties of the EBG structure (surface impedance, reflection phase, dispersion diagram).

11.3.1 Microstrip Patch Antenna on Mushroom-Like-EBG Structure

There are many forms of periodic EBGs such as mushroom-like EBG, uniplanar EBG (UC-EBG), fork-like EBG, and spiral-like EBG. The mushroom-like EBGs known as high impedance surface (HIS) were initially proposed by Sievenpiper [12] and they are made up of metal patches, dielectric substrate, and vias connected with patches and ground plane, as shown in Figure 11.4. The mushroom-like EBG has a frequency bandgap and very high surface impedance characteristics, which is generally called the in-phase reflection band [13]. This conventional mushroom-like EBG structure can be used in antenna design to suppress the propagating of surface wave, but the cell size is still too large to be used in some practical applications. Thus, it is necessary to investigate the smaller EBGs for wide use in antenna fields.

The mushroom-like EBG structure is a 2-D EBG surface that consists of a metallic patch and a via that connects the patch to the ground as shown in Figure 11.4. The operating mechanism of the structure can be explained using an effective medium model with equivalent lumped LC elements. The shorting vias have an equivalent inductive effect, while the gaps between the patches have an equivalent capacitive effect. At its resonant frequency, the impedance is very high, and the structure impedes any surface waves, resulting in a frequency bandgap.

A unit cell EBG structure is composed of a metallic patch, a substrate, a ground plane, and a metallic via connecting the patch to the ground plane. The gap between adjacent patches has the effect of capacitance while the current flowing through the metallic via has an inductive effect. For an EBG structure with patch

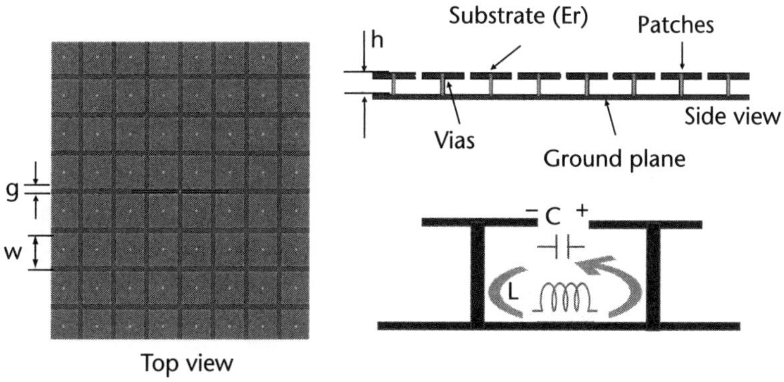

Figure 11.4 Mushroom-like EBG structure and equivalent circuit model of the conventional EBG unit cell [15].

width W, patch gap width g, dielectric constant ε_r, and substrate thickness h, the values of the inductor L and the capacitor C can be predicted with the following equations [13].

$$L = \mu h \tag{11.1}$$

$$C = W\varepsilon_0 \frac{1+\varepsilon_r}{\pi} \cosh^{-1} \frac{2W+g}{g} \tag{11.2}$$

$$f_0 = \frac{1}{2\pi\sqrt{LC}} \tag{11.3}$$

$$f = \frac{1}{\eta} f_0 \sqrt{\frac{L}{C}} \tag{11.4}$$

where μ is the free space permeability, ε_r is the substrate dielectric constant, and ε_0 is the free-space permittivity. The bandgap of such a structure can be determined by f_0, which is the central bandgap frequency, and Δf, which is the bandgap width. η is the free-space impedance, which is equivalent to 120π.

The surface impedance and reflection coefficient for the uniplanar compact EBG structure can be estimated based on this parallel LC equivalent circuit configuration. Hence, the surface impedance may be represented as

$$Z_s = \frac{j\omega L}{1-(\omega/\omega_0)^2} \tag{11.5}$$

where ω_0 is given by

$$\omega_0 = \frac{1}{\sqrt{LC}} \tag{11.6}$$

The bandgap of a periodic structure is calculated from its dispersion characteristics [14], which can be obtained by an eigenmode solver. In a structure, the eigenmode solver determines the resonant frequency for a given wavenumber, which is an important parameter to describe the propagation property of EM waves. There may be several different propagation constants at the same frequency. Each propagation constant is known as a specific mode with its own phase velocity, group velocity, and field distribution. The relation between phase constant and resonant frequency is referred to as the dispersion diagram.

A patch antenna design with an inset feed on the 1.6-mm-thick ($h = 0.026\ \lambda_{4.9\ \mathrm{GHz}}$ FR-4 substrate with a dielectric constant of 4.5 and a loss tangent of 0.025 is shown in Figure 11.5. The dimensions of the patch are 18.5 mm × 14.4 mm ($L = 0.29\lambda_{4.9\ \mathrm{GHz}}$, $W = 0.230\lambda_{4.9\ \mathrm{GHz}}$). The feed has a length of 5.3 mm from the edge and a notch width of 0.3 mm. The microstrip line width has a 1.5-mm EBG structure. The bandgap of the EBG structure was adjusted by tuning its width to 7.5 mm and its gap to 0.5 mm to cover the frequency of 4.9 GHz. The via diameter connecting the EBG patch and the ground plane is 1 mm. The number of columns of mushroom-like EBG structures placed around the patch antenna was varied to determine the number of columns required to effectively suppress the surface waves. The EBG structures were placed at least one period away from the edges of the patch antenna. Simulations were done for a different number of EBG columns. The patch antenna of the same ground size without EBG is also simulated for comparison purposes.

The simulation results are tabulated in Table 11.1. A patch antenna with 11 columns exhibited noticeable improvements in terms of the antenna bandwidth and gain of the antenna.

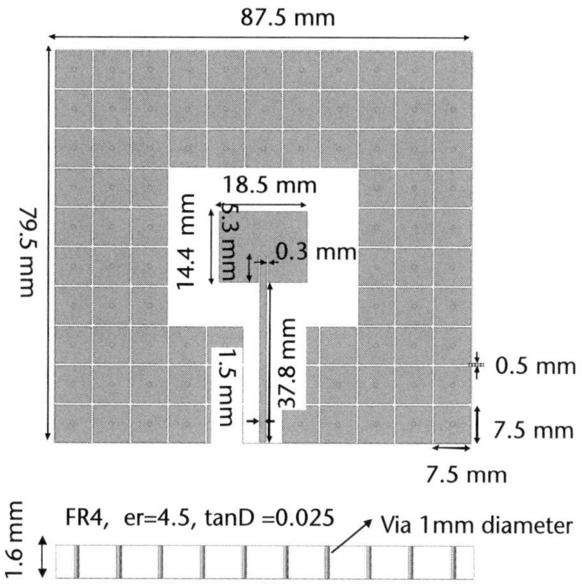

Figure 11.5 Microstrip patch antenna design using mushroom-like EBG structure.

Table 11.1 Antenna Performance with and without EBG

EBG Structure Size (Columns)	*Resonant Frequency (GHz)*	*Bandwidth (MHz)*	*Bandwidth Percentage*	*Gain (dBi)*
Without EBG	4.86	56	1.15%	6.20
7	4.88	185	3.79%	6.05
9	4.89	205	4.19%	6.76
11	4.91	227	4.60%	7.24

The S-parameter comparison of the antenna with the same size of the ground plane only and with the EBG structure is shown in Figure 11.6. An antenna with an EBG structure has around 4 times the bandwidth as compared to an antenna without any EBG structure. Unlike uniplanar and other EBGs, mushroom-like structures do not have any interconnecting microstrip lines that restrict their placement periodically. The electric field distribution over the patch antenna and the radiation pattern is shown in Figure 11.7. The gain improved from 6.2 dBi to 7.24 dBi, with the EBG structure.

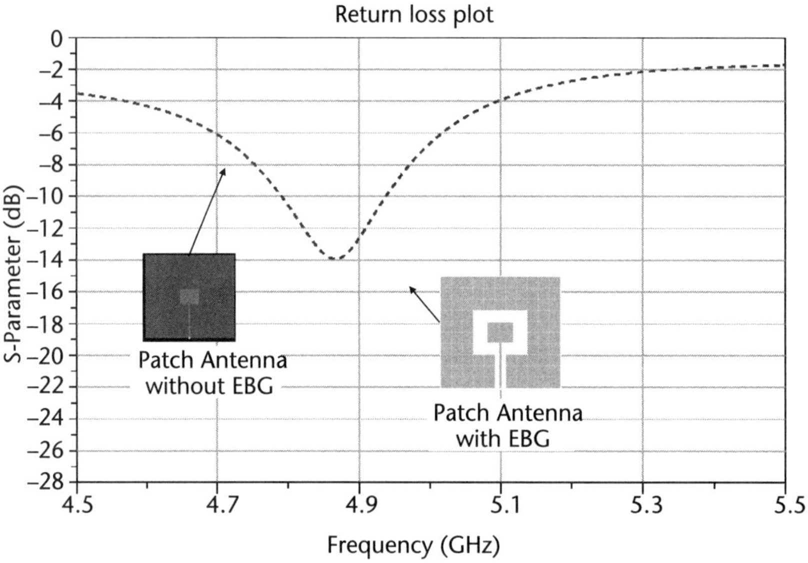

Figure 11.6 Microstrip patch antenna result with and without EBG.

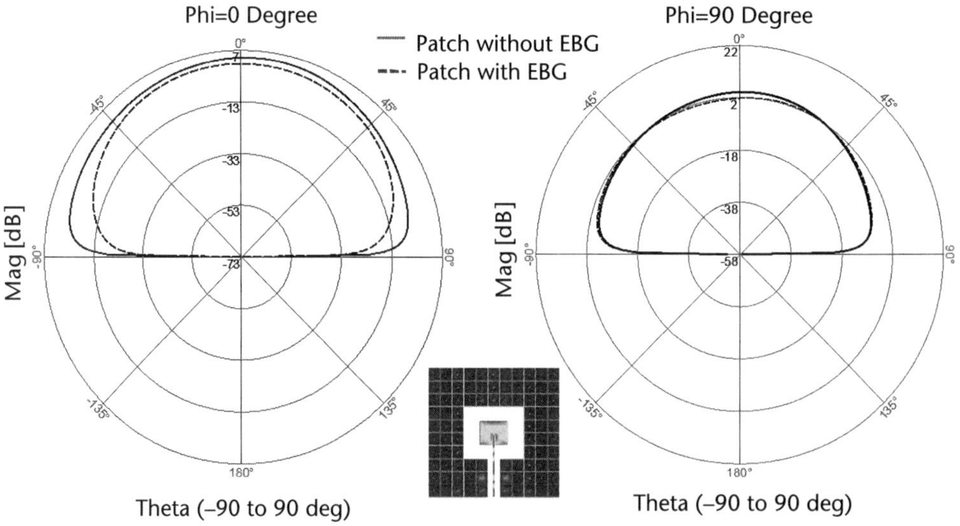

Figure 11.7 Radiation pattern of a mushroom-like EBG patch antenna.

11.3.2 Uniplanar EBG Cells with a Rectangular Microstrip Antenna

The design of a rectangular microstrip antenna based on the EBG cell-loading technique is shown in Figure 11.8. By incorporating the U-EBG cells on the ground plane of the rectangular microstrip antenna, the performance of the antenna can be enhanced. An important feature in the U-EBG design is the removal of vertical vias. Thus, it simplifies the antenna fabrication process and is compatible with microwave and millimeter-wave circuits. There are numerous configurations of EBG structures according to their application in the antenna. In this design, a square EBG periodic structure with stripline is used in the ground plane of the microstrip patch antenna. In the microstrip antenna design, an EBG structure can be used to design a high impedance ground plane that acts as a thin AMC metamaterial capable of reducing surface waves near resonance.

The zero-order bandwidth for an antenna over a magneto-dielectric substrate with thickness can be approximated by [18]:

$$BW \approx \frac{96\sqrt{\mu_r t / \varepsilon_r \lambda_0}}{\sqrt{2}\left[4+17\sqrt{\mu_r \varepsilon_r}\right]} \tag{11.7}$$

Thus, for a given miniaturization factor (constant $\sqrt{\mu_r \varepsilon_r}$) the antenna bandwidth can be enhanced by increasing μ_r/ε_r.

The rectangular microstrip antenna is designed and fabricated on low-cost glass epoxy substrate of relative permittivity of 4.4 and a substrate height of 1.6

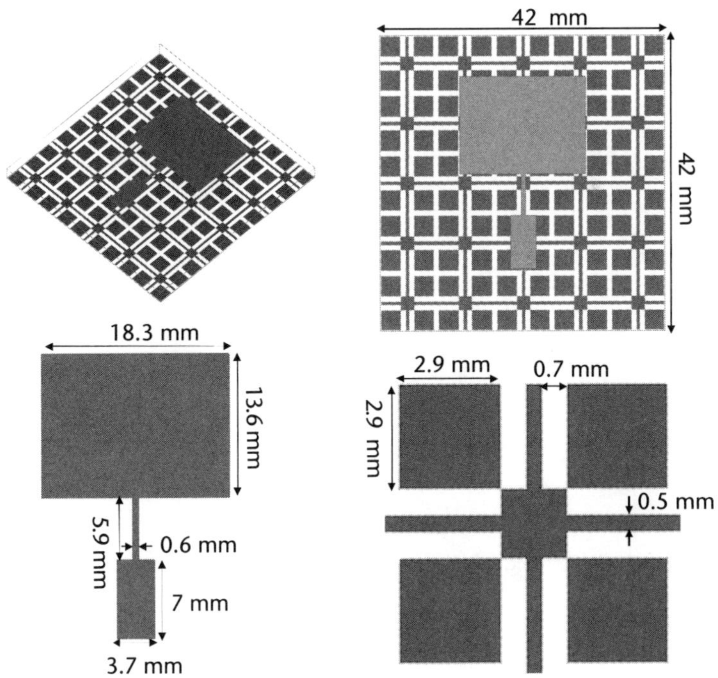

Figure 11.8 Design of a rectangular microstrip antenna based on the EBG cells as a ground plane.

mm. The rectangular microstrip patch length is 13.6 mm and the width is 18.3 mm. The enlarged view of the EBG cell is also shown in Figure 11.8. There are 5 × 5 EBG cells with a dimension of 7.7 mm × 7.7 mm. The total size of the ground plane is 42mm × 42 mm for the design. U-EBG microstrip antenna performance is compared with microstrip patch antenna with metallic ground plane without EBG structure. The return loss versus frequency characteristics of the antenna is shown in Figure 11.9.

The antenna bandwidth is 34% with a U-EBG structure that is almost 4 times more than the bandwidth of patch antenna without EBG as listed in Table 11.2.

11.3.3 Bowtie Antenna with EBG Cells

The ultrawideband (UWB) antenna designs may be broadly divided as traveling-wave structures such as the Vivaldi antenna [19, 20], frequency-independent structures such as the biconical antenna or the bowtie antenna [21, 22], self-complementary antennas that are characterized by a self-complementary metallization such as the logarithmic spiral antenna and fractal antennas [23], and combinations of these such as the log-periodic antenna [24–26]. The bowtie antenna, first proposed by Lodge and later reexamined by Brown and Woodward, demonstrates its benefits. There are various shapes of bowtie antenna such as the triangular shape, rounded edge, or slotted shape.

A bowtie antenna design is shown in Figure 11.10. This is printed on a 0.64-mm-thick dielectric substrate with a 10.2 (Rogers RT6010) dielectric constant. The feeding structure consists of a microstrip line continued by the transmission line formed by two parallel strips. One side of the feed consists of a microstrip

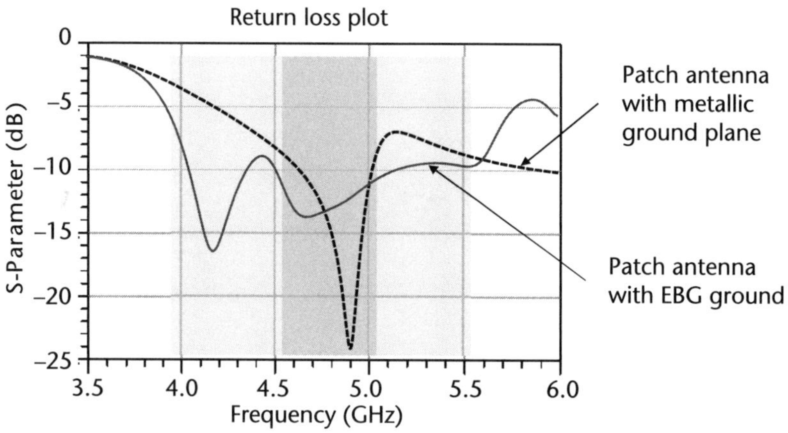

Figure 11.9 Antenna return loss plot with metallic and with EBG ground plane.

Table 11.2 Antenna Bandwidth with and without EBG

Antenna	*Bandwidth (–9 dB)*	*Bandwidth in %*
Rectangular microstrip patch antenna with ground plane	425 MHz	8%
Patch antenna with EBG	1,575 MHz	34%

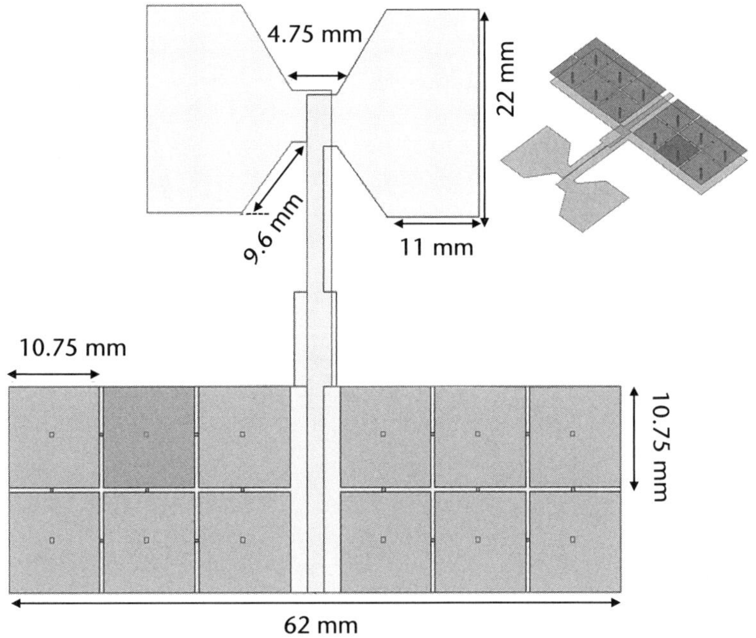

Figure 11.10 Double bowtie antenna design over a rectangular EBG structure.

line, one of the parallel lines, and one arm of the printed bowtie antenna. The other side consists of a truncated ground plane, the other line of the parallel strip transmission line, and the second arm of the bowtie antenna. The microstrip line is designed with a characteristic impedance of 50Ω. The parallel strips are selected based on some parametric studies and optimization procedures for the best antenna performance.

The truncated ground plane acts as a reflector; thus, the radiation patterns are expected to peak in the opposite direction of the ground plane edge. The dimensions used for this printed bowtie antenna are shown in Figure 11.10. The total area of the substrate is approximately 1,300 mm^2 at the central frequency. The antenna was built and tested. The computed and measured reflection coefficients and 3-D radiation at 2.5 GHz are shown in Figure 11.11. An antenna with a –10-dB bandwidth is 21%.

11.3.4 Array Antenna Design Using EBG

There are many frequency bands for different communication systems, and these are allocated to serve different purposes in modern life. Many of these applications and devices mainly operate in Ka-band, which is allocated in the microwave radio region of the EM spectrum.

For high gain, printed array antennas are used in these applications. Mutual coupling is one major issue in printed array antenna design. This coupling becomes especially critical in arrays of patch antennas where coupling comes from two paths. The first path is due to the free-space radiation, which is present in all types of array antennas. The second path comes from surface waves and it constitutes a very important factor in patch antennas. In these types of antennas, surface waves

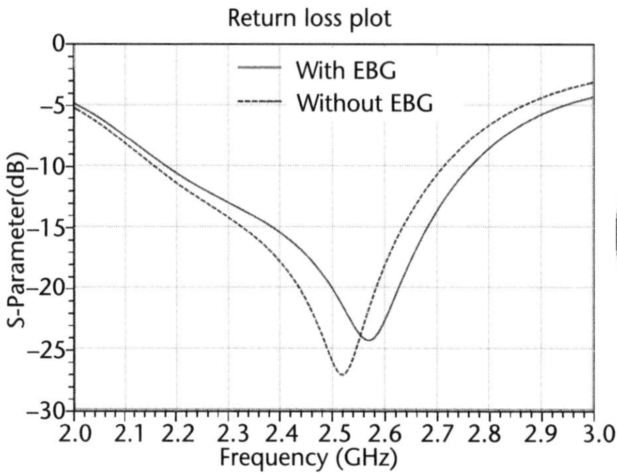

Figure 11.11 Bowtie antenna return loss and 3-D radiation pattern plot.

are strongly excited in E-plane when the antenna is operated in the fundamental mode (TM_{10} for rectangular patches). In this mode, the field distribution inevitably excites the first propagating mode of surface waves in E plane, given that this mode has no cutoff frequency.

Array antennas with a periodic structure like EBG, as shown in Figure 11.12, have the ability to suppress surface-wave propagation in a frequency band. The EBG structure is made of rectangular metallic elements (patch type) whose size to suppress surface-wave propagation is related to (usually smaller than) $0.5\lambda_{er}$. The simulated antenna return loss plot is shown in Figure 11.13. The antenna resonates at 28.8 GHz.

The lower cutoff frequency of this structure to behave as an EBG is related to the periodicity (here, 1.65 mm) in the direction of propagation (E-plane). Simulations were carried out using the FEM. The mutual coupling in the E-plane is shown

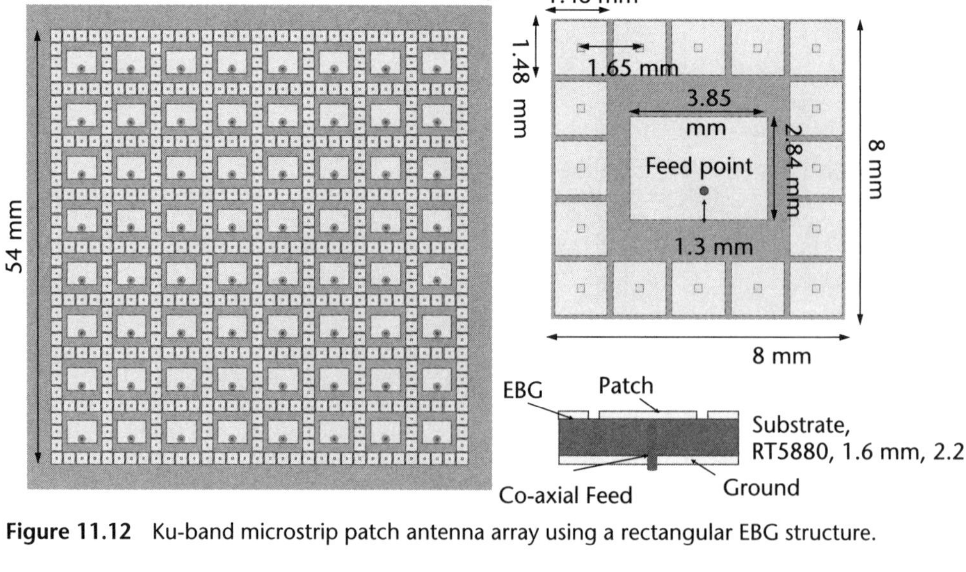

Figure 11.12 Ku-band microstrip patch antenna array using a rectangular EBG structure.

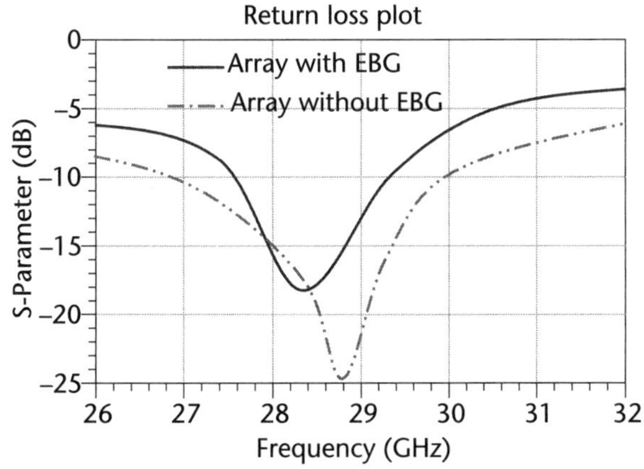

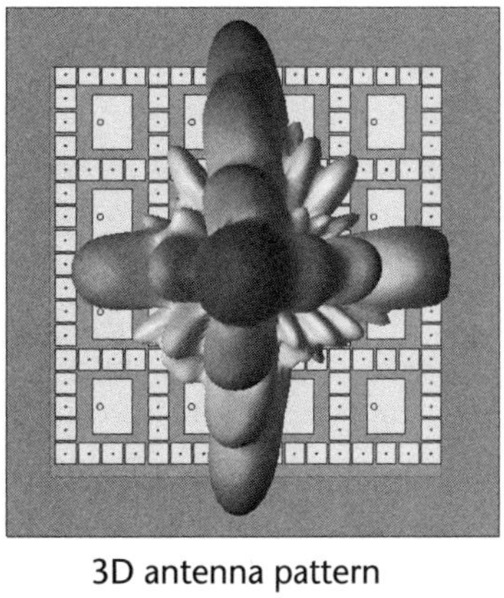

Figure 11.13 A 4 × 4 antenna array return loss and radiation pattern (top view).

in Figure 11.14. In the frequency band where the antenna is matched (with an S_{11} smaller than 10 dB), the maximum coupling level is –22 dB.

11.4 FSS

FSS is another area of interest to broadband antenna researchers. The FSS comprises periodically arranged metallic patch elements or aperture elements within a metallic screen and exhibits total reflection (patches) or transmission (apertures) about the element resonance. A periodic surface is basically an assembly of identical elements arranged in a 1-D or 2-D infinite array. Its basic characteristic is to scatter or reflect certain frequencies of EM waves incident on it. The FSS is used in

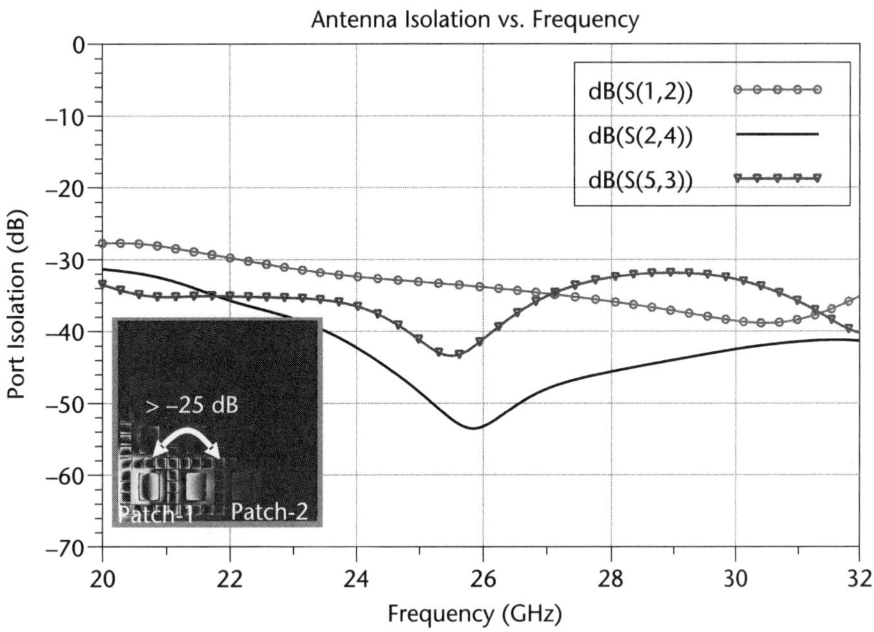

Figure 11.14 Simulated mutual coupling between antenna elements is greater than –25 dB.

various applications such as aerospace, antenna design, medical, and microwave devices. In microwave and medical applications, FSS technology is used to reduce the leakage of power and to prevent EMI. In aerospace applications, it is used to reduce the radar cross-section (RCS) of various devices such as radome, antennas, and microwave circuits. Sometime FSS is placed on top of the microstrip patches to reduce the cross-polarization level of the antenna [27].

High-performance, low observable antennas can be designed using the FSS structure to the planar antenna, where the FSS is employed either as superstrate or high-impedance ground plane of the antenna. The incorporation of the FSS to the antenna mainly enhances its gain and bandwidth. In printed antennas, to reduce the scattering component outside the operating bandwidth, the conventional metallic ground plane can be replaced with a stopband FFS of the antenna, because it largely reflects the energy of incident waves.

There are four different types of possible element-type FSS arrays: plate, center connected, loop, and combination. Figure 11.15 shows some examples of each element type. These elements vary from narrowband to wideband structures. In the first group, the plate group, the FSS elements in general do not have very desirable characteristics. Primarily, the elements have x and y dimensions around a half-wavelength, which limits how closely these elements can be placed next to one another. In addition, plates are highly inductive elements with small capacitances between them, leading to issues regarding achieving resonance. It is impossible to achieve perfect reflection if the FSS does not resonate. This is because, at resonance, the FSS becomes a short circuit and thus acts as a PEC ground plane.

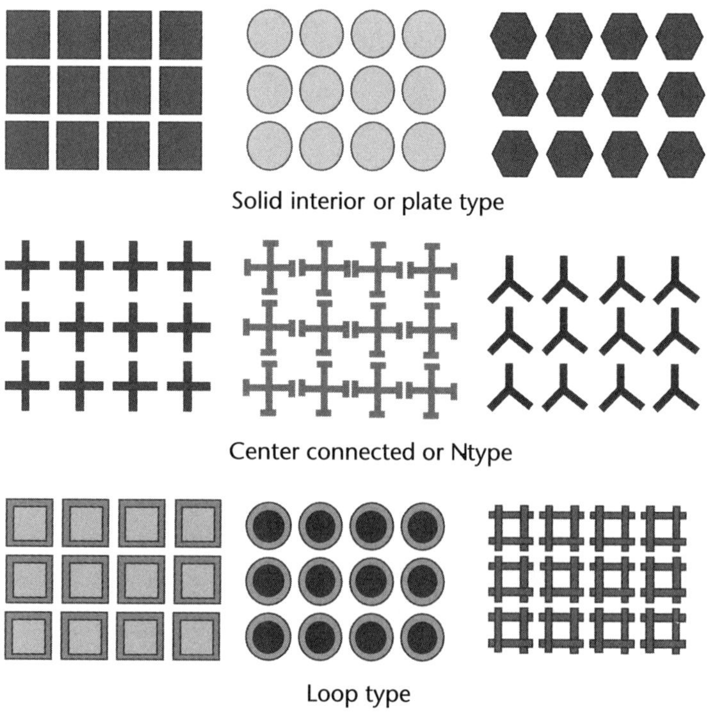

Figure 11.15 The three major groups of FSS elements.

The second group, the center-connected elements, is a viable candidate for both radiating and nonradiating arrays. The most recognizable from the center-connected group is the dipole. This is by nature a narrowband element, but when it is placed in a well-designed array this element can reach a bandwidth ratio of 4:1 without the addition of a dielectric matching section added in front of the array. The third group, the loop elements, is, in general, an excellent choice for nonradiating element FSS arrays. These FSS elements are typically smaller in the x and y directions than the center-connected elements with respect to a wavelength and thus can be spaced close to one another. As will be discussed in more detail later, this leads to a delayed onset of grating lobes and generally wider bandwidth.

An FSS pattern on a 2-mm-thick RO3003 substrate is shown in Figure 11.16. Periodic boundary conditions are used on four sides of the unit cell to simulate the infinite 2-D array. Perfectly matched layers (PMLs) on the top and bottom of the unit cell absorb the excited mode from the input port and any higher-order modes generated by the periodic structure. Unit cell simulation setup is shown in Figure 11.17.

An EM simulation model of a unit cell is shown in Figure 11.18. This structure acts as a passive resonant absorber. The FSSs have found applications in multiband reflector antennas and radome such as the Cassegrainian subreflectors in parabolic dish antennas, especially in communications and defense applications. The simulation transmission loss and the electric field are shown in Figure 11.19.

11.5 Antennas on PBG

Microstrip planar antennas fabricated on a substrate with a high dielectric constant (GaAs, Si, and InP) are strongly preferred for easy integration with the MMIC RF front-end circuitry. However, it is well known that patch antennas on high dielectric constant substrates are inefficient radiators due to surface wave losses and have very narrow (1%–3%) frequency bandwidth. This situation becomes very severe

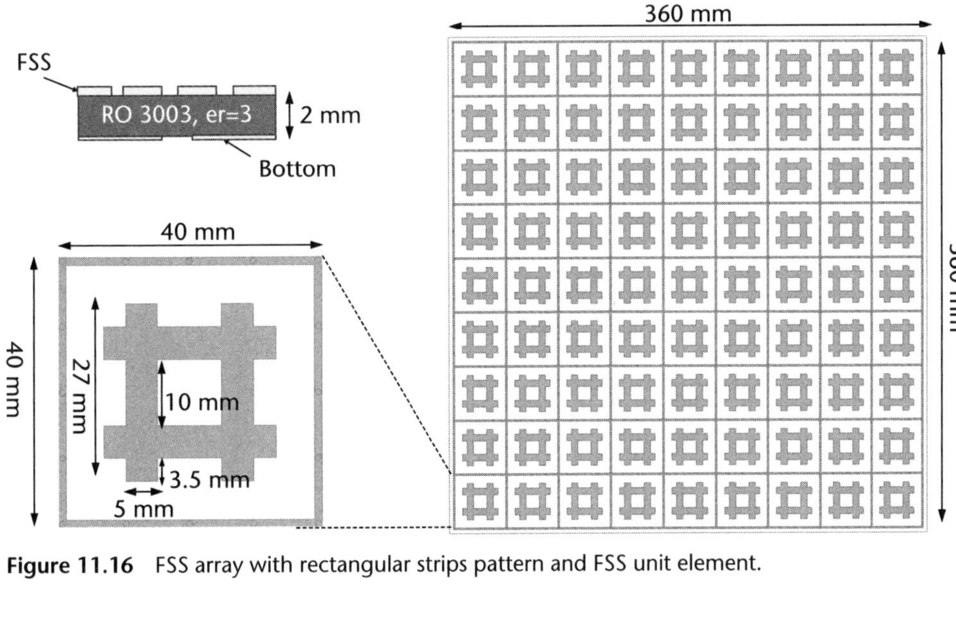

Figure 11.16 FSS array with rectangular strips pattern and FSS unit element.

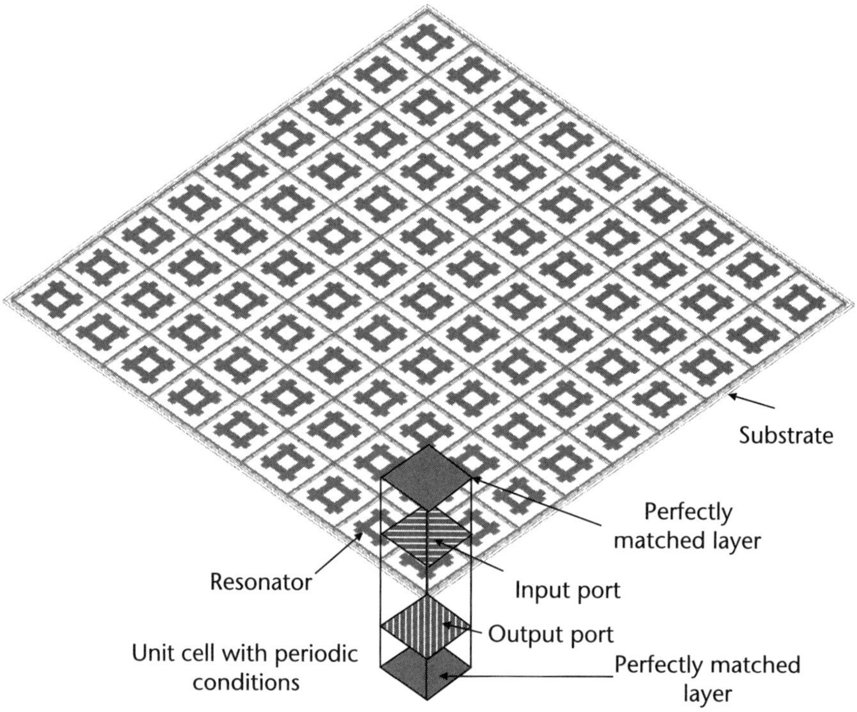

Figure 11.17 Layout geometry of the FSS and EM simulation setup for a unit FSS cell.

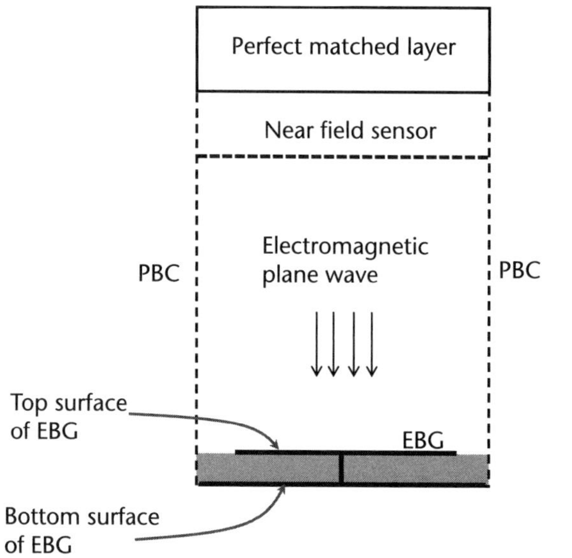

Figure 11.18 Unit cell model of FSS with boundary conditions and plane-wave excitation for EM simulation.

as applications move to higher frequencies, resulting in microstrip patch antennas with reduced efficiency and gain as well as an unacceptably high level of mutual coupling and cross-polarization within an array environment. PBG technology has been applied as a new promising solution to these problems. Another way to get

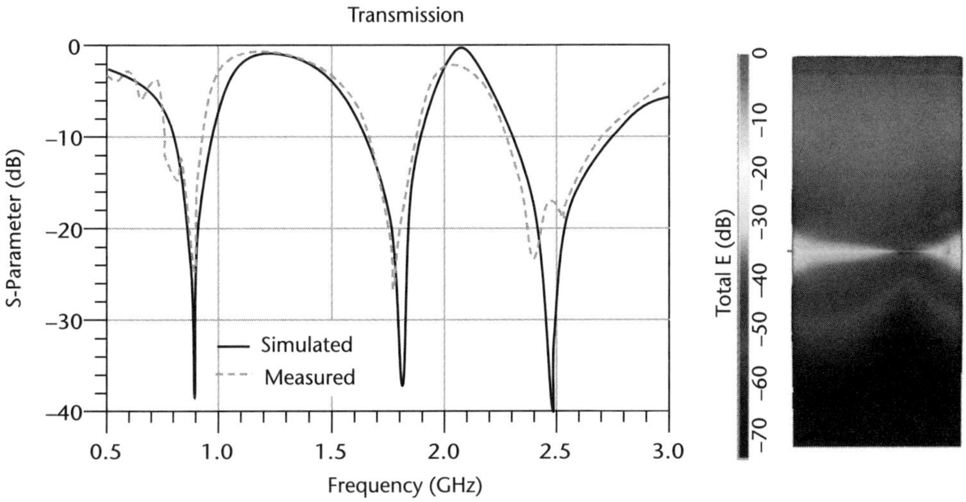

Figure 11.19 Simulated and measured reflection coefficient and E-field profiles in the xz-plane, the unit cell cavity.

broadband antenna performance is increasing the substrate thickness that introduces the low Q-factor of the cavity below the patch. However, this technique is restricted by the surface-wave generation leading to low gain and low efficiency of the antenna.

The PBG structure is basically a periodic metallic pattern printed on a dielectric substrate for microwave and millimeter-wave applications as shown in Figure 11.20. This technique provides a stopband of EM waves propagating through it, called the bandgap. The frequency range of the stopband depends on the geometry dimensions and its pattern. The wave is attenuated during propagating through the substrate if the antenna operating frequency falls within this stopband. In this way, the generation and propagation of the surface wave are stopped, which increases antenna efficiency. The PBG structure can be formed by using metallic, dielectric, ferroelectric, or ferromagnetic implants. Dielectric PBG structures have been used

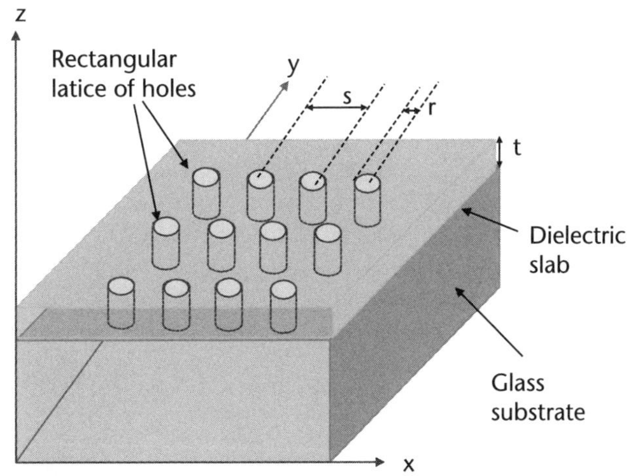

Figure 11.20 A PBG structure with a periodic metallic pattern printed on a dielectric substrate.

for microstrip circuits [10]. These require drilling of a periodic pattern through the substrate. A PBG structure that requires only partial etching of the ground plane is compatible with monolithic circuit technology. Photonic structures have been applied to the study of the microwave frequency ranges [10, 11].

A microstrip planar antenna design that uses photonic crystals as a substrate to enhance the antenna performance at the terahertz frequency regime is shown in Figure 11.21. The dominant surface wave in microstrip patch antennas on a dielectric substrate with the ground is the TM0 mode, which has no cutoff frequency. A 2-D PBG lattice can prohibit the propagation of the TM0 mode along the dielectric substrate. The incorporation of PBG unit cells in the dielectric substrate volume reduces the effective dielectric constant and refractive index and improves the gain and radiating efficiency of the antenna. The effective dielectric constant and refractive index of the PBG material depends on the dimension and periodicity of the PBG unit cell. The microstrip patch antenna basic structure consists of a conducting patch and a ground plane separated by 5-μm-thin glass substrate. This is the terahertz antenna with thin substrate and high dielectric constants, which results in tightly bound fields within the dielectric material. The end of the microstrip feed line is connected with a coaxial line for impedance transition. The antenna input impedance can be matched by tuning the length, width, and position of the microstrip feed line.

The PBG is used to control the refractive index of the substrate material to be less than 1. The refractive index of the PBG material depends on the lattice factor (k), where k is the ratio of the radius of the cylindrical air cavity unit cells to the PBG unit cell interspacing distance. PBG is also used for filtering, as they block certain bands; these structures can be used to block undesired frequencies. The antenna is simulated using the FEM solver. Antenna return loss plot and mesh data are shown in Figure 11.22. Antenna resonance frequencies are 2.24 THz and 2.61 THz.

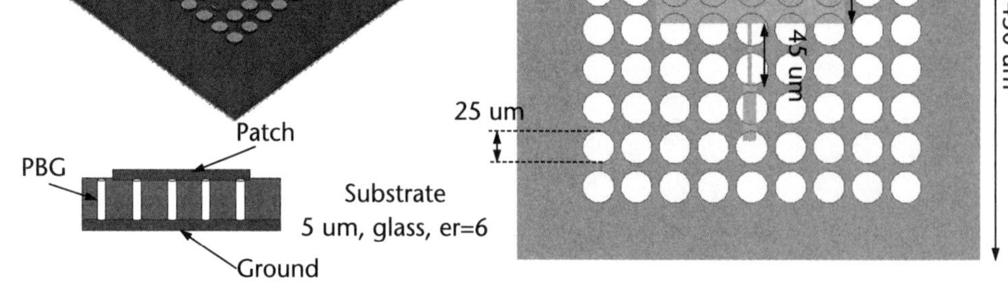

Figure 11.21 Microstrip patch antenna design with photonic crystals in the ground plane. The PBG is constructed on a grounded dielectric substrate by drilling holes in the substrate.

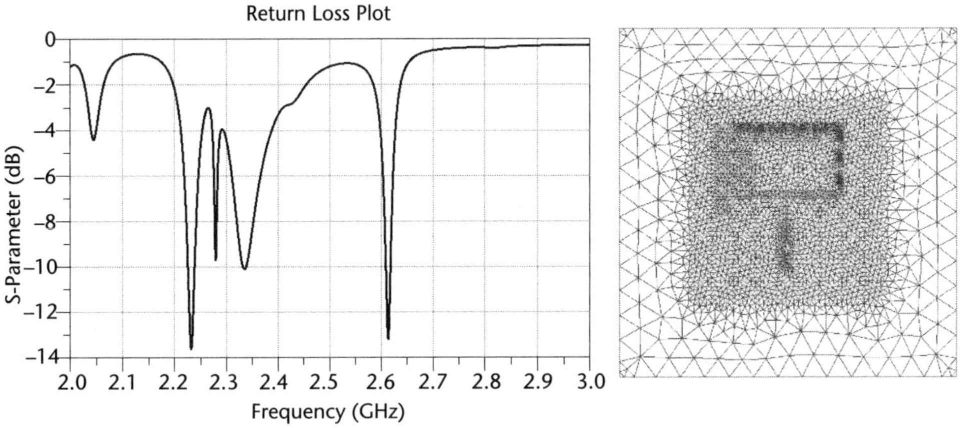

Figure 11.22 Computed return loss plot using the FEM technique and FEM mesh.

11.6 Fractal Antennas

Miniature antennas are of prime importance due to the available space limitation on the user equipment and the oncoming deployment of diversity and MIMO systems in many applications. The major antenna miniaturization techniques include material loading, lumped element loading, the use of ground planes, short circuits, and the antenna's environment and geometry optimization. Among these miniaturization techniques, geometry optimization has gained much interest with the introduction of fractals in antenna engineering due to their inherent miniaturization ability. The term fractal, which means irregular fragments or broken, was originally coined by Mandelbrot [28] to describe a family of complex shapes that possess an inherent self-similarity or self-affinity in their geometrical structure. The original motivation for the development of fractal geometry came mainly from the study of the patterns of nature. For example, fractals have been effectively used to model complex natural objects such as galaxies, cloud boundaries, mountain ranges, snowflakes, coastlines, ferns, and much more. By using fractals in antennas, some of the unique properties can be achieved:

- Reduction of physical radiator size;
- Multiband behavior is a result of self-similarity;
- Self-similar radiation pattern;
- Consistence performance over the huge frequency range;
- Added inductance and capacitance without components.

To overcome narrow bandwidth and low gain limitation of conventional microstrip patch antennas, these antennas are designed with fractal geometries. Broadband antennas, logarithmic spiral, and log-periodic structures can be considered as fractal antennas, and these antennas can be also analyzed based on fractal antenna theory. Frequency independence in antennas starts with Rumsey's principle [29]. As per Rumsey's principle, the impedance and pattern properties of an antenna will be frequency-independent if the shape is specified only in terms of angles. The spiral antenna can be considered as a smoothly self-symmetric structure. Similarly, in the log-periodic antenna, various performance properties repeat in the same geometric ratio as its element size and spacing.

11.6.1 Fractal Antenna Geometries

There are some common fractal geometries that have been found to be useful in developing new and innovative designs for antennas such as the Sierpinski gasket, Sierpinski carpet fractal structure, Koch snowflake, fractal rectangular curve (FRC), and Hilbert curve as shown in Figure 11.23. A few of the other fractal geometry types are Cantor's comb, the Mandelbrot set, and the Lorenz attractor.

Nathan Cohen, a radio astronomer at Boston University, was a fractal antenna innovator who experimented with wire fractal antennas (von Koch curves) and fractal planar arrays (Sierpinski triangles). In 1988, he built the first known fractal antenna. Fractal antenna geometries have two common properties: self-similarity and space-filling. While the self-similarity property is successfully applied to design multiple band fractal antennas, the space-filling property can be used to compact the antenna size.

11.6.2 Sierpinski Gasket Fractal Antenna

The Sierpinski gasket is named after the Polish mathematician Sierpinski, who described some of the main properties of this fractal shape in 1916 [30]. The original gasket is constructed by subtracting a central inverted triangle from the main triangle shape as shown in Figure 11.24. After the removal, three equal triangles remain on the structure, each one being half of the size of the original one. The same

Figure 11.23 Common fractal antenna geometries, Sierpinski gasket, Koch snowflake, Hilbert curve, and FRC.

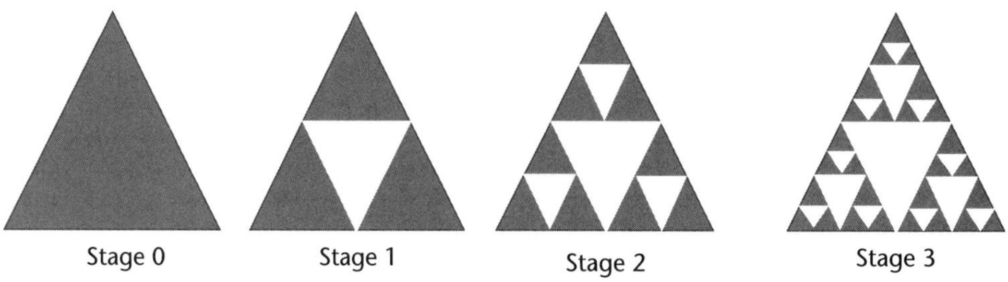

Figure 11.24 Several stages in the construction of a Sierpinski gasket fractal.

subtraction procedure can be iterated on the remaining triangles and, if the iteration is carried out an infinite number of times, the ideal fractal Sierpinski gasket is formed. In this ideal structure, each shape of its three main parts is exactly equal to the whole object but scaled by a factor of 2 and so is each of the three gaskets that compose any of those parts.

The first few stages in the construction of the Sierpinski gasket are shown in Figure 11.24. The Sierpinski gasket is an example of a self-similar fractal. The procedure for geometrically developing this fractal begins with an equilateral triangle contained in the plane, as shown in stage 0. The next step is to remove the central triangle with vertices that are located at the midpoints of the sides of the original triangle, shown in stage 1. The same process is then repeated for the three remaining triangles, as illustrated in stage 2. The next two stages (i.e., stages 3 and 4) in the construction of the Sierpinski gasket are also shown in Figure 11.24. A Sierpinski triangle antenna design is shown in Figure 11.25.

The antenna is printed on a 1.6-mm-thick RT Duroid substrate with a dielectric constant of 2.3. The structure is fed through a 1.5-mm diameter, 50Ω coaxial probe with a SubMiniature version A (SMA) connector on the bottom side of the plane. The size of the antenna is 10 cm × 10 cm. The sides of the initiator equilateral triangle are 6 cm. The antenna has been constructed through four iterations in this design, so four-scaled versions of the Sierpinski gasket are found on the antenna (circled regions), the smallest one being a single triangle. The antenna is analyzed using the MoM. The return loss against frequency and 2-D radiation patter=n plots at two frequencies are shown in Figure 11.26 for different iterations. For the first iterations of the antenna, the resonant frequencies observed are 4.3 GHz, 7 GHz, 8.3 GHz, and 9.8 GHz, respectively, with the prominent frequencies being 4.3 GHz and 7 GHz. The second iteration resonates at 3.1 GHz, 6.6 GHz, and 9.2 GHz.

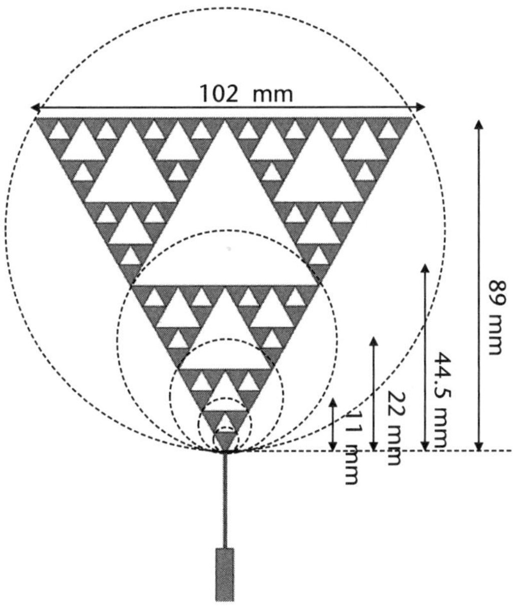

Figure 11.25 Sierpinski monopole antenna and design dimensions.

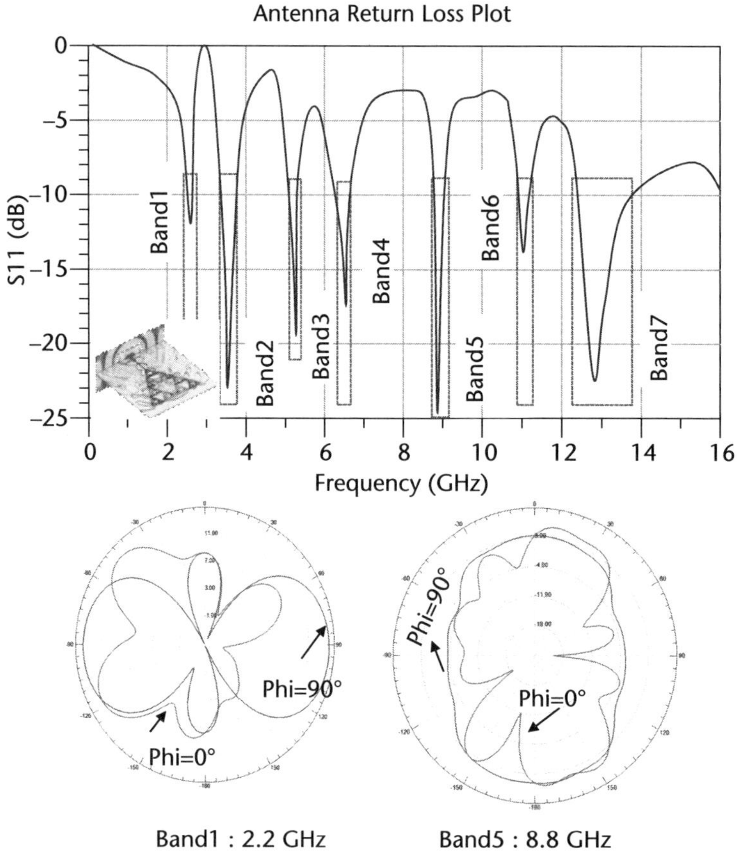

Figure 11.26 Sierpinski monopole antenna return loss and 2-D radiation pattern plot at 2.2 GHz and 8.8 GHz.

It is seen that the lower resonant frequencies are less emphasized in case of the single-layered structures of the Sierpinski geometry. Thus, the true multiband character is not so emphasized by this single-layered design approach.

11.6.3 Square Shape Fractal Antenna Design

The UWB fractal antennas with a notch are used in many broadband systems. In applications such as Wi-Fi and GPS, operation at two or more subbands is necessary. A valid alternative to the broadening of total bandwidth is the use of a multifrequency fractal antenna in a single radiator. A square shape fractal geometry with four stages is shown in Figure 11.27.

A square shape fractal antenna design and dimensions are shown in Figure 11.28. At the initial stage, antenna geometry looks like a simple square patch designed on the RT Duroid substrate with $\varepsilon_r = 2.2$ and a substrate thickness of 1.57 mm.

The square patch has a dimension of 25 mm × 25 mm. The first iteration of the fractal antenna has been constructed by inscribing the circular patch of diameter of 24 mm and subtracting it from the square. The second iteration has been

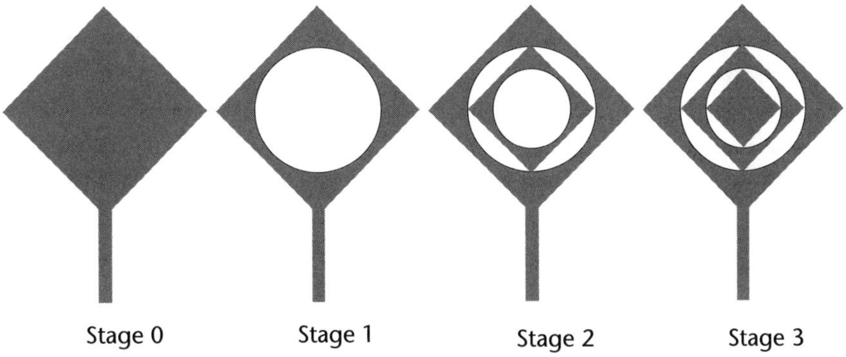

Figure 11.27 Square shape fractal geometry with four stages in the construction.

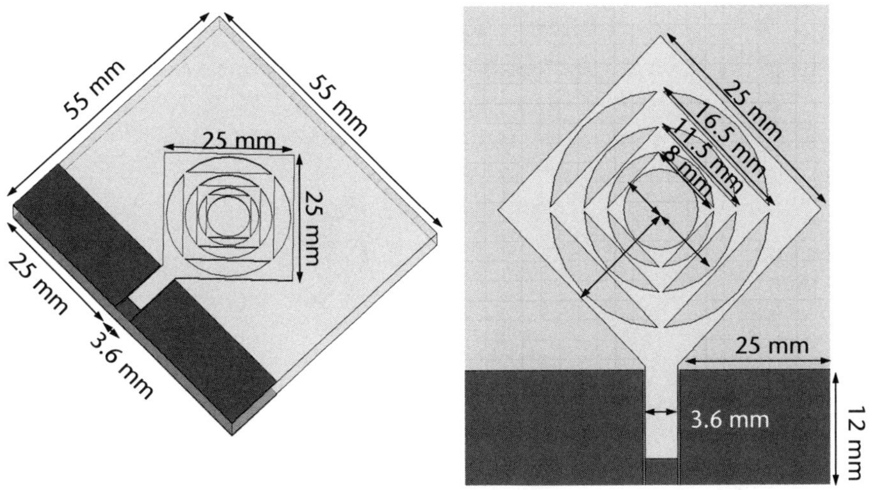

Figure 11.28 Square shape microstrip fractal antenna geometry and design dimensions.

constructed by making the square of the dimension 16.5 mm × 16.5 mm and an inscribed circle with a diameter of 15.5 mm and subtracting it from the square. Likewise, the third iteration is constructed by making a square of 11.5 mm × 11.5 mm and inscribing the circle with a diameter of 10.5 mm and subtracting it from the square. In the fourth iteration, a square of dimension of 8 mm × 8 mm is made, and an inscribed circle of a diameter of 7.5 mm is subtracted from it. This fractal antenna has been fed with a CPW feed line.

After simulating this fractal antenna, it is noticed that the operating resonant frequency of the antenna depends on patch size and substrate thickness. So, for better performance of the antenna, these parameters should be optimized. The simulated return loss of the antenna is shown in Figure 11.29. Antenna 2-D radiation patterns at 4 GHz and 8.5 GHz are shown in Figure 11.30.

The antenna has an operational resonant frequency at 2.1 GHz, 4.4 GHz, 5.4 GHz, 6.4 GHz, and 7.7 GHz normalized to a 10-dB return loss value, which is covered in WLAN and Wi-Max bands for communication.

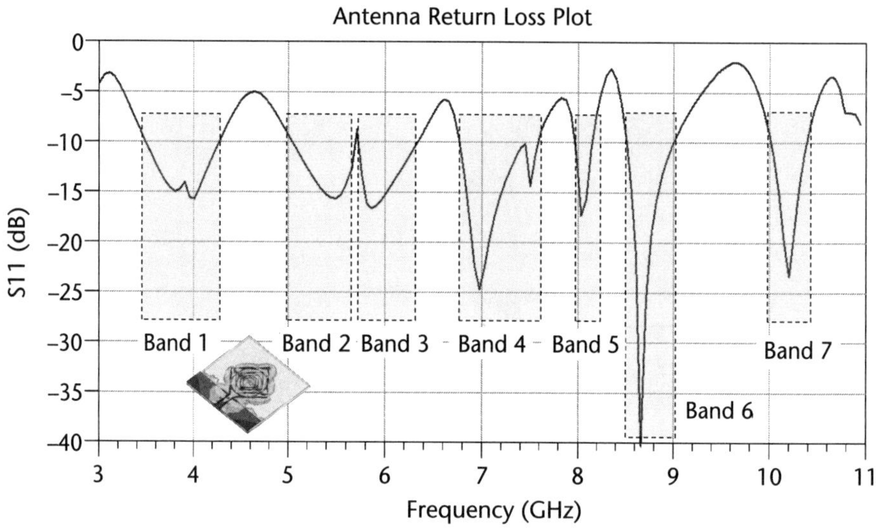

Figure 11.29 Return loss plot of the square shape fractal antenna.

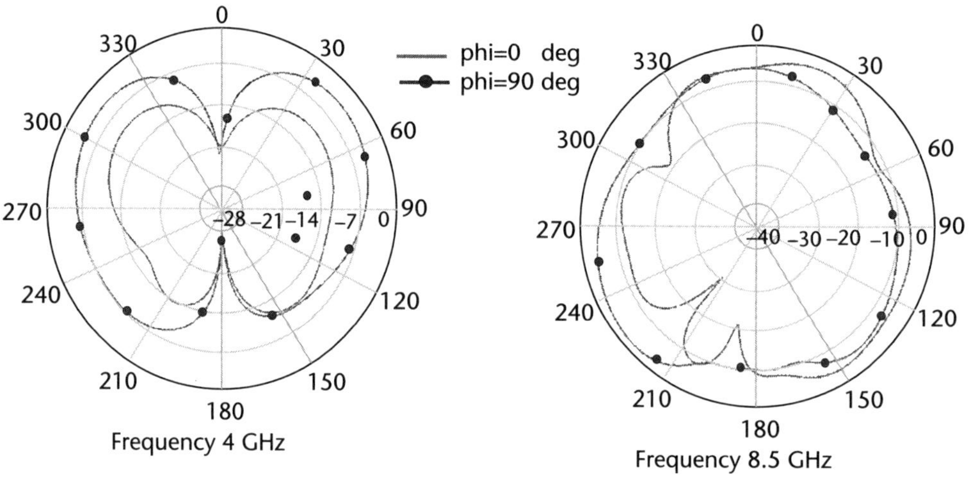

Figure 11.30 Antenna radiation pattern at 4 GHz and 8.5 GHz in the antenna operating frequency band.

11.6.4 Koch Snowflake Fractal Antenna

The Koch snowflake is another popular fractal antenna. The creation of the Koch fractal starts with a straight segment with a length L, and then this initiator line segment is divided into three parts of equal length (i.e., $L/3$ each). Next, the middle segment is replaced with other two segments of the same length, with 60° as intersection angle. This geometry is called the generator and is the first iterated version of fractal geometry. By repeating this process further, higher iterations are generated. This fractal also starts out as a solid equilateral triangle in the plane, as illustrated in stage 0 of Figure 11.31. However, unlike the Sierpinski gasket fractal, which is formed by systematically removing smaller triangles from the original structure,

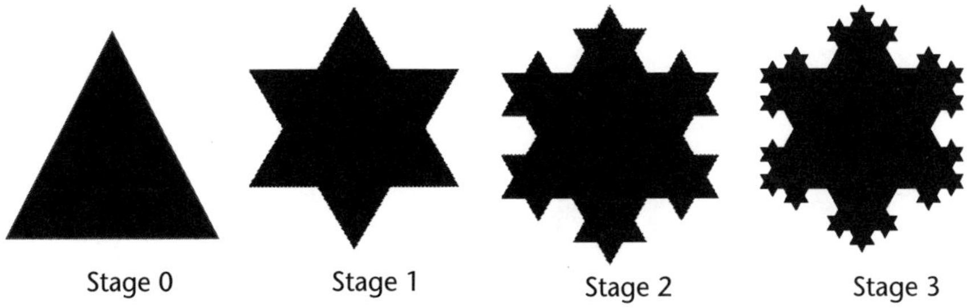

Figure 11.31 Four stages in the construction of a Koch snowflake fractal.

the Koch snowflake is constructed by adding smaller triangles to the structure in an iterative fashion. This process is clearly represented in Figure 11.31, where the first four stages in the geometrical construction of a Koch snowflake are shown.

For the design of triangular patch antenna, there are three important parameters: resonant frequency (f_r), dielectric material of the substrate (ε_r), and thickness of the substrate (d). The resonant frequency of the fundamental mode of such an antenna is given as:

$$f_r = \frac{2c}{3a\sqrt{\epsilon_r}} \tag{11.8}$$

where c is the speed of light, ε_r is the relative permittivity of the substrate, and a is the patch side length. The first iteration of the Koch fractal antenna is obtained by replacing the sides of an equilateral triangle by a Koch curve. After the first iteration of the Koch fractal, the resulting shape is a star shape patch, as shown in Figure 11.31. After the second iteration of the Koch fractal antenna, the resulting shape is as shown in stage 2. In each new iteration, the area of the geometry increases.

If $A_k$is the area at one iteration, then the area A_{k+1} of the next iteration can be computed:

$$A_{k+1} = A_k + \frac{\sqrt{3}}{12}\left(\frac{4}{9}\right)^{k-1} a^2 \tag{11.9}$$

where a is the side length of the initial triangle that has an area of:

$$A_0 = \frac{\sqrt{3}}{4}a^2 \tag{11.10}$$

Figure 11.32 shows a microstrip Koch snowflake fractal antenna operating at multiple frequency bands between 2 GHz and 16 GHz. Some fractal shapes are not broadband structures like Koch is narrowband, but this is good for miniaturization.

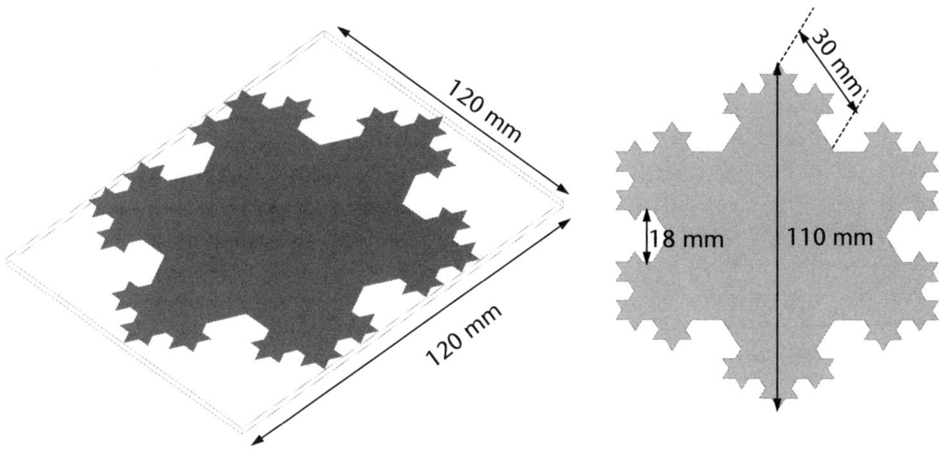

Figure 11.32 Koch snowflake fractal microstrip antenna geometry and design dimensions.

11.6.5 UHF Hilbert Antenna Array for Partial Discharge (PD) Detection

PD is a common local dielectric breakdown phenomenon of an electrical insulation system, and it often takes place in gas-insulated switchgears, power transformers, switches cabinets, and power cables [31, 32]. A PD is normally accompanied by EM and acoustic emissions [31]. A PD online monitoring system, as shown in Figure 11.33, is an effective method in inspecting insulation defects and fault diagnosis in electrical apparatus. Compared with ultrasonic detection and pulse current detection, the UHF method possesses the advantage of high sensitivity and strong anti-interference capability. The design of the UHF antenna is the key technology in the UHF detection approaches because the performance of UHF antenna determines the ability of the PD online detection systems. In general, the frequency of EM waves arising from PD varies from 300 MHz to 3 GHz, and the ratio bandwidth reaches up to 10:1. To cover the entire frequency band, several broadband antennas (e.g., planar equiangular spiral antenna, Hilbert fractal antenna, horn antenna, and biconical log-periodic antenna) are employed. However, these antennas suffer from

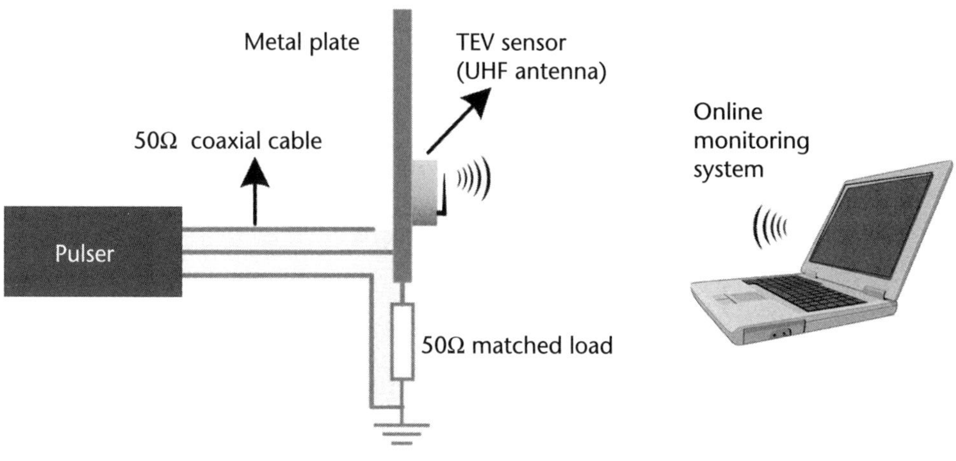

Figure 11.33 PD detection system and detection process.

excessive size, and it is inconvenient for the internal installation. The space-filling property of fractal antenna helps to lengthen the effective current path within the limited area and the self-similar property aids the generation of multiple resonant modes [33].

A UHF Hilbert antenna array for PD monitoring is designed. The first four steps in the construction of the Hilbert curve are shown in Figure 11.34. The Hilbert curve is a space-filling fractal curve that is self-avoiding (i.e., has no intersection points). The performance of the Hilbert fractal antenna can be affected by numerous factors, such as the order, side dimension, width of the conductor, and thickness of the substrate and feed point. The increasing number of order corresponds to the increasing number of resonant frequencies of a Hilbert fractal antenna and decreasing first resonant frequency.

Considering the size and resonant frequencies of the Hilbert fractal antenna, the optimal design is based on the third-order Hilbert curve, as shown in Figure 11.35. The Hilbert fractal antenna is designed synthetically through simulations. The simulation model includes conductor patch, substrate, ground plane, and probe feed. The conductor patch is filled with the third Hilbert curves, and the material of the substrate is FR4 epoxy with a dielectric constant of 4.4. The total

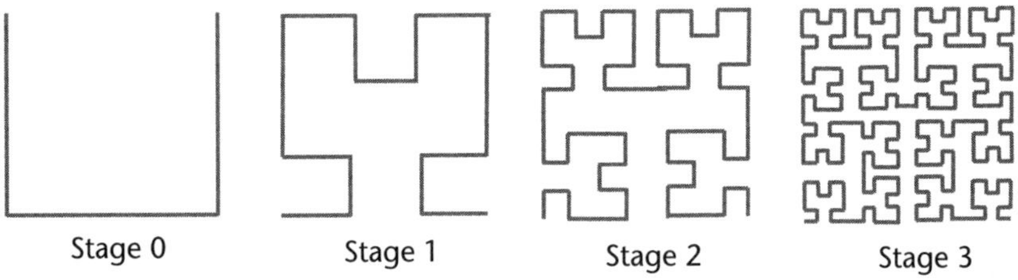

Figure 11.34 The first four stages in the construction of a Hilbert curve.

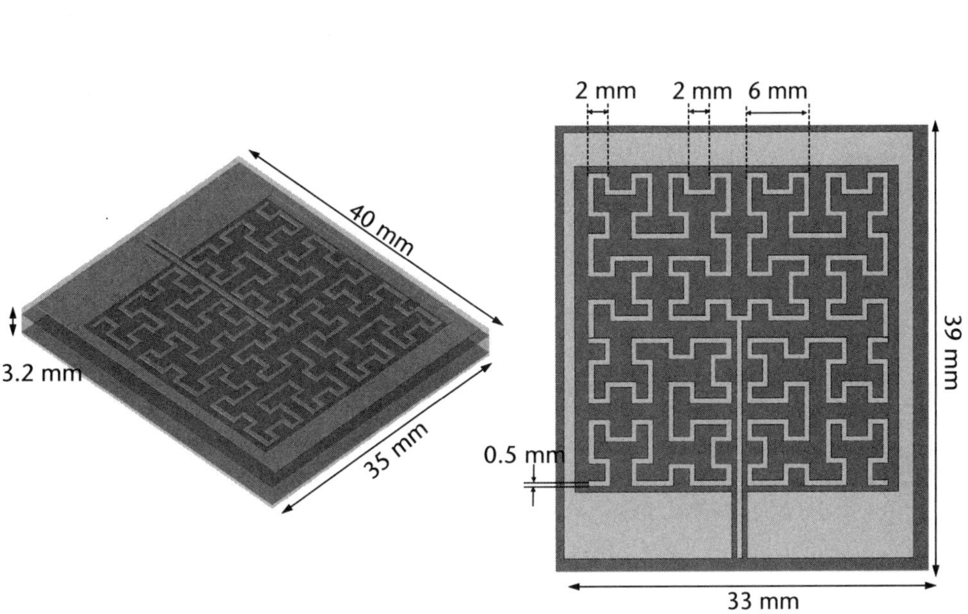

Figure 11.35 Microstrip Hilbert antenna structure and design dimensions.

antenna size is 40 mm × 35 mm, which is very compact in the 2.4-GHz frequency band. The impedance bandwidth of the Hilbert fractal antenna is narrow in the range of 2 GHz to 2.41 GHz as shown in Figure 11.36. The gain of the antenna at 2.4 GHz is 3 dBi.

11.7 UWB Antenna

The commonly used frequency band allocated for UWB applications is 3.1–10.6 GHz, although lately there appear to be systems with operation bands started from 300 MHz and sometimes up to 20 GHz, depending on the fields of use and the applications. Due to the FCC and ITU regulations, the UWB assigned an unlicensed band in the range of 3.1–10.6 GHz with the transmitted power emission limitation of –41.3 dBm/MHz, and the other frequency range with as low power as –75 dBm/MHz as shown in Figure 11.37.

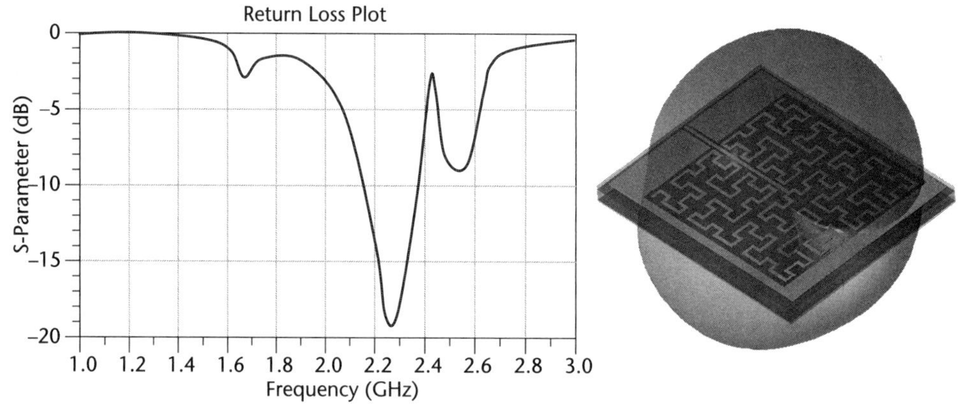

Figure 11.36 Antenna return loss and radiation pattern plot.

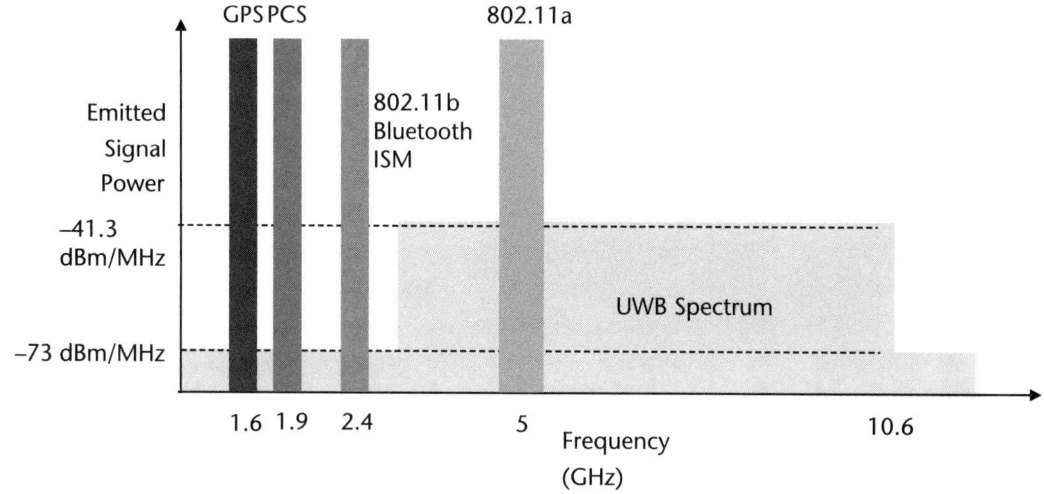

Figure 11.37 UWB spectrum.

11.7.1 UWB Hexagonal Printed Monopole Antenna for GPR Systems

Ground-penetrating radar (GPR) systems are used for subsurface investigations. These are used for the detection of objects buried beneath the Earth's surface such as landmines, pipes, cables, and hidden tunnels. GPR uses a high-frequency microwave signal that is transmitted into the ground, and reflected signals are returned to the receiver as shown in Figure 11.38. The reflected signals are processed to get information. The transmit and receive antennas are one important component in any GPR system [34].

The desired features of a GPR antenna include UWB operation capability and good impedance matching. One of the problems found in the impulse radar system is the generation of short pulses. Stepped frequency continuous wave (SFCW) ground penetration radar (SFCW-GPR) normally use inverse discrete Fourier transform (IDFT) to transform the frequency-domain data into time-domain pulse data. For UWB operation, some of the required properties of an antenna are compact size, wide impedance bandwidth, unidirectional radiation pattern, low-power operation, ease of manufacturing, and unipolar configuration. Printed monopole antennas with a very large bandwidth are a popular antenna design choice as they meet most of the above requirements. For SFCW-GPR applications, it is also vital that the antenna has a flat transfer function in the frequency domain.

A hexagonal UWB antenna is designed for transmitting a monocycle pulse generated by a 100–1,000-MHz SFCW system. As per requirement, the antenna has a low VSWR (≤ 2), the flat transfer function in the entire bandwidth, and exhibits very small late-time ringing (in the time domain).

A hexagonal monopole UWB antenna for a 3–10-GHz frequency band is designed as shown in Figure 11.39. For UWB antennas, instead of resonance or operating frequency, the lower band-edge frequency and total bandwidth become the design parameters. The lower frequency depends mainly on the maximum height of the substrate, while the bandwidth of the antenna depends on how the impedance

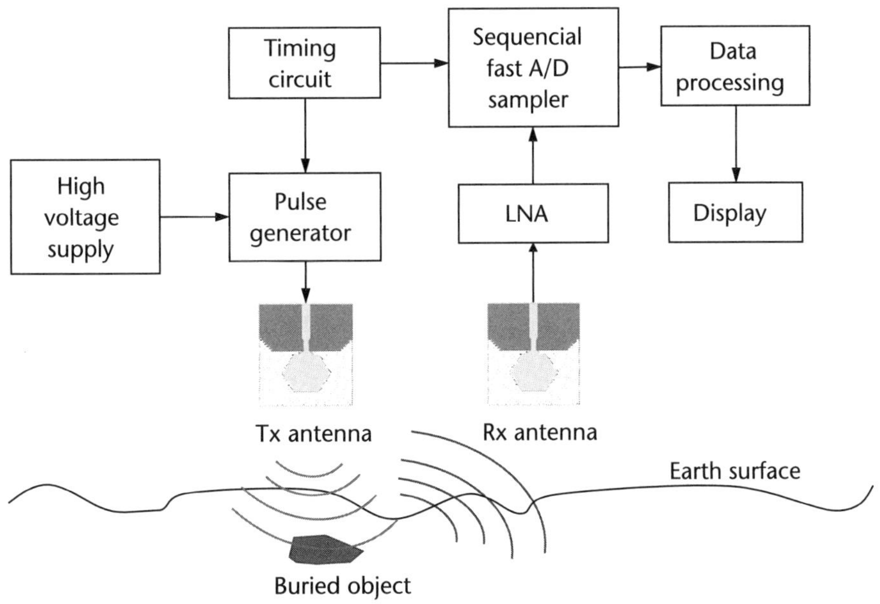

Figure 11.38 SFCW-GPR system.

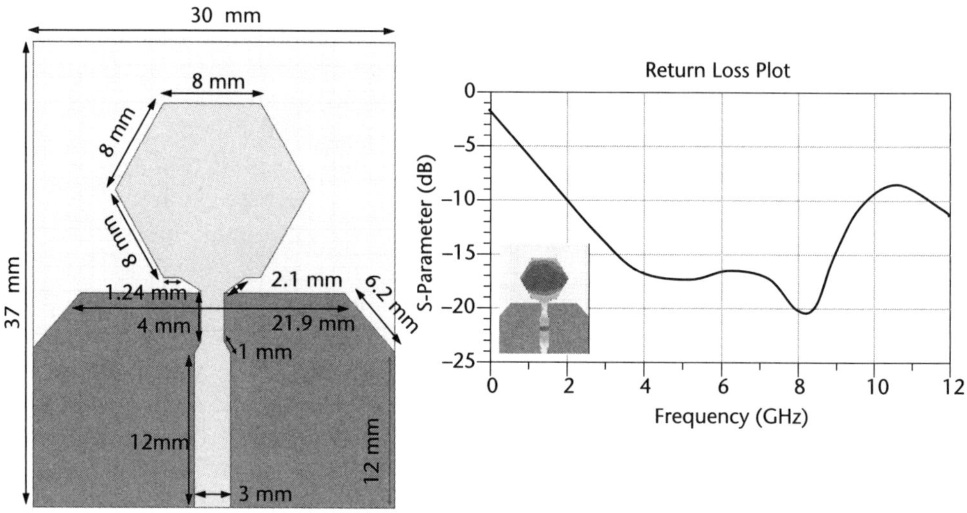

Figure 11.39 UWB hexagonal patch antenna design and return loss plot.

of various modes of the patch is matched with the feed line. The substrate used is FR4 of a 1.6-mm thickness and a dielectric constant (ε_r) of 4.4. The antenna size is 30 mm × 37 mm. In this design, truncated ground structures are used. Usually to reduce the size of the antenna without affecting the bandwidth, gain, and other antenna parameters, various methods such as using high-permittivity dielectric material and magneto-inductive waveguide loaded slots on the patch can be used. The antenna performance is analyzed using the FEM simulation technique. The simulation results are shown in Figure 11.39. Antenna return losses are lower than –10 dB for the entire frequency band (2.95–10 GHz) and the VSWR is also less than 2 for the entire frequency. The bandwidth of the antenna meets the requirement of the 100–1,000-MHz SFCW-GPR application.

11.7.2 Broadband Rectangular Patch

To improve the antenna bandwidth, several techniques are used such as partial ground plane feeding structure, inserting a capacitor between the patch and the ground plane, inserting a chip resistor between the patch and the ground plane, and inserting a chip inductor that broadens impedance bandwidth. A broadband monopole rectangular patch antenna with the partial ground plane is designed as shown in Figure 11.40. The antenna design is optimized for the frequency range from 1.9 GHz to 9 GHz. This design can be used for various applications operating in this frequency range.

A Teflon-based dielectric substrate with a relative permittivity ε_r of 2.65, a thickness of 0.8 mm, a length of 15 mm, and a width of 13 mm is used. The radiating patch element and ground plane are on the top and bottom of the dielectric substrate, respectively, and are made of copper material with a thickness t of 0.035 mm and a conductivity σ of 5.96e 7 s/m. The ground plane length and width are 28 mm × 9 mm. To achieve 50Ω output impedance matching with the SMA connector, a transmission line fed with a width of 2 mm and a length of 13 mm is used. The

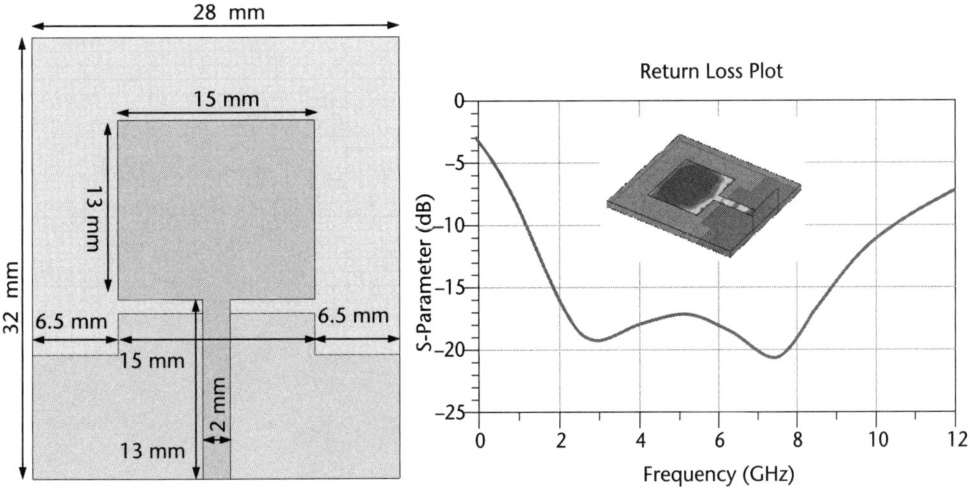

Figure 11.40 Broadband rectangular patch antenna design and simulated return loss and E-field plots.

antenna 2-D radiation pattern and gain versus frequency plots are shown in Figure 11.41.

11.7.3 Slot-Loaded Patch

One technique to improve the bandwidth of the microstrip patch antenna is to cut a half-wavelength-long slot in the patch at a desired resonant frequency as shown in Figure 11.42. Slots are embedded in the printed patch. Their dimensions and positions are properly selected to the first two broadside-radiation modes of the patch to be perturbed such that their resonance frequencies get close to each other to form a wide impedance bandwidth [1, 2]. Slots cut into the patch are of different types (e.g., commonly used slots are the E-slot, U-slot, and H-slot). The antenna performance of the U-slot patch antenna is shown in Figure 11.43.

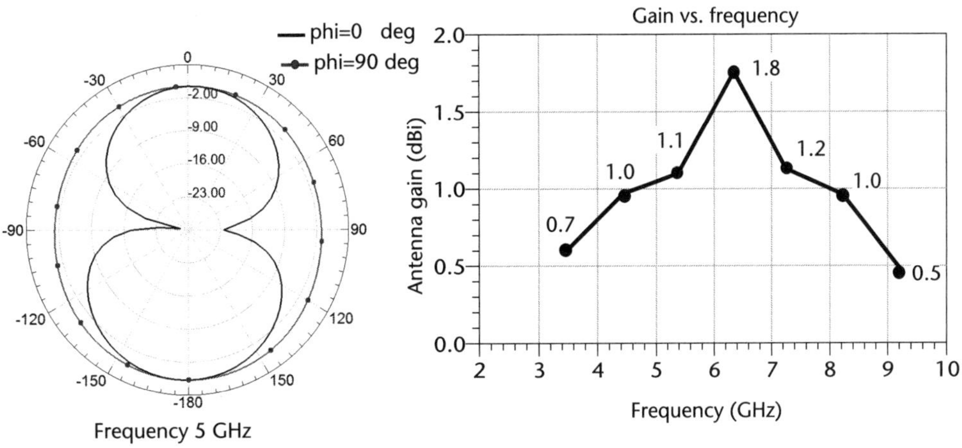

Figure 11.41 The 2-D radiation pattern at 5 GHz and gain versus frequency plot of the broadband rectangular patch.

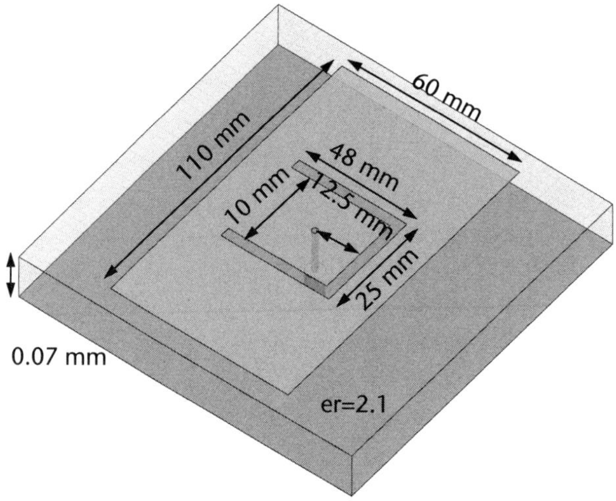

Figure 11.42 U-Slot patch antenna structure and design dimensions.

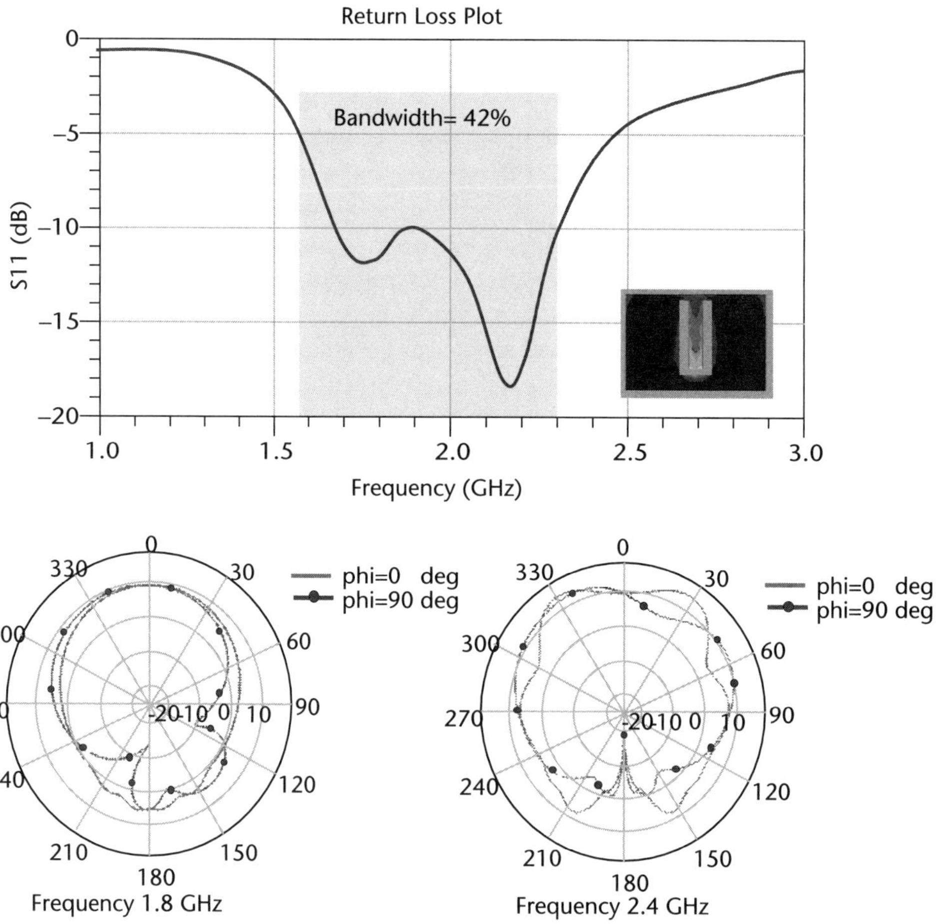

Figure 11.43 U-slot patch antenna simulated return loss and 2-D cut pattern plot at 1.8 GHz and 2.4 GHz.

11.7.4 Wideband Spiral Antenna for Software-Defined Radio-Based Radar

Software-defined radio (SDR)-based radar finds many applications in modern electronics systems such as air traffic radar controls, real-time processing FMCW radar for target detection, and even space communications. SDR has many advantages such as a multipurpose system with the ability to reuse the hardware and signal processing algorithms, low cost, and faster development. In this system, a wideband antenna is applied to scan over all the frequencies. The spiral antenna belongs to the class of frequency-independent antennas and has a radiation pattern, impedance, and polarization that remain virtually unchanged over a large bandwidth. A perfect spiral antenna that radiates a circularly polarized wave will not receive the wave reflected from the interface, because it is predominantly cross-polarized. A broadband two-arm logarithmic spiral antenna as shown in Figure 11.44 is designed for 3–6-GHz frequency band.

The spiral antenna consists of two identical arms that are shifted by 180° with respect to each other. The shape of the spiral arm is like a filling of two identical curves shifted by 90°. The curve is described by the function [29]:

$$r = r_0 e^{a(\varnothing + \varphi)} \tag{11.11}$$

where r_0 is the starting distance at $\varphi = 0°$, a determines the increasing rate of radius r, $\varnothing$ is the variable angle, and φ determines rotation of the curve. The outer antenna curve is set up for $\varphi = 0°$ and inner antenna curve is set up for $\varphi = 90°$.

11.7.5 UWB Vivaldi Antennas

The important parameters for the UWB antennas are the impedance bandwidth, the radiation efficiency, usable gain, the directivity, the beamwidth of the main lobe, and the minimization of the sidelobes and back lobes. Vivaldi antennas are the microstrip type of traveling-wave UWB antennas, which carries all the benefits

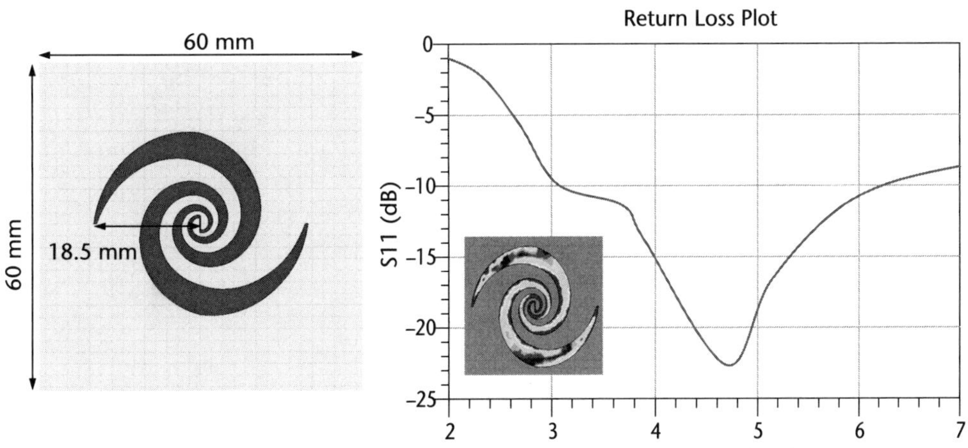

Figure 11.44 Two-arm logarithmic spiral antenna with self-complementary pattern and antenna return loss plot.

that microstrip designs have. Like all other microstrip antennas, the Vivaldi antenna design consists of two parts: the radiator (the part of the antenna body shape responsible for generating radiation by the currents flowing through it), and the matching element (the part of antenna that makes impedance transition from the radiator to the system impedance for which the antenna is designed as shown in Figure 11.45). The radiator part also acts as matching between the transmission line and the free space. The radiator part of antenna is made of copper layers or any other conducting material. The radiator shape is limited by the surrounding curves. One curve, called the inner or radiation curve, looks inside towards the body of the substrate, and the second one is placed close to the side edge of the substrate and can be called the outer or directivity curve. The radiator plates are followed with the transmission line, which provides impedance matching between radiator plates and the system. Radiation curves are the main contributors of antenna radiation; they act as a load in an antenna circuit and provide radiation resistance R_R. Directivity curves are responsible for radiation beam formation, controlling the main beam width and its direction.

The antenna current distribution at the edges of the conductor plate is much higher than the current in the middle; thus, the curves through the edges take most of the responsibilities for the radiation. The shapes of inner curves are crucial for antenna performance, because they are the main contributors of the radiation. The current strength through the inner curve is the highest.

11.7.5.1 Antipodal Vivaldi Antenna for the GB-SAR System

Subsurface radar including GPR is used to observe and detect anomalies beyond the surface. One of the applications is the through-the-wall radar, where the key objective is the detection of moving targets behind the wall. The monitoring of moving targets as well as static objects behind the wall can be done by ground-based SAR (GBSAR). These radars find applications to inspect the structure of building walls. Damages on building walls are sometimes obvious from the outside; however, in

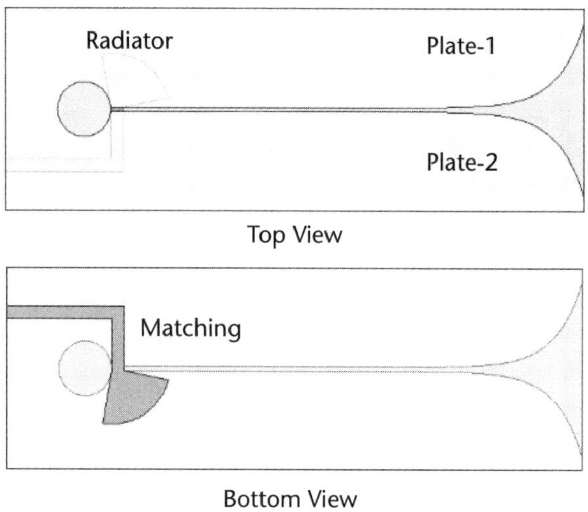

Figure 11.45 Vivaldi antenna design (top and bottom views).

many cases, damages are hidden inside the structures of the walls, and they cannot be found from the outside.

One of the main features of the GB-SAR system is radar polarimetry. To have a higher resolution of the radar image, a very wide frequency range for improving the range resolution is used. However, even with higher resolution, it is not easy to observe very small details of the structures. Radar polarimetry is a technique that can overcome this specific problem. To achicve a wide-frequency range and full polarimetric radar data acquisition, two antenna types, antipodal Vivaldi and spiral antennas, are used. Both these antennas are for wide frequency range operation, but the Vivaldi antenna is for linear polarization, and the spiral antenna is for circular polarization.

A wideband antipodal Vivaldi antenna design to achieve UWB performance is shown in Figure 11.46. The total size of the antenna was 12.4 mm × 12.4 mm. Between different substrate materials, different thicknesses of the substrate, and different shapes of antenna flare, the widest usable bandwidth was achieved for the 3–18-GHz frequency range. The high dielectric material of 6.15 is used for the substrate (RO3006). The main purpose is to improve the usable gain in a frequency. For constructing antenna design, an elliptical flare is used.

11.8 Substrate-Integrated Waveguide Microstrip Patch Antenna for Radar Applications

In the early 2000s, SIW was invented, and it offers an approach of the highly efficient driving patch antenna and array for linear polarizations [35–38] and circular polarizations [39]. In high-frequency applications, microstrip devices are not very efficient because of losses; also, microstrip device manufacturing requires very tight tolerances. Waveguide devices are less lossy at high frequency and less lossy compared to microstrip; however, these are bulky, and their manufacturing process is difficult. Therefore, a new concept emerged: SIW, which is a transition between dielectric-filled waveguide (DFW) and microstrip. The DFW is converted to SIW with the help of vias for the side walls of the waveguide. Another potential use case of

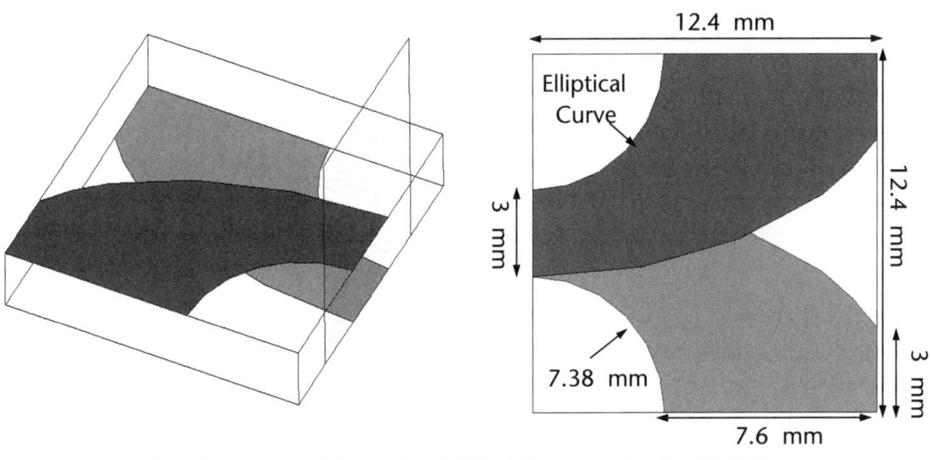

Figure 11.46 Design dimensions of the antipodal Vivaldi antenna for the GB-SAR system.

SIW-based antenna is the integration of microwave active devices with an antenna on the same board to reduce the overall physical volume and cost in the RF front-end subsystems. There has been growing interest in implementing SIW technology in active circuits and complete systems, including active integrated antennas.

SIW technology is basically a hybrid of microstrip and DFW technologies. Starting with a PCB substrate, the top and bottom metal layers provide two of the waveguide walls. Then two parallel rows of vias are placed, forming the side walls of the waveguide. Microstrip antennas designed with SIW technology have excellent performance because they suppress the propagation of surface waves, increase the bandwidth, and decrease both end-fire radiation and cross-polarization radiation. SIW-based microstrip cavity-backed antenna design is shown in Figure 11.47. The cavity-backed antenna has better performance because of the suppression of the propagation of the surface wave. This increases the bandwidth, decreases end-fire radiation, decreases cross-polarization radiation, and increases the applicable frequency range.

Table 11.3 shows the dimensions of the antenna. Arlon 25N material with a dielectric constant of 3.38 and a loss tangent of 0.025 is used as antenna substrate. To design and analyze an antenna, a full-wave FEM solver is used before fabrication. Antenna length and slot length control the resonance frequency. The physical length and width of the SIW square patch antenna is 6.6 mm, while the via-hole diameter is 1 mm and the inter-via distance is 2 mm. A patch with a 2-mm slot length resonates at 10.15 GHz as shown in Figure 11.48. Resonance frequency shifted at a lower frequency if there was an increase in antenna length and slot length.

The electric field for the TE10 mode on the SIW antenna at 10.1 GHz is shown in Figure 11.48. The simulation result shows that the magnitude of E-field is concentrated in the center of the SIW cavity. In this condition, the array of via-holes of the SIW is used as a boundary to prevent the EM fields from escaping from the SIW cavity.

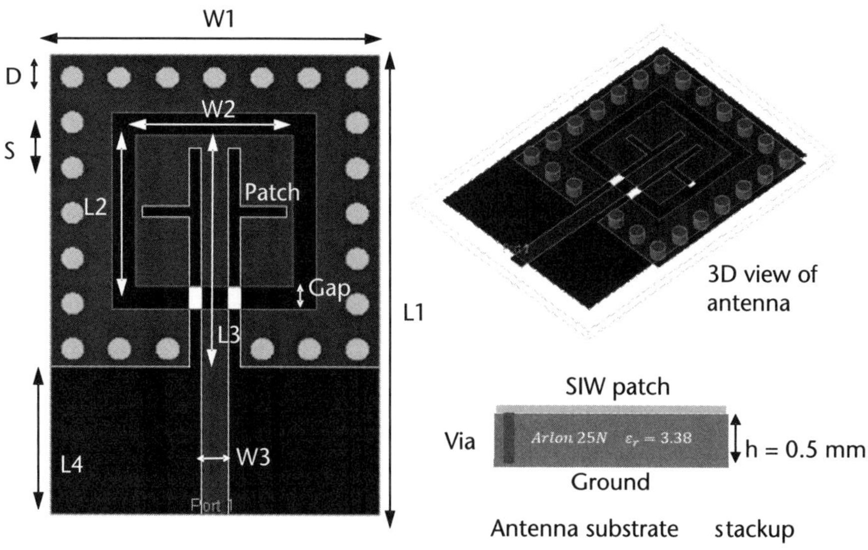

Figure 11.47 SIW patch antenna structure, substrate stack-up, and design dimensions.

Table 11.3 Antenna Dimensions

Symbol	*Value (mm)*	*Symbol*	*Value (mm)*
W1	14	W3	1.15
L1	21	L4	7
D	1.1	Slot L	2
S	2	Slot W	0.5
L3	9.6	W2	6.6
Gap	1	L2	6.6

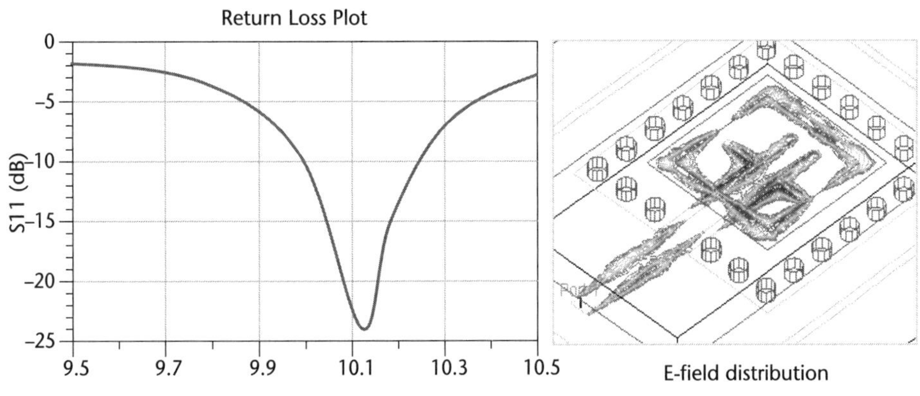

Figure 11.48 SIW patch antenna return loss plot and E-field for the TE10 mode.

Figure 11.49 shows the radiation pattern for the single-mode antenna at 10.15 GHz. The pattern represents the main lobe magnitude of 4.6 dBi at a 5.0° direction from the origin point. The $\varphi = 0°$ cut radiation pattern at 10.15 GHz of the fabricated and simulated design is shown is Figure 11.49. SIW structures show propagation characteristics like the ones of classical rectangular waveguides, including the field pattern and the dispersion characteristics.

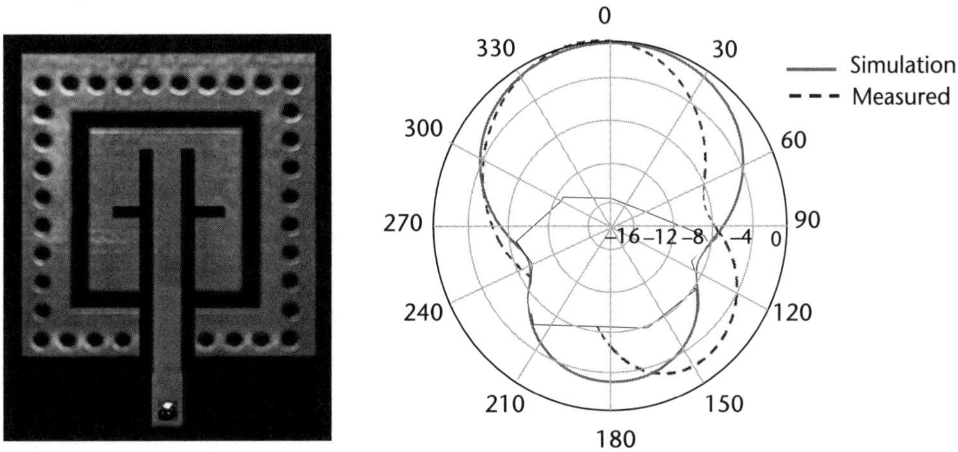

Figure 11.49 Fabricated SIW patch antenna and 3-D and $\varphi = 0°$ cut patterns at 10.15 GHz.

11.9 Summary

In this chapter, various broadband and UWB designs and applications were discussed. The design of broadband antennas based on EBG, PBG, FSS fractal, and many other latest technologies were presented with antenna performance results. Antenna geometries, materials, design solutions, and numerical tools adopted for the analysis and design have been explained. Several classes of broadband antennas have been identified. The EM performances, the dimensions, and the materials, as well as the design solutions useful to improve the antenna bandwidth of printed monopoles, metamaterials, and fractals, have been covered. The advent of the broadband system in the wireless communication area has demanded the design of antennas that must operate effectively over a wide range of frequencies. The widespread commercial deployment of UWB systems has sparked a new interest in UWB antennas.

References

[1] Kumar, G., and K. P. Ray, *Broadband Microstrip Antennas*, Norwood, MA: Artech House, 2002.

[2] Gardiol, F. E., and J. -F. Zurcher, *Broadband Patch Antennas*, Norwood, MA: Artech House, 1995.

[3] Wong, K. L., *Compact and Broadband Microstrip Antennas*, New York: Wiley, 2002.

[4] Pozar, D. M., and D. H. Schaubert, *Microstrip Antennas: The Analysis and Design of Microstrip Antennas and Arrays,* New York: IEEE Press, 1995.

[5] Bhartia, P., et al., *Microstrip Antenna Design Handbook*, Norwood, MA: Artech House, 2000.

[6] Veselago, V. G., "The Electrodynamics of Substances with Simultaneously Negative Values of ε and μ," *Soviet Physics Uspekh.*, Vol. 10, No. 4, January-February 1968, p. 509.

[7] Pendry, J. B., et al., "Low Frequency Plasmons in Thin Wire Structures," *J. Phys.: Condens. Matter*, Vol. 10, March 1998, p. 4785.

[8] Smith, D. R., and N. Kroll, "Negative Refractive Index in Left Handed Materials," *Physical Review Letters*, Vol. 85, No. 14, October 2000, pp. 2933–2936.

[9] Ziolkowski, R. W., "Design, Fabrication, and Testing of Double Negative Metamaterials," *IEEE Transactions on Antennas and Propagation*, Vol. 51, No. 7, July 2003, pp. 1516–1529.

[10] Radisic, V., Y. Qian, and T. Itoh, "Broadband Power Amplifier Using Dielectric Photonic Bandgap Structure," *IEEE Microwave Guided Wave Lett.*, Vol. 8, January 1998, pp. 13–14.

[11] Kesler, M. P., J. G. Maloney, and B. L. Shirley, "Antenna Design with the Use of Photonic Bandgap Materials as All Dielectric Planar Reflectors," *Microwave Opt. Tech. Lett.*, Vol. 11, No. 4, March 1996, pp. 169–174.

[12] Sievenpiper, D., et al., "High-Impedance Electromagnetic Surfaces with a Forbidden Frequency Band," *IEEE Transactions on Microwave Theory and Techniques*, Vol. 47, 1999, pp. 2059–2074.

[13] Sievenpiper, D. F., "High-Impedance Electromagnetic Surfaces," UCLA, Ph.D. dissertation, 1999.

[14] Toyota, Y., et al., "Stopband Analysis Using Dispersion Diagram for Two-Dimensional Electromagnetic Bandgap Structures in Printed Circuit Boards," *IEEE Microwave and Wireless Components Letters*, Vol. 16, No. 12, 2006, pp. 645–647.

[15] Yang, F., and Y. Rahmat-Samii, "Microstrip Antennas Integrated with Electromagnetic Band Gap (EBG) Structures: A Low Mutual Coupling Design for Array Applications," *IEEE Transactions on Antennas and Propagation*, Vol. 51, No. 10, October 2003.

[16] Yang, L., et al., "A Novel Compact Electromagnetic Band Gap (EBG) Structure and Its Application for Microwave Circuits," *IEEE Transactions on Microwave Theory and Techniques*, Vol. 53, No. 1, January 2005.

[17] Yang, F., and Y. Rahmat-Samii, "Reflection Phase Characterizations of the EBG Ground Plane for Low Profile Wire Antenna Applications," *IEEE Transactions on Antennas and Propagation*, Vol. 51, No. 10, October 2003, pp. 2691–2703.

[18] Hansen, R. C., and M. Burke, "Antennas with Magneto-Dielectrics," *Microwave Opt. Tech. Lett.*, Vol. 26, No. 2, July 2000, pp. 75–78.

[19] Hood, A. Z., T. Karacolak, and E. Topsakal, "A Small Antipodal Vivaldi Antenna for Ultrawide-Band Applications," *IEEE Antennas and Wireless Propagation Lett.*, Vol. 7, 2008, pp. 656–660.

[20] Ying, L., and A. -X. Chen, "Design and Application of Vivaldi Antenna Array Antennas," *8th International Symposium Propagation and EM Theory 2008 (ISAPE2008)*, November 2008, pp. 267–270.

[21] Kiminami, K., A. Hirata, and T. Shiozawa, "Double-Sided Printed Bow-Tie Antenna for UWB Communications," *IEEE Antennas and Wireless Propagation Lett.*, Vol. 3, No. 1, 2004, pp. 152–153.

[22] Ito, Y., et al., "Unidirectional UWB Array Antenna Using Leaf-Shaped Bowtie Elements and Flat Reflector," *IEE Electron Lett.*, Vol. 44, No. 1, January 2008, pp. 9–11.

[23] Naghshvarian-Jahromi, M., "Novel Wideband Planar Fractal Monopole Antenna," *IEEE Transactions on Antennas and Propagation*, Vol. 56, No. 12, December 2008, pp. 3844–3849.

[24] Chen, S. Y., P. H. Wang, and P. Hsu, "Uniplanar Log-Periodic Slot Antenna Fed by a CPW for UWB Applications," *IEEE Antennas and Wireless Propagation Lett.*, Vol. 5, No. 1, December 2006, pp. 256–259.

[25] Pantoja, R., A. Sapienza, and F. M. Filho, "A Microwave Printed Planar Log-Periodic Dipole Array Antenna," *IEEE Transactions on Antennas and Propagation*, Vol. AP-35, October 1987, pp. 1176–1178.

[26] Calmon, A., G. Pacheco, and M. Terada, "A Novel Reconfigurable UWB Log-Periodic Antenna," *IEEE Antennas and Propagation Society International Symposium 2006*, July 2006, pp. 213–216.

[27] Sung, Y. J., "Bandwidth Enhancement of a Wide Slot Using Fractal-Shaped Sierpinski," *IEEE Transactions on Antennas and Propagation*, Vol. 59, No. 8, August 2011, pp. 3076–3079.

[28] Mandelbrot, B. B., *The Fractal Geometry of Nature*, New York, W. H. Freeman, 1983.

[29] Rumsey, V. H., *Frequency Independent Antennas*, New York: Academic Press, 1966.

[30] Sierpinski, W., "Sur une courbe dont tout point est un point de ramification," *C. R. Acad.*, Paris, Vol. 160, No. 302, 1915.

[31] Li, J., et al., "UHF Stacked Hilbert Antenna Array for Partial Discharge Detection," *IEEE Transactions on Antennas and Propagation*, Vol. 61, No. 11, November 2013, pp. 5798–5801.

[32] Hai-Feng, Y., et al., "Development of Multi-Band Ultra-High-Frequency Sensor for Partial Discharge Monitoring Based on the Meandering Technique," *IET Sci., Meas., Technol.*, Vol. 8, No. 5, September 2014, pp. 327–335.

[33] Li, T., et al., "Development Simulation and Experiment Study on UHF Partial Discharge Sensor in GIS," *IEEE Transactions on Dielectr. Electr. Insul.*, Vol. 19, No. 4, August 2012, pp. 1421–1430.

[34] Daniel, D. J., *Ground Penetrating Radar*, 2nd ed., IEE Radar Sonar, Navigation and Avionics Series, London, UK: Institution of Engineering and Technology, 2004.

[35] Wahab, W. M. A., S. Safavi-Naeini, and D. Busuioc, "Wide-Bandwidth 60-GHz Aperture-Coupled Microstrip Patch Antennas (MPAs) Fed by Substrate Integrated Waveguide (SIW)," *IEEE Antennas Wireless Propag. Lett.*, Vol. 10, 2011, pp. 1003–1005.

[36] Mikulasek, T., et al., "2 × 2 Microstrip Patch Antenna Array Fed by Substrate Integrated Waveguide for Radar Applications," *IEEE Antennas Wireless Propag. Lett.*, Vol. 12, 2013, pp. 1287–1290.

[37] Ghassemi, N., and K. Wu, "High-Efficient Patch Antenna Array for E-Band Gigabyte Point-to-Point Wireless Services," *IEEE Antennas Wireless Propag. Lett.*, Vol. 11, 2012, pp. 1261–1264.

[38] Li, Y., and K. M. Luk, "60-GHz Substrate Integrated Waveguide Fed Cavity Backed Aperture-Coupled Microstrip Patch Antenna Arrays," *IEEE Transactions on Antennas and Propagation*, Vol. 63, No. 3, March 2015.

[39] Li, Y., et al., "Axial Ratio Bandwidth Enhancement of 60-GHz Substrate Integrated Waveguide-Fed Circularly Polarized LTCC Antenna Array," *IEEE Transactions on Antennas and Propagation*, Vol. 60, No. 10, October 2012, pp. 4619–4626.

Selected Bibliography

Caloz, C., and T. Itoh, *Electromagnetic Metamaterials: Transmission Line Theory and Microwave Applications*, New York: Wiley-IEEE Press, 2005.

Cheype, C., et al., "An Electromagnetic Bandgap Resonator Antenna," *IEEE Transactions on Antennas and Propagation*, Vol. 50, No. 9, September 2002, pp. 1285–1290.

Chung, K. B., and S. W. Hong, "Wavelength Demultiplexers Based on the Superprism Phenomena in Photonic Crystals," *IEEE Transactions on Antennas and Propagation*, Vol. 81, August 2002, pp. 1549–1551.

El Hamdouni, A., et al., "A Novel Design of a CPW-Fed Printed Fractal Antenna for UWB Applications," *2015 Third International Workshop on RFID and Adaptive Wireless Sensor Networks (RAWSN)*, Agadir, 2015, pp. 38–41.

Engheta, N., and R. W. Ziolkowski, *Metamaterials Physics and Engineering Explorations*, New York: Wiley, 2006.

FCC 02-48, Revision of Part 15 of the Commission's Rules Regarding Ultra-Wideband Transmission Systems, First Report & Order, Washington DC, Adopted February 14, 2002, Released April 22, 2002.

Jang, Y. W., "Characteristics of a Large Bandwidth Rectangular Microstrip-Fed Inserted Triangular Patch in a Circular Slot Antenna," *Microwave Journal*, Vol. 45, No. 5, May 2002, pp. 288–298.

Kaneko, S., et al., "Detecting Characteristics of Various Type Antennas on Partial Discharge Electromagnetic Wave Radiating Through Insulating Spacer in Gas Insulated Switchgear," *IEEE Transactions on Dielectrics and Electrical Insulation*, Vol. 16, No. 5, October 2009, pp. 1462–1472.

Amiri, N., and K. Forooraghi, "Dual-Band and Dual-Polarized SIW-Fed Microstrip Patch Antenna," *IEEE Antenna and Wireless Propagation Letters*, Vol. 13, 2014, p. 1605.

Park, J. S., et al., "A Novel Resonant Microstrip RF Phase Shifter Using Defected Ground Structure," *30th European Microwave Conf.*, France, 2000, pp. 72–75.

Radisic, V., et al., "Novel 2-D Photonic Bandgap Structure for Microstrip Lines," *IEEE Microw. and Guided Wave Lett.*, Vol. 8, No. 2, 1998, pp. 69–71.

Sabban, A., "New Fractal Compact Printed Antennas," *2017 IEEE International Symposium on Antennas and Propagation & USNC/URSI National Radio Science Meeting*, San Diego, CA, 2017, pp. 2197–2198.

Yang, L., et al., "A Novel Compact Electromagnetic Band Gap (EBG) Structure and Its Application for Microwave Circuits," *IEEE Transactions on Microwave Theory and Techniques*, Vol. 53, No. 1, January 2005.

CHAPTER 12

Modern Printed Antennas Design Trends and Applications

12.1 Introduction

The world is undergoing a major wireless revolution in terms of both mobile technology and wireless systems that are also leading new trends and technologies in antenna design world. The wireless communication industry is bringing new products with advanced features to support wireless services anytime and anywhere. The fast development of the communication technology such as wearable devices, cell phone technology, 5G, and personal area networks (PANs) for remote retrieval and monitoring of surrounding information has demanded that antennas should be able to operate with multiband characteristics in wireless communication devices. There is tremendous research interest in designing the compact size of antennas to offer more space to integrate other electronic components. All these needed attributes push the antenna researcher into challenges in designing new broadband, multifrequency, compact, low-cost, and reconfigurable antennas for the new generation of mobile communication devices, as well as for UWB communication systems. This chapter provides an idea of the latest antenna design trends for modern wireless communications.

In this chapter, printed antennas used in many wireless communication devices are discussed to facilitate the idea of current research and development trends towards different antenna applications. RFID antennas and its applications are covered in Section 12.2. Low-profile RFID tag and reader antenna design is presented to cover the operation bands of low-frequency and high-frequency bands. For navigation applications, antenna designs are presented in Section 12.3. Section 12.4 describes the features of the most recent antennas for medical applications. In addition, this chapter also describes wearable antennas. The wearable antennas find potential applications in healthcare, entertainment, identification systems, sport, smart home, and space.

12.2 RFID Antennas

RFID technology is one of the major methods for object identification purposes. It provides a unique identifier for objects and, just like a bar code or magnetic strip, the RFID device must be scanned to retrieve the identifying information [1–6]. RFID systems are also used in health monitoring with wearable devices [2]. It is also an important technology in the Internet of Things (IoT) that communicate with different sensors. The integration of RFID systems in production chains and process management is common. Antenna designs for RFID tags and readers are very specific and are a challenging topic. This section presents some advanced antenna design methods that can help RFID designers to reach good performance in many cases. Several realistic examples are considered with design and antenna performance data.

12.2.1 RFID System Working Principle

The principle of operation of a passive RFID system is illustrated in Figure 12.1. It typically contains three parts: a scanning antenna, also called the RFID reader, a transponder that is an RFID tag, and a transceiver with a decoder to interpret the data. The tag consists of an antenna and an application specific integrated circuit (ASIC) chip.

Both tag and reader exhibit complex impedances Z_a and Z_c, respectively. The RFID tag can operate in the interrogation zone of the reader. When it receives the modulated signal with sufficient power from the reader (downlink), the tag switches the chip impedance between two states Z_c^0 and Z_c^1. At these two impedance states, the RFID tag presents a specific complex RCS. The tag generates the information back by varying its input impedance and thus modulating the backscattered signal (uplink).

In most of the RFID systems, tags are attached to all items that need to be tracked. These tags are made from a small tag chip that is connected to an antenna. The tag chip contains a memory module that stores the electronic product code (EPC) and other variable information so that it can be read and tracked by RFID readers anywhere. An RFID reader system is a network-connected device (fixed or mobile) with an antenna that sends power as well as data and commands to the tags.

Fundamentally, RFID devices are categorized as either active or passive. Active RFID uses tags that need a power source for operation. They are either connected to a battery or some other power source. Having a battery embedded on the RFID system makes the product bulky, which is why active tags are impractical for retail trade. As compared to the active ones, passive tags do not require a power source or battery for operation. They absorb the EM energy propagated from a reader's

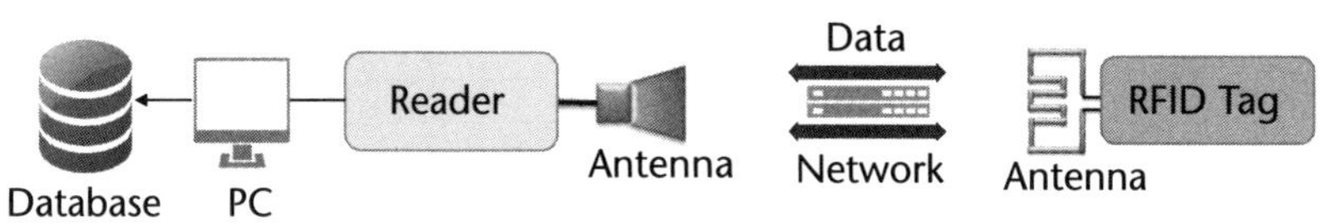

Figure 12.1 Block diagram of a typical RFID system.

RF field to supply all the needed power to operate. Passive tags are commonly used because these are cost-effective and small enough to fit into a practical adhesive label.

12.2.2 RFID Frequency Band Allocation

There are many different RFID frequency bands or RFID frequencies that are used around the globe listed in Table 12.1.

12.2.3 RFID Antennas

For any RFID antenna, there are two operating regions: near field and far field depending on the communication distance between the reader and the tags. Normally, if an antenna has a small loop-shaped antenna in the middle, it usually has near-field capabilities. Many RFID antennas use a near-field coupling mechanism (i.e., inductive coupling). When RFID antennas work based on near-field effects, the reader and the tag are relatively closely coupled to each other, which affects the bandwidth and performance of the antenna. In the RFID design, coupling must be considered when designing the antennas. For the RFID antenna design, parameters such as bandwidth, the radiation resistance, efficiency, and Q all need to be considered to

Table 12.1 RFID Frequency Band/Spectrum Allocation

Frequency Band	*Description*	*Typical Range*	*Typical RFID Application*
125–134.2 kHz and 140–148.5 kHz	Low frequency	Up to 0.5m	These frequencies can be used globally without a license; often used for vehicle identification
6.765–6.795 MHz	Medium frequency	Up to 0.8m	Inductive coupling is used
13.535–13.567 MHz	High frequency also called 13.56 MHz	Up to 1m	These RFID frequencies are typically used for electronic ticketing, contactless payment, access control, garment tracking, and so forth
26.957–27.283 MHz	High frequency	Up to 1m	Inductive coupling only, and used for special applications
433 MHz	Ultrahigh frequency (UHF)		These RFID frequencies are used with backscatter coupling for applications such as remote-control car keys
858–930 MHz	UHF	1 to 20 meters	These RFID frequencies cannot be accessed globally, and these are significantly restricted on their use; when they are used, they are often used for asset management, container tracking, and baggage tracking and often in conjunction with Wi-Fi system
2.400–2.483 GHz	SHF		Backscatter coupling (only available in the United States and Canada)
2.446–2.454 GHz	SHF	3m upwards	These RFID frequencies are used for long-range tracking and with active tags, RFID, and automatic vehicle identification (AVI)
5.725–5.875 GHz	SHF		Backscatter coupling, not widely used for RFID

meet the system requirements. RFID antennas are designed to resonate only to a narrow frequency band that is centered on the designated RFID system frequency.

12.2.4 RFID Tag Antenna Design

The role of the tag antenna and its impact on the RFID system performance is very critical. A passive tag is activated shortly by the RF scan of the reader. The generated small electrical current is just enough for transmission of an ID number. Passive RFID tags use an induced antenna coil voltage for operation. The tag antenna not only transmits the radio wave carrying the information stored in the tag, but it also needs to catch the EM wave from the reader to generate energy for the tag operation. Tag antennas gather energy and channel it to the chip to turn it on. Usually, the larger the tag antenna's area, the more energy it will be able to collect and channel toward the tag chip. Tag antennas can be made from a variety of materials; these can be printed, stamped with conductive ink, etched, or even vapor-deposited onto labels. A tag antenna should be compact, low-cost, and easy to fabricate for mass production. In most cases, the tag antenna should have hemispherical coverage or omnidirectional radiation. Usually, the impedance of the tag chip is not 50Ω, so the antenna impedance is realized for the conjugate match with the tag chip impedance to supply the maximum power to the tag chip.

Passive UHF RFID tag antennas are widely used for inventory tracking and sensing in various applications. The tag cost is an important factor in antenna design and it should be as low as possible. From tag cost point of view, chipless approach is an interesting option. Chipless RFID tags, also called Beacon Range Factor (BRF) barcodes, are an emerging area of RFID technology for ultralow-cost RFID applications. Cost and application also play an important role in material selection for tag antenna design.

The meandered dipole antenna is most used tag antenna type in RFID because of the size and tunability requirements. A thin flexible polyester substrate with a dielectric permittivity of 3.5 and a thickness of 2 mils (0.051 mm) is used for the tag antenna presented in this section. The top antenna trace is made of copper and had a thickness of 0.7 mil (0.018 mm). The antenna geometry and design parameters are shown in Figure 12.2 at 915 MHz (860–960 MHz UHF band). The typical tag antenna design specifications are listed in Table 12.2.

A traditional printed dipole antenna size is large, and it will exceed the maximum allowed dimension of the tag. Thus, a meander-formed dipole antenna is designed to decrease the antenna's size as much as possible with respect to the matching impedance. Meandering allows the antenna to be compact and to provide omnidirectional performance in the plane perpendicular to the axis of the meander. A loading bar with the same width as the meander trace is added to have

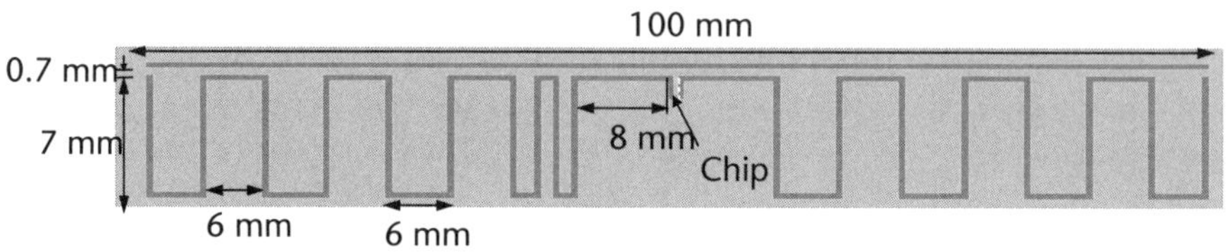

Figure 12.2 UHF RFID tag antenna design dimensions.

Table 12.2 Typical Specifications of RFID Tag Antenna

Parameters	*Typical Value*
Frequency	915 MHz
Maximum dimension	110 mm × 15 mm × 0.05 mm
Maximum area	16.5 cm^2
Gain	1.2 dBi
Range	10m

a better control over the antenna resistance. To provide a better match for the tag chip capacitive impedance, one meandered section is further meandered to obtain additional inductance.

The loaded meander antenna has several key variable parameters: loading bar width, length, spacing, meander step width, and meander step height. These parameters influence antenna gain and impedance, which determine tag resonance, peak range, and bandwidth. The maximum allowed antenna length is decided by the size requirements. The read range is calculated using the Friis free-space formula:

$$R = \frac{\lambda}{4\pi}\sqrt{\frac{P_t G_t G_r \tau}{P_{th}}} \tag{12.1}$$

where λ is the wavelength, P_t is the power transmitted by the reader, G_t is the gain of the transmitting antenna, G_r is the gain of the receiving tag antenna, P_{th} is the minimum threshold power necessary to provide enough power to the RFID tag chip, and τ is the power transmission coefficient given by

$$\tau = \frac{4R_c R_a}{\left|Z_c + Z_a\right|^2}, 0 \le \tau \le 1 \tag{12.2}$$

where $Z_c = R_c + jX_c$ is the chip impedance and $Z_a = R_a + jX_a$ is the antenna impedance.

Thin planar tag antenna is optimized using the MoM-based EM solver. The tag range is computed from (12.1) using simulated values of antenna gain and impedance and known values of chip impedance at the threshold power. A figure of merit (FoM) of RFID tag is calculated, with P_{min} as the minimum power necessary to activate the chip and the area defined as the substrate area of the entire design:

$$FoM = -P_{min}(dBm)10.\log_{10}\left(\frac{Area(cm^2)}{20}\right) \tag{12.3}$$

The simulated result of the antenna for different dipole lengths is shown in Figure 12.3. This antenna can be easily tuned by trimming. Loading bar length and lengths of the meander trace can be varied to obtain optimum reactance and resistance matching. The trimming is obtained by punching holes through the antenna

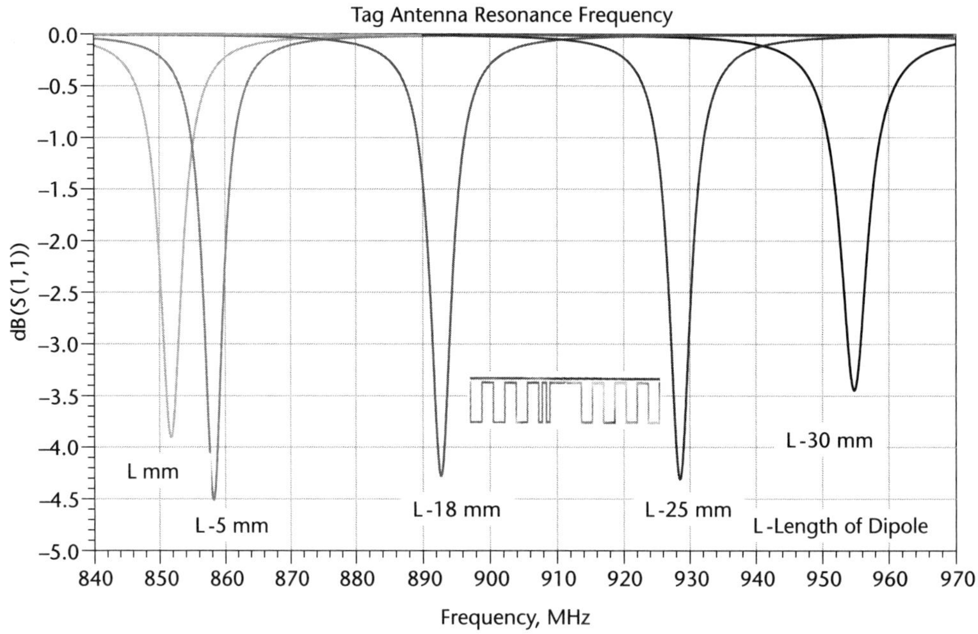

Figure 12.3 Simulated result of a tag antenna for different dipole lengths.

trace at defined locations. Such a tunable design is desirable where it is often needed to provide a solution for an application with a minimum lead time. The antenna resonance frequency shifts at a higher value with a reduced dipole length.

12.2.5 Fat-Tag or Dog-Bone RFID Antenna

Fat-tag RFID dipoles, also known as broadband dipoles, are characterized by having the maximum amount of metal on the tag face. Increasing the amount of metal on the tag is achieved by the etching process or with silver ink printing. By increasing the wire thickness, inductance value is reduced, and the increased metal surface produces a large capacitance.

Since these tags have a lot of metal, they are often best used on high dielectric surfaces like wood, glass, and plastic. Figure 12.4 is showing one such RFID tag antenna designed in UHF (860–960 MHz) band. In this tag, the EPC C1G1 memory chip is used. EPC is used for RFID in today's global, information-hungry trading networks. Gen2 or EPC global Class1 Generation2 defines the physical and logical requirements for a passive-backscatter, RFID system operating in the 860–960-MHz frequency range. EPC C1 Gen2 represents a major step in performance, standardization, and quality. This tag is modeled and simulated using the FEM EM solver. The antenna radiation plot is shown in Figure 12.5.

12.2.6 RFID Printed Tag Antenna on Hydrophobic Paper

Paper-based electronics find application in RFID tag antenna design. Paper is considered one of the best organic substrate candidates for UHF and microwave applications. Inkjet-printing using conductive silver ink is used to print electronics on the

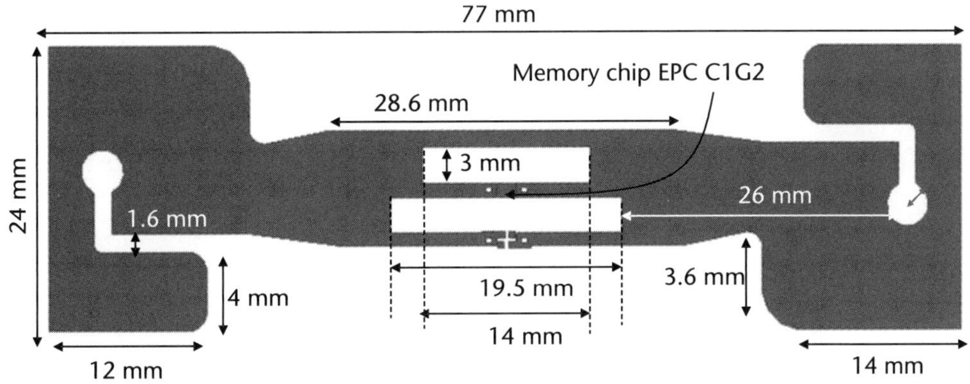

Figure 12.4 Fat-tag RFID antenna design in the UHF band.

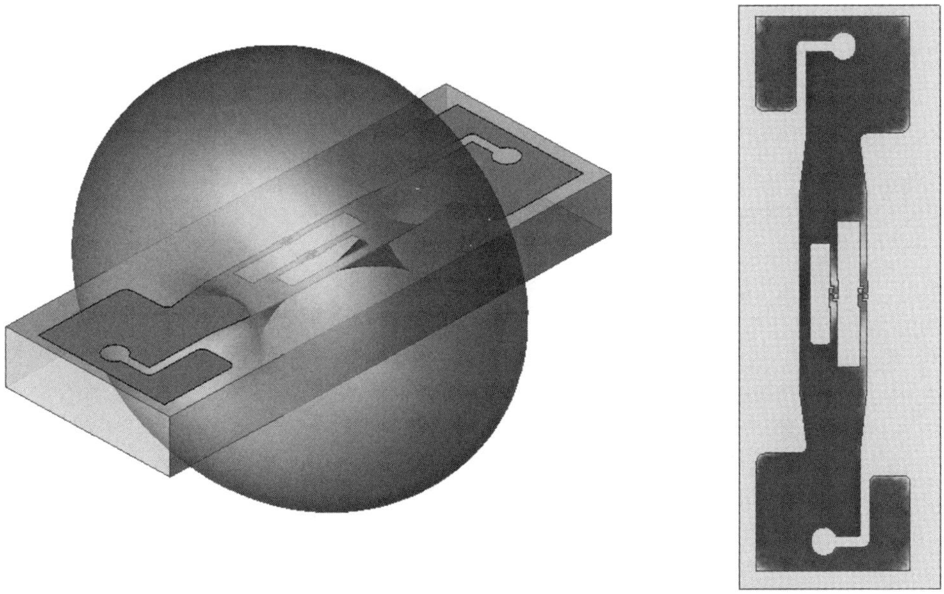

Figure 12.5 Fat-tag RFID antenna radiation pattern and E-field plot.

surface of a paper substrate. In addition, paper can be made hydrophobic by adding certain textiles; thus, mass-fabricating RFID inlays on paper becomes more feasible.

12.2.7 Reader Antenna

The reader antenna produces RF energy to activate the RFID tag, which modulates the information and reflects the remaining energy back to the reader antenna; this is called backscatter coupling. These antennas convert electrical current into EM waves that are then radiated into space where they can be received by a tag antenna and converted back to electrical current. In a near-field RFID system, the performance of the RFID system depends on the type of the reader antenna employed, which is viewed as a near-field reader antenna. Designing a reliable near-field reader antenna is an engineering challenge. In the UHF RFID near-field systems, two coupling mechanisms are employed between tags and reader antennas and capacitive and magnetic coupling. In most of the near-field RFID applications, the interaction between the reader and tag antennas is based on magnetic coupling, in which the magnetic field stores most of the reactive energy. Usually, electrically small loops can produce a strong magnetic field over the near-field region [7, 8]. These magnetic coupling-based antennas can detect large numbers of loop-like tags, but their short reading distance limits the scope of their applications.

The second mechanism is capacitive coupling, in which most of the energy of reader antenna is stored in the electric field. The electric coupling reader antennas can be assessed using the leaking wave of a microstrip transmission line. The reader antenna can be a signal turn or a multiple turn as shown in Figure 12.6. The RFID near-field reader antennas can be designed based on loop structure. An EM E-field plot on a loop-based RFID reader antenna at 12.5 MHz is shown in Figure 12.7.

12.2.7.1 Far-Field Reader Antenna Design

In some actual application scenarios, UHF near-field RFID reader antennas with given radiation patterns are desired. Far-field patterns are often used to achieve a

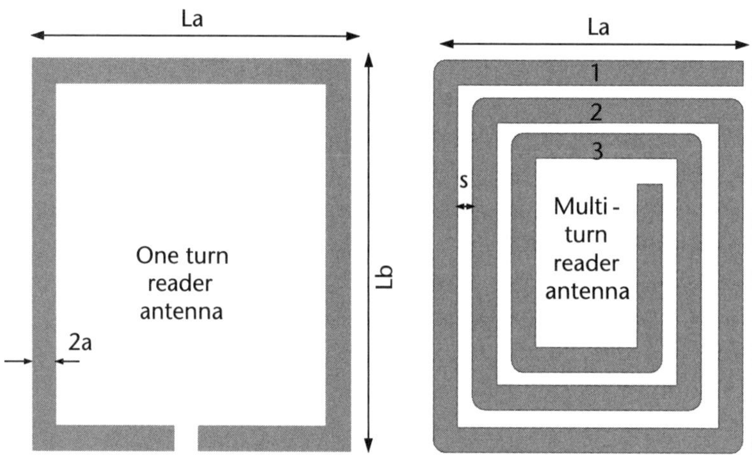

Figure 12.6 Signal-turn and multiple-turn RFID reader antennas.

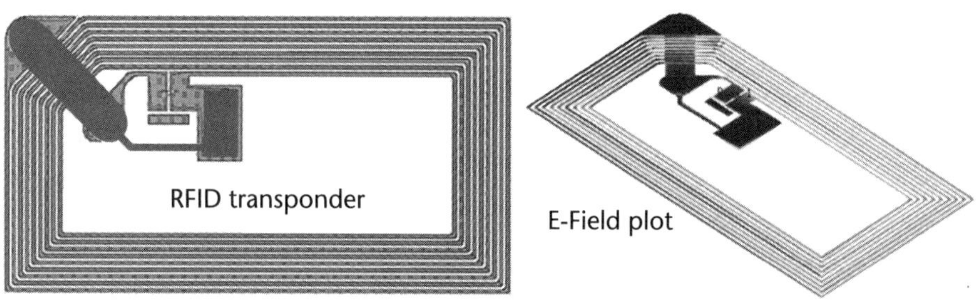

Figure 12.7 RFID reader antenna design at 13.5 MHz.

long-read range; for example, in a medication-monitoring management system, a reader antenna needs to detect near-field tags to realize medical product tracking while reading a far-field tag. Here, the far-field tag might serve as an environment monitor through offering the environment parameters (e.g., temperature) collected by sensors embedded in the far-field tag antenna or work as an identification of the operators.

A far-field-based reader antenna is designed for UHF RFID application at 915 MHz (902–928 MHz frequency range). The design parameters (all dimensions are in millimeters) of this planar antenna is shown in Figure 12.8.

The antenna consists of a microstrip-to-coplanar strip line, Marchand balun, a meandered driven dipole, a folded finite-size ground plane, and a closely coupled parasitic element. This antenna is appropriate for RFID handheld readers [9]. A microstrip-to-coplanar stripline (CPS) Marchand balun, which works as a matching network, is inserted between the feed line and the dipole element. The lengths of the parasitic element and driven dipole are optimized for simultaneously achieving excellent input impedance matching and high antenna front-to-back ratio. The dipole arms are meandered to reduce the antenna size. A tuning stub is added near the microstrip feed line, which is electrically connected to the top ground plane and provides a capacitive loading between the feed line and the ground plane. It further improves the antenna input impedance matching. The antenna is designed on a 1-mm-thick FR4 epoxy substrate with a dielectric constant of 4.4 and a loss

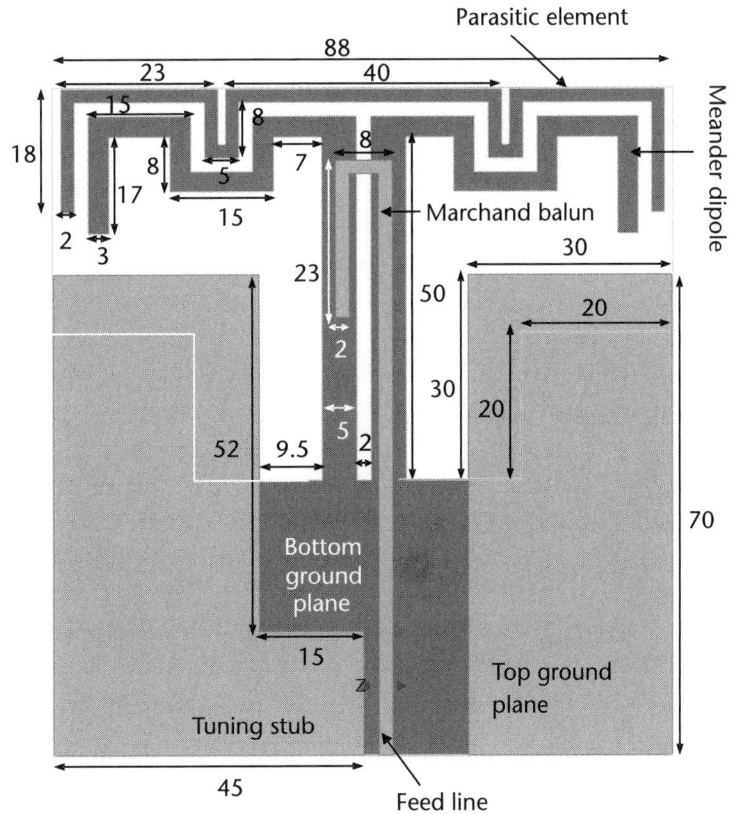

Figure 12.8 Far-field RFID reader antenna dimension (in millimeters) at 915 MHz [9].

tangent of 0.02. Normally, the conductors used in tags are copper, silver, or aluminum. The simulated antenna return loss, E-field plot, and radiation pattern are shown in Figure 12.9. The simulation was performed using the FEM.

The simulated center frequency is at 915 MHz. The slight frequency shift between the results can be mostly attributed to the fabrication tolerance, especially the electrical connection between the top and bottom ground planes. The simulated 10-dB return loss bandwidths are from 885 to 985 MHz. Antenna gain is 3.7 dBi at 915 MHz. Strong near-field coupling between the driven dipole and the parasitic element gives rise to the relatively high current densities along both elements and, in turn, helps further broaden the impedance bandwidth of the reader antenna. As the parasitic element is placed further apart from the driven dipole, the magnitude of the near-field coupling decays dramatically and has less influence on the impedance matching.

12.2.8 RFID Read Range Calculation

Operating range or RFID read range is one of the most significant criteria for evaluating the performance of RFID systems, especially UHF RFID systems. A longer operating range system finds more potential application opportunities and ensures a more reliable performance. The read range is the maximum distance at which the RFID reader can detect the backscattered signal from the tag. Because reader

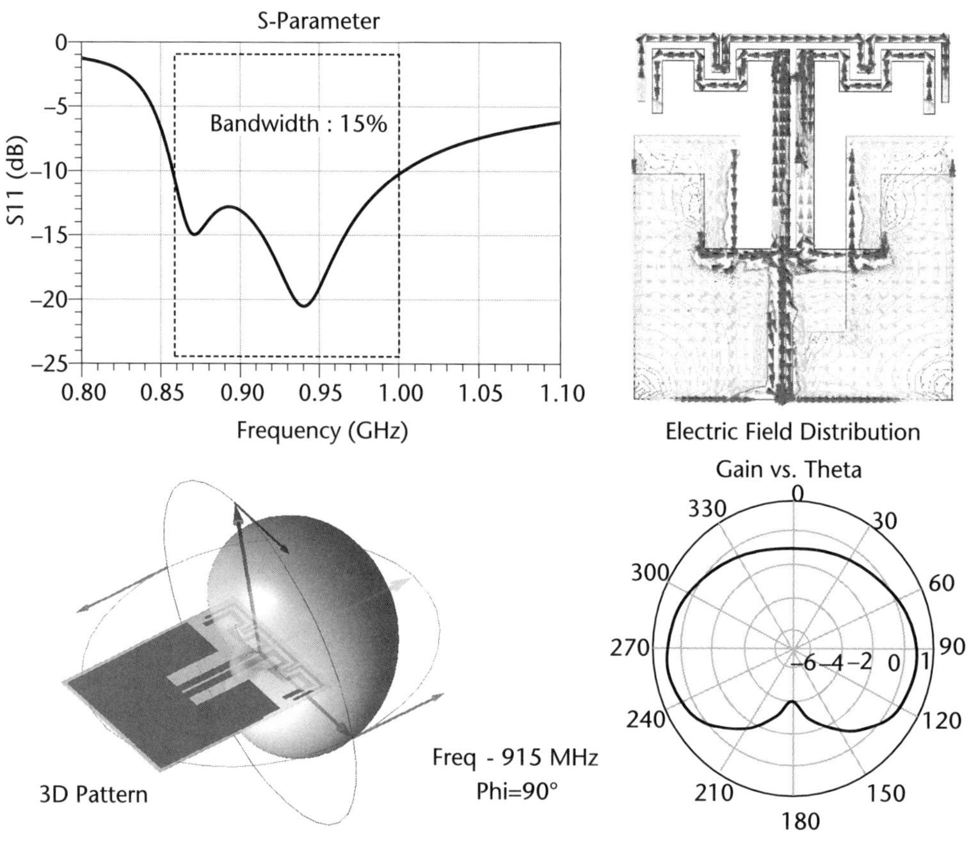

Figure 12.9 Reader antenna return loss, E-field, and radiation pattern plots at 915 MHz.

sensitivity is typically high as compared to tag, the read range is defined by the tag response threshold. Read range is also sensitive to the material on which the tag is placed, the tag orientation, and the propagation environment. The read range measurement is conducted in a controlled environment, such as anechoic chamber or transverse EM (TEM) cell.

In both methods, the tag position is fixed, and transmitter output power is varied by controlled attenuation. This setup allows one to carry accurate tag range characterization and avoid using large and prohibitively expensive chambers or cells. A compact TEM cell is a suitable tool for measuring small tags while a larger anechoic chamber can be used to measure tag performance on various objects. In the anechoic chamber, the tag is placed at a fixed distance from the reader antenna as illustrated in Figure 12.10. The operating range is determined by the whole RFID system design including RFID reader and tag antennas. Theoretically, read range can be calculated using the Friis equation (12.1). Experimental or simulation setup to calculate read range is shown in Figure 12.10. Complete simulation is carried out using FEM to find electric field strength at a different distance at the tag antenna when the reader is excited.

12.3 Radio Navigation Satellite System Antennas

Radio navigation satellite system antennas (RNSS) is a system that uses satellites to provide autonomous geospatial positioning. It permits small electronic receivers to determine their location (longitude, latitude, and altitude) to high precision (within a few meters) using time signals transmitted from satellites. A satellite navigation system with global coverage is called GNSS. At present, the United States' GPS, the European Union's Galileo, and Russia's GLONASS are global operational GNSSs. Navstar GPS of the United States is an operational, 31-satellite GNSS with

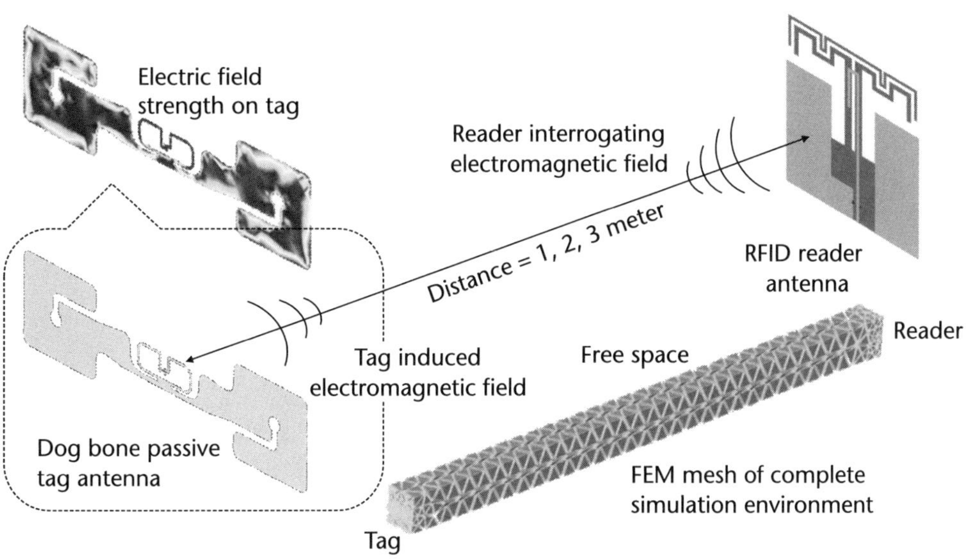

Figure 12.10 RFID read range simulation environment.

downlinks in the L1, L2, and L5 bands. The typical position accuracy available to a user is better than 30m. The Russian GLONASS has 24 satellites currently, and it provides a civilian radio navigation-satellite service. The Galileo system has been under planning by the European Commission since 2008. The Galileo system created by the European Union (EU) went live in 2016. Galileo is intended to provide horizontal and vertical position measurements within 1m precision and better positioning services at higher latitudes than other positioning systems. The Chinese COMPASS navigation system consists of a constellation of 30 nongeostationary satellites and five geostationary satellites. The integration between GNSS and other related technologies, such as telecommunications (GSM, GPRS, UMTS), the Geographic Information System (GIS), and the Inertial Navigation System (INS), has created numerous applications.

GNSS receivers include frequency bands such as GPS L5, Galileo E5/E6, and the GLONASS bands in addition to the legacy GPS bands, and the antenna feeding a receiver may need to cover some or all these bands as shown in Table 12.3.

12.3.1 RNSS Frequency Bands

Figure 12.11 shows the different frequency bands for modern RNSS [10]. The two primary Aeronautical Radio Navigation Service (ARNS) bands are allocated for safety applications, which includes the upper L-band (1,559–1,610 MHz) with GPS L1, Galileo E1, BeiDou B1, and GLONASS G1 and the lower L-band (1,164–1,214 MHz) with GLONASS G3, GPS L5, and Galileo E5. The remaining GPS L2, GLONASS G2, and Galileo E6 signals are in the bands allocated for radiolocation services and RNSS on a primary basis.

12.3.2 Antennas for RNSS

The antenna is a critical component of GNSS receiver system. The signals received from GNSS satellites are very weak and they can arrive from virtually any direction with signals from different satellites arriving simultaneously. GNSS antennas are typically GPS antennas as GPS has been the navigation system for years, but other

Table 12.3 GNSS Frequency Band Allocations

RNNS	*No. of Satellites*	*Frequency Band*	*Country*	*Band*
GPS	32	1.57542 GHz (L1 signal), 1.2276 GHz (L2 signal), 1.164 GHz (L5)	United States	L1, L2, L5
GLONASS	24	Around 1.602 GHz (SP), around 1.246 GHz (SP)	Russia	L1, L2, L3, L5
Galileo	30	1.164–1.215 GHz (E5a and E5b), 1.260–1.300 GHz (E6), 1.559–1.592 GHz (E2-L1-E11)	Europe	L1, L2
COMPASS (Beidou)	30	1.561098 GHz (L1), 1.589742 GHz (B1-2), 1.20714 GHz (B2), 1.26852 GHz (B3)	China	L1, L2, L5
NAVIC	8	1,176.45 MHz (L5 band) 2,492.028 MHz (S-band)	India	L5, S

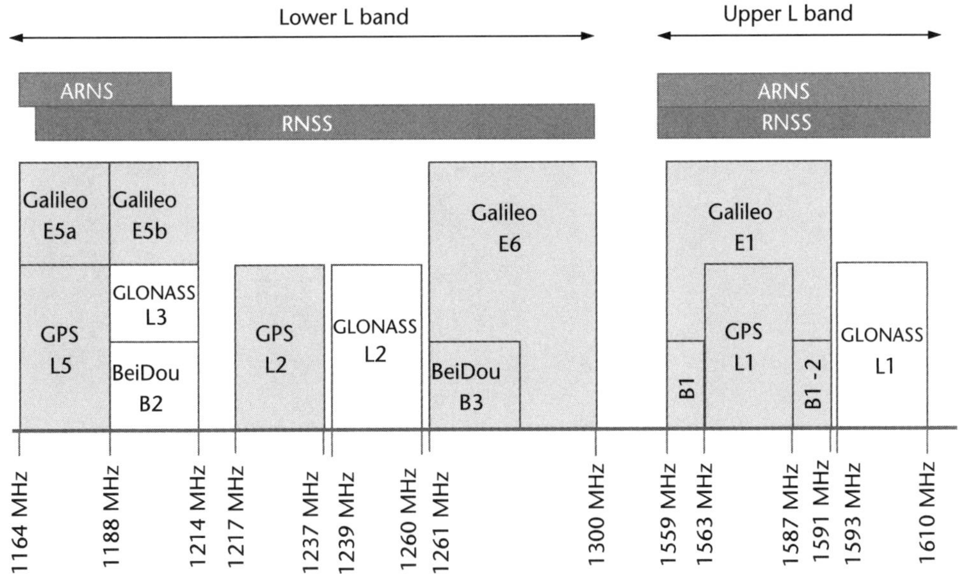

Figure 12.11 RNNS frequency bands.

systems have been and are being developed. Many important properties of GNSS antennas affect functionality and performance, including frequency coverage, gain pattern, circular polarization, and interference handling.

Spaceborne systems at L-band normally use circular polarization (CP) signals for transmitting and receiving. The changing relative orientation of the transmitting and receiving CP antennas as the satellites orbit the Earth does not cause polarization fading as it does with linearly polarized signals and antennas. Also, CP does not suffer from the effects of Faraday rotation caused by the ionosphere in the Earth's environment. Due to the Faraday rotation, an EM wave from space arrives at the Earth's surface with a different polarization angle than it would have if the ionosphere was absent. This results in signal fading and potentially poor reception of linearly polarized signals.

The antenna gain pattern optimizes gain and coverage for satellites over $0° < \theta < \theta_c$ as shown in Figure 12.12 and mitigates problems due to multipath from satellite platform, ionosphere, troposphere, and platform environment, as well as interferences. The ideal gain pattern and spatial coverage vary from application to application. Before 2010, conventional GPS generally called for $70° < \theta_c < 80°$ for terrestrial applications and $70° < \theta_c < 100°$ for airborne applications. After 2010, GNSS applications were expected to move toward a higher cutoff elevation angle, probably about $50° < \theta_c < 70°$ for terrestrial applications and $60° < \theta_c < 80°$ for airborne applications. A GNSS antenna with a higher cutoff angle θ_c above the horizon and good axial ration (or low cross-polarization) would have a higher antenna gain over noise temperature to detect the low line-of-sight CDMA signal desired.

GNSS receive antenna design challenges basically come from two major areas: the large bandwidth requirement and the constraints by the platform on which the antenna operates. Feed network and cost are other major challenges. Because the GNSS antenna needs to detect extremely low GNSS signals, about 25 dB below

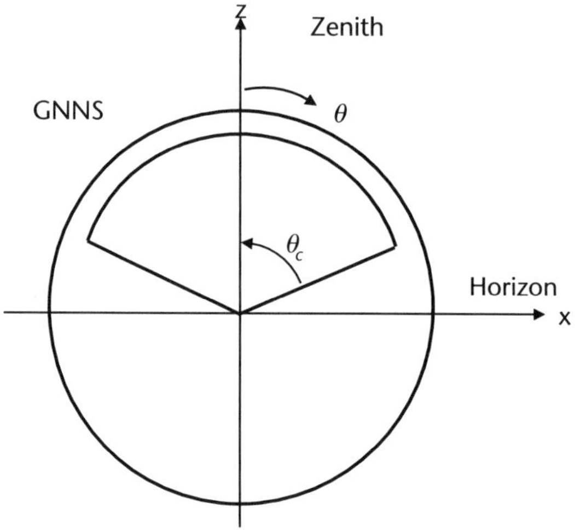

Figure 12.12 Ideal antenna gain patterns for GNSS antennas.

the noise level on Earth, the use of LNA close to antenna is highly desirable. The design cost of a GNSS antenna is a fundamental design consideration driven by the market. A testimony is the printed antennas, whose widespread use is due not only to its low profile but also for its low cost enabled by its simple feed.

GNSS satellites use right-hand circular polarization (RHCP), and therefore a GNSS antenna receiving the signals must also be designed for RHCP. The quality of the circular polarization is defined by specifying the axial ratio (AR). AR is the measure of the polarization ellipticity of an antenna designed to receive circularly polarized signals. An AR close to 1 (or 0 dB) is best (indicating a good CP). AR< 3 dB or cross-polarization (LHCP) of <10 dB is desired over $\theta < \theta_c$ for a medium-quality GNSS antenna. A minimum antenna gain $G_{\min}$ in RHCP is desired over the band. Generally, $G_{\min} > -10$ dBic in RHCP is needed to provide sufficient SNR for detection. However, this low gain requirement can be further relaxed by up to 10 dB or so by using various software and hardware techniques for applications with lower precision requirement.

12.3.3 GPS Spiral Antenna Design

Spiral antennas operate over a wide range of frequencies, and these belong to the class of frequency independent antennas. The main characteristics of spiral antennas are CP, broadband, and low directivity. The antenna impedance, polarization, and radiation pattern of such antennas remain unchanged over a large bandwidth. Spiral antennas are compact antennas because of their windings. A bandwidth of 5:1 or 10:1 comfortably can be obtained, and a stable input impedance is achieved through a self-complementary geometry. This wideband characteristic of the spiral antenna makes it an attractive choice for GNSS where a single antenna is required to transmit and receive over multiple channels.

Figure 12.13 shows a single-arm rectangular spiral antenna. It includes a conductive arm extending from the center outwards. This antenna radiates an axial beam with its outer rectangular periphery loop length being approximately one guide wavelength. The antenna is designed and simulated using the FEM for the L1

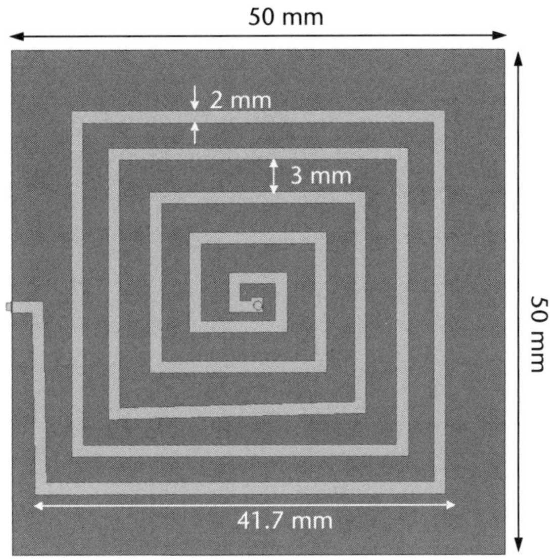

Figure 12.13 Rectangular spiral antenna for the GPS L1 band.

band of GPS. The length of spiral (L), width (W), spacing between turns, and width of stripline (w) parameters control the spiral geometry and its performance. The antenna is designed on a 1.1-mm-thick FR4 substrate with a dielectric constant of 4.24 and a loss tangent of 0.015. This is a bidirectional antenna and here substrate thickness plays a major role in controlling antenna behavior.

Important antenna performance parameters such as gain, radiation patterns, and impedance of the antenna are shown in Figure 12.14. The antenna has dual resonance frequency at 1.508 GHz and 1.576 GHz. The antenna gain is 1.2 dBi. This antenna is covering only the L1 band; hence, this has a very limited application for GPS.

For broadband GPS covering multiple bands, the Archimedean spiral antenna as shown in Figure 12.15 is used. The spectra of GPS, GLONASS, Beidou, and Galileo spread densely across 1.15–1.62 GHz, covering a frequency bandwidth of 446 MHz or an octave bandwidth of 1.383:1. For performance and redundancy, it

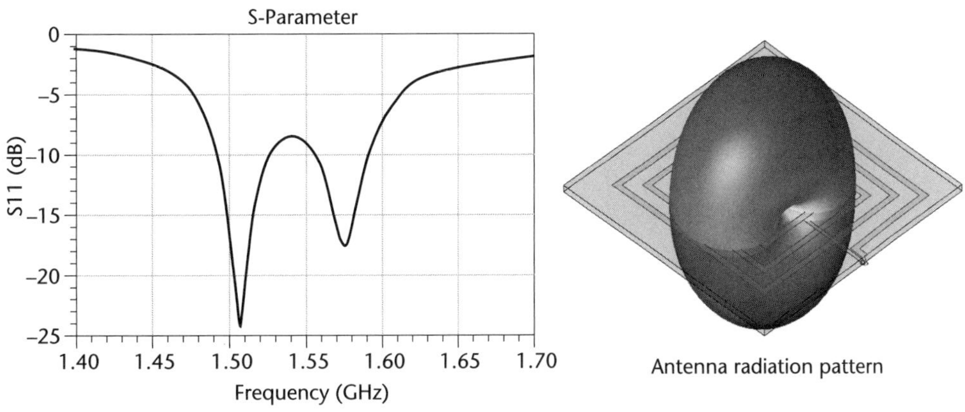

Figure 12.14 GPS L1 band rectangular spiral antenna performance results.

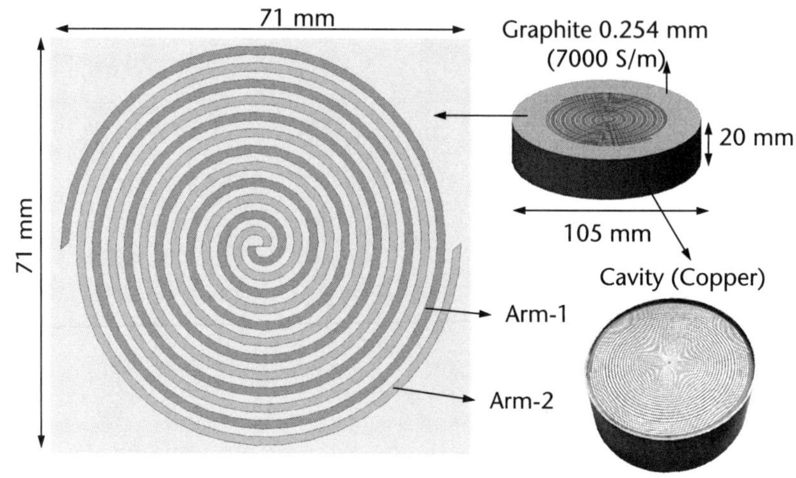

Figure 12.15 Two-arm Archimedean spiral antenna design.

is highly desirable for a GNSS receiver to cover more than one of these GNSSs. The Archimedean spiral antenna [11, 12] can be classified as a frequency-independent antenna in the sense that its input impedance and gain remain almost constant throughout the bandwidth. At low frequencies, the radiation zone is close to the outermost part of the spiral, while at high frequencies it is near the center. Therefore, the lowest cutoff frequency of the spiral antenna is associated with its outer radius and the highest cutoff frequency is related to its inner radius. This is the reason that spiral antenna bandwidth can be very large, only depending on size and printing accuracy. Figure 12.15 shows a two-arm Archimedean spiral antenna.

The arm of the Archimedean spiral is defined by the equation:

$$r = a\varphi \tag{12.4}$$

Equation (12.4) states that the radius r of the antenna increases linearly with the angle θ. The rate at which the spiral flares out is controlled by the parameter a. Arm 2 of the spiral is the same as arm 1 but rotated at 180°. Both arms are 1.4 mm wide. The direction of rotation of the spiral arms defines the direction of antenna polarization. Extra arms may be included to form a multiband spiral antenna. These antennas operate in three ways: traveling wave, leaky wave, and fast wave. The traveling wave, formed on spiral arms, allows for broadband performance. Fast wave is due to the mutual coupling phenomenon occurring between the arms of the spiral. The leaky wave leaks the energy during propagation through the spiral arms to produce radiation. Typically, the spiral is cavity-backed, and the conductive cavity changes the antenna pattern to a unidirectional pattern.

The spiral antenna shown in Figure 12.15 is excited in a balanced mode. In this mode excitation amplitudes are same but there is 180° phase difference between the two arms. This spiral antenna radiates a circularly polarized wave in the antenna-axis direction normal to the antenna plane over a wide frequency range. For exciting the antenna, a coaxial line is used with a wideband balun circuit, which transforms the unbalanced mode of the coaxial line into the balanced mode required for the spiral antenna. The simulated antenna result using the FDTD solver is shown in Figure 12.16. The AR of this GNSS antenna is an important parameter in enhancing SNR and suppression of interference such as those due to multipath scattering and propagation. The result shows that its maximum AR is about 0.4 dB on boresight, slowly increasing to 3.8 dB at 70°, over 1,100–1,700 MHz.

12.3.4 Sine-Wave Meander Archimedean Spiral Antenna

The size of the Archimedean spiral antennas is determined by the lowest operating frequency, the depth of the cavity, and the length of the feed balun. Many techniques have been applied to overcome these size limitations by reducing the aperture and/or by decreasing the profile [13, 14]. An efficient way to miniaturize the antenna aperture is by meandering the spiral antenna arms, which extends the electrical length of arms and lowers the operating frequency [15]. A sinuous meander Archimedean spiral line is used to miniaturize the aperture and a controlled-width exponential spiral line that helps to minimize the input impedance [16]. Such antennas have been used for circularly polarized broadband communications and can be easily flush-mounted in many systems. They are characterized by radiation characteristics and stable input impedance over several octaves.

This section presents a sine-wave meander Archimedean spiral antenna (SMARSA) operating over a frequency range from 1.05 GHz to 1.95 GHz, as shown in Figure 2.17, where the ARSA lines are inserted with a sine wave. With the same amplitude, the sine curve is longer than both the rectangle curve and the sawtooth curve in the same period; additionally, the sine wave is continuous and derivable in the whole period. So it is better for a SMARSA to hold the fine radiation characteristics of a typical ARSA. For getting a unidirectional operation, the antenna is backed by an absorber-loaded cavity. This eliminates back-side radiation without decreasing the frequency band of operation, but at the cost of an efficiency reduction. The exponentially tapered microstrip balun is used as a wideband balun necessary for the balanced feeding of the antenna.

A SMARSA operating over a frequency range from 1.05 GHz to 1.9 GHz is shown in Figure 12.17. With the same amplitude sine curve is longer than both the rectangle curve and the sawtooth curve in the same period. This eliminates back-side radiation without decreasing the frequency band of operation, but at the cost of an efficiency reduction.

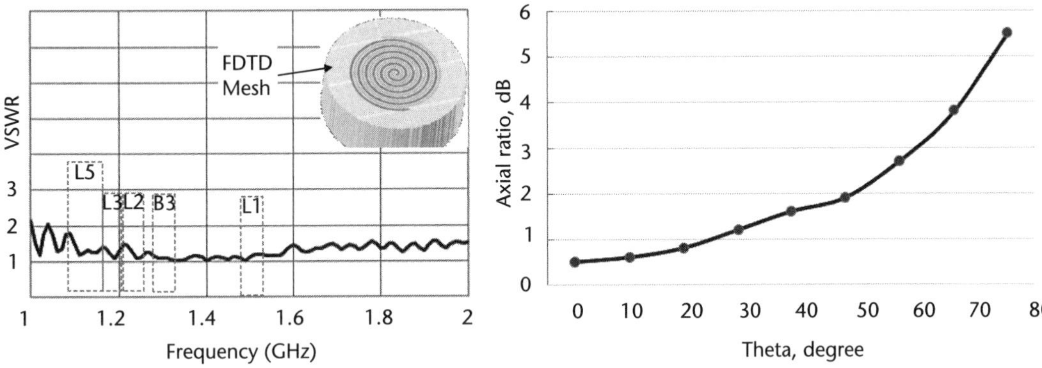

Figure 12.16 Archimedean spiral antenna VSWR and AR plot.

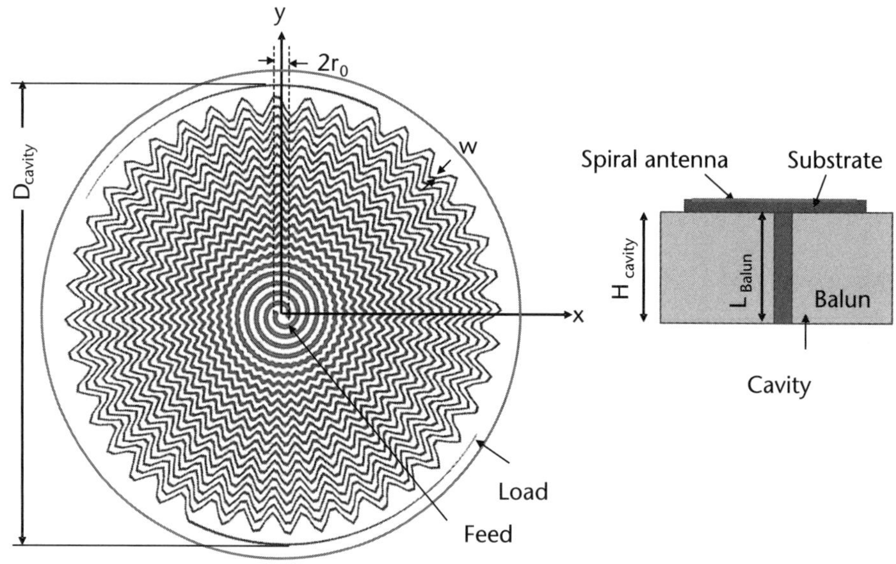

Figure 12.17 Configuration of SMARSA.

The spiral line is defined by the function

$$r = r_0 + a\varnothing + m.\left(\frac{\varnothing - \varnothing_{start}}{\varnothing_{end} - \varnothing_{start}}\right).\sin(n\varnothing) \tag{12.5}$$

where r_0 is the original radius of the spiral line, a is the spiral constant, and $\varnothing$ is the winding angle rising from the starting angle $\varnothing_{start}$ and the ending angle $\varnothing_{end}$. In addition, m and n are the amplitude and the angular frequency of the sine wave, respectively. The antenna is backed by an absorber-loaded cavity, whose depth H_{cavity} is equal to the length of the balun (L_{balun}) decided by the lowest frequency of the band. R is the radius of the final antenna and R_{inner} is the inner radius of the ring-shaped parasitic antenna. The cavity diameter D_{cavity} is 105 mm while antenna diameter is 95 mm. The cavity height H_{cavity} is 30 mm.

The antenna is printed on RT5880 with a dielectric constant of 2.2 (thickness of 1 mm). The simulated antenna return loss plot is shown in Figure 12.18. The RHCP antenna AR is from 0.22 dB (L5 band) to 4.9 dB (L1 band). The antenna gain also varies from 1.12 dB to 1.9 dB. This antenna covers all frequency bands of GNSS. A tapered microstrip is placed in the end of the antenna arm to weaken the terminal reflection.

12.4 Printed Antennas for Medical Applications

Biomedical engineering is involved for improving medical diagnosis and treatment. Insulin pumps, glucose monitoring, deep brain simulations, and endoscopy are a few examples of the medical applications that are based on a remote monitoring system and a body implantable unit. Body implantable devices are generally used in applications such as monitoring temperature and blood pressure, wirelessly transferring

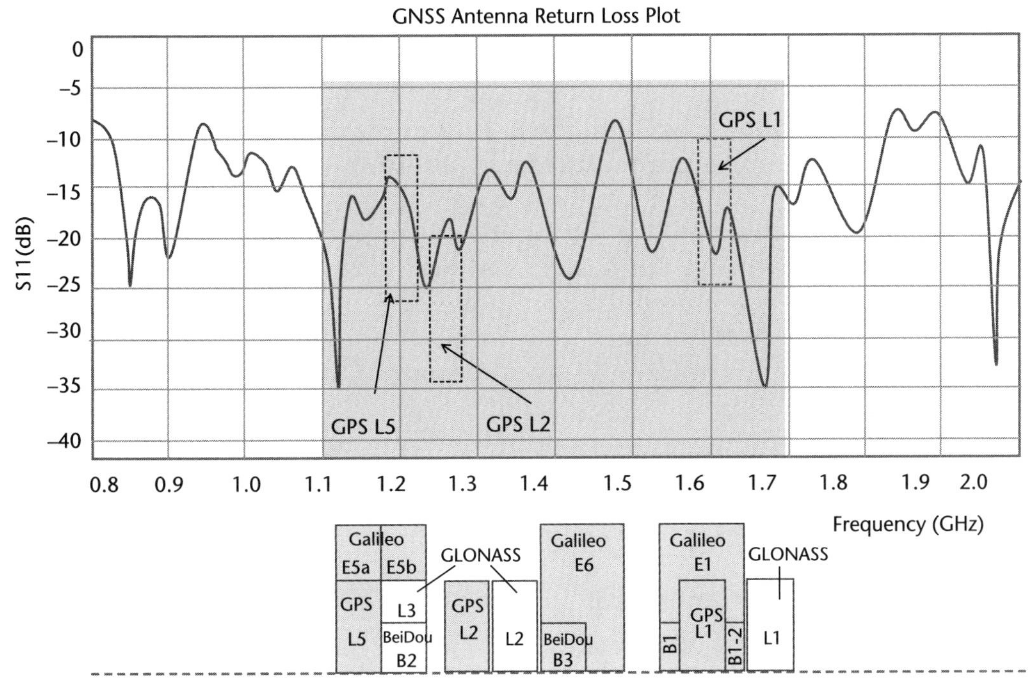

Figure 12.18 SMARSA return loss plot.

diagnostic data from an electronic device implanted in the human body for human safety, and care such as a pacemaker to an external RF receiver and tracking nonindependent people or lost pets. In medical therapy, local and/or whole-body hyperthermia provides successful modality in treatment of some malignant tumors. Microwave energy is one of the operative ways of inducing rapid hyperthermia, but there are certain issues like heating of a relatively large volume of tissue and deep laying tissues. In general, the required characteristics of a microwave radiator include an effective deposition of the energy in a defined tissue volume, minimum leakage of microwave energy into the outside of the treated area, proper impedance matching, lightweight system, and easy-to-handle design. The antenna plays a key role in such system designs to meet these requirements.

Because of different electrical properties of human tissue, designing an antenna for an implanted application is challenging. Implantable medical devices can communicate wirelessly with an external monitoring device. Biomedical telemetry is either in real time or stored physical signals are communicated to the receiver. As per the recommendation of FCC, the Medical Implant Communication Services (MICS) band of 402–406 MHz is suggested for implantable antennas. The MICS frequency band has replaced a previously used low-frequency inductive link, which suffers from short-range communication and slow data rate. The maximum transmit power requirement at this band is very low, about 25 μW. This minimizes the risk of interference with other users of the same band. The maximum bandwidth utilization at a time is 300 kHz, which makes it a low bit rate system compared with Bluetooth or Wi-Fi. Implanted devices are placed into human body and implantable antenna ensures wireless biotelemetry. These antennas can be implanted either into human bodies or can just be mounted over the torso (skin-fat-muscle) to form a biocommunication system between medical devices and exterior instruments for short-range biotelemetry applications.

Printed and microstrip radiators offer the advantage of being small, lightweight, and capable of conforming to the shape of the human body, but the electrical performance of antenna is altered significantly in vicinity to the human body. These facts make the biomedical antenna design complicated. The electrical performance of a new class of wideband wearable printed antennas for medical applications is presented in this section. Low-profile, compact tunable antennas are important in the development of wearable human biomedical systems. The antenna resonant frequency can be tuned by using a varactor to compensate variations in antenna resonant frequency at different locations on the body. A conformal applicator employing a radiator consisting of a few printed dipoles was previously used for inducing hyperthermia at 2.45 GHz [17]. RF transmission properties of human tissues have been investigated in several articles [18, 19]. Several wearable antennas have been designed in the past decade [20–23].

12.4.1 Antenna Design for Wireless Body Area Network (WBAN)

WBAN is widely utilized in the fields of medical and motion monitoring, personal entertainment, modern wars, and virtual reality games [18]. Wearable antennas are crucial for WBAN applications [19]. These can be divided into three common modes: in-body mode [20–23], on-body mode, and off-body mode [22, 23]. In-body mode, the antenna is implanted into human body, which needs to consider the body tissue's influence on antenna performance. On-body mode shows that the wearable antenna can communicate with other wearable antennas located at other positions along the body surface, which demands maximum directivity along the body surface. Off-body mode represents that the antenna can communicate with a medical base station or other devices far away from the body, which requires maximum radiation in the broadside direction.

Figure 12.19 shows the design of a wearable antenna at ISM frequency band at 2.45 GHz for wireless vital signs monitors (VSMs). VSMs generally measure temperature, respiration, heart rate, and, more recently, blood oxygen saturation levels. To do this, the sensor device mainly incorporates an integrated antenna, a low-power digital processing technology and RF circuitry. A repeatable simulation phantom model (human body model) is required to overcome predictable variations in human test subjects and to comparatively analyze the performance of different VSM systems. An ideal phantom has a variable tissue model. In this design, a tissue phantom model as shown Figure 12.19 is used that is electromagnetically representative of the spread in tissue properties of the thorax region of a wider range of the human population [24].

The design of this antenna is discussed in Chapter 3 (Section 3.5) in detail. The antenna is printed on flexible substrate, which conforms to the surface of the human body. The EBG structure is used as shown in Figure 12.20 to increase the impedance bandwidth of the antenna to be relatively large, which will mitigate against resonant frequency shifts and detuning when mounted on variable tissue morphologies. Furthermore, to fully characterize the performance of the antenna, it is wrapped and placed on the chest. The return loss ($|S_{11}|$) and the radiation efficiency of the antennas are simulated on the physical tissue phantom.

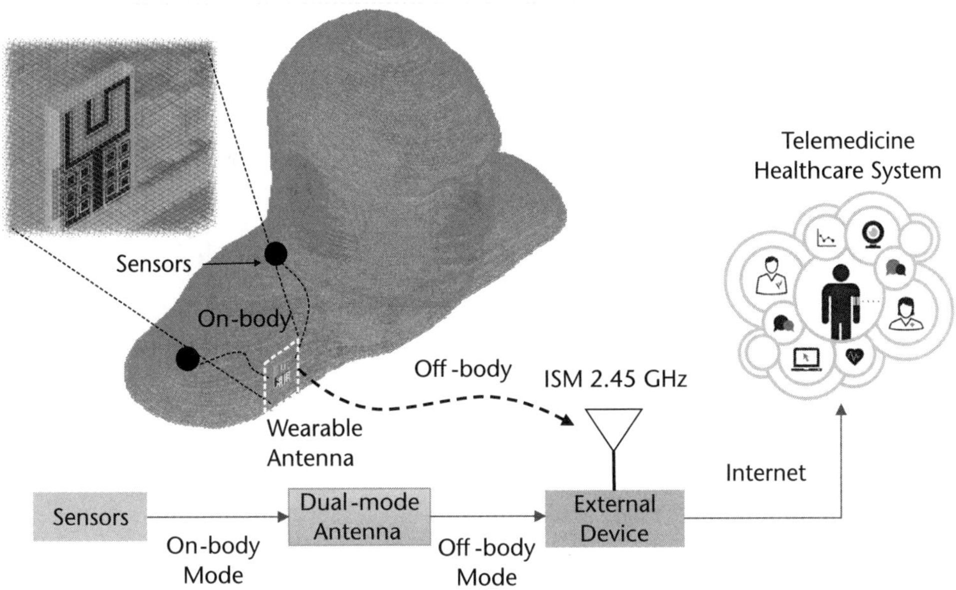

Figure 12.19 Dual-band wearable antenna design and on-body and off-body systems.

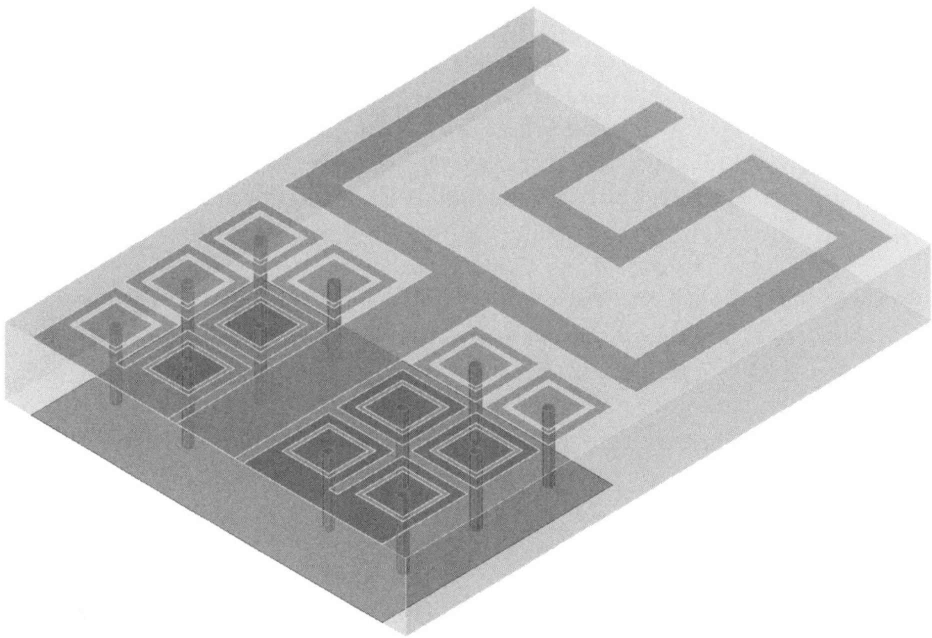

Figure 12.20 Dual-band wearable antenna with EBG structure.

The return loss plots for the antenna are shown in Figure 12.21. The total radiation efficiency of the antennas is measured on each side of the tissue phantom, composed of a thick side with a relatively larger tissue layer and a lean side with a thin tissue layer. The measured radiation efficiency is between 8% and 30%. The maximum total radiation efficiency is achieved on the phantom side with the thickest tissue layer. The minimum performance is achieved where the tissue layer

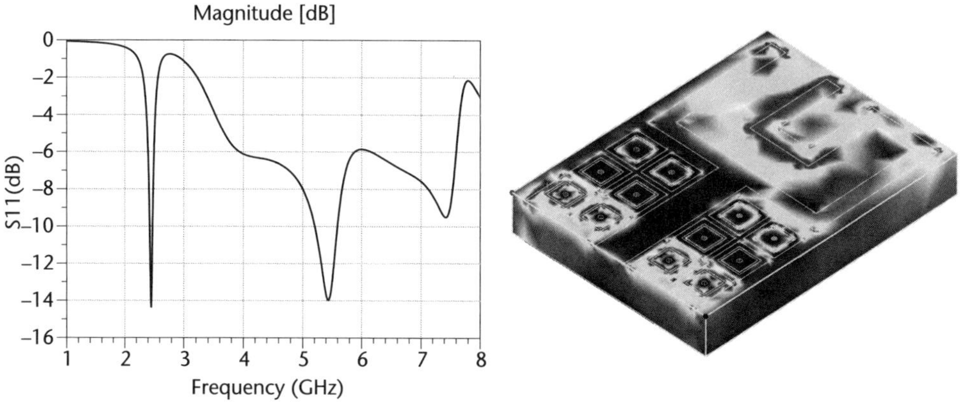

Figure 12.21 Dual-band wearable antenna return loss and E-field plots.

is thinnest, which means that the antenna is position closer to the inner phantom tissue layer, which has a relatively larger electrical conductivity.

12.5 IoT Antennas

IoT applications are extensively used for consumer and industrial markets for various purposes such as computations and sensing as well as connecting things within the region. IoT is a booming market; gradually, IoT modules are becoming compact and demand for incorporating more than one wireless technology is increasing. These requirements have increased the demand for IoT antennas compliant to IoT modules. IoT devices are commented using wireless communications links to other devices. Driven by the demands of the IoT, more antennas are required to be integrated into devices to make all wireless interconnectivity possible. IoT systems shall be able to incorporate many different and heterogeneous end systems for the digital services including everything from smart health, smart homes, smart retails, smart wearables, smart agriculture, smart factories, smart cities, and the smart grid. As fundamental units of machine-to-machine (M2M) communications which is the basis for IoT, the performance of microwave antennas directly affects the performance of IoT devices.

Antenna design challenges related to the IoT include providing reliable connectivity and maintaining a reasonable performance within an ever-smaller footprint. Many devices will need to operate in an increasingly crowded radio spectrum with the possibility of co-site conditions, operation near the human body, and other challenging installed environments. Antenna placement is crucial in these cases.

The choice of an antenna system for is a critical component of all these node-end smart devices. Designing an antenna with an effective performance in IoT devices requires engineers to examine several factors including antenna size, from what is needed to what is possible, antenna shape, and placement. Therefore, IoT module antenna designers face the restrictions of maintaining good antenna performance in ever-shrinking volumes and under extreme interference conditions. The major types of IoT antennas are chip, wire, whip, PCB, and proprietary, as shown in Figure 12.22.

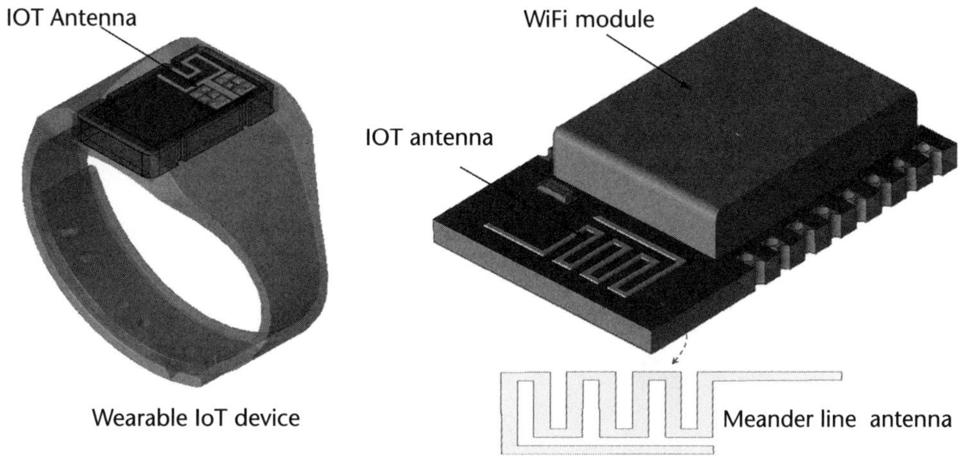

Figure 12.22 IoT device and integrated printed antennas.

12.5.1 Low-Power Wide Area Network (LPWAN) IoT Antenna Design

IoT systems are characterized by the transmission of small amounts of data and low transmission rate, which drastically reduce the bandwidth requirements. Usually, less than 1-MHz bandwidth is enough for most of the IoT applications. New wireless standards for LPWANs, such as LoRa or Sigfox, well match the low data rate needed by IoT applications. Compact antennas allow for easier integration within IoT devices, but miniaturization is strictly connected to the antenna radiation efficiency η_r through the lower bound limit on the antenna quality factor Q [25]

$$Q_{lb} = \eta_r \left(\frac{1}{ka} + \frac{1}{(ka)^3} \right) \tag{12.6}$$

where k is the wavenumber and a is the radius of the sphere enclosing the antenna.

Figure 12.23 is showing IoT antenna design covering 2.2–6-GHz frequency band. The printed antenna is designed using a flexible substrate of ULTRALAM 3850 by Rogers consisting of 100-μm-thick LCP substrate [26]. The substrate has a dielectric constant of 3.14 and a loss tangent of 0.0025. These materials satisfy robustness, flexibility, and thermal conditions in which the antenna can perform effectively under extreme bending.

The antenna design is initiated from a circular monopole radiator with the addition of a circular slot ground that surrounds the radiator. The rectangular slots present on the radiator and partially filled gaps in the ground slot help the generation of CP radiation for a certain frequency range. The antenna is fed with a 50Ω CPW transmission line. It has a compact size with a total area of 46 × 46 mm^2. The resulting reflection coefficient confirms that the antenna maintains less than –10 dB within bandwidth starting from 2.2–4.5 GHz and 5.1–5.9 GHz. At a lower frequency, the antenna has an omnidirectional pattern that tends to deform as the frequency increases, which is observed at 5.8 GHz.

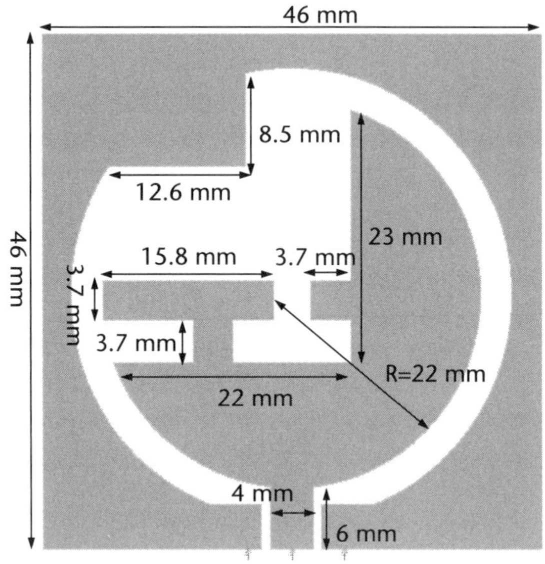

Figure 12.23 IoT antenna design dimensions.

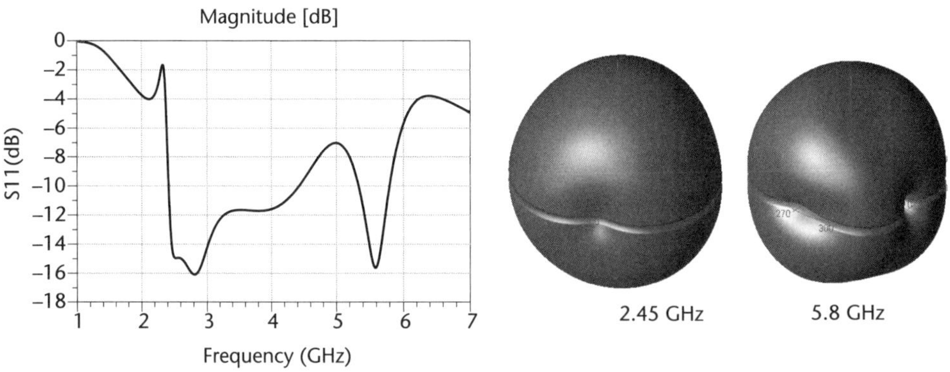

Figure 12.24 IoT antenna return loss and radiation pattern plots.

12.5.2 Quasi-Yagi IoT Antenna Design

Quasi-Yagi antennas are used for IoT device communications for their properties, such as compact size, easy fabrication process, light weight, and end-fire radiation pattern [26–31]. A traditional Yagi-Uda antenna has a dipole driver in the center, a single reflector on one side, one or more directors on the other side, and a feeding structure. The directivity of a microstrip Yagi antenna significantly depends on the number of directors. There is always a trade-off between the bandwidth and antenna gain. Bandwidth gets narrow as more elements are used to increase the gain. The design parameters of a quasi-Yagi antenna for X-band IoT applications are shown in Figure 12.25.

In a microstrip Yagi-Uda antenna, EM energy is coupled from the driven element dipole through space into the parasitic dipoles and then gets reradiated to form a directional beam. This process is analogous to the conventional Yagi antenna, but the slight difference is that the substrate also plays some role in coupling by

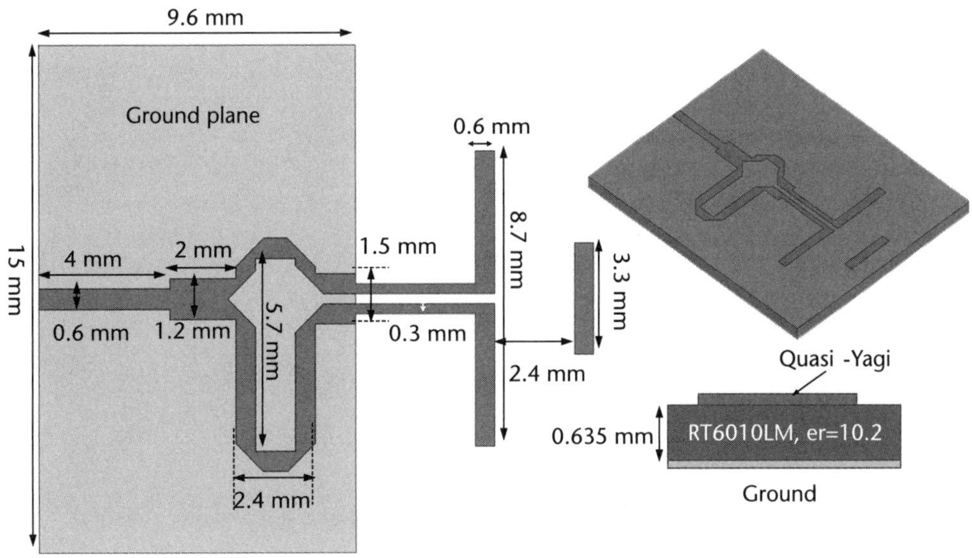

Figure 12.25 Quasi-Yagi antenna design parameters.

surface waves. The microstrip patch antenna normally couples very little energy as it radiates largely in broadside and slightly along the direction of the ground plane. The field radiated by the driven element is reflected by the long reflector element positioned at the reverse end-fire direction. The parasitic director elements help to improve further the radiation in the end-fire direction. The basic design parameters are the length of driven element (0.45λ–0.49λ), the length of directors (0.4λ–0.45λ), and the separation between directors (0.3λ–0.4λ). The separation between the directors and the length of the directors need not be uniform.

The antenna is fabricated on a 0.635-mm-thick RT6010LM substrate with metallization on both sides. The top metal layer consists of a microstrip feed, a broadband microstrip-to-coplanar stripline (CPS) balun, and two dipole elements, one of which is the driver element fed by the CPS and the other of which is the parasitic director. The bottom metal plane is a truncated microstrip ground, which serves as the reflector element for the antenna. The parasitic director element on the top plane simultaneously acts as an impedance-matching parasitic element and directs the antenna propagation toward the end-fire direction. One of the most effective features of this antenna is the use of the truncated ground plane as the reflector element.

In the quasi-Yagi antenna, the simplest method to enhancement the bandwidth is to increase the distance between the ground plane and radiator, but it will increase substrate thickness. The main limiting factor of bandwidth in the quasi-Yagi antenna is its feed. A broadband microstrip-to-CPS transition structures is used to enhance the impedance bandwidth of the quasi-Yagi antenna in [28], although the delay line used in the balun structures still restricts the antenna's bandwidth. UWB balun or coplanar waveguide feeding has been used in many designs to enhance the bandwidth [29, 30]. Approximately 40% bandwidth is achieved using this technique as shown in Figure 12.26.

The FEM-based simulation results of the antenna are shown in Figure 12.26. The simulated bandwidth (−10 dB) is 40%. The radiation patterns are measured

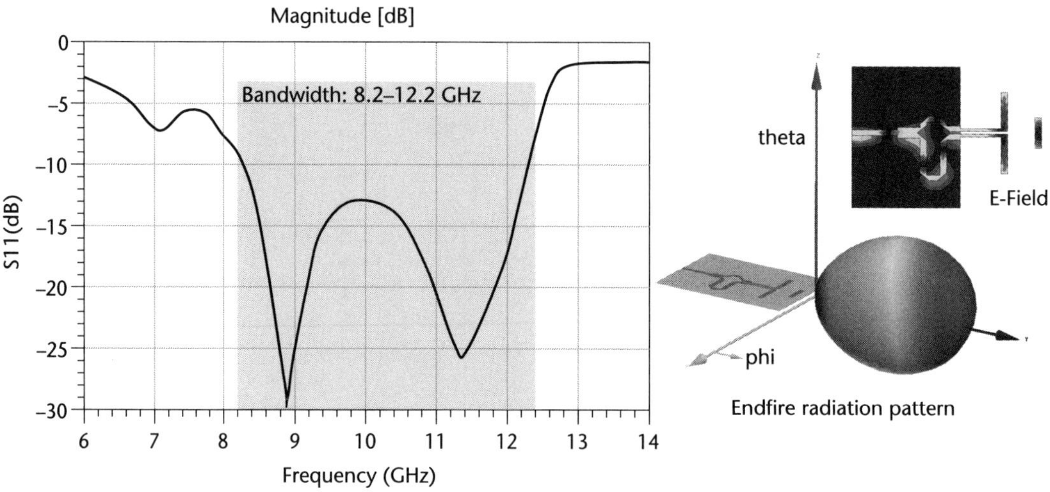

Figure 12.26 Quasi-Yagi antenna return loss, E-field, and radiation pattern plots.

across the bandwidth, the antenna gain is better than 5.6 dB, and the measured front-to-back ratio is better than 8.1 dB within the operating band. The front-to-back ratio is a critical parameter for a unidirectional antenna. The directivity of the one director Yagi antenna can be increased by adding extra metallic strips acting as directors, but it will reduce the bandwidth of antenna.

12.6 Summary

The fast development in the field of wireless communications demands that compact devices be capable of operating in more than one frequency band. The design of innovative compact and high-performance printed antennas with multiple communications protocols, for example, ISM band (2.4 GHz), the IEEE 802.11b/g, and the IEEE 802.11a, is a challenge for the designers. In this chapter, modern application-based printed antenna design was discussed covering antennas for RFID, GPS, IoT, and printed antennas for medical applications. Printed antenna design techniques were presented that make them suitable for modern communications applications. Printed antennas are also used as a part of the power supply block in sensor systems. Low-power sensor systems work by harvesting energy from the ambient in different environments. These systems are usually called rectennas. Printed antennas are good candidates for this purpose due to the availability of broadband design and ease of integration with the required electronics.

References

[1] Azzouzi, S., et al., "New Measurement Results for the Localization of UHF RFID Transponders Using an Angle of Arrival (AoA) Approach," *Proc. IEEE RFID*, April 2011, pp. 91–97.

[2] Amendola, S., L. Biachi, and G. Marrocco, "Movement Detection of Human Body Segments," *IEEE Antennas and Propagation Magazine*, Vol. 57, No. 3, June 2015, pp. 23–37.

[3] Finkenzeller, K., *RFID Handbook: Fundamentals and Applications in Contactless Smart Cards, Radio Frequency Identification and Near-Field Communication*. New York: Wiley, 2010.

[4] Perret, E., et al., "BRF and THz Identification Using a New Generation of Chipless RFID Tags," *Radio Engineering*, Vol. 20, Special Issue: Emerging Materials, Methods, and Technologies in Antenna & Propagation, No. 2, June 2011, pp. 380–386.

[5] Marrocco, G., L. Mattioni, and C. Calabrese, "Multiport Sensor RFIDs for Wireless Passive Sensing of Objects-Basic Theory and Early Results," *IEEE Transactions on Antennas and Propagation*, Vol. 56, No. 8, August 2008, pp. 2691–2702.

[6] Lee, B., and B. Yu, "Compact Structure of UHF Band RFID Tag Antenna Mounted on Metallic Objects," *Microw. Opt. Technol. Lett.*, Vol. 50, No. 1, January 2008, pp. 232–234.

[7] Qing, X. M., C. K. Goh, and Z. N. Chen, "A Broadband UHF Nearfield RFID Antenna," *IEEE Transactions on Antennas and Propagation*, Vol. 58, No. 12, December 2010, pp. 3829–3838.

[8] Shi, J., X. Qing, and Z. N. Chen, "Electrically Large Zero-Phase-Shift Line Grid-Array UHF Near-Field RFID Reader Antenna," *IEEE Transactions on Antennas and Propagation*, Vol. 62, No. 4, April 2014, pp. 2201–2208.

[9] Hua, R., and T. Ma, "A Printed Dipole Antenna for Ultra High Frequency (UHF) Radio Frequency Identification (RFID) Handheld Reader," IEEE Transactions on Antennas and Propagation, Vol. 55, No. 12, December 2007, pp. 3742–3745..

[10] Subirana, J. S., J. J. Zornoza, and M. Hernandez-Pajares, *GNSS Signal*, University of Catalunia, Spain, 2011, https://gssc.esa.int/navipedia/index.php/GNSS_signal.

[11] Kaiser, J., "The Archimedean Two-Wire Spiral Antenna," in *IRE Transactions on Antennas and Propagation*, Vol. 8, No. 3, May 1960, pp. 312–323.

[12] Wang, Y. -W., G. -M. Wang, and H. -Y. Zeng, "Design of a New Meander Archimedean Spiral Antenna," *Microwave and Optical Technology Letters*, Vol. 52, No. 10, 2010, pp. 2384–2387.

[13] Filipovic, D. S., and J. L. Volakis, "Broadband Meanderline Slot Spiral Antenna," *IEE Proc. Microw. Antennas Propag.*, Vol. 149, No. 2, April 2002, pp. 98–105.

[14] Kramer, B. A., C. -C. Chen, and J. L. Volakis, "Size Reduction of a Low-Profile Spiral Antenna Using Inductive and Dielectric Loading," *IEEE Antennas Wireless Propag. Lett.*, Vol. 7, February 2008, pp. 22–25.

[15] O'Brien, J. M., et al., "Miniaturization of a Spiral Antenna Using Periodic Z-Plane Meandering," *IEEE Transactions on Antennas and Propagation*, Vol. 63, No. 4, April 2015, p. 1843.

[16] Gao, X., and Z. Shen, "UHF/UWB Tag Antenna of Circular Polarization," *IEEE Transactions on Antennas and Propagation*, Vol. 64, No. 9, September 2016, pp. 3794–3802.

[17] Sterzer, F., et al., "Microwave Apparatus for the Treatment of Cancer," *Microwave J.*, Vol. 23, 1980, pp. 39–44.

[18] Hall, P. S., and Y. Hao, *Antennas and Propagation for Body-Centric Wireless Communications*, Norwood, MA: Artech House, 2006.

[19] Xiaomu, H., S. Yan, and G. A. E. Vandenbosch, "Wearable Button Antenna for Dual-Band WLAN Applications with Combined On and Offbody Radiation Patterns," *IEEE Transactions on Antennas and Propagation*, Vol. 65, No. 3, March 2017, pp. 1384–1387.

[20] Liu, C., Y. -X. Guo, and S. Xiao, "Capacitively Loaded Circularly Polarized Implantable Patch Antenna for ISM Band Biomedical Applications," *IEEE Transactions on Antennas and Propagation*, Vol. 62, No. 5, May 2014, pp. 2407–2417.

[21] Liu, C., Y. X. Guo, and S. Xiao, "Compact Dual-Band Antenna for Implantable Devices," *IEEE Antennas Wireless Propag. Lett.*, Vol. 11, 2012, pp. 1508–1511.

[22] Liu, Z. G., and Y. X. Guo, "Dual Band Low Profile Antenna for Body Centric Communications," *IEEE Transactions on Antennas and Propagation*, Vol. 61, No. 4, April 2013, pp. 2282–2285.

[23] Liu, Z. G., and Y. X. Guo, "Compact Low-Profile Dual Band Metamaterial Antenna for Body Centric Communications," *IEEE Antennas Wireless Propag. Lett.*, Vol. 14, December 2015, pp. 863–866.

[24] Conway, G. A., et al., "Layered RF Phantom Characterisation for Wireless Medical Vital Sign Monitors," *IEEE International Symposium on Antennas and Propagation*, 2013.

[25] Best, S. R., "Low Q Electrically Small Linear and Elliptical Polarized Spherical Dipole Antennas," *IEEE Transactions on Antennas and Propagation*, Vol. 53, No. 3, 2005, pp. 1047–1053.

[26] Zahran, S. R., Z. Hu, and M. A. Abdalla, "A Flexible Circular Polarized Wide Band Slot Antenna for Indoor IoT Applications," *2017 IEEE International Symposium on Antennas and Propagation & USNC/URSI National Radio Science Meeting*, San Diego, CA, 2017, pp. 1163–1164.

[27] Yen, C. C., and P. Nysen, "Broadband Quasi-Yagi Antenna with Enhanced Radiation Pattern for MIMO Applications," *IEEE International Symposium on Antennas and Propagation Society International Symposium*, 2013, pp. 880–881.

[28] Sor, J., et al., "A Broadband Quasi-Yagi Antenna Array," *European Microwave Conference,* 1999, pp. 255–258.

[29] Tzyh-Ghuang, M., et al., "A Modified Quasi-Yagi Antenna with a New Compact Microstrip-to-Coplanar Strip Transition Using Artificial Transmission Lines," *IEEE Transactions on Antennas and Propagation*, 2009, pp. 2469–2474.

[30] Kan, H. K., et al., "Simple Broadband Planar CPW-Fed Quasi-Yagi Antenna," *IEEE Antennas and Wireless Propagation Letters*, Vol. 6, 2007, pp. 18–20.

[31] Kwangseok, H., et al., "Broadband CPS-Fed Yagi-Uda Antenna," *Electronics Letters*, Vol. 45, No. 24, 2009, pp. 1207–1209.

Selected Bibliography

Fujita, K., et al, "UWB Circularly Polarized Planar Antenna on Flexible Substrate," *Microwave Workshop Series on RF and Wireless*, 2013.

Jeong, S. H., and H.-W. Son, "UHF RFID Tag Antenna for Embedded Use in a Concrete Floor," *IEEE Antennas Wireless Propag. Lett.*, Vol. 10, 2011, pp. 1158–1161.

Liang, W., et al., "A Novel Simple and Compact Microstrip-Fed Circularly Polarized Wide Slot Antenna with Wide Axial Ratio Bandwidth for C-Band Applications," *IEEE Transactions on Antennas and Propagation*, Vol. 64, No. 4, 2016, pp. 1552–1555.

Lin, K. H., S. L. Chen, and R. Mittra, "A Capacitively Coupling Multifeed Slot Antenna for Metallic RFID Tag Design," *IEEE Antennas Wireless Propag. Lett.*, Vol. 9, 2010, pp. 447–450.

Tran, H. H., S. X. Ta, and I. Park, "A Compact Circularly Polarized Crossed-Dipole Antenna for an RFID Tag," *IEEE Antennas Wireless Propag. Lett.*, Vol. 14, 2015, pp. 674–677.

Zahran, S. R., A. Gaafar, and M. Abdalla, "A Flexible UWB Low Profile Antenna for Wearable Applications," *IEEE Int. Ant. & Prop. Symp.*, 2016, pp. 1931–1932.

CHAPTER 13

Automotive Antennas

13.1 Introduction

Modern vehicles are fitted with multiple antennas to enable communication and allow a moving vehicle to connect with the outside world. In automobiles, many electronic devices are usually designed to increase safety or enhance the driving experience, and most of them depend on wireless communication to perform their task to the roadside environment. Antennas on a vehicle form the basis of navigation, telemetry, and all kinds of onboard communication. Systems such as Bluetooth, satellite navigation, radio, vehicle tracking, collision notification, and remote diagnostics all depend on high-performance antennas that perform as intended in challenging environments. Automobile manufacturers have been looking to create cost-effective, fuel-efficient vehicles with attractive designs. This leads to a focus on lightweight vehicles with reduced aerodynamic drag, and sleek and improved styling. Many significant innovations in the world of vehicular antennas are driven by these market preferences along with the technological factors.

Automotive communication has witnessed a tremendous growth in recent years due to modern communication. In addition to the traditional amplitude and frequency modulation (AM/FM), the present-day automobiles are equipped with multitude of information and entertainment systems such as navigation, GSM, Bluetooth, satellite radio, RKES, and TPMS. Apart from these services, higher-end cars are equipped with DAB, satellite in-vehicle TV, SDARS, ETC, short-range and long-range radar system for ACC, parking assistance, and collision detection and avoidance. In the present scenario, premium cars are integrated with at least 10 to 30 dedicated antennas.

Driverless cars will rely on radar to a much higher degree. These cars will have to autonomously keep a safe driving distance from other vehicles, exit and enter highways, change lanes safely, and make many other driving decisions that are now handled by human drivers. Vehicles are becoming ever more connected to other cars using V2X (vehicle to everything) communication, to transport infrastructure, to pedestrians, and to data centers. V2X is the umbrella term for the car's communication system, where information from sensors and other sources travels via low-latency, high-reliability links and high bandwidth, paving the way to fully autonomous driving.

Simulation techniques are critical in automotive antenna design because the range of antenna types and principles of operation, as well as the range of geometric scale involved, means that no single numerical technique will be suitable for all simulation cases. Because of antenna shape and operating frequency, simulating an FM windscreen antenna should use a different approach than simulating a roof-mounted integrated shark-fin antenna, which is again different from simulating a 77-GHz bumper-integrated collision avoidance system antenna. The choice of numerical method depends on the frequency of operation, geometrical complexity, materials used, and overall electric size of the model being studied. A strong trend currently is to combine multiple numerical approaches to provide a hybrid solution. With EM simulators such as MoM, FDTD and FEM automotive engineers can analyze radar system performance in complex, dynamic environment with improved accuracy. EM and system-level simulations are important in designing, testing, and validating many aspects of automotive radar systems. High reliable parametric antenna arrays design and their performance analysis by mounting them on vehicle bumpers are possible with EM simulations. Strong vehicle scattering can cause antenna radiation pattern sidelobes to rise and multiply; such a problem is solved by modeling the array interaction with the vehicle and predicting the installed antenna pattern. The main antenna beam shift, gain loss, shape distortion, depolarization, and sidelobe levels are all critical metrics in modeling the performance of an LRR and can be analyzed and studied thoroughly using simulations. In this chapter, antenna designs and the frequencies that are commonly used in the automotive environment are discussed.

13.2 Frequency-Modulated Continuous-Wave Automotive Radar System

Modern automobiles are equipped with radar and proximity sensors. Radar is prevalent in safety, navigation, and driver assistance systems. The radar systems in the 24-GHz and 77-GHz frequency bands are used for functions such as autonomous cruise control, collision avoidance, blind spot detection, cross-traffic alerts, parking assistance, pedestrian detection, and emergency braking. Automotive frequency-modulated continuous-wave (FMCW) radar needs target elevation angle information not only in azimuth angle but also in range and velocity. The automotive radar can detect moving objects including vehicles and pedestrians or some static objects as shown in Figure 13.1.

In an FMCW radar, the transmitted signal is a linear frequency-modulated continuous-wave (L-FMCW) chirp sequence, whose frequency versus time characteristic follows a sawtooth pattern, as shown in Figure 13.2. A local oscillator generates the L-FMCW signal, also known as the chirp signal, for amplification and transmission from the transmit antenna. Any object presents in the coverage of the radar signal, reflect the transmitted signal back. Mixing the reflected signal with the local oscillator signal produces an intermediate frequency (IF) output, which is digitized and subsequently processed in a digital signal processing. In the case of moving objects, the IF signal also has a Doppler component that depends on the relative velocity between the radar and the target. Looking at the phase shift of the IF signal from one chirp to the next chirp provides an estimate of the Doppler and

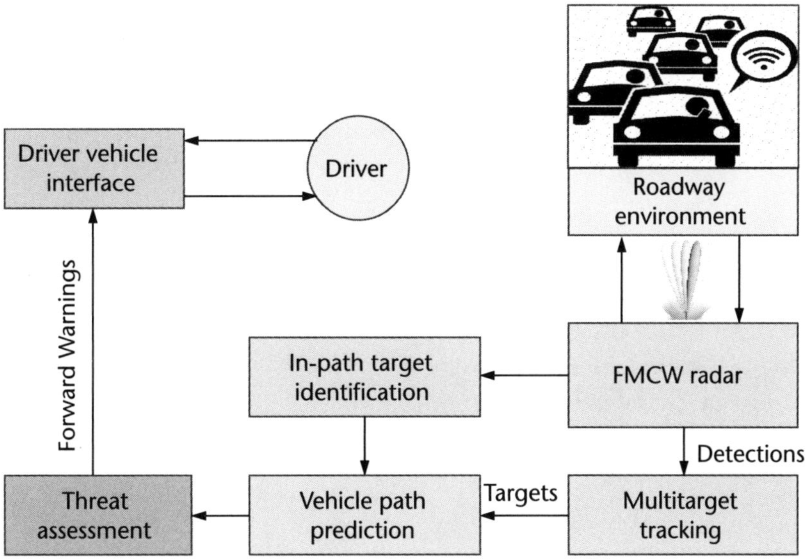

Figure 13.1 A basic FMCW radar system.

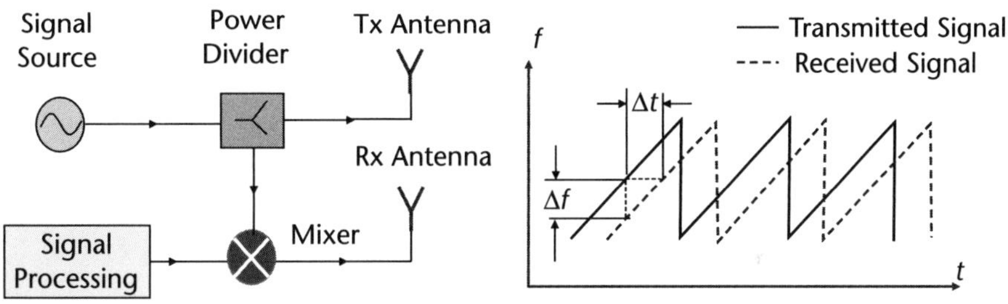

Figure 13.2 Block diagram of the FMCW automotive radar and the frequency-time relationships of transmitted and received signals.

hence the relative velocity. This is typically accomplished by performing a second FFT across chirps.

Figure 13.2 is the general block diagram of a FMCW automotive radar, and it is capable of measuring both the distance and velocity of a moving object. The time delay can be calculated by

$$\Delta t = \frac{\Delta f}{k} \tag{13.1}$$

where Δf is the frequency difference between the transmitted and received signals, which could be measured from the mixer output and k is the frequency sweep rate. The distance between the observer and target is then given by

$$d = \frac{c.\Delta t}{2} \tag{13.2}$$

Frequency bands 24 GHz and 77 GHz are two specific frequency bands for automotive radar applications. The high-frequency 77-GHz solution offers advantages such as broader bandwidth and a smaller dimension, but there are more challenges in design and implementation. The 77-GHz band could be divided into two subbands: 76–77 GHz and 77–81 GHz (also called the 79-GHz band). The former has been approved by most countries, while the latter is only available in Europe so far but has been under discussion in other countries. The functions of automotive radar sensors vary with their maximum ranges. Three major classes of radar systems are typically employed in automotive safety systems:

- *Short-range radar (SRR):* SRR is mainly used for collision proximity warning and safety and to support parking assist features.
- *Medium-range radar (MRR):* MRR is used to watch the corners of the vehicle, perform blind spot detection, observe other-vehicle lane crossover, and avoid side/corner collisions.
- *Long-range radar (LRR):* LRR is for forward-looking sensors, ACC, and early collision detection functions.

LRR is usually mounted in the front grill of the vehicle to measure the distance of objects ahead (up to 200m) and it has a narrow beam. SRR offers a broader beam and is used to monitor the vicinity of a vehicle (within 30m). MRR, which can be installed on the front, the rear, or the side area of the vehicle, is used for different applications as shown in Figure 13.3. Detailed comparisons of the three radar types are given in Table 13.1.

13.3 Automotive Antenna Design Requirements

The antennas used for automotive communication and its radiation performance depend on the type of application. Table 13.2 gives the list of services available in

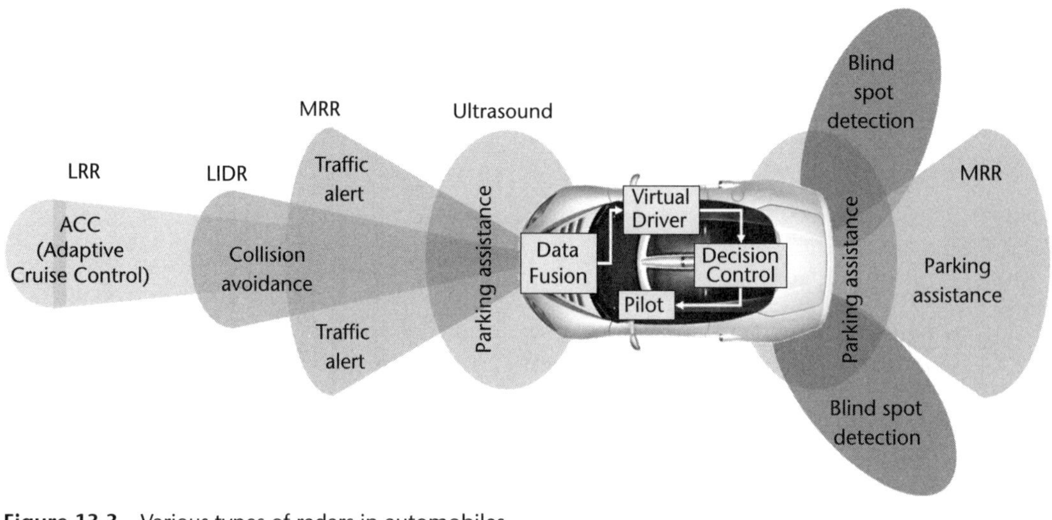

Figure 13.3 Various types of radars in automobiles.

Table 13.1 Automotive Radar Classifications

Radar Type	*LRR*	*MRR*	*SRR*
Range (m)	100–200	60–150	30
Applications	ACC	Cross-traffic alert (CTA), ACC	Blind spot detection, parking aid
EIRP	55 dBm	–9 dBm/MHz	–9 dBm/MHz
Frequency band	76–77 GHz	77–81 GHz	77–81 GHz
Distance resolution	0.5m	0.5m	0.1m
Velocity resolution	0.6 m/s	0.6 m/s	0.6 m/s
Velocity accuracy	0.1 m/s	0.1 m/s	0.1 m/s
3-dB beamwidth (Az)	±15°	±40°	±80°
3-dB beamwidth (El)	±5°	±5°	±10°

Table 13.2 Automotive Applications and Antenna Requirements

Serial Number	*Application*	*Frequency*	*Type of Radiation Pattern*
1	AM	1 MHz	Omnidirectional
2	FM	88–110 MHz	Omnidirectional
3	In-vehicle TV	50–400 MHz	Omnidirectional
4	DAB	100–400 MHz	Omnidirectional
5	GSM, 3G, and LTE	800/900 MHz	Omnidirectional
		1,800/1,900 MHz, 2,600 MHz	Omnidirectional
6	Wi-Fi and WiMAX	2.4 GHz and 2.3, 2.5, and 2.7 GHz and 3.5 GHz	Omnidirectional
7	V2V	5.9 GHz	Omnidirectional
8	RKES	315 MHz/413 MHz/434 MHz	Omnidirectional
9	Navigation system (GPS)	1.176 GHz, 1.382 GHz, 1.575 GHz	Directional
10	Satellite radio	1.472 GHz	Directional
11	SDARS	2.3 GHz	Directional
12	ETC	5.8 GHz	Directional
14	Automotive radar	24 GHz and 77 GHz	Directional
15	TPMS	315 MHz/413 MHz/434 MHz	Isotropic
16	Bluetooth	2.45 GHz	Isotropic
17	Wi-Fi (internal)	2.45 GHz	Isotropic

modern automobiles and the corresponding antenna radiation requirements. The radiators used in vehicles are grouped into three major categories, namely, omnidirectional, directional and isotropic. The choice of antenna technology depends on the antenna's performance with respect to the intended application and the location where the antenna is to be placed in the vehicle. Some of the technologies are rod antennas, printed antennas, on-glass antennas, and glued foil antennas. Automobile antennas should have following characteristics:

- Multiband capability and good spatial coverage;
- Compact design;
- Fits into the overall picture of the car body;
- High flexibility with a minimum number of antenna modules;
- Diversity and MIMO capability;
- Extremely cost-effective;
- Low in-car SAR exposure.

13.4 Antennas for Automotive Radars

As an indispensable part of the automotive radar system, the antenna is the main component in radar. The structure of antenna should be as compact as possible for easy integration into the vehicle and it must be suitable for mass production to minimize the average cost per unit. An automotive system, performance-wise, is required to maintain a low sidelobe level, so that misdetections could be avoided. All these facets pose challenges to the antenna design for automotive radars. In the earliest days of automotive radar, the horn antenna or parabolic antenna was often chosen for the system [1, 2]. However, these antennas are too large to fit in the modern vehicle; also due to the high cost, they are very impractical for commercial use. Revolution happens when the development of printed technology makes planar antennas available even at millimeter-wave frequencies. Planar antennas could be mass-produced at a very low cost because these are more compact and more appropriate for commercial applications. Based on the vehicle design, these antennas are placed at various locations in cars as shown in Figure 13.4.

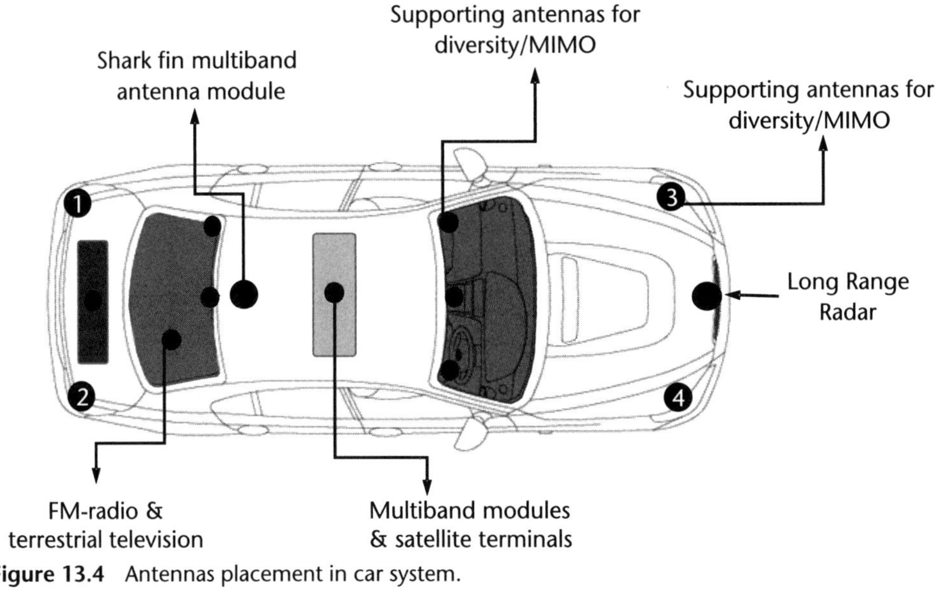

Figure 13.4 Antennas placement in car system.

13.4.1 Low-Frequency Automobile Antennas

AM/FM antennas used in automobiles are low-frequency antennas. AM radio works in the frequency band between 535–1,605 kHz, while FM radios function from 88–108 MHz [3]. This makes AM/FM radio communications the two lowest-frequency forms of wireless communications in vehicles. Because many antenna dimensions correspond to the electrical length at the desired frequency of operation, antennas that operate at lower frequencies tend to be larger than those that operate at higher frequencies. Blade, or shark-fin, antennas represent most of the radio antennas in modern vehicles [4, 5]. These are composed of vertically oriented PCBs covered with a radome. Their appearance and small size allow them to be mounted on the roof of the vehicle, which limits the interference that the antenna experiences from the vehicle itself. A vehicle designer wanting to avoid a shark-fin antenna may go with antennas printed onto the windshield and/or the rear windshield. These on-glass antennas have received much interest in recent years due to the immense number of potential designs that can be implemented on the large glass surfaces found in automobiles [6, 7].

DAB is another low-frequency antenna used in vehicles. These antennas may be like AM/FM antennas, as the DAB band (100–400 MHz) overlaps the FM band (88.0–108.0 MHz). Just as the DAB band overlaps the AM/FM radio at the lower frequencies, it also overlaps RKES and TPMS at the higher frequencies of the band. RKE is a wireless communication system in which a driver utilizes a transmitter attached to their key ring to communicate with a receiver on their vehicle. This can provide several convenient functions, such as locking and unlocking doors, opening automatic doors from a distance, and starting the vehicle. The TPMS comprises air pressure sensors on all four tires paired with wireless transmitters. These communicate with a receiver mounted near the center of the vehicle, which relays the information to an electronic control unit (ECU) mounted in the instrument panel. The ECU uses the data from the receivers to calculate the tire pressure, and it alerts the driver if an irregular tire pressure is detected. The TPMS and RKES both operate between 300 and 400 MHz. Also, both systems are designed to work only at a short range in and around the vehicle. Because of this feature, these antennas are different from dedicated short-range communications (DSRC) antennas that must operate at medium to long ranges strictly outside the vehicle.

13.4.2 High-Frequency Automobile Antennas

Satellite navigation antennas communicate with GPS satellites to calculate the vehicle's location and velocity and provide the user with directions to their destination. These systems operate with a center frequency of 1.575 GHz [7]. When GPS antennas are discussed, they are often mentioned in conjecture with Satellite Digital Audio Radio Service (SDARS) antennas. Although there is a sizable gap between the center frequency of GPS systems (1.575 GHz) and that of SDARS antennas (2.3 GHz), their operation requirements are similar [8]. Bluetooth technology (2.4 GHz) is often used inside vehicles to create a link between passengers' mobile phones and the vehicle's onboard audio and navigation system. Some vehicles employ Wi-Fi to communicate with passengers' wireless devices within the vehicle cabin, although this is less common (2.4 GHz). Vehicle detection radar systems typically use antenna arrays to detect large metallic objects. The system detects not only the presence of the object, but also the speed of the object relative to the speed of the vehicle to which the radar is attached. Radar antenna arrays operate at very high frequencies (~74 GHz) and are attached to the front and rear bumpers of the host vehicle. Because of this, they have minimal interference from other wireless.

13.5 Automotive Glass-Mounted FM/DAB Antennas

The lowest frequencies used in vehicles are often for AM and FM radio as shown in Figure 13.5. At the beginning of car radio history, a quarter-wavelength FM rod antenna played the major role in automotive radio reception. Over the years, new antenna concepts, such as a windscreen-printed antenna, have emerged [9]. These antenna structures, inconspicuously embedded into the windscreen of a car, offer several significant advantages over standard rod antennas on a vehicle. Another trend is represented by the integration of as many services as possible in a single antenna. Nowadays, there is a steadily increasing demand towards DAB (1,452–1,492 MHz), which offers better audio quality than FM and is more suitable for mobile scenarios.

Many glass-mounted antennas are based on wire geometry, although the antenna may or may not be an actual wire. A glass wire antenna is formed by using the wire of a very thin diameter or a silk-screened film that is laminated between layers of glass in the vehicle windows. No additional aerodynamic drag is provided by glass-mounted antennas and it also creates no wind noise, which is a significant advantage over mast-type designs. Glass antennas are more directional than mast antennas, which can lead to nulls in the reception on certain angles around the vehicle. Different shapes are used for wire glass antennas, often forming a grid or

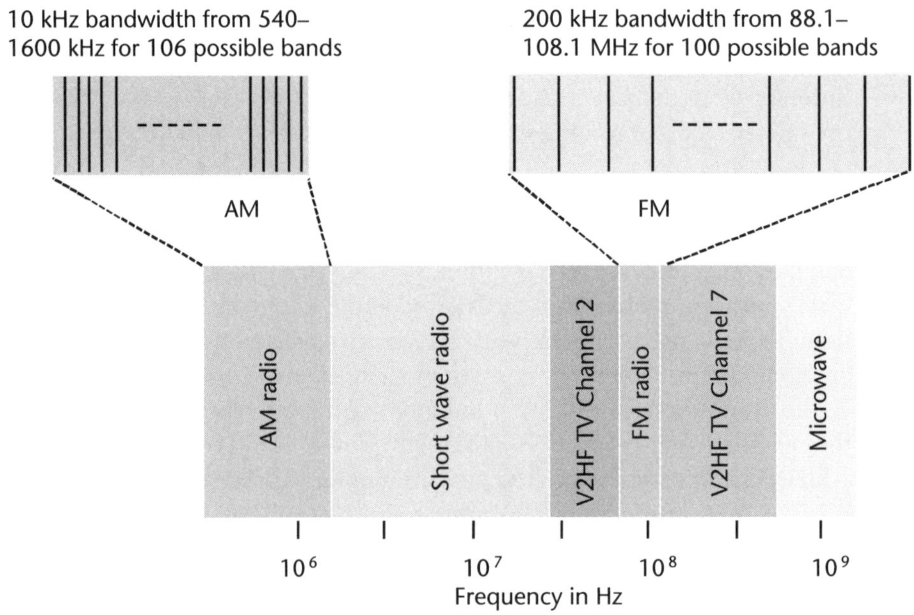

Figure 13.5 AM and FM frequency bands in radio communication.

meandering geometries, with a shape that works well on one vehicle not necessarily performing well on other vehicles. No universal glass-mounted antenna has yet been discovered. This is because of the vehicle body on the antenna's impedance and radiation, which is significant for on-glass antennas. Antennas oriented vertically may provide better reception of vertically polarized signals.

A windscreen antenna is designed for the services FM 88–108 MHz and DAB in the VHF-band (174–240 MHz) as shown in Figure 13.6 with a 1.5-mm metallic copper strip.

The heater-grid antenna is more effective at AM frequencies than the standard whip monopole antenna. The reason that the heater-grid antenna is more effective is that it has a larger effective height than the whip antenna. A preamplifier is used to compensate for poor radiation performance of the heater-grid antenna. The distance between the horizontal lines on the heater-grid is electrically small (at AM/FM frequencies), such that the RF current flow in the horizontal direction is the same as it would be on an equivalent conducting plate. To allow the RF current to flow vertically, as it would on a conducting plate, 1 to 3 additional vertical lines must be added to the heater grid. The antenna copper structures are placed on the windscreen of the car. This adaptive windscreen antenna consists of multiple FM and DAB antenna structures. The different frequency bands are combined by means of filter circuits. Glass thickness is 3.2 mm, and permittivity is 7.4. Figure 13.7 shows the simulated result of the antenna using the MoM.

13.6 Shark-Fin Antenna

Shark-fin antennas have gained considerable attention, to meet the demands of built-in system to support Wi-Fi and LTE inside the automobile. Low-profile design is necessary as several dedicated antennas are placed inside a small shark-fin module [10–12]. As a result, antenna performance can be largely affected by the

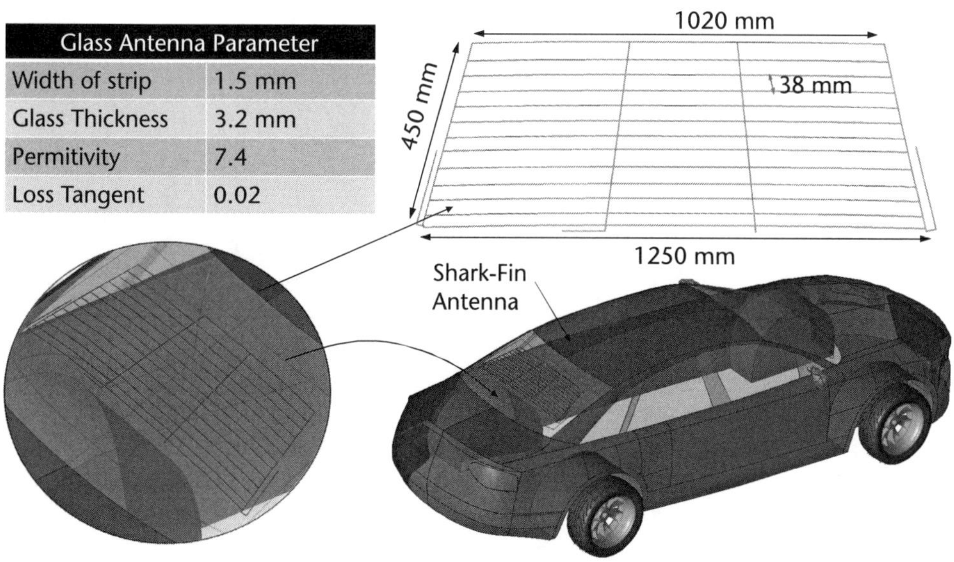

Glass Antenna Parameter	
Width of strip	1.5 mm
Glass Thickness	3.2 mm
Permitivity	7.4
Loss Tangent	0.02

Figure 13.6 Antenna placement on the car and glass antenna design parameters.

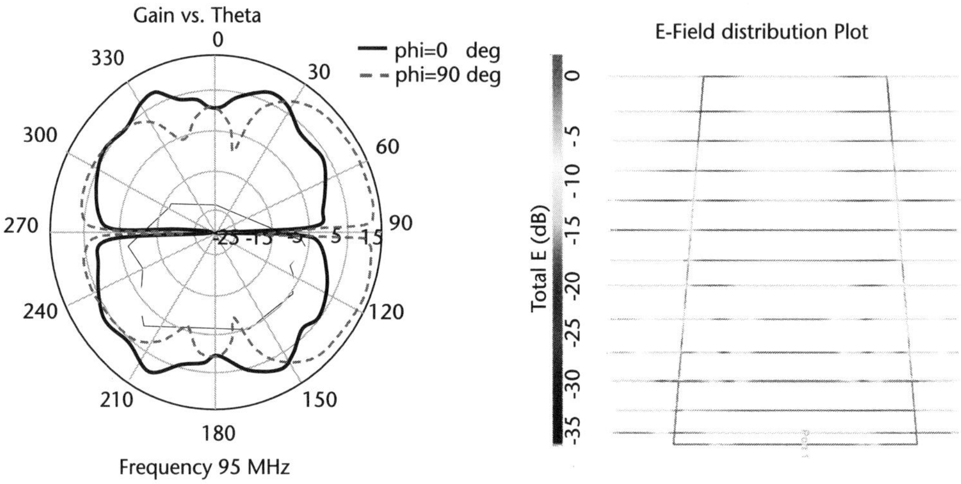

Figure 13.7 Antenna radiation patterns and current distribution over the antenna.

limited space available in typical shark-fin modules. This module is typically shaped like a blade or dorsal fin and is usually located on the roof towards the rear of the vehicle as shown in Figure 13.8. One of the main challenges is the antenna integration in a small-form factor such as the shark-fin case on the roof of a vehicle and its multiband behavior between antenna elements. Antenna elements placed closely should be highly isolated within a small volume for covering simultaneous multiband operations. In automobiles, several cellular and ad hoc-based communication standards are under consideration for offering a safe, efficient, and comfortable driving environment. For LTE, GPS, WLAN, and car-to-car (C2C) communications, antenna design becomes challenging in terms of compactness, bandwidth, efficiency, and mutual coupling. Cellular bands such as LTE700, LTE750, LTE800 MHz, LTE1800 MHz (1,710–1,880 MHz), LTE2600 MHz (2,520–2,690 MHz),

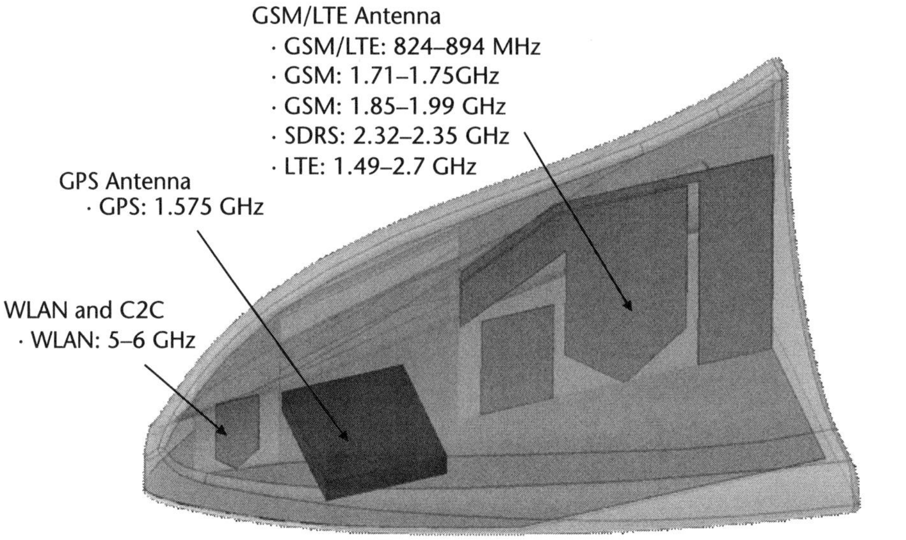

Figure 13.8 Designed shark-fin antenna covering several cellular and ad-hoc-based communication standards.

and GSM900 require the operation along with many other standards, for example, DCS1800 and PCS1900 (from 1,710–2,700 MHz). The DSRC system working in 5,850–5,925-MHz frequency band is a new standard intended for intervehicular communications to improve traffic safety for Intelligent Transportation Systems (ITS). A Wireless Access in the Vehicular Environment (WAVE) protocol, the IEEE 802.11p standard at 5.9-GHz band, can be a great candidate for vehicle-to-vehicle communication due to its real-time operation when anticollision messages and safety information must be provided to vehicles.

A shark fin antenna as shown in Figure 13.8 is designed to cover the frequency range of 698–2,700 MHz (VSWR < 3), which would serve major GSM/LTE standards. A GPS antenna for 1.575 GHz and one-tab monopole antenna covering the frequency range of 5–6 GHz is integrated into the shark-fin module and provide services for WLAN and C2C. Three antenna elements are used to cover all frequency bands and applications: one element for the low LTE band, the second element for GPS, and the third element for the WLAN band. In the case of multiple antennas placed side-by-side, high mutual coupling between the two antenna elements is a serious issue. Additional filters are needed to mitigate mutual coupling but with extra fabrication costs. In this design, low mutual coupling (lower than –22 dB) between the two antennas is achieved in the frequency range of 698–6,000 MHz without any additional filters.

13.6.1 LTE/GSM Antenna Design

A monopole antenna has an omnidirectional radiation pattern and it radiates with equal power in all azimuthal directions perpendicular to the antenna, which is why it is a suitable antenna choice for LTE frequency bands. It is possible to reduce the antenna height using an L-shaped monopole as shown in Figure 13.9. For conformal placement of the LTE antenna, the tail of the monopole is bended as per shark-fin geometry [13]. The length of the L-monopole determines the lower resonance. The first resonance of the monopole antenna is mainly determined by the length of

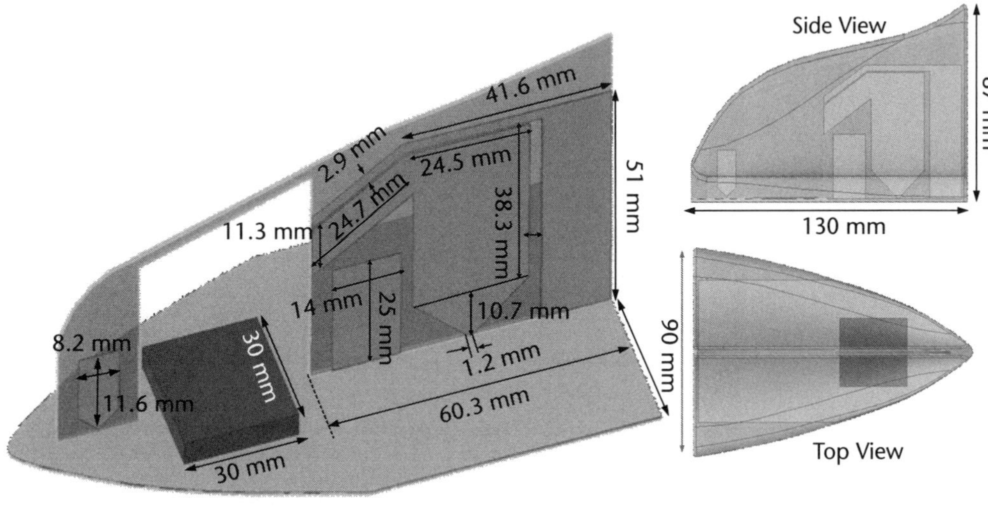

Figure 13.9 Design dimensions of LTE/GSM and WLAN shark-fin antenna.

the right-side sleeve, and the second resonance is attributed mostly to the length of the main central monopole. As both the left sleeve and the monopole have similar L-shape structures, their lengths are about 0.3λ. The VSWR plot is shown in Figure 13.10 for all antenna resonance frequency in GSM/LTE band. Bluetooth is incorporated into many present-day vehicles to allow hands-free calling and the operation of an equipped mobile through the vehicle's multimedia system.

To depict the working principles of the LTE and WLAN antennas, its E-field distributions are shown in Figure 13.11. This antenna can be used for the 2.4-GHz ISM band, which has seen enormous growth in the past decade due to the ubiquitous application and implementation of Wi-Fi and Bluetooth, which occupy part of this band.

13.6.2 GPS Antenna

Guidance and navigation facilities are becoming more cost-effective and are seeing a large uptake in the modern market. These navigation systems usually rely on the constellation of approximately 30 GPS satellites to determine the location of the vehicle before plotting it on a map. The GPS L1 band is received in a narrow 20-MHz channel centered at 1.575 GHz. The GPS antenna is placed between LTE and WLAN antennas. A square patch of 30 mm × 30 mm with a diagonal probe feeding as shown in Figure 13.12 is used as radiating elements. The patch is printed on an Arlon1000 substrate (with a relative permittivity of 10.2, a loss tangent of 0.003, and a thickness of 1 mm).

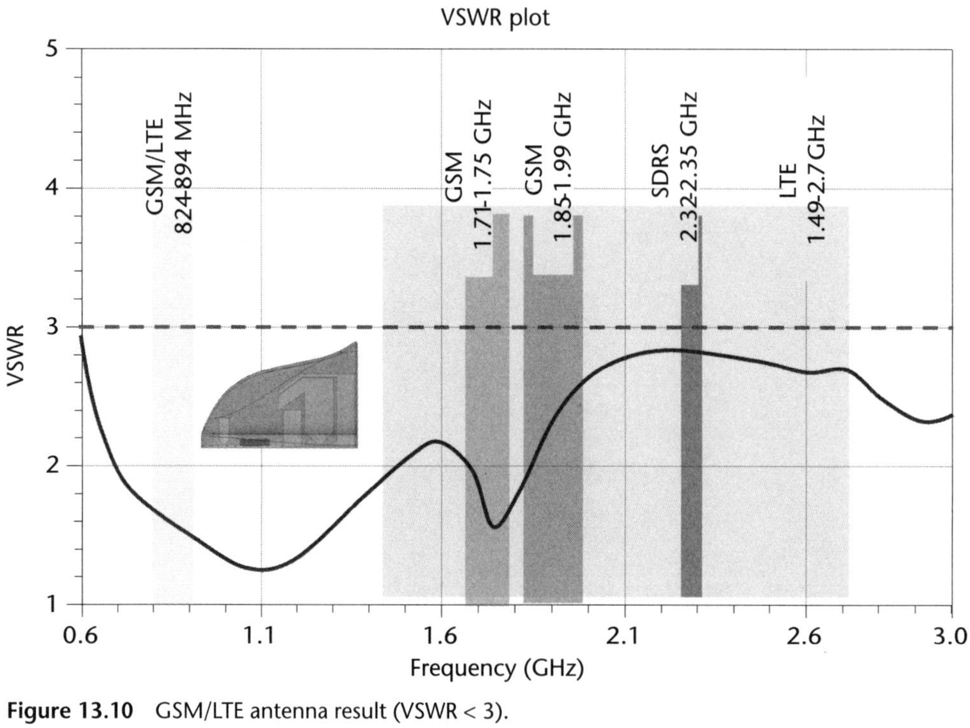

Figure 13.10 GSM/LTE antenna result (VSWR < 3).

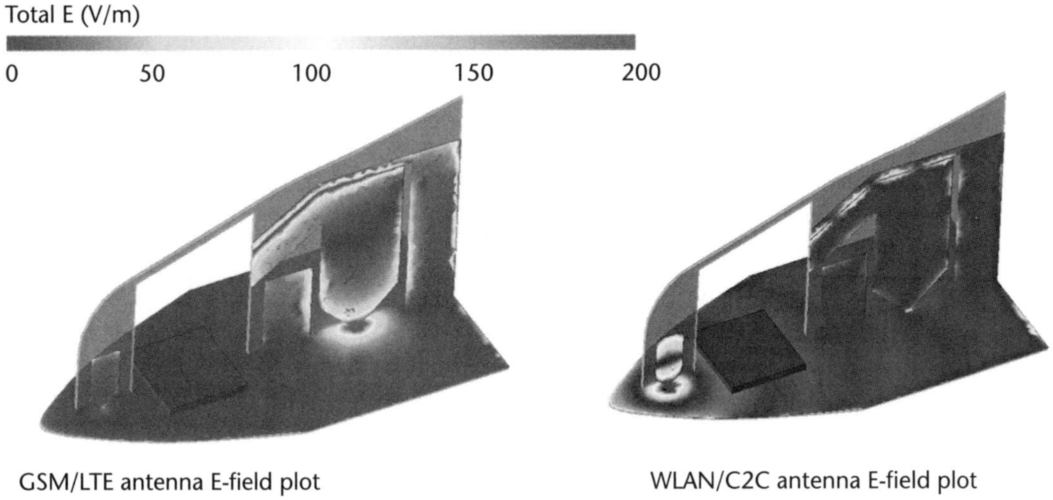

Figure 13.11 E-field distribution on LTE and WLAN antenna.

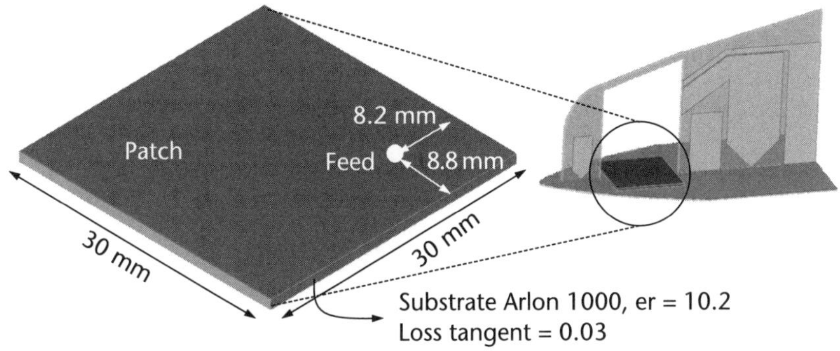

Figure 13.12 GPS microstrip patch antenna design dimensions.

The simulated gain of the antenna is 2.1 dBi as shown in Figure 13.13 at 1,575 MHz, a reasonable value for the square patch on a high permittivity dielectric substrate.

Sometimes GPS antennas are designed for circular polarization by truncating a corner of a square patch or with one or two feed points. Based on application and design, a circular polarization can also be obtained from a single-point square or a circular patch.

13.6.3 WLAN Antenna

A WLAN antenna provides services to high-band WLAN (IEEE 802.11n, 802.11ac) and C2C communications (IEEE 802.11p). For the WLAN antenna, a tab monopole with dimensions of 11.6 mm × 8.2 mm is used to cover the frequency range of 5–6 GHz [14]. It fits the shark-fin module. As it is electrically small and impedance=mismatched in the working frequency range of the LTE antenna (i.e. 698–2,700 MHz), low mutual coupling levels between them are expected in this

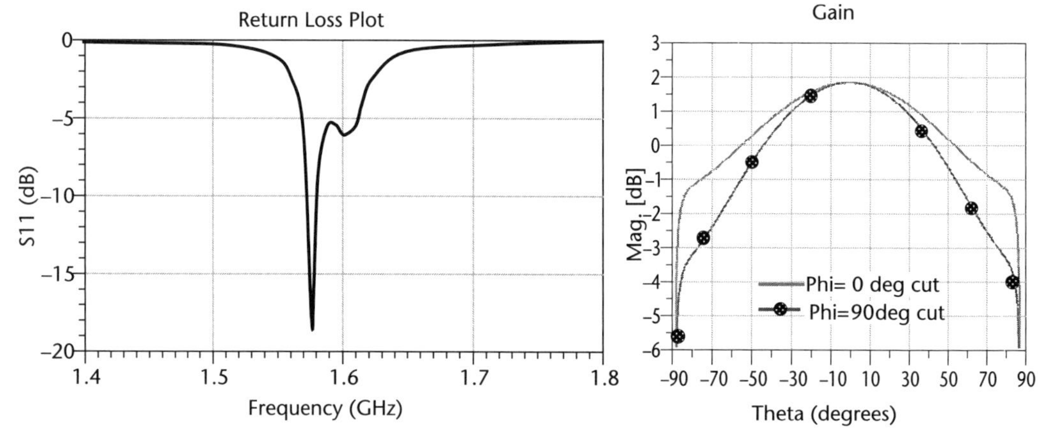

Figure 13.13 Simulated return loss and antenna gain plot of the GPS antenna.

range. The antenna resonating frequency range is between 5 GHz and 6.5 GHz, as shown in Figure 13.14.

13.7 Grid Array Antenna for 24-GHz Automotive SRR

The two most popular millimeter-wave automotive radars are LRR and SRR systems. The LRR operating at the 24/77-GHz frequency spectrum is characterized by narrow beams to control the driving path in the front of the car and to estimate the relative distance between the car and the vehicle ahead of it. This ensures a minimum safety distance between the vehicles and hence reduces the accident risk. The LRR can range up to 250m and requires a bandwidth of less than 1 GHz with a typical spatial resolution of 0.5m [15]. However, the SRR gives a wide field of view for target detection near the car, typically around 30m. The modern vehicles are fitted with at least 10 SRRs to ensure the safety aspects of the vehicle. In this section, an antenna design is discussed for the SRR that requires a wideband operation for

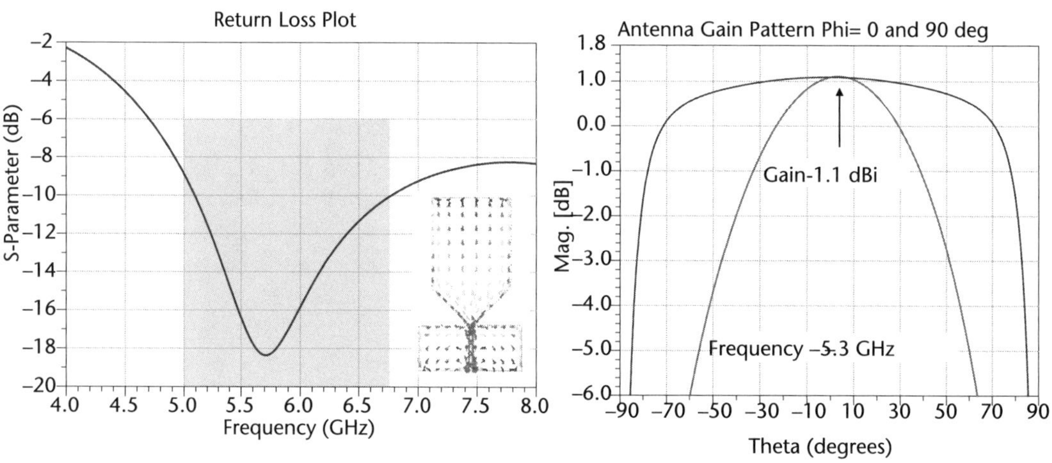

Figure 13.14 WLAN and C2C antenna return loss and radiation pattern.

interference-free target detection. A very specific type of antenna array called the grid array antenna (GAA), which has gained recent research interest for automotive radar antenna development, is designed.

GAA was proposed by Kraus [16] in 1964. He found that the performance of the GAA depends largely on the synchronization of surface currents. In 1981, Conti et al. [17] adopted it as a monopulse antenna. The GAA for 24-GHz automotive radar sensor applications was first studied by Zhang et al. [18, 19]. A GAA, which is backed by many grid cells, is a low-profile, flat antenna consisting of numerous radiation and transmission-line elements. The microstrip GAA has all the usual benefits of conventional printed radiators along with broad bandwidth, high gain, adequate cross-polarization control, and beam scan ability with frequency. The grid is a periodic structure of rectangular frames with dimensions equal to about one wavelength by one-half wavelength.

The GAA consists of horizontal transmission lines and vertical radiating elements. The radiation originates from the short edges, while long edges act as transmission-line feed elements. The width and length of the transmission line control the impedance and frequency specifications, respectively. There is no beam scanning in GAA, but it is designed to radiate a narrow beam along the elevation plane and a broad beam along the azimuth plane to enhance the lateral coverage of the automotive radar system. In the GAA, the current on the vertical radiating sections are aligned in phase and constructively interfere to produce radiation at the far-field distance, whereas the current in the transmission line, which is separated by a full wavelength, changes phase and interferes destructively. Therefore, the horizontal sections contribute only to a small radiation. Hence, the GAA produces a narrow beam along the elevation plane and a broad beam along the azimuth plane. This antenna is positioned vertically in the front side of the car beneath the plastic bumper to achieve the lateral coverage.

Figure 13.15 shows geometries of the 24-GHz microstrip GAA. It comprises a radiating grid, a substrate, a metalized ground, and one feed probe. The radiating grid with 61 meshes is etched on the top surface of the substrate. The length of the short side is 4.6 mm, which is approximate to half of the guided wavelength at 2 GHz, and the length of the long side is 9.23 mm (7 mm +2.23 mm), which is approximate to one of the guided wavelengths. The width of the short side is 2.23 mm, while the width of the long side is 0.38 mm. The substrate is a 65 mm × 65 mm square. A coaxial probe is drilled through the ground and the substrate to feed the radiating element. Rogers 5880 with an ε_r of 2.2 and a tangent loss of 0.001 with a thickness of 0.787 mm is chosen as the board material.

To monitor blind spots, 24-GHz automotive radars are used that require the maximum directivity of the antenna to be larger than 20 dBi, the impedance and radiation bandwidth to be larger than 0.25 GHz [18]. The radiation characteristics are evaluated using MoM based on the obtained current distribution along the GAA. Figure 13.16 shows $|S_{11}|$ as a function of frequency. The input impedance bandwidth is 2.4 GHz. The simulated 3-dB gain bandwidth is 1.54 GHz from 22.8 GHz to 25.2 GHz with a maximal directivity of 20.21 dBi at 24 GHz. The radiation patterns at 24 GHz in phi = 0° and 90° cut-planes are shown in Figure 13.16.

Another configuration of the 24-GHz microstrip GAA is shown in Figure 13.17. It is composed of 25 grids. This antenna can fit in the circular aperture radar.

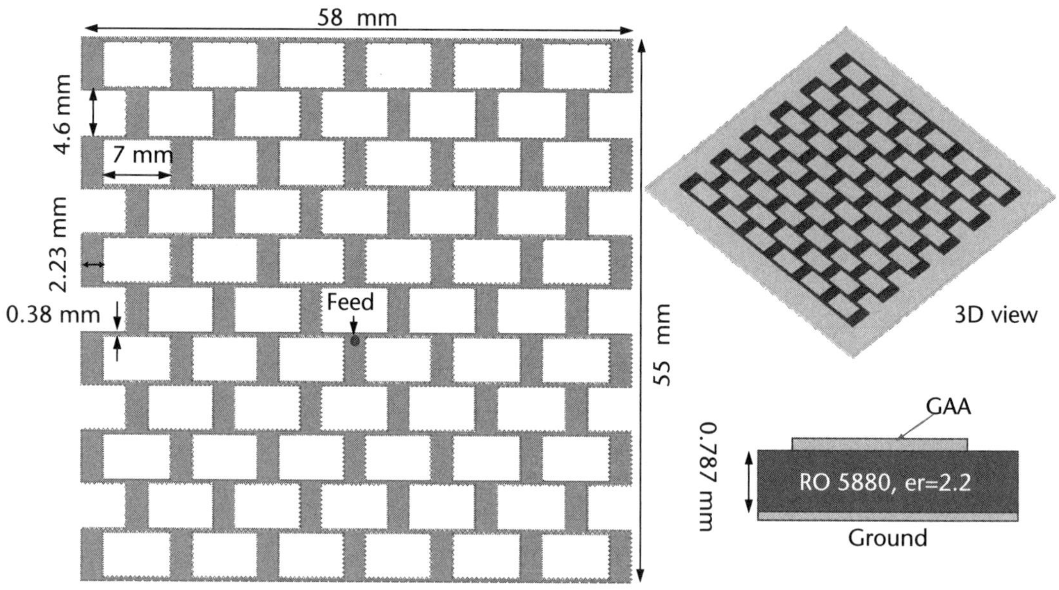

Figure 13.15 Design dimensions of a 24-GHz microstrip GAA.

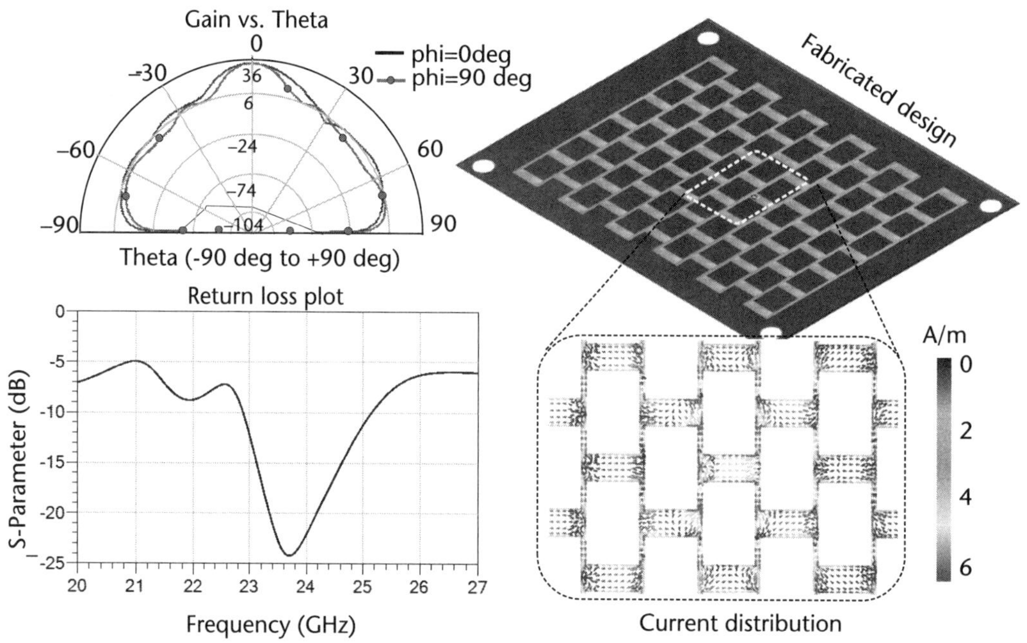

Figure 13.16 Fabricated 61-grid microstrip GAA and results.

13.7.1 Microstrip Comb-Line Antenna Arrays for LRR

The microstrip comb-line antenna array is a form of GAA discussed previously. As shown in Figure 13.18, if we remove the top and bottom conducting lines from the GAA, then it becomes a comb-line antenna.

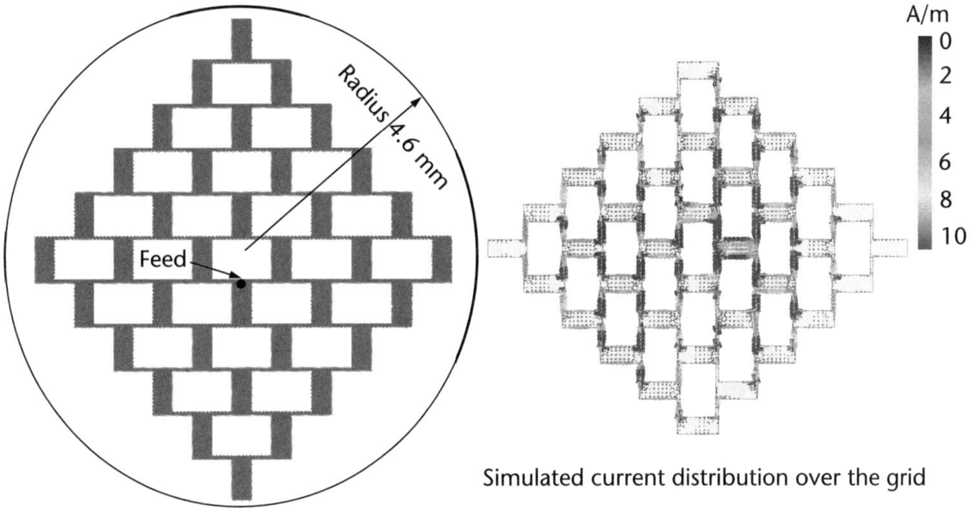

Figure 13.17 GAA for circular aperture.

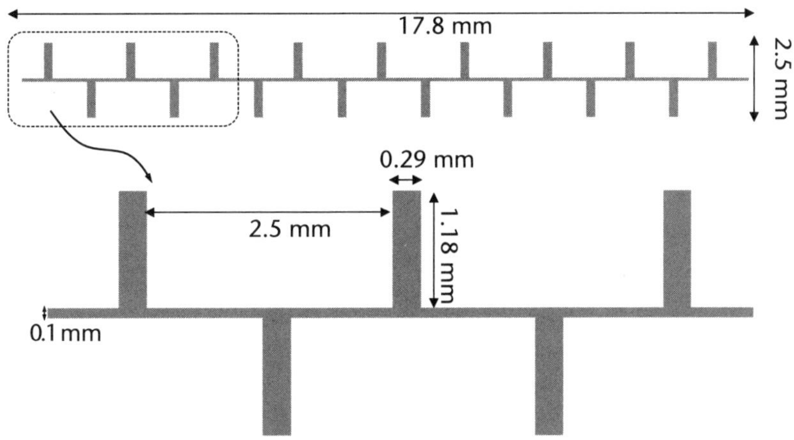

Figure 13.18 Microstrip comb-line antenna array for LRR.

The comb-line array should be realized in a standing-wave configuration with 45° polarization orientation for an automotive LRR application. To begin with, 90° linearly polarized arrays are first designed. Figure 13.18 shows a 90° uniform comb-line antenna array in a standing-wave configuration. The length of open stub and the spacing between adjacent stubs both equal a half-wavelength; the amplitude distribution of the array is controlled by the open stub width. In the simplest case of uniform array, the width of each stub is the same, so power will be evenly spread over the aperture. A 5-mil-thick Rogers RO3003 substrate with an ε_r of 3.2 and a loss tangent of 0.016 is used for antenna design. The length and width of a 50Ω half-wavelength stub are 1.18 mm and 0.29 mm, respectively. The feedline of 0.1-mm width with the distance between adjacent stubs of 1.48 mm is used to feed the signal.

The simulation results of the reflection coefficient from 74 to 84 GHz and radiation patterns at 77 GHz are presented in Figures 13.19 for 17 elements. The return loss is better than 10 dB from 75.5 to 78.2 GHz. The 3-D radiation pattern demonstrates a broadside beam with a fan-shaped main lobe. Peak radiation occurs and the maximum gain is 13.1 dBi. In the $\theta = 0°$ plane radiation pattern, the half-power beamwidth (HPBW) is approximately 12° and the sidelobe level is –12.5 dB.

13.7.2 Amplitude Tapered Antenna Array

Theoretically, the best sidelobe level that a uniform linear array could obtain is about –13.46 dB. To accomplish better sidelobe suppression, nonuniform amplitude distribution is formed on the array by applying several amplitude tapering methods. In this design, Taylor distribution is selected as this technique offers an optimal balance between beamwidth and sidelobe level. A compromise between the Dolph-Chebyshev and the Taylor one-parameter distribution was developed by Taylor and called the $\bar{n}$ distribution [20]. This method offers a modest improvement in efficiency over the one-parameter distribution by making the first few sidelobes at an equal level. The radiation pattern of a Taylor weighted array exhibits the following properties: the first $\bar{n}$ sidelobes near the main lobe are at the same level $1/R_0$, while the rest decrease monotonically. Both parameters $\bar{n}$ and $1/R_0$ are specified by the designer. The value of $\bar{n}$ should be carefully determined, as it will influence the peak positions of amplitude distribution on the array. Generally, for –25 dB, the sidelobe level should be no less than 3 [21]. To improve the sidelobe suppression, the uniform array is modified with a 20-dB ($\bar{n}$=3) Taylor amplitude taper [22]. The corresponding voltage ratio is

$$R_0 = \cosh(\pi.X) = 10 \tag{13.3}$$

The constant X can be calculated by

$$X = \frac{1}{\pi}\text{sech}^{-1}(R_0) = 0.95 \tag{13.4}$$

The scaling factor σ is given by

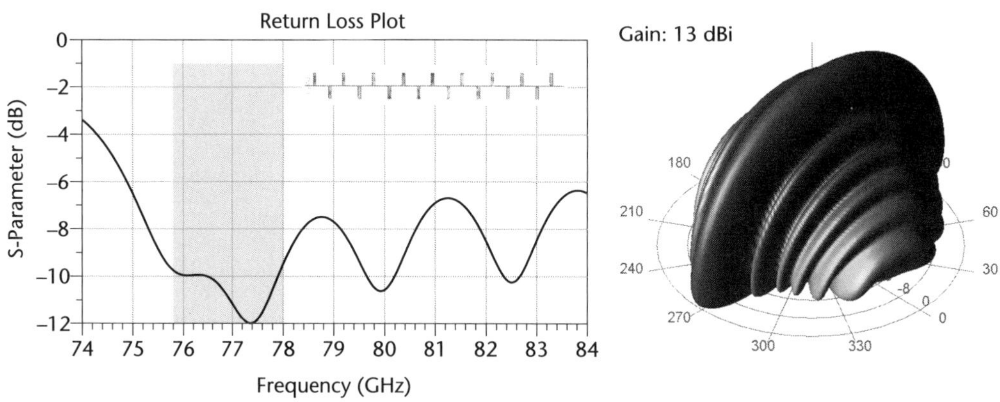

Figure 13.19 Antenna performance results of the comb-line array.

$$\sigma = \frac{\bar{n}}{\sqrt{X^2 + \left(\bar{n} - \frac{1}{2}\right)^2}} \tag{13.5}$$

and the normalized current distribution over the array can be expressed as

$$I(z) = \frac{\lambda}{l}\left[1 + 2\sum_{p=1}^{\bar{n}-1} SF(p, X, \bar{n})\cos\left(2\pi p \frac{z}{l}\right)\right] \tag{13.6}$$

where Taylor space factor SF(p, X, ($\bar{n}$) is given as

$$SF(p, X, \bar{n}) = \begin{cases} \dfrac{\left[(\bar{n}-1)!\right]^2}{(\bar{n}-1+p)!(\bar{n}-1-p)!} \displaystyle\prod_{m=1}^{\bar{n}-1}\left[1 - \left(\frac{\pi p}{z_m}\right)^2\right], |p| < \bar{n} \\ 0, |p| \geq \bar{n} \end{cases} \tag{13.7}$$

where zeros z_m are

$$z_m = \pi \frac{l}{\lambda}\cos\theta_m = \begin{cases} \pm\pi\sigma\sqrt{X^2 + \left(m - \dfrac{1}{2}\right)^2}, 1 \leq m < \bar{n} \\ \pm m\pi, \bar{n} \leq m \leq \infty \end{cases} \tag{13.8}$$

and θ_m is the location of the mth sidelobe in the radiation pattern. The $\bar{n}$ distribution gives a narrower beamwidth and higher efficiency, without significantly degrading its robustness. Normalized current levels on the 90° comb-line antenna array with a 20-dB ($\bar{n}$ = 3) Taylor amplitude taper are given in Figure 13.20. Only half of the values need to be calculated because the distribution is symmetric. Once the current distribution is established, the following task is to decide the corresponding stub widths.

The stubs at two ends of the array (stub 1 and stub 13) will carry the lowest currents, so their characteristic impedances are the highest among all stubs. Stub 1 and stub 13 are assigned the minimum manufacturable line width of 0.11 mm, which corresponds to the characteristic impedance of 87Ω. The characteristic impedances of other stubs can be obtained by

$$Z_{0,i} = Z_{14-i} = \frac{I_1 Z_1}{I_i} = \frac{0.53x87}{I_i} = \frac{46.11}{I_i} \tag{13.9}$$

The –10-dB impedance bandwidth is from 74 to 78.4 GHz. As shown in Figure 13.21, a 77-GHz elevation plane radiation pattern has a –4-dB less sidelobe label compared to an array with equal amplitude distribution, the HPBW is about 12°, and the sidelobe suppression is better than 16.3 dB. The sidelobe suppression has been improved 4 dB after applying the amplitude taper, and the differences in the

Element No.	Normalized Current	Z_0 (Ohm)	Line Width (mm)
1 & 13	0.53	87.00	0.11
2 & 12	0.56	82.33	0.13
3 & 11	0.64	72.04	0.17
4 & 10	0.76	60.67	0.23
5 & 9	0.88	52.39	0.29
6 & 8	0.97	47.53	0.34
7	1.00	46.11	0.36

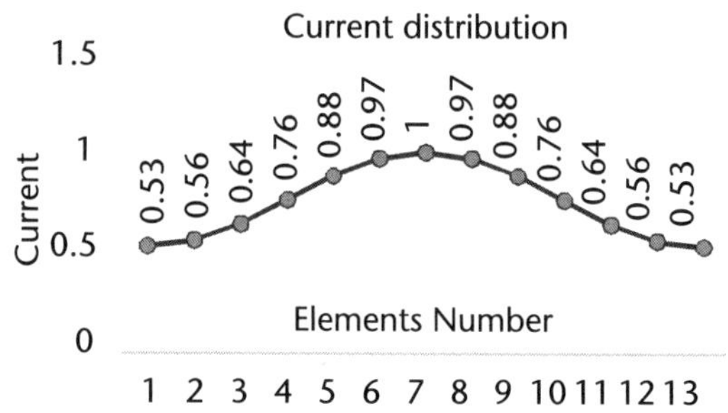

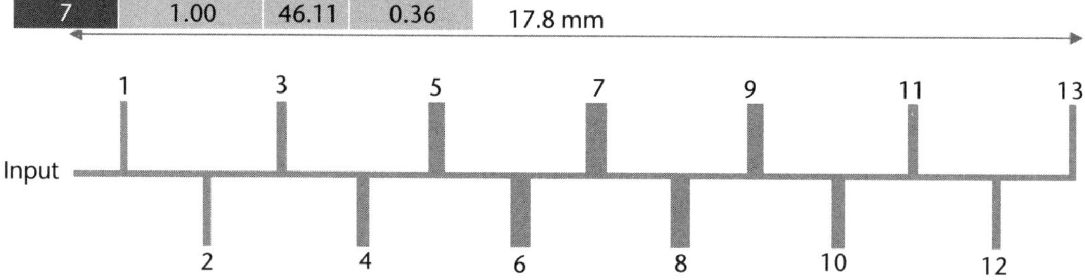

Figure 13.20 Normalized current distribution for the Taylor method with $N = 13$ and the sidelobe level 18 dB and ($\overline{n} = 3$) and the line width of each element's stub.

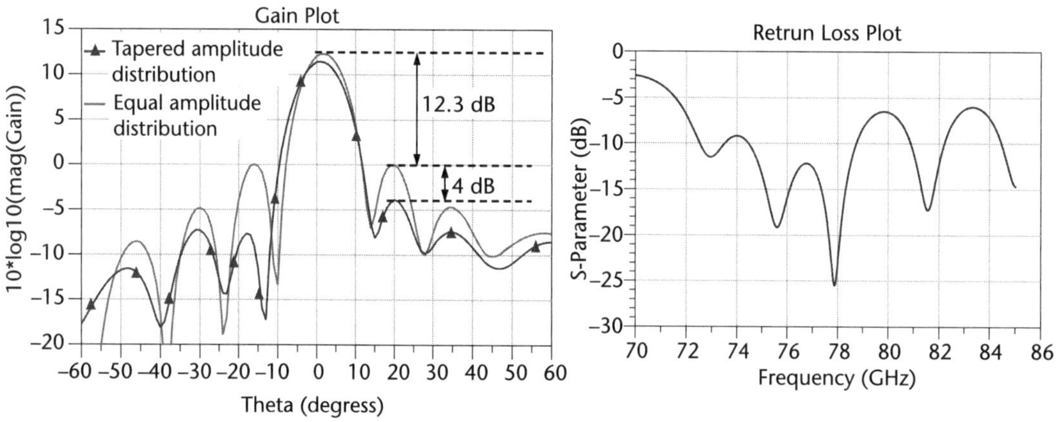

Figure 13.21 Phi = 0° plane radiation pattern (at 77 GHz) and return loss plot.

first three sidelobes' power levels of the Taylor array are much smaller than those of the uniform array, as it is expected.

13.7.3 Amplitude Tapered 45° Antenna Array

In a 90° comb-line antenna, all open-circuit stubs are perpendicular to the main feed line, so those arrays are 90° linearly polarized. The disadvantage of this configuration is that automotive radars equipped with such antennas cannot distinguish between the reflected signals and signals transmitted by vehicles from the opposite directions.

One solution is to create a 45 linearly polarized array by tilting the orientation of stubs to 45°, and then signals from vehicles traveling in opposite directions are orthogonal (assuming that 45° antennas are installed on both vehicles) and the interferences will be minimized. The design principle of 45° comb-line antenna array is very similar to that of a 90° array.

Figure 13.22 shows a 45° 13-element comb-line array with an 18-dB ($\bar{n}$ =3) Taylor amplitude taper. The stub widths are the same as the 90° array. The antenna results of the 45° amplitude tapered array are shown in Figure 13.22. The reflection coefficient is below 10 dB from 76.4 to 79 GHz. Using 8 linear arrays, a phased array as shown in Figure 13.23 is realized. This phased array antenna can communicate with multiple targets on the road. An antenna simulated radiation pattern is shown with a 20-dBi gain.

13.8 Planar Array for Millimeter-Wave Automotive Radar

Planar antennas are commonly used in sensing applications at frequencies as high as 77 GHz and above, for example, in automotive sensing. Early vehicular radar development was in the microwave frequency range of 10–24 GHz [23–30]. To reduce the physical size of the radar sensors for use on automobiles, the frequency of operation had to be increased to the millimeter-wave range. At those frequencies, the radar antenna can be made electrically large to produce the required high directivity and narrow beamwidth, while keeping the physical size small enough for installation on an automobile. Typical antenna requirements for millimeter-wave automotive radar are multibeam or scanning capability, high gain up to 35 dBi, and moderate to low sidelobes.

Radar technology has been investigated for use on automobiles since the 1970s and has been employed for various functions on automobiles since the early 1980s [31, 32]. The initial usage of microwave radar was for collision warning applications on commercial vehicles, such as ambulances, buses, and trucks. Now these

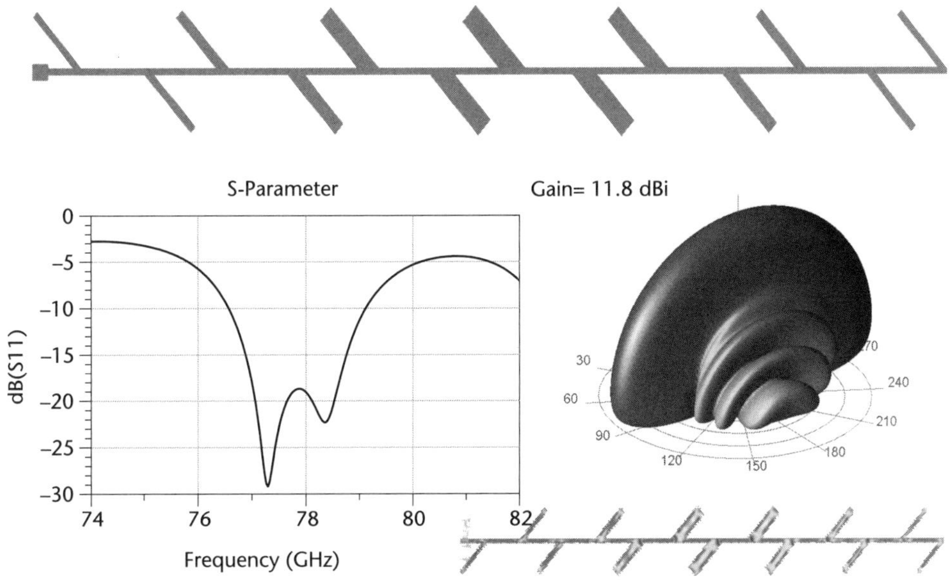

Figure 13.22 Layout design, reflection coefficient, and radiation pattern of the 45 amplitude tapered comb-line array.

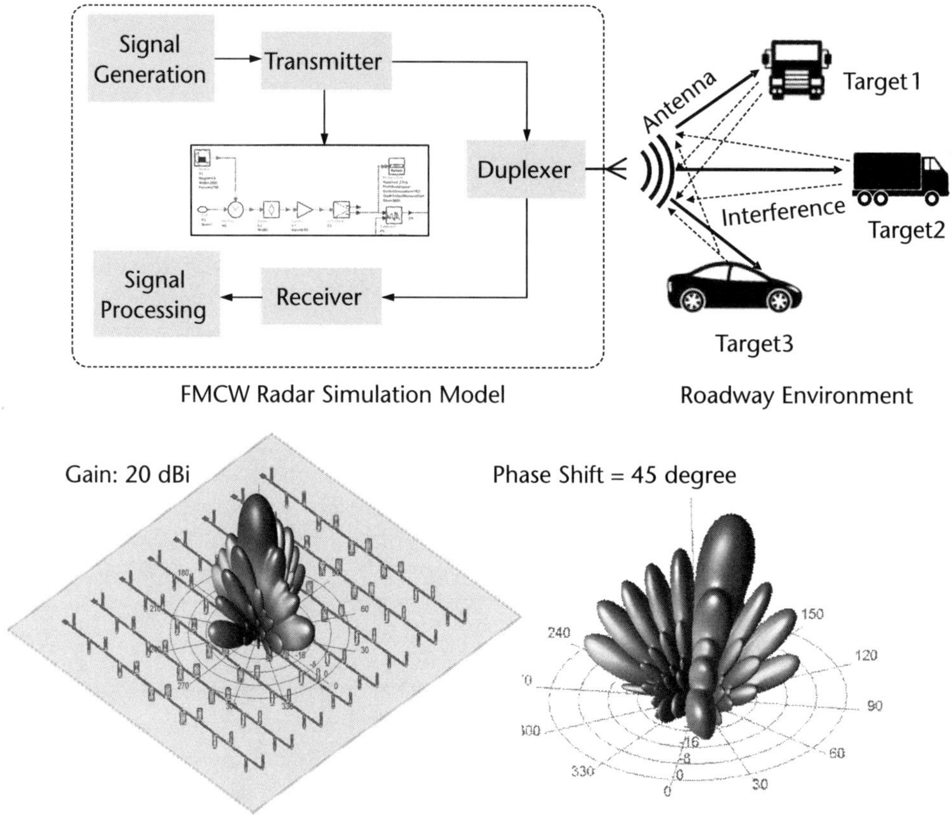

Figure 13.23 FMCW radar simulation model and antenna array 3-D radiation pattern.

radars are aimed to improve vehicle safety, efficiency, and economy. Radar can also be used for target identification and for detecting road conditions by making use of scattering signature information.

Automotive radar sensors are making it to the broader automobile market as part of ACC. ACC radar sensors provide range and closing-rate information to the cruise control system, which can control the brakes, throttle, and automatic transmission of the vehicle. The ACC system adapts the speed of the vehicle according to the speed of the vehicle ahead, to maintain a separation time interval between the two vehicles. The driver sets the minimum separation desired and the maximum speed. The radar system can also be used to locate and track multiple targets on the road ahead to anticipate traffic conditions in the driving lane. The same radar sensors used for ACC can also be used in pre-crash sensing, parking aid, and collision warning/avoidance (CW/A) systems, as shown in Figure 13.24. When used in a CW/A application, the radar sensors allow for warning signals to be given and for air bags to be activated. CW/A systems can also be designed to take control of a vehicle when a collision is anticipated.

13.8.1 ACC Radar System Requirements

The ACC radar must be able to detect targets of varying sizes up to 200m ahead. These targets could be moving relative to the radar at speeds of up to 250 km/h.

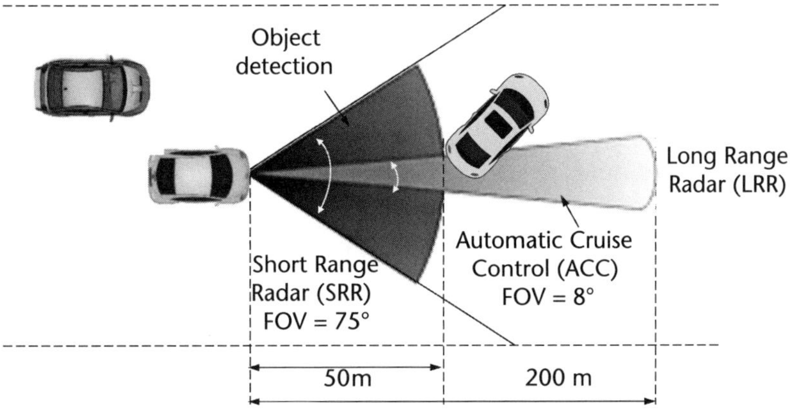

Figure 13.24 Precollision systems in vehicular radar to detect potentially dangerous crashes.

The radar must have an angular coverage of at least ±8° in the azimuth plane. Table 13.3 summarizes the typical performance specifications of ACC radar modules.

13.8.2 ACC Radar Antenna Design

A multimodal radar for an ACC system based on FMCW radar driving multiple antenna arrays is discussed in this section. This multibeam radar with an electronic antenna scanning technique operates at 77 GHz for long-range, narrow-angle coverage (200m, ±10°). This example illustrates the use of tapered series fed antenna array (15 × 16 elements) and Rotman lens for beam scanning. One of the primary requirements of ACC radar is low sidelobe levels. In a microstrip antenna array, the most commonly used approach for reducing the sidelobe level of a uniform array involves amplitude tapering, in which the excitation amplitudes of the array elements generally decrease with distance from the center of the array. To obtain the amplitude shading coefficients for a linear array of uniformly spaced point sources, several techniques [33, 34] have been developed.

The series-fed taper antenna array shown in Figure 13.25. The array mounted on the top of RT5880 dielectric substrate of 0.787-mm thickness and a relative

Table 13.3 Typical Specifications of ACC Radar

Parameters	*Typical Value*
Transmit frequency	76–77 GHz
Transmit frequency	>10 dBm
Target detection range	2m to 200m
Relative velocity	Up to 250 km/h
Angular coverage, azimuth	8.5° wide coverage with 3° minimum resolution
Angular coverage, elevation	3° to 4° single beam
Antenna gain	26–34 dBi
Antenna sidelobe level	>20 dBc

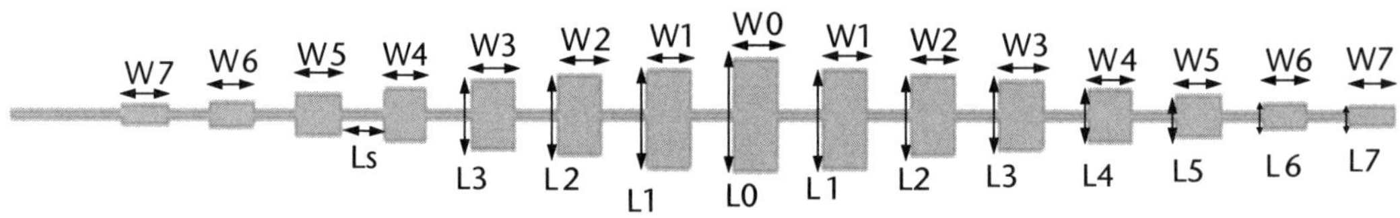

Figure 13.25 Tapered series fed microstrip antenna array design at 77 GHz.

dielectric constant $\varepsilon_r = 2.2$. The design parameter of antenna array is given in Table 13.4.

With end-fed arrays, the elements nearest the feed couple only a small amount of power and therefore must be narrow. The feed line (0.155 mm) is small compared to the narrowest patch. The radiating patches themselves are resonant so that the input line to a patch is matched. Each patch element is tapered according to the excitation current amplitude obtained from the Chebyshev synthesis method.

The antenna is simulated using the FEM full-wave analysis technique. The 1-D linear array has a –10-dB bandwidth of about 0.5 GHz, and the return loss at 77 GHz is –20 dB, the gain is 17.2 dB, the sidelobe level is approximately –24 dB, which meets the antenna sidelobe-level requirements for ACC radar, and the width of the E-plane is 7.2°. The planar antenna array is designed by using 16 linear antenna arrays as shown in Figure 13.26. The element spacing between the antenna elements is 0.5λ. The size of a 15 × 16 element antenna array is 41 × 43 mm.

The maximum array scan angle is ± 30°. The simulation result of its pattern is shown in Figure 13.27 for E-plane and H-plane. The sidelobe reduction technique is applied only in the E-plane direction. By comparing the 1-D linear array pattern, we can find that the E-plane pattern has not changed substantially, the beamwidth was compressed to 6.8° in the H-plane pattern, and the sidelobe level is –13.1dB.

13.8.3 Rotman Lens Design for Electronic Scanning of the Antenna Beam

The electronic scanning antenna (ESA) can position the antenna beam instantaneously to any position, but phase shifters for the millimeter-wave frequency range are pricey or for many frequencies not available. Rotman lenses offer a compact alternative for antenna scanning technologies. Multibeam concepts used the Rotman

Table 13.4 Tapered Series Fed Microstrip Antenna Dimensions

Parameter	*Value (mm)*	*Parameter*	*Value (mm)*
L0	2.205	L5	0.830
W0	1.255	W5	1.332
L1	1.874	L6	0.488
W1	1.306	W6	1.289
L2	1.546	L7	0.415
W2	1.264	W6	1.342
L3	1.314	Ls (spacing)	1.297
W3	1.131	Feed line W	0.155
L4	0.9776	W4	1.281

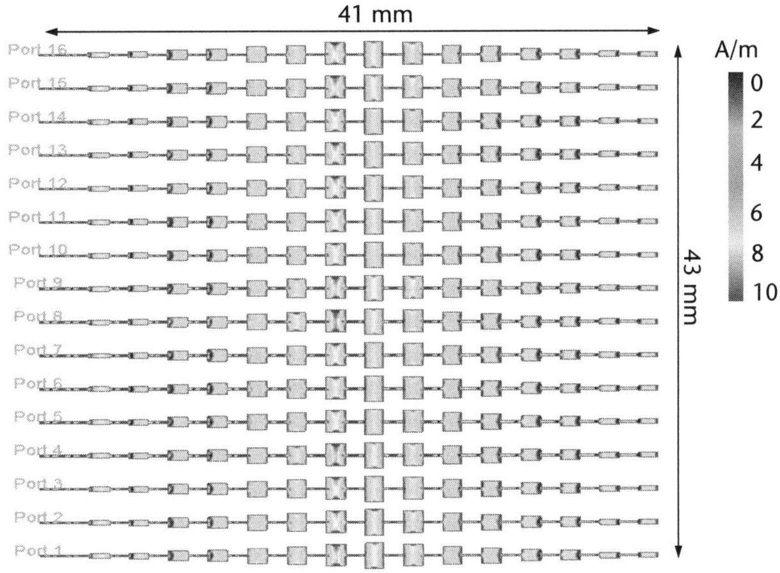

Figure 13.26 Current distribution in tapered series fed microstrip antenna array.

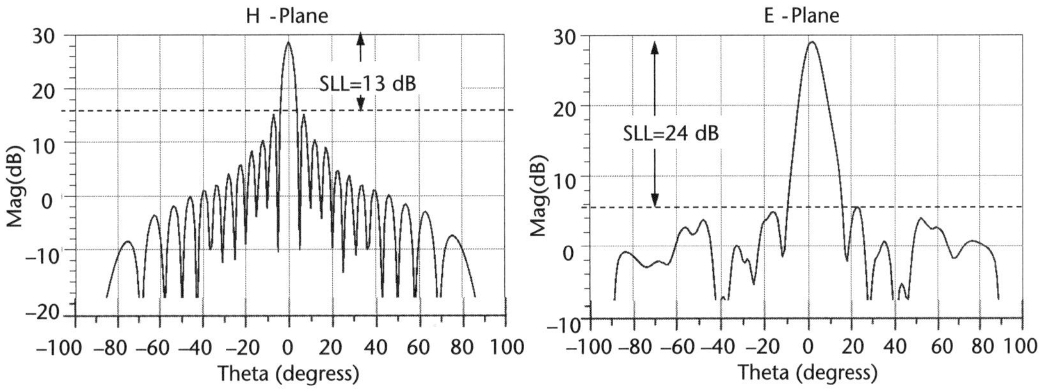

Figure 13.27 Tapered series fed antenna array radiation pattern at 77 GHz.

lens for the frequency range between 6 GHz and 94 GHz [35–38] and are used in many applications. A microstrip line connects each antenna input port with one antenna element as shown in Figure 13.28.

Based on the different lengths inside the Rotman lens, the radiated energy from the antenna elements will produce a beam along a defined direction. The Rotman lens equations include the design parameter for the parallel plate region and the length differences for the connecting lines between antenna port and antenna element. For a lens design based on solving Rotman's design equations, the output parameters contain the x, y-coordinates for the positions of the array and focal ports and the different length between the array ports and the antenna elements.

For the realized lens designs, nine main input parameters are used: focal angle, focal ratio, focal length, beam angle to ray angle ratio, maximum beam angle, number of beams, number of array elements, array element spacing, and design

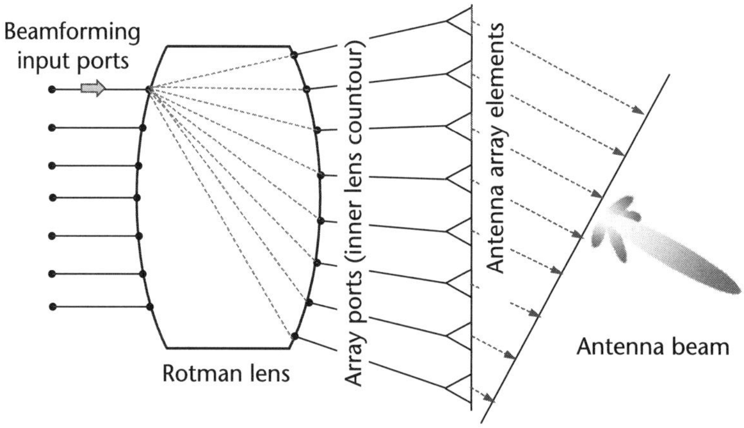

Figure 13.28 Sketch of the functional principle of a Rotman lens.

frequency. Figure 13.29 shows the complete antenna system along with a Rotman lens with 7 beam ports for the scan range ±25° and 15 × 16 antenna elements. As a compromise between the overall size of the lens and the phase aberrations of the beamforming system, a focal length of 8λ was chosen. By optimizing the ratio between the on-axis focal length and the symmetrical off-axis length f/g = 1.13, the residual phase errors remain <1° over the array antenna elements. Rotating the beam ports pointing towards the center of the output arc avoids asymmetrical illumination of the array ports.

Figure 13.30 shows a pattern of scanning beam at ±25° in the H-plane of the phased array antenna. As can be seen from the figure, when the scanning angle is 0, the gain is 29.5 dB, the sidelobe level is –24.2 dB, and the beamwidth is 6.8°. With the increase of the scanning angle, the sidelobe level increased gradually, and the gain also decreased slightly, as shown in Figure 13.31.

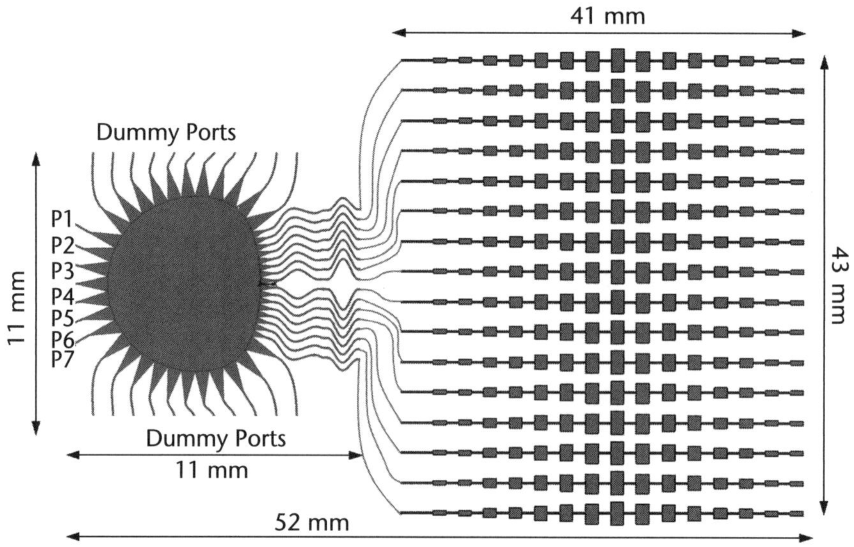

Figure 13.29 Microstrip antenna array design with Rotman lens.

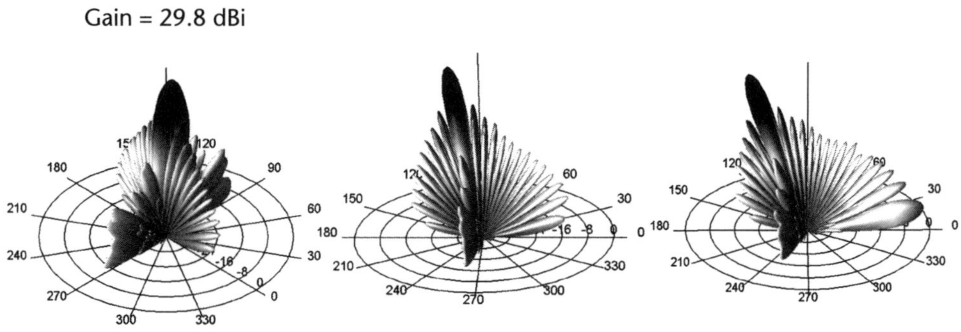

Figure 13.30 Antenna beam direction for phase differences 0°, 45°, and 90°.

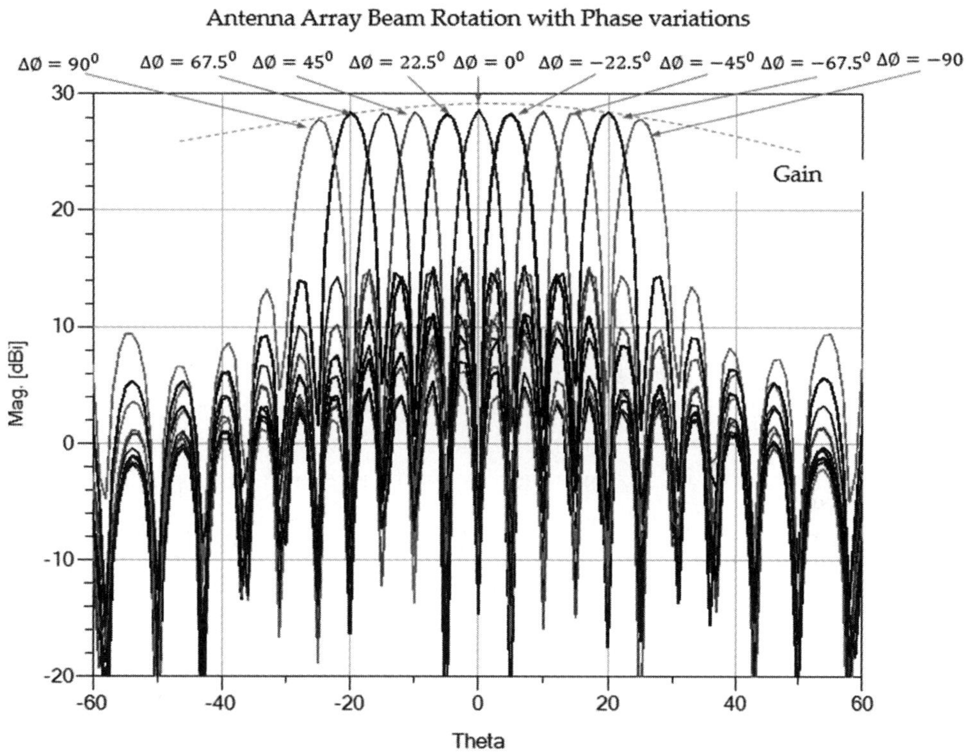

Figure 13.31 A 2-D antenna beam direction for different phase angle.

13.9 Automotive System-Level Simulation

Automotive system-level simulation includes comprehensive tools of FMCW radar waveform generation, signal modulation, antenna modeling, channel simulation, and signal processing to establish the unique automotive radar simulation scenarios. System-level simulation is important to validate the automotive radar at microwave and millimeter-wave frequency bands with various simulation scenarios such as evaluating the effect of the transmit and receive antenna channel number, weather, and clutter and verifying the unique radar waveform design, direction of angle (DOA) degree calculation, micro-Doppler effect detection, and more as shown in Figure 13.32.

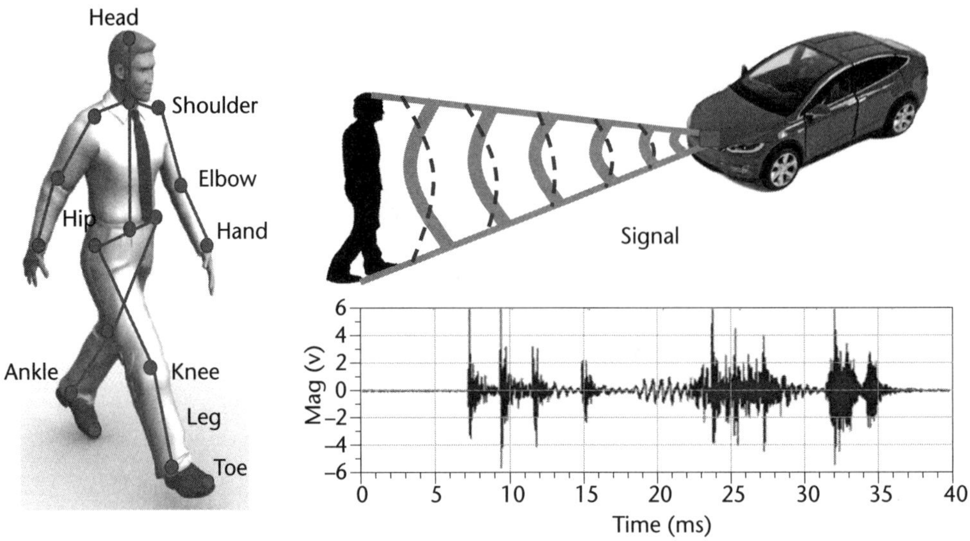

Figure 13.32 Walking passenger modeling and detection simulation.

The system-level simulation integrates critical, safety-conscious, and complicated scenarios into a reliable, leading-edge design of automotive radar. It provides solution under cost and time-constrained projects to design and develop an algorithm such as DOA and phase comparison. System simulations have proven to be a valuable aid during the development process, keeping costs for prototyping small and enabling development engineers to gain substantial insight into the circuits and investigate different conceptual approaches. The radar system-level simulator has been created in ADS to simulate the RF front end, including all nonlinear effects, target model, and antenna effect as shown in Figure 13.33.

Simulation setup consists of three main building blocks: the transmitter and receiver subcircuits and the channel cosimulation (including the antenna characteristics) that are all embedded in a dataflow simulator called Ptolemy, which is steering the whole simulation process from ramp generation to the back-end signal processing. All simulations are performed in a technique called Circuit Envelope (CE) in ADS. The equivalent simulation model of system-level simulation is shown in Figure 13.33.

13.10 Antenna Array for the ETC System

ETC, which enables the electronic collection of toll payments, is applied in different contexts (e.g., highways, bridges, tunnels) [39, 40]. ETC systems are usually developed by means of RFID technology. Passive tag systems are usually preferred due to their main advantage of being cheaper. The reader is placed at a certain height above the lane and irradiates energy toward the tag (placed on a vehicle) to enable a communication link.

A unique identification number that is embedded on the tag is read by the roadside RFID reader, as the vehicle reaches the toll plaza. This unique number is sent to a central computer. An applicable toll amount is deducted from a prepaid account

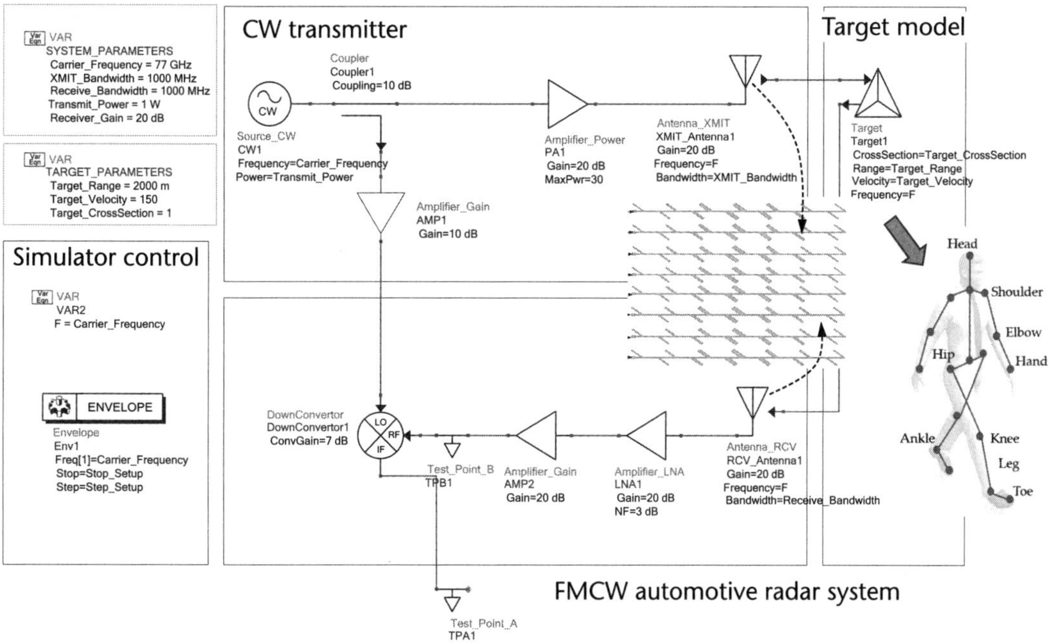

Figure 13.33 System-level simulation setup of radar including target model and antenna.

that is linked to that tag. The main objective of ETC is to eliminate the delay on toll roads. In 1986, ETC was first introduced in Bergen, operating together with traditional tollbooths. The world's first use of completely unaided full-speed electronic tolling was introduced by Trondheim in 1991. Any ETC system must address four aspects: vehicle identification, classification, transaction processing, and violation handling. Vehicle identification is usually implemented by means of opening a communication channel between the vehicle and the roadside unit (RSU), leading to an automatic vehicle identification (AVI) system. There are different AVI systems that can be categorized according to the employed communication technology: GPS, GSM, Wi-Fi, and DSRC in the UHF and microwave frequency spectra, as well as infrared [2]. Among them, DSRC is probably the most consolidated technology for road toll applications, as also demonstrated by the proliferation of numerous communication standards (IEEE 802.11p, ETSI, and so forth).

13.10.1 Phased Array Antenna for ETC Application at 5.8 GHz

In automatic toll collection and access control systems, an exact allocation between the transmitted data and the position of the communicating vehicles is important for an enforcement of unauthorized ones. A phased array antenna is a more flexible solution and is well suited to substitute several fixed beam antennas. The phased array antenna with a steerable narrow main lobe to cover four lanes for 5.8-GHz DSRC is shown in Figure 13.34. By using the microwave DSRC between the tag or the onboard unit (OBU), which is installed on the front window of vehicles and the reader or roadside unit (RSU) on the ETC road of toll collection station, the passing vehicles' toll could be collected automatically without stopping.

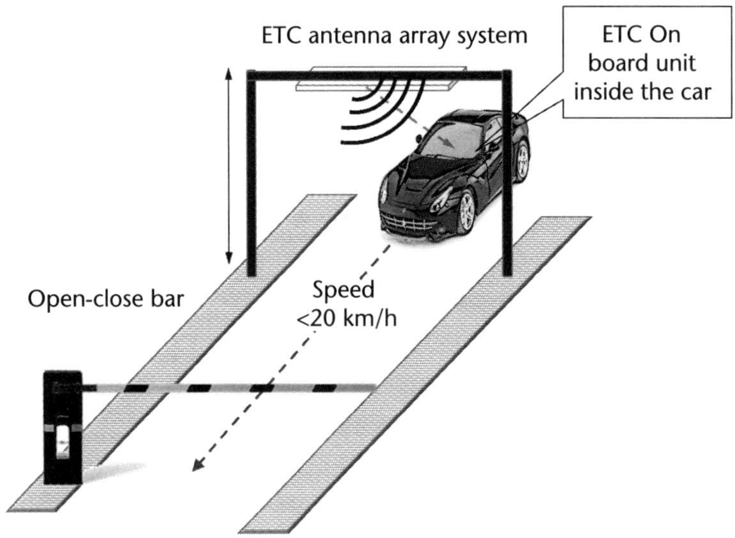

Figure 13.34 Antenna array system for RFID-based ETC.

For 5.8-GHz DSRC, an antenna must have a high gain, a low profile, a compatibility with a monolithic microwave integrated circuit (MMIC) as well as being simple and low-cost to manufacture. Planar microstrip patch arrays for left-handed circular polarization (LHCP), with 1 × 4 elements, is suitable for this application as shown in Figure 13.35. One of the advantages in the adoption of CP is its capability to achieve a good immunity to multipath propagation. In a vehicular environment, particularly rich of scatterers, this is an important advantage. Phase-shifter modules are designed using a 4 × 4 Butler matrix connected to each element of the antenna.

Microstrip array antenna resonating at 5.8 GHz provides a peak gain of about 13.2 dBi at the center frequency. It radiates a beam with a half-power lobe width of 18° in the horizontal plane and 34° in the vertical plane. The patch is printed on the Rogers 4003 substrate with the thickness h of 0.8 mm and ε_r of 3.38. It is easy to get CP of this type of antenna by trimming the opposite corners of a square patch and exciting at the feed point as in Figure 13.35. This single patch antenna has the dimensions 14.1 × 14.1 mm^2. A common measure for the quality of the achieved circular polarization is the AR.

$$\mathrm{AR}(\text{Axial ratio}) = {E_{max}}/{E_{min}} \tag{13.10}$$

This feed of antenna arrays is well matched with the array of 4 patch antennas at 5.8 GHz. The gain of this antenna arrays is 13.2 dB as shown in Figure 13.36.

The phased array antenna (reader antenna) and tag antenna installed in car are shown in Figure 13.37. The advent of autonomous vehicles will drive the need for multiple sensors, spanning microwave and millimeter-wave radar, as well as visible and infrared spectra.

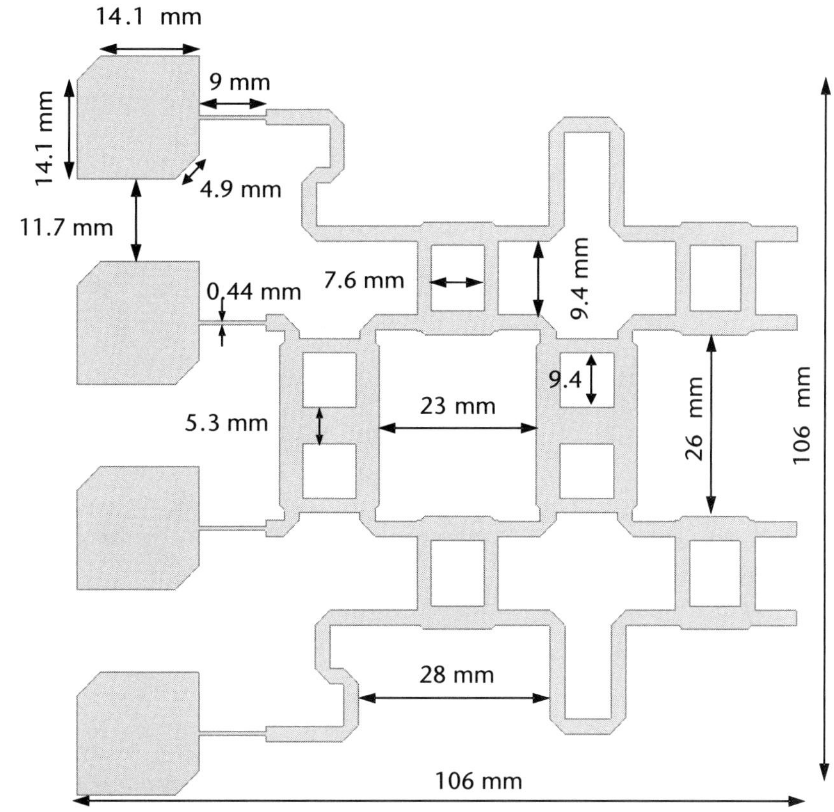

Figure 13.35 Phased array antenna design for ETC.

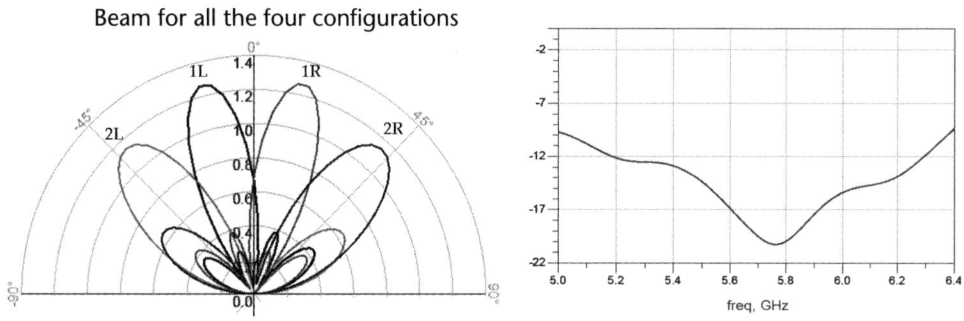

Figure 13.36 Antenna beam scanning and return loss plot.

In addition, radar sensors in the future are likely to have a significantly higher degree of functionality, with the ability to image the road scene, detect and track objects and obstructions, and locate safe paths through unfolding hazards. These radars will be able to monitor road conditions, even locating and reporting potholes and debris on the road. Modeling the installation effects of the radar antenna, including the defocusing effect of bumpers, fascia, and grills, is important. Analysis of the radar module and its installation location can affect the radar's ability to detect the direction of an object, as well as to reduce the radar's sensitivity to objects

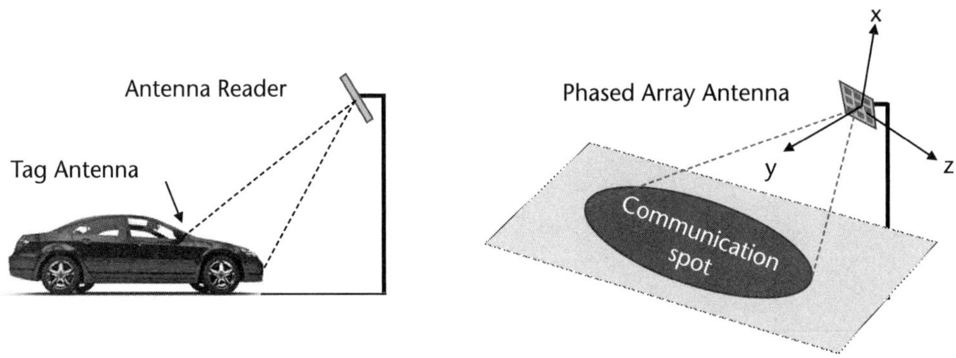

Figure 13.37 Antenna array and propagation model with a car.

that are more difficult to detect. The antenna design must tightly control the distribution of radiation energy, focusing it on the areas of the road under observation and minimizing energy broadcast to locations that are not of interest to the sensor. The antenna array is then placed on a virtual vehicle in a virtual world and evaluated as a complete radar system.

13.11 Summary

In this chapter, various antenna designs for automotive applications were discussed. Automotive antennas have seen significant advancements in recent years. Various services and requirements in automotive applications demand multiband antennas. In the near future, MIMO diversity and beamforming will become more important due to bandwidth-hungry multimedia applications. Future automotive developments will likely focus on two key areas: further integration of the antenna components with the vehicle structure, and new antenna configurations for radar and vehicle-to-vehicle communication services.

References

[1] Meinel, H. H., "Evolving Automotive Radar – From the Very Beginnings into the Future," *8th European Conf. Antennas and Propagation*, 2014, pp. 3107–3114.

[2] Rohling, H., "Milestones in Radar and the Success Story of Automotive Radar Systems," *11th Int. Radar Symp.*, 2010.

[3] "United States Frequency Allocations: The Radio Spectrum," *U.S. Department of Commerce, National Telecommunication and Allocations Information Administration*, Office of Spectrum Management, October 2003.

[4] Jairam, H., "Novel Blade Antenna Using Microstrip Patch Elements," *Microwave Conf., 19th European*, London, UK, 1989, pp. 1118–1122.

[5] Walbeoff, A., and R. Langley, "Multiband PCB Antenna," *Proceedings of IEE Microwaves*, Antennas and Propag., Centre, Sittingbourne, UK, December 2005, pp. 471–475.

[6] Lindenmeier, H., J. Hopf, and L. Reiter, "Active AM-FM Windshield Antenna with Equivalent Performance to the Whip Now as Standard Equipment in Car Production," *IEEE AP-S Int. S.*, Munich, Germany, June 1985, pp. 621–624.

[7] Murakami, et al., "Glass Antenna for Automobile," U.S. Patent No. 5,334,988, 1994.

[8] Heidari, A. A., M. Heyrani, and M. Nakhkash, "A Dual-Band Circularly Polarized Stub Loaded Microstrip Patch Antenna for GPS Applications," *Progress in Electromagnetics Research*, Vol. PIER 92, 2009, pp. 195–208.

[9] Fujimoto, K., and J. James, *Mobile Antenna Systems Handbook*, Norwood, MA: Artech House, 1994.

[10] Geissler, M., K. Scharwies, and J. Christ, "Intelligent Antenna Systems for Cars," *Proc. German Microw. Conf.*, 2014, pp. 1–3.

[11] Ghafari, E., A. Fuchs, and D. Eblenkamp, "A Vehicular Rooftop, Shark-Fin, Multiband Antenna for the GPS/LTE/Cellular/DSRC Systems," *Proc. IEEE-APS Top. Conf. Ant. Prop. Wirel. Comm.*, August 2014, pp. 237–240.

[12] Gallo, M., S. Bruni, and D. Zamberlan, "A Novel Fully Integrated Fin Antenna for Automotive Application," *Proc. Europ. Conf. Antennas Propag. (EuCAP)*, 2014, pp. 2986–2989.

[13] Gavilanes, G., et al., "Comparative Characterization of Four Antennas for VANETs by On-Field Measurements," *Proc. Int. Conf. Antennas Electron. Syst.*, Tel Aviv, Israel, 2013, pp. 1–5.

[14] Johnson, J. M., and Y. Rahmat-Samii, "The Tab Monopole," *IEEE Transactions on Antennas and Propagation*, Vol. 45, No. 1, January 1997, pp. 187–188.

[15] Malley, O., et al., "Rear-Lamp Vehicle Detection and Tracking in Low-Exposure Color Video for Night Conditions," *IEEE Transactions on Intelligent Transportation Systems*, Vol. 11, 2010.

[16] Kraus, J. D., "A Backward Angle-Fire Array Antenna," *IEEE Transactions on Antennas and Propagatation*, Vol. 12, January 1964.

[17] Conti, R., et al., "The Wire-Grid Microstrip Antenna," *IEEE Transactions on Antennas and Propagation*, Vol. 29, January 1981, pp. 157–166.

[18] Zhang, L., W. Zhang, and Y. P. Zhang, "Microstrip Grid and Comb Array Antennas," *IEEE Transactions on Antennas and Propagation*, Vol. 59, pp. 4077–4084, November 2011.

[19] Zhang, B., L. Zhang, and Y. P. Zhang, "Design of Low-Cost Linearly-Polarized Microstrip Grid Array Antenna for 24 GHz Doppler Sensors," *IEEE Topical Meeting on Microwave Photonics*, Singapore, October 18–21, 2013.

[20] Taylor, T. T., "Design of Line-Source Antennas for Narrow Beamwidth and Low Side Lobes," *Trans. IRE*, Vol. AP-3, 1955, pp. 16–28.

[21] Balanis, C. A., *Antenna Theory: Analysis and Design*, 3rd ed., New York: Wiley-Interscience, 2005.

[22] Wu, D., et al., "A 76.5 GHz Microstrip Comb-Line Antenna Array for Automotive Radar System," *2015 9th European Conference on Antennas and Propagation (EuCAP)*, Lisbon, 2015, pp. 1–3.

[23] IEEE Standard for Information Technology – Telecommunications and Information Exchange Between Systems – Local and Metropolitan Area Networks – Specific Requirements, Amendment 6: Wireless Access in Vehicular Environments, 2010.

[24] Mariottini, F., et al., "Design of a Compact GPS and SDARS Integrated Antenna for Automotive Applications," *IEEE Antennas and Wireless Propag. Lett.*, Vol. 9, May 2010, pp. 405–408.

[25] Eom, S. Y., et al., "Design and Test of a Mobile Antenna System with Tri-Band Operation for Broadband Satellite Communications and DBS Reception," *IEEE Transactions on Antennas and Propagation*, Vol. 55, 2007.

[26] Mousavi, P., et al., "A Low-Cost Ultra-Low Profile Phased Array System for Mobile Satellite Reception Using Zero-Knowledge Beamforming Algorithm," *IEEE Transactions on Antennas and Propagation*, Vol. 56, No. 12, 2008, pp. 3667–3679.

[27] Kilian, A., and L. -P. Schmidt, "Conformal Superstrate Loaded Automotive Radar Antenna for Integration in Polymer Car Body Panels," *2010 Proceedings of the Fourth European Conference on Antennas and Propagation (EuCAP)*, 2010, pp. 1–5.

[28] Mariottini, F., et al., "Design of a Compact GPS and SDARS Integrated Antenna for Automotive Applications," *IEEE Antennas and Wireless Propagation Letters,* Vol. 9, 2010, pp. 405–408.

[29] Baek, S., and S. Lim, "Low-Profile Internal Automotive Antenna for WiBro Vertical Polarized Signal Reception," *2010 IEEE Antennas and Propagation Society International Symposium (APSURSI)*, 2010, pp. 1–4.

[30] Koch, N., "Triple-Service GPS-DAB/DMB-WiMAX Patch Antenna for Automotive Applications," *Loughborough Antennas & Propagation Conference 2009 (LAPC 2009)*, 2009, pp. 605–608.

[31] Belohoubek, E. F., "Radar Control for Automotive Collision Mitigation and Headway Spacing," *IEEE Transactions on Vehicle Technology*, Vol. 31, May 1982, pp. 89–99.

[32] Brus, E., "Vehicular Radar: The Ultimate Aid for Defensive Driving?" *Microwaves & RF*, September 1987, pp. 53–58.

[33] Riblet, H. J., "Discussion on 'A Current Distribution for Broadside Arrays Which Optimizes the Relationship Between Beamwidth and Side-Lobe Level'," *Proc. IRE*, Vol. 35, June 1947, p. 489.

[34] Pritchard, R. L., "Optimum Directivity Patterns for Linear Point Arrays," *J. Acoust. Soc. Amer.,* Vol. 25, September 1953, pp. 879–891.

[35] Rotman, W., and Turner R. F, "Wide-Angle Microwave Lens for Line Source Applications," *IEEE Transactions on Antennas and Propagation*, November 1963, pp. 623–632.

[36] Rausch, E. O., A. F. Peterson, and W. Wiebach, "Electronically Scanned Millimeter Wave Antenna Using a Rotman Lens," *Radar 97*, Publication No. 449, October 1997, pp. 374–379.

[37] Fuchs H. -H., and D. Nüßler, "Design of Rotman Lens for Beam-Steering of 94 GHz Antenna Array," *Electronic Letters,* Vol. 35, No. 11, May 27, 1999.

[38] Metz, C., et al., "Fully Integrated Automotive Radar Sensor with Versatile Resolution," *IEEE Transactions on Microwave Theory and Techniques*, Vol. 49, No. 12, December 2001, pp. 2560–2566.

[39] Li, M. W., et al., "On the Feasibility of Using 802.11p for Communication of Electronic Toll Collection Systems," *Proc. International Conference on Network-Based Information Systems (NBiS),* 2011.

[40] Fuxjaeger, P., et al., "Towards Privacy Preserving Wi-Fi Monitoring for Road Traffic Analysis," *IEEE Intell. Transp. Syst. Mag.*, Vol. 8, No. 3, 2016, pp. 63–74.

Selected Bibliography

Eibert, T. F., et al., "Electromagnetic Field Transformations for Measurements and Simulations (Invited Paper)," *Progress in Electromagnetics Research*, Vol. 151, 2015, pp. 127–150.

Guan, N., et al., "A Roof Automobile Module for LTE-MIMO Antennas," *Proc. IEEE-APS Topical Conf. Antennas Propag. Wireless Commun.*, Turin, Italy, 2015, pp. 387–391.

Gulam, M., N. Alsath, and M. Kanagasabai, "Ultra-Wideband Grid Array Antenna for Automotive Radar Sensors," *IET Microwaves, Antennas & Propagation*, Vol. 10, No. 15, October 12, 2016, pp. 1613–1617.

Ku, B., et al., "A 77–81-GHz 16-Element Phased-Array Receiver with ±50° Beam Scanning for Advanced Automotive Radars," *IEEE Transactions on Microwave Theory and Techniques*, Vol. 62, No. 11, November 2014, pp. 2823–2832.

Ku, B., et al., "A High-Linearity 76–85-GHz 16-Element 8-Transmit/8-Receive Phased-Array Chip with High Isolation and Flip-Chip Packaging," *IEEE Transactions on Microwave Theory and Techniques*, Vol. 62, No. 10, October 2014, pp. 2337–2356.

Mailloux, R. J., *Phased Array Antenna Handbook*, 3rd ed., Norwood, MA: Artech House, 2017.

Rabinovich, V., N. Alexandrov, and B. Alkhateeb, *Automotive Antenna Design and Applications*, Boca Raton, FL: CRC Press, 2010.

Rabinovich, V., and N. Alexandrov, *Antenna Arrays and Automotive Applications*, New York: Springer, 2013.

Xu, J., Z. N. Chen, and X. Qing, "CPW Center-Fed Single-Layer SIW Slot Antenna Array for Automotive Radars," *IEEE Transactions on Antennas and Propagation*, Vol. 62, No. 9, September 2014, pp. 4528–4536.

Yeap, S. B., X. Qing, and Z. N. Chen, "77-GHz Dual-Layer Transmit-Array for Automotive Radar Applications," *IEEE Transactions on Antennas and Propagation*, Vol. 63, No. 6, June 2015, pp. 2833–2837.

CHAPTER 14

Transceiver for Phased Array Radar

14.1 Introduction

The antenna is one of the most complicated aspects of RF design, but it is probably also the most ignored part of an RF system design. The performance, range, and legality of an RF link are critically dependent upon the antenna. However, antenna placement and analysis in RF system are often left until the end of the design and are expected to fit into whatever space is left, no matter how that location may affect the antenna performance. Many of these RF system designs ultimately go with degraded performance or multiple redesigns are carried out. Engineers designing an RF transceiver for the first time can quite easily confuse or misunderstand the meaning of antenna specifications and how to apply them in system-level analysis. For example, the gain of an antenna is very different from the gain of an amplifier.

This chapter presents the antenna design consideration in a complete transceiver system for the active phased array radar at C-band. The heart of an active phased array radar is the TRM. The TRMs provide power amplification during transmit and LNA during receive, as well as amplitude and phase control for beam steering and sidelobe reduction. Generally, complex RF and microwave transceiver chips are based on different fabrication technologies such as MMIC, LTCC, microwave integrated circuit (MIC), RF integrated circuit (RFIC), RF MEMS, and various packaging technologies on same RF board. Design and performance optimization of a complete transceiver system, along with an antenna system, are challenging due to the presence of a component based on multiple technologies in a single TRM. A low-power phased array transceiver that includes LNA, PA, single-pole-double-through (SPDT) switch, and lowpass filter (LPF), along with an antenna array is designed at 5.4 GHz to explain the design process of a transceiver integrated with an antenna system. This system-level design uncovers the key design challenges and techniques in implementing an integrated phased array transceiver.

14.2 TRM Architecture

TRMs are key elements in any active antenna aperture system. The transmit power is distributed at the aperture, eliminating the central high-power transmitter needed in passive phased array antennas [1]. A system-level block diagram of the transceiver (T/R) module is shown in Figure 14.1.

A transceiver module contains a transmit path and a receive path. The transmit path consists of a phase shifter, a driver amplifier, and a power amplifier (PA). A SPDT switch is used to provide the duplexing function between transmit and receive paths. The receive path consists of a limiter, an LNA, and a phase shifter. Active phased arrays use an individual solid-state T/R microwave module at each of their radiating elements. This way avoids the distribution and phase shifter losses encountered in the passive array design. Additionally, the active aperture array techniques help to generate very large power for long-range surveillance and tracking radar systems.

The concept of an active phased array is highly dependent on the availability of compact, low power consumption, and high-reliable TRMs. The large number of individual TRMs integrated with the respective radiating elements of the active array ensures a great degree of redundancy in case of the failure of the elements (graceful degradation). Compared to passive array systems, due to the close connection of the TRMs to the radiating elements, the losses are low. This results in a low receive noise figure and a high transmit efficiency. The main functions of a TRM are the generation of transmit power, LNA of received signals, and phase shift in the transmit and receive chain for beam steering.

To fulfill the demands on bandwidth, power, robustness, weight, and requirement of the TRM, new emerging semiconductor and packaging technologies are implemented for the next generation TRMs. The fast development of technology, especially in MMIC, SiGe, CMOS, and GaN, has enabled the integration of most of the components of TRMs into one small chip commonly called a multicore chips (MCC). Small chips (<20 mm^2) enable the possibility of integrating the electronics in the back of the antenna and reduce the cost of the phased array antenna system. As active phased array antennas consist of hundreds, thousands, or even tens of

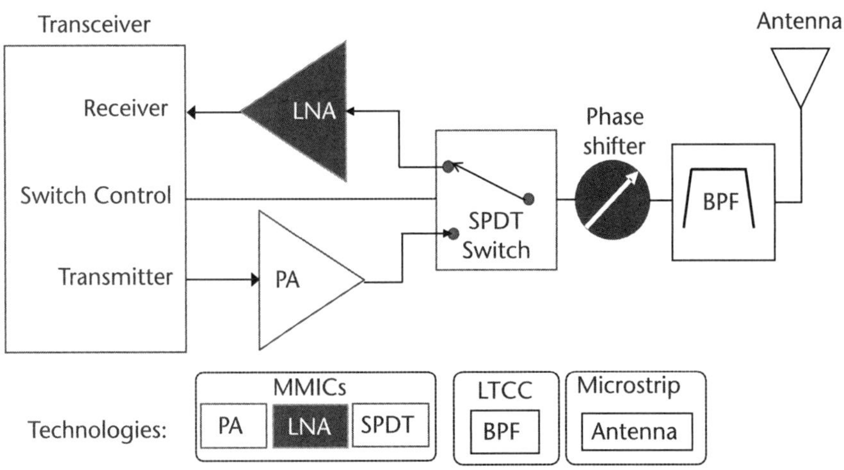

Figure 14.1 Block diagram of active phased array TRM.

thousands of elements per aperture, the TRM has a big impact on the cost and weight of the entire antenna system. The constraint aiming for the cost and weight reduction is the trend towards higher integration both on the module level and on the chip. TRM components for active phased array radar, as shown in Figure 14.1, are designed using three different major technologies listed here.

- MMIC technology: SPDT switch, LNA, and PA;
- LTCC technology: LPF, 3-pole LPF based on a 6-layer LTCC technology;
- Microstrip technology: Antenna, C-band microstrip patch antenna array.

TRM requirements are derived from the phased array antenna requirements and can vary significantly depending on the application. The following key parameters are typically specified for a TRM:

- Operating frequency and bandwidth;
- Beamwidth in azimuth and elevation;
- Maximum output power;
- Power-added efficiency (PAE);
- Spurious and harmonic output;
- Receive noise figure;
- Receive gain and third-order intercept;
- Number of amplitude and phase bits;
- Amplitude and phase root-mean-square (rms) errors;
- Mean time between failure (MTBF);
- Azimuth and elevation steering angles;
- Size of TRM.

These modules are sized to fit within the lattice of a phase array, which is a function of operating wavelength. A good rule of thumb is that, within the plane of the array, the modules must stack together to meet a half-wavelength spacing. Typically, the spacing between modules is chosen to be around $0.5–0.8\lambda_0$ for wide antenna scanning beam performance. At C-band (5.4 GHz), the spacing between modules is 2.5 cm. All these modules are integrated into a single system-on-package (SoP) chip as shown in Figure 14.2.

The SoP module has a highly integrated transceiver MMIC mounted on thick anodized aluminum. Using a selectively anodized aluminum substrate, high-power Tx MMICs are mounted on the bottom aluminum layer. With this arrangement, heat can be easily dissipated from bare chips. The Rx MMICs are mounted on the top aluminum layer with embedded passives such as a microstrip line, a meander line inductor, and a metal-insulator-metal (MIM) capacitor for the bias circuit. To provide isolation and minimize the possibility of oscillation, the LNA is gated off in the transmit mode while the driver amplifier and PA are off in the receive mode.

PAE is a vital parameter for minimizing the prime power requirements and cooling load of the active array. Depending on the technology used, transmit PAEs on the order of 20% to 25% are typically achieved. For a high-power TRM package,

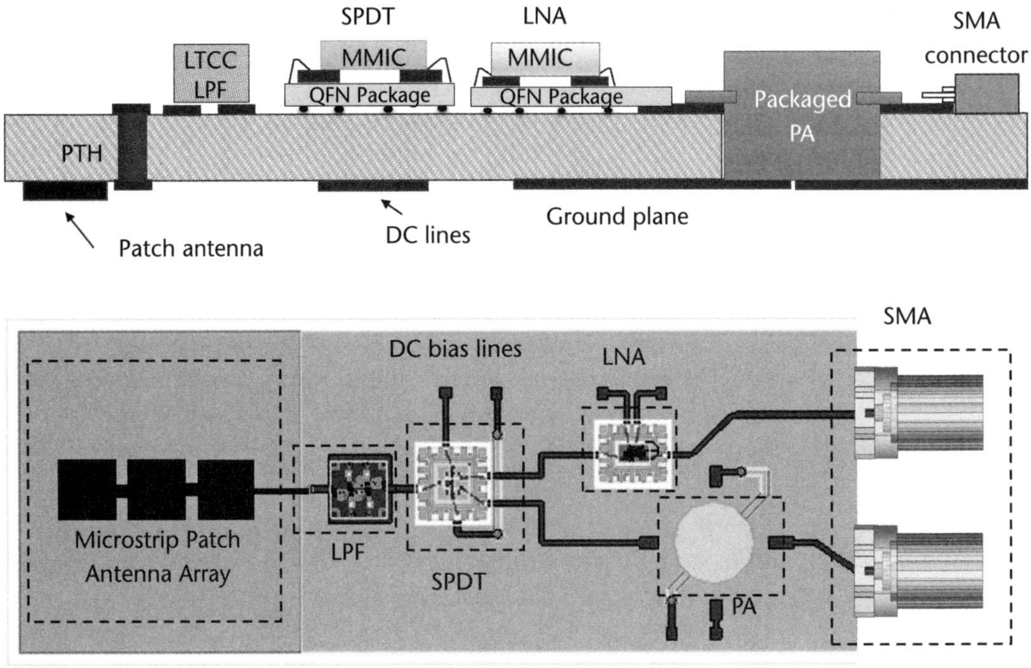

Figure 14.2 Cross-section (side and top) view of a transceiver system.

thermal conduction is a major parameter because it determines the operating temperature of the GaAs, SiGe, and GaN MMICs [2]. The amplitude weighting is essential to control the sidelobe performance of the beams, whereas proper phase shifting enhances the reception from the desired direction and rejects emissions from other directions, increasing SNR at the receiver output [3].

Because all the layers of the TRM include RF components, there is a need for efficient via-interconnects between the components placed on both side of the substrates. The important properties of the vertical via interconnects are the low insertion loss, sufficient electrical shielding, high mechanical stability, compactness, compatibility with multilayer structures, and easy assembly. Over the years, various researchers and companies developed several TRMs at different frequency bands and using multiple technologies.

14.3 TRM Design and Simulation

In the first phase of design, all individual transceiver components are modeled and optimized to meet design specifications separately. Individual components of the T/R model are simulated using circuit simulations, and then these chips are integrated on a PCB. The FEM solver has been used for complete chip and board-level simulation of the layout design to extract the couplings and parasitic parameters of TRM. EM simulation is carried out without active components (like FETs) for a complete system. The design is then simulated for the S-parameter using circuit simulators along with active components. The typical design specifications of a TRM are listed in Table 14.1.

Table 14.1 Design Specifications of Transmit and Receive Channels

	Specifications	
Parameters	*Tx*	*RX*
Frequency	5.2–5.4 GHz	5.2–5.4 GHz
Gain	20 dB	19 dB
Power output per channel	30 dBm	2.4 dB
Beamforming	Electronics	
Efficiency	>40%	
Insertion loss	3 to 6 dB	
Phase shift	360°	
Antenna gain	16–20 dBi	3–10 dBi
Power (T/R)	30 dBm	–50 dBm
P1dB	29 dBm	
OIP3	35 dBm	

A complete design flow diagram of the designed transceiver is shown in Figure 14.3. The design process of the individual components of PA, LNA, LTCC filter, and antenna module is discussed in this section.

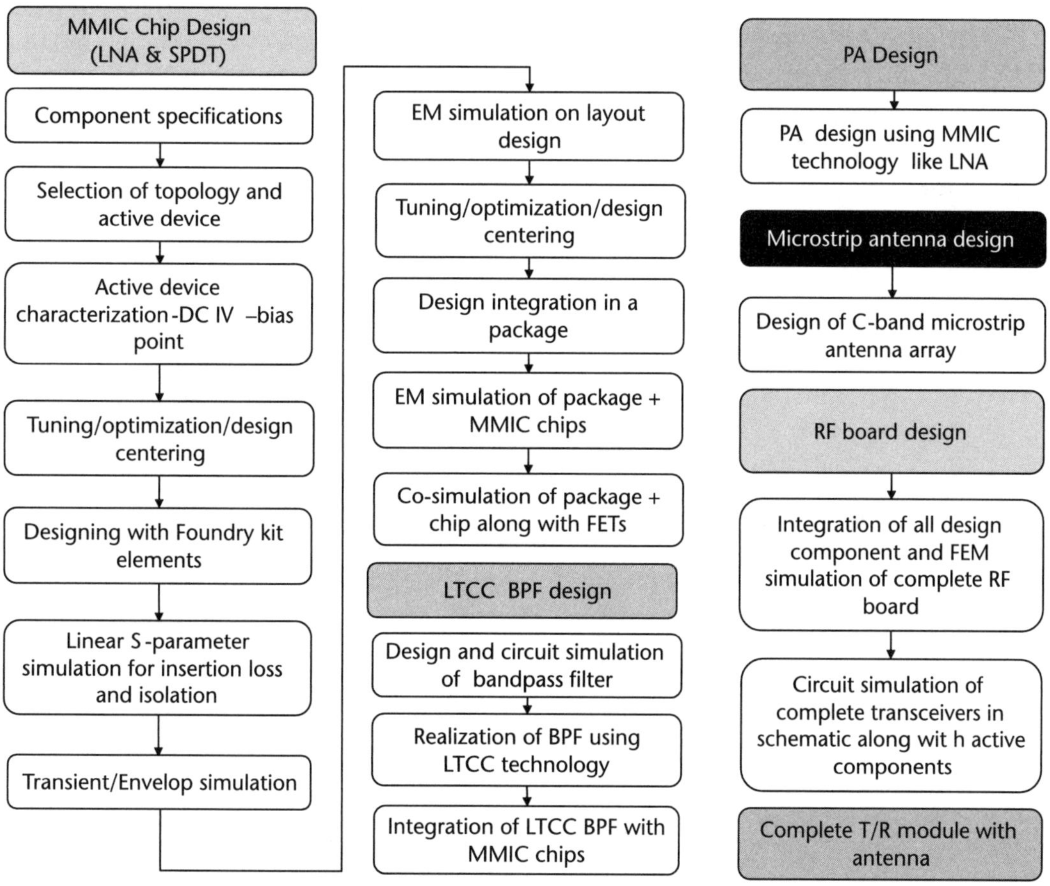

Figure 14.3 Design flow diagram of the complete transceiver system.

14.3.1 MMIC LNA Design

An LNA is used to amplify signals of very low strength, usually from an antenna where signals are barely recognizable and should be amplified without adding any noise. The LNA is a key component of any receiver system. The main function of a LNA is to amplify the small signal received from the antenna without adding too much noise signal. By using an LNA, the effect of noise from subsequent stages of the receiving chain is diminished by the gain of the amplifier. It is placed at the front end of a receiving system to improve the noise figure (NF) of the total system as well as providing the initial boosting of signal. LNAs with higher gain, lower NF, and wider bandwidth are desirable in most of the communication systems.

In the LNA design, the goal is to realize an NF as low as possible while surviving a large input signal. The first step in designing the LNA is to decide which profile provides the best combination of features and performance. The next step is to choose the device size. The selection of active device depends on the requirement of gain and NF of the amplifier. The progress in physics of semiconductors have been the origin of a new generation of components, called heterojunction transistors such as high electron mobility transistors (HEMTs). Another family of HEMTs, named pseudomorphic HEMT (pHEMT), offers better noise performance. The pHEMT process is selected based on high maximum available gain (MAG) and low NF. Device size affects the performance of LNAs such as bandwidth, DC power consumption, NF, and nonlinear performance [4].

In this section, the design and development of a single-stage MMIC LNA using 0.1-μm GaAs pHEMT process in microstrip architecture as shown in Figure 14.4 are discussed. A wideband bias decoupling circuit is designed at C-band, which also suppresses low-frequency bias-induced oscillation to stabilize the amplifier from dc to 8 GHz.

The operating point must provide a compromise between a minimal noise factor (NF_{min}) and a high gain (MaxGain). The operation point (V_{ds} = 3V, V_{gs} = –0.5V) on a biased curve is selected as the biasing point. $\lambda/4$ tramsmission lines are

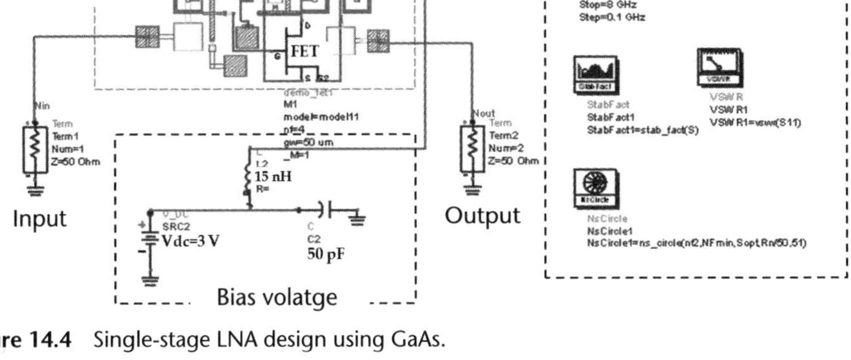

Figure 14.4 Single-stage LNA design using GaAs.

implement and folded in the biasing circuit to minimize the resistive loss adding to the signal.

In a multistage LNA design, a good compromise between each stage's noise and gain performance is made to meet the NF and gain specifications of the LNA. According to Friis' formula [5]:

$$N_f = N_{f1} + \frac{N_{f2} - 1}{G_1} + \frac{N_{f3} - 1}{G_1 G_2} + \ldots \tag{14.1}$$

So, to obtain low noise behavior from the overall amplifier not only the noise figure of the single stage must be considered but also its available gain, as shown in (14.1). When an amplifier is unstable, there is a risk of oscillations, due to reflection coefficients at the input and at the output of the transistor that are greater than unity. Stability is the most important factor that should be considered when designing LNA; the K-B factor is always used to determine if the amplifier is unconditionally stable [6]:

$$K = \frac{1 - |S_{11}|^2 - |S_{22}|^2 + |\Delta|^2}{2|S_{12}S_{21}|} \tag{14.2}$$

$$B = 1 + |S_{11}|^2 - |S_{22}|^2 - |\Delta|^2 \tag{14.3}$$

$$\Delta = S_{11}S_{22} - S_{12}S_{21} \tag{14.4}$$

When considering the stability of the amplifier, not only the stability in the operating frequency range should be checked, but also stability in the whole band (from DC to f_{max}) should be checked to make sure that it is unconditionally stable. Conditional stability (Stability Factor below 1) is when the stability of the quadruple is checked only for certain source impedance and load values at the operating frequency. If the stability conditions are satisfied whatever the impedance of the source and the load, the quadruple is then unconditionally stable (stability factor over 1).

In the design and optimization process, first LNA is realized using ideal components based on behavioral models. After design optimization for the desired performance, these idle components are replaced with actual MMIC component models provided by foundries. EM simulation is the performance along with trace routing and a chip-package model. Figure 14.5 shows the LNA design and matching networks.

Simulated results from 1 to 8 GHz are shown in Figure 14.6. LNA gain is 6.7 dB and the NF value is 2.64 at 5 GHz. The NFs in the microwave regime typically range from 1 to 2 dB. Including losses and other effects, module NFs in the 3 to 4-dB range are generally good.

In Figure 14.6, NF and gain of a GaN-based MMIC low noise amplifier are shown. The obtained NF is very close to the expected one over the whole frequency

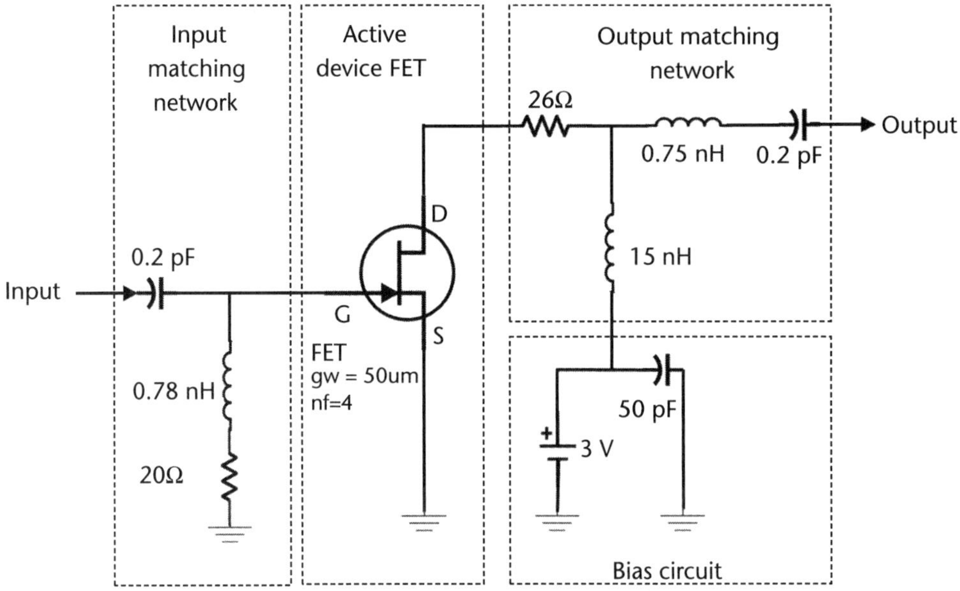

Figure 14.5 LNA layout design and matching networks.

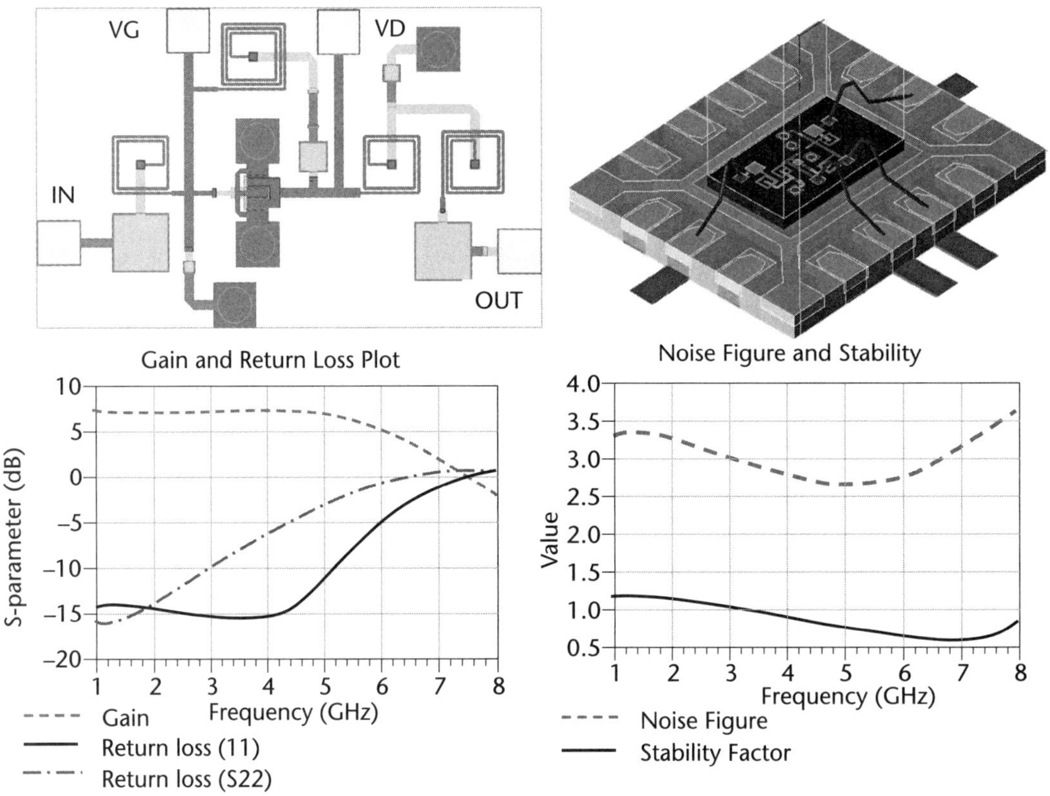

Figure 14.6 Layout design and LNA with QFN package and simulated results.

band. For frequencies from 3 GHz up to 7 GHz, the NF is lower than 3 dB. The overall performance summary of the designed LNA is given in Table 14.2.

14.3.2 PA Design

In early phased array radar, microwave tubes were key component for RF power amplification. Even today in some radar applications, microwave tubes, mainly miniature traveling-wave tubes (TWTs), are used as the RF PA for active phased arrays. However, it is not practical to apply TWT in active phased array antennas for each radiating element because of the large volume, weight, and cost of TWTs. Generally below X-band, silicon RFIC meets many phased array system requirements, but for efficient, linear power amplification, GaAs MMICs are much preferred. In most TRMs, GaAs MMIC amplifiers are used with typical output power levels in the range of 5W to 10W [7, 8]. Broader bandwidth, higher output power levels, increased PAE values, and higher operating voltages are some advantages of GaAs MMIC. For high-power applications, the use of amplifiers based on AlGaN/GaN is a very promising approach [9, 10].

The design of a broadband PA is considerably more complex than that of a linear amplifier. In a PA, an active device operates in a nonlinear regime (i.e., its properties depend not only on the terminations at the signal frequency but also on the signal power level and terminations at harmonics). The design discussed in this section is based on GaAs pHEMT. The starting point for the HEMT model is the drain current formulation. A common feature in the drain formulation [11, 12] is the drain voltage saturation parameter:

$$I_D \sim \tanh(\alpha V_D) \tag{14.5}$$

The drain current variations near pinch-off are given by [12]:

$$I_D \sim \beta(V_D - V_{DSO}) \tag{14.6}$$

The load-pull technique [13] is adapted to optimize the impedance for getting the maximum values of power efficiency and output power of the amplifier. The approximate value for the load impedance can be estimated as:

Table 14.2 Performance Summary of Designed LNA

Parameters	*Value*
Frequency	5 GHz
Voltage gain	6.7 dB
Return loss (S_{11})	−11 dB
NF	2.7
Input IP3 in-band	12 dBm
DC	4 mA
Supply voltage	3V

$$R_{opt} = \frac{V_{dc}}{\left(\frac{I_{max}}{2}\right)} = \frac{V_{dc}}{I_{dc}} \tag{14.7}$$

With the help of the load-pull technique, load and source pull analysis is carried out on the design. In this technique, PAE contours and output power contours are drawn in the Smith chart as shown in Figure 14.7. An optimum output impedance of the device is chosen using these contours. A similar procedure is used to analyze the effect of source-pull. The load impedance R_{opt} for optimum output power is determined based on the load-pull simualtion with a large-signal model. A matching network is then designed to match the impedance R_{opt}.

In the case of multistage PA, if the gate and drain-line inductors are matched, and the drain capacitance is made equal to the gate capacitance, then the input and output currents will be phase synchronized [14]. Assuming this condition and using image-impedance-matched terminations, the voltage gain can be derived as

$$\frac{V_{out}}{V_{in}} = \frac{-N \cdot g_m}{2 \cdot \sqrt{1 - \left(\omega/\omega_c\right)^2}} \sqrt{\frac{L}{C}} e^{-N\theta} \tag{14.8}$$

where N is the number of stages, g_m is stage transconductance, ω_c is cutoff frequency of the lines, and θ is the propagation constant of the lines. The performance of a multistage amplifier is limited by the weakest link in the cascade of stages from input to output. The amplifier can be optimized by recognizing the impact of gain, output power, and PAE from each stage on the overall performance. The overall PAE of a three-stage amplifier is given by [15]:

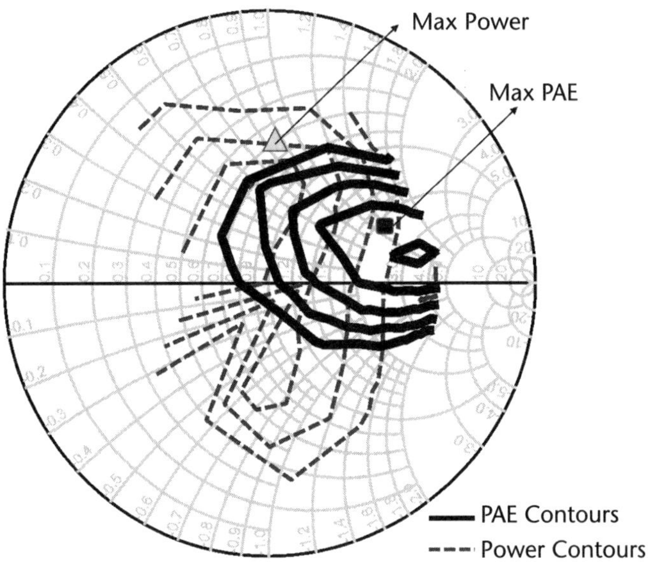

Figure 14.7 PAE contours and output power contours.

$$PAE \approx \frac{1}{1/PAE_3 + 1/(G_3 \cdot PAE_2) + 1/(G_2 \cdot G_3 \cdot PAE_1)} \tag{14.9}$$

where G_{1-3} and PAE_{1-3} are the respective gain and PAEs of each stage.

A two-stage MMIC balanced amplifier is designed to meet the requirements of a transmit phased array operating at 5 GHz with an output power in the range of 1W. The complete design consists of 2 two-stage amplifiers in parallel as shown in Figure 14.8. The input signal to the pair of amplifiers is split via a branch line coupler, which is implemented with microstrip lines. The outputs from the two amplifiers are combined via a branch line coupler identical to the one at the input. The design process includes evaluation of the HEMT device used in this amplifier, including determination of device transconductance g_m versus bias and load-pull analysis. This allows the designer to pick a bias point where the g_m is maximum, which also maximizes the gain. For a minimum NF, the optimal source impedance is calculated, as well as the corresponding load impedance for maximum gain, both as a function of bias voltage. Usually, the first-stage device is quite unstable by itself, and stabilization networks are added.

Input matching network is designed using shunt-C, series-L matching network to generate the desired source impedance for the input stage. An interstage matching network transforms the impedance seen looking into the second stage to the desired impedance to present to the first stage. Output matching is designed by using series-C, shunt-L or series-L, shunt-C networks for generating the desired load impedance for the second stage. Large signal and small signal FET performance is simulated using the circuit simulation models.

The PA matching networks are designed to provide a wide small signal and power bandwidth simultaneously. The PA is unconditionally stable and includes on-chip power supply bypassing capacitors. Lumped element capacitors and spiral inductors are used in design to reduce the size of the chip. Layout design, circuit simulation test bench, and PA results are shown in Figure 14.9. The maximum measured output power is 15W, when biased for class-AB operation with V_{DS} = 30V. The gain of the PA is 25 dB.

The overall performance summary of the designed LNA is given in Table 14.3.

14.3.3 MMIC SPDT Switch Design

SPDT switches are extensively used in phased array radars for T/R applications. SPDT switches are used to selectively couple signals between two RF ports, where

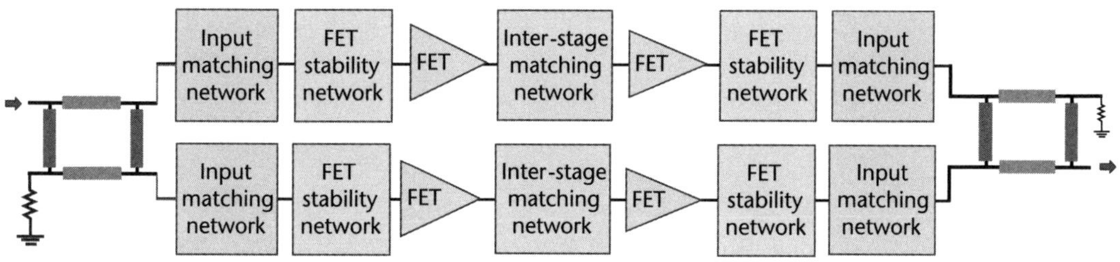

Figure 14.8 Design MMIC balanced PA topology.

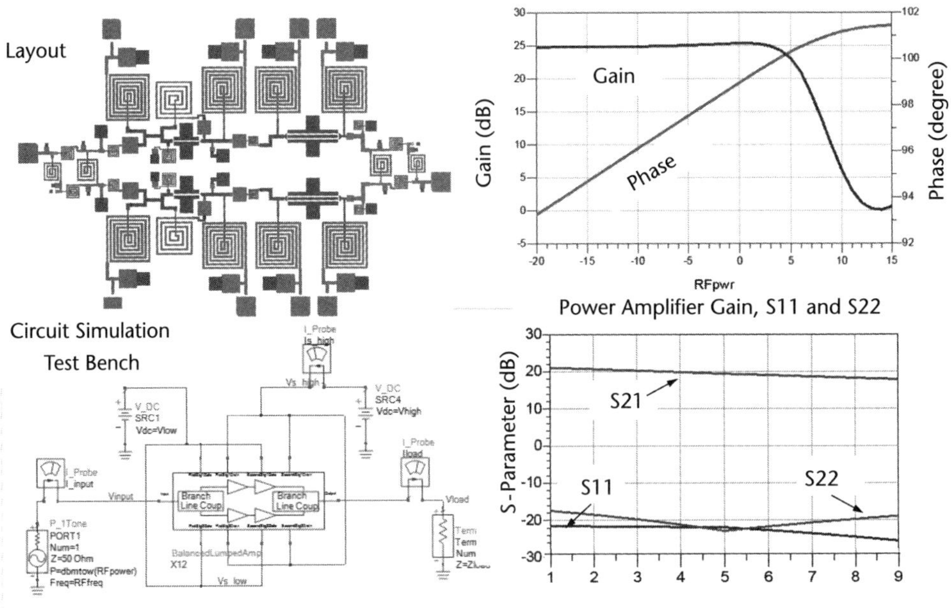

Figure 14.9 MMIC PA design, circuit simulation model, and results.

Table 14.3 Performance Summary of Designed PA

Parameters	*Value*
Frequency	5 GHz
Tx output power level	26.5 dBm
Vcc supply voltage	5V
PA gain	24 dB
PAE	11.80%
Package	QFN
1-dB compression point	25.7 dBm
NF (minimum)	2.8

it provides a simultaneous high isolation to the inactive path and low insertion loss for the active path. GaN HEMT is appropriate for high-power MMIC switch designs due to its high thermal conductivity, high breakdown voltage, and high electron mobility. High-power microwave SPDT switches are needed for high-power handling in phased array systems. Although a mechanical switch can handle high power, it is large, slow in switching time, heavy, and costly. Solid-state switches based on the pin diode can also handle high power, but it draws significant DC when the diodes conduct [16]. This is unacceptable for the low DC power systems. Another disadvantage of the PIN diode is that it is a current-controlled device; therefore, it requires more complex driver circuitry than voltage-controlled FETs. The GaAs FET switches draw very low DC compared to pin diode, but these have limited RF power-handling capability [17]. Design specifications of a typical SPDT switch are listed in Table 14.4.

Table 14.4 MMIC SPDT Switch Design Specifications

Parameters	*Value*
Frequency range	C-band, 5% bandwidth
On-state insertion loss	≤0.6 dB
Off-state isolation	≥30 dB
0.1-dB compression	≥4W (+36 dBm)
Switching time	≥25 ns
Return loss	≥10 dB

Rise time and fall time are fundamental to switch design. This is the time required for switching to take place between two states in the switch response. Rise time is defined as the period between the off state to the on state, typically from 10% of this condition to 90% of the square-law-detected RF power and can be expressed in terms of circuitry resistance ($R_{circuit}$) and load capacitance (C_{load}) as shown in (14.10) [18]:

$$T_r = 2.2R_{circuit}C_{load} \tag{14.10}$$

Conversely, fall time is the period between 90% on state to 10% off state and can be calculated using (14.11):

$$T_f = 2.2R_{circuit}C_{load} \tag{14.11}$$

FOM is used to rate the switching characteristics of different switching elements, which is based on capacitance realization when FET is in the off state (C_{off}) and R_{on} values. The FOM to rate SPDT switches can be represented by [19].

$$FOM = \frac{1}{2\pi C_{off} R_{on}} \tag{14.12}$$

In SPDT switch design, the three most common configurations are series diodes, shunt diodes, and a combination of series and shunt diodes. Many commercially available SPDT switches make use of both series and shunt devices in combination. The SPDT switch discussed in this section has been designed using standard series-shunt configuration. The MMIC design process kit contains metal semiconductor field effect transistors (MESFETs) with nonlinear models, which are used as basic switching elements. SPDT switch has one input and two output arms as shown in Figure 14.10.

RF power is directed into one of the arms by switching on the required arm and switching off the other arm. The active switch provides a low insertion loss and good return loss usually matched with 50Ω. The inactive switch should provide high isolation and high input impedance, ideally providing a perfect open circuit. With perfect isolation, ideally, no power should be detected in the off arm. However, due to the parasitic and coupling effects, some RF power leaks into the

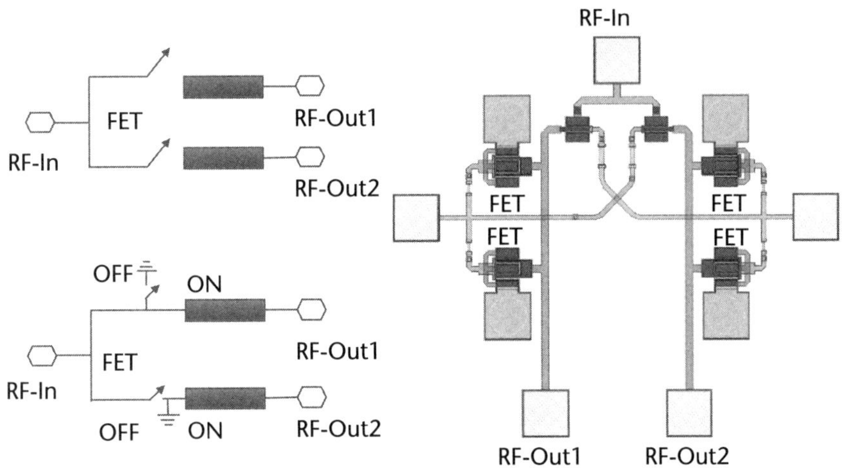

Figure 14.10 Basic topology of SPDT switch and layout design of SPDT switch.

off arm. The leaked power is grounded by placing a shunt MESFET in the on state as shown in Figure 14.10. This switch is designed based on GaAs MMIC for low insertion loss and high isolation from DC to 10 GHz. The use of FETs as switching elements is well recognized [20]. When the gate is biased at 0V (the on state), a small resistance is present between the source and drain. When the gate is biased beyond pinch-off (the off state), the source and drain are capacitively coupled.

The topology of the DC to 10 GHz SPDT switch is indicated in Figure 14.11. Switching FETs are used in a shunt configuration in conjunction with quarter-wave transformers. The circuit comprises two equal but independently biased arms; in a

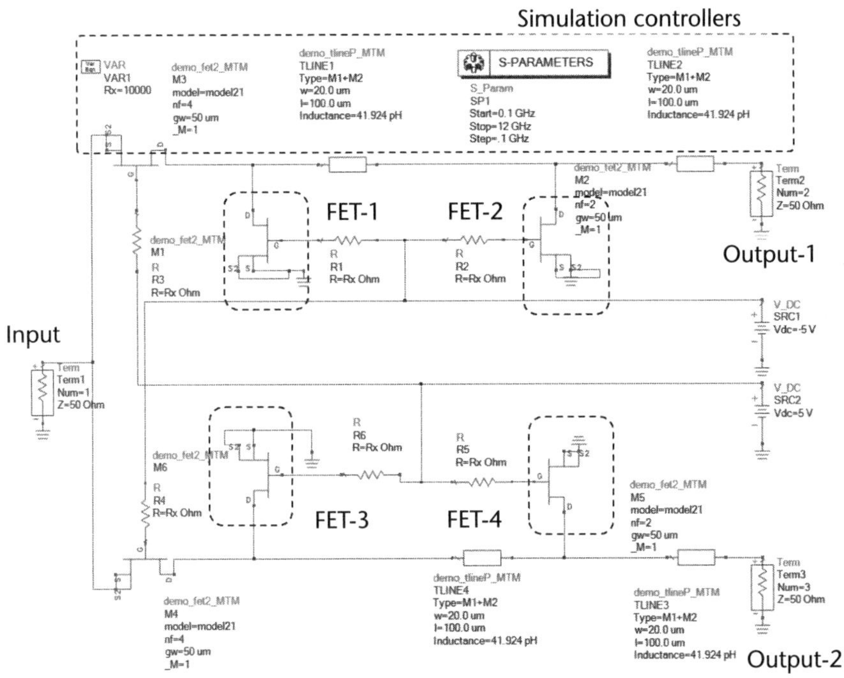

Figure 14.11 Circuit design of SPDT switch using MESFET.

normal operation, one arm passes the signal (the on arm) and the other arm isolates the signal (the off arm). The FETs are connected through series inductive lines. This improves the isolation by reducing the power detected in the off arm.

RF flows from source to drain when the gate is at 0V. Hence, series FETs have input and output at the source and drain, respectively. One single DC bias is used to connect all on devices and another single DC bias is used to connect all off devices. After verifying the basic switch action with idle components, these are replaced with MMIC design foundry components. The chip is packaged using QFN package. QFN chip size is 3.2 mm × 3 mm as shown in Figure 14.12. The small signal performance of the DC to 8-GHz switch is shown in Figure 14.12. Insertion loss is between 2 dB and 3 dB up to 10 GHz. Isolation is more than 30 dB in entire frequency band.

14.3.4 LTCC LPF

LTCC is a multilayer ceramic technology that has the ability to embed the passive elements, such as resistors, capacitors, and inductors, into a ceramic interconnect package. In this, the active elements are mounted in the top layer. As a multilayer

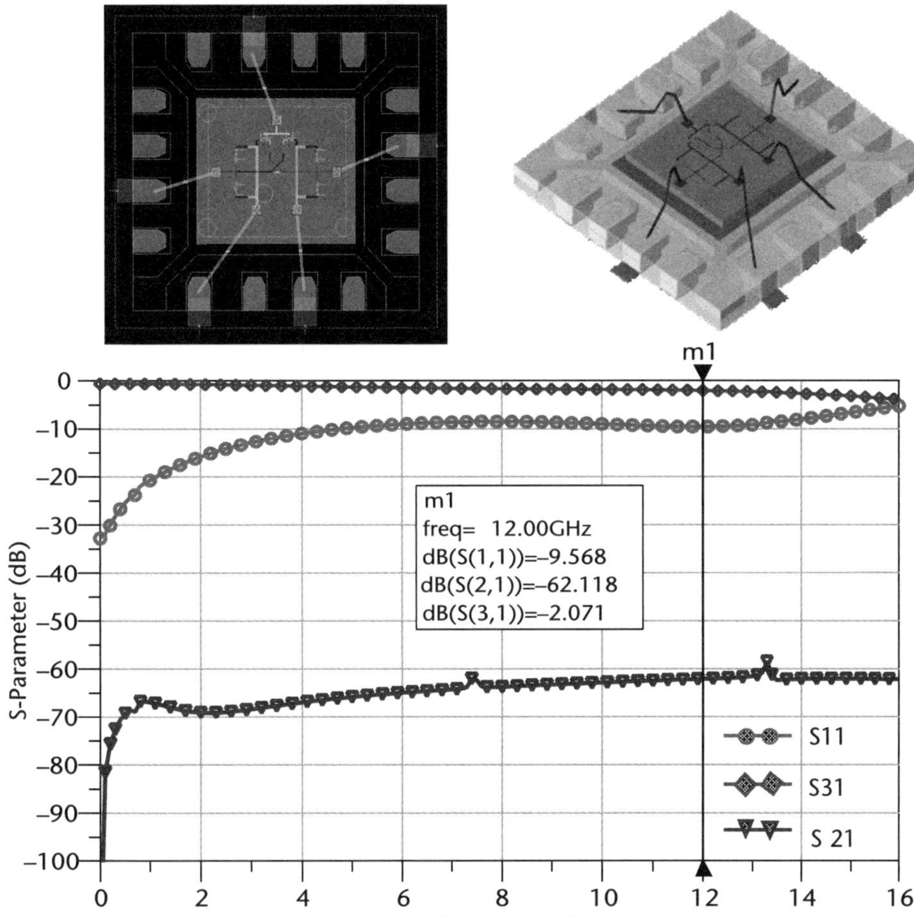

Figure 14.12 SPDT switch inside QFN package and S-parameter results.

ceramic technology, LTCC technology has an ability to print complicated circuits at each layer. Then all layers are punched and dried in parallel. These layers are then aligned, stacked, and co-fired into a highly compact structure at a temperature between 800°C and 900°C [21]. In LTCC, the substrate serves both as carrier for components and lines and as central part of the housing. The multilayer capability of LTCC is the enabler for a very compact design with nearly all control and supply lines buried in inner layers. One major drawback of LTCC is its poor thermal conductivity, which is usually in the order of 3–4 W/mK. Another disadvantage is the difficulty of managing the shrinkage control during the firing process.

The LPF is an important passive component in RF front-end sections. A good LPF has low passband insertion loss and high suppression at the rejection area. High suppression can be provided by generating finite transmission zeros at the rejection area. LTCC technology plays an important role in filter design. The size of the LTCC filter mostly depends upon the realized form of the inductance implemented in the LTCC structure. The inductances can be realized by the grounding transmission lines in LTCC structure, for the self-induction of the lines [22] or by the coupling effect between two transmission lines [23]. The filter, first optimized by the modern network synthesis method, is based on the lumped lowpass prototype filter. The LC circuit of common lowpass prototype filter is shown in Figure 14.13. LPF design has many different solutions, in which the most commonly used are the Butterworth filter and the Chebyshev filter. The Butterworth filter has flatter response characteristics, but the Chebyshev filter has steeper attenuation characteristics. A 3-pole LPF up to 6 GHz has been designed by using a spiral inductor and capacitor. The locations of the transmission zeros are estimated by the network analysis method. The filter is realized in the LTCC process and simulated by a full-wave FEM simulator.

The basic components' spiral inductor and capacitor have been designed with parameterized value. LPF results, as shown in Figure 14.14, have been optimizing at 5 GHz by varying parameters that change the value of the inductor (L) and

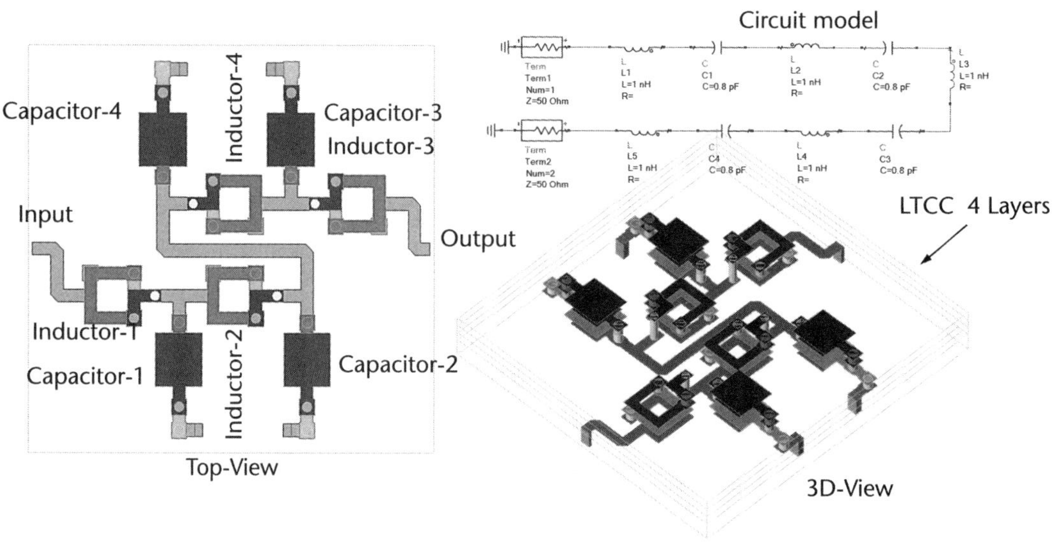

Figure 14.13 LTCC LPF circuit and layout design.

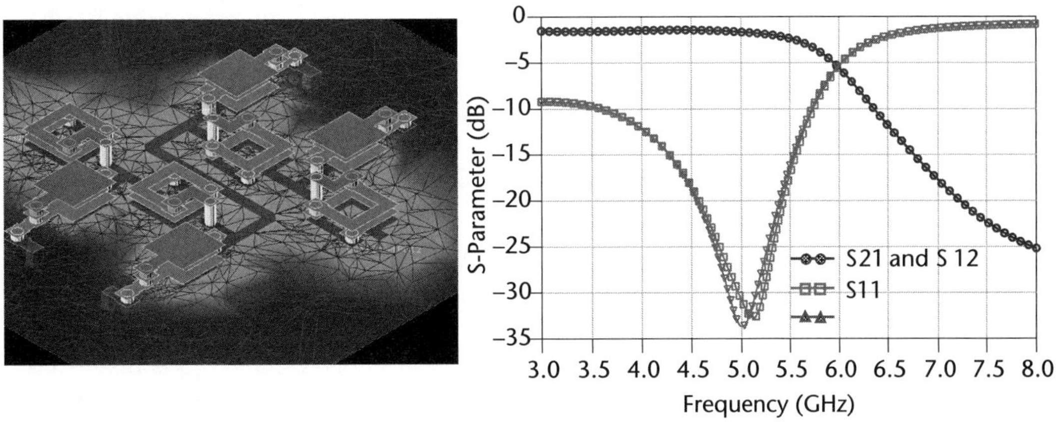

Figure 14.14 E-field on LPF design and simulation S-parameter data.

capacitor (C). The filter has a 3-dB passband from DC to 5.5 GHz. The insertion loss is less than 1.2 dB from DC to 5 GHz. The return loss is better than 10 dB from DC to 5.5 GHz. The volume of the filter is $2.5 \times 2.5 \times 0.5$ mm^3.

14.3.5 A 6-Bit MMIC Phase Shifter

Phase shifters are an essential component of the phased array system that provides the necessary insertion phase to each element in the antenna array. Phase shifters can be implemented as active or passive phase shifters. Active phase shifters also known as amplifier type phase shifters provide phase shift by continuously changing the device gain using gate control voltage [24]. They provide 0° to 360° phase shift with gain to compensate for losses from other components. However, active phase shifters suffer from large DC power consumption and have complex circuitry. In comparison, passive phase shifters have higher linearity with lower power consumption. They can further be classified as analog and digital phase shifters. Analog phase shifters provide continuous phase variation by changing the capacitance of the device using a control voltage, but require complex biasing networks and consume moderate DC power. Digital phase shifters provide a discrete set of phase shift states and consist of a cascade of several phase bits that increment the phase in binary steps. They are immune to noise on their voltage control lines and have negligible power consumption. Digital phase shifters find extensive applications in phased array antennas [25] and their resolution, along with the phase error, decides the precision of the BFN.

A 6-bit digital phase shifter provides 0°–360° phase shift in 64 discrete steps with 5.625° phase resolution. The phase shifter is designed using the pHEMT GaAs MMIC process. The process includes two metal interconnect layers, precision resistors, MIM capacitors, and inductors. The design of the phase shifter using this process enables on-chip integration with other active stages leading to a single-chip solution for the transmit module of the BFN. Each individual bit of the phase shifter needs to be designed using optimum topology to provide low insertion loss, flat phase shift response over the desired band of operation, and good tolerance to process variations. The line length L1 required for a phase shift of $\Delta\varnothing$ is calculated as:

$$\Delta\varnothing = \varnothing_{FET} - \beta L1 \tag{14.13}$$

$$L1 = \frac{(\varnothing_{FET} - \Delta\varnothing)\lambda_g}{2\pi} \tag{14.14}$$

where β, $\varnothing_{FET}$, and λ_g represent phase propagation constant, insertion phase of FET Q1, and guide wavelength of the signal in microstrip line, respectively.

Phase bit topology is shown in Figure 14.15. For 11.25° and 22.5° phase bits, the reference state is obtained when Q1, Q2 are turned on and Q3 is pinched off. The inductor L_p is used to resonate the off-state capacitance of Q3. In the phase shifting state, Q1 and Q2 are turned off and Q3 is turned on. For the 45° phase bit, the reference state is obtained when Q4 is turned on and Q5 is pinched off. In the phase-shifting state, Q4 is turned off and Q5 is switched on. The 90° and 180° phase shift designs are implemented using switched filter topology. The implementation of higher phase shift values using this topology provides a low amplitude imbalance compared to the embedded device type topology. The required phase shift is obtained by switching the signal between highpass filter and LPF networks matched for their insertion losses at the center frequency of operation. The filter

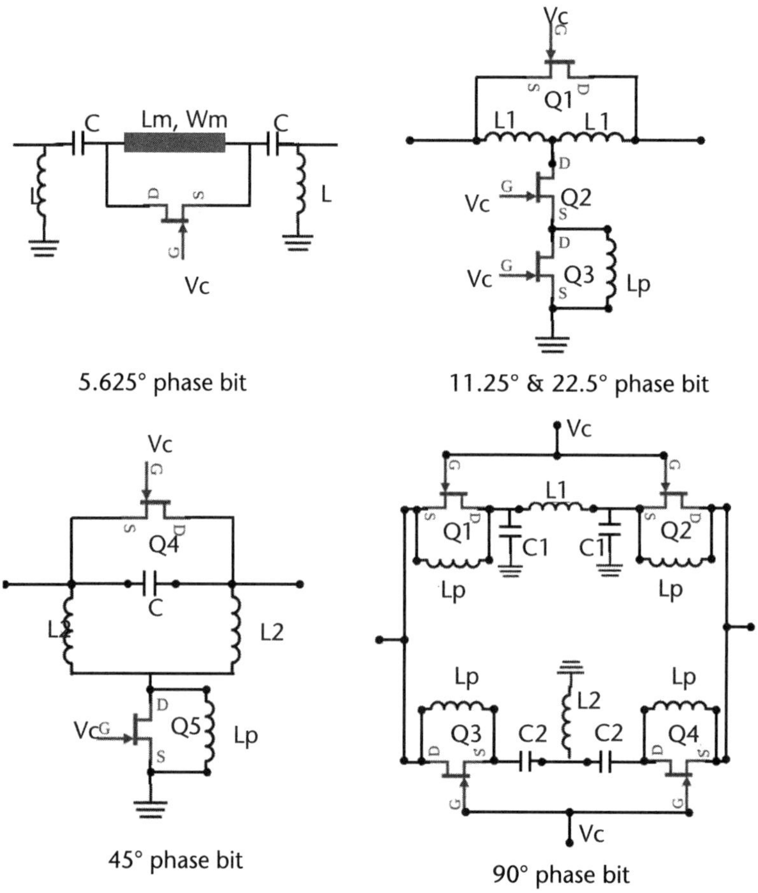

Figure 14.15 Circuit topologies of the phase shifter bits.

networks are connected between two SPDT switches as shown. When transistors Q1 and Q2 are switched on and Q3 and Q4 are off, the signal passes through the LPF network, providing a phase lag of (∅/2) to the signal. When the transistor biases are reversed, the signal passes through the highpass filter network, providing a phase lead of (∅/2) to the signal. Figure 14.16 shows the phase variation at the output of a 1:8 network using a 6-bit phase shifter.

All designed chips are placed in QFN package and then integrated onboard as shown in Figure 14.17. After the integration, TRM performance is validated by circuit design simulations. Component placing and trace routing connecting all the components on the board are other important aspects that are optimized.

14.3.6 Antenna with TRM

Any printed antenna element or array antenna discussed in this book can be integrated with T/R for phase array application. Phased arrays consist of arrays of radiating elements that combine energy in free space to form a single equivalent

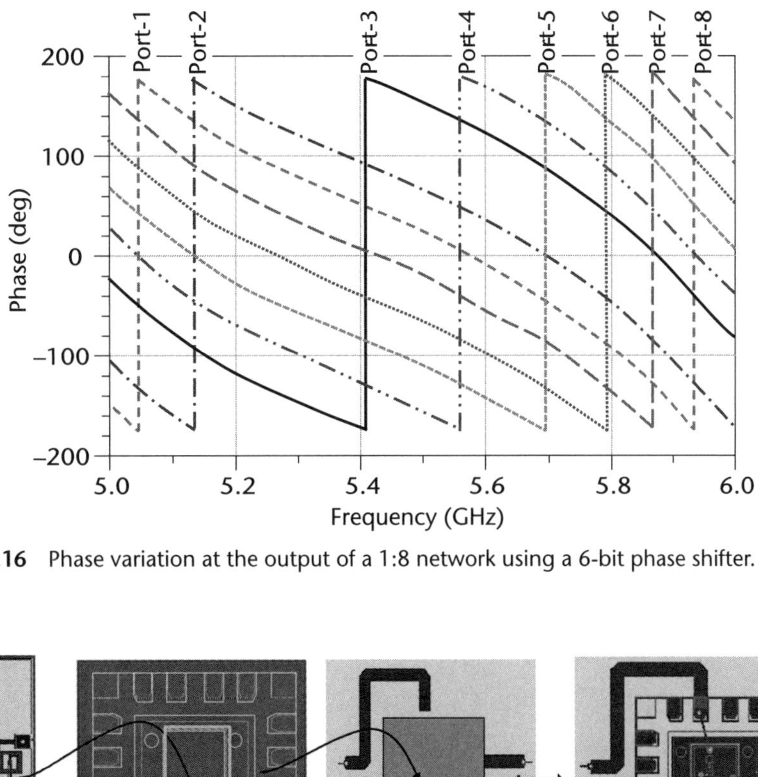

Figure 14.16 Phase variation at the output of a 1:8 network using a 6-bit phase shifter.

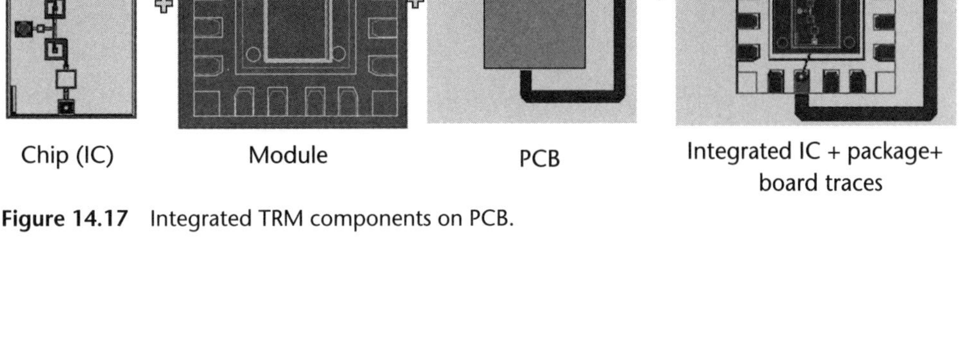

Figure 14.17 Integrated TRM components on PCB.

aperture. Sometime phased arrays are specifically designed to be uniform linear arrays, where the elements are equally spaced, so optimum phase adjustments result in constructive interference.

14.4 Design of Phased Array Radar Using the Transceiver Module

In the realization of a complete phase array system, TRM, along with an antenna system as shown in Figure 14.18, is designed and simulated. Simulation is undertaken to show the antenna pattern produced by an array of 8 linear arrays, each having 4 elements (12-dB gain) driven by the 8 channels of TRM. Phase shifters can be placed either after the TRM or before the TRM. Based on the phase shifter placement in which phase shifting is performed, phased arrays can be categorized into four district types: RF phase shifting, LO phase shifting, IF phase shifting, and digital phased arrays.

The hybrid beamforming technique is used in this design. Hybrid beamforming is a technique that can be used to screen beamforming between the digital and RF domains. RF system designers implement hybrid beamforming to balance flexibility and cost trade-offs while still maintaining a system that meets the required performance parameters. Hybrid beamforming designs are developed by combining multiple array elements into subarray modules, for example, in this design, four antenna elements are combined to form a subarray. A TRM is dedicated to a subarray in the array and therefore fewer TRMs are required in the system. The number of antenna elements in the subarray, and the positioning, can be selected to ensure that system-level performance is met across a range of steering angles.

Figure 14.19 shows a simulated return loss and antenna beam scanning of complete system. The return loss of each linear array is below −20 dB in the operating frequency band. The antenna beam scanning results are presented in Figure 14.19 in the form of rectangular plots showing the antenna responses versus

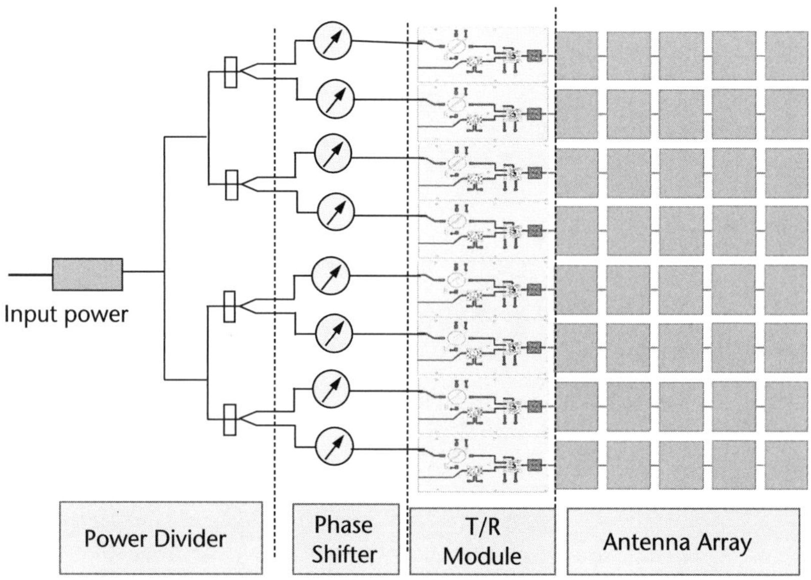

Figure 14.18 Phase array antenna along with designed TRM.

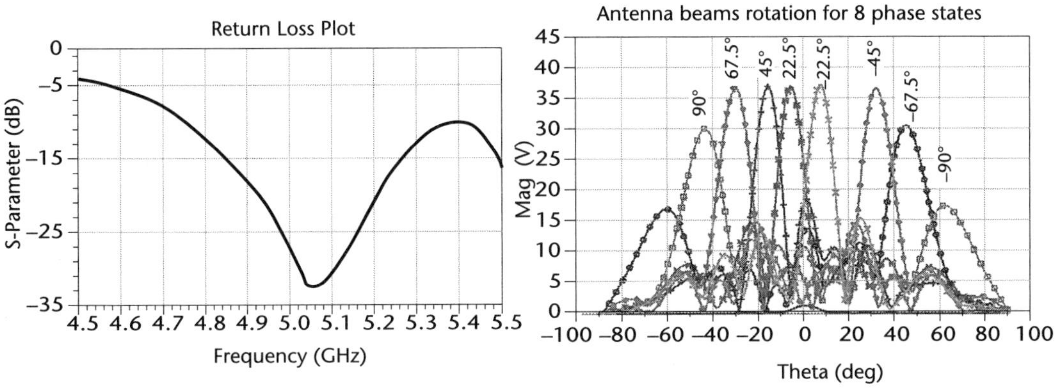

Figure 14.19 Return loss and squinting of antenna beam with variation of phase.

azimuth angle. The array beam is controlled by a single bias voltage applied at the input to the phased array.

A scan angle of 65° has been achieved. The 3-dB beamwidth is 17° at scan angle of 15° and 14° at the scan angle of 30° while the sidelobes are maintained at approximately below –12 dB. As shown in Figure 14.19, the antenna array beam is sequented with different phase values in the feeder network. Simulated return loss, insertion loss, and phase array patterns are given in Table 14.5.

14.5 Summary

The quest for increased bandwidth in the latest generation wireless communication systems, radar, phased arrays, and other high-performance RF applications imposes challenging demands on system integration, power consumption, and thermal management. In this chapter, C-band TRMs for phased array application was discussed.

Table 14.5 Results of a Complete Phased Array System

Design Name	*Parameters*	*Specification*
MMIC-SPDT Switch	Frequency range	DC to 8 GHz
	Insertion loss	<1 dB
	I/O return loss	<–15 dB
	Isolation	>30 dB
MMIC-LNA	Gain	6.7 dB
	NF	2.64 at 5.08 GHz
	Stability factor	1.43 at 5.08 GHz
MMIC-PA	Gain	18.2 dB
LTCC-LPF	Pass bandwidth	Up to 5.6 GHz
	Input/output return loss	<–15 dB
Microstrip antenna	Frequency range	5–5.3 GHz (±300 MHz)
	Polarization	Linear polarization
	Gain	7 dB
Transceiver system	Supply voltage	Single bias supply (3V)
	Overall, I/O return loss	<–20 dB

Various components of T/R models have been designed and simulated. The performance of a complete phased array system has been analyzed using the FEM and circuit simulation using active components. The transceiver system has been applied in an active phase array to demonstrate beam sequencing with variation of phase values. As an antenna designer only, antenna design is not sufficient; it is important to analyze antenna performance and behavior to check with all other components of TRM.

References

[1] Brookner, E., "Phase Arrays and Radars Past, Present and Future," *IEEE Microwave J.*, Vol. 49, No. 1, January 2006, pp. 24–46.

[2] Koriyama, S., et al., "Millimeterwave Ceramic Package for a Surface Mount," *IEEE MTT-S Int. Microw. Symp. Dig.*, Boston, MA, June 2000, pp. 61–64.

[3] Agrawal, A., and E. Holzman, "Beamformer Architectures for Active Phased-Array Radar Antennas," *IEEE Transactions on Antennas and Propagation*, Vol. 47, No. 3, March 1999, pp. 432–442.

[4] Del Alamo, J. A., "The High-Electron Mobility Transistor at 30: Impressive Accomplishments and Exciting Prospects," *2011 International Conference on Compound Semiconductor Manufacturing Technology*, Indian Wells, CA, May 16–19, 2011.

[5] Marsh, S., *Practical MMIC Design*, Norwood, MA: Artech House, 2006.

[6] Pozar, D. M., *Microwave Engineering*, 3rd ed., New York: John Wiley & Sons, 2005.

[7] Mancuso, Y., P. Gremillet, and P. Lacomme, "T/R-Modules Technological and Technical Trends for Phased Array Antennas," *IEEE Microwave Theory and Techniques Symposium Digest,* San Francisco, CA, June 2006, pp. 614–617.

[8] Hommel, H., and H. -P. Feldle, "Current Status of Airborne Active Phased Array (AESA) Radar Systems and Future Trends," *European Radar Conference Digest*, Amsterdam, 2004, pp. 121–124.

[9] Kinghorn, A. M., "Where Next for Airborne AESA Technology?" *IEEE Radar Conference Proceedings,* Rome, May 2008, pp. 287–290.

[10] Edwards, T., "Semiconductor Technology Trends for Phased Array Antenna Power Amplifiers," *European Radar Conference Proceedings*, Manchester, October 2006, pp. 269–272.

[11] Materka, A., and T. Kacprzak, "Computer Calculation of Large-Signal GaAs FET Amplifier Characteristics," *IEEE Transactions on Microwave Theory and Techniques*, Vol. MTT-33, No. 2, February 1985, pp. 129–135.

[12] Curtice, W., and M. Ettenberg, "A Nonlinear GaAs FET Model for Use in the Design of Output Circuits for Power Amplifiers," *IEEE Transactions on Microwave Theory and Techniques*, Vol. MTT-33, No. 12, December 1985, pp. 1383–1393.

[13] Cripps, S. C., *Advanced Techniques in RF Power Amplifier Design*, Norwood, MA: Artech House, 2006.

[14] Ballweber, B. M., R. Gupta, and D. J. Allstot, "A Fully Integrated 0.5-5.5 GHz CMOS Distributed Amplifier," *IEEE Journal of Solid-State Circuits*, Vol. 35, No. 2, February 2000, pp. 231–239.

[15] Cheung, T. S. D., and J. R. Long, "A 21-26-GHz SiGe Bipolar Power Amplifier MMIC," *IEEE Journal of Solid-State Circuits*, Vol. 40, No. 12, December 2005, pp. 2583–2597.

[16] Triquint Data Sheet, TGS4302, High Power Ka-Band SPDT Switch.

[17] Triquint Data Sheet, TGS2306, High Power DC-18GHz SPDT FET Switch.

[18] Viranjay, M. S., and G. Singh, *MOSFET Technologies for Double-Pole Four-Throw Radio-Frequency Switch,* New York: Springer, 2014.

[19] Tipirneni, N., "High Power III-Nitride Transistors for Energy Conversion," Ph.D. dissertation, Dept. Elec. Eng., Univ. of South Carolina: ProQuest, 2007.

[20] Ayasli, Y., "Microwave Switching with GaAs FETs," *Microwave Journal*, Vol. 25, No. 11, 1982, pp. 61–74.

[21] Scrantom, C. Q., and J. C. Lawson, "Where We Are and Where We're Going," *IEEE International Microwave Symposium Digest*, 1999, pp. 193–200.

[22] Tang, C. -W., "Harmonic-Suppression LTCC Filter with the Step-Impedance Quarter-Wavelength Open Stub," *IEEE Transactions on Microwave Theory and Techniques*, Vol. 52, No. 2, February 2004, pp. 617–624.

[23] Rambabu, K., and J. Bornemann, "Simplified Analysis Technique for the Initial Design of LTCC Filters with All-Capacitive Coupling," *IEEE Transactions on Microwave Theory and Techiques*, Vol. 53, No. 5, May 2005, pp. 1787–1791.

[24] Bahl, I. J., *Control Components Using Si, GaAs, and GaN Technologies*, Norwood, MA: Artech House, 2014.

[25] Xiaofeng, Y., and S. Jiangyi, "C-Band 6-Bit Phase Shifter for a Phase Array Antenna," *Journal of Semiconductors*, Vol. 34, No. 4, 2013, pp. 1–4.

Selected Bibliography

Andricos, C., I. J. Bahl, and E. L. Griffin, "C-Band 6-Bit GaAs Monolithic Phase Shifter," *IEEE Microwave Theory and Techniques*, Vol. MTT-33, December 1985.

Bessemoulin, A., et al., "A 1-Watt Ku-Band Power Amplifier MMIC Using Cost-Effective Organic SMD Package," *Proc. 34th Eur. Microw. Conf.*, October 2004, pp. 349–352.

Bettidi, A., et al., "X-Band Transmit/Receive Module MMIC Chipset on Emerging GaN and SiGe Technologies," *Proc. IEEE Phased Array Syst. Tech. Symp. Dig.* October 2010, pp. 250–255.

Cohen, E., et al., "A Bidirectional TX/RX Four-Element Phased Array at 60 GHz with RF-IF Conversion Block in 90-nm CMOS Process," *IEEE Transactions on Microwave Theory and Techniques,* Vol. 58, No. 5, May 2010, pp. 1438–1446.

Kang, D. W., et al., "Single and Four-Element Ka-Band Transmit/Receive Phased-Array Silicon RFICs with 5-Bit Amplitude and Phase Control," *IEEE Transactions on Microwave Theory and Techniques,* Vol. 57, No. 12, December 2009, pp. 3534–3543.

Ku, B. H., D. W. Kang, and S. C. Hong, "CMOS Integrated Circuits for X-Band Phased Array Systems," *Proc. 54th MWSCAS Symp.*, August 2011, pp. 1–5.

Mangraviti, G., et al., "A 4-Antenna-Path Beamforming Transceiver for 60 GHz Multi-Gb/s Communication in 28 nm CMOS," *IEEE Int. Solid-State Circuits Conf. (ISSCC) Dig. Tech. Papers*, February 2016, pp. 246–247.

Poon, A. S. Y., and M. Taghivand, "Supporting and Enabling Circuits for Antenna Arrays in Wireless Communications," *Proc. IEEE,* Vol. 100, No. 7, July 2012, pp. 2207–2218.

CHAPTER 15

Phased Array and BFN for 5G Communication Systems

15.1 Introduction

Phased array systems operating at millimeter wave are already a focused area of research by the emerging 5G communication market. Based on 5G applications requirement, there is an increasing interest in making broader use of the millimeter-wave frequency band for communications on small mobile platforms where narrow antenna beams from low-volume radiating apertures provide enhanced communication security. However, currently, millimeter-wave systems are not very user-friendly and are designed to be platform-specific, lacking interoperability, and are thus reserved for only the most complex platforms. To increase the use of millimeter-wave phased arrays and make them broadly applicable for 5G systems, many technical challenges must be addressed, including wideband frequency coverage, precision beam pointing, mesh networking, and antenna systems. Multibeam phased arrays have driven technology to use the millimeter-wave frequency spectrum for communication systems.

The 5G wireless communication is the next major phase of mobile communications standards beyond the current LTE-Advanced. To cater to 5G's faster speed (1 to 10 Gbps), the millimeter-wave frequency band has been recommended for 5G wireless communication. The unlicensed 60-GHz band has been targeted for a short-range high data rate communication system. This band is useful for many reasons, including a large amount of spectrum availability (7 GHz), suitability for frequency reuse, and high FCC effective isotropic radiated power (EIRP) limit (40 dBm) [1]. One major challenge in the implementation of 60-GHz wireless network is high path loss, implying a need for high-gain beamforming antennas.

Due to the 5G demand for high data rate communication, phased array systems are found to be significant in wireless communications. Phased arrays can adapt their radiation patterns and cancel out information in unwanted directions depending on the operating environment. This feature significantly contributes to the reception quality and data capacity in communication systems. Mobile communications networks use MIMO technology to achieve high data rates. With MIMO beamforming, it is possible to reduce transmit power and still get better channel performance.

The BFN is the key component of multibeams antennas. BFNs are complex networks used to precisely control the phase and amplitude of the RF energy passing through them, which is conveyed to the radiating elements of an antenna [2–4]. In the 5G system, the devices will use a smart antenna array to beam-shape its own receiver sensitivity and transmit energy. The Butler matrix has been used widely over the years in the beamforming systems. The Butler matrix consists of many 90° hybrid couplers and fixed phase shifters. It has N input ports and N output ports to drive an array of N antenna elements [5–8]. It can produce N orthogonally space beams with the full gain of the array.

15.2 5G Communication System

In the past 10 to 15 years, the speed at which communication standards have developed has been very rapid. Cellular generation differs in four major aspects: radio access, bandwidth, data rates, and switching schemes. Initially, the cellular concept was introduced in the 1G technology, which made the large-scale mobile wireless communication possible. The 1G cellular system, mainly the analog system, had a bandwidth ranging from 10 to 30 kHz depending on system type and services. The 1G was only suitable for voice services. Digital communication has replaced the analog technology in the 2G, which significantly improved the wireless communication quality. The first release of the Third Generation Partnership Project (3GPP) LTE system was standardized in December 2008 timeframe, with commercial deployments in 2001. The first release of LTE-Advanced was in March 2011, which made LTE formally compliant with the International Telecommunication Union (ITU) definition of the 4G wireless technology, called IMT-Advanced. The 4G wireless networks can support data rates of up to 1 Gbps for low mobility, such as roaming/local wireless access, and up to 100 Mbps for high mobility such as mobile access. LTE and its extension, LTE-Advanced systems, are practical 4G systems. However, the industries and standards bodies are now organizing a timeframe to standardize 5G technology, which is expected to be around 2020.

The 5G use cases and their requirements have already been studied by many 5G research organizations, academic institutes, and telecommunications firms, as shown in Figure 15.1. The 5G can deliver up to 100 times the speeds that customers are currently getting with 4G LTE. It can also accommodate many more devices expected to attach to networks due to the expansion of the IoT.

These are some major advantages of the 5G technology:

- *Better education system:* A student sitting in any part of the world can attend the class.
- *Parallel multiple services:* A user can know the weather and location while talking with another person.
- *Easier and economical medical treatment:* A doctor can treat the patient located in another part of the world.

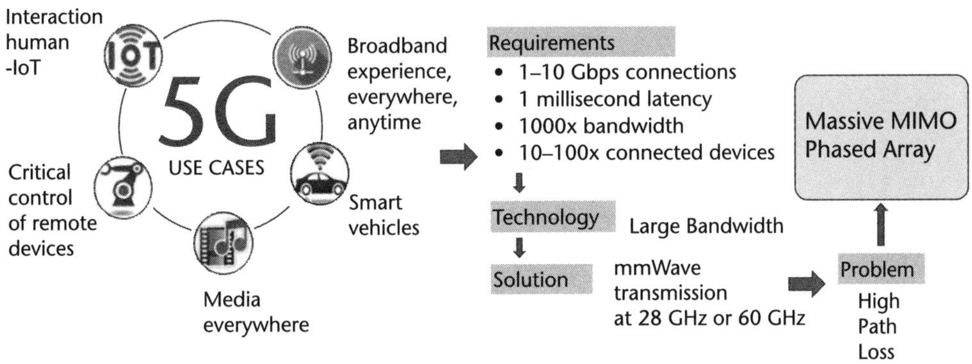

Figure 15.1 Applications of the 5G system and technology driven by the 5G requirements.

- *Easy monitoring and tracking:* A governmental organization and investigating agencies can monitor any part of the world.
- *Better disaster management:* Possible natural disasters including tsunamis and earthquakes can be detected faster.
- *Always connected and responsive mobile network:* Mobile data downloads will be much faster, and always on and responsive mobile Internet offers a power mobile experience. The 5G networks will enable secure access to cloud storage and access to enterprise applications.

15.2.1 Massive MIMO and Antenna Technologies

In 5G communication, a massive number of antenna elements in phased array enable beamforming technology, which combines radio waves and forms a single sharp beam. The beam compensates the propagation loss at higher frequencies and ensures a transmission area with radii of several hundred meters. This technology utilizes multiple propagation paths available in a radio communication channel between a mobile station and a base station. MIMO is a wireless network that allows the transmitting and receiving of more than one data signal simultaneously over the same radio channel, as discussed in Chapter 4. The usual MIMO networks tend to use two, four, or eight antennas as shown in Figure 15.2. However, massive MIMO is a system with a very high number of antennas (>100). The advantage of a MIMO network over a standard one is that it multiplies the capacity of a wireless connection without requiring more spectra. The more antennas with which the RF system is equipped, the greater number of possible signal paths and the better the performance in terms of data rate and link reliability.

The large number of antennas in a massive MIMO network will also make it far more resistant to interference and intentional jamming than current systems that only utilize a few antennas. In the 5G system, a large number of antennas increase the multiplexing capability for higher data rates with improved system capacity. As the 5G system will use millimeter waves or submillimeter waves, applying multi-antenna technologies in these higher frequency spectra have an advantage because of compact antenna sizes. Massive MIMO allows tens to thousands of arrays to be configured at the base station as shown in Figure 15.2. This enables more accurate

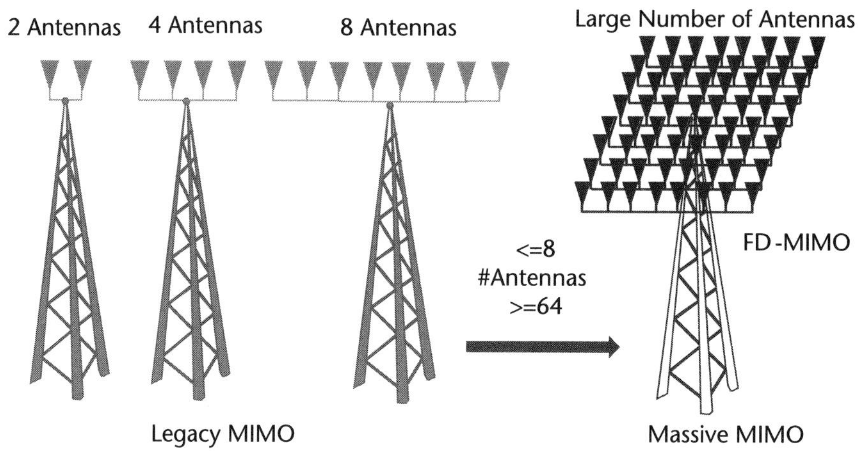

Figure 15.2 Massive MIMO and antenna array system.

beam control and higher spectral efficiency than legacy MIMO. Massive MIMO also includes the use of spatial multiplexing and interference mitigation to increase system capacity. These antennas provide substantial interference suppression for enhanced performance.

15.2.2 MIMO Antennas for Picocells

In the 5G communication system, the complete area will be dividing in picocell as shown in Figure 15.3. A picocell is a small cellular base terminal (CBT) typically covering a small area, such as offices, shopping malls, train stations, and stock exchanges. The size of a cell is from 10m to hundreds of meters. The same concept is used in 3G and 4G network also, but cell size is large in that, typically from hundreds of meters to some kilometers [9]. This infrastructure fulfills the capacity demand of a 5G system. In the 4G system, only a few antennas are needed that radiates the signal in all the direction, but in 5G this will be replaced with a massive phased array antenna system to cover both vertically and horizontally placed devices in a picocell.

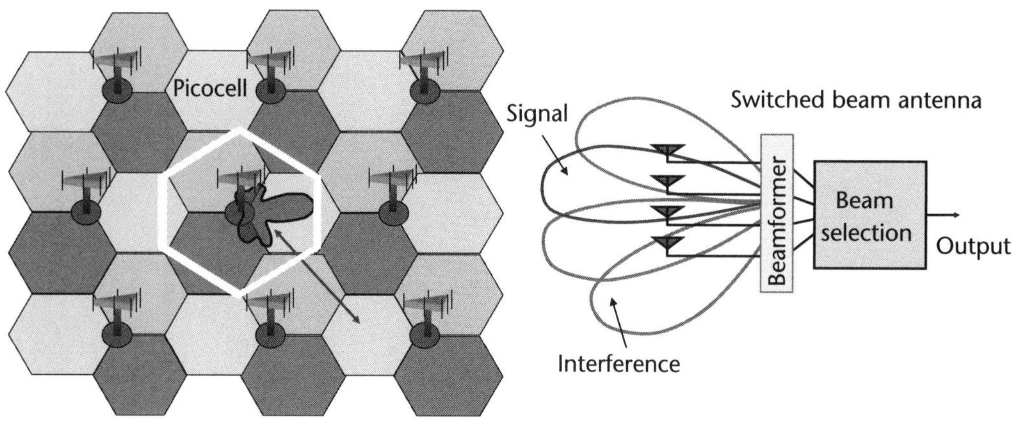

Figure 15.3 The cellular picocell concept and antenna BFN covering these cells.

15.2.3 Spectrum for 5G Communication

Currently, there has been a lot of discussion about the frequency spectrum that 5G technology will use. Cellular network operators all over the world are conducting trials with the objective to deploy the technology commercially. For 5G communication, different countries have proposed and are working on different frequency bands that range all the way from 2.4 to 71 GHz. Countries all over the world are using both low-frequency and high-frequency bands for the initial 5G trials. The 5G spectrums in some countries are shown in Table 15.1; around the world, these bands have been allocated or targeted.

15.2.4 Path Loss Model in a Cellular Cell

In a communication system, the path loss models describe the signal attenuation between the receive and transmit antennas as a function of the propagation distance and other parameters. Path loss is a key factor in the design of the 5G communications system. The channel path loss determines many elements of the communications system design such as the antenna gain, height, and transmitter power. Generally, path loss, which is scaled in decibels relative to a reference power or transmitted power, is assumed to have a linear dependence with logarithmic distance, expressed [9] as:

$$\overline{PL(d)}dB = \alpha + \overline{\beta}\ .10\log_{10}(d) \tag{15.1}$$

where $\overline{PL(d)}$ is a mean path loss over the distance, α is the floating intercept, $\overline{\beta}$ is a linear slope, and d is the distance between transmitter and receiver. Usually, the linear slope $\overline{\beta}$ is extracted by a best-fit linear regression to (15.1) using path loss values in the decibel scale versus the logarithm of measurement distances. The esti-

Table 15.1 5G Spectrum Consideration/Allocations in Various Countries

Country	*Low-Frequency Band*	*High-Frequency Band*
United States	2.5 GHz, 3.7–4.2 GHz, 3.55–3.7 GHz, 5.9–7.1 GHz	24.25–24.45 GHz, 24.75–25.25 GHz, 27.5–28.35 GHz, 38.6–40 GHz, 60–71 GHz
Finland	3.4–3.8 GHz	26.5–27.5 GHz
France	3.46–3.8 GHz	26 GHz
Germany	3.4–3.8 GHz	26–27.5 GHz
Ireland	3.4–3.8 GHz	26 GHz
Italy	3.6–3.8 GHz	26.5–27.5 GHz
Russia	3.4–3.8 GHz	26 GHz
Spain	3.4–3.8 GHz	26.5–27.5 GHz
United Kingdom	3.4–3.6 GHz, 3.6–3.8 GHz	26.5–27.5 GHz
China	3.3–3.6 GHz, 4.4–4.5 GHz, 4.8–5 GHz	24.25–27.5 GHz, 37–43.5 GHz
Japan	3.6–4.2 GHz, 4.4–4.9 GHz	27.5–28.28 GHz, 60–71 GHz
Korea	3.42–3.7 GHz	26.5–29.5 GHz
Australia	3.4–3.8 GHz	24–30 GHz
New Zealand	3.4–3.485 GHz, 3.51–3.587 GHz	24.25–28.35 GHz

mate approach employed is the least-square linear regression fit, in which the linear slope β can be derived by:

$$\overline{\beta} = \frac{\sum_i^n \left(d_i - \overline{d}\right) \mathrm{x} \left(PL_i - \overline{PL}\right)}{\sum_i^n \left(d_i - \overline{d}\right)^2} \tag{15.2}$$

where d_i is the distance of the ith measurement power delay profile (PDP) for a given antenna pointing angle and receiver location, d is the average distance in decibels of all d_i in decibels from the measurement snapshot, PL_i is the path loss value of the ith measurement snapshot in decibels, and $\overline{PL}$ is the average path loss of the entire data set in decibels, respectively. The constant α in decibels is the floating intercept of the linear regression fit in (15.1) that fits the data empirically, and can be thought of as a globally optimum reference attenuation set point that determines the tilt of the path loss model and is found as [9]:

$$\alpha(dB) = \overline{PL}(dB) - \beta.10\log_{10}(d) \tag{15.3}$$

15.2.5 The 60-GHz Spectrum for 5G Application

Currently, the portion of the EM spectrum that is used for mass wireless communications (Wi-Fi, cellular, Bluetooth, and so forth) falls under 0.5–6-GHz region, which is a fraction of the large spectrum, and is considerably crowded. This spectrum is also called the sub-6-GHz band. Engineers and researchers have been working for a long time studying other portions of the spectrum that can be used to develop wireless technologies. The unlicensed 60-GHz band is one of these options. Previously, the 60-GHz band was used only for point-to-point radio links like satellite-to-satellite communications, but now this band currently has several short-range, high-throughput wireless technologies. The rain attenuation is high in this frequency band and, because of that, the EM losses are significant. More importantly, at this frequency, there is a peak in oxygen absorption. Figure 15.4 shows the atmospheric attenuation at millimeter-wave frequencies. There is around 9 GHz of bandwidth in this unlicensed band, depending on the country. An antenna compatible with these applications needs to have a large bandwidth that covers the entire band.

The power ratio between the main lobe and sidelobes is about 13.2 dB in typical array antennas. By developing an antenna power-supply circuit that sends more power to antenna elements close to the center and diminishingly less power towards the periphery, it is possible to achieve low sidelobes with a power ratio of approximately 20 dB. This technique is also known as amplitude tapering feeding, which gives an improvement of 7 dB (or one-fifth, compared to the conventional amount). By applying this technology and arranging multiple antennas array so that each faces a slightly different direction, multiple users can simultaneously communicate without interfering with each other. Even if multiple users concentrate in one location and simultaneously communicate, the reduction of each user's transmission speeds can be minimized, which enables smooth, high-speed communications.

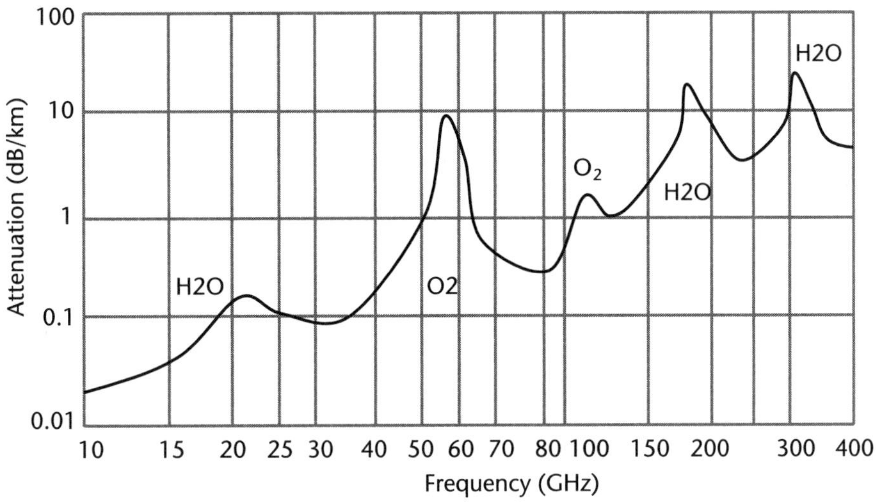

Figure 15.4 Atmospheric absorption of millimeter waves.

This allows the provision of a comfortable communications environment, even when watching high-resolution videos or uploading videos to the cloud.

15.3 Millimeter-Wave 5G BFN

Compared to the linear array, a planar array provides higher resolution of beam steering and more symmetrical patterns with lower sidelobes. Butler matrix-based BFN has been used extensively over the years. It is a cost-effective passive beamforming feed network for phased array antenna applications without needing active devices such as mixers, switches, phase shifters, and transceivers. However, the typical Butler matrix designs need to deploy many air bridges, vias, and bent cables that result in a complicated prototyping process. Planar and printented technology has enabled compact and cost-effective Buttler matrix-based BFNs. As shown in Figure 15.5, an 8-way Butler matrix is composed of twelve 3-dB quadrature couplers, 8

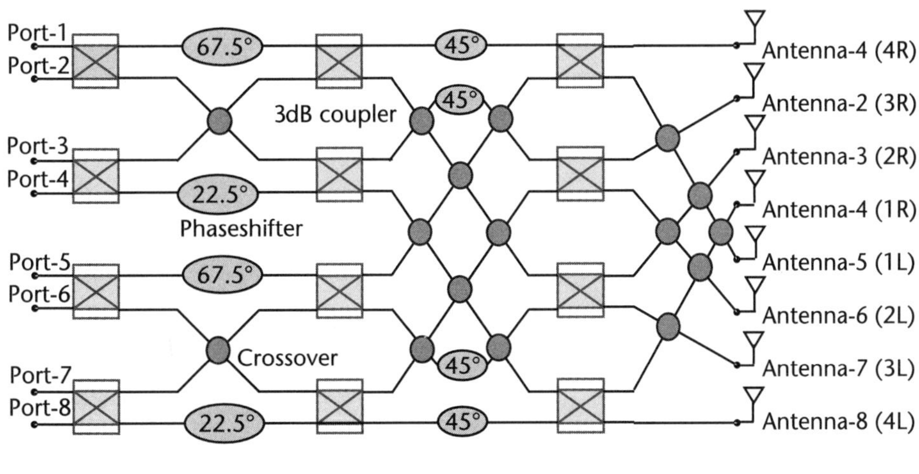

Figure 15.5 Block diagram of an 8 × 8 Butler matrix-based BFN and phased array system.

phase shifters, and 16 crossovers. The microstrip line-based 3-dB coupler and phase shifter are designed individually before integrating into a Butler matrix.

A planar Butler matrix is realized on a 4-mil-thick LCP substrate material and a dielectric constant of 3.2. The LCP substrate is used as a low-cost dielectric material for high-frequency antenna and passive networks design that provides a very reliable performance at high frequencies. It has very reliable electrical properties up to millimeter waves like the low dielectric constant of 3.2 and the low loss tangent of 0.002–0.004 at 60 GHz. Its permeability is comparable to that of glass and very close to that of ceramic. It has a low coefficient of thermal expansion (CTE) as low as 6.9×10^{-6}/K. These material properties make LCP a suitable candidate for multichip module (MCM) technology in millimeter-wave applications.

The design process block diagram of a complete 3-D full-dimension (FD)-MIMO phased array system is shown in Figure 15.6. In the first phase of the design, a single 8 × 8 Butler matrix beamformer is designed that can scan the antenna beam either horizontally or vertically at a time. To scan the antenna beam in all directions, an FD BFN is required. To make an FD beamformer, 8 such Butler matrix BFNs are stacked horizontally and 8 vertically to form a 3-D 64-port network for 64 elements of microstrip planar antenna array. There are three building blocks in this BFN: the microstrip-based 3-dB coupler, 0-dB crossover, and phase shifters. All these components are designed and optimized individually based on microstrip technology. The individual components, as well as the entire Butler matrix structure, are analyzed using the FEM EM solver. The hybrid coupler is used to generate signals 90° out of phase at its outputs. Crossover is used for crossing two RF transmission-line signals with a minimal coupling between them. By cascading two 90° hybrids, a crossover can be implemented. Three different 67.5°, 45°, and 22.5° phase shifters are designed and placed in a BFN.

The next phase in the design is to place all these designed components to form a Butler matrix BFN. It is designed by placing twelve 3-dB hybrid couplers, sixteen 0-dB crossovers, and eight phase shifters. A total of 36 components are used to design a single 8-port BFN. Using the principle of reciprocity, this structure can be

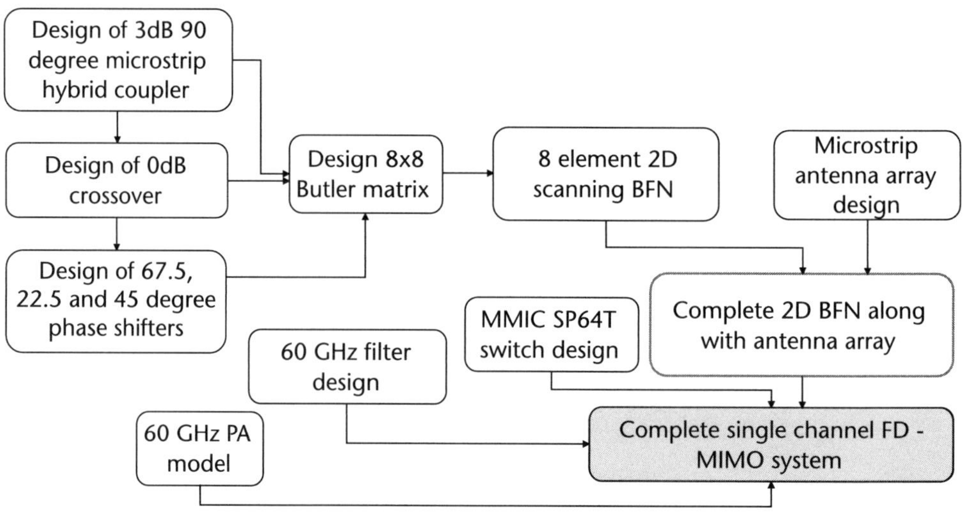

Figure 15.6 Design flow diagram of an 8 × 8 Butler matrix-based phased array system.

utilized in the same manner for receiving or transmitting antenna applications. At 60 GHz, the overall size of a beamformer is about 25 × 18 mm^2. The progressive phase delays at all antenna ports (from P1 to P8) for eight different input signals are given in Table 15.2. In this design, the MMIC SP64T is used to switch beams from one port to another in 64 antenna arrays.

15.3.1 A 3-dB Hybrid Coupler Design

The hybrid couplers are used to generate signals 90° out of phase at its outputs. The port closer to the input port is leading in phase by 90°. Port 4, located on the same side as input port 1, is isolated because there is no power reaching it. A quadrature hybrid or 90° hybrid is used to generate 90° out-of-phase signals at its outputs. The 3-dB hybrid coupler is designed using microstrip-based technology as shown in Figure 15.7. The size of a 3-dB coupler is 70 × 48 mils.

15.3.2 The 0-dB Crossover Design

The crossovers are an effective means of crossing two RF transmission-line signals with minimal coupling between them. It is also known as a 0-dB coupler. By cascading two 90° hybrids, a crossover can be implemented. The geometry used to

Table 15.2 Phase Distribution of 8 × 8 Butler Matrix

	Input Beam Port (in Degrees)							
Output	*4L (P2)*	*3L (P6)*	*2L (P4)*	*1L (P8)*	*1R (P1)*	*2R (P5)*	*3R (P3)*	*4R (P7)*
A1	157.5	157.5	135	90	–112.5	112.5	–135	–180
A2	–45	–90	–157.5	112.5	–135	–180	112.5	22.5
A3	112.5	22.5	–90	135	–157.5	112.5	0	–135
A4	–90	135	22.5	157.5	–180	45	–112.5	67.5
A5	67.5	–112.5	45	–180	157.5	–22.5	135	–90
A6	–135	0	112.5	–157.5	135	–90	22.5	112.5
A7	22.5	112.5	–180	–135	112.5	–157.5	–90	–45
A8	–180	–135	–112.5	–112.5	90	135	157.5	157.5
Δ Phase	157.5	112.5	67.5	22.5	–22.5	–67.5	–112.5	–157.5

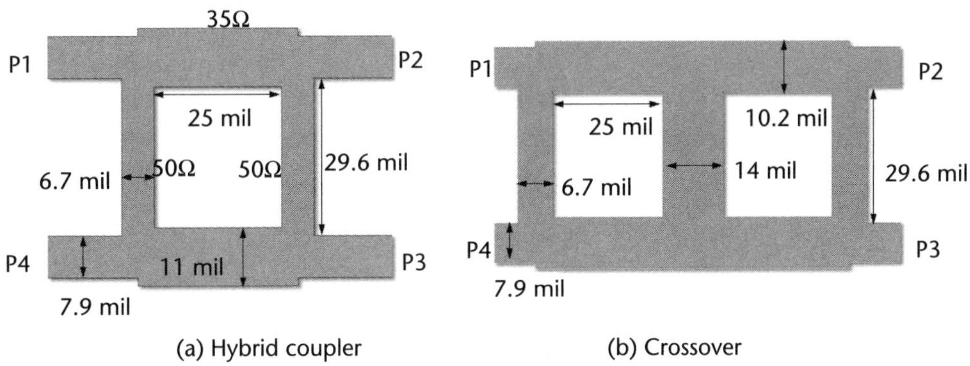

Figure 15.7 The 3-dB hybrid coupler and 0-dB crossover designs and dimensions.

implement the crossover is shown in Figure 15.7(b). The size of a 0-dB crossover is 90×48 mils.

15.3.3 The 22.5°, 45°, and 67.5° Phase Shifter Designs

The phase shifter is implemented using microstrip lines of different length. The length of the section corresponding to the ∅ radian phase shift is given by

$$\varnothing = (2\pi)/\lambda_L \tag{15.4}$$

where λ_L is the length of the line in fraction of wavelength, ∅ is the phase shift in radians, and λ is the wavelength.

15.3.4 The 8 × 8 BFN Design

The 8 × 8 BFN consists of eight RF inputs that independently feed eight RF outputs with varying phase delays as shown in Figure 15.8. The purpose of this network is to uniformly feed the eight antenna elements with progressive phase delays of 22.5°. This arrangement creates eight fixed radiating beams accessed by eight independent

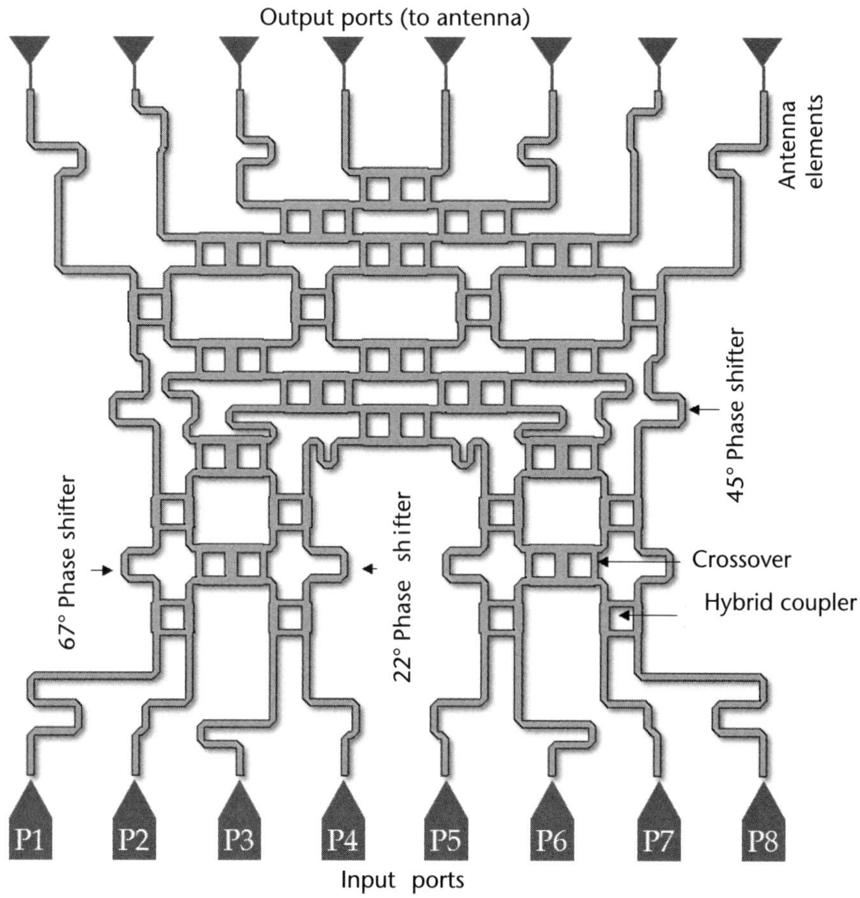

Figure 15.8 Layout design of an 8 × 8 Butler matrix.

inputs. The use of 90° hybrid couplers provides high isolation between each input port, which allows a switch network to toggle between these ports without affecting the antenna performance. The S-parameter of the Butler matrix is shown in Figure 15.9(a). These results correspond to the assigned ports shows the reflection coefficients when considering each port. Return loss is better than 10 dB across the frequency band of 58.3–61.5 GHz. Figure 15.9(b) shows the simulated phase shift at adjacent antenna elements for excitations.

15.4 A 60-GHz Microstrip Antenna Array Design

Millimeter-wave microstrip antenna, particularly around the center frequency of 60 GHz with the advantages of small volume, light weight, thin section, low cost, easy integration, and easy conformance, has attracted lots of interest in the antenna design for the 5G communication systems [10–16]. Considering the layout request of the 5G transceiver, normally the antenna should be in the same plane with the transceiver. The antenna array should be placed at one side of the transceiver in a limited space. At 60 GHz, the antenna design starts with the accurate determination of the single antenna element impedance characteristics allowing tuning the elementary antenna element over the desired frequency band. Table 15.3 provides an overview of the antenna array design specifications.

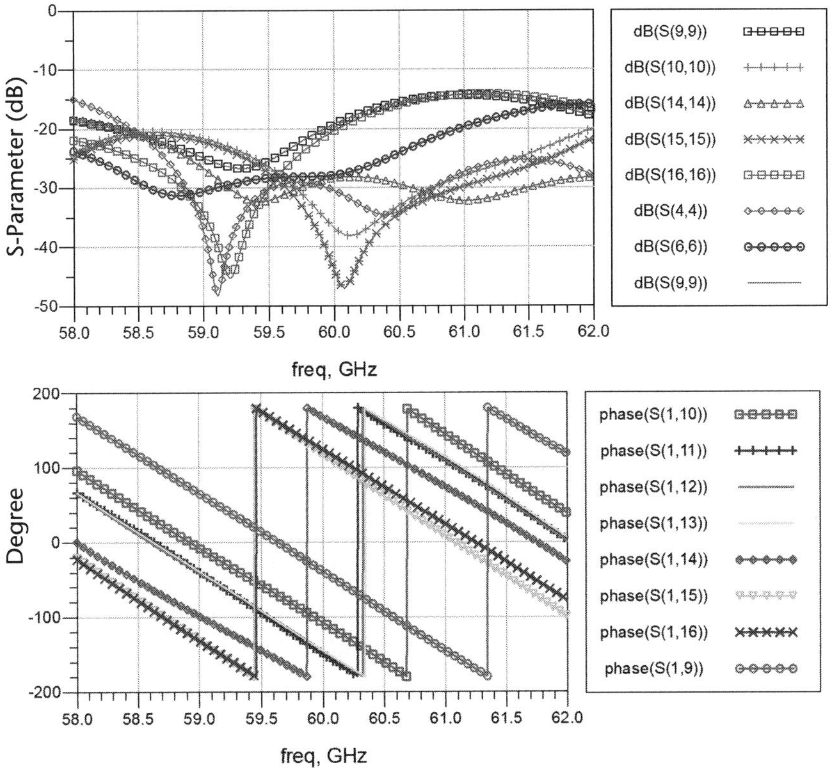

Figure 15.9 Simulation result of the Butler matrix: (a) reflection coefficient and (b) phase plot at output ports.

Table 15.3 Design Parameter of Antenna Array

Antenna Parameter	*Value*
Operating frequency	60 GHz
Substrate thickness	$h = 4$ mil
Substrate permittivity	$\varepsilon_r = 3.2$
Loss tangent	$\tan\delta = 0.004$
Bandwidth (–10-dB return loss)	1 GHz
Maximum gain (planar array)	22 dBi

A patch of 54 mils in size has been designed to operate at 60 GHz with a bandwidth of 1 GHz (10-dB return loss) and a peak gain of 6 dBi. The substrate has a thickness of 4 mils with a dielectric constant of 3.2. The design parameters of eight elements series-fed linear array are shown in Figure 15.10(a). Length and width of patch element are 54 mils and the spacing between array elements is $0.75\lambda_0$. A quarter-wavelength transformer of 3.7 mils is used to match patch impedance with feedline impedance. This transformer has a significant effect on the return losses due to changes in its characteristic impedance.

Eight linear antenna arrays are fed with eight outputs of BFN as shown in Figure 15.10(b). The excitation of each linear array is done by the center feeding along the microstrip feed line. The return loss is optimized by adjusting the patch dimensions and impedance transformer at the feeding point. As shown in Figure 15.11, a bandwidth of 1 GHz for –10-dB return losses has been achieved. The half-power

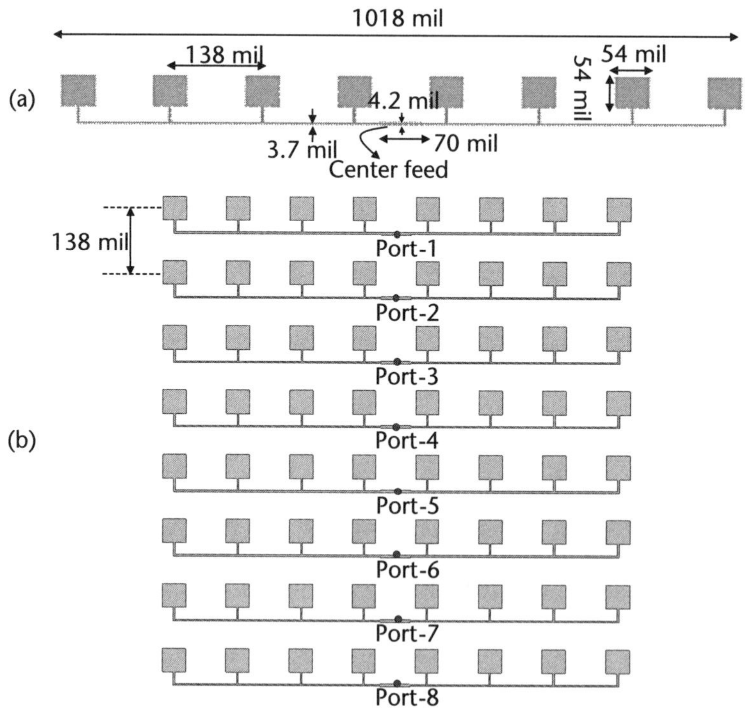

Figure 15.10 (a) Linear array of 8 elements and (b) planar array designs of 8 × 8 elements (all linear arrays are fed individually with the output of the Butler BFN).

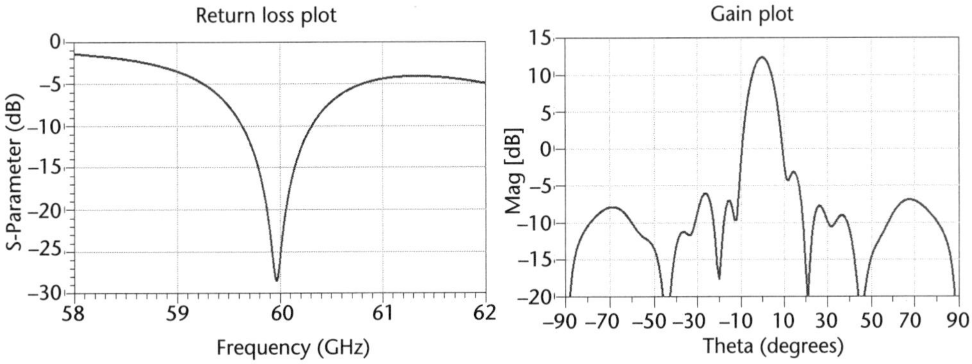

Figure 15.11 Linear array simulated antenna performance results: (a) return loss plot and (b) radiation pattern (phi = 0° cut) at 60 GHz.

beamwidth (HPBW) of the array is 8° and it undergoes negligible fluctuations over the frequency range from 59.5 to 60.5 GHz. In this frequency range, the maximum gain and sidelobe levels fluctuate ±0.7 dB and ±3.5 dB, respectively. Linear array antenna gain is 13 dBi as shown in Figure 15.11(b).

15.5 Phase Array System

The planar antenna array is integrated with an 8 × 8 Butler matrix in a single chip. As shown in Figure 15.12, the antenna is placed on a top layer of a chip stack-up, while the Butler matrix network is at the bottom layer of the chip.

Both antenna and BFN share a common ground plane. The output of BFN is connecting through via to linear microstrip array element. The complete system is

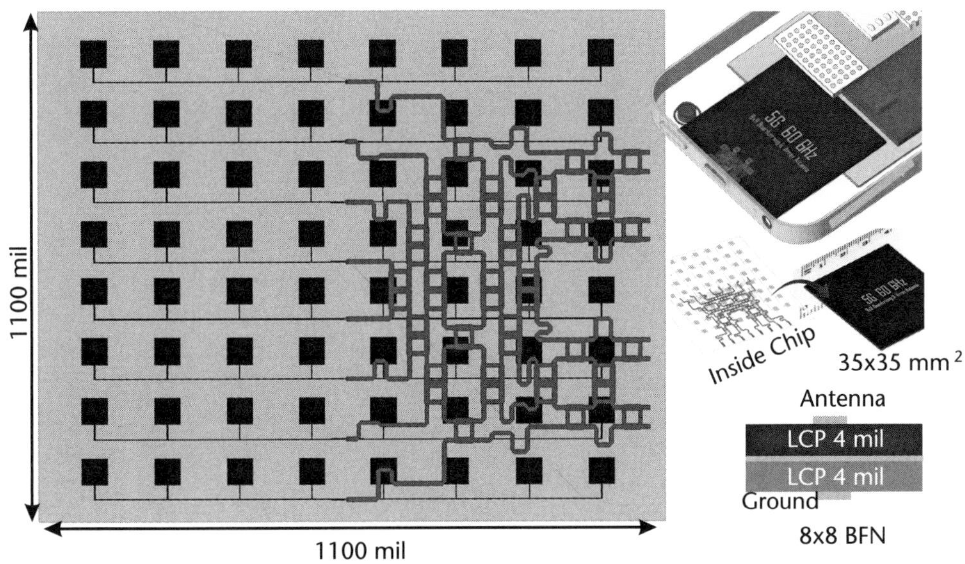

Figure 15.12 Integrations of BFN with planar antenna array.

simulated, and antenna beam scanning is plotted in Figure 15.13 with the RF signal to eight input ports of the Butler matrix.

The gain of 8 × 8 elements array antenna is 22 dBi. This antenna supports the 5G mobile communication to operate at the millimeter-wave frequency with beam scanning capability. The integrated array antenna and BFN are placed in a single 35 × 35 mm^2 chip. This chip can be easily integrated with other circuits, thereby making fabrication more cost-effective and suitable for 5G. For MIMO, the antenna array can be easily integrated with RF systems. Steerable beam ranging from −30° to +30° is achieved with this network.

15.6 FD-MIMO

In communications networks, MIMO technology is used to achieve high data rates. The MIMO approach implements a spatial diversity technique, using multiple transmit and receive antennas. The technique compensates for the signal fading effect, achieving increased channel capacity. FD-MIMO is one of the key technologies currently investigated for 5G. FD-MIMO utilizes many antennas (≥64) placed in a planar antenna array panel to realize high-order multi-user (MU)-MIMO transmissions. It provides link reliability and significant gains in wireless data rates by using large numbers of antennas at the base transceiver station (BTS). In an FD-MIMO system, a base station with a 2-D active array supports multi-user joint elevation and azimuth beamforming, which results in much higher cell capacity compared to conventional systems. As shown in Figure 15.14, the BTS antenna based on the FD-MIMO concept can scan in both the azimuth and elevation planes. With the acceptance of advanced digital signal processing schemes, an FD-MIMO antenna system can support higher-order MU-MIMO, which delivers a multifold improvement in system performance compared to conventional MIMO systems.

The base stations may contain hundreds of antennas using beamforming to focus their power on a tight beam toward the handset. Mobile handsets have room for six to ten antennas, but power and cost limitations on the handset may limit the number of antennas to one or two. As millimeter-wave spectrum has insufficient

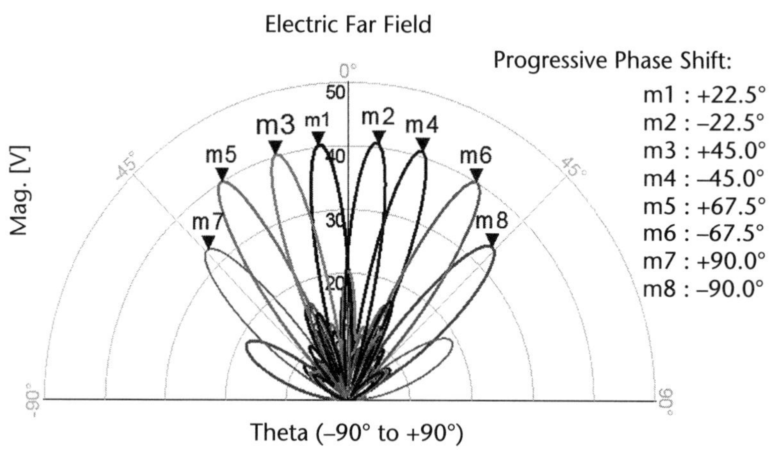

Figure 15.13 Phased array antenna beam scanning for eight RF inputs.

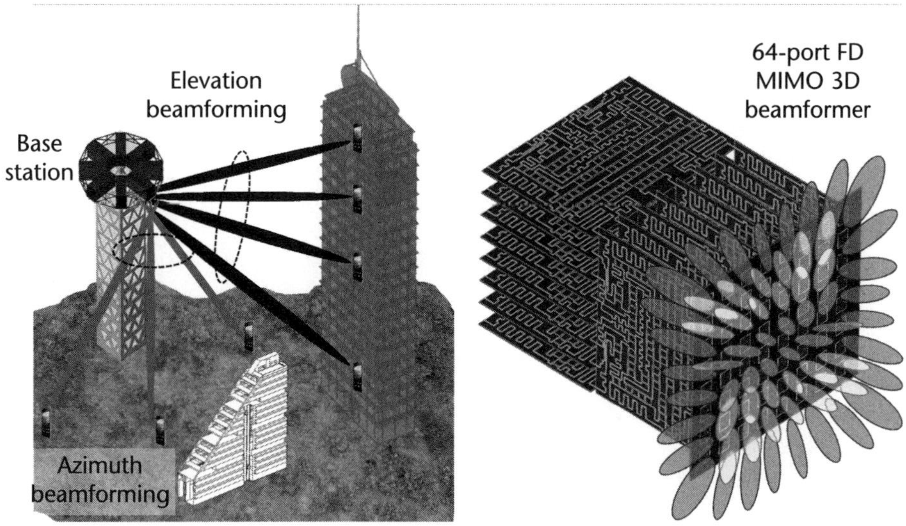

Figure 15.14 Beam scanning in a 5G FD-MIMO system and volumetric phased array BFN.

propagations characteristics, beamforming is thought to be an important element to control the direction of transmitted signals. As seen in Figure 15.14, there are multiple inputs that are communicating devices and multiple outputs that are antenna beams covering all the dimensions of space. That is why this technique is called the FD-MIMO.

15.7 FD-MIMO Phased Array System

A 3-D beamforming phase array is developed by stacking eight units of 8×8 BFN horizontally and eight units vertically and then these are combined as shown in Figure 15.15. There are a total of 64 input ports and 64 output ports fed to the antenna system.

To evaluate the performance of a 3-D phased array system, simulation is carried out using 64 switching circuits as shown in Figure 15.16 for 2-D scanning. The antenna scan angle is measured using this setup. In this setup, complete transmit chain including a PA, SP64T switch, and RF filter is used. The SP64T switch can be turned on for any specific port by turning biasing voltages (S1 port to S6 port) on and off as shown in Figure 15.16.

For each input port, there are a total of 64 antenna scan angles in both the vertical and horizontal directions as shown in Figure 15.17. This multibeam antenna has ±60° coverage utilizing eight individual narrow beam patterns each with a half-power beamwidth of 7°.

15.8 5G Communication System Design

The 5G wireless technology will enable gigabit data, ultralow latency, and massive connectivity for machine-to-machine (M2M), connected vehicles (V2X), and IoT applications. To accomplish these goals, 5G technology developers need to

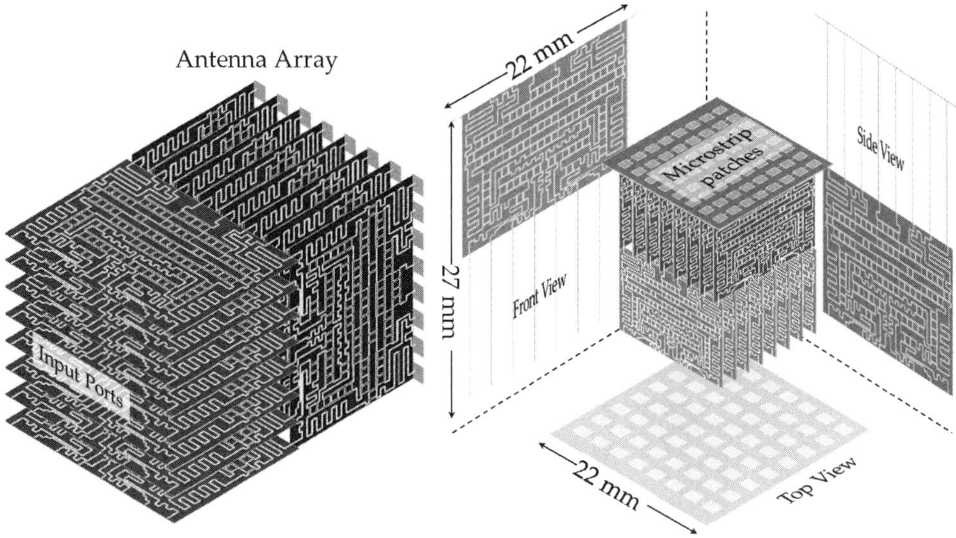

Figure 15.15 The 3-D BFN design and dimensions.

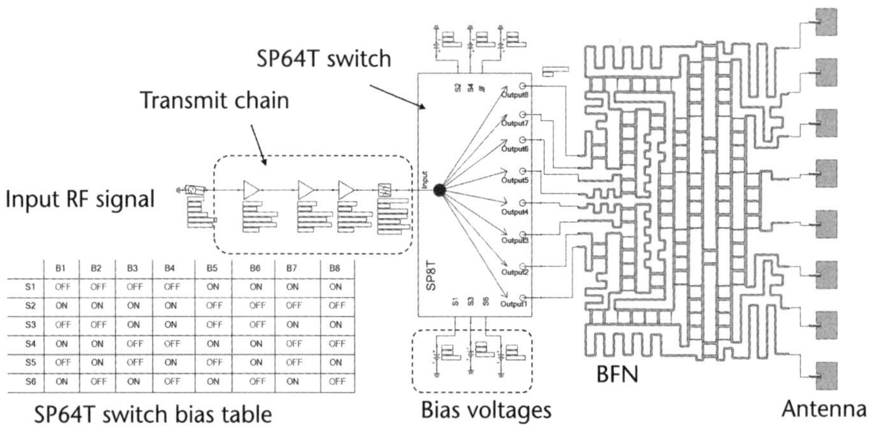

Figure 15.16 BFN simulation setup along with the RF transmit chain.

perform many integrated and system-level simulation, such as beamforming antenna arrays along with RF system to achieve higher data rates by using millimeter-wave frequencies. To evaluate the performance, the BFN must be integrated into a system-level model and evaluated over a collection of the parameter, steering, and channel combinations [17, 18]. Because of the complexities of 5G antenna design, RF system engineers need to include the antenna characteristics in their simulations to correctly design the entire T/R system. All these activities are best done early in the design process.

As the industry moving toward highly integrated 5G phased-array systems, it is critical for in-house systems experts to work closely with hardware developers, with both fully exploring the capabilities and trade-offs among possible architectures and integration technologies [19–22]. An ideal phased array system design flow handles the start-to-finish front-end development, embedding RF/microwave

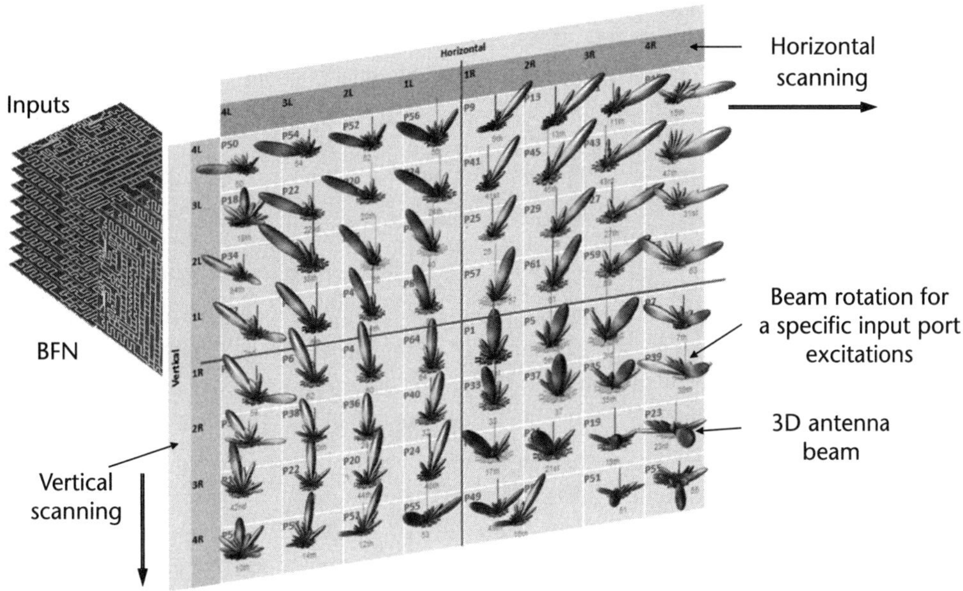

Figure 15.17 Volumetric beam rotation of phased array antenna for 64 RF inputs.

circuit simulation and/or measured data of components models within a phased-array system hierarchy. These analysis and simulations enable the system designer to select the optimum solution, ranging from hybrid modules through a fully integrated RFIC device, addressing the specific requirements of the targeted application. The inability of high-level system tools to accurately model the interactions between the RF components, which are typically specified and characterized individually, results in design failure, and high costs of development. Because the overall phased array performance is neither driven purely by the antenna nor by the microwave electronics in the feed network, combined analysis and simulation capture their interaction to accurately predict true system behavior. Circuit, system, and EM cosimulation together enable verification throughout the design process.

MIMO phased arrays and the corresponding RF and digital circuits are critical components of the 5G communication system. A balance must be made in these system designs to meet overall system performance goals and system-level cost objectives. The modeling of a complete system including RF, antenna, and signal processing components can help to verify design choices at the earliest phases of the project and reduce the associated challenges. In a phased array, system-level simulation can account for RF effects such as the S-parameters of off-the-shelf phase shifters and attenuators and X or S-parameters of nonlinear amplifiers and mixers. With this ability, users can characterize the radiation of spurious intermodulation signals from the array, which helps to avoid interference with nearby antennas.

In this section, system-level design and simulations method for incorporating 5G phased array antenna that is coupled to RF circuits is discussed. This method provides great accuracy with electrical modeling of a phased array antenna in the presence of the TRM and other subsystems of the 5G MIMO. Furthermore, the system simulator models the antenna and RF interaction to obtain a more realistic performance of the overall system. In a 5G system, more devices are being integrated into packages that need different types of design and analysis tools. Innovative design flow and simulation methods are needed from the component level implementation all the way through to system-level performance verification.

The block diagram concept of 5G transceiver system is shown in Figure 15.18. It can be divided into three sections, the first section is the base-band system, the second part is the RF design block and the last section is the radiating element. In complete phased array design there are many design considerations for all these three sections:

- *Phased array system-level design considerations:* Drive system with modulated signals; simulation with various array geometries, tapers, imperfections, array mutual coupling, and element radiation patterns; and perform system-level digital signal processing simulation.
- *Phased array RF system design considerations:* RF circuit design for of individual elements (PA, LNA, mixer) of the RF link; and antenna arrays excited from the amplifier and feed network (EM excitation) antenna elements mismatch effects (main lobe gain, beamwidth, sidelobe suppression).
- *Phased array antenna design considerations:* Number of antenna elements, the shape of the phased array; and the spacing between the elements and the arrangement of the antenna elements.

A typical phased array antenna design includes parameters such as array geometry, element spacing, the lattice structure of the elements, and element tapering. Moreover, the effects of mutual coupling among antenna elements are important to characterize before the final design is implemented. After completing an initial configuration of the array design, architectural partitioning can be iteratively evaluated against the overall system performance goals. The 5G network will be very complex compared to other generation networks because of the requirements and technology as shown in Table 15.4, and that is a reason to evaluate system performance along with antenna design. Among these specifications, high data rate and data capacity are the driving force in 5G. To meet these requirements, the phased array system introduces the beam-steering function and enables the system to achieve a good link when incoming signals are coming from different angles, but the steering range will still be limited. The 5G network design starts

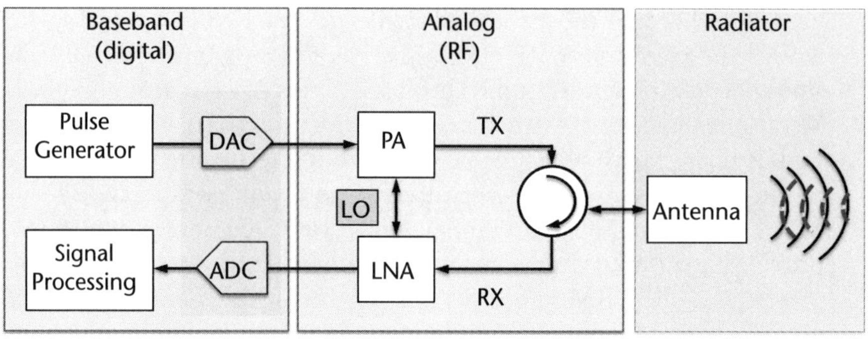

Figure 15.18 Block diagram of a typical communication system.

Table 15.4 All Generation Comparison

Parameters	*1G*	*2G*	*3G*	*4G*	*5G*
Development	1970–1980	1990–2001	2001–2010	2011–2015	2020
Data capacity	2 Kbps	Up to 64 Kbps	2 Mbps	Up to 1 Gbps	Up to 10 Gbps
Technology	Analog wireless	Digital wireless	FDD and TDD CDMA, WCDMA, UMTS	WLAN, OFDMA, MC-CDMA, network LMPS	Unified IP, broadband LAN/ WAN/PAN/ WLAN, OFDM modulation
Multiplexing	FDMA	TDMA	CDMA	CDMA	CDMA, BDMA
Switching type	Circuit	Circuit	Packet, Ckt	All packet	All packet
Handoff	H	H	H	H & V	H &V
Frequency	800–900 MHz	850–1,900 MHz	1.6–2.5 GHz	2–8 GHz	3 to 300 GHz
Core network	PSTN	PSTN	PSTN	Internet	Internet

with system-level simulation. In the first stage, the design blocks are replaced with behavior models to see the complete system performance and other parameters. In the real world, these behavior models of systems are replaced with actual designed components or hardware. The 5G communications will the user the perception of near infinite capacity. This requires a significant change in data rates that will be facilitated by a move to higher transmission frequencies where wider bandwidths are more readily available. System-level design and analysis of the 5G communication system along with phased array antenna array is discussed in the next section.

15.9 The 5G Phased Array System Design

Two critical technologies in the delivery of 5G services are the phased array antenna and hybrid BFN. Beamforming is at the heart of modern wireless communication systems such as 5G, LTE, and WLAN and is used in antenna arrays in radar, sonar, medical imaging, communication and audio systems [23, 24]. Beamforming controls the direction of the transmission or reception of signals at the antenna array and it is a key technique that is used to improve the SNR of received signals, eliminate undesirable interference sources, and focus transmitted signals on a specific location in 5G communication. To achieve a controlled and flexible beamforming in an active array design, it is desirable to have independent weighting control over each antenna array element. This requires separate TRMs for each antenna element.

15.9.1 System-Level Design and Simulation

Millimeter-wave front ends require substantial antenna and beamforming gain to overcome lossy propagation loss as compared to sub-6-GHz systems. The 27.5–29.5-GHz band is a strong candidate for the 5G system and much of the research undertaken to date has considered this band. Figure 15.19 shows a typical architecture of the millimeter-wave front-end communication system. The key parameters

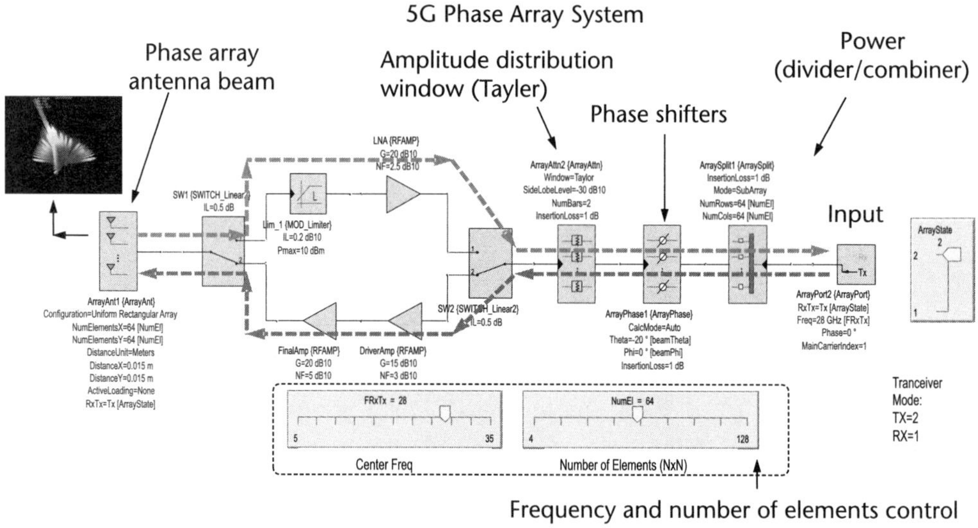

Figure 15.19 5G BFN system level simulation setup.

and specifications for the designed millimeter-wave MIMO transceiver system are listed in Table 15.5.

The system design supports all required phase shifting, path switching, and amplification to drive each antenna array element appropriately. The system engineer performs such simulations to generate design specifications and parameter values for each component used in this communication chain. These specifications are passed on to RF and antenna design engineers. RF engineers design millimeter-wave component circuits and characterize their models to capture the fluctuation of the individual device. Then the component level design is verified for the system-level performance evaluation.

Table 15.5 Key System Parameters of the 64-Channel 5G MIMO Transceiver System

Key System Specifications	*Values*
Receiver sensitivity	28.1 GHz
Signal bandwidth	1 GHz
Antenna array configuration	64 elements
Element gain	7.2 dBi
Array gain	22.8 dBi
Number of RF TRMs	64
Max. transmit power	27 dBm (10-dB backoff)
Max. EIRP for single stream	42 dBm (Tx power + element gain + array gain)
Transmit power dynamic range	27.7 dB
Receive power dynamic range	62 dB
Beam-steering range	Azimuth: ± 60°, Elevation: ± 60°
Duplexing scheme	Time division multiplexing (TDM)
OFDM subcarrier spacing	260 kHz

15.9.2 Antenna Design and Simulation

Printed antennas are very suitable antenna types for the 5G application. A 4 × 4 planar antenna as shown in Figure 15.20 is designed based on antenna specifications received from the system engineer. The antenna array is designed on RT Duroid 5880 substrate (ε_r of 2.2 and thickness of 0.79 mm). This antenna covers a 28-GHz 5G frequency band (25.5 GHz to 30 GHz) and the peak gain is 16 dBi. To achieve sufficient directivity, a beamwidth of 6° in the scanned direction is realized. Element spacing of 5.2 mm is chosen to prevent grating lobes, which are unintended beams (arising from spatial aliasing) that can degrade sensitivity. Because the wavelengths are small at millimeter-wave frequencies, this allows such arrays to be incorporated into a small chip form factor. Base stations are likely going to require hundreds of elements to meet the EIRP requirements for increased coverage, whereas end-user equipment may only require 16–32 elements.

The array EIRP must be chosen and the output power required per element at P1dB (P_{out}) is determined based on the number of array elements. Given an EIRP, P_{out} is calculated using:

$$P_{out} = EIRP - 10\log N - G_{ANT} + L_M \tag{15.5}$$

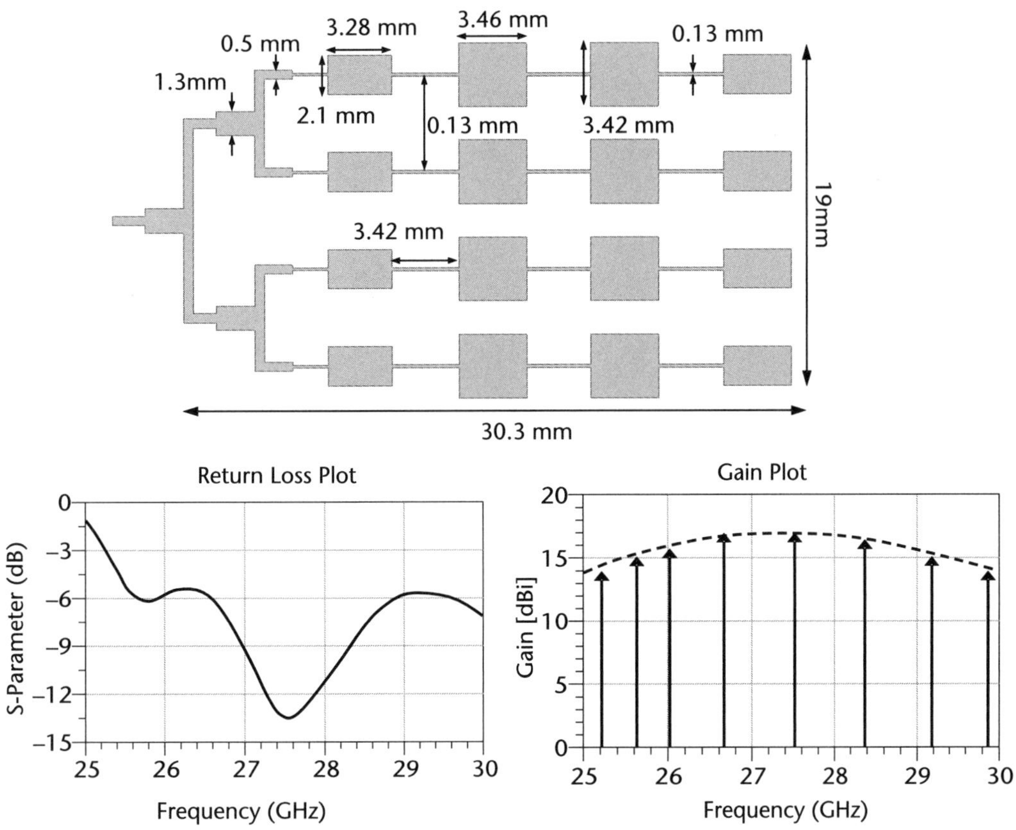

Figure 15.20 The 28-GHz 4 × 4 planar antenna array and simulated return loss and gain plot.

where N is the number of array elements, G_{ANT} is the N-element antenna gain assuming uniform distribution across the array and includes the antenna ohmic, mismatch loss, and feed line loss, and L_M is the power loss due to any amplitude and phase mismatch between the elements. Figure 15.21 presents the required P_{out} for different EIRP values and array sizes. In this design P_{out} is chosen around 10 dBm for low power consumption per chip and to mitigate the thermal issues. An EIRP of 43 dBm can be achieved with an array size of $N = 64$ (16×4 elements).

The array factor of a 2-D ($N_x \times M_y$) uniform planar array can be written as [25]

$$AF(\varnothing,\theta) = \sum_{n=0}^{N_x-1}\sum_{m=0}^{M_y-1} a_{nm} e^{-j2\pi\sin\varnothing\left(n\cos\theta d_x + m\sin\theta d_y\right)} / \lambda_0 \tag{15.6}$$

where $\varnothing$ is the azimuth angle, θ is the elevation angle, a_{nm} is the complex coefficient at each element, λ_0 is the wavelength in vacuum, and (d_x, d_y) represent the horizontal and vertical element spacings. For N_x-element uniform linear array at horizontal direction, the horizontal far-field radiation pattern is:

$$P(\varnothing) = E(\varnothing)\sum_{n=0}^{N_x-1} a_n e^{-j2\pi nd\sin\varnothing} / \lambda_0 \tag{15.7}$$

where $E(\varnothing)$ is the horizontal pattern of element, d is the element spacing, and a_n is the complex excitation at the nth element.

15.9.3 RF Circuit Design and Simulation

Normally, the design of an RF system is performed in a top-down approach, where the application requirements and constraints drive component design decisions. After getting the design specification of individual components from system engineers used in the transceiver chain, the RF engineer starts a realization of these components using circuit simulation. The circuit simulation setup in the ADS for a 4-channel transmit chain is shown in Figure 15.22. All major RF components

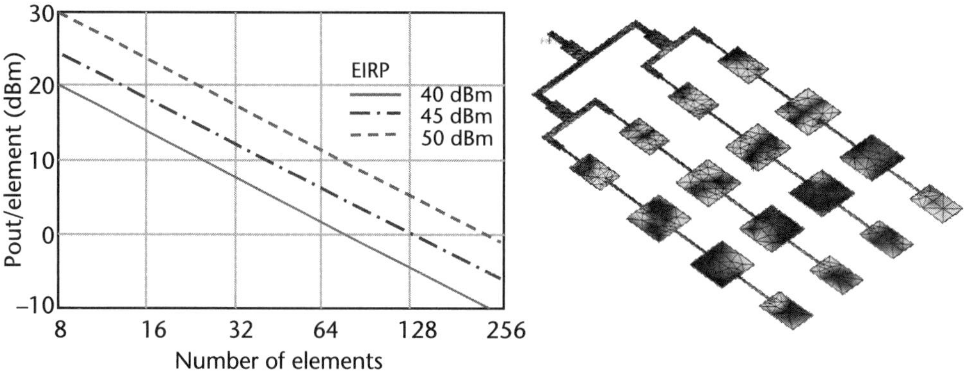

Figure 15.21 Required output power per element for different EIRP values versus number of array elements.

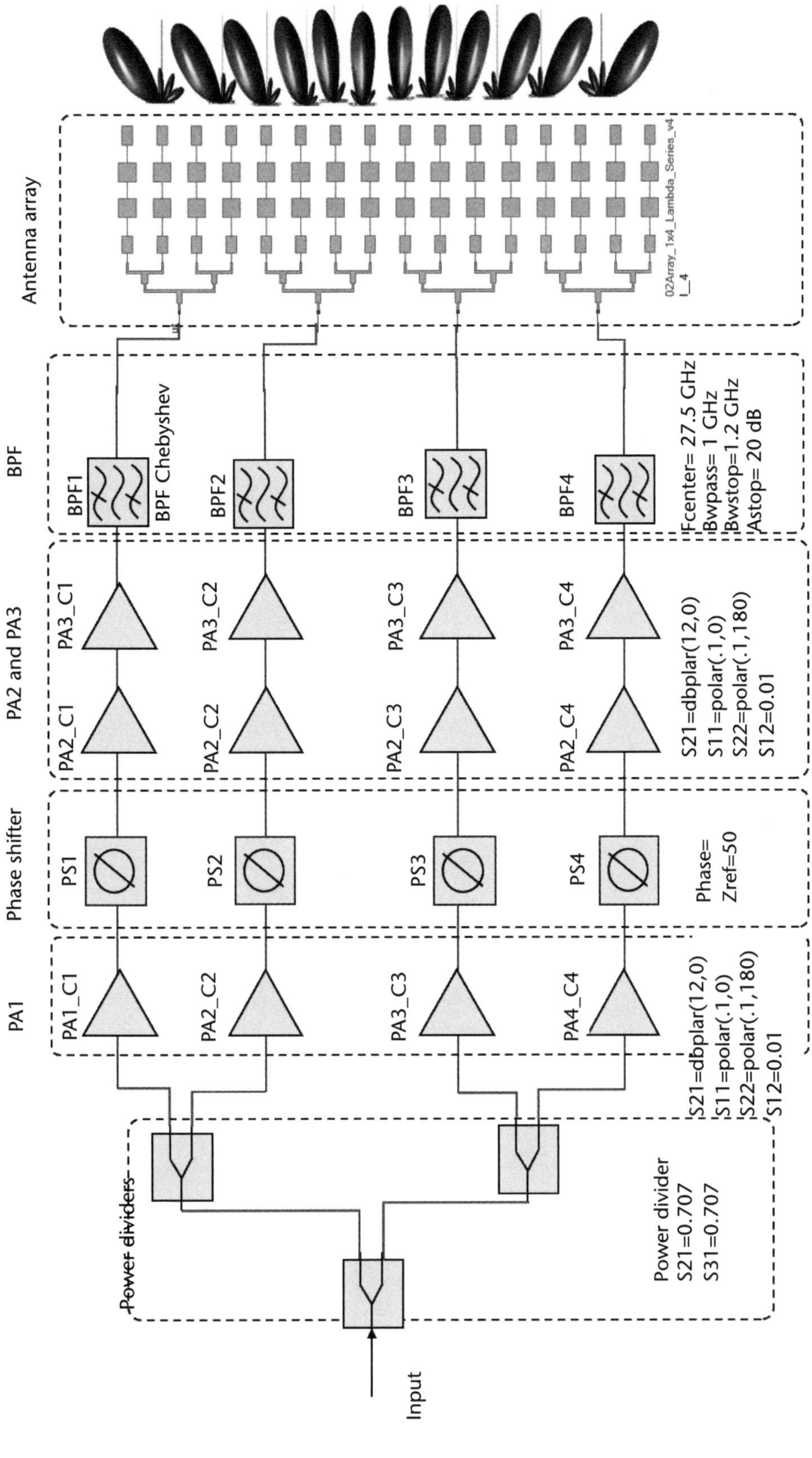

Figure 15.22 Four-channel 5G phased array transmit module schematic simulation setup and rotated antenna beams for different phase values.

(LNA, PA, filter, and phase shifter) of the transceiver chain are designed based on MMIC technology, as discussed in Chapter 14.

This communication system is built on a low-cost PCB that only consumed approximately 10W of DC power in either its transmit (Tx) or receive (Rx) modes. It interfaces to an array of a 4-antenna subarray, where each subarray is driven by one of four parallel PAs. Some degree of filtering is required immediately after each PA, for harmonic rejection and suppression of receive band noise and unwanted spurious outputs. The inputs to all communication channels are driven in-phase from a common RF input via a 4-way splitter as shown in Figure 15.22.

The performance of an RF PA dominate the overall transmitter (Tx) performance, as its PAE dictates the power and heat dissipation for the entire Tx. For enhanced user experience and massive MIMO antennas, the 5G system will require more PAs to be integrated into the RF front-end modules, making the design of a 5G PA more critical than that of a 4G PA design. The output power (P_{out}), linearity, reliability, cost, and form factors of a PA are all very important. SiGe BiCMOS processes are used for PA design. The 4-bit phase shifter is designed with pHEMT process as shown in Figure 15.23. This phase shifter allows the insertion phase of each channel to be set with a 22.5° resolution. The lower bits (22.5° and 45°) are implemented using switched-filter networks. Higher bits (90° and 180°) are implemented by switching between highpass filter and LPF networks.

15.10 System Validation and Test

Once the individual components have been designed and meet their individual performance requirements, it is needed to assess how they will perform when deployed in the actual operational environment for which they are intended. After completing design of individual circuit designs, the next step is the verification and test of the prototype hardware.

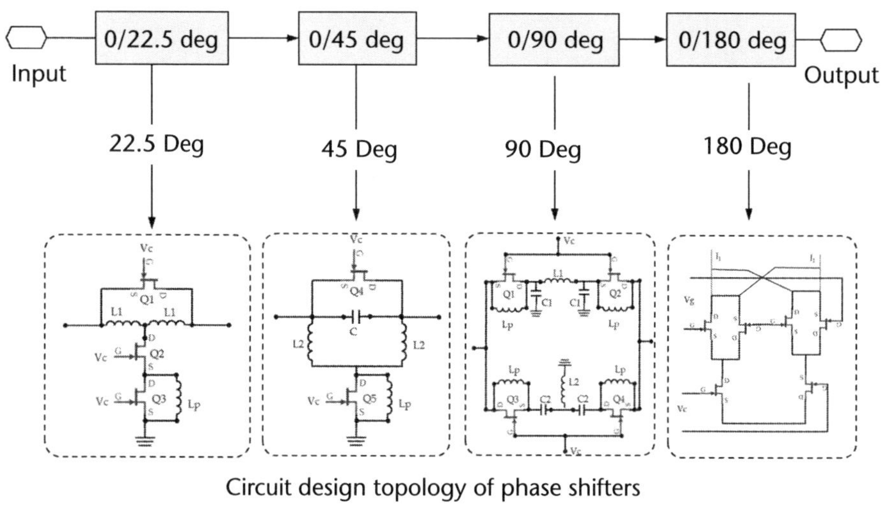

Figure 15.23 Four-bit phase shifter design.

As explained earlier, the millimeter-wave RF front-end components and phased array antenna are integrated into the package. Each component performance data is measured (like S-parameter, X-parameter, antenna patterns), and this data is used in system-level validation. In a system simulation, behavior models (like PA, LNA, antenna) are replaced with actual measured data. System-level simulation is carried out to see the actual performance of the complete system. There may be some degradation in the system performance with actual components. Based on system requirements, design components may need reoptimization of system-level changes. System-level validation is a kind of virtual test and it saves on the design cost and production cycle. Designed RF components of the communication system and antenna array are integrated in a 5G communication system as shown in Figure 15.24 along with digital baseband blocks. The local oscillator (LO) performance is critical in the millimeter-wave 5G communication system. One major contribution of the degradation of the modulation precision of the millimeter-wave transceiver system is the phase noise of LO. The influence of the phase noise on the error vector magnitude (EVM) can be expressed as [26]:

$$EVM_{PN} = \sqrt{\left(\int_{-\infty}^{\infty} S_{PN}(f)df \right)} = \sqrt{(PN_{INT})} \tag{15.8}$$

where $S_{PN}(f)$ is the phase noise spectrum density and PN_{INT} is the integral phase noise.

The simulated phase array 5G communication system antenna beams are shown in Figure 15.25. The main beam of array antenna always points in the direction of

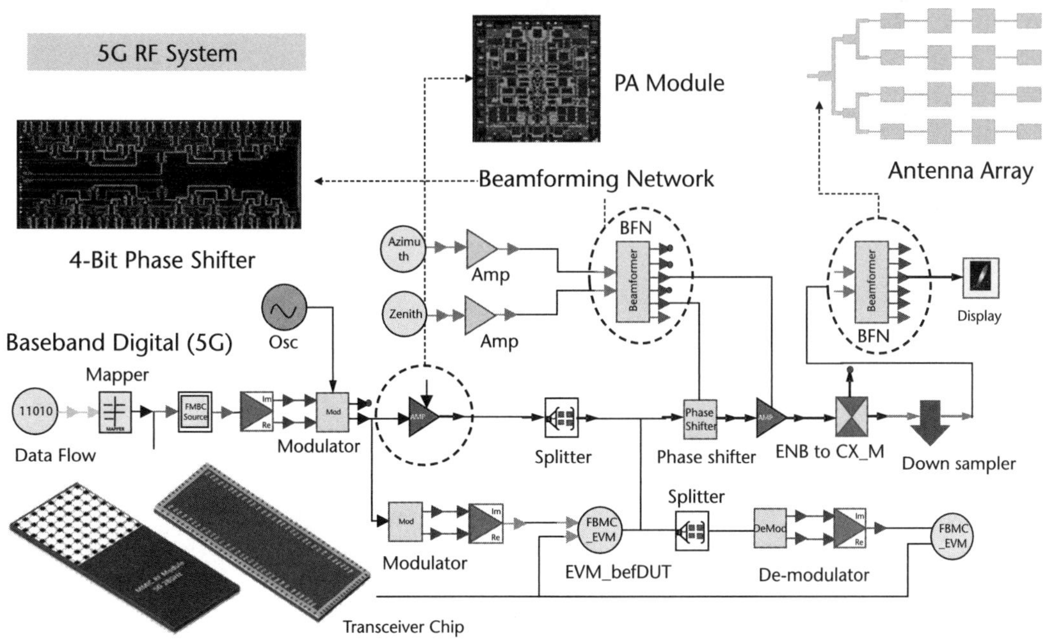

Figure 15.24 Complete 5G communication system validation along with designed components.

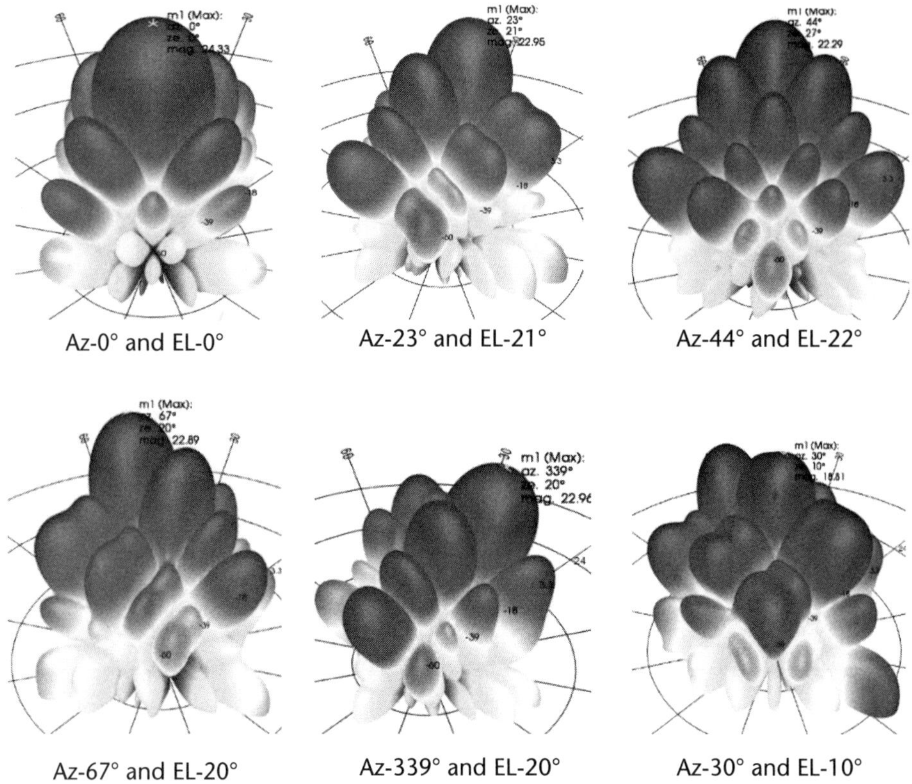

Figure 15.25 Antenna beam patterns for different phase shifts in the complete 5G system.

the increasing phase shift. The complete phased array network provides 64 beams with gains between 28 dBi and 29 dBi. It maintains a return loss less than –10 dB within the frequency band of 27.9–29.4 GHz. This multibeam antenna has ±60° beam coverage. The maximum value, which can be achieved for the field of view (FOV) of a phased array antenna, is 120° (60° left and 60° right). Grating lobes are generated beyond the FOV scan angle. Amplitude tapering can be applied to reduce sidelobe levels in the antenna array.

15.11 Massive MIMO Based on Digital Beamforming

In previous section, FD-MIMO phased array system is based on analog beamforming technique. The antenna array beam of massive MIMO system can be concentrated in a narrow range based on digital beamforming (DBF) technology also. The digital beamforming performs the phase shifting in the digital domain. The RF signal is first downconverted and then digitized using an analog-to-digital converter (ADC) whose digital output is passed to a digital baseband processing block where the beamforming operation is performed. The key technology of the massive is the active multibeam antenna system, which consists of multiple (≥ 64 antenna elements) and transceivers. Massive MIMO relies on spatial multiplexing and a simple method for spatial multiplexing is to deploy linear precoding like beamforming. In practical systems, DBF technology has the advantages of flexibility, adaptive control, and accuracy. DBF technology is established with the development of advanced digital signal processing, by which transmitting and receiving the useful signal can be implemented with a high-gain narrow beam generated in the direction of the useful signal and null steering generated in the direction of interference. The conceptual diagram of DBF is shown in Figure 15.26.

The received signal vector of the antenna array with N elements is defined as X,

$$X = [x\_1 x\_2 x\_3 \ldots x\_N]^T \tag{15.9}$$

where x_i refers to the complex signal received by the ith element.

$$x_i = I_i + jQ_i \tag{15.10}$$

The amplitude and phase of the ith received signal are expressed as [27–29]

$$|x_i| = \sqrt{(I_i^2 + Q_i^2)} \tag{15.11}$$

$$\varnothing_i = \arctan\left(Q_i / I_i \right) \tag{15.12}$$

To form the kth beam, the complex weighted coefficient W_{ik} is given by:

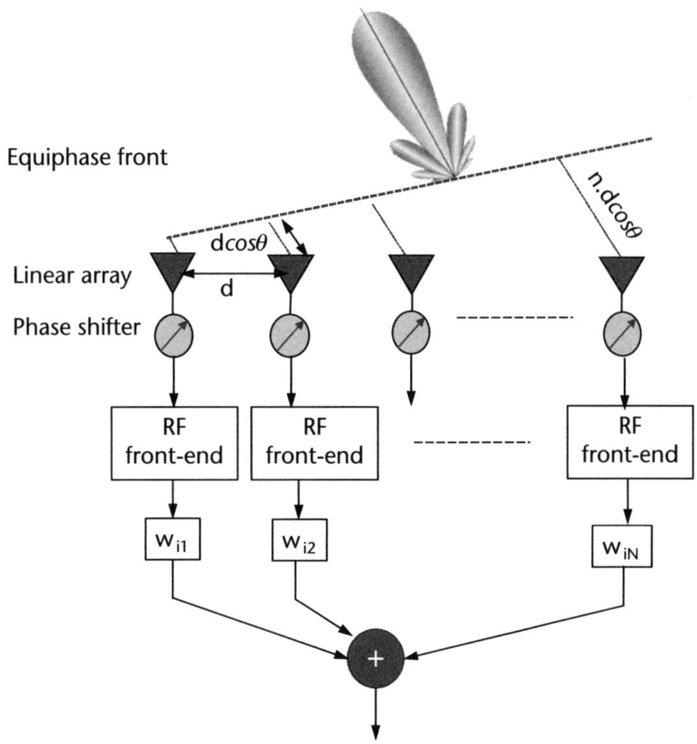

Figure 15.26 The conceptual diagram of a digital BFN.

$$W_{ik} = a_i \exp\left[-ji\Delta\varnothing_{Bk}\right] \tag{15.13}$$

Then the weight of W_k of the kth beam received signal vector is defined as:

$$W_k = \left[W_{1k}\, W_{2k} \,\ldots\, W_{Nk}\right] \tag{15.14}$$

The output $D_k(\theta)$ of the DBF system is described by:

$$D_k(\theta) = W^T X \tag{15.15}$$

where $|D_k(\theta)|$ is the radiation pattern of the kth beam.

In massive MIMO system based on digital beamforming, there are many challenges associated with implementation [26]. Each RF channel requires a high-speed, high-resolution ADC, which is challenging to implement and will increase the overall power consumption, especially in the case of large arrays.

15.12 Summary

Active millimeter-wave antennas and BFNs are planned to deploy in unprecedented volumes over the next few years, fueled by the rapidly emerging 5G wireless communications systems. The combination of high-performance silicon processes together with advanced analog and digital design techniques has accelerated the adoption of millimeter-wave ICs to the extent that advanced 5G and SATCOM networks are now anticipated several years earlier than the industry originally projected. The phased array antenna is one of the most important components in the 5G system. Emerging 5G radio access technologies will drastically change the landscape of not only the cellular communication ecosystem, but more specifically the way in which the 5G system and antennas are designed. As per 5G requirement for more data bandwidth, the 5G new radio (NR) will push into new frequency ranges traditionally reserved for satellite and defense applications. The associated adaptive antenna system technologies required to accomplish this and increase overall spectral efficiency, namely beamforming and massive MIMO. Where usually most of the radio functionality could be evaluated independent of the antenna system, the use of adaptive antenna system technologies makes it impossible to dissociate the radio performance from the antenna performance.

In this chapter, a beam-steerable phased array antenna at 60 GHz for a 5G FD-MIMO system was discussed, designed, and modeled and the results were presented. The designed phased array system has 3-D Butler matrix as its feed to achieve a steerable beam ranging from –60° to +60°. Both the feed and the array antenna are designed using waveguide technology resulting in low loss and high-power network. The design and simulation of a completed 5G communication system, BFN, and antenna design of a millimeter-wave 28-GHz transceiver with a beam-steered patch antenna array were also covered in this chapter. Design techniques are established to achieve the desired level of performance using the system, circuit, and EM simulation technologies.

References

[1] Doan, C., et al., "Design Considerations for 60 GHz CMOS Radios," *IEEE Communications Magazine*, Vol. 42, No. 12, December 2004, pp. 132–140.

[2] Lee, J., et al., "A Switched Array Antenna Module for Millimeter-Wave Wireless Communications," *Proc. Global Symp. Millim. Waves*, 2008, pp. 161–163.

[3] Thompson, J., et al., "5G Wireless Communication Systems: Prospects and Challenges," *IEEE Communications Magazine*, February 2014, pp. 62–65.

[4] Roh, W., et al, "Millimeter-wave Beamforming as an Enabling Technology for 5G Cellular Communications: Theoretical Feasibility and Prototype Results," *IEEE Communications Magazine*, Vol. 52, No. 2, 2014, pp. 106–113.

[5] Pattan, B., "The Versatile Butler Matrix," *Microwave Journal*, November 1, 2004.

[6] Gruszczynski, S., and K. Wincza, "Broadband 4x4 Butler Matrices as a Connection of Symmetrical Multisection Coupled-Line 3-dB Directional Couplers and Phase Correction Networks," *IEEE Transactions on Microwave Theory and Techniques*, Vol. 57, No. 1, 2009.

[7] Butler, J., and R. Howe, "Beamforming Matrix Simplify Design of Electronically Scanned Antennas," *Electron. Design*, No. 9, April 1966.

[8] Liang, J. W., and A. J. Paulraj, "On Optimizing Base Station Array Topology for Coverage Extension in Cellular Radio Networks," *45th IEEE Vehicular Technology Conference*, 1995, pp. 866–870.

[9] MacCartney, G. R., et al., "Path Loss Models for 5G Millimeter Wave Propagation Channels in Urban Microcells," *2013 IEEE Global Communications Conference (GLOBECOM)*, Atlanta, GA, 2013, pp. 3948–3953.

[10] Pandey, A. K., "Design of a Compact High Power Phased Array for 5G FD-MIMO System at 29 GHz," *2016 Asia-Pacific Microwave Conference (APMC)*, New Delhi, 2016, pp. 1–4.

[11] Hong, W., et al., "Design and Analysis of a Low-Profile 28 GHz Beam Steering Antenna Solution for Future 5G Cellular Applications," *IEEE MTT-S International Microwave Symposium (IMS2014)*, Tampa, FL, 2014, pp. 1–4.

[12] Khalily, M., et al., "Design of Phased Arrays of Series-Fed Patch Antennas with Reduced Number of the Controllers for 28-GHz mm-Wave Applications," *IEEE Antennas and Wireless Propagation Letters*, Vol. 15, 2016, pp. 1305–1308.

[13] Brookner, E., *Practical Phased-Array Antenna Systems*, Norwood, MA: Artech House, 1991.

[14] Cheon, Y., and Y. Kim, "Millimeter-Wave Phased Array Antenna with Wide Beam Coverage," *2016 10th European Conference on Antennas and Propagation (EuCAP)*, Davos, 2016, pp. 1–3.

[15] Helander, J., et al., "Performance Analysis of Phased Array Antennas in Cellular Handsets for 5G," *IEEE Antennas Wireless Propag. Lett.*, 2015.

[16] Zhang, J., et al., "5G Millimeter-Wave Antenna Array: Design and Challenges," *IEEE Wireless Communications*, Vol. 24, No. 2, April 2017, pp. 106–112.

[17] Rappaport, T. S., "Millimeter Wave Mobile Communications for 5G Cellular: It Will Work!" *IEEE Access*, May 30, 2013.

[18] Roh, W., et al., "Millimeter-Wave Beamforming as an Enabling Technology for 5G Cellular Communications: Theoretical Feasibility and Prototype Results," *IEEE Communications Magazine*, Vol. 52, No. 2, February 2014, pp. 106–113.

[19] Hong, W., "Design and Analysis of a Low-Profile 28GHz Beam Steering Antenna Solution for Future 5G Cellular Applications," *IEEE MTT-S International*, June 2014.

[20] Chang, S., W. Hong, and J. Oh, "Quantization Effects of Phase Shifters on 5G mmWave Antenna Arrays," *Proc. IEEE Int. Symp. Antennas Propag. USNC/URSI Nat. Radio Sci. Meet.*, Vancouver, BC, Canada, July 2015, pp. 2119–2120.

[21] Kibaroglu, K., M. Sayginer, and G. M. Rebeiz, "An Ultra-Low-Cost 32-Element 28 GHz Phased-Array Transceiver with 41 dBm EIRP and 1.0–1.6 Gbps 16-QAM Link at 300 Meters," *Proc. IEEE RFIC, Honolulu*, HI, June 2017, pp. 73–76.

[22] Valdes-Garcia, A., et al., "A Fully-Integrated Dual-Polarization 16-Element W-Band Phased-Array Transceiver in SiGe BiCMOS," *Proc. IEEE Radio Freq. Integr. Circuits Symp. (RFIC),* June 2013, pp. 375–378.

[23] Guo, C., et al., "Advances on Exploiting Polarization in Wireless Communications: Channels, Technologies, and Applications," *IEEE Commun. Surveys Tuts.*, Vol. 19, No. 1, First Quarter, 2017, pp. 125–166.

[24] Kildal, P. S., et al., "MIMO Characterization on System Level of 5G Microbase Stations Subject to Randomness in LOS," *IEEE Access*, Vol. 2, 2014, pp. 1062–1075.

[25] Bhattacharyya, A. K., *Phased Array Antennas: Floquet Analysis, Synthesis, BFNs and Active Array Systems*, New York: Wiley, 2006.

[26] Gu, Q., *RF System Design of Transceivers for Wireless Communications*, New York: Springer, 2005.

[27] Guangyi, Z., and Z. Yujie, *Phased Array Technology*, Beijing, China: Electronic Industry Press, 2006.

[28] Roh, W., et al, "Millimeter-Wave Beamforming as an Enabling Technology for 5G Cellular Communications: Theoretical Feasibility and Prototype Results," *IEEE Communications Magazine*, Vol. 52, February 2014, pp. 106–113.

[29] Yang, B., et al., "Digital Beamforming-Based Massive MIMO Transceiver for 5G Millimeter-Wave Communications," *IEEE Transactions on Microwave Theory and Techniques*, Vol. 66, No. 7, July 2018, pp. 3403–3418.

Selected Bibliography

Daftari, N., et al., "A Wideband 25-35 GHz 5-Bit Low Power 2x2 CMOS Beam Forming Network IC for Reconfigurable Phased Arrays," *IEEE CSICS*, New Orleans, LA, 2015, pp. 1–4.

Fettweis, G., and S. Alamouti, "5G: Personal Mobile Internet Beyond What Cellular Did to Telephony," *IEEE Communications Magazine*, Vol. 52, No. 2, February 2014, pp. 140–145.

Glazunov, A. A., et al., "Impact of the Spatial User Distribution on the Coverage Antenna Pattern of Maximum Ratio Combining in Random Line-of-Sight," *Proc. Eur. Conf. Antennas Propag.*, Lisbon, Portugal, 2015, pp. 1–5.

Golcuk, F., T. Kanar, and G. M. Rebeiz, "A 90–100-GHz 4×4 SiGe BiCMOS Polarimetric Transmit/Receive Phased Array with Simultaneous Receive-Beams Capabilities," *IEEE Transactions on Microwave Theory and Techniques*, Vol. 61, No. 8, August 2013, pp. 3099–3114.

Gu, X., et al., "W-Band Scalable Phased Arrays for Imaging and Communications," *IEEE Communications Magazine*, Vol. 53, No. 4, April 2015, pp. 196–204.

Jacomb-Hood, A., and E. Lier, "Multibeam Active Phased Arrays for Communications Satellites," *IEEE Microwave Magazine*, Vol. 1, No. 4, December 2000, pp. 40–47.

Kant, G. W., et al., "EMBRACE: A Multi Beam 20,000-Element Radio Astronomical Phased Array Antenna Demonstrator," *IEEE Transactions on Antennas and Propagation*, Vol. 59, No. 6, June 2011, pp. 1990–2003.

Kim, H. T., et al., "A 28 GHz CMOS Direct Conversion Transceiver with Packaged Antenna Arrays for 5G Cellular System," *Proc. IEEE Radio Freq. Integr. Circuits Symp. (RFIC),* June 2017, pp. 69–72.

Kim, S. Y., et al., "A Compact Antenna-in-Package 60-GHz SiGe BiCMOS Radio," *Proc. IEEE RFIC*, Tampa, FL, June 2014, pp. 287–288.

Liu, D., et al., *Advanced Millimeter Wave Technologies*, New York: John Wiley and Sons, 2009.

Liu, D., et al., "Antenna-in-Package Design Considerations for Ka-band 5G Communication Applications," *IEEE Transactions on Antennas and Propagation*, 2017.

Ojaroudiparchin, N., et al., "A Switchable 3D-Coverage Phased Array Antenna Package for 5G Mobile Terminals," *IEEE Antennas and Wireless Propagation Letters*, Vol. 15, 2016.

Ojaroudiparchin, N., M. Shen, and G. Pedersen, "Multi-Layer 5G Mobile Phone Antenna for Multi-User MIMO Communications," *Proc 23rd Telecommun. Forum Telfor*, November 2015, pp. 559–562.

Puglielli A., et al., "Design of Energy- and Cost-Efficient Massive MIMO Arrays," *Proc. IEEE*, Vol. 104, No. 3, March 2016, pp. 586–606.

Rajagopal, S., et al., "Antenna Array Design for Multi-Gbps mmWave Mobile Broadband Communication," *Proc. IEEE GLOBECOM'2011*, Houston, TX, 2011, pp. 1–6.

Sayginer, M. and G. M. Rebeiz, "An Eight-Element 2–16-GHz Programmable Phased Array Receiver with One, Two, or Four Simultaneous Beams in SiGe BiCMOS," *IEEE Transactions on Microwave Theory and Techniques*, Vol. 64, No. 12, December 2016, pp. 4585–4597.

Shahramian, S., M. J. Holyoak, and Y. Baeyens, "A 16-Element W-Band Phased Array Transceiver Chipset with Flip-Chip PCB Integrated Antennas for Multi-Gigabit Data Links," *Proc. IEEE RFIC*, Phoenix, AZ, May 2015, pp. 27–30.

Shakib, S., et al., "A Highly Efficient and Linear Power Amplifier for 28-GHz 5G Phased Array Radios in 28-nm CMOS," *IEEE J. Solid-State Circuits*, Vol. 51, No. 12, December 2016, pp. 3020–3036.

Tomkins, A., et al., "A 60 GHz, 802.11ad/WiGig-Compliant Transceiver for Infrastructure and Mobile Applications in 130 nm SiGe BiCMOS," *IEEE J. Solid-State Circuits*, Vol. 50, No. 10, October 2015, pp. 2239–2255.

APPENDIX A

Important Antenna and Electromagnetics Formulas

A.1 Maxwell's Equations

$$\nabla . E = \frac{\rho_v}{\epsilon} \text{ (Gauss's Law)} \tag{A.1}$$

$$\nabla . H = 0 \text{ (Gauss's Law for magnetism)} \tag{A.2}$$

$$\nabla \mathrm{x} E = -\mu \frac{\partial H}{\partial t} \text{ (Faraday's Law)} \tag{A.3}$$

$$\nabla \mathrm{x} H = J + \varepsilon \frac{\partial E}{\partial t} \text{ (Ampere's Law)} \tag{A.4}$$

A.2 Microstrip Antenna Radiation Pattern and Directivity

A measurement antenna radiation pattern is represented by a graphically surrounded sphere as shown in Figure A.1. Electric and magnetic fields (far-field radiation fields) at a distance equal to the radius of the sphere are measured in the E-plane and the H-plane.

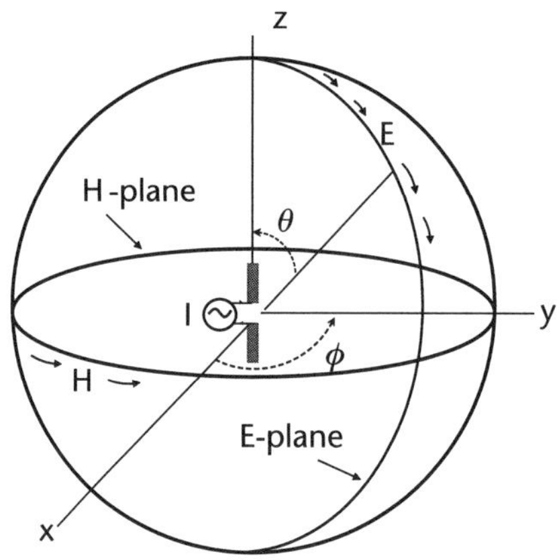

Figure A.1 Antenna-radiated fields evaluated on an imaginary sphere surrounding a dipole

For a rectangular microstrip patch antenna with length L and width W, the normalized radiation pattern (far-field intensity) is at a point $E(\theta, \varnothing)$ given by:

$$E_\theta = \frac{\sin\left(\frac{kW\sin\theta\sin\varnothing}{2}\right)}{\frac{kW\sin\theta\sin\varnothing}{2}}\cos\left(\frac{kL\sin\theta\cos\varnothing}{2}\right)\cos\varnothing \tag{A.5}$$

$$E_\varnothing = -\frac{\sin\left(\frac{kW\sin\theta\sin\varnothing}{2}\right)}{\frac{kW\sin\theta\sin\varnothing}{2}}\cos\left(\frac{kL\sin\theta\cos\varnothing}{2}\right)\sin\varnothing\cos\theta \tag{A.6}$$

where k is the wavenumber given by $k = 2\pi/\lambda$. The magnitude of the field is calculated by:

$$A(\theta,\varnothing) = \sqrt{E_\theta^{\,2} + E_\varnothing^{\,2}} \tag{A.7}$$

The directivity of antenna is calculated by

$$D(\theta,\varnothing) = \frac{4\pi R^2 P_{avg}(\theta,\varnothing)}{P_{rad}} \tag{A.8}$$

where $P_{avg} = \frac{1}{2}Re\left[\vec{E}\,x\,\vec{H}\right]$ is the radiation pointing vector (power per unit area) and P_{rad} is the total radiated power.

A.3 Antenna Bandwidth Calculation

The impedance bandwidth (BW) of the microstrip antenna is defined as the frequency range over which it is matched with that of the feed line within specified limits. The bandwidth is inversely proportional to its quality factor Q and is given as

$$BW = \frac{VSWR - 1}{Q\sqrt{VSWR}} \tag{A.9}$$

where VSWR is defined in terms of the input reflection coefficient τ

$$VSWR = \frac{1+|\tau|}{1-|\tau|} \tag{A.10}$$

The reflection coefficient τ is defined in terms of antenna input impedance Z_{in} and the characteristic impedance Z_0 of the feed line:

Table A.1 Important Parameters and Units

Parameter	*Symbol*	*Unit/Value*
Capacitance	C	F
Conductivity	σ	$1/\Omega \cdot$ m
Current	I	A
Current density	J	A/m^2
Dielectric constant	ε	F/m
Dielectric constant in space	ε_0	F/m
Electric field	E	V/m
Electric field displacement	D	V/m
Magnetic field	B	Tesla, Wb/m^2
Magnetic field strength	H	A/m
Magnetic flux	ψ	Wb
Magnetic potential	A	Wb/m
Mutual-inductance	M	H
Permeability	μ	H/m
Permeability in space	μ_0	H/m
Power	P	W
Relative dielectric constant	ϵ_r	
Resistance	R	Ω
Self-inductance	L	H
Skin depth	δ_s	M

$$\tau = \frac{Z_{in} - Z_0}{Z_{in} + Z_0} \tag{A.11}$$

Table A.1 shows a list of parameters and their units.

A.4 Constants

The following are constants:

- $\varepsilon_0 = 8.85419 \times 10^{-12}$ F/m
- $\mu_0 = 4\pi \times 10^{-7}$ H/m
- $\eta_0 \approx 377\Omega$

APPENDIX B

Designed Antenna Project List

No.	*Project Name*	*Application*	*Chapter*
1	Microstrip Patch Antenna	2.4-GHz ISM Band	2
2	Coaxial-Fed Patch Antenna	Coal Mine	2
3	Circular Polarized Millimeter-Wave Antenna	5G	2
4	Dual-Orthogonal Fed Square CP Antenna	GNSS and GPS	2
5	CPW-Fed Dual-Band WLAN Antenna	WLAN	3
6	Inverted-F Bluetooth Antenna	Bluetooth	3
7	Planar Inverted-F Antenna (PIFA)	LTE	3
8	MIMO Antenna Design	LTE	3
9	Mobile MIMO Antenna Design	Mobile Antenna	4
10	Reconfigurable Frequency and Polarization Antenna	Wireless Devices	5
11	Wilkinson Power Divider	High Port Isolation	6
12	Printed Antenna Array	Weather Radar	6
13	Circular Aperture Radar Microstrip Antenna Array	Radar	6
14	Multilayer Antenna Design	Broadband Applications	6
15	Millimeter Wave Antenna Array for Automotive Radar	Automotive Radar	6
16	Cylindrical Microstrip Antenna Analyses for Large Radii	Mobile Phone and Other Designs with Curvature	6
17	Cylindrical Microstrip Antenna Analyses for Small Radii	Mobile Phone and Other Designs with Curvature	6
18	Conformal Antenna on Missile Surface	Missile Application	6
19	Conformal Microstrip Antenna Array Fed with Waveguide	Airborne Remote Sensing	6
20	Dual-Polarized SAR Antennas Design	Microwave Imaging	7
21	Beamforming network using microstrip line	Microwave Imaging	8
22	Cosec square Shaped Pattern Array Antenna	Microwave Imaging	8
23	Square Coaxial Line Beamforming Network	High Power Feed	8
24	Dual-Frequency Shared Aperture Antenna Array	Microwave Imaging	9
25	Dual Polarized Feed Network	Microwave Imaging	10
26	Dual Polarized X-Band Array	Microwave Imaging	10
27	Microstrip Patch Antenna on Mushroom-Like EBG Structure	Broadband Applications	11
28	Uniplanar EBG Cells with Rectangular Microstrip Antenna	Broadband Applications	11

No.	Project Name	Application	Chapter
29	Bowtie Antenna with EBG Cells	Broadband Applications	11
30	Array Antenna Design Using EBG	Broadband Applications	11
31	Frequency Selective Surface (FSS)	FSS	11
32	Antennas on Photonic Bandgap (PBG)	High-Frequency Applications	11
33	Sierpinski Gasket Fractal Antenna	Compact Antennas	11
34	Square Shape Fractal Antenna	Compact Antennas	11
35	Koch Snowflake Fractal Antenna	Compact Antennas	11
36	Hilbert Antenna Array	Partial Discharge Detection	11
37	Hexagonal Monopole Antenna	Broadband Applications	11
38	Broadband Rectangular Patch	Broadband System	11
39	Slot Loaded Patch	Broadband System	11
40	Spiral Vivaldi Antenna	Broadband System	11
41	Wideband Antipodal Vivaldi Antenna	Broadband Applications	11
42	SIW Microstrip Patch Antenna	Healthcare	11
43	RFID Tag Antenna	RFID	12
44	Fat Tag or Dog Bone RFID Antenna	RFID	12
45	RFID Reader Antenna	RFID	12
46	GPS Spiral Antenna Design	GPS, GNSS	12
47	GPS Circular Spiral Antenna Design	GPS	12
48	Sine-Wave Meander Archimedean Spiral Antenna	GPS	12
49	Wireless Body Area Network (WBAN) Antenna	Healthcare	12
50	Dual-Band Wearable Antenna with EBG Structure	Healthcare	12
51	LPWAN IoT Antenna	IoT	12
52	Quasi-Yagi IoT Antenna Design	IoT	12
53	Automotive Glass Mounted FM/DAB Antennas	Automotive	13
54	Shark Fin Antenna	Automotive	13
55	LTE/GSM Antenna Design	Automotive	13
56	Grid Array Antenna for 24 GHz Automotive SRR	Automotive	13
57	Microstrip Comb-Line Antenna Arrays for LRR	Automotive	13
58	45° Amplitude Tapered Comb-Line Antenna Array	Automotive	13
59	ACC Radar Antenna Design	Automotive	13
60	Rotman Lens Design for Electronic Scanning of Antenna Beam	Automotive	13
61	Phased Array Antenna for ETC Application at 5.8 GHz	Automotive	13
62	MMIC Low-Noise Amplifier (LNA)	Phased Array	14
63	Power Amplifier Design	Phased Array	14
64	MMIC SPDT Switch	Phased Array	14
65	LTCC LPF Filter	Phased Array	14
66	Six-Bit MMIC Phase Shifter	Phased Array	14
67	Phased Array Radar Using Transceiver Module	Phased Array	14
68	Millimeter-Wave 5G 8 × 8 Butler Matrix-Based Beamforming Network	5G BFN	15
69	60-GHz Microstrip Antenna Array Design for 5G	5G Antenna Array	15
70	FD-MIMO Phased Array System	5G Phased Array	15
71	5G Phased Array	5G	15

Acronyms and Abbreviations

ACC	Adaptive cruise control
ADC	Analog-to-digital converter
ADS	Advanced Design System
AESA	Active electronic scanning array
AM	Amplitude modulation
AMC	Artificial magnetic conductor
APN	Airborne Doppler Navigator
AR	Axial ratio
ARNS	Aeronautical Radio Navigation Service
ASIC	Application specific integrated circuit
ASP	Aperture-stacked patch
AUT	Antenna Under Test
AVI	Automatic vehicle identification
AWACS	Airborne warning and control
AWS	Advanced Wireless Service
BFN	Beamforming network
BW	Bandwidth
CAD	Computer-aided design
CDMA	Code division multiple access
CEM	Computational electromagnetics
CISPR	Comité International Spécial des Perturbations Radioélectriques
CP	Circular polarization
CPW	Coplanar waveguide
CTA	Cross-traffic alert
CTR	Current taper ratio
CW/A	Collision warning/avoidance systems
DAB	Digital audio broadcasting
DBF	Digital beamforming
DBS	Direct broadcast satellite
DC	Direct current
DCS	Digital communications system
DCU	Digital control unit
DFW	Dielectric-filled waveguide
DMB	Digital multimedia broadcasting
DSRC	Dedicated short-range communications

EBG	Electromagnetic bandgap
ECU	Electronic control unit
EDA	Electronic design automation
EFIE	Electric field integral equation
EIRP	Effective isotropic radiative power
EM	Electromagnetic
EMC	Electromagnetic compatibility
EMI	Electromagnetic interference
ESA	Electronic scanning antenna
ETC	Electronic toll collection
EVM	Error vector magnitude
FAA	Frequency agile antenna
FDTD	Finite difference time domain
FEM	Finite element method
FET	Field effect transistor
FFT	Fast Fourier transform
FM	Frequency modulation
FMCW	Frequency-modulated continuous wave
FOV	Field of view
FRC	Fractal rectangular curve
FSS	Frequency selective surfaces
GAA	Grid array antenna
GHz	Gigahertz
GIS	Geographic Information Systems
GLONASS	Russian Global Navigation Satellite System
GNSS	Global Navigation Satellite System
GPS	Global Positioning System
GSM	Global System for Mobile Communications
HAC	Hearing aid compatibility
HFSS	High-frequency structure simulator
HIS	High impedance surface
HPBW	Half-power beamwidth
Hz	Hertz
IDFT	Inverse discrete Fourier transform
IFA	Inverted-F antenna
INS	Inertial Navigation System
IoT	Internet of Things
ISM	Industrial, Scientific, and Medical Band
ITS	Intelligent transport system
LCP	Liquid crystal polymer
LHCP	Left-hand circular polarization
LNA	Low-noise amplifier
LPWAN	Low-power wide area network
LTCC	Low temperature co-fired ceramic
LTE	Long Term Evolution
MCM	Multichip module
MEG	Mean effective gain
MFIE	Magnetic field integral equation

MHz	Megahertz
MICS	Medical Implant Communication Services
MIMO	Multiple-input multiple-output
MISO	Multiple-input single-output
MMIC	Microwave monolithic integrated circuits
MoM	Method of Moments
MSS	Mobile satellite service
MU-MIMO	Multi-user MIMO
NF	Noise figure
NRI	Negative Refractive Index
OBU	Onboard unit
PA	Power amplifier
PAA	Phased array antenna
PAE	Power-added efficiency
PAN	Personal area network
PBG	Photonic bandgap
PCB	Printed circuit board
PD	Partial discharge
PDF	Probability density function
PDP	Power display profile
PEC	Perfect electric conductor
PIFA	Planar inverted-F antenna
PMC	Perfect magnetic conductor
PNF	Planar near test facility
PTFE	Polytetrafluoroethylene
QFN	Quad-flat no-leads
RCS	Radar cross-section
RF	Radio frequency
RFID	Radio frequency identification
RHCP	Right-hand circular polarization
RL	Return loss
RKES	Remote keyless entry service
RNNS	Radio Navigation Satellite System
RSU	Roadside unit
SAM	Standard anthropomorphic model
SAR	Synthetic aperture radar
SCL	Square coaxial line
SDARS	Satellite Digital Audio Radio Service
SFCW	Stepped frequency continuous wave
SISO	Single-input single-output
SIW	Substrate-integrated waveguide
SLL	Sidelobe level
SMARSA	Sine-wave meander Archimedean spiral antenna
SNR	Signal-to-noise ratio
TCM	Transparent conducting material
TE	Transverse electric (mode/field)
TEM	Transverse electromagnetic (mode/field)
TM	Transverse magnetic (mode/field)

TPMS	Tire-pressuring monitoring system
T/R	Transmit/receive
TRM	Transmit-receive module
TTD	True time delay
UC-EBG	Uniplanar EBG
UHF	Ultrahigh frequency
UMTS	Universal Mobile Telecommunications System
UWB	Ultrawideband
VNA	Vector network analyzer
VSWR	Voltage standing wave ratio
WBAN	Wireless body area network
WCDMA	Wideband code division multiple access
WiGig	Wireless Gigabit Alliance
WiMax	Worldwide Interoperability for Microwave Access
XPD	Cross-polarization decoupling

About the Author

Anil Pandey received a B.E. in electronic and communication engineering from KEC in 2001, and master of technology (M. Tech) in RF microwave engineering from the Indian Institute of Technology (IIT) in Varanasi, India, in 2003. His M. Tech thesis was "Broadband Multilayer Printed Antennas." From 2003 to 2006, he was a scientist at the Indian Space Research Organization (ISRO) in Ahmedabad, India. Since 2007, he has been in the Design Engineering Software (DES) area of the research and development division at Keysight Technologies, India, where currently he is a principal engineer. He has conducted over 16 years of research in the area of printed antennas. He has worked on various projects involving a wide range of antenna technology, including printed antennas, broadband microstrip antennas, phased array systems, 5G, mobile phone antennas, automotive radar, SAR antennas, and multilayer microstrip arrays. He has successfully completed many research projects and also managed several projects in these fields.

Mr. Pandey's research interests include microwave and antenna engineering, automobile antennas, biomedical engineering, and communication and system engineering. During his work in ISRO, he gained experience in antenna designs, electronic design automation (EDA) tools, project management, and training. He developed microstrip phased array antennas and a high-power feed network for satellite remote-sensing applications. At Keysight Technologies, he managed and led groups and developed various active and passive microwave components by employing MMIC, LTCC, printed, and waveguide-based technologies. In addition, he developed wideband microstrip, MIMO antennas, mobile phone antennas, a 5G communication antenna system, radar antenna arrays, couplers, power dividers, and wideband monopulse comparators. Mr. Pandey has published more than 40 papers in journals and conference proceedings. He has two patents on RF microwave and antenna design. Additionally, he has served on the technical and organizational committees of several international conferences.

Index

Q

R

S

X

Y

Recent Titles in the Artech House Antennas and Electromagnetics Analysis Library

Christos Christodoulou, Series Editor

Adaptive Array Measurements in Communications, M. A. Halim

Advanced Computational Electromagnetic Methods and Applications, Wenhua Yu, Wenxing Li, Atef Elsherbeni, Yahya Rahmat-Samii, editors

Advances in Computational Electrodynamics: The Finite-Difference Time-Domain Method, Allen Taflove, editor

Advances in FDTD Computational Electrodynamics: Photonics and Nanotechnology, Allen Taflove, editor; Ardavan Oskooi and Steven G. Johnson, coeditors

Analysis Methods for Electromagnetic Wave Problems, Volume 2, Eikichi Yamashita, editor

Antenna Design for Cognitive Radio, Youssef Tawk, Joseph Costantine, and Christos Christodoulou

Antenna Design with Fiber Optics, A. Kumar

Antenna Engineering Using Physical Optics: Practical CAD Techniques and Software, Leo Diaz and Thomas Milligan

Antennas and Propagation for Body-Centric Wireless Communications, Second Edition, Peter S. Hall and Yang Hao, editors

Antennas and Site Engineering for Mobile Radio Networks, Bruno Delorme

Antennas for Small Mobile Terminals, Kyohei Fujimoto and Koichi Ito, editors

Analysis of Radome-Enclosed Antennas, Second Edition, Dennis J. Kozakoff

Applications of Neural Networks in Electromagnetics, Christos Christodoulou and Michael Georgiopoulos

The Art and Science of Ultrawideband Antennas, Second Edition, Hans G. Schantz

AWAS for Windows Version 2.0: Analysis of Wire Antennas and Scatterers, Antonije R. Djordjević, et al.

Broadband Microstrip Antennas, Girsh Kumar and K. P. Ray

Broadband Patch Antennas, Jean-François Zürcher and Fred E. Gardiol

CAD of Microstrip Antennas for Wireless Applications, Robert A. Sainati

The CG-FFT Method: Application of Signal Processing Techniques to Electromagnetics, Manuel F. Cátedra, et al.

Computational Electrodynamics: The Finite-Difference Time-Domain Method, Third Edition, Allen Taflove and Susan C. Hagness

Design and Applications of Active Integrated Antennas, Mohammad S. Sharawi and Oualid Hammi

Electromagnetics and Antenna Technology, Alan J. Fenn

Electromagnetic Modeling of Composite Metallic and Dielectric Structures, Branko M. Kolundzija and Antonije R. Djordjević

Electromagnetic Waves in Chiral and Bi-Isotropic Media, I. V. Lindell, et al.

Electromagnetics, Microwave Circuit and Antenna Design for Communications Engineering, Peter Russer

Engineering Applications of the Modulated Scatterer Technique, Jean-Charles Bolomey and Fred E. Gardiol

Fast and Efficient Algorithms in Computational Electromagnetics, Weng Cho Chew, et al., editors

Frequency-Agile Antennas for Wireless Communications, Aldo Petosa

Fresnel Zones in Wireless Links, Zone Plate Lenses and Antennas, Hristo D. Hristov

Handbook of Antennas for EMC, Second Edition, Thereza Macnamara with John McAuley

Handbook of Reflector Antennas and Feed Systems, Volume I: Theory and Design of Reflectors, Satish Sharma, Sudhakar Rao, and Lotfollah Shafai, editors

Handbook of Reflector Antennas and Feed Systems, Volume II: Feed Systems, Lotfollah Shafai, Satish Sharma, and Sudhakar Rao, editors

Handbook of Reflector Antennas and Feed Systems, Volume III: Applications of Reflectors, Sudhakar Rao, Lotfollah Shafai, and Satish Sharma, editors

Introduction to Antenna Analysis Using EM Simulators, Hiroaki Kogure, Yoshie Kogure, and James C. Rautio

Iterative and Self-Adaptive Finite-Elements in Electromagnetic Modeling, Magdalena Salazar-Palma, et al.

LONRS: Low-Noise Receiving Systems Performance and Analysis Toolkit, Charles T. Stelzried, Macgregor S. Reid, and Arthur J. Freiley

Measurement of Mobile Antenna Systems, Second Edition, Hiroyuki Arai

Microstrip Antenna Design Handbook, Ramesh Garg, et al.

Microwave and Millimeter-Wave Remote Sensing for Security Applications, Jeffrey A. Nanzer

Mobile Antenna Systems Handbook, Third Edition, Kyohei Fujimoto, editor

Multiband Integrated Antennas for 4G Terminals, David A. Sánchez-Hernández, editor

New Foundations for Applied Electromagnetics: The Spatial Structure of Fields, Said Mikki and Yahia Antar

Noise Temperature Theory and Applications for Deep Space Communications Antenna Systems, Tom Y. Otoshi

Phased Array Antenna Handbook, Third Edition, Robert J. Mailloux

Phased Array Antennas with Optimized Element Patterns, Sergei P. Skobelev

Plasma Antennas, Theodore Anderson

Polarization in Electromagnetic Systems, Second Edition, Warren Stutzman

Practical Microstrip and Printed Antenna Design, Anil Pandey

Printed MIMO Antenna Engineering, Mohammad S. Sharawi

Quick Finite Elements for Electromagnetic Waves, Giuseppe Pelosi, Roberto Coccioli, and Stefano Selleri

Radiowave Propagation and Antennas for Personal Communications, Third Edition, Kazimierz Siwiak

Reflectarray Antennas: Analysis, Design, Fabrication and Measurement, Jafar Shaker, Mohammad Reza Chaharmir, and Jonathan Ethier

RF Coaxial Slot Radiators: Modeling, Measurements, and Applications, Kok Yeow You

Solid Dielectric Horn Antennas, Carlos Salema, Carlos Fernandes, and Rama Kant Jha

Switched Parasitic Antennas for Cellular Communications, David V. Thiel and Stephanie Smith

Ultrawideband Antennas for Microwave Imaging Systems, Tayeb A. Denidni and Gijo Augustin

Ultrawideband Short-Pulse Radio Systems, V. I. Koshelev, Yu. I. Buyanov, and V. P. Belichenko

Understanding Electromagnetic Scattering Using the Moment Method: A Practical Approach, Randy Bancroft

Wavelet Applications in Engineering Electromagnetics, Tapan Sarkar, Magdalena Salazar Palma, and Michael C. Wicks

For further information on these and other Artech House titles, including previously considered out-of-print books now available through our In-Print-Forever® (IPF®) program, contact:

Artech House
685 Canton Street
Norwood, MA 02062
Phone: 781-769-9750
Fax: 781-769-6334
e-mail: artech@artechhouse.com

Artech House
16 Sussex Street
London SW1V HRW UK
Phone: +44 (0)20 7596-8750
Fax: +44 (0)20 7630 0166
e-mail: artech-uk@artechhouse.com

Find us on the World Wide Web at: www.artechhouse.com

RAW MATERIALS FOR GLASS AND CERAMICS

RAW MATERIALS FOR GLASS AND CERAMICS

Sources, Processes, and Quality Control

Christopher W. Sinton

WILEY

John Wiley & Sons, Inc.

Published by John Wiley & Sons, Inc., Hoboken, New Jersey
Published simultaneously in Canada

Library of Congress Cataloging-in-Publication Data:
Sinton, Christopher W.
Raw materials for glass and ceramics : sources, processes, and quality control / Christopher W. Sinton.
p. cm.
Includes bibliographical references and index.
ISBN-13 978-0-471-47942-0
ISBN-10 0-471-47942-X
1. Ceramic materials. 2. Glass manufacture. 3. Glass. I. Title.

TP810.5.S56 2006
666—dc22

2005021295

10 9 8 7 6 5 4 3 2 1

CONTENTS

PREFACE

Like most industries, commercial glass and ceramic manufacturers strive to make high-quality products at the lowest possible cost. Efficiency is found with consistency and control, by being able to repeat the process over and over again without changes or surprises. In order to attain the goal of steady production, a consistent raw material batch is required, following the tired but accurate adage of "garbage in, garbage out." Of course, there are other variables that affect the manufacturing process, but raw materials are a major factor. Unfortunately, many glass and ceramic engineering students graduate from an academic background without a basic understanding of where raw materials come from and how they are processed. If the only experience of raw materials is scooping powders out of a bag during laboratory exercises, the individual will not be suitably equipped to enter into the industrial world.

This book is designed for the undergraduate student enrolled in a ceramic or glass engineering program. It can be used as the primary text for a course in raw materials or it can be used as a supplemental text for a course in glass and ceramic processing. It is hoped that it also will provide a reference for individuals already in the commercial industry and for artists that want a better understanding of raw materials. This book cites many informative sources, and the reader is encouraged to refer to these sources for more detailed and comprehensive information regarding each topic that is covered.

I am indebted to many people who have helped with the preparation of the book. I am grateful for reviews of sections by Bill Carty, Peter Ciullo, David Hall, Charles Merivale, Mark Noirot, Konrad Rieger, John W. Sinton, and Peter Van Stone. I benefited greatly from conversations with John Buckner, Seth Buzby, Juan Carlos Factorovich, Ernie Guter, Scott Misture, Nik Ninos, Jim Shelby, and Phil Tucker. I would like to extend my thanks to the cumulative effects of my advisors: Ray Coish, Bob Duncan, Dave Christie, Haraldur Sigurdsson, and Bill LaCourse. I am grateful for the resources of the Scholes Library at Alfred University, the Bailey Howe Library at the University of Vermont, and the libraries of Middlebury College.

RAW MATERIALS FOR GLASS AND CERAMICS

1

INTRODUCTION

The history of human exploitation of materials is the history of the advancement and growth of civilizations. With each discovery of a process to transform a natural raw material into a new tool or energy source, humans were better able to control and construct their environment. This is evident by the names of anthropogenic epochs: the Stone Age, the Bronze Age, and the Iron Age.

With each new discovery, the primary components for the process or tool immediately became valuable. These raw materials and other mineral-based wealth have both historical and current significance. Wars have been—and will continue to be—fought over the ownership of metal ores, petroleum, and other resources required to keep modern industrialized society moving. These materials are the basis of all industry, and today we use virtually all of the elements found on the periodic table.

There is no official Ceramic or Glass Age, although objects made of glass and ceramics have long been part of human civilization. Perhaps this omission is related to the historic lack of a relationship between battlefield advantage and glass and ceramics. Nevertheless, these two products have long been invaluable in domestic applications and, therefore, the raw materials to make them have had value.

Until relatively recently, the materials used to make most glass and ceramics have generally been naturally occurring inorganic minerals such as silicates and oxides of metals. Clay and sand perhaps come to mind when one considers the dominant raw materials in these industries. These are examples of relatively crude materials removed from the ground and minimally processed. Although clay and sand continue to be important components, the progressive introduction of complex, rapid fabrication processing of traditional products and the growth of advanced glass and ceramics has forced the business of raw materials to evolve and become more sophisticated. Both the glass and ceramic industry are similar in the need for efficient and economic procurement, beneficiation, and handling of natural and synthesized materials to produce a final product. Modern manufacturing may further require higher purity specifications, the use of waste or recycled materials, and the compliance with environmental and human regulations involved in mining and processing.

For our purposes, we can define a raw material as the precursor ingredients used at the beginning of the manufacturing process for a given ceramic or glass product. Reed (1995) classified ceramic raw materials into the following categories:

- Nonuniform crude from natural deposits
- Refined industrial minerals that have been beneficiated to remove mineral impurities to significantly increase the mineral purity and physical consistency
- High-tonnage industrial inorganic chemicals that have undergone significant chemical processing and refinement to significantly upgrade the chemical purity and improve the physical characteristics

Under each of these categories, a specific mineral or compound can be further classified by mineralogy, chemical composition, and purity, and a variety of physical characteristics, such as particle size distribution and bulk density. Because most raw materials are powders, many of the characteristics pertain to submillimeter-sized particles and granules. As glass and ceramic materials are placed into higher-performance applications, the raw material powders must adhere to higher standards of purity and uniformity. Not only are the initial components of the raw materials important (e.g., the composition of the bauxite used to make alumina), but processes such as grinding, mixing, and classification can also impart properties, both desirable and undesirable (Naito et al., 2003). The desired characteristics of the starting components are defined by the quality and performance of the final manufactured product and these characteristics directly affect the price of the raw material. In general, highly processed materials are more costly due to higher energy and labor requirements needed to produce them.

Because of the inherent link between raw material characteristics and processing behavior, an understanding of raw materials is of utmost importance in economically and efficiently producing glass and ceramic products. In this book we begin with a description of the origin and distribution of elements and minerals in the earth, which, in turn, are related to the discovery of ore deposits and subsequent extraction of material from the ground. By understanding the geological processes that concentrate compounds and minerals, one can better understand the issues specific to each material. In addition, industrial separation and beneficiation techniques often simulate the natural processes that concentrate minerals. For example, heavy minerals such as zircon can be deposited in a river where the water velocity slows, just as an industrial hydrocyclone will separate two minerals based on particle density or size.

The general processing routes that are used to beneficiate, purify, and control particle size are described in the third chapter. Most raw materials are processed to varying degrees, depending on the ore deposit and the desired characteristics and purity levels. Each additional processing step adds to the final cost of the final product; therefore, the manufacturer needs to continually balance purity with cost. Advances in processing have led to the availability of higher purity materials, critical for making high performance products. In addition, better processing has allowed for the exploitation of previously uneconomical ore deposits. As with any industry that relies on natural materials, the extraction and processing activities can result in environmental impacts that, depending on the country, are subject to regulations. Complying with these regulations can be costly and will affect the prices of the product.

In the fourth chapter, we will describe the structure of the raw materials industry. Many of the materials used in glass and ceramics can be classified as industrial minerals while the more refined components can be considered chemicals. The concept of location is presented, as the geography of primary sources (mines) as related to the location of the manufacturer is extremely important, particularly for the low-cost, high-tonnage materials where transportation costs may equal or exceed the cost of the material itself. In other cases, the proximity of a raw material source may even be more important than its composition.

In chapters 5 and 6, we will learn about individual raw materials in terms of mineralogy, distribution, general processing routes, uses, and specifications. First are the naturally occurring minerals, or those that are removed from the ground and processed without substantial alteration to the composition or structure of the material. Some natural materials are feedstocks that are used to produce synthetic materials that are either not found in nature, such as carbides and nitrides, or those that are rare in nature, such as mullite and alumina. Chapter 7 addresses the growing pressure to use materials that do not come from virgin minerals. These include recycled glass cullet and the inorganic slags and ashes that are byproducts of metal smelting and combustion of coal and wood.

The last chapter is dedicated to material characterization and quality control. As with any modern manufacturing process, physical and compositional characteristics of the raw materials need to be consistent. Consistency allows the manufacturer to predict the behavior of the fabrication process and therefore to increase overall efficiency. Another important issue is awareness and compliance with environmental and health issues associated with handling. This is very important because the handling of dry powders and the need for high operating temperatures can release harmful particulate or volatile emissions into the air (from burning fossil fuels or from fine raw materials). The salient problems for each material are discussed in the last part of each specific material section.

Please note that all percentages of components are reported as weight percent unless otherwise stated.

REFERENCES

Naito, M., N. Shinohara, and K. Uematsu (2003). Raw Materials. In *Handbook of Advanced Ceramics,* ed. S. Somiya, 81–129. Oxford: Elsevier.

Reed, J. S. (1995). *Principles of Ceramics Processing,* 2nd ed. New York: Wiley-Interscience.

2

THE SOURCE—EARTH, MINERALS, AND MINERAL DEPOSITS

2.1 FORMATION OF THE EARTH AND DISTRIBUTION OF THE ELEMENTS

Almost all of the elements in the periodic table are found within the Earth in varying amounts, some very common and some extremely rare. Most are bound within compound silicate and oxide materials but several, such as gold and silver, can be found in elemental form. The distribution of the elements is heterogeneous in all three dimensions, across the surface and within the interior of the Earth. The current dispersion of the elements is the result of more than four billion years of geologic activity, and the procedure of finding and extracting these materials for industrial use begins with an understanding of these processes.

It is generally accepted that the solar system evolved from the cooling of a rotating gaseous cloud, or solar nebulae. Assuming that the cloud was relatively homogenous, the variation in the compositions of the sun and the different planets within the solar system demonstrates that considerable fractionation of the elements occurred during its coalescence. Despite the large-scale heterogeneity between the planets, the Earth may have originally been a somewhat homogenous mass of molten material and gases. The composition of this primitive, undifferentiated earth is thought to be represented by the composition of certain stony meteorites from nearby asteroid belts. A primitive Earth may have consisted of the accretion of these meteorites into a protoplanet.

Over time, the Earth evolved from a homogeneous body to the vertically and horizontally stratified planet we see today. As it cooled, density-driven fractionation (i.e., separation into components) formed five general layers: the core, mantle, crust, hydrosphere, and atmosphere (Figure 2.1). Metallic iron segregated and sank to the core, bringing with it the *siderophile* elements, or those that readily combine with iron such as nickel. The *lithophile,* or elements that readily combine with oxygen, such as silicon, concentrated in the mantle and the crust. Water collected in the hydrosphere and the volatile components collected in the atmosphere. While the hydrosphere, atmosphere, and the surface of the crust can be directly observed by humans, the structure of the earth below the surface of the crust must be inferred indirectly from the behavior of seismic waves (i.e., wave energy from earthquakes) that propagate through the earth. The core is probably composed predominantly of nickel and iron and is subdivided into a solid inner core and a liquid outer core. The mantle is composed of iron- and magnesium-rich silicate minerals such as olivine and enstatite pyroxene. The crust, which is separated from the mantle by an abrupt change in mineralogy and composition, is a mix of silicate minerals that are, on average, richer in lighter elements, such as silicon and aluminum, relative to the mantle. The crust is further subdivided into oceanic crust, which is, on average, 5 to 10 km thick and lies beneath

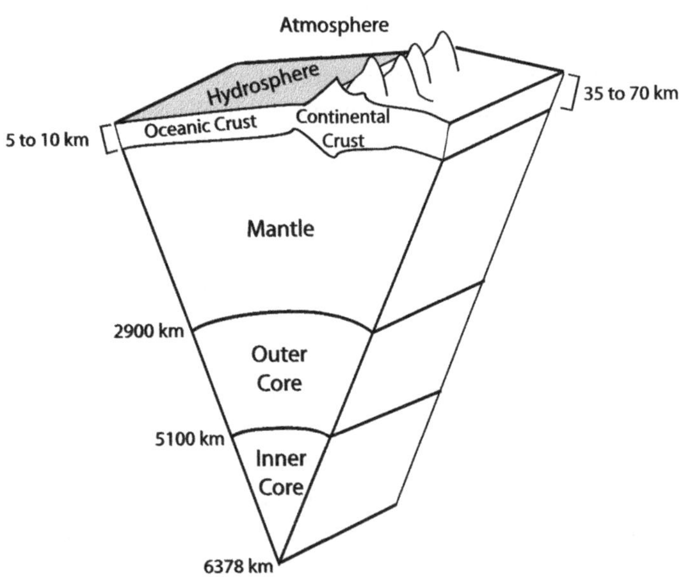

Figure 2.1 Schematic showing the layering of the solid earth.

the oceans, and continental crust, which is about 35 to 70 km thick and makes up the crust under the continents and the shallow continental shelf areas. Oceanic crust continually forms at the middle of the volcanic, mid-ocean ridges and consumed at subduction zones (see section 2.3)

Table 2.1 gives the average elemental abundances of the elements in both the crust and seawater. These values are approximate, and there can be variations between different studies. However, it is generally agreed that the most abundant elements in the Earth's crust are oxygen (46.6 percent), silicon (27 percent), aluminum (8.2 percent), iron (6.3 percent), calcium (5.0 percent), magnesium (2.9 percent), sodium (2.3 percent), and potassium (1.5 percent). These eight elements make up 98 percent of the Earth's crust by weight. It is no accident that most traditional glass and ceramic products are made from these materials. Nevertheless, there are other elements of interest that exist in much smaller amounts. Luckily, the elements are not distributed homogeneously within the crust. The billions of years of geological movement at the Earth's crust—coupled with its interactions with the hydrosphere and atmosphere—have provided mechanisms to mobilize the elements. This mobilization allowed the elements and their compounds to become concentrated, leading to the formation of deposits enriched with specific elements. These elements are contained within the minerals and rocks that make up the deposit.

2.2 ROCKS, MINERALS, AND MINERALOGY

Mineral Categories

Elements can combine to form *minerals,* which are naturally occurring solids that have characteristic crystal structures and defined compositions. There are approximately 2,000 known minerals on Earth, although only about 30 of them make up most of the common rocks. The study of mineralogy is an exhaustive topic on its own (e.g., Klein and Hurlbut, 1993; Wenk and Bulakh, 2004), but here we give a basic review of the aspects of mineralogy that are relevant to this book.

Minerals can be described and categorized by crystal structure and composition. The composition is one of the main factors that controls the crystal structure of a mineral (the others are temperature and pressure). Crystals are formed by linking polyhedra made from the constituent cations (positively charged ions such as Mg^{2+}) and anions (negatively charged ions such as O^{2-}). Each ion tends to attract or

Table 2.1 Abundance of elements in the crust and seawater

Atomic #	Element	Average Continental Crust ppm	Seawater ppm	Atomic #	Element	Average Continental Crust ppm	Seawater ppm
1	Hydrogen	1,400	108,000	22	Titanium	6,600	0.001
2	Helium	0.0055	—	23	Vanadium	190	0.0015
3	Lithium	20	0.17	24	Chromium	140	0.0006
4	Beryllium	1.9	$6.00\ e^{-7}$	25	Manganese	1,100	0.002
5	Boron	8.7	4.44	26	Iron	63,000	0.03
6	Carbon	1,800	28	27	Cobalt	30	0.0001
7	Nitrogen	20	0.5	28	Nickel	90	0.002
8	Oxygen	464,000	857,000	29	Copper	68	0.003
9	Fluorine	540	1.3	30	Zinc	79	0.05
10	Neon	0.003	0.00012	31	Gallium	19	$3.00\ e^{-5}$
11	Sodium	23,000	11,500	32	Germanium	1.4	$6.00\ e^{-5}$
12	Magnesium	29,000	1,326	33	Arsenic	2.1	0.023
13	Aluminum	82,000	0.05	34	Selenium	0.05	0.0004
14	Silicon	270,000	1	35	Bromine	3	67.3
15	Phosphorus	1,000	0.07	36	Krypton	0.00015	0.00021
16	Sulfur	420	928	37	Rubidium	60	0.12
17	Chlorine	170	19,870	38	Strontium	360	8.1
18	Argon	1.5	0.45	39	Yttrium	29	$1.30\ e^{-5}$
19	Potassium	15,000	416	40	Zirconium	130	$2.60\ e^{-5}$
20	Calcium	50,000	4.2	41	Niobium	17	$1.00\ e^{-5}$
21	Scandium	26	$1.50\ e^{-6}$	42	Molybdenum	1.5	0.01

44	Ruthenium	0.0001	7.00 e^{-7}	66	Dysprosium	6.2	9.10 e^{-7}
45	Rhodium	0.0007	1.00 e^{-5}	67	Holmium	1.2	2.20 e^{-7}
46	Palladium	0.0063	0.001	68	Erbium	3	9.00 e^{-7}
47	Silver	0.08	0.0001	69	Thulium	0.45	2.00 e^{-7}
48	Cadmium	0.15	5.00 e^{-5}	70	Ytterbium	2.8	8.00 e^{-7}
49	Indium	0.16	1.00 e^{-7}	71	Lutetium	0.56	1.50 e^{-7}
50	Tin	2.2	1.00 e^{-5}	72	Hafnium	3.3	8.00 e^{-6}
51	Antimony	0.2	0.0002	73	Tantalum	1.7	2.00 e^{-6}
52	Tellurium	0.001	—	74	Tungsten	1.1	1.20 e^{-5}
53	Iodine	0.49	0.06	75	Rhenium	0.0026	1.00 e^{-6}
54	Xenon	2.0 e^{-5}	1.00 e^{-5}	76	Osmium	0.0018	—
55	Cesium	1.9	0.0004	77	Iridium	4.0 e^{-5}	—
56	Barium	340	0.03	78	Platinum	0.037	—
57	Lanthanum	34	3.40 e^{-6}	79	Gold	0.0031	5.00 e^{-5}
58	Cerium	60	1.20 e^{-6}	80	Mercury	0.067	5.00 e^{-5}
59	Praseodymium	8.7	6.00 e^{-7}	81	Thallium	0.53	1.00 e^{-5}
60	Neodymium	33	2.80 e^{-6}	82	Lead	10	3.00 e^{-5}
62	Samarium	6	4.50 e^{-7}	83	Bismuth	0.025	0.0002
63	Europium	1.8	1.30 e^{-7}	90	Thorium	6	—
64	Gadolinium	5.2	7.00 e^{-7}	91	Protactinium		
65	Terbium	0.94	1.40 e^{-7}	92	Uranium	1.8	0.0033

Values are from www.webelements.com

coordinate itself with as many ions of opposite charge as their sizes permit. This is known as ionic coordination. The coordination number refers to the number of oppositely charged ions that can surround an ion. This number depends on the size ratio and charges of the cations and anions. Electrical neutrality must be kept, that is, the total charges of the anions must equal that of the cations. Table 2.2 lists the coordination numbers, range of size ratios, and examples for each. A common example is the combination of silicon and oxygen. The Si^{4+} cation has an ionic radius of 0.04 nm and the O^{2-} anion has an ionic radius of 0.14 nm, giving a cation to anion ratio of 0.286. According to Table 2.2, the combination of silicon and oxygen will result in a coordination number of 4, that is, a silicon cation will surround itself with four oxygen anions. This arrangement is also referred to as tetrahedral coordination as the polyhedron that is formed is a tetrahedron.

When these polyhedra are repeated in a three dimensional pattern, a crystal is made. Crystals are categorized into seven systems defined by three or four imaginary axes of equal or unequal length. The most simple is cubic, which consists of three mutually perpendicular axes of equal length. Other crystal systems are hexagonal, tetragonal, rhombohedral, orthorhombic, monoclinic, and triclinic. For a more detail on crystals systems, see Callister (2002) or Klein and Hurlbut (1993).

The two most abundant elements in the Earth's crust are oxygen and silicon, so naturally the most common minerals contain these two elements in the form of the SiO_4^{-4} (silica) tetrahedron, which is the basic building block of all silicate minerals. Silica tetrahedra can link or polymerize by sharing one oxygen atom at the corners of the tetrahedra, but never more (that is, edges or faces), as there is too much repulsion between two silicon atoms to be that close. Silicate minerals can be found in a variety of compositions and structural configurations. These are categorized as the following:

Table 2.2 Coordination number and size ratios of cations and anions

Coordination #	Cation/Anion Ratio	Coordination	Example
2	0–0.155	linear	HF
3	0.155–0.225	triangular	BO_3
4	0.225–0.414	tetrahedral	SiO_4
6	0.414–0.732	octahedral	spinel
8	0.732–1	cubic	NaCl
12	1	hexagonal, CCP	metals

- *Nesosilicates or orthosilicates*—These are independent SiO_4 tetrahedra. Examples include olivine, zircon, and garnet.
- *Sorosilicates*—Epidote is an example of this mineral, composed of two SiO_4 groups.
- *Cyclosilicates*—Ring structures of linked SiO_4 tetrahedra form such minerals as cordierite and beryl.
- *Inosilicates*—Chains and double chains of SiO_4 tetrahedra form such minerals as pyroxene and wollastonite.
- *Phyllosilicates*—Planar sheets of SiO_4 tetrahedra are the basis of clay minerals and mica.
- *Tectosilicates*—These are three-dimensional frameworks of linked SiO_4 tetrahedra. Examples include quartz, feldspar, and nepheline

In addition to the silicate minerals, there are several other important rock-forming and ore mineral groups:

- *Carbonates*—These consist of one or more metals combined with the carbonate ion (CO_3^{2-}). Examples of carbonate minerals are calcite ($CaCO_3$) and magnesite ($MgCO_3$).
- *Metal oxides and hydroxides*—Many metals in contact with the atmosphere will combine with oxygen or oxygen and water to form oxides and hydroxides. Some common examples are hematite (Fe_2O_3) and gibbsite ($Al(OH)_3$).
- *Sulfides*—In reducing conditions, some metals may combine with sulfur to form sulfide minerals, such as pyrite (Fe_2S). Certain metals (iron, copper, zinc, and nickel) have a strong affinity to sulfur.
- *Sulfates*—In oxidizing conditions, sulfur can combine with metals to form sulfate minerals, such as gypsum ($CaSO_4 \cdot 2H_2O$).
- *Phosphates*—Phosphate ions (PO_4^{3-}) can combine with metals to form phosphate minerals. An example is the calcium phosphate mineral apatite ($Ca_5[PO_4]_3[OH,F,Cl]$).
- *Halides*—Metals can combine with the halide anions (most commonly with chlorine) to form minerals such as halite (NaCl).
- *Borates*—Boron naturally occurs in combination with oxygen and metal cations to form one of the 150 known borate minerals.

Ionic Substitution and Mineral Impurities

There can be considerable variation from the nominal formula of a given mineral due to the incorporation of other ions. This variability is of great concern in ceramic and glass manufacturing because of the potential for contamination of undesirable elements, such as iron and titanium, and because they can lead to inconsistencies in the batch. Deviation from the nominal composition of a mineral can be attributed to ionic substitution within the crystal lattice. For many minerals, there exists a *solid solution* series where the composition can vary over a specific range between two end-member compositions. An example of this is the mineral olivine, which has a solid solution series between Mg_2SiO_4 to Fe_2SiO_4, with Fe^{2+} and Mg^{2+} occupying the same site within the structure. This is an example of simple ionic substitution, where X^{2+} is replaced by Y^{2+}. A coupled substitution occurs when a pair of ions replaces another pair of ions, while retaining a charge balance. An example of this is the plagioclase feldspar solid solution series, where Na^+Si^{4+} substitutes for $Ca^{2+}Al^{3+}$ within the crystal lattice.

Ionic substitution can also occur when a minor ion replaces the major ion without a solid solution. Factors that influence ionic substitution include: (1) ionic radius; and (2) the temperature at which crystal formed. In general, ions can readily substitute into a mineral if ionic radii differ by less than 15 percent. Higher-temperature minerals have greater disorder and a more open structure, and therefore can accept larger or smaller ions more readily. The substitution may involve two species to retain charge balance. For instance, if Al^{3+} replaces Si^{4+} in a silicate mineral, a monovalent cation such as Na^+ will need to be retained close to the exchanged lattice site to maintain charge neutrality.

The bulk composition of a mineral grain may also diverge from its nominal formula due to the presence of other minerals or materials, either as discrete inclusions within the particle or as coatings on the surface. It is common to see inclusions of other solid phases within a larger crystal. Large mineral inclusions may be separated during processing if the mineral grains are pulverized to a size where the inclusions are liberated (see Chapter 3). Fine, disseminated particles are essentially impossible to remove, and their presence can prevent a mineral from being used.

Rocks

Rocks can be defined as a cohesive collection of one or more minerals that have crystallized or have been cemented together. Rocks that form from the solidification and crystallization of molten magma are called *igneous*. Magma that reaches the surface and is expelled as lava forms extrusive rocks, which usually are cooled quickly and therefore have small mineral grains, or crystallites, and an amorphous glassy matrix. Magma that is injected into the Earth's crust and slowly cooled forms intrusive rock bodies, such as granite batholiths. The slow cooling allows the growth of minerals, and as a result, these rocks have much larger mineral grain sizes and no glass. The types of minerals and their relative proportions in the rock depend on the original composition of the melt and, to a lesser extent, on the pressure at which crystallization occurred. Magmatic rocks are generally classified by the amount of silica in their bulk composition and by the size of the crystal grains. *Basalt* is a general term for a silica-poor volcanic magma that is relatively high in iron and magnesium. It is produced by partially melting the mantle, which is composed predominantly of iron and magnesium silicate minerals. If a basaltic magma does not reach the surface and is allowed to cool slowly, it forms a *gabbro,* which has the same bulk composition as the basalt but a coarser grain size. *Granite* is a general term for a coarse-grained igneous rock that has high silica and alumina and low iron and magnesium. Granitic melts are generally produced by melting the silica-rich continental crust. When a granitic magma erupts at the surface, it may form a glass (obsidian) or a highly porous, fine grained volcanic rock (rhyolite). If the eruption is explosive, a fine particulate ash is expelled. Table 2.3 shows the average bulk composition of basalt and granite.

Rocks that are derived from material that was eroded at the Earth's surface and deposited by water, wind, ice, or biologic activity are *sedimentary.* Clasts, or pieces of rock that are cemented together (often the cement is made of silica and/or calcium carbonate precipitated from groundwater), can form sandstones or, if the particles are clay, a shale. Rocks that are deposited after the evaporation and subsequent precipitation of dissolved ions are *evaporites,* such as salt or gypsum. The

Table 2.3 Average major element bulk composition of granite and basalt (weight %)

Element oxide	Granite/Rhyolite	Gabbro/Basalt
SiO_2	74.0	51.0
Al_2O_3	13.8	14.2
CaO	1.2	10.5
MgO	0.3	6.3
FeO/Fe_2O_3	2.0	13.0
Na_2O	3.0	2.2
K_2O	5.4	0.8
TiO_2	0.2	2.0

assemblage of evaporite minerals in a deposit depends on the dissolved load in the water from which they were derived. Sediments that are derived from organic material, such as shells, are biogenic sediments. An example of a biogenic sedimentary rock is limestone.

If a rock is buried and subjected to fluids, heat, and pressure, it may recrystallize new minerals and form a *metamorphic* rock. For example, if a limestone is metamorphosed, it can form marble. The composition of the original rocks may be altered because hot water or vapor can remove or add material during metamorphism.

Many of these rocks and their constituent minerals are very desirable from an economic and industrial view. If found with the appropriate composition and physical characteristics, and if they are relatively accessible, these rocks can become economical ore deposits. In the next section, we discuss how, on a large scale, geologic processes concentrate the elements into these mineral deposits.

2.3 GEOLOGIC PROCESSES AND FORMATION OF MINERAL DEPOSITS

Plate Tectonics

The millennia of geologic activity on the Earth's crust, which continues today in the form of volcanism, hot springs, and earthquakes, is due to the heat engine within the Earth. Unlike the moon, which is more or less solidified and shows no volcanic or seismic activity, the Earth is constantly moving, albeit slowly. This movement is driven by the energy generated predominantly by the decay of radioactive materials within the Earth, such as isotopes of uranium, thorium, and potassium. The heat causes convective currents within the Earth's mantle, which in turn causes movement and the release of this energy through the crust at the Earth's surface.

The Earth's crust is not made of a continuous hard shell but is rather made of discrete rigid plates that float on top of the mantle. *Plate tectonics* is the theory that the Earth's lithosphere, or rigid part of the crust and upper mantle, is divided into a dozen major plates that slide over the mantle in various directions. Because of the motion of the plates, they will collide, pull apart, or scrape against each other. Each type of interaction causes a characteristic set of geologic processes and resulting structures.

The plate margins can be generally divided into three types: *divergent, convergent,* and *transform.* At divergent boundaries, two plates separate or break apart and move away from each other. Continental rifts occur where a continent breaks apart. In the early stages of breakup, the continental crust is thinned and uplifted, causing faulting and earthquakes. The mantle beneath the rift rises to fill the void, bringing heat and molten magma. An example is the Central Africa rift valley, which is marked by high heat flow, hydrothermal activity (hot springs), and volcanism. As divergence continues, the continental crust on the upper part of the lithospheric plate is filled with basalt to form oceanic crust. Divergent margins in the oceanic crust are called mid-ocean ridges. These ridges form the longest mountain chain in the world and are constantly active, producing new crust that progressively moves away from the ridge.

A boundary where two plates collide is called a *convergent margin.* Where an oceanic plate converges with a continental plate, the denser oceanic crust is subducted or forced under the continental lithosphere, resulting in earthquakes. The subducting crust is subjected to heat and pressure, which drives off water and other volatiles that promote melting of the mantle and the base of the continental crust. The resulting silica-rich magmas rise from behind the subduction zone to form linear chain of volcanoes on the continental crust. For example, the Cascade Mountain range of northwestern North America is a magmatic arc that formed as the oceanic Juan de Fuca plate subducts below the North American continental plate. When two continental plates collide, there is no subduction (and therefore no volcanism) because there is no density difference between the two masses. Rather, the edges of the plates weld together to produce a high mountain range, such as the Himalaya Mountains.

Transform boundaries are found where plates are sliding past each other along one fault or a group of faults. The motion of the two plates is parallel but in opposite directions so that the boundary is a shear zone. The San Andreas fault in California is a transform boundary between the Pacific and North American Plates. In general, there is localized deformation and metamorphism associated with movement along the fault.

The movement of heat and material to the Earth's surface that takes place at these plate margins—coupled with the interaction between the crust, hydrosphere, and atmosphere—result in a continuous change in the composition and structure of the crust over geological timescales (millions of years). The transfer of energy and material provide an opportunity to mobilize, fractionate, and concentrate elements and minerals. For example, the average crustal abundance of boron is 8.7 ppm

(Table 2.1). Without concentrated deposits of boron, it would take, on average, more than 1,100 tons of rock to produce one ton of boron. The geological processes that form ore deposits can generally be subdivided into subsurface and surface processes (Kesler, 1994). Subsurface processes involve magmas and hydrothermal systems, while surface processes include organic, chemical, and physical activities associated with water and wind. Table 2.4 summarizes the types of ore-forming geological processes and the associated mineral deposits.

Magmatic Processes

Some ore-forming processes that occur below the surface can involve magmas. Magmas are formed by partially melting the minerals in the mantle or crust, so their composition will reflect the composition of the source. The thermally buoyant magma segregates from its source and then forces its way up toward the surface of the crust. If the magma makes it to the surface and erupts, it will quickly cool and there will be little to no compositional differentiation from the primary magma composition. However, if the magma does not make it to the surface and slowly cools in a subsurface magma chamber, there is the possi-

Table 2.4 Geologic processes that can form mineral deposits (after Kesler, 1994)

Type of Process	Deposit or Mineral Deposited
	Subsurface Processes
Involving water	Groundwater—uranium, sulfur Basinal brines Seawater
Involving magma	Crystal fractionation—chromium, vanadium Immiscible liquid—nickel, copper, cobalt, platinum group elements Magmatic water—pegmatites
	Surface Processes
Weathering	Laterite—nickel, bauxite, gold, clay
Physical sedimentation	Placer—gold, platinum, diamond, ilmenite, rutile, zircon, sand, and gravel
Chemical sedimentation	Evaporite—halite, sylvite, borax, trona Chemical—iron, sulfide minerals,
Organic sedimentation	Hydrocarbons—oil, natural gas, coal Minerals—limestone, dolomite, phosphate

bility of separating the elemental components of the magma through crystallization and physical separation of minerals of different compositions. It is these slowly cooled magmas that have a higher potential of forming ore deposits.

As magma slowly cools and the temperature falls below the liquidus of the melt, mineral crystals will begin to form or *nucleate,* with the order of crystallization and composition of the minerals depending largely on the initial composition of the magma and pressure within the crust. Phase diagrams can be used to estimate the minerals that are in equilibrium with a given bulk magma composition. As the temperature and composition of the residual magma in the chamber changes with time, so does the mineral assemblage. This is a simplistic version of what may actually occur in nature, as recharge of the magma chamber with fresh molten material or assimilation of the rock around the chamber may occur.

When a mineral nucleates and grows, it incorporates specific elements into its crystal lattice and in the process removes them from the melt. This process results in the differentiation or fractionation of the elements from a relatively homogeneous melt. For instance, olivine ($[Mg,Fe]SiO_4$) commonly is the first mineral to crystallize in a cooling basaltic or gabbroic magma, which depletes the residual melt in magnesium and iron. Many elements that are present in minor or trace amounts may not form their own minerals and will partition into other minerals, substituting for major element ions in the lattice. The partitioning of elements into phases in a cooling magma can be described using partition coefficients (k_d):

$$k_d = [M]_{\text{mineral}}/[M]_{\text{magma}}$$

where M is the concentration of a given element in the mineral and the magma, respectively. If the $k_d > 1$, then the element is compatible in mineral. If $k_d < 1$, then the element is incompatible. For example, in a basaltic magma, the k_d for Cr in the mineral olivine ranges from 1 to 3, so Cr is *compatible* in olivine. However, k_d values for sodium in olivine are in the range of 0.01 to 0.02, so sodium is *incompatible* in olivine. Some general observations of element compatibility include the following (after Henderson, 1982):

- The transition minerals in the periodic table between scandium and copper are enriched in magnesium- and iron-rich minerals such as olivine and pyroxene.

- Elements with atomic number 39 (yttrium) and higher (except the platinum group elements) are highly incompatible in most rock-forming minerals and prefer to reside in the magma.
- Sulfide magmas will be enriched in copper, iron, nickel, cobalt, silver, gold, selenium, tellurium, indium, and rhenium.
- Iron-oxide magmas are enriched in copper, uranium, and fluorine.

As a result of slow cooling and crystallization, a gabbroic igneous intrusion can form a layered deposit, in which bands of minerals collect at the bottom of the magma chamber as they crystallize and settle. One example of a layered intrusion is the Bushveld Complex of South Africa, which is up to 7 km thick in some parts and is a major source of strategically important metals such as chromium, vanadium, and platinum group elements (including palladium, rhodium, osmium, and iridium).

Higher-silica magmas, such as granites, generally do not form layered intrusions due their higher viscosity, which hinders the gravitational settling of crystals. Nevertheless, fractionation of elements in granites can occur through the formation of *pegmatites*. Many of the less-common elements that are incompatible in most common rock-forming minerals will remain in the residual liquid of a cooling granitic magma body. This residual liquid, which is also rich in volatiles such as water, collects at the top of the intrusion. Pressure from this volatile-rich body may cause sudden cracking of the surrounding rock, and the rapid depressurization will allow quick crystallization of minerals. It is thought that the depolymerization of the silica-rich melt by high water content increases the diffusion rates of elements, allowing rapid crystallization. The resulting pegmatite is characterized by minerals with very large crystals sizes, with some commonly up to 1 m in diameter. Pegmatitic mineral assemblages usually contain quartz, feldspars, and hydrous minerals such as mica and/or hornblende. Other minerals include tourmaline, beryl, and lithium minerals such as lepidolite and spodumene. Pegmatites are important sources of less common elements, such as beryllium, cesium, niobium, tantalum, tungsten, uranium, and thorium. A rock often associated with pegmatites is aplite, which is a mix of very low iron quartz and feldspar. Aplite can be used as a source of alumina in glass manufacturing.

Another mechanism for magmatic differentiation occurs when a cooling magma divides into two immiscible liquids, a silicate melt and sulfide or iron-oxide melt. Some iron oxide melts may have trapped gasses that allow it to reach the surface and erupt. These deposits are often associated with copper, uranium, and fluorine compounds (Kesler, 1994).

Some rare magmas are very rich in carbon dioxide. One type that is rich in carbon dioxide and calcite are referred to as *carbonatites*. These are of interest because they can host elements such as vanadium, niobium, tantalum, and the rare earth elements.

Subsurface Aqueous Processes

In both subsurface and surface aqueous processes, the solubility of an ion in water or fluid is greatly determined by its *ionic potential,* which is defined as the ionic charge divided by the ionic radius (Z/r). Ions that have a low charge and relatively large ionic radii, such as the alkali element sodium ($Na^{+}Z/r = 1.0$), have lower ionic potentials than transition metals such as titanium ($Ti^{4+}Z/r = 5.9$). Cations can be subdivided into those with low ionic potential ($Z/r < 3$ to 4), intermediate ionic potential (3 to $4 < Z/r < 10$ to 12), and high ionic potential (10 to $12 < Z/r < 50$).

Ionic potential influences an ion's interaction with polar molecules such as water (Railsback, 2003). In aqueous solutions, low ionic potential cations usually hydrate or form ion pairs. The more concentrated charge of intermediate ionic potential cations can repel some of the protons of water molecules to the point of expelling the proton and forming hydroxide compounds that may be insoluble, such as $Al(OH)_3$. For multivalent metals, oxidation state of the cation plays an important role in solubility. For example, Fe^{2+} ($Z/r = 2.7$) may be soluble in a given solution, but upon oxidation to Fe^{3+} ($Z/r = 4.7$), it may become insoluble and precipitate out of solution as iron hydroxide. Cations of highest ionic potential will repel most protons of complexing species and thus form soluble hydroxo- or oxo-complexes such as SO_4^{2-} or HCO_3^{-}. Thus, the ionic potential of cations play an important part in their behavior when a rock is exposed to solutions in subsurface processes.

The exposure of minerals to water is common below the Earth's surface because rocks are, to varying degrees, porous materials with voids and fractures between crystal grains that allow water to pass through. The ability of water to flow through rock is referred to as *permeability*. The basic mechanism of subsurface, aqueous ore-forming processes is the dissolution, transportation, and redeposition of materials that can take place when water permeates through and is in contact with rock. Aqueous geochemistry is concerned with the varying ability of water to dissolve minerals under different conditions, such as temperature and pH. The water involved in these processes can be groundwater (which originates as atmospheric water), seawater, or hydrous fluids released during metamorphism. The changes that take place between both water and rock are referred to as *hydrothermal alteration* reactions. Preexisting minerals and solutions can change in composition and structure during alteration, either gaining or losing material. The material dissolved in the hydrothermal solution can be transported to another location and then precipitated in open pores, fractures, and veins. Precipitation occurs when the solution is subjected to changes in its physical or chemical environment, such as temperature, pH, and oxidation state.

Hydrothermal springs are one example of this process. Groundwater or seawater is heated by the proximity to a heat source, such as a magma chamber, and the hot fluid has a greater capacity to dissolve and carry minerals compared to cold water. The thermally buoyant solution is forced to the surface, where it is quickly cooled and forces the precipitation of dissolved material. Such springs are found on land in volcanically active regions, such as the continental volcanic arcs (e.g., the Cascade Mountains, the Andes, etc.) or at mid-ocean ridge environments, where circulating seawater, up to 350°C, exits into 2°C water, resulting in the precipitation of metal sulfide and oxide minerals that collect at the base of the vents or within cracks near the surface of the ocean floor.

In addition to causing hydrothermal circulation, the heat from a magmatic intrusion can cause localized metamorphism that can form potential ore deposits. One class of metamorphic deposits of particular interest is the *skarn*. Skarns are calcium-bearing silicates formed by the introduction of silicon, aluminum, iron, and magnesium into a carbonate rock, such as limestone or dolomite, by the intrusion of a silicic magma (e.g., granite). Skarns are a major source of wollastonite ($CaSiO_3$) and metal oxide and sulfide ores.

Surface Processes

Unlike the subsurface processes, which are driven in large part by energy from the Earth's interior, the movements of water and air at the Earth's surface are ultimately driven by energy from the sun. This energy continually changes the surface of the Earth by cycles of weathering, eroding, mobilizing, and depositing material. Weathering is the decomposition process of rocks and minerals that are exposed to the atmosphere and hydrosphere. Physical weathering is the mechanical breakdown of rocks caused by the action of plants, freeze/thaw cycles, and moving water. Chemical weathering involves the dissolution of rock forming mineral by the removal of soluble elements. The degree of chemical weathering depends on the pH of water, which can be affected by dissolved CO_2, SO_2, and organic acids, the presence of which enhance the decomposition of minerals.

Some elements are more susceptible to chemical weathering. Those elements that are readily soluble in ground and surface water may eventually be transported by rivers into the oceans while the less soluble elements will remain in place (Table 2.5). As discussed previously in this section, the ionic potential of cations determines their behavior in solutions. Ions of low ionic potential ($Z/r < 4$), such as the alkali and alkaline earth elements, are leached from rocks and soils by weathering processes and collect in seawater. Intermediate ionic potential ions ($4 < Z/r < 16$), such as Al^{3+}, Si^{4+}, Ti^{4+}, and Fe^{3+}, will remain in the soil as either an oxide or hydroxide. These ions also tend to have very low concentrations in seawater. Table 2.1 shows the composition of dissolved elements in seawater, demonstrating the relative deficiency of intermediate ionic potential metals compared to the abundance of alkali and alkaline earth elements.

A simplified model of weathering at the Earth's surface starts with a rock, such as granite, that is composed of quartz (silicon dioxide), feldspars (calcium, sodium, and potassium aluminosilicates) and mica (a hydrous alkali aluminosilicate). The less-stable feldspars and mica will initially break down into clays (see section 5.2), while the quartz will remain relatively unaltered. The size of the clay particles will depend greatly on the initial size of the minerals that were weathered (Pruett and Murray, 1993; Dombrowski, 2000). Further weathering can dissolve some of the silica in the clays to form aluminum hydroxides and, if iron is present, iron hydroxides. These residual materials, of

Table 2.5 Relative susceptibility of major elements to mobilization during weathering

Element	Relative Mobility
Aluminum	essentially immobile
Silicon	relatively immobile, but removed under intense weathering
Iron	immobile
Magnesium	mobile
Calcium	very mobile
Sodium	very mobile
Potassium	very mobile

which all except the most insoluble elements have been removed, are called *laterites*. Laterites are the ore deposits from which aluminum and aluminum oxide can be derived (see section 6.1). Laterites derived from weathered ultramafic rocks (consisting of olivine and pyroxene) can become rich in nickel. Weathered deposits vary greatly in mineralogy and chemical composition because the process of weathering depends on a broad range of variables, including the composition of the primary minerals, the pH of the water, temperature, dissolved load in the water, and time.

Physical weathering can mobilize solid particles, which will collect to form sedimentary deposits. Like a modern industrial cyclone used in particle classification (Chapter 3), currents of air and water can physically sort minerals based on their density and particle size. An example of a water-mediated sedimentary deposit is a beach sand. The relatively insoluble quartz is transported by rivers and collected on beaches, where the action of the waves sorts and removes fine or soluble particles. In moving water, such as a river, deposits can form where water velocity decreases enough to allow the high-density grains, such as zircon and gold, to settle while keeping the less-dense quartz grains in suspension.

Sediments that are deposited and remain loosely combined are referred to as *unconsolidated*. The process of changing unconsolidated sedimentary deposits to rock is called *diagenesis*. During this process, subsequent deposition of sediments on top of the pre-existing material will compact it. At the same time, groundwater that flows through the pores of the sediments may deposit material on the surface of the grains. The deposition of material may be sufficient to bind individual particles together to form a rock. For instance, beach sand deposits may eventually be cemented together to become quartz sandstones. The cement usually consists of either silica or calcium carbonate, with or without iron oxides and clay. For high-purity silica sandstones, the cement must be composed predominantly of silica.

In arid regions that are subjected to high surface temperatures, bodies of water can evaporate, leaving behind the dissolved load as an evaporite deposit. In restricted seas such as the Red Sea between Africa and Asia, marine evaporites can form deposits of salt, gypsum, and minerals containing potassium and bromine. In arid continental regions, evaporation of intermittent lakes can form nonmarine evaporites. The minerals that form in these areas depend on the local geology, which, in turn, determines what elements will dissolve into the water that eventually collects in the lakes. The massive trona (soda ash mineral) mines of Wyoming were derived from nonmarine evaporites that precipitated from ancient lakes.

Organic activity at the Earth's surface over the millennia has produced considerable accumulations of material. In most oceanic regions, the organic remains of animals are oxidized into carbon dioxide and water, leaving the inorganic shells and skeletons of sea animals and plankton to be deposited. These remains can be made of calcium carbonate (either as calcite or aragonite), amorphous silica, or apatite. Ancient reefs can eventually form limestone. Some limestones can be converted to dolomite ($CaMg[CO_3]_2$) after deposition as a result of the penetration of magnesium-rich waters through the rock, which can exchange calcium in the limestone for magnesium to create a mixed carbonate. Silica deposits, such as diatomite and chert, can form from the accumulation of the internal skeletons of diatoms (a type of plankton). Apatite deposits, which are important phosphate ores, come from the accumulation of skeletons and teeth of marine fauna.

Most carbonaceous organic matter that accumulates in water eventually oxidizes to carbon dioxide. However, in areas of low oxygen conditions, organic remains will be preserved. Some of these materials were buried and subjected to heat and pressure, resulting in the formation of fossil fuels such as coal, petroleum, and natural gas.

2.4 EXPLORATION AND DEPOSIT EVALUATION

The exploration geologist uses a variety of tools to discover potentially useful ore deposits. The first step is to conduct initial field work to determine the type of bedrock that is exposed and to use deductive reasoning to place constraints on the geological environment of formation. For example, if one were searching for an alkali feldspar deposit, one would look at silicic rocks of igneous origin such as granite. To find uncommon elements that normally exist in low abundance, geochemical studies of the soil, sediment in streams, and water can be conducted to look for anomalous enrichments that can indicate the proximity of an ore deposit.

In order to explore regions below the surface, indirect geophysical methods are used. These methods include studies of the magnetic field and propagation of seismic waves, which are particularly useful in areas covered by thick soil or sand deposits. Magnetometers can be used to detect concentrations of magnetic minerals, such as magnetite. Seismic waves travel through rock at speeds that vary with density, porosity, and other physical parameters. Seismic waves are produced naturally by earthquakes, but exploration geologists use explosives or physical thumping near the surface. By measuring differences in the speed of seismic waves through the Earth, an image is generated to show the presence or absence of potential deposits. Seismic imaging is used often in gas and petroleum exploration.

After a potential deposit is identified, evaluation is done by drilling, which is the most expensive part of the exploration process and can cost millions of dollars for a single deposit (Kesler, 1994). Drilling is done with a drill rig consisting of a superstructure that holds the motors, pumps, and drilling pipe. A carbide or diamond-embedded bit is attached to lengths of pipe that, when connected, make up the drill string. Rotating the drill string from the superstructure cuts through the rock while drilling mud pumped through the pipe removes rock fragments. These rock cuttings are evaluated to determine the mineralogy of the rock. Drilling has become highly developed and can be done on both land and sea.

A final feasibility study is conducted based on information from the deposit evaluation, expected cost of extraction, market value of the product, and distance from the market. Environmental regulations, which can vary between different countries and regions, also plays a large part in determining if a deposit can be economically exploited. If costs are significantly lower than the estimated value, the deposit may be placed into production.

REFERENCES

Callister, W. D. (2002). *Materials Science and Engineering: An Introduction,* 6th ed. New York: John Wiley and Sons.

Dombrowski, T. (2000). The Origins of Kaolinite—Implications for Utilization. In *Science of Whitewares II,* ed. W. M. Carty and C. W. Sinton, 3–12. Westerville, OH: American Ceramic Society.

Henderson, P. (1982). *Inorganic Geochemistry.* Oxford: Pergamon Press.

Kesler, S. E. (1994). *Mineral Resources, Economics, and the Environment.* New York: Macmillan Publishing.

Klein, C., and C. S. Hurlbut, Jr. (1993). *Manual of Mineralogy,* 21st ed. New York: John Wiley and Sons.

Pruett, R. J., and H. H. Murray (1993). The Mineralogical and Geochemical Controls That the Source Rocks Impose on Sedimentary Kaolins. In *Kaolin*

Genesis and Utilization, ed. H. H. Murray, W. M. Bundy, and C. C. Harvey, 149–170. Boulder, CO: The Clay Minerals Society.

Railsback, L. B. (2003). An Earth Scientist's Periodic Table of the Elements and Their Ions. *Geology*, v. 31(9), pp. 737–740.

Wenk, H., and A. Bulakh (2004). *Minerals: Their Constitution and Origin*. Cambridge: Cambridge University Press.

3

MINERAL EXTRACTION AND PROCESSING

When demand for a particular raw material is recognized by the glass and ceramic industry and a suitable deposit has been established (see Chapter 4), the product is removed from the earth and processed to the desired specifications. These steps need to be accomplished with minimized expense and in compliance with all national and local human health and environmental regulations. Knowledge of these processes is essential to understanding the cost, quality, and availability of each raw material that the manufacturer requires. It also enables the consumer to communicate better with the vendors and suppliers.

The minerals in a deposit that can be economically extracted are referred to as an *ore.* Virtually all ores are a mix of desired minerals as well as *gangue,* which are the undesired materials. The entire process of obtaining a marketable product can be broken into unit operations that are conducted in a logical order. The process begins with extraction from the ground, or mining, followed by *comminution,* which is the procedure of breaking the material into a size where the desired minerals can be separated from the gangue. The material then undergoes beneficiation to clean and concentrate the desired minerals, followed by size classification. Some materials may require other steps or modifications, such as dewatering, to produce a finished product. Refined materials are produced by chemical purification of the natural minerals. In all, the path from the ground to the industrial user can be complex and each additional step adds to the cost of the final product. This chapter describes the general procedure for raw material extraction and processing. Much of this subject comes under the broader topic of industrial mineral processing. For more detailed work on this subject, the reader is referred to such texts as Kelly and Spottiswood (1982), Wills (1997), and Fuerstenau and Han (2003).

3.1 MINING

After an ore deposit has been identified for exploitation, the material is removed through mining. The overall goal of mining is to first expose and then break up the desired rock, usually through blasting, mechanical, and hydraulic methods, to a size that is small enough so that it can be removed from the mine. There are two general classes of mine: open pit (or surface) and underground. Open-pit mines are normally used if the deposit is within 100 to 200 m of the surface and the desired mineral and the gangue are intimately mixed (Kesler, 1994). Underground mines are used to extract ores that lie below 200 m from the surface or to follow veins of high-grade ores. In general, the cost and difficulty of mining increases with depth, and in the United States, 97 percent of mined material comes from open pits while, on average, 50 percent of mines in the rest of the world are open pit. This discrepancy can be attributed to the greater manual labor required in underground mines, which increases costs greatly where wages are higher.

An open-pit mine begins with the removal of overburden, which is the undesirable soil and rock that lies over the deposit. When the ore body is exposed, explosives are used to loosen the material enough so that it can be removed by mechanical equipment. As with digging any hole, the slope of the pit walls must be shallow enough to retain its integrity. Therefore, the mine is constructed as a series of benches, which requires the top to be widened as the pit deepens (Figure 3.1). The final depth of the mine is limited by the amount of overburden that must be removed and the increasing cost of moving ore up and out.

Another type of surface mining is dredging or alluvial mining (Figure 3.2). This is one of the least expensive methods of mining and is used for unconsolidated ores that lie underwater. A slurry of water and material is pumped to the surface and subsequently processed. An example of this would be mining sands for aragonite, quartz, and heavy minerals such as zircon and ilmenite.

Underground mines are accessed either by a vertical shaft which reaches the desired strata of the deposit or by an inclined shaft that follows the ore strata (Figure 3.3). Material can be loosened by explo-

Figure 3.1 Photograph of an open-pit mine pyrophyllite mine (printed with permission from R.T. Vanderbilt).

sives and/or mechanical machinery and then moved to the surface by a conveyor system. A common underground mining pattern is the room and pillar, in which pillars of material are left to support the mine and keep it from collapsing. The depth of the mine can be limited by such variables as water intrusion and increasing temperature. The deepest mine in the world is the 3,466 m-deep Western Deep Levels gold mine in South Africa (Kesler, 1994).

Another method of subterranean mining is solution or brine mining. For this, solid minerals are dissolved in place by pumping a solvent—usually water—into the deposit, allowing the material to dissolve, then pumping the solvent and its dissolved load back to surface. This method is used to mine relatively soluble minerals, such as halite and other salts that formed in evaporite deposits and were subsequently buried. For example, a horizontal bed of halite (sodium chloride) or nahcolite (sodium bicarbonate) can be accessed by two vertical drill holes that are connected with a horizontal passage made by either horizontal drilling or hydraulic fracturing. Hot water pumped down one hole will progressively dissolve the horizontal bed, and the brine will emerge from the other vertical hole. Brine mining has the advantage of relatively lower costs and decreased environmental impact as no

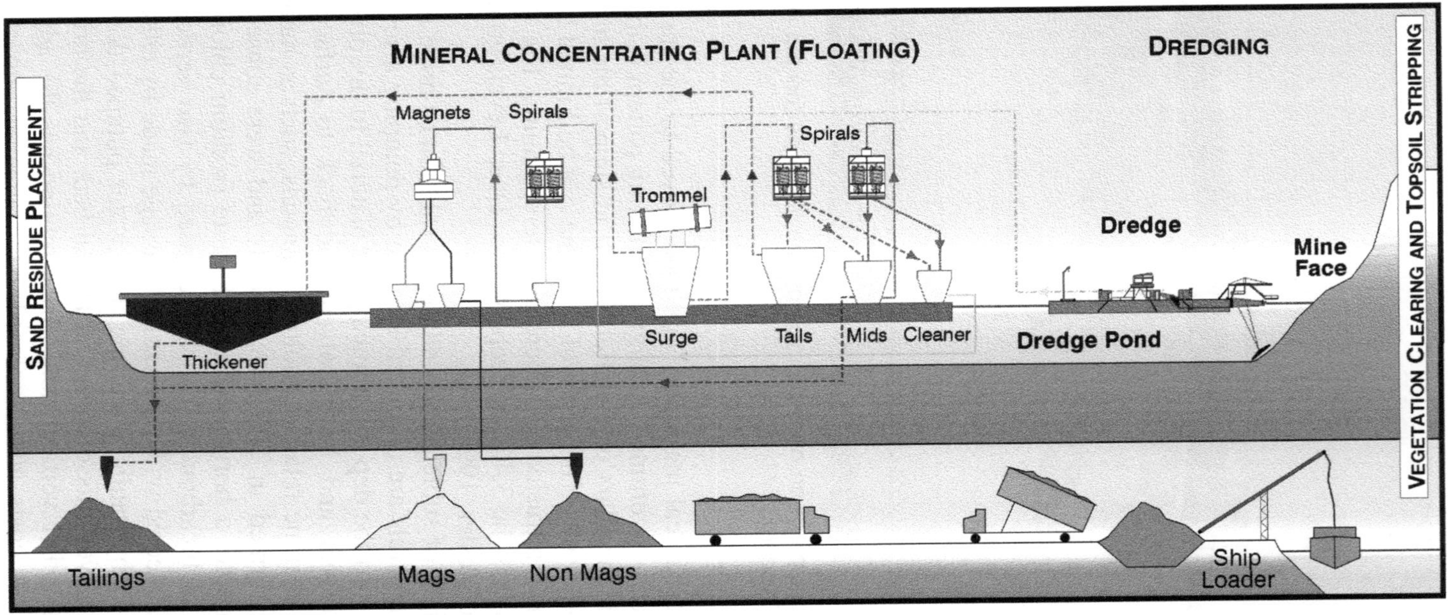

Figure 3.2 Photograph of a floating dredge and concentrator collecting unconsolidated sand in an artificial pond. The concentrator removes much of the unwanted quartz sand and sends the heavy mineral concentrate to the shore (printed with permission of Consolidated Rutile Limited).

Figure 3.3 Photograph of an underground wollastonite mine in New York (printed with permission from R.T. Vanderbilt).

overburden is removed. However, leaking solution has the potential to cause water contamination or undesirable changes in groundwater patterns.

The environmental impact of the mine needs to be minimized and the operations need to comply with all applicable regulations. The dominant intermediary for environmental problems is water. Most mines reach below the water table, and therefore intruding groundwater needs to be removed by continuous pumping. The gangue produced from the mine and from the processing plant may be stacked in tailings piles and will be exposed to precipitation. The water from both the mines and the tailings may contain suspended sediment or may have leached potentially hazardous materials from the deposit, which can pose a hazard to environmental and human health if released. For instance, mining can expose sulfide minerals to the atmosphere, which will chemically weather and release sulfuric acid. The increased acidity will, in turn, increase the ability of the water to leach more metals.

3.2 COMMINUTION

After the removal of ore from a mine, the *as-mined* or *run-of-mine* material is usually a mix of the desired minerals and gangue. An exception is where the material is hand-sorted at the mine, which can be the case for large, pegmatitic minerals. For a mixed ore, the desired minerals must separated from the gangue to produce a concentrate. This process of separation begins with comminution, which is the progressive size reduction of the ore to a point that the desired minerals are liberated from the gangue. Comminution is an important and expensive aspect of mineral processing. It is very energy intensive and can account for 50 percent of a plant's operating cost (Herbst et al., 2003).

In order to determine the most-efficient comminution route, the texture of the as-mined ore must be known. Texture is a general term that refers to the size, distribution, and shape characteristics of the ore minerals and their relationship to the gangue minerals. This is determined by examined cut and polished specimens of the ore using optical microscopy or electron beam microscopy. Figure 3.4 is a photomicrograph

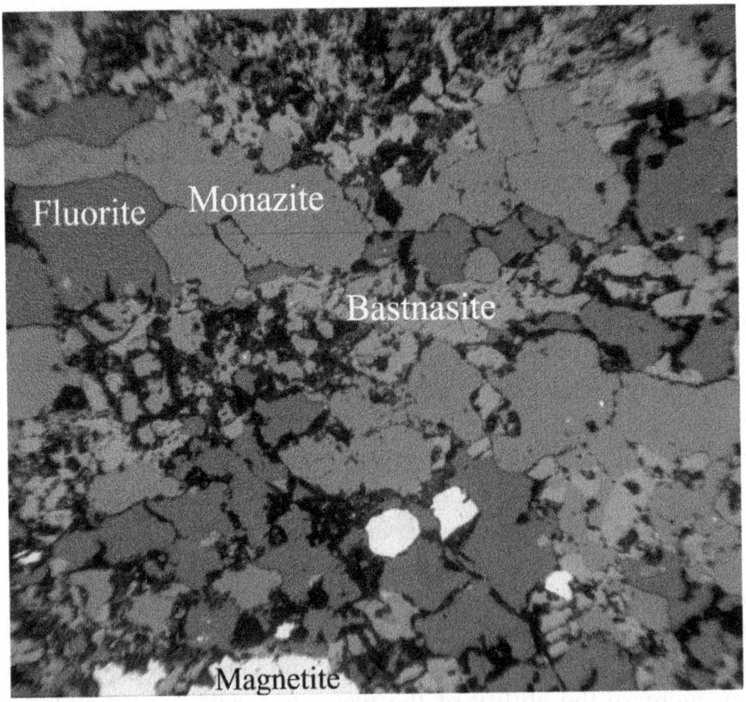

Figure 3.4 A photomicrograph that shows the textural characteristics of a rock ore taken in reflected light at 100× magnification. The rock is a rare-earth mineral-rich magnetite-hematite iron ore from Bayan Obo, Inner Mongolia, China. The minerals include magnetite (lightest color); bastnaesite (light gray; locally prismatic); monazite (medium gray); and fluorite (dark gray). The magnetite is used as an iron ore and the bastnaesite and monazite are used for rare-earth element extraction (courtesy of R. Hagni).

of a mixed iron and rare-earth mineral ore showing the size, shape, and dissemination characteristics of the crystals. One needs to determine the relative sizes of all minerals and ensure that sufficient sampling of the entire ore body has been conducted to assess any heterogeneities in composition and texture throughout the deposit. Knowing the texture, mineral phases, and other factors, a plan is made for the processing route of the ore. This is most simply displayed as a flow diagram, such as that shown in Figure 3.5.

Comminution involves crushing and grinding through a single or multistage process, depending on textural observations. For some ores, such as sandstone, only a small percentage of the rock is gangue, while for others, such as zircon, the majority of the ore is gangue. For unconsolidated sands and clays, mechanical size reduction is usually not

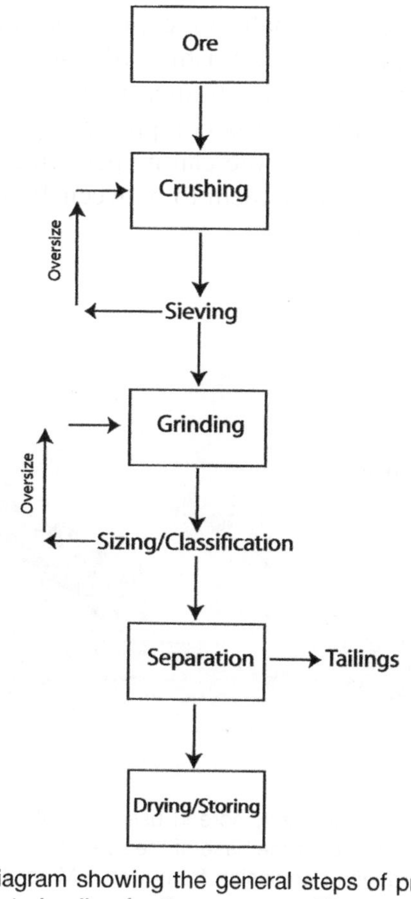

Figure 3.5 A flow diagram showing the general steps of processing an ore from extraction at the mine to loading for the consumer. There will be detailed subprocesses for each different material, but the general path is similar for all raw materials.

needed. Initial comminution is usually done at the mine to reduce the particle size to a point that it can be conveyed and transported to a mill. Initial sizing of the material to a size that can be transported is often done by a set of stationary parallel bars called a grizzly screen (Figure 3.6).

Further comminution must result in the liberation of individual particles that are either wholly mineral or wholly gangue. However, during comminution, particles that are composed of both ore and gangue, or middlings, are produced (Figure 3.7a). Middlings are particles that contain both ore and gangue in which the ore is considered to be locked into the particle and distributed in various patterns and textures. Figure 3.7b shows some of the basic textures of middlings. The processor strives to reduce the amount of middlings in order to increase liberation, though this must be balanced by the grain size. Very small particle sizes are expensive to produce, can cause problems in the separation processes, and may be too small to meet the specifications of the user.

Comminution is divided into crushing (down to 6- to 14-mesh) and grinding (down to micron and sub-micron sizes). Crushing is usually a dry process in which the ore is compressed against a rigid surface. Primary crushing of as-mined ore can be performed on pieces up to 3 m in diameter and can reduce them to between 10 and 20 cm diameter

Figure 3.6 Photograph of a grizzly screen used for initial sizing at an aggregate quarry. Oversize pieces collect at the base of the screen and undersize are conveyed to a separate pile (used with permission from Metso Minerals).

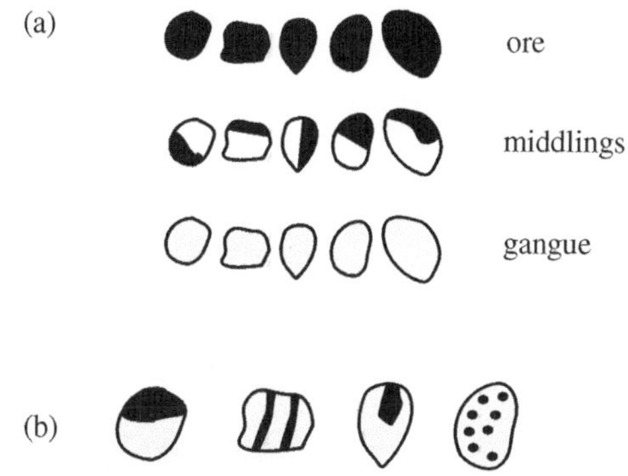

Figure 3.7 Schematic diagram showing (a) particle sets after crushing of desired ore (black), gangue (white), and middlings; and (b) the four basic types of middlings or locked particles (based on Kelly and Spottiswood, 1982).

particles (Wills, 1997; Hayes, 1993). This is performed with jaw crushers, which use the movement of a pivoting plate against a stationary plate to cyclically compress and release material (Figures 3.8a and b). Some primary crushing is also done with gyratory crushers, which use a conical grinding spindle that rotates on an eccentric within a hardened conical shell. Crushing takes place between the cone and the shell. Gyratory crushers are preferred to jaw crushers when higher crushing rates are required. Primary crushing generally uses 0.5 kWh/ton of material (Herbst et al., 2003).

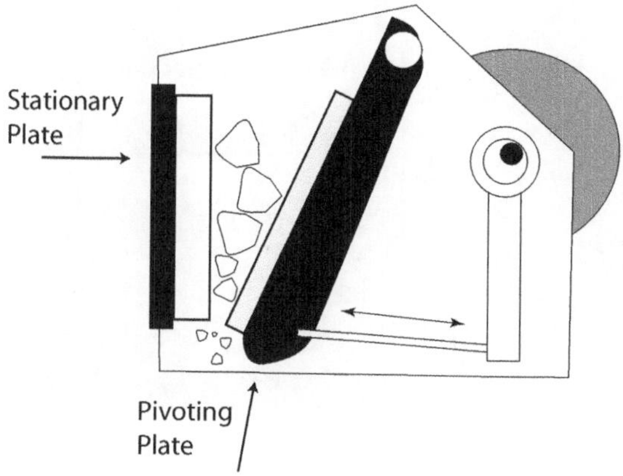

(a)

Figure 3.8 (a) Schematic diagram of a jaw crusher; (b) jaw crusher used for primary crushing of feldspar at the Piedra Grande plant in San Luis, Argentina.

(b)

Figure 3.8 (*Continued*)

After primary crushing, particles that are generally less than 15 cm diameter are sent to secondary crushers, which reduce the particles to a size suitable for grinding (2.0 to 0.5 cm diameter) (Wills, 1997). Cone crushers are commonly used for secondary crushing. These are small, modified gyratory crushers that operate at higher speeds and hammer the particles rather than slowly compressing them (Figures 3.9a and b).

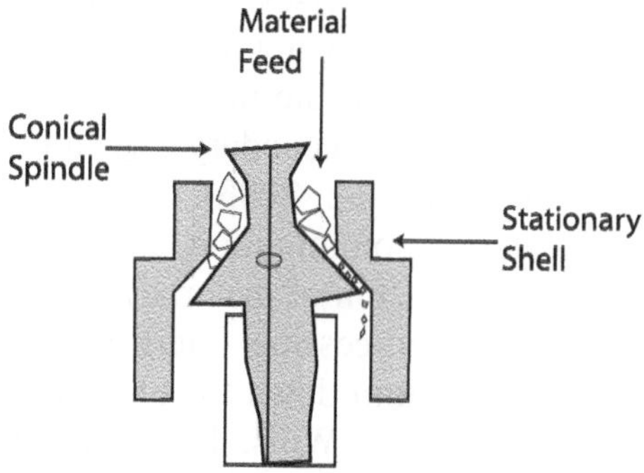

(a)

(b)

Figure 3.9 (a) Schematic cutaway diagram of a cone crusher; (b) stationary cone crusher that can be used for secondary crushing (used with permission from Metso Minerals).

More friable ores, such as limestone and phosphate ores, may by crushed using roll crushers. These consist of two horizontal, parallel cylindrical rollers that rotate toward each other. The desired particle size is set by the space between the rollers. Another relevant type of secondary crushing device is the impact crusher, such as the hammer mill. Rather than using compression to break up particles, these use the impact of metal beaters. An advantage to this is that the impacted particle will have no residual stresses, which may be the case for a particle broken by pressure as in a jaw or gyratory crusher. These concerns are relevant to the aggregate industry, where residual stresses may cause cracks in concrete. Secondary crushing generally uses 1 kWh/ton of material (Herbst et al., 2003).

Grinding is the last step of comminution, and it can also be the most energy intensive (5 to 25 kWh/ton of material). therefore, it is the most expensive stage of mineral processing (Herbst et al., 2003). Particle size is reduced from up to 5 mm down to sub-micron sizes by a combination of impact and abrasion (Wills, 1997), depending on the optimum size for liberation of the particles. *Volume grinding* is a term used when size is reduced by impact, and *surface grinding* is used for techniques that abrade the surface of the larger particle to make smaller particles. Volume grinding is performed in tumbling mills with material that is either dry or in an aqueous suspension.

Grinding mills are large drums or trunnions that rotate on a horizontal axis. Ore is continuously fed through one end of the trunnion, and the crushed material exits from the other end or from the bottom (Figure 3.10). Rod mills use steel rods as the grinding media, and ball mills use steel or ceramic spheres. The grinding media wear at a rate of 0.1 to 1 kg per metric ton of ore (Wills, 1997), depending on the hardness of the minerals being processed. The particles that are shed by the grinding media are considered contaminants and therefore must be removed. This is particularly relevant for products that require low iron specifications, such as flint glass. This particular problem of contamination can be solved by grinding without media, or self-autogeneous grinding, in which comminution is achieved by particle-particle impact. Semi-autogeneous milling uses a small charge of grinding media.

Different ores have different grindability characteristics, depending primarily on hardness of the constituent minerals, the fracture habit of the crystals (platy, prismatic, etc.) and the strength of the boundaries between individual grains. Therefore, each material requires its own set of grinding parameters. The optimum final grain size needed to liberate

Figure 3.10 Example of a rod mill with steel rods that run the length of the drum. A ball mill is the same, except that it is charged with spherical crushing media (used with permission from Metso Minerals).

the ore minerals is a function of texture, initial mineral grain size, and separation method. Some ore separation systems require particle sizes below 100 μm. Decreasing particle size increases energy costs because crushing and grinding equipment require large amounts of electrical power. For particle sizes below 500 μm, the energy requirements increase dramatically.

Submicron-sized particles can be made by physically breaking down larger particles or synthesizing or building up the particles from the atomic and molecular scale (Naito et al., 2003). In the past, the minimum grinding size was about 1 μm, while the maximum size for a built-up particle was 0.1 μm. This gap has been bridged by recent advances in submicron grinding techniques. Surface grinding methods, such as friction mills, appear to be the most effective for submicron particle production.

3.3 PARTICLE SELECTION, CLASSIFICATION, AND SEPARATION

After individual particles are liberated by comminution, they then follow different pathways ultimately leading to either the loading dock or the tailings (waste) pile. The separation of desirable and undesirable materials is done by taking advantage of differences in physical and chemical properties of different particles. The most commonly used properties follow (after Hayes, 1993):

- Size
- Density
- Magnetic susceptibility
- Electrical conductivity
- Surface chemistry

Methods of particle separation using these properties generally include the following (Miller, 2002):

- Selective barriers such as a screens or filters
- Separators based on particle density
- Separators based on fluid and particle interactions
- Separators based on particle surface or electrical characteristics

Screening

Selection on the basis of size (which refers to both the diameter of an idealized sphere and the actual particle shape) can be done using screens or the differential movement of particles through fluids. Screening is used throughout the crushing and grinding circuits to ensure that the feed size is compatible for individual steps. Specifically, particles are selected by screens either to continue through the process (undersize) or to return to a previous step (oversize) for continued crushing or grinding. Screening involves passing the particle stream over a perforated surface, such as wire mesh and punched plates of different opening sizes. Wire mesh can be made from any metal or alloy that can be drawn into wire, though abrasion of the mesh can introduce undesirable elements such as metal flakes. In order for screens to function, they must be agitated either by shaking, vibrating, or rotating in a drum (Figure 3.11). In general, the efficiency of screens decreases with decreasing mesh size and therefore screening is generally not performed on particles smaller than 200 μm (Hayes, 1993). A discussion of different types of screens and screening theory can be found in Kelly and Spottiswood (1982), Wills (1997), and Schmidt (2002).

Mesh sizes are described by several classification systems that can vary by country. The number of wires used per inch defines the mesh number in a standard sieve, though the aperture or opening size de-

Figure 3.11 Screens are used throughout the crushing and grinding circuits for particle-size classification. This is an example of a large, inclined screen that would be used during primary or secondary crushing (used with permission from Metso Minerals).

pends on the gauge of wire used. Table 3.1 shows standard international mesh sizes. In the United States, the two common standard sieve schemes are Tyler Standard Screen Scale and the USA series. The Tyler sieve series was introduced in 1910 by W.S. Tyler Inc, with numbers referring to the number of wires per inch. A modified version was adopted by the American Society for Testing and Materials (ASTM). The German Institute for Standardization (DIN) and the British Standardisation Institute (BS) have their own sieve sizing schemes. The International Standards Organization (ISO) uses a scale based on the actual square aperture size, and this is the most accepted scheme globally.

Classification

The sizing of particles by hydraulic classification results in the separation of a feedstock into two size fractions: oversize or sands and

Table 3.1 Standard sieve sizes

Mesh Size (μm)	Tyler Mesh	ASTM-E11 Mesh	BS-410 Mesh	DIN-4188 mm
5	2,500		2,500	0.005
10	1,250		1,250	0.010
15	800		800	0.015
20	625		625	0.020
22				0.022
25	500		500	0.025
28				0.028
32				0.032
36				0.036
38	400	400	400	
40				0.040
45	325	325	350	0.045
50				0.050
53	270	270	300	
56				0.056
63	250	230	240	0.063
71				0.071
75	200	200	200	
80				0.080
90	170	170	170	0.090
100				0.100
106	150	140	150	
112				0.112
125	115	120	120	0.125
140				0.140
150	100	100	100	
160				0.160
180	80	80	85	0.180
200				0.200
212	65	70	72	
250	60	60	60	0.250
280				0.280
300	48	50	52	
315				0.315
355	42	45	44	0.355
400				0.400
425	35	40	36	
450				0.450
500	32	35	30	0.500
560				0.560
600	28	30	25	

Table 3.1 ***(Continued)***

Mesh Size (μm)	Tyler Mesh	ASTM-E11 Mesh	BS-410 Mesh	DIN-4188 mm
630				0.630
710	24	25	22	0.710
800				0.800
850	20	20	18	
900				0.900
1000	16	18	16	1.0
1120				1.12
1180	14	16	14	
1250				1.25
1400	12	14	12	1.4
1600				1.6
1700	10	12	10	
1800				1.8
2000	9	10	8	2.0
2240				2.24
2360	8	8	7	
2500				2.5
2800	7	7	6	2.8
3150				3.15
3350	6	6	5	
3550				3.55
4000	5	5	4	4.0
4500				4.5
4750	4	4	3.5	
5000				5.0

undersize or slimes. The separation is done by the differential settling behavior of particles in a moving fluid, such as water or air. It is the same mechanism used in nature to sort particles in sedimentary deposits. For instance, when a river carrying a suspended load meets a body of water, the decrease in velocity will force the coarse sands to deposit while the clays and silts will remain suspended and settle further from the confluence. The outcome of hydraulic classification is the same as screening, but hydraulic classification is better suited and more economical for separating relatively small particles (<10 mm). Air classification is used predominantly in the fine and very fine range of 300 μm down to 1 μm (Nied and Horlamus, 2002). Hydraulic classification is based on the dependence of settling rates on particle size. Particles are accelerated into a given trajectory, either through gravity or from introduced centrifugal forces. Assuming laminar flow, particle settling within the medium can be described using Stokes Law:

$$v = \frac{(2gr^2)(\rho_1 - \rho_2)}{9\eta}$$

where v is the settling velocity (m/s), g is the gravitational constant (m/s^2), r is the radius of the particle (m), ρ_1 is the density of the particle (g/m^3), ρ_2 is the density of the medium, and η is the dynamic viscosity of the fluid (poise or g/cm-sec). It is apparent from this equation that settling velocity is dependent on particle radius. In practice, the movements of particles are more complex due to nonspherical shapes, particle-particle interactions, and turbulent flow around the particle. Nevertheless, this is the general basis for wet and air classifiers, tables, spiral concentrators, jigs, and cyclones.

Gravity-based classifiers use gravity along with the resistance exerted upon the particles by the medium. Simple gravity sedimentation classifiers are large pools into which a slurry of particles suspended in a liquid is introduced. However, fines that are less than 10 μm can increase the viscosity of the slurry and negatively affect the process (Wills, 1997). The larger particles are deposited while the smaller particles remain in suspension and are removed by the flow of liquid. *Jigging* is a gravity method that is used to separate particles of different sizes and specific gravity. It can be very efficient, despite the fact that the principles of the method are not well understood (Wills, 1997). In a jig, a slurry that is placed on a screen is cyclically disturbed by a reciprocating plunger that forces water through the screen. The repeated dilation causes the heavier particles to work through the interstices to the bottom of the collection while the lighter particles move to the top and are discharged. *Hydrosizers* use a current of water directed up from the bottom of a vessel. The fine material will remain suspended while the coarser material settles to the bottom, where it is discharged. *Spiral classifiers* use a revolving helical screw to both agitate the slurry and to move settled coarse particles up an incline. The fines remain suspended in the liquid medium, normally water, and are thus separated from the coarse fraction.

Hydrocyclones and air cyclones are widely used in mineral processing to dewater a slurry, to remove slimes, or to classify the final product sizes. Cyclones may be used as an integral part of a closed grinding circuit, in which the coarse underflow is sent back through the grinding process and the fine overflow continues on to the next process. For precisely sizing the final products, multiple stages of cyclones are often used. In hydrocyclones, a mixture of particles and water (pulp) or air enters the cylindrical portion (top) of the cyclone under pressure through the feed pipe and rotation of the pulp begins inside the cyclone. Centrifugal forces accelerate the particles toward the outer wall as it passes into the conical section, while drag forces try to move a particle inward. The magnitude of centrifugal forces that act on the mixed particles depends on the angular velocity of the pulp and the weight of the particle, which, in turn, depends on its size and specific gravity. The design geometry of the cyclone forces the coarser particles to collect near the inner surface of the cyclone and move downward to the underflow discharge. The lighter particles migrate toward the center and, with most of the water, are carried in an upward spiraling column near the axis of rotation to the vortex finder and through the overflow discharge. By controlling the diameter of the underflow apex relative to the diameter of the overflow, it is possible to vary the size of the particles in the overflow and underflow. Hydrocyclones work well when the particle size is in the range of 5 to 100 μm (Trawinski, 2002). Air cyclones operate in a manner similar to hydrocyclones, except that the particles are suspended by air. In addition, particle agglomeration needs to be carefully monitored, as this can cause misclassified particles. Agglomeration depends partly on particle size (with smaller particles having a greater tendency to adhere), and partly on material properties such as conductivity, surface roughness, and moisture or oil content (Nied and Horlamus, 2003).

Relevant parameters in cyclone design include chamber diameter; diameter of the vortex finder (generally the smaller the diameter, the finer the separation); chamber length; inlet area; and angle of the conical section. Important system parameters affecting performance include: temperature; pressure drop; percent solids per volume; particle shape; solids specific gravity; and volumetric flow rate. The feed volume is extremely important because it greatly influences the apparent viscosity of the slurry. If the viscosity is too high, then the fluid may tend to drag particles into the overflow stream. When operating at a feed density above 50 percent solids (by weight), small changes in the feed density can lead to dramatic changes in the separation process.

An important measurement used in cyclones is the cutpoint or separation size. This is usually reported as the D50 and is defined as the particle size in which the solids have a 50 percent chance of reporting to the underflow. Another term that is often used refers the percentage of material that passes into the overflow. For instance, P80 is the particle size in which 80 percent of the hydrocyclone overflow is finer than that size. The P80 depends on the D50, the sharpness of the separation and the size distribution of the pulp fed into the cyclone.

Magnetic Separation

Mixed particles can be separated on the basis of magnetic properties. Materials can be classified as either *paramagnetic,* which have magnetic permeability greater than air, or *diamagnetic,* which have magnetic permeability lower than air. Paramagnetic materials are attracted to a magnet, while diamagnetic substances are repelled. Ferromagnetic materials are a class of strongly paramagnetic materials, such as magnetite and metallic iron. Paramagnetic minerals that are commonly separated in commercial operations include ilmenite ($FeTiO_3$), rutile (TiO_2), wolframite ($[Fe,Mn]WO_4$), monazite ($[Ce,La,Th]PO_4$), siderite ($FeCO_3$), pyrrhotite (FeS), chromite ($FeCr_2O_4$), hematite (Fe_2O_3), and manganese minerals (Wills, 1997).

The magnets used in mineral processing are either permanent magnets or electromagnets. Ferrites are the least expensive type of permanent magnets and generate magnetic fields up to 0.2 Tesla. In order to generate the higher magnetic strengths required to separate weakly paramagnetic minerals, such as rutile, rare-earth permanent magnets are used since they can generate magnetic fields up to 2.4 Tesla (Norrgran, 2003). Solenoid coil electromagnets can operate up to 2 Tesla. Commercial magnetic separators operate continuously in the process route as the particles pass by (Bronkala, 2002). They can be classified as low-intensity and high-intensity and use wet or dry feed. Typical dry feed designs use rotating drums or belts (Figure 3.12a). Dry, low-intensity separators are commonly used for coarse material to separate highly magnetic particles from material with low magnetic susceptibility (e.g., ilmenite from quartz sand). A wet feed magnetic filter is a magnetized steel grid through which a slurry passes, collecting any ferromagnetic particles.

Suspended magnets are permanent or electromagnets placed over a material stream and are generally used for removing magnetic materials, such as tramp iron, which are the iron and steel particles that have been abraded during processing (Figure 3.12b). These particles pose a hazard to equipment and are highly undesirable in low-iron specification materials, such as those used for flint glass.

Electrical Conductivity

Due to differences in atomic bonding type and strength, all materials have characteristic electrical conductivity properties, which can be the basis for a separation process. High-tension separators are often used in processing sands for heavy minerals. Grains are charged with static electricity and separated by passing through an electric field. Separation is based on different electrostatic affinities. Systems can use a falling stream of material through an electric field or onto a rotating drum with a controlled electrostatic charge so that grains of a certain affinity will stick to the drum and others will fall. This system is somewhat similar to the belts and drums used in magnetic separation (Figure 3.12a). In practice, high tension separators can be used for particles in the range of 50 to 500 μm diameter (Hayes, 1993) and must be completely dry.

Froth Flotation

The separation of particles using the differences in the surface chemistry properties is the basis of froth flotation separation (Yarar, 2002). In this method, a feedstock of mineral particles, optimally with sizes between 10 to 500 μm, is mixed with additives that are designed to

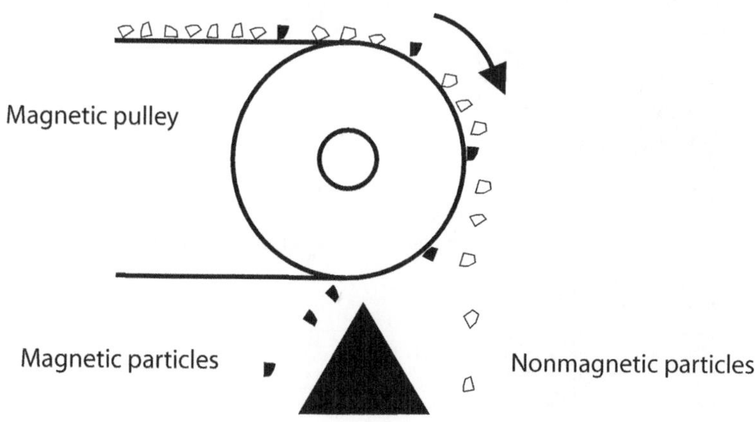

(a)

Figure 3.12 (a) Schematic of a magnetic pulley. The pulley drum is a permanent magnet, allowing the magnetic particles to remain attached to the belt slightly later than the nonmagnetic particles. (b) A flat plate magnet is used to remove tramp iron in the feldspar crushing circuit of the Piedra Grande plant in San Luis, Argentina.

(b)

Figure 3.12 (*Continued*)

modify the surface characteristics of the minerals. This "pulp" is placed in a vessel through which air bubbles are introduced from the bottom. Minerals that are rendered hydrophobic by the additives attach themselves to the bubbles and are lifted through the suspension to form a froth at the top (Figure 3.13). Hydrophilic minerals are wetted by water and remain in the vessel. With direct flotation, the froth contains the desired mineral and is skimmed off and collected while the unfloated gangue minerals remain in the pulp. In reverse flotation, the undesired gangue mineral is floated and removed.

The attachment of a solid particle to an air bubble can only be done if the surface energy is reduced by that attachment. This is represented by the following:

$$\gamma_{sg} - \gamma_{sl} < \gamma_{lg}$$

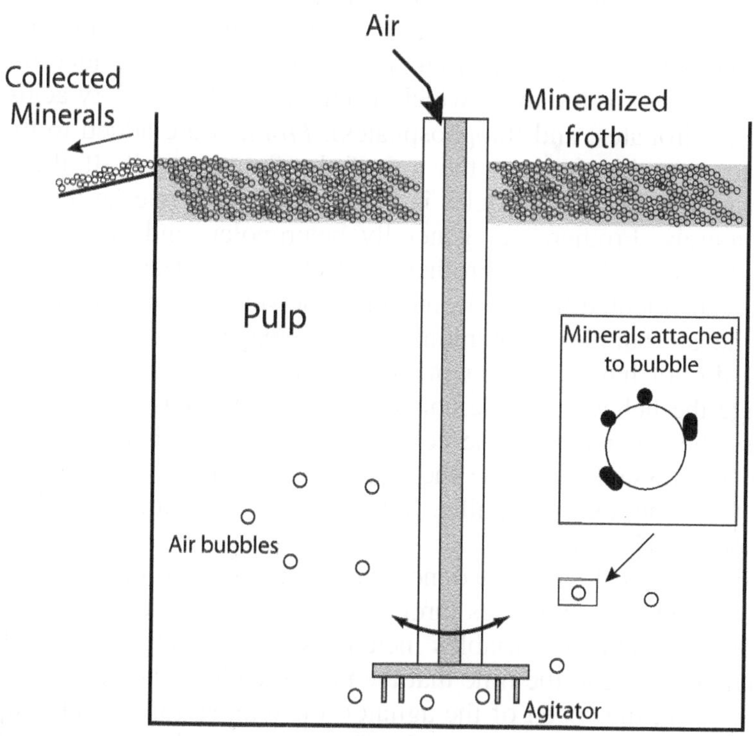

Figure 3.13 Schematic describing the froth flotation process. The inset shows how the floated particles are attached to individual bubbles that then rise to the surface.

where γ_{sg} is the interfacial energy between a solid particle and air (gas), γ_{sl} is the interfacial energy between the particle and the liquid, and γ_{lg} is the interfacial energy between the liquid and the air bubble (Wills, 1997). If this requirement is fulfilled, then the particle can be considered hydrophobic, and will tend to stick to the bubbles. Hydrophobic minerals are nonpolar with surfaces that do not readily attach to dipolar water molecules. Examples of hydrophobic minerals are talc, diamond, coal, and graphite. Most minerals of interest are polar and have strong covalent or ionic bonding at the surfaces that readily react with water molecules, and are therefore hydrophilic. Such minerals can be classified by the degree of hydrophobicity, with the following general mineral groups listed in increasing hydrophobicity: sulfides, sulfates, carbonates, oxides/hydroxides, silicates.

The reagents used in flotation can be classified as collectors, frothers, and auxiliary reagents (Yarar, 2002). *Collectors* are organic surfactants that allow minerals to become hydrophobic by adsorbing to the surface of the mineral. Collector molecules can be ionizing (dissociate in water) or nonionizing. Ionizing collectors are heteropolar molecules, having both a polar end that attaches to the mineral surface and a nonpolar end that is water repellant. Ionizing collectors are subdivided into cationic (amines) or the more widely used anionic types (fatty acids, xanthates, sulfonates, and thiophosphates). *Frothers* are added to the pulp to keep a stable froth at the top of the flotation cell. If the bubble carrying the mineral particles is not stable, it will release the particles prematurely. Frothers are generally heteropolar, and some collectors can also act as frothers. Frothers collect at the surface, with the nonpolar end oriented to the air, reducing the surface tension and stabilizing the air bubbles. Common collector reagents are alcohols, since these do not have collector properties. *Auxiliary reagents* are used to change the behavior of the collector. Activators may be needed to increase the efficiency of the collector for minerals that are difficult to float. Depressants increase the selectivity of the collector by making selected minerals hydrophilic. Substances such as lime may be added to modify pH of the pulp.

Commercial flotation is done using a series of continuously operating cells. The slurry passes through gates from one cell to the other by gravity. Operational variables include: solids loading of the pulp, the amount of reagent, the time that the pulp spends in the cells, the height of the froth, the speed of the agitator, the temperature of the pulp, and the size of the solids in the pulp. The latter variable is extremely important, and flotation processes are closely linked with grinding and classifying circuits to ensure the proper particle size is attained.

3.4 DEWATERING AND DRYING

For nearly all of the processing systems, there is a need to remove the water from a slurry of minerals before being shipped. Exceptions are clays that are shipped as a slurry in tanks or some sands that remain damp during shipping. Dewatering can be an expensive and time-consuming part of material processing, with the cost generally increasing with decreasing particle size and increased surface area (Dahlstrom, 2003).

The dewatering process is decided by several factors (Dahlstrom, 2003):

- Particle size and shape
- Weight and volume percent of solids
- Fluid viscosity and temperature
- pH and composition of the feed
- Variation and range of feed quality
- Specific gravity of the solids and liquid
- Quality requirement of discharge

Dewatering processes can be classified into three general methods, all of which may be used to a certain extent in a single processing route (Wills, 1997; Hayes, 1993):

1. Sedimentation
2. Filtration
3. Drying

Sedimentation is the easiest and usually the initial step of dewatering. Thickeners and clarifiers, circular tanks with conical bottoms and a central slurry feed, are the types of equipment that are used for sedimentation. Both use the gravitational force on a particle suspended in a liquid medium to collect in the bottom of the tank. The settled particles are removed from the tank bottom and the water exits from the top as the overflow. For very fine particles, such as clays and other particles less than 10 μm, flocculants can be used to cause the particles to coagulate or flocculate. Flocculants compress the electrical double layer that surrounds each particle, allowing the attraction of particles though van der Waal's forces. Simple coagulants can be inorganic salts with highly charged cations, such as calcium and aluminum salts, or polymers such as polyacrylamides, gums, and starches. Flocs are compound particles with a more open structure, where particles are separated by a flocculant molecule.

Forced sedimentation can be accomplished through centrifugal separation, using either a hydrocyclone or a centrifuge. Coagulants or flocculants cannot be used with a hydrocyclone. Hydrocyclones may also be limited by the particle size and generally cannot be used if the particle size is less than 10 μm.

Filtration uses a physical barrier that allows the liquid through but retains the solids. In this process, the solids form a filter cake behind the filter, which requires the liquid to move through the filter cake before it passes through the filter. Wide-particle-size distributions tend to block capillaries that provide conduits for the water. This tends to decrease the filtering rate, and often either pressure from the feed end or a vacuum from the filtrate end is applied to enhance filtration. Filtering can be done as a continuous, batch, or semicontinuous process (Dahlstrom, 2003). Continuous filters are constructed of fabric belts, disks, or drums that use gravity or an applied vacuum to remove water. The filter cake is removed by scraping or by an application of compressed air. Batch filters use fabric to retain the solid material while applied pressure squeezes water out of the slurry. The plate and frame filter press is a common batch-type that is used in dewatering of whiteware bodies.

Drying with heat is the last process to be performed before the final product is stockpiled, bagged or loaded, and shipped. Thermal drying is not used for slurried materials (with the exception of spray drying) because it takes approximately 1,800 Btu/lb. (4,187 kJ/kg) (Dahlstrom, 2003). Final drying reduces weight and, therefore, transportation costs, and it improves conveying for most materials. At the same time, a product that is too dry may have dusting problems, so for each product there is an optimum moisture content. Dryers remove water primarily by forcing hot air into a drying chamber. A common dryer is the rotary dryer, which is an inclined rotating cylinder into which the moist material is continuously fed along with hot air. In direct rotary dryers, hot air is in direct contact with the material. Indirect rotary dryers have heat applied outside of the cylinder and are used when dust needs to be minimized or when processing fine material.

REFERENCES

Bronkala, W. J. (2002). Magnetic Separation. In *Ullmann's Encyclopedia of Industrial Chemistry*, 6th ed. Weinheim, Germany: Wiley-VCH.

Dahlstrom, D. A. (2003). Liquid-Solid Separation. In *Principles of Mineral Processing,* ed. M. C. Fuerstenau and K. N. Han, 307–362. Littleton, CO: Society of Mining, Metallurgy, and Exploration.

Fuerstenau, M. C., and K. N. Han, eds. (2003). *Principles of Mineral Processing.* Littleton, CO: Society of Mining, Metallurgy, and Exploration.

Hayes, P. (1993). *Process Principles in Minerals and Materials Production,* 2nd ed. Brisbane: Hayes Publishing.

Herbst, J. A., Y. C. Lo, and B. Flintoff (2003). Size Reduction and Liberation. In *Principles of Mineral Processing,* ed. M. C. Fuerstenau and K. N. Han, 61–118. Littleton, CO: Society of Mining, Metallurgy, and Exploration.

Kelly, E. G., and D. J. Spottiswood (1982). *Introduction to Mineral Processing.* New York: John Wiley and Sons.

Kesler, S. E. (1994). *Mineral Resources, Economics, and the Environment.* New York: Macmillan Publishing.

Miller, S. A. (2002). Mechanical Separation Techniques. In *McGraw-Hill Encyclopedia of Science & Technology,* 9th ed. New York: McGraw-Hill.

Naito, M., N. Shinohara, and K. Uematsu (2003). Raw Materials. In *Handbook of Advanced Ceramics,* ed. S. Somiya, 81–129. Oxford: Elsevier.

Nied, R., and H. Horlamus (2002). Air Classifying. In *Ullmann's Encyclopedia of Industrial Chemistry,* 6th ed. Weinheim, Germany: Wiley-VCH.

Norrgran, D. A. (2003). Cleaning Ceramic Feedstocks with Magnets. *Ceramic Industry,* v. 153(8), pp. 11–14.

Schmidt, P. (2002). Screening. In *Ullmann's Encyclopedia of Industrial Chemistry,* 6th ed. Weinheim, Germany: Wiley-VCH.

Trawinski, H. (2002). Hydrocyclones. In *Ullmann's Encyclopedia of Industrial Chemistry,* 6th ed. Weinheim, Germany: Wiley-VCH..

Venkatraman, P., F. S. Knoll, and J. E. Lawver (2003). Magnetic and Electrostatic Separation. In *Principles of Mineral Processing,* ed. M. C. Fuerstenau and K. N. Han, 221–244. Littleton, CO: Society of Mining, Metallurgy, and Exploration.

Wills, B. A. (1997). *Mineral Processing Technology,* 6th ed. Oxford: Butterworth and Heinemann.

Yarar, B. (2002). Flotation. In *Ullmann's Encyclopedia of Industrial Chemistry,* 6th ed. Weinheim, Germany: Wiley-VCH.

4

THE RAW MATERIALS INDUSTRY

It has been pointed out that the glass industry is necessarily linked to economic geology "from the inescapable fact that everything we put in the glass furnace is either a mineral or a mineral fuel, or derived from one of them" (Edwards, 1988). The same can be said for the ceramic industry. The business of mining and supplying materials to the glass and ceramic manufacturers is part of the commerce of industrial minerals, which incorporates aspects of economic geology, chemical engineering, and industrial economics. The term *industrial mineral* is not solidly defined, but generally includes all useful nonmetallic minerals with the exception of fossil fuels and gemstones (Bates, 1994). Much of what applies to the industrial minerals sector can be applied to the subsectors of glass and ceramic raw materials. In fact, the processing techniques described in the previous chapter can be applied to many materials in the broad field of extractive metallurgy and industrial minerals.

The cost of raw materials is obviously an important factor in all manufacturing operations, and the ceramic and glass sectors are no exception. Depending on the product, materials costs can be from 15 percent for high-volume glass and ceramic ware to a majority of the cost for specialty products made from highly refined and/or rare materials. For glass production, about 55 percent of the cost can be attributed to materials (DOE, 1996). There are many factors that are used to determine the value and eventual price. In this chapter, we present an overview of the business of evaluating, selling, and transporting raw materials.

4.1 DEMAND AND RESOURCE AVAILABILITY

Raw material suppliers in the free-market world respond predominantly to economic forces of supply and demand. Demand for minerals and industrial chemicals, which is the range of prices and amounts that industrial consumers are willing to buy, stimulates supply (Wagner et al., 2002). Raw materials companies then respond to the demand by providing a supply of the goods at the desired specifications. When demand increases, the price may increase or producers may supply more materials to the market at a stable price. If demand increases faster than supply, then shortages may occur until the supplies are replenished. The balance of supply and demand will then determine the price.

Because the ultimate source of the raw materials is the ground, it is the price determined by the market for a particular mineral that dictates whether a deposit is economically viable. High demand may spur the development of new deposits or reopen operations that had ceased. Alternatively, decreasing mineral demand can suppress prices, even to the point that some producers cease production. For some materials, the glass and ceramic industries are the largest users and therefore are the dominant drivers for demand. For instance, commercial glassmaking is the largest user of soda ash, so it dominates the demand for this material. Alternatively, some materials are used by other industries, and therefore, demand and price for these materials are divided between more than one consumer. An example is kaolin, which, in addition to being a major component of whitewares and fiberglass, is widely used as a filler in paper, plastics, and paints.

The demand for a raw material by an industry gives that material some intrinsic value within a market economy. Another important concept of mineral value is *place value* (Manning, 1995), which is analogous to the three real estate rules of "location, location, and location." Place value is a qualitative assessment that takes into account the monetary value for the commodity and the costs of transporting it to market. A high-place-value mineral has low intrinsic value and is expensive to transport; place dominates the delivered cost. Examples are sand and gravel. Low-place-value commodities, such as diamonds and high-purity kaolin, have high intrinsic value and the location of origin has little impact on cost. Geologic providence decides where a deposit is located, while market price and costs associated with mining, processing, and transportation determine if it is economically feasible to exploit.

There are several terms used to define the relative availability of mineral wealth (adapted from Wagner et al., 2002):

- *Resource.* A concentration of naturally occurring solid, liquid, or gaseous material in or on the Earth's crust in such form and amount that economic extraction of a commodity from the concentration is currently or potentially feasible.
- *Reserve base.* That part of an identified resource that meets specified minimum physical and chemical criteria related to current mining and production practices, including those for grade, quality, thickness, and depth. It may encompass those parts of the resources that have a reasonable potential for becoming economically available within planning horizons beyond those that assume proven technology and current economics. The reserve base includes those resources that are currently economic (reserves), resources that are marginally economic (marginal reserves), and some resources that are currently subeconomic (subeconomic resources).
- *Reserves.* That part of the reserve base that could be economically extracted or produced at the time of determination. The term *reserves* need not signify that extraction facilities are in place and operative. Reserves include only recoverable materials; thus, terms such as *extractable reserves* and *recoverable reserves* are redundant and are not a part of this classification system.

These definitions change with variation in value of a particular material, extraction and processing technologies, and new discoveries. In particular, processing technologies that can separate previously unusable resources can dramatically change the estimated reserves of a given material.

The dominant pricing factors are competitive economic forces, but geopolitical pressures can also affect availability. Due to the geographically heterogeneous distribution of mineral deposits throughout the world, some countries and regions are endowed with more mineral wealth than others, and more than one war has been fought for them, though not necessarily for most bulk glass and ceramic raw materials. Such drastic measures are kept for strategically important materials such as petroleum, which is greatly concentrated in the Middle East, and platinum group metals, of which almost 90 percent reside in South Africa. Other important geopolitical factors are tariffs and subsidies that can be used to protect some mining and chemical industries of a given country.

Environmental concerns can influence the availability of a resource. For example, if a feldspar deposit resides in an environmentally sensitive area, it may not be available for exploitation even if the deposit could be mined economically. This would be a situation where society values the wilderness over the economic worth.

4.2 RELATIONSHIP BETWEEN SUPPLIER AND MANUFACTURER

Historically, the manufacture of ceramic and glass products was intimately related to the industry of gathering raw materials to the extent that the producer located operations close to the raw material source. As methods of rapid transportation evolved, such as trains, highways, and large ships, manufacturing operations could be placed further away from the raw material sources. This has been particularly beneficial where the raw material batch requires several components from geographically distant sources.

There has been a great deal of change in the minerals industry over the past century. In the earlier part of the twentieth century, most manufacturers owned the mines to supply their operations. In some cases, some of the smaller entities became larger corporations through consolidation or the deposits were mined to exhaustion. Due to the latter, some nations found that they needed to find new resources outside of their borders. As a result, multinational corporations were formed to exploit the mineral wealth of other nations that had favorable conditions for mining and exporting. Thus began the movement of the industrial minerals industry to a more globalized market dominated by large companies.

This is not always the case, as manufacturers of lower-value ceramic products in both developed and developing countries may still operate their own in-house mines. In the recent past, several glass and ceramic manufacturing companies have chosen, at some point, to acquire their own mines and raw-material processing facilities. The advantages are control over cost, supply, and specifications and the ability to make long-term planning decisions. However, it necessarily increases the complexity and size of the company, and, in fact, this trend toward *backward integration* in North America seems to have ended in the mid-1990s as companies decided to concentrate on their core business and allow raw materials suppliers to take over the mineral business (Edwards and Harben, 1998).

Currently, most of the companies that supply the ceramic and glass industry are involved in the broader business of industrial minerals, that is, the mining, refining, and distribution of mineral products to manufacturers. Other suppliers involved in the ceramic and glass industry include distributors of industrial process gases, fuels, polymers, and recycled or waste materials.

4.3 PURCHASING

The manufacturer's decision of purchasing a raw material is based on several variables, including material price, transportation costs, and product specifications. The driving factor is usually price, and most companies would like to "obtain the lowest price material . . . from as close to the rock face as possible" (Sugden, 2002). In addition to costs, consistency of delivery and product are essential. Specifically, the material must be available on a regular basis, and the physical and chemical characteristics must not deviate from the desired specifications over time.

The price of a material often depends on where it is purchased, and there are several terms that should be defined. *Ex works* refers to the cost of the goods at the seller's site (factory, warehouse, etc.). Title to the material then passes at the factory door, and the purchaser is responsible for loading the goods onto the carrier and pays all freight and insurance charges from that point. *FOB* is an acronym for *Free on Board,* which refers to the price for which the seller undertakes the risk and expense to load the goods onto a carrier at a specified point and title passes after loading. *FOB destination* means that title passes when the material arrives at the factory and the seller is responsible for delivery. *FOB origin* means that ownership of the material is transferred to the buyer from the seller at the moment the goods are transferred to the carrier. *CIF* is an acronym for *cost insurance and freight,* which is an export term in which the seller pays all expenses involved in the delivery of material to an agreed port. The buyer pays for customs clearance and subsequent transportation.

The change in the glass and ceramic industries toward consolidation by larger, multinational companies has produced changes in raw material purchasing practices (Edwards and Harben, 1998). An example is the movement from local purchasing at the plant level to centralized purchasing, where a single main office negotiates the deal for the entire company with the individual plant managing the deliveries. There are positive and negative aspects to this practice. Single sourcing can result in lower prices, improved data sharing and quality control, and better strategic planning. The negative aspects of single sourcing include a reduction in flexibility in both price and supply.

The relationship between the manufacturer and the supplier is key to ensuring a smooth raw material program. Communication from the manufacturer lets the supplier know the desired specifications and pricing. At the same time, the supplier must communicate with the manufacturer any problems or changes that may occur. As in any commodity, raw materials can be purchased using a variety of mechanisms. Long-term contracts with a company can lock in a given price, and therefore protect the buyer from potential volatility in prices. This may be important for materials prone to short-term fluctuations in price, such as fuels.

4.4 TRANSPORTATION

A manufacturer needs to have all raw materials delivered on a prescribed schedule in order to maintain stocks and to avoid stoppages in works. This is particularly important to the large-volume manufacturer, such as a commercial glass plant, where it is not acceptable to stop production. Transportation contributes significantly to the cost of raw materials. In some instances, the transportation costs for large-volume, low-cost materials such as dolomite and sand can exceed the FOB or ex-works cost of the material.

Commodity materials are moved by an extensive network of rail, waterway, air, and road systems. *Intermodal transportation* is the combination of two or more methods of moving freight in order to reach distant market and distribution points. For obvious reasons, airplanes are used only for small-volume, very-high-value materials. Trucks can carry up to forty short tons of material in the United States, depending on state regulations and the material being carried. For bulk granular material such as sand and soda ash, rail delivery usually comes in covered hopper cars that discharge by gravity through an opening at the bottom of the car or with pneumatic attachments. Pneumatic discharging is done using either a vacuum system that sucks the material out or using positive pressure that blows the material out of the car. Individual railcars or groups of railcars may be owned by the supplier or may be leased from the rail company (Moore, 2003). Minerals may move in small groups of railcars in larger trains or, using soda ash as an example, involve the entire train. Barges and ocean-going ships can carry the largest amounts of material, but these are limited to delivering in ports or in navigable rivers.

Traditionally, rail is the more cost effective land-based method when travel distances exceed 300 miles and it is often the lowest-cost way to move high-volume materials, with each car capable of handling 100 to 120 short tons. Deregulation has been particularly important for the railroad industry in the United States. The Staggers Act of 1980 allowed railroads to negotiate directly with shippers for services, to set rates more readily, and to enter and exit markets with more freedom. This has resulted in an increase in overall efficiency of rail service and a substantial decrease in average rail rates.

Although higher in cost per ton delivered, trucking has several advantages over rail, at least in countries with well-maintained systems of highways. In fact, in some countries, such as some in South America, the rail systems are not well maintained and trucking is the only option. One particular advantage is an increase in competition among carriers, which can lead to reduced costs and better service. However, trucking costs are susceptible to fluctuations in the cost of diesel fuel and carriers will apply a fuel surcharge when prices increase. Other variables associated with truck transportation are the relative locations of the source and the manufacturing plant, the number of competing transportation companies, and the volume of back haul (i.e., the amount the truck can carry on its return trip) from the delivery location.

Increases in the cost of truck and rail and decreases in the cost of oceanic shipping, coupled with changes in international trade policies have made overseas products increasingly attractive (Grahl, 2003). However, the distances can have a negative effect in terms of the relationship between the supplier and buyer. In addition, unless the manufacturing site is close to a maritime port, the materials must still be shipped overland by truck or rail.

In many instances, railroad lines have a monopoly that can affect delivery, such that rail deliveries are on average 80 percent to 90 percent on time, requiring customers to increase their storage capacity (Buzby, 2003). Other costs associated with rail include the leasing of the cars and demurrage. *Demurrage* is a fee levied by the shipping company if the buyer does not unload the shipment by a specified date agreed upon by contract. Charges are usually assessed on a daily basis after the deadline. If the purchaser is not organized or inexperienced, charges can be in the thousands of dollars per month at a single plant. However, unloading quickly and before the negotiated time is usually not compensated for by a reduction in charges (although this may be the case if the material is being unloaded from a ship).

Transportation systems need to provide reliable service, but one must always have a contingency plan in the event that raw materials cannot be delivered by the usual route. For instance, if a company receives sand by rail and there is a delay due to congestion or a labor strike, the company may need to negotiate with a trucking company for short-term delivery. This may significantly increase the short-term transportation costs, but non-delivery is never acceptable.

REFERENCES

Bates, R. L. (1994). Overview of the Industrial Minerals. In *Industrial Minerals and Rocks,* 6th ed., ed. D. D. Carr, 3–5. Littleton, CO: Society for Mining, Metallurgy, and Exploration, Inc.

Buzby, S. (2003). Personal communication.

DOE (1996). Glass: A Clear Vision for a Bright Future. Washington, DC: U.S. Department of Energy. www.eere.energy.gov/industry/glass/pdfs/glass_vision.pdf.

Edwards, G. H., and P. W. Harben (1998). The North American Glass Industry—Fewer but Larger Players. In *Raw Materials for the Glass Industry,* ed. J. B. Griffiths, 4–10. London: Industrial Minerals. 10.

Edwards, G. H. (1988). The Application of Economic Geology in Glass Raw Materials: A Case History. In *Advances in the Fusion of Glass,* ed. D. F Bickford, 44.1–44.9. American Ceramic Society, Westerville, Ohio.

Grahl, C. L. (2003). Inside the Raw Materials Industry, *Ceramic Industry,* v. 153(10), pp. 25–28.

Manning, D. A. C. (1994). *Introduction to Industrial Minerals.* London: Chapman and Hall.

Moore, P. (2003). In Line with Minerals. *Industrial Minerals,* July, pp. 68–69.

Sugden, A. (2002). Changes and Developments of Non-Plastic Raw Materials for Whitewares. *International Ceramics,* Issue 1, pp. 9–16.

Wagner, L. A., D. E. Sullivan, and J. L. Sznopek (2002). Economic Drivers of Mineral Supply. U.S. Geological Survey Open-File Report 02-335, Reston, Virginia.

5

NATURALLY OCCURRING MINERALS

5.1 SILICA

Introduction

Silica (SiO_2) is one of the most widely used industrial compounds, and it can be considered the most important material in the manufacture of glass and ceramics. It is composed of the two most abundant elements on Earth, silicon and oxygen. It also happens to be the basic building block for all silicate minerals that are the major constituents of the solid earth. Free silica—that is, silica not bound within other silicate minerals—can be found in several crystallographic forms, depending on temperature and pressure. At atmospheric pressure, pure silica will crystallize into three mineral structures with increasing temperature: quartz, tridymite, and cristobalite. Quartz is the thermodynamically stable form found in natural deposits and is a major rock-forming mineral in many igneous, metamorphic, and sedimentary rocks. Tridymite and cristobalite are metastable under ambient conditions, and minor amounts are found in natural deposits.

Most commercial silica comes from sands that are mined as unconsolidated beach deposits, cemented sandstones, or a type of metamorphosed sandstone called quartzite. Lesser amounts of high-purity quartz come from pegmatites and hydrothermally deposited natural quartz crystals. Commercial silica is commonly called sand, quartz, flint, or, when ground to 200 to 325 mesh, it is referred to as flour. In all soda-lime-silicate and most commercial borate glasses, quartz is the major batch ingredient. In traditional ceramics, quartz is added as filler in a triaxial porcelain body. Silica flour is used in glazes as a glass-forming oxide. The resistance of silica minerals and silica glass to thermal shock, chemical attack, and mechanical wear allows them to be used in a variety of refractory products such as bricks, kiln and furnace liners to foundry sand. Silica sand and natural quartz crystal can be beneficiated to make a feedstock for refined silica products such as elemental silicon metal, silicon carbide, silica glass, and gels. Highly refined vitreous silica is used in optical and photonic applications. Synthetic quartz crystal used in electronic and optical applications can be made from high-purity quartz feedstocks.

Composition and Mineral Structure

Silicon (Si) is a semimetallic element that was first isolated in 1823 by the Swedish chemist Baron Jöns Jakob Berzelius. In its elemental form, silicon forms grayish-black crystals, but it is never found naturally in this state. Rather, it occurs in combination with oxygen, which, under the relatively low pressures in the earth's crust, form SiO_4 tetrahedra, the basic building block and fundamental structural element for all silicate minerals (Figure 5.1). The bonding between the silicon and oxygen atoms is partially ionic and partially covalent. Almost all silicate minerals are formed by the linking of SiO_4 tetrahedra through the sharing of the oxygen from one tetrahedron to the next. Most igneous rocks and metamorphic and, with the exception of carbonates, sedimentary rocks contain silicate minerals.

Minerals that are composed of pure silica at atmospheric pressure include the polymorphs (i.e., minerals with the same composition but different structure) quartz, cristobalite, and tridymite and their respective low (α) and high (β) subpolymorphs. The stability of these minerals is dependent on temperature and pressure (Figure 5.2). Table

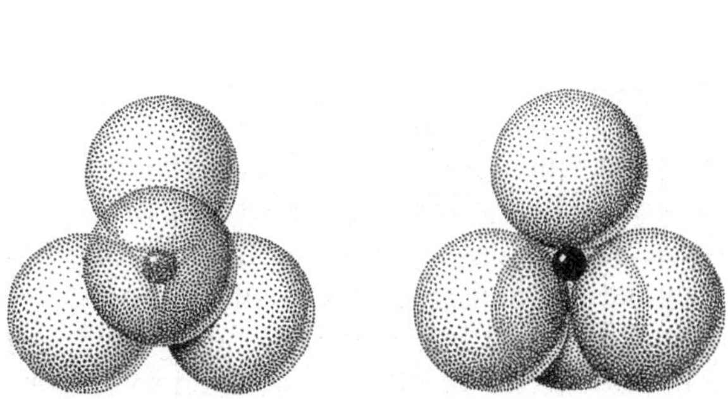

Figure 5.1 Schematic of silica tetrahedra with silicon atom at the center (from Klein and Hurlbut, 1993; used with permission from John Wiley & Sons).

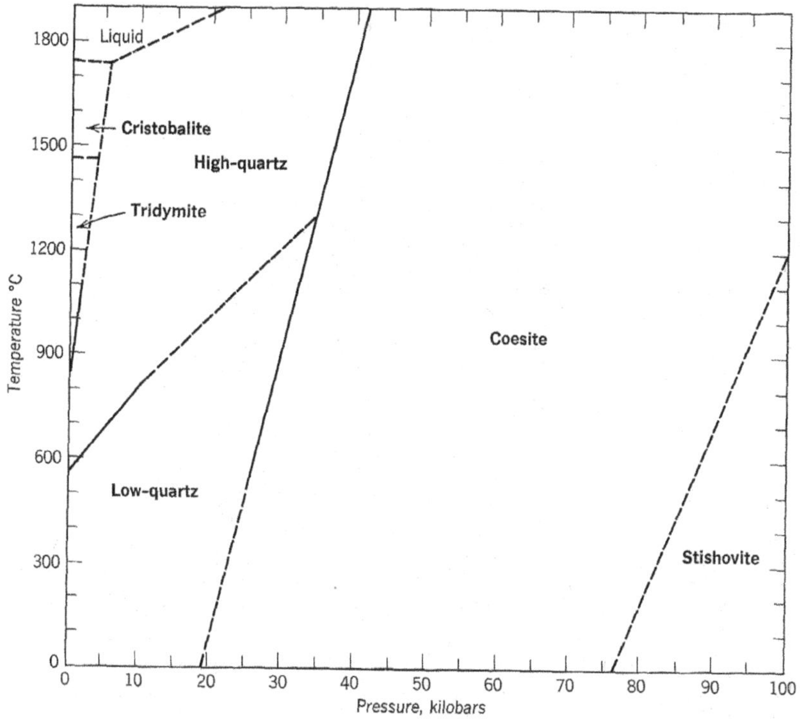

Figure 5.2 Stability field for silica polymorphs (from Klein and Hurlbut, 1993; used with permission from John Wiley & Sons).

5.1 lists the stability ranges and characteristics for each polymorph. Silica polymorphs generally have low thermal expansion coefficients (Table 5.2). Low quartz, which is stable under ambient conditions and the most common naturally occurring silica mineral, has a rhombohedral structure, while high quartz has a hexagonal structure (Figure 5.3). Both quartz structures are relatively complex with a spiral arrangement along the c-axis. The spiral can be either right-handed or left-handed, which is tested by the rotation of light propagating along the c-axis. The spiral structure does not allow the formation of a cleavage plane, so when quartz fractures, it has a characteristic conchoidal pattern (see Figure 5.4).

The higher-temperature polymorphs have greater lattice energies with more expanded structures, resulting in lower specific gravity and refractive indices (Klein and Hurlbut, 1993). High tridymite is made of sheets of silica tetrahedra that form a hexagonal ring structure with the tetrahedral apices alternately pointing up and down in an AB-AB-

Table 5.1 Silica minerals stable at atmospheric pressure

Mineral/Material	Stability Range	Specific Gravity	Hardness	Cell
Low quartz (α)	Up to 573°C	2.65	7	rhombohedral
High quartz (β)	573°C–870°C	2.53	7	hexagonal
Low tridymite (α)	Metastable 190°C–380°C	2.28	6–7	orthorhombic
High tridymite (β)	870°C–1470°C	2.22	NA	hexagonal
Low cristobalite (α)	Metastable 200°C–275°C	2.33	6–7	tetragonal
High cristobalite (β)	1470°C–1713°C melting point	2.20	NA	cubic
Vitreous silica	metastable	2.20	5.5–6.5	NA

Table 5.2 Thermal expansion of silica minerals: $\alpha_T = \alpha_o + \alpha_1 T + \alpha_2 T^{-1} + \alpha_3 T^{-2}$

Polymorph	$\alpha_o \times 10^5$	$\alpha_1 \times 10^8$	$\alpha_2 \times 10^4$	$\alpha_3 \times 10^2$
α quartz	2.7513	2.9868	0.05572	9.1181
β quartz	2.0604	3.4694	1.30754	−163.76
Tridymite	2.1797	−1.53	0	0
Cristobalite	0.5	0	0	0

Compiled by Ottonello (1997)

AB sequence (Figure 5.5a). High cristobalite, like tridymite, is made of sheets of silica tetrahedra that form a cubic structure with a stacking of silica tetrahedral layers in an ABC-ABC sequence (Figure 5.5b). Low cristobalite is formed during the cooling of high cristobalite.

The nature of the conversions between the three low-pressure polymorphs is very important. The reconstruction process is slow, and the actual temperatures at which reconstruction occurs can vary, depending on the rate of temperature change and whether the mineral is being heated or cooled. The sluggish nature of the polymorphic changes allows both tridymite and cristobalite to remain metastable at temperatures below their equilibrium temperature range. This is due to the need of breaking the Si-O bonds in order to complete the reconstruction.

It is important to recognize that the transformation temperatures are equilibrium temperatures, and as such, may not be applicable to in-

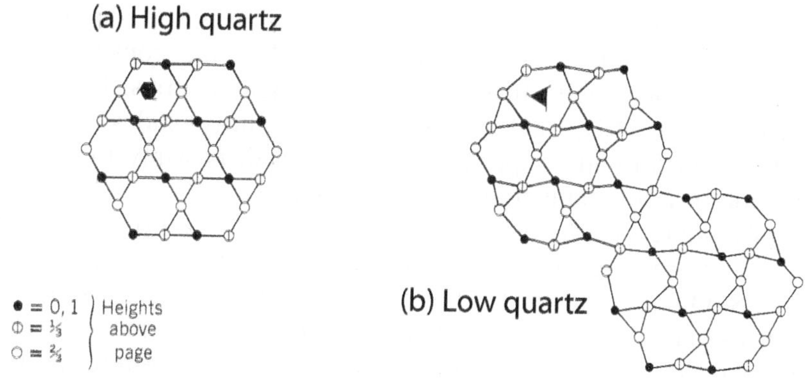

Figure 5.3 Framework of silica tetrahedral for (a) high quartz and (b) low quartz projected parallel to the c-axis (from Klein and Hurlbut, 1993; used with permission from John Wiley & Sons).

Figure 5.4 Secondary emission microscopy (SEM) photomicrograph of ground quartz. The scale bar is 10 μm. Note the conchoidal fracture surface (courtesy of B. Pinto).

dustrial situations where the system may never be or is rarely in equilibrium. For example, it is not uncommon to observe β-quartz convert directly to β-cristobalite at 1000°C. The reaction is still sluggish and it may take as long as 24 hours to complete the conversion, but tridymite is completely missing from the reaction series. If a flux is present, such as H_2O, CaO, Na_2O, or other impurity oxides, the reconstructive transformation appears to be hastened, possibly through the formation of a low-temperature glass phase.

Unlike the reconstructive polymorphic changes, the inversions within each polymorph are rapid and reversible as the process is *displacive,* that is, Si-O bonds are rotated but not broken. For instance, on heating, low or α-quartz undergoes an abrupt displacive transformation to high or β-quartz at 573°C. This transformation is commonly referred to as the *quartz inversion*. During this process, the Si-O-Si bond angle in α-quartz increases from 158° to 180° at about 573°C (see Figure 5.3). The O-Si-O bond angle within each SiO_4 tetrahedron remains constant at 105.8°, since it remains intact and does not change. Compared to

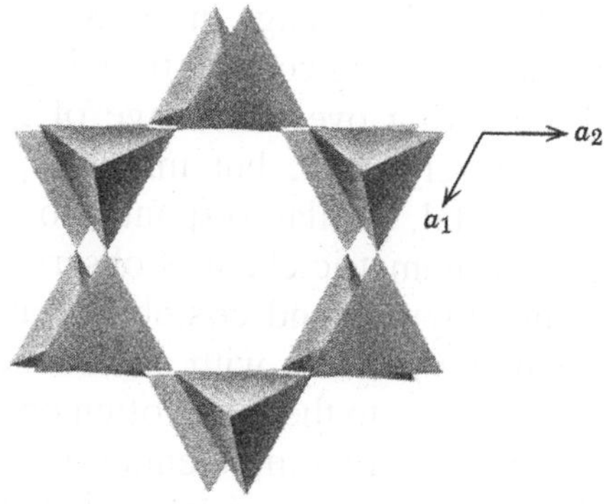

(a)

(b)

Figure 5.5 (a) Layers of silica tetrahedra of high tridymite projected onto (0001); (b) high-cristobalite structure projected onto (111) (from Klein and Hurlbut, 1993; used with permission from John Wiley & Sons).

α-quartz, β-quartz has greater symmetry and lower specific gravity, so the α-quartz to β-quartz inversion results in a volumetric increase. The volumetric change associated with the quartz inversion is sufficiently great to cause cracking of the quartz crystal on cooling provided the crystal is sufficiently large. As an example, a quartz crystal that was crack-free at room temperature can be heated above 573°C and, upon cooling, the displacive transformation causes the surface of the quartz crystal to contract (the volume is reduced on cooling) at exactly 573°C, but the interior of the crystal is still above 573°C (and therefore, still β-quartz), thus placing the surface in sufficient tension to cause spontaneous cracking. If the particle is cooled rapidly, the cracks can be observed to grow into the crystal as the sample cools (Carty, 2004). Tridymite and cristobalite exhibit displacive transformations between the high and low forms that are similar to quartz in that they are rapid and reversible and only entail a change in the Si-O-Si bond angle. The transformation temperature during cooling from high to low cristobalite is not well defined, occurring over the range of 225°C to 250°C for reasons that are not entirely clear, but may, in part, be due to the thermal history of the crystal. Similar to spontaneous cracking of quartz crystals on cooling, the volumetric changes observed during displacive transformations within tridymite and cristobalite are sufficiently large to cause cracking. A major concern with cristobalite is that the transformation temperature is close to the range often observed in household and commercial cooking, resulting in potential problems for cookware.

There are very few structural impurities within the crystal lattice of quartz with only limited substitution of Al^{3+} with an alkali ion such as sodium to balance the charge. Total impurity concentrations in natural quartz crystals are low at around 1000 ppm (Heaney, 1994). The more open structures of cristobalite and tridymite can allow slightly more opportunities for ionic substitution relative to quartz (Deer et al., 1992). Impurities in all polymorphs can be present as discrete inclusions that are not part of the crystal structure. Inclusions of some minerals, such as iron and titanium oxides, will give color to quartz, which is normally clear to colorless.

In addition to crystalline silica, there are several microcrystalline (cryptocrystalline) and amorphous forms of silica that are found in nature. These include diatomite, opal, chalcedony, and chert. *Diatomite* is a chalklike, friable sedimentary rock that is formed by the sedimentary accumulation of the tests or internal skeletons of diatoms, a type of plankton. *Opal* is a soft, hydrated amorphous silica material ($SiO_2 \cdot nH_2O$) with 5 percent to 10 percent water. *Chalcedony* is a brown to gray microcrystalline quartz deposited from aqueous solutions. *Chert* and flint are hard, sedimentary rocks consisting of chalcedony and opal.

Geological Occurrence and Distribution

The most common silica polymorph in the Earth's crust is α-quartz, and it is a primary rock-forming mineral, particularly in continental rocks such as granite (Heaney, 1994). During crystallization of a granitic magma, quartz will crystallize along with feldspar and a hydrous silicate mineral such as mica or amphibole. Exceptionally large and often pure quartz crystals can crystallize in pegmatites, so these features are a source of lump quartz. Quartz may recrystallize during metamorphism either within a mineral assemblage or as discrete veins.

Quartz is relatively insoluble in water under ambient conditions at the earth's surface and therefore silica is found in low concentrations in surface water and seawater (~1 ppm; Table 2.1). However, silica is much more soluble in subterranean or hydrothermal fluids that are subjected to high temperature and pressure. When these fluids are subjected to a drop in pressure and temperature, such as when they approach the surface, high-purity, single-crystal quartz can precipitate in veins and cavities (Figure 5.6a). Some of these are used as gems, such as the violet-colored amethysts. Clear quartz crystals that may have some imperfections are often called *lascas* and are used as a source of high-purity silica and seed crystals for synthesized quartz crystal (Figure 5.6b).

(a)

Figure 5.6 (a) Photograph of a clear quartz crystal from Brazil. The scale bar is one inch. (b) Synthetic quartz crystals are produced using a pressurized hydrothermal process (reprinted with permission from Sawyer Technical Materials, LLC).

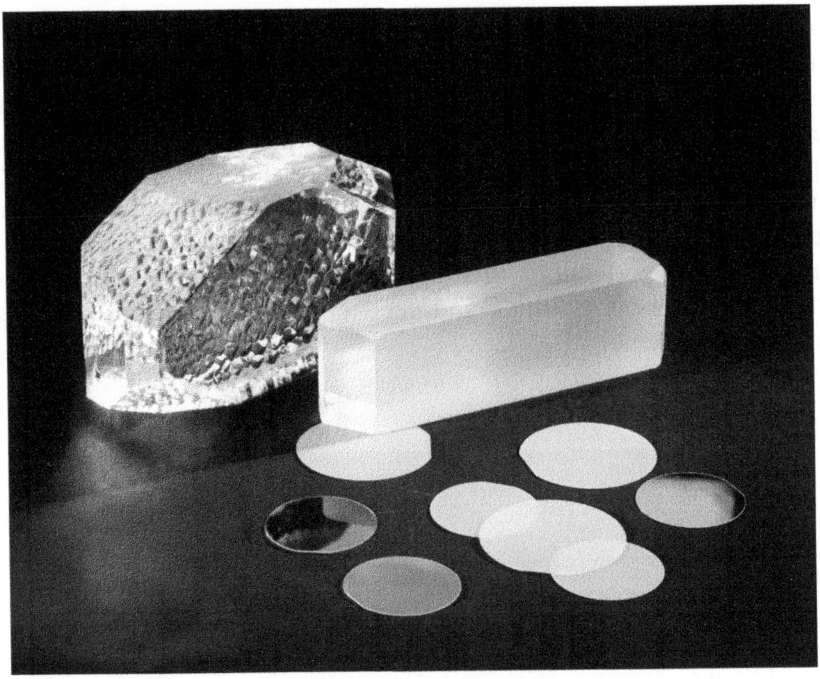

(b)

Figure 5.6 (*Continued*)

The relative insolubility of quartz at the Earth's surface and its strong chemical bonds make it very resistant to weathering. During the decomposition of continental rocks that contain free quartz, such as granite, quartz will remain intact while the other minerals (feldspars and mica) will breakdown. If the quartz is eroded and transported, it will collect in sand deposits such as beaches or river deltas and channels. Repeated reworking of the sand by water breaks up and removes any less-resistant minerals. Over time, this purifies the sand, although other resistant minerals will remain, such as ilmenite, zircon, chromite, and the aluminosilicate minerals sillimanite, andalusite, and kyanite. In general, sand deposits that are the result of several cycles of erosion and deposition have the highest purity. Quartz grain morphology differs between and often within sand deposits, depending on the environmental conditions under which the deposit was formed. Grains may be sharp and angular, subangular, subrounded, or rounded.

If the sand is buried by successive layers of sand or other sediments, the sand will be compacted by the weight of the overlying material. As groundwater moves through the voids between the sand grains, a cementitious material such as silica or calcite will precipitate. When cemented (lithified), the deposit becomes a sandstone. In order to be used commercially, sandstones must be cemented by silica, as those cemented by calcite are undesirable. Sandstones are classified by the content of silica and impurities such as feldspar and mica. Orthoquartzite is composed of at least 95 percent quartz. Arkosic sandstones contain feldspar, and greywacke refers to a sandstone with feldspar, clay, and rock pieces. Orthoquartzite is the only variety that is useful as a commercial product (Harben and Bates, 1984). *Gannister* is a term used for a type of fine-grained sandstone that is associated with coal deposits (Worrall, 1982). *Tripoli* is formed by leaching calcareous cherts or siliceous limestones, which removes the calcia and leaves the silica.

If a sandstone deposit is buried further by continued sedimentation or tectonic processes, it may be subjected to heat and pressure, which will convert the quartz sandstone to a dense *quartzite*. The quartz grains may be strained from metamorphism, and therefore when heated, the quartz grains from a quartzite may transform to cristobalite more rapidly than quartz sand (Baumgart et al., 1984). Unlithified beach and glacial deposits can be used as a source of quartz, particularly where quartz sandstone is not readily available. Although deposits are very common in all countries, sands should be at least 95 percent silica and free of refractory iron or titanium minerals in order to be of value. Lower-purity deposits require more beneficiation to remove the unwanted minerals, thereby adding to the cost. However, feldspathic sands that contain few iron-bearing minerals can be processed to obtain both feldspar and quartz where suitable local feldspar and sandstone deposits are not available. The processing of feldspar mined from pegmatites (see section 5.4) produces very high purity quartz as a byproduct. This material is generally more expensive than sandstone or quartzite, but it is useful where very high purity is desired.

Silica sand is found all over the globe (Figure 5.7), and the United States is the largest producer and consumer (Dolley, 2002). Glassmaking is one of the larger consumers of the more pure quartz sands. The bulk of quartz sand used in the United States glass industry comes from blanket sand deposits that were formed hundreds of millions of years ago. These were originally marine beaches and dunes that formed after the erosion of older sandstones, and the constant reworking has resulting in exceptionally pure sandstones. The two most valuable are the Oriskany Sandstone (West Virginia and Pennsylvania) and the St. Peter Sandstone (Illinois and Missouri). The St. Peter sandstone is of early Middle Ordovician age and is exposed from Minnesota as far

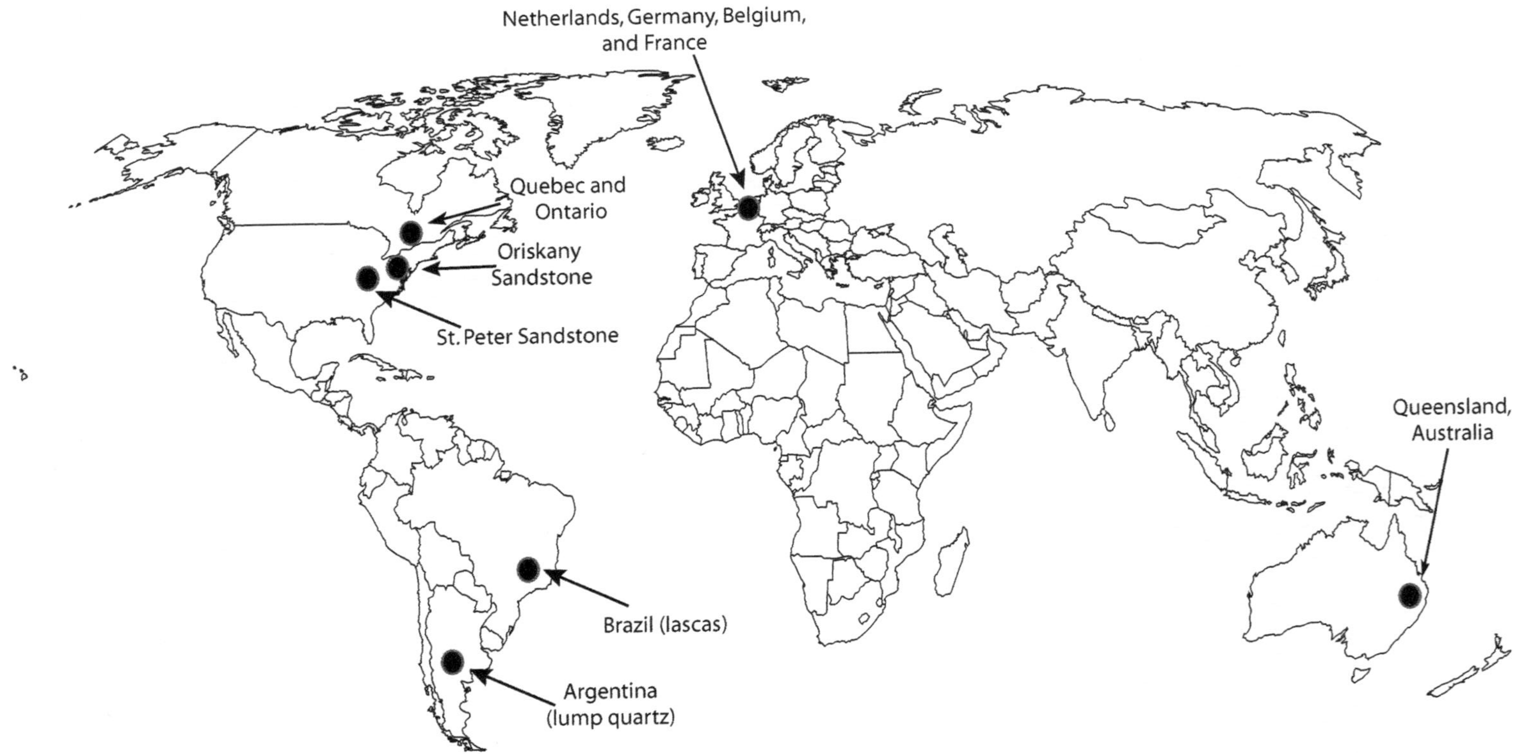

Figure 5.7 Global map showing some of the larger deposits of silica used in the glass and ceramic industry. Because of its general low value, bulk sand is not shipped great distances.

south as Missouri, with an average thickness of 20 m (Chang, 2002; Zdunczyk and Linkous, 1994). The Oriskany sandstone is the premier sandstone of the Appalachian Belt. It is a very hard marine deposit that is exposed from New York to Virginia (Zdunczyk and Linkous, 1994). Younger, unconsolidated sand deposits are found along the Atlantic coastal plain. For instance, the Miocene age Cohansey Formation of southern New Jersey is a 10 to 25 meter-thick deposit of subangular sand mixed with minor amounts of clay, coarse sand, and heavy minerals (Zdunczyk and Linkous, 1994). Feldspathic sands mined along the California coast are normally upgraded with sand transported from Nevada, such as the Eureka Quartzite, or other sandstone deposits in the central United States.

In other regions of North America, silica sand is mined in Mexico from marine sand deposits along the coast of the Gulf of Mexico. Most sand for use in glass comes from the states of Veracruz and Tamaulipas. In Canada, silica sand is produced in most provinces (Dumont, 2001). In Quebec, silica is mined from a quartzite deposit at Saint-Donat-de-Montcalm and from a sandstone deposit at Saint-Canut. Other sites include Bécancour and La Mabaie. In Ontario, quartzite is mined from Badgeley Island in northern Georgian Bay and is processed to the south at Midland. In Alberta, sand is mined from sand dunes in the Bruderheim region. In British Columbia, a high purity silica is found at the Moberley mine in the Golden area.

Sand deposits are found all over Europe, although the highest-quality deposits are mined in Belgium, France, the Netherlands, and Germany (Loughbrough, 1993; Flörke and Graetsch, 2002). In the Netherlands, sand is dredged from the bottom of the North Sea (Kesler, 1994). In the United Kingdom, silica sand comes from both unconsolidated sand and hard sandstone deposits (Burrows and Broad, 2000). Flint is found in the Upper and Middle Chalk of the United Kingdom (Worrall, 1982).

In South America, the majority of industrial sand comes from unconsolidated beach and river deposits. In Argentina, high-purity lump quartz comes from the processing of a group of pegmatites in the province of San Luis. Brazil is known as a major source of quartz-crystal lascas and gems such as amethyst. It also has river sand deposits to supply its own glass industries.

In Asia, much of the silica is provided by local sources. An exception is Japan, where only two-thirds of consumption is satisfied within the country. The major source of high-purity sand imports is Australia (Loughbrough, 1993). In that country, unconsolidated beach sand deposits are found along the Queensland coastline north of Brisbane. The sand occurs in Pleistocene to Holocene sand deposits that have been heavily weathered and leached of alkali and iron-bearing materials. The major deposits include the southern Queensland islands of North Stradbroke, Bribie, Moreton, and Fraser Island and the coastal dune systems of Beachmere, Peregian-Noosa, Cooloola, Byfield-Shoalwater Bay, Cape Flattery and Shelburne Bay, and Mourilyan (Bruvel, 1999). Almost 90 percent of the Queensland sand production is exported to Asian countries such as Japan, Korea, Taiwan, and the Philippines.

Mining and Processing

Quartz Sand Quartz sand has a high place value (see Chapter 4); that is, it is a relatively low-value commodity so that the source of sand is close to the end user to minimize transportation costs. Because each deposit has a unique set of physical and compositional characteristics, the mining and processing steps for each site are unique (Paige, 2004). Aside from the nature of the deposit (i.e., sandstone, unconsolidated, etc.), some of the characteristics that determine the beneficiation process include the particle-size distribution and particle morphology, the presence of coatings on the sand particles, and the occurrence of any undesirable minerals. It is important to know if the undesirable minerals are present as discrete particles or as coatings and inclusions in the quartz. Typical gangue consists of minerals that are resistant to weathering and corrosion and, unfortunately, many of these contain iron and titanium that can contribute to unwanted coloring of glass or contribute to unmelted stones. These minerals include limonite (a hydrated iron oxide), ilmenite ($FeO\text{-}TiO_2$), rutile (TiO_2), and magnetite (Fe_2O_3). Clays can also be present, although small amounts are not as much of a concern because aluminum and silicon in clay can be part of the gross batch composition. Refractory minerals such as zircon ($ZrSiO_4$), chromite (a Cr-Mg-Fe-Al spinel oxide), corundum (Al_2O_3), and the Al_2SiO_5 minerals andalusite, kyanite, and sillimanite are extremely undesirable because they will not melt and can cause significant defects in a glass. Refractory minerals are measured as the number of grains per unit mass of sand. For instance, if a sand shipment contains one grain of refractory mineral per 500 grams, a typical glass plant using 350 tpd of sand would have in excess of 700,000 defects per day (Sadowski, 2003).

Virtually all silica sand is mined by open pit or by dredging using standard mining equipment. Some high-purity quartz veins and crystals may be mined by underground methods. Unconsolidated deposits below water are gathered using a suction dredge and less often with a dragline. Dry, unconsolidated sands may be dug with power shovels or loosened with high-pressure water hoses. Sandstones are drilled and blasted to initially loosen the material so that it can be transported to a crushing facility, usually located close to the mine.

The overall goal of processing quartz sand is to clean the grains, increase the overall purity of the product by removing gangue, and classify into the desired grain size distribution. Consolidated sandstones need to be crushed to free the quartz particles. For very pure deposits, simply washing followed by sizing is sufficient to produce a marketable product. Attrition scrubbing is used to remove clay or silt particles that have adhered to the sand grains (Sadowski, 2003). In this process, the sand is combined with water to make a slurry of up to 75 percent solids and then agitated by an impeller in a tank for several minutes. This causes the particles to collide which abrades the clays and silts off the quartz. The process may be done in a series of tanks, depending on the amount of clay.

After washing and scrubbing, the fines ($<100\ \mu m$) need to be removed from the slurry, a process that is referred to as desliming. The coarse fraction (>0.5 mm) is removed using screens while the finer particles are removed using a hydrocyclone. For sands that have more than 5 percent fines, a hydrosizer may be used (Sadowski, 2003). For particles with surface coatings and staining, an acid wash of sulfuric acid added into a heated, rotating drum can be used (Zdunczyk and Linkous, 1994; Eurosil, 2003). In British sands, this extra step can reduce iron levels from 0.2 percent to 0.030 percent (Burrows and Broad, 2000).

The physical separation of impurity minerals can be accomplished through several processing steps. Magnetic and paramagnetic minerals such as iron oxides can be removed using magnetic separation techniques. Mica, feldspar, and other nonmagnetic materials can be removed by gravity separation or froth flotation.

The drying of quartz sand is a two step process of first removing the excess water by storing in a drain shed or passing the material over a vacuum table followed by thermal drying (Paige, 2004). Thermal drying is one of the most expensive steps in sand processing. It is done using either a rotary drier or a fluidized bed drier. After drying, sand is sized using screens or cyclones, then conveyed to storage silos.

For the manufacture of silica flour, final grinding is done in a rod or ball mill, with some products ground with ceramic media to reduce the potential for metal contamination.

Quartz Crystal Natural quartz crystal can be found where it precipitates in veins and cavities from hydrothermal fluids. Natural crystals are usually mined and sorted by hand from both open pit and underground mines. However, synthetic, high-purity crystals made using a pressurized hydrothermal process have begun to replace natural quartz. Below 300°C, increased pressure moderately increases silica solubility in water and added salts have little effect (Fournier, 1985). However, above 300°C, increased pressure and the addition of salts can greatly increase the solubility of silica. This behavior is taken advantage of to make synthetic quartz crystals. In this process, natural quartz lascas (the nutrient) are placed in the bottom of a steel autoclave that is partially filled with a solution of deionized water and a mineralizer, either sodium hydroxide, sodium carbonate, and lithium salts (Figure 5.8).

Figure 5.8 A technician unloads quartz crystals grown using a pressurized hydrothermal process (reprinted with permission from Sawyer Technical Materials, LLC).

The mineralizer helps increase the solubility of silica into the fluid. Only a small amount of mineralizer is required. For example, lithium is added in concentrations of 0.01 to 0.10 moles/liter (Sawyer Technical Materials, 1999). An initial crystal is needed on which the growing crystal can nucleate, so thin seed crystals of about 1 mm are suspended in the upper part of vessel. The autoclave is brought up to 300°C to 400°C and 140 MPa, with heat transferred from the bottom of the vessel. A vertical temperature gradient is maintained where the top is 40°C to 50°C lower than the bottom (Wenk and Bulakh, 2004). The solution becomes saturated in silica at the bottom and rises by convection to the cooler top, causing crystallization on the seed crystals. Conditions are set to allow crystal growth of about 1 mm per day to ensure that the crystal is free of defects, with crystal growth monitored using X-rays. The growing process requires weeks to months for completion. The resulting crystal should be free of inclusions and defects that will reduce its performance and, therefore, value. Incorrect growth rates can cause structural defects while impurity inclusions can result from reactions between the dissolved quartz, the steel vessel surfaces and the mineral salts in the solution that form complex alkali ferro-silicate phases (Sawyer Technical Materials, 1999).

Synthesized Silica Products

Sodium silicate is made by reacting sand with soda ash ($NaCO_3$) or sodium hydroxide (NaOH). Sodium silicate is used as a feedstock for the production for precipitated silica and silica gels. Precipitated silica is made by reacting sodium silicate with either hydrochloric acid (HCl) or sulfuric acid (H_2SO_4) to precipitate minute silica particles (Harben, 2002). Upon drying, the material has a particle-size distribution between 100 and 1 μm. Silica gel or colloidal silica can also be made by reacting sodium silicate with an acid to produce a suspension of fine silica particles. An alternative process is to pass sodium silicate through an ion exchange system to remove the sodium. In either case, the colloidal suspension is set and washed to make a hydrogel. If it is dried and ground, a silica gel is produced. If the hydrogel structure remains intact on drying, it is an aerogel. If the aerogel is partially collapsed, it becomes a xerogel (Harben, 2002).

Synthesized cristobalite is produced by calcining high-purity quartz at high temperature in a rotary kiln. This converts the quartz to its high-temperature polymorph, which is quenched and the cristobalite remains in a metastable state. Synthesized cristobalite consists of small octahedral crystals, which aggregate into rounded particles. In the United Kingdom, flint is calcined to about 1100°C to remove organic impurities and to convert some of the quartz to cristobalite for a final product that is 8 percent to 26 percent cristobalite (Sugden, 2001).

Fused quartz glass is optically transparent, extremely durable with a Mohs hardness of 5.5 to 6.5, and has a low thermal expansion coefficient of 5.4 to 5.6 × 10^{-7} °C^{-1}. It can be doped with titania to give it a zero expansion coefficient. There are several manufacturing methods that use beneficiated quartz sand, lascas, or chemically purified silicon tetrachloride ($SiCl_4$) as the starting component. There are four general methods using sand or lascas: carbon electrode, carbon arc, plasma arc, or gas-fired continual extrusion (Harben, 2002). In the carbon electrode method, which is the most common, silica is placed in a furnace with two carbon electrodes. An electrical current is introduced and the center of the silica batch is raised to 2300°C, which forms an egg-shaped center of fused silica surrounded by cristobalite. The mass is cooled and the fused silica is separated. Approximately 1,500 kWh of power is required per ton of fused silica produced. High-purity fused silica can be made by feeding quartz into an oxygen-hydrogen flame along with chlorine. The chlorine will react with impurity metals to make volatile chlorides, which are subsequently removed (Beall, 1994). Another method for manufacturing very-high-purity vitreous silica is through vapor phase hydrolysis. In this process, silicon tetrachloride is oxidized in a flame to produce SiO_2.

Uses and Specifications

Silica is used in a wide variety of glass and ceramic applications. Most commercial soda-lime-silicate glasses and ceramic glazes have silica as the major network former, and silica is a common filler in porcelain bodies. Because of the variety of applications, there is a wide spectrum of chemical and physical specifications for silica raw materials. Purity, particle size and shape, and particle-size distribution are all important. Increases in purity and decreases in grain size will both cause a significant increase in raw material cost.

Glass Silica makes up more than 70 percent by weight of typical soda-lime-silicate glasses and 55 percent by weight of reinforcement fiberglass. As a result, glass manufacturing is one of the largest users of high-purity silica sand and accounts for about 40 percent of sand produced in the United States. It is also the most exacting in terms of purity and consistency of product (Paige, 2004). Purity and consistency are the most important factors and specifications refer to composition and size distribution of the grain. Iron, titanium, and chromium are the impurity elements of most concern because of their ability to color the glass. The levels of aluminum, magnesium, calcium, and the alkalis, which are present in most container and flat glasses, should be known and kept consistent because they can affect the melting properties and final composition of the glass. Refractory minerals such as those mentioned in the previous section should be kept as low as possible to prevent defects.

The particle-size distribution is critical to ensure consistent batching and melting. The desired particle-size distribution depends on the glass product and the design of the melter. Larger or oversize grain sizes may remain unmelted and may produce stones in the final glass. Finer particles smaller than 140 U.S. mesh, referred to as fines, may be carried into the regenerators or lost as dust. Most vendors try to keep fines to less than 5 percent.

Grain size can affect the composition and melting behavior of the batch as well. For example, iron is often associated with the finer particle-size fraction relative to the coarse fraction (Paige, 2004). Particle morphology can have some effect on melting rate. Angular or irregularly shaped grains have more surface area than rounded grains, allowing for faster reaction and melting rates.

There are several national standard specifications for glass sand, although most manufacturers set their own standard sizes and purity levels in response to client demand. Published specifications for the United States, United Kingdom, and Brazil are shown in Table 5.3. Purity for silica used in glass manufacture can be tested using the ASTM C146-94a standard, which includes the determination of silica and of total alkali and alkaline earth oxides, and the separate determination of iron, titanium, chromium, zirconium, and loss on ignition (ASTM, 2002). Examples of commercial glass sand compositions are given in Table 5.4.

For flat glass, the purity level of sand should be at least 97.5 percent SiO_2 but is normally greater than 99.7 percent. It must not vary by more than 0.3 percent from batch to batch. Fe_2O_3 must be less than 1.4 percent (though most commercial sands are <0.5 percent) and Al_2O_3 should be less than 0.5 percent and not vary more than 0.1 percent (Loughbrough, 1993). The grain size of flat glass sand should be between 40 and 140 U.S. mesh (420–103 μm) (IMI, 2000). Average particle sizes are in the range of 250 μm to 325 μm. Particles between

Table 5.3 Specifications for silica used in glass manufacturing

	SiO_2	Al_2O_3	Fe_2O_3
General Specifications			
Flat glass	>99.5%	<0.3%	<0.04%
Container flint glass	>98.5%	<0.5%	<0.035%
E-glass	>99.0%		0.1%
Insulation fiberglass	>95.5%	<2.2%	<0.3%
Porcelain	>97.5%	<0.55%	<0.2%
Enamels	>97.5%	<0.55%	<0.02%
UK Specifications (BS 2975)			
Grade A Optical and ophthalmic glass	>99.7%		<0.013%
Grade B Tableware and lead crystal glass	>99.6%		<0.010%
Grade C Borosilicate glass	>99.6%		<0.010%
Grade D Colorless container glass	>99.8%		<0.03%
Grade E Clear flat glass	>99.0%		<0.10%
Grade F Colored container glass	>97.0%		<0.25%
Grade G Insulating fiberglass	>94.5%		<0.3%
Brazil Standards (ATBIAV)			
Type A: Special glass, including optical	>99.50%	<0.20%	<0.002%
Type B: High-quality clear glass, including crystal	>99.50%	<0.20%	<0.015%
Type C: Clear glass, including container and flat	>99.40%	<0.30%	<0.03%
Type D: Colored glass	>99.0%	<0.50%	<0.15%

Harben, 2002; Zdunczyk and Linkous, 1993; British Standard 2975, 1988

500 μm and 750 μm should be less than 5 percent, and none should be greater than 750 μm

Container glass specifications for purity can be less stringent than flat glass, particularly for colored containers. Sands with minimum of 95 percent SiO_2 can be used, but generally higher purity levels of greater than 99.5 percent are desirable for most applications. Fe_2O_3 must not exceed 0.06 percent for a clear flint glass (Chang, 2002), but can be present in amounts up to 0.25 percent to 0.35 percent for amber glass. Al_2O_3 should be less than 0.5 percent. The particle-size specifications are similar to those for flat glass.

In continuous fiberglass production, specifications for size and composition are very strict. Ground silica flour is used with all particles

Table 5.4 Specifications for some commercial silica products

	UK (wt.%)[1]	USA (wt.%)[1]	Fused Silica[3]
SiO_2	97.30	99.70	>99.7 wt.%
Fe_2O_3	0.101	0.031	<280 ppm
Al_2O_3	1.46	0.11	<1700 ppm
K_2O	0.88	0.02	<120 ppm
Na_2O	0.11	<0.01	<50 ppm
CaO		0.01	
Cr_2O	0.0003		
TiO_2	0.04	0.02	<250 ppm
LOI	0.28	0.1	
Morphology	rounded	subangular	

[1] Chelford high silica sand from Pleistocene glacial sediments—WBB Minerals
[2] Mapleton #1 dry silica—U.S. Silica
[3] Minco Min-Sil

smaller than 75 μm and the majority of particles less than 53 μm. The grinding should be done with ceramic grinding media to ensure that no steel is introduced as a contaminant. For E glass, which is the most common fiberglass composition used for textile or reinforcement fibers, SiO_2 should be greater than 99 percent and Fe_2O_3 should be less than 0.1 percent (Loughbrough, 1993). The requirements for discontinuous mineral wools such as those used in thermal insulation are less stringent, and often silica sources are from natural basalt and diabase or blast furnace slag (Dolley, 2002).

Silica Glass Pure silica glass is often referred to as fused quartz or fused silica. In the past, these two terms were used to distinguish between clear (fused quartz) and opaque (fused silica) silica glass. Now, fused quartz refers to glass that is produced by melting naturally occurring crystalline quartz, and fused silica refers to glass made from chemically purified silica.

Glasses made from pure silica are used in a variety of optical and electronic applications (refractories are discussed in the following sections). These uses take advantage of the many properties of pure silica, such as corrosion resistance and low thermal expansion. Vitreous silica has very low thermal expansion up to 1000°C, although above 1050°C it can convert to cristobalite (Routschka, 2002). The low thermal expansion of silica glass makes it particularly useful in telescope mirrors and other optical devices where dimensional stability is critical.

Optical-fiber-grade vitreous silica must be ultra-pure and completely free of any transition metal contamination that may attenuate the light signal. Fused silica is used in the manufacture of halogen and high-intensity discharge lamp tubes. The high-temperature environment attained during the operation of these light sources requires an optically transparent material that can withstand heat and thermal shock.

In the semiconductor industry, fused quartz tubing is used for furnace chambers in the manufacturing of silicon wafers. High-purity silicon is melted in a fused silica crucible, and a single crystal is pulled from the melt. Grades of silica used in this application can be greater than 99.999 percent SiO_2. The very stringent purity requirements call for total contaminant levels of less than 25 ppm and total alkali levels less than 1 ppm. Low hydroxyl and alkali contents are critical because these are extremely detrimental to the performance of silicon devices. Hydroxyl can lead to nucleation of cristobalite in the crucible and the subsequent devitrification of the fused quartz. Devitrification will reduce strength and possibly lead to mechanical failure.

Quartz Crystal Although many optical devices use silica glass, there are several key applications for quartz crystal in optical and electronic products. The high electromagnetic radiation transparency of quartz makes it useful for optical devices such as Brewster windows and prisms, polarizers, ultraviolet windows, birefringent filters, and tuning elements for laser optics. Quartz is a piezoelectric material; that is, when exposed to an electric field, the crystal undergoes strain. Conversely, when the crystal is placed under strain it produces a weak electric field. It is this property that allows the use of quartz crystal in oscillator mechanisms in watches, radio receivers, and electronic filters.

There are several grades of quartz crystal, depending on the number of inclusions and crystal dislocations. Quartz used in optical applications can be found in laser grade, optical grade, and near optical grade based on its homogeneity and inclusion content. The International Electrotechnical Commission sets standards for grading quartz crystal (standard EIC 60122 and others) used in electronic devices. Quartz resonators can be characterized by an electrical quality factor (Q) which is a measure of the efficiency of the resonator in converting between electrical and mechanical energy (Sawyer Technical Materials, 1997). The number of impurities in the quartz is a major factor of the overall resonator Q, and this component may be referred to as *material Q*. Inclusion counts per unit volume may be reported. The presence of dislocations within the crystal structure can be measured by etch channel density, in which a representative slice of quartz crystal is etched in ammonium biflouride and inspected with a microscope. The number of etch channels per unit area relates directly to the grade of product.

Refractories The low thermal expansion and high melting temperature of silica, coupled with its low cost, make it an excellent material in a variety of refractory applications. However, the temperature dependent polymorphism and its associated volume changes need to be kept in mind. In foundry applications, quartz sand is used in molds and cores for casting. The refractoriness, thermal shock resistance, cohesiveness when moistened, and high-temperature strength of the sand are the most important properties in casting (Flörke and Graetsch, 2002). The refractoriness, or the temperature at which grain sintering begins, is very dependent on the mineralogy with the higher quartz content materials having the highest level of refractoriness. For example, the presence of feldspar, which acts as a flux, can dramatically lower the sintering temperature. Particle-size distribution, grain shape, and surface area are other important factors. Coarse sand has a lower specific surface area and requires lower binder levels, but this must be balanced with the fact that fine sand gives a smoother surface to the casting, thus reducing the amount of machining required for the finished product. Well-rounded grains increase permeability and allow the escape of gases that build up during the casting process. Foundry sand grades are based on weight percent of quartz (with the rest as clay and feldspar), which relates to temperature service: more than 99 percent quartz for temperatures higher than 1500°C; 95 percent to 99 percent for 1500° to 1300°C; and less than 95 percent quartz for lower than 1300°C. As examples, cast-iron molds must contain more than 85 percent SiO_2 and steel casting molds require 95 percent SiO_2 (Flörke and Graetsch, 2002).

Quartz is not suitable for refractory brick that is subjected to heating and cooling cycles due to the quartz inversion. Therefore, refractory brick is made of either cristobalite or fused silica. Cristobalite brick is made by firing quartz sand mixed with 1.5 percent to 2.0 percent CaO at 1450°C for up to 14 days (Harben, 2002). The desired transformation of the α-quartz to cristobalite in the brick is strongly affected by the source of silica. Quartzite transforms much more rapidly to cristobalite than sand or sandstone (Baumgart, 1984). However, the rate of transformation can differ among quartzites, depending on crystal size, impurities, straining of crystals as a result of metamorphism, and amount of cryptocrystalline cementitious matrix. The ideal quartzite would have fine grains of strained crystals cemented with cryptocrystalline silica. Grain boundary impurities may cause the formation of lower-density tridymite between 870°C and 1470°C.

The specifications of silica brick depend on the service environment and in general, sandstone or quartzite of 96 to 98 percent SiO_2 is used. The combined total of Al_2O_3 and TiO_2 should not exceed 2.5 percent or, for high grade, 1 percent. Alkali elements, which would significantly lower the refractory service, should not exceed 0.1 percent. Ideally, the total alumina and alkalis should be less than 5 percent (Keegan, 1998). CaO should be below 0.3 percent and MgO below 0.1 percent. The amount of iron oxides should be low, particularly for those used in carbonizing industries, such as coke ovens, because iron acts as a catalyst that will affect the brick (Baumgart, 1984).

Fused quartz or fused silica is used in investment casting, kiln walls, and glass tank crowns. In these application, the SiO_2 content should be 99.4 percent to 99.9 percent (Table 5.4), with commercial material available in particle sizes of 4 mesh to smaller than 5 μm (Keegan, 1998). Fused silica brick has extremely low thermal expansion and is very resistance to acids. It can be used in applications with stable temperatures up to 1650°C and in applications with cyclical temperatures up to 1094°C. Above 1094°C, fused silica may devitrify and begin to form cristobalite, which will reduce the structural integrity of the brick.

Porcelain In traditional porcelain ceramic manufacturing, silica is used in both the body and glazes. In a typical whiteware body, silica is used as filler in amounts of 10 percent to 40 percent of the batch. The silica reacts with the flux to form a glassy phase that provides strength and durability to the ware. The silica is finely ground silica sand, calcined silica, or calcined flint (potter's flint). For glazes, silica flour, used in amounts of up to 25 percent to 45 percent of the glaze formulation, serves to increase the viscosity and durability and decrease the thermal expansion. The silica flour should contain more than 97.5 percent SiO_2, less than 0.55 percent Al_2O_3, and less than 0.2 percent Fe_2O_3 (Loughbrough, 1993).

Quartz in porcelain can be a source of mechanical weakness due to the presence of residual quartz in the fired bodies (Carty and Pinto, 2002). Some studies relate the weakness to the β- to α-quartz inversion at 573°C that occurs on cooling, leading to residual stresses in the matrix and ultimately to cracking within the microstructure (e.g., Warshaw and Seider, 1967). There appears to be a link between the particle size of the quartz used in the body and the strength of quartz-based porcelain with a maximum strength in samples containing 26 μm quartz particles (Carty and Pinto, 2002). The problem of residual quartz can also be addressed in part by substituting some of the silica with alumina (e.g., Austin et al., 1946; Mostafa, 2003).

Health and Environmental Concerns

In the United States, silica is regulated under the Occupational Safety and Health Administration (OSHA) and handling of it must comply with all regulations concerning hazard communication and worker training. This is due to the hazard posed by exposure to airborne crystalline silica. Prolonged and excessive exposure to crystalline silica can cause silicosis, a noncancerous but nonreversible and sometimes fatal lung disease. More than one million U.S. workers are exposed to crystalline silica, and each year more than 250 die from silicosis (OSHA, 2004). Crystalline silica may also be carcinogenic and increase the chances of scleroderma and other connective tissue disorders. Inhaled silica may also increase the risk of tuberculosis and chronic kidney disease. Most of the OSHA regulations are oriented toward sandblasting and foundry operations, but can be applicable to the glass and ceramic industries where silica is handled in batch preparations. Measures that should be in place to reduce exposure to inhalation of silica include a reduction in dust using such methods as wetting silica materials and working with adequate ventilation. For detailed information on the health dangers posed by respirable silica, the reader is directed to the OSHA Web site (OSHA, 2004).

5.2 CLAY

Introduction

Clay is the backbone of traditional ceramic manufacturing and is an integral raw material for some refractories and fiberglass. Its characteristic plasticity, durability, low cost, and ubiquity in virtually all areas of the globe have made it a historically important commodity. However, clay is perhaps one of the most complex materials used in industry and, therefore, can be one of the most difficult to control during processing. A thorough understanding of clay is needed to successfully and efficiently manufacture clay-based products.

The term *clay* is used rather generally, referring to a mineral group, a sedimentary deposit or rock, or a grain size of less than 2 μm. Clay minerals are all hydrous aluminosilicate sheet minerals that characteristically have fine particle size and plasticity. They are found within a wide range of compositions and degrees of crystallinity and can occur in various particle-size distributions and morphologies. Commercial clay products can be classified into the following six categories: kaolin, ball or plastic clay, fire clay, common clay/shale, Fuller's earth, and bentonite. These are distinguished by composition, plasticity, color, absorption qualities, and firing characteristics. The first three have kaolinite as the major clay mineral with minor amounts of the clay minerals illite and smectite as well as nonclay minerals such as quartz. The value of clays can be directly related to the mineralogical and chemical purity of the product. Clays used in high-quality ceramic products require beneficiation to remove unwanted minerals and compounds that affect the properties of the clay during forming and firing.

In addition to ceramic manufacturing, clays are an important industrial mineral used in a wide range of other applications such as paper (which consumes more high-quality clays than the ceramic industry), paint, rubber reinforcement, drilling muds, and pharmaceuticals. There is a large body of literature on clays and clay minerals, with entire textbooks and professional organizations (such as the Clay Minerals Society) dedicated to the subject.

Composition and Mineral Structure

All clay minerals are hydrous sheet silicates (phyllosilicates) based on layers of Si^{4+} tetrahedrally coordinated with oxygen and sheets of Al^{4+} or Mg^{2+} octahedrally coordinated with hydroxyl. Clay minerals can be described using end member sheet structures of $Si_2O_5^{2-}$ or silica layers, combined with an octahedral layer of either $Al_2(OH)_4^{2+}$(gibbsite layer), or $Mg_3(OH)_4^{2+}$ (brucite layer) (Worrall, 1986). Within each unit of the octahedral layer, there are three potential cation sites. Clay minerals with two of these sites occupied by Al^{3+} or other cations are referred to as *dioctahedral*. Clays with all three sites filled by Mg^{2+} or other cations are referred to as *trioctahedral*. Different clay minerals are made using several combinations of these sheet structures as well as isomorphous substitution of the Si^{4+}, Al^{3+}, or Mg^{2+} by other cations. Most of the tetrahedral sites are occupied by Si^{4+} with limited substitution by Al^{3+} or Fe^{3+}. The octahedral sites are normally occupied by Al^{3+} or Mg^{2+} but they can accommodate Fe^{2+}, Fe^{3+}, Ti^{4+}, Ni^{2+}, or Cr^{+3}. F^- may substitute for hydroxyl ions in some clays, which is of concern because fluorine is released during firing of the clay.

Clay minerals can be classified by the layered sequence of tetrahedral and octahedral sheets:

- 1 : 1 are clays that have one tetrahedral and one octahedral sheet per clay layer.
- 2 : 1 are clays with an octahedral sheet sandwiched between two tetrahedral sheets.
- 2 : 1 : 1 are 2 : 1 clays that contain an additional octahedral sheet between individual clay mineral layers.

Individual sheets of 1 : 1 clays can be linked together through hydrogen bonding of the hydroxyl ions within the octahedral layers and the oxygen ions of the tetrahedral layer (Figure 5.9). The 2 : 1 clays cannot be hydrogen bonded because only the tetrahedral layer is exposed at the surface of the crystal, although individual crystals may be linked by weaker Van der Waals bonds. The spacing between individual layers, or basal spacing, is influenced by this bonding, and clay minerals can be described physically by the basal plane spacing.

Substitutions of cations with lower valence numbers in the tetrahedral and octahedral sites (for example, Al^{3+} for Si^{4+} in the tetrahedral layer) will lead to a net negative charge on the clay particle. This charge can be balanced by large cations, such as Ca^{2+}, K^{+}, and Na^{+}, or attract

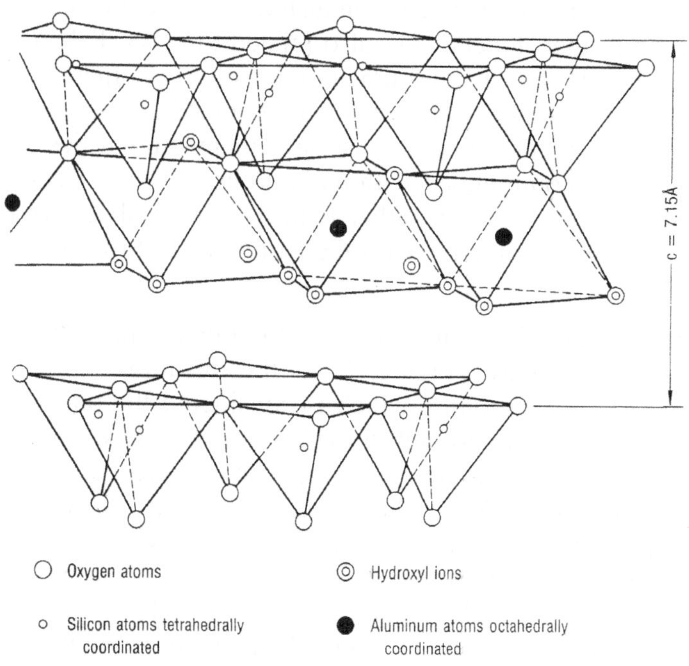

Figure 5.9 Idealized structure of kaolinite (from Pickering and Murray, 1994; courtesy of the Society for Mining, Metallurgy, and Exploration (SME), www.smenet.org).

polar molecules, such as water, that reside outside of the crystal structure and occupy the space between the layers (interlayer cations). These interlayer species are responsible for properties such as cation exchange capacity (CEC) and swelling ability.

Clays can be described as swelling and nonswelling when exposed to water. Swelling ability and CEC are both related to negative charges associated with oxygen anions that are not balanced by cations in the tetrahedral or octrahedral sites. The charges can be balanced by the adsorption of exchangeable cations in a solution. The CEC refers to the total amount of cations that can be retained by a given amount of material and is reported either as moles per unit mass, or more often in milliequivalents per 100 grams (meq/100g). CEC for a given mineral is often reported as a range of values, as it can depend on the surface area of the particles, the degree of crystallinity, and, for some clay minerals, the pH of the solution. An equivalent is defined as the amount of substance it takes to combine with one mole of hydrogen ions.

Clay minerals are categorized based on the different mineral structures into four general groups: kaolinite, smectite, illite, and chlorite (Table 5.5). These have characteristic basal spacings of 7 Å, 10 Å, 15 Å, and 14.5 Å, although the spacing can change with degree of hydration (Deer et al., 1992). An additional clay group is vermiculite, which is of little concern in the ceramic industry, so it is not discussed further here. Several other minerals are also related to clays. Palygorskite and sepiolite are fibrous, magnesium aluminosilicate minerals similar to clays, but with a chain structure for the octahedral sites. Chlorite and mica are sheet silicate minerals that have many similarities to clay and are often found commingled with natural clays. The 1:1 clay minerals that contain a brucite sheet rather than a gibbsite sheet are the trioctahedral serpentine minerals with the formula $Mg_3Si_2O_5(OH)_4$.

Kaolinite Group If a silica sheet is combined with a gibbsite sheet to make a 1:1 clay (Figure 5.9), the resulting composition is $Al_2Si_2O_5(OH)_4$, or *kaolinite,* which is considered the purest clay mineral and is the most important in the ceramic industry. There is essentially no substitution of Si^{4+} in the tetrahedral sites, but there can be minor substitution of Mg^{2+}, Fe^{3+}, or Ti^{4+} for Al^{3+} in the octahedral site (Geise, 1988). Excess SiO_2 or Al_2O_3 that is present in amounts higher than expected from the stoichiometry can occur as coatings on individual grains or agglomerates (Prasad et al., 1991).

The dioctahedral kaolinite minerals include kaolinite, halloysite, dickite, and nacrite. These all have the same nominal composition, but differ with respect to the way that individual layers are stacked. The

Table 5.5 Properties of clay mineral groups

	Kaolinite/Halloysite	Illite	Smectite	Chlorite
Formula	$Al_2Si_2O_5(OH)_4$	$K_{0.8}Al_2(Si_{3.2}Al_{0.8})O_{10}(OH)_2$	$Si_8Al_4O_{20}(OH)_4 \cdot nH_2O$	$(Mg,Fe,Mn,Al)_{12}(Si,Al)_8O_{20}(OH)_{16}$
Structure	1:1 dioctahedral	2:1 mostly dioctahedral	2:1 di- and trioctahedral	2:1:1 trioctahedral
Spacing (Å)	7.1/10	10	variable, mostly 15	14.2
Interlayer cations	none	K^+	Ca^{2+} and Na^+	none
Interlayer water	none for kaolinite 1 layer for halloysite		variable	none
Glycol	halloysite only	no effect	two layers, 17 Å	No effect
CEC (meq/100g)[1]	3–15 (kaolinite) 40–50 (halloysite)	10–40	70–130	10–40
Specific Surface Area (m^2/g)[1]	5–40 (kaolinite) 1100 (halloysite)[2]	10–100	40–800	10–55
Hardness	2.0–2.5	1–2	1.5–2.0	1.0–2.0
Paragenesis	Alteration of silicic rocks and feldspars in acidic conditions	Alteration of micas and feldspars under alkaline conditions with high Al and K concentrations	Alteration of mafic/basic and volcanic rocks in alkaline conditions; deficiency of K but available Mg and Ca	Hydrothermal alteration of primary ferromagnesian minerals in igneous rocks

After Deer et al. (1994)

[1] Encyclopedia Britannica Online (2004)

[2] Upper limit of estimated values

layers of kaolinite, dickite, and nacrite all are oriented such that layers are stacked and held together with hydrogen bonds, giving a basal spacing of 7.2 Å. An individual kaolinite particle is a thin, flat, hexagonal platelet (Figure 5.10) with triclinic symmetry. Dickite and nacrite differ from kaolinite in that the individual layers are stacked to give monoclinic symmetry. All three kaolinite minerals are nonswelling in water.

Halloysite is similar in composition to the other kaolinite minerals, but it can incorporate water in the interlayer sites. The random stacking of halloysite leads to an absence of hydrogen bonding, which allows water molecules to enter between successive layers and increases the basal spacing from about 7 to 10 Å. The hydrated form is unstable in most ambient conditions, and if not kept in a moist environment it will lose water and convert to a kaolinite-like metahalloysite (Geise, 1988). Halloysite often occurs as coils, tubes, and fibers within kaolinite plates.

When heated, the kaolinite minerals and halloysite will initially release adsorbed water at temperatures between 110° to 400°C (Deer et al., 1992). At 550° to 600°C, they will lose the structural water and form metakaolin. When metakaolin is heated beyond 800°C, the metakaolin structure will be irreversibly disrupted and between 900° to 1000°C form an aluminosilicate spinel that will then convert to mullite and possibly cristobalite (Carty and Senapati, 1998).

Figure 5.10 SEM photograph showing the flat, hexagonal morphology of individual kaolinite platelets (courtesy of W.M. Carty).

Smectite Group Smectites occur as aggregates of very small platelets that are 1 to 3 μm and may be irregular, elongate, fibrous, or hexagonal. They are difficult to view in optical microscopes and must be seen using electron microscopy (Figure 5.11). Smectites (Figure 5.12) are 2:1 dioctahedral or trioctahedral clays, which can be described by isomorphous substitution of cations between the end members of pyrophyllite $Al_2Si_4O_{10}(OH)_2$ and talc $Mg_3Si_4O_{10}(OH)_2$. Pyrophyllite, which is made of a gibbsite layer between two silica layers, and talc, which consists of a brucite layer between two silica layers, are not considered to be clay minerals and are separately described in section 5.3. The substitutions in the tetrahedral and octahedral sites, rarely more than one-third atom per unit formula, lead to a net negative charge that must be compensated for by cations attached outside of the crystal lattice. Therefore, smectites have cations that occupy some of the interlayer sites. The weak van der Waals bonds allow the smectites to be relatively easily separated by charge-compensating cations and water. The result is that these clays swell in the presence of water and have a great capacity for ion exchange with

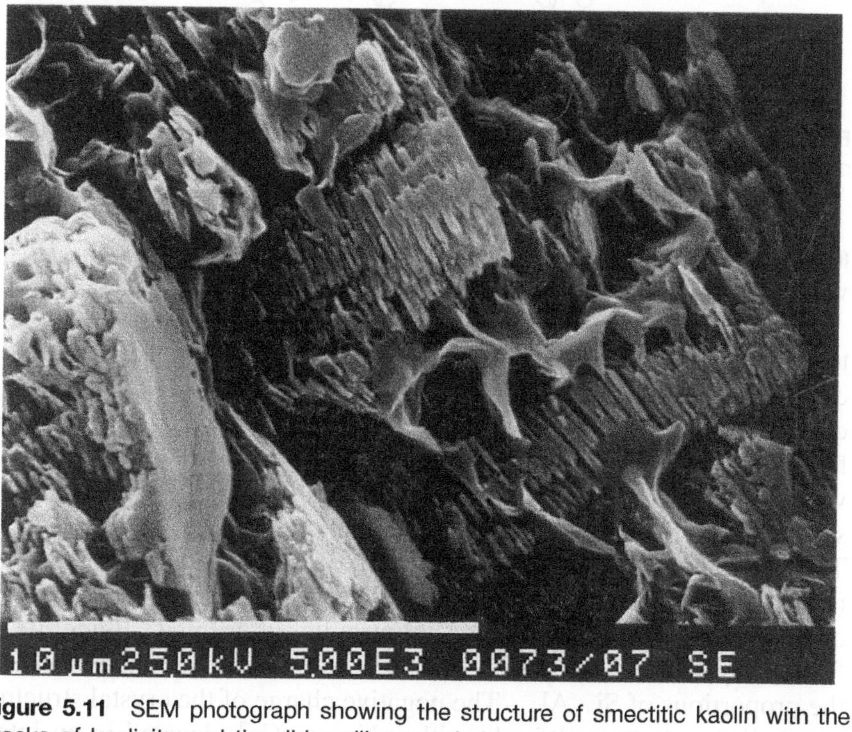

Figure 5.11 SEM photograph showing the structure of smectitic kaolin with the stacks of kaolinite and the ribbon-like morphology of smectite (courtesy of W.M. Carty).

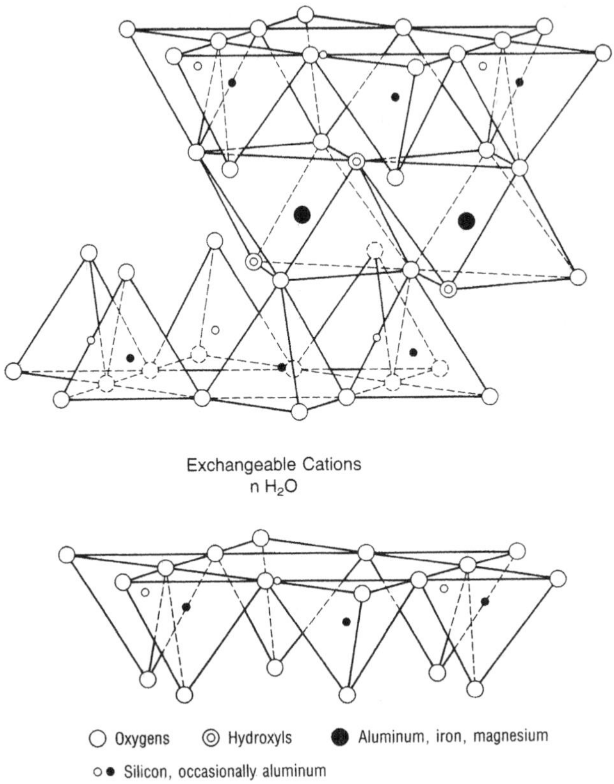

Figure 5.12 Idealized structure of smectite (from Elzea and Murray, 1994; courtesy of the Society for Mining, Metallurgy, and Exploration (SME), www.smenet.org).

a CEC in the range of 70 to 130 meq/100 g (Gaines et al., 1997), varying with the grain size and cation in the interlayer site.

There is a range of smectite mineralogies, and the reader is directed to Güven (1988) for a review of this subject. For ceramics, the most common smectite mineral of interest is montmorillonite. This mineral derives its name from the Montmorillon region of France where it was first classified. Montomorillonite is a dioctahedral smectite mineral in which some of the Al^{3+} of the octahedral sites in the pyrophyllite structure is occupied by a divalent cation such as Mg^{2+} (Figure 5.12). The crystals have a nominal formula of $(\frac{1}{2}Ca,Na)_{0.7}(Al,Mg,Fe)_4(Si,Al)_8O_{20}(OH)_4 \cdot nH_2O$ (Deer et al., 1992). If the octahedral sites are occupied by Fe^{3+}, the mineral nontronite is formed. Beidellite is formed by the substitution of Al^{3+} for Si^{4+} in the tetrahedral site in the proportions of $Si_{7.3}Al_{0.7}$. The negative charge of the crystal structure is usually compensated for by either Na^+ or Ca^{2+} located in an interlayer site, although other alkali and alkaline earth cations can substitute.

The amount of water that is adsorbed into the interlayer sites of montmorillonite (that is, its swelling ability) is greatly affected by the type of interlayer cation and the type of mineral. The calcium montmorillonites have Ca^{2+}occupying the interlayer site with two or three layers of water while the sodium montmorillonites have Na^+ and a continuously variable number of layers. As a result, the sodium smectites have a greater swelling ability relative to the calcium smectite. In the presence of ethylene glycol, smectites will swell to 17 Å, and this property is a commonly used test to identify smectite.

Upon heating, smectites lose adsorbed and interlayer water between 100°C to 250°C, causing a reduction of the basal spacing to about 10 Å. Upon further heating, the minerals lose hydroxyl ions at 450°C to 700°C. Dehydration can be reversible even after heating to 600°C (Deer et al., 1992). The final fusion temperature is dependent on the composition.

Illite and Mica Illite is the most common type of clay in sediments and has a crystal structure similar to the mica minerals because of the presence of K^+ as the interlayer cation (Figure 5.13). The mica structure is 2:1 and can be derived from the end-member structures of

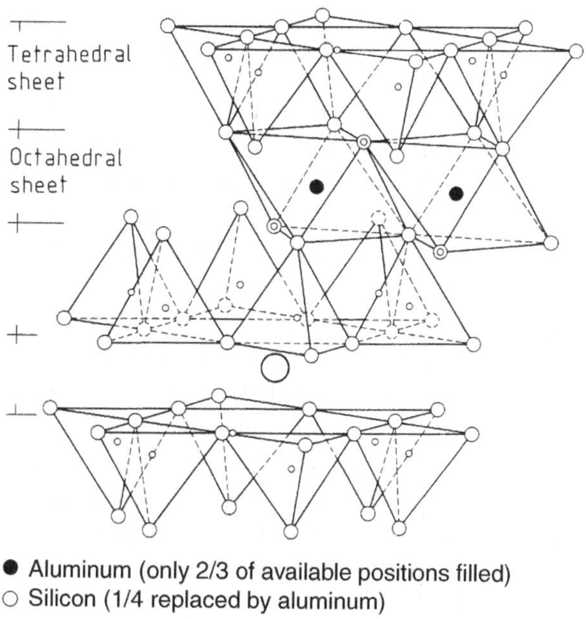

Figure 5.13 Idealized structure of illite (from Murray, 2002, used with permission from Wiley-VCH).

pyrophyllite and talc, with substitutions of Al^{3+} for Si^{4+} in the tetrahedral site. If one atom of aluminum substitutes for a Si^{4+} cation in the pyrophyllite structure, the resulting negatively charged unit is $Al_2(Si_3Al)O_{10}(OH)_2^-$. If the charge is balanced by an interlayer K^+, this forms muscovite mica $KAl_3Si_3O_{10}(OH)_2$. Muscovite occurs as a coarse, platy mineral that is common in granites and is the third most common mineral in the Earth's crust (after feldspar and quartz). Illite has a structure similar to muscovite, but generally has fewer potassium ions and more silicon and magnesium ions and water molecules in the lattice than mica. Illite occurs as fine particles that can be as small as 0.05 μm (Worrall, 1982). Illites are found with a range of compositions, as either dioctahedral or trioctahedral minerals, with the dioctahedral group more common. Illite is often associated with smectite (smectite can weather to illite), and the two types can occur as intimately mixed layers. The cation exchange capacity for pure illite is 10 to 40 meq/100g (Gaines et al., 1997), although higher observed values may be due to the presence of smectite.

Upon heating, illite loses adsorbed water between 110°C and 200°C. Hydroxyl molecules within the structure are driven off between 550°C and 600°C. At temperatures between 850°C and 950°C the crystal structure is disrupted and will convert to spinel at 1000°C (Deer et al., 1992).

Chlorite Chlorite is a mineral that resembles mica and can have a wide variation in composition, given as $(Mg,Fe,Mn,Al)_{12}(Si,Al)_8O_{20}(OH)_{16}$. It is a product of the low-temperature alteration of ferromagnesian igneous rocks such as basalt. The wide variety of compositions of chlorite has led to several classification schemes that are reviewed in Bailey (1988). Chlorite has a 14 Å 2:1 layer structure with substitutions in both the tetrahedral sites and octahedral sites that result in a net negative charge (Figure 5.14). The charge is balanced by cations in the interlayer space. The interlayer space may be occupied by a brucite-like sheet of magnesium and hydroxyl ions. The differences between the various minerals within the chlorite group are based on ionic substitutions in the brucite-like interlayer sheet and the 2:1 layers. Most chlorites are trioctahedral and have crystal symmetry that is usually triclinic, although sometimes monoclinic (Bailey, 1988). These clays are highly undesirable for white-firing porcelain ceramics and fiberglasses because they typically contain appreciable amounts of iron. However, chloritic clays are desirable for glazed tile.

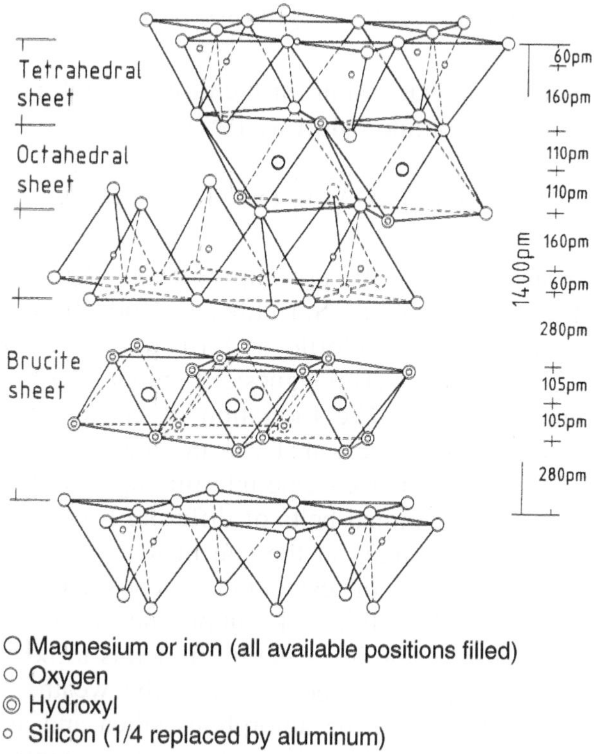

Figure 5.14 The structure of chlorite (from Murray, 2002; used with permission from Wiley-VCH).

Geological Occurrence and Distribution

Clays are formed as weathering and alteration products of primary aluminosilicate minerals such as feldspar and mica. A simplified equation to display this process is the alteration of orthoclase (potassium feldspar) to muscovite mica or illite is as follows:

$$3KAlSi_3O_8 + 2H^+ \rightarrow KAl_3Si_3O_{10}(OH)_2 + 6SiO_2 + 2K^+$$

Continued weathering and the removal of some of the potassium can alter the muscovite to form illite. The illite is further weathered to form kaolinite minerals. Note that the process causes the progressive loss of potassium into the weathering solution:

$$2KAl_3Si_3O_{10}(OH)_2 + 2H^+ + 3H_2O \rightarrow 3Al_2Si_2O_5(OH)_4 + 2K^+$$

There is an alternative expression for the direct alteration of potassium feldspar directly to kaolinite (Murray, 1989):

$$2KAlSi_3O_8 + 3H_2O \rightarrow Al_2Si_2O_5(OH)_4 + 4SiO_2 + 2KOH$$

Similar reactions occur for the decomposition of the sodium feldspar albite ($NaAlSi_3O_8$) to kaolinite. Smectites can form from the alteration of volcanic ash deposits. Chlorite is formed by the weathering or metamorphism of basic (ferromagnesian) minerals such as olivine, pyroxene, and amphibole. The formation of clays is a variable and spatially heterogeneous process that can produce intermixed clays on several scales, from the micron-scale intermixing of illite-smectite and kaolinite-halloysite to regional belts of differing kaolins.

If the clay minerals remain at the site in which they were formed, they can become a *primary* or residual clay deposit. The *in situ* weathering of a granitic or feldspathic rock can produce a very clean kaolinite if the original rock had few iron-bearing minerals to begin with. *Bentonites* are another primary deposit, where the weathering of silicic volcanic ash deposits alters the amorphous silica and alkali rich ash particles into a mix of clays that is predominantly montmorillonite.

Clays can erode from the location of formation and be transported by streams and rivers, eventually being deposited when the water velocity decreases in a body of water (for instance, in a river delta or lake). The settling clays will form tabular lenses and discontinuous beds, which are *secondary* or sedimentary deposits. Settling may be enhanced if the suspended clays enter into salt water because the increased cation concentration of the water forces the coagulation of the clays, increasing the effective particle size and accelerating deposition. Hard kaolins, with the denser packing, probably formed in marine environments. In many sedimentary deposits, it appears that the clays were weathered further after deposition. Post-depositional processes can increase the percentage of kaolin through more weathering and through oxidation that can remove reduced metal species, such as pyrite, and organic material (Pickering and Hirsh, 1989). Post-depositional processes may also cause coarsening of kaolin particles through the growth of vermicular stacks (Manning, 1995). Significant impurity minerals associated with secondary clay deposits include unwanted clay minerals (smectite, illite, and chlorite), feldspar, quartz, mica, and iron oxides.

If a sedimentary clay deposit is lithified after being buried and subjected to pressure and temperature, it may become *shale*. Shale is used as a material for terra cotta ceramics such as drainage and roofing tiles. If the shale deposit is metamorphosed further, it may become a slate, which is not necessarily of interest to the ceramic industry but is used to make durable roofing material.

Kaolin Kaolins or china clays are high-grade white clays consisting of primarily kaolinite. Kaolin can also refer generally to a white-firing, low-iron clay. Mined kaolins generally consist of 85 percent to 95 percent kaolinite minerals (including halloysite). Kaolins are white in color, have very fine particle size, and have variable plasticity. The grades of kaolin that are commercially available generally are based on particle size and on color, or brightness. Subtypes of kaolins, specifically Georgia kaolins (Elzea et al., 2000), include the following:

- *Soft.* These tend to break easily and are made of fine to coarse kaolinite platelets that have been recrystallized and have low iron.
- *Hard.* These are tightly packed with submicron platelets, and are slightly higher in iron.
- *Flints.* These are very hard, with clay grains cemented by silica that precipitated from interstitial fluids after clay deposition.

Kaolin world reserves are estimated at 14.2 billion tons, with almost 87 percent of this residing in the United States and Brazil (Figure 5.15). Add to this the reserves of Ukraine, United Kingdom, and China, and these five countries host up to 96 percent of the total estimated kaolin reserves (Virta, 2002a). However, it is difficult to compare production between countries because reported numbers may be combined with other clays or may be reported as slurry weight rather than a dry weight basis.

There are a number of kaolin deposits in North America (Patterson and Murray, 1984). Major commercial kaolin deposits in North America are found in what is referred to as the *Kaolin Belt* of South Carolina, Georgia, and Alabama. This area accounts for approximately 35 percent of the world's kaolin production (Chang, 2002). The belt is 30 km wide and 400 km long and occurs along the fall line where the sedimentary coastal plain meets the older, crystalline Appalachian Mountains. The deposits are Cretaceous and Lower Tertiary (70 to 50 million years) sediments forming lenticular beds associated with quartz sands (Patterson and Murray, 1994). The sediments are thought to come from

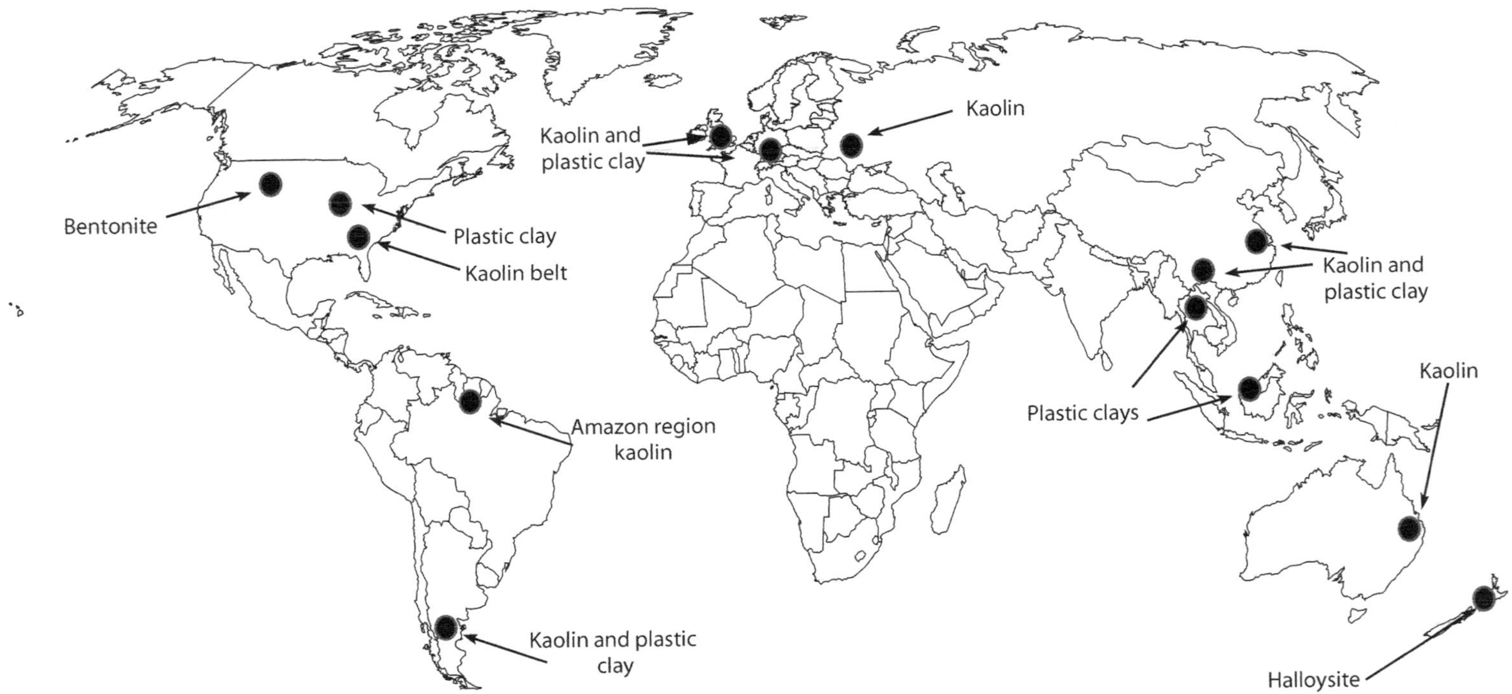

Figure 5.15 Map showing some of the important global deposits of clays.

the weathering of feldspathic crystalline rocks in the Appalachians during the Cretaceous uplift of the Piedmont (Austin, 1998). Deposition occurred in deltas, offshore bars, and muddy lagoons similar to those now existing along the coast of Georgia and South Carolina (Murray, 2002). The clays were subsequently altered by groundwater after deposition. Individual deposits range in thickness from 3 to 15 meters and can extend laterally from a few hundred meters to kilometers in length and width.

The Georgia and South Carolina kaolins generally contain about 85 percent to 95 percent kaolinite. The Cretaceous kaolins are relatively coarse with about 60 percent to 70 percent of the particles finer than 2 μm and the Tertiary kaolins range from about 85 percent to 95 percent finer than 2 μm (Murray, 2002). Other minerals associated with the clays include quartz, muscovite, biotite, smectite, ilmenite, anatase, rutile, leucoxene, goethite, and traces of zircon, tourmaline, kyanite, and graphite. In Mexico, there are kaolin mines in most states (Patterson and Murray, 1984). These deposits are both primary and secondary with a range in mineralogy.

In Europe, some of the highest-quality kaolins come from southwestern England, specifically in the Cornwall region (Figure 5.16). These deposits, first discovered in 1748, are primary deposits formed by the hydrothermal alteration of granitic intrusions which altered much of the feldspar to kaolinite, but left the quartz, mica, tourmaline,

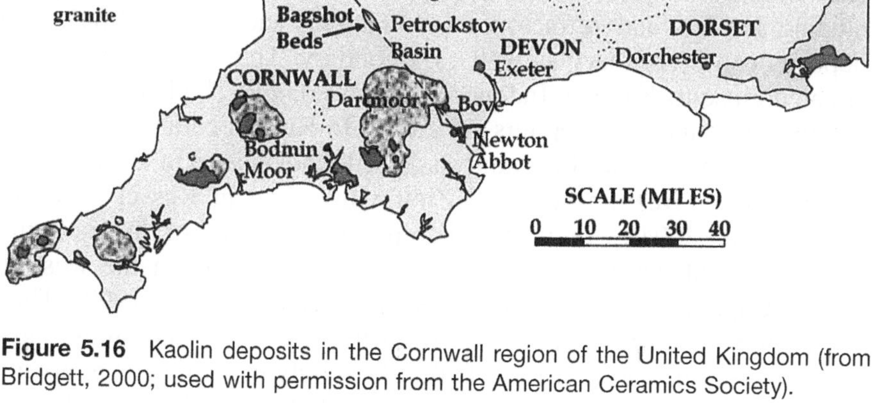

Figure 5.16 Kaolin deposits in the Cornwall region of the United Kingdom (from Bridgett, 2000; used with permission from the American Ceramics Society).

and some of the feldspar intact. The degree of alteration of the granite varies with depth (up to 300 m from the surface) and length, leading to heterogeneities within the deposits. In general, a typical Cornwall deposit is made of a mix of high-purity kaolin (15 percent), quartz sand (70 percent) and unweathered rock (15 percent) (Bridgett, 2000; Pickering and Murray, 1994). The other significant producers in Europe are Germany, the Czech Republic, and southern Ukraine. In Germany, kaolin was first discovered and used for porcelain in Meissen. Now, mining occurs in primary deposits near Dresden and in secondary deposits in Bavaria in the Oberfalz region (Newman, 2002). The Bavarian deposits consist on average of 76 percent quartz, 12 percent feldspar, and 12 percent kaolin (Pickering and Murray, 1994).

Brazil has the second largest world kaolin resources with the majority occurring in the Amazon Region states of Amapá and Pará (Pereira da Silva, 2002). A Pliocene age secondary deposit 40 m thick and extending over an area of 60 km^2 is mined along the Jari River. The kaolin is made of 80 percent to 98 percent kaolinite and has a fine particle size similar to the Tertiary deposits of Georgia and South Carolina (Murray, 2002). A large, secondary kaolin deposit of coarser particle size is mined in the Rio Capim area. Argentina mines sedimentary kaolins in the Patagonian provinces of Chubut, Rio Negro, and San Luis (SEGEMAR, 2005). Colombia is a very large producer of low to medium quality kaolins that are used in ceramics (Harben, 2002).

In Asia and Australia, China is one of the largest producers of kaolin. Unlike the western countries, where kaolin production is dominated by a few large mines and companies, Chinese kaolin is produced in up to 700 small mines (Dickson and Harben, 2000). The largest kaolin mine in China is a primary deposit located southwest of Shanghai near Suzhou (Zheng et al., 1996). Japan produces kaolinite-based clays including kaolins and refractory clay (Harben and Bates, 1984). The deposits can be primary as well as secondary. The major deposits are situated close to Nagoya. Indonesia, Malaysia, and Thailand are growing in importance as suppliers of kaolin. In Russia, ceramic grade kaolins are mined at the Prosyanovskoe, Glukhovetskoe in Ukraine and Kyshtymskoe, Yeleninskoe, and Zhuravlinyi Log in Russia (Maslennikova et al., 2004). Commercial kaolin deposits in Australia are found in Queensland, with the two largest deposits located on the western side of Cape York at Weipa and at Skardon River (Wallis, 1999). They are primary deposits associated with a heavily weathered, aluminum-rich laterite. The kaolin is mined after the overlying bauxite is removed. New Zealand is known for its high-purity halloysite deposits (Taylor,

2000), which are primary deposits that lie north of Auckland. The deposits are believed to originate from the hydrothermal alteration of silica-rich rhyolites (Murray et al., 1977). The crude ore consists of about 50 percent halloysite and 50 percent silica minerals (Keegan, 2000).

Ball or Plastic Clay The term *ball clay* comes from the manner in which it was originally mined in England, where it was cut into cubes that became rounded. The term is slowly being phased out and replaced by the term *plastic clay*. Plastic clays are composed primarily of kaolinite (50–70 percent), with the remainder consisting of illite, mica, smectite (usually montmorillonite), quartz (5–30 percent), and organic material such as lignite. Other nonclay contaminants include pyrite, siderite, anatase, gypsum, and dolomite (Echlin, 2002). In general, these clays are more plastic and more variable in color than kaolins. The increased plasticity can be attributed to the finer particle size relative to kaolins, ranging between 50 and 90 percent smaller than 2 μm, and increased amounts of organic matter. In suspensions, plastic clays have higher viscosities than kaolins because the finer particle size and therefore higher surface area result in a higher dispersant demand. Plastic clays generally fire to a light cream color.

All plastic clays are sedimentary in origin and can be found all over the world. However, there are only a few large deposits that are of the quality suitable for modern whiteware manufacturing, which is the largest consumer of plastic clay. In the United States, commercial clays are mined from 38 quarries in Tennessee, Texas, Kentucky, and Mississippi (Virta, 2002a). The major production quarries are in western Tennessee and western Kentucky. These deposits are Upper Cretaceous to Middle Eocene sediments that appeared to have been deposited in broad river delta environments (Murray, 2000). The thickness of the deposits can vary greatly, although most exploited deposits are 3 to 6 m thick. In South America, better quality plastic clays come from deposits in the Argentine Patagonian provinces of Chubut, Santa Cruz, and Neuquén (SEGEMAR, 2005).

In Europe, high-quality ball clays are mined in the southwest of England in the Bovey and Petrockstowe Basins in Devon and the Wareham district in Dorset. These are mixed sediments composed of kaolinite, illite, quartz, lignite, and minor amounts of smectite that were deposited in small basins of late Eocene to middle Oligocene (36–28 million years) (Manning, 1995). Within South Devon, there are several subtypes of ball clays: black, dark blue, light blue, and high silica.

Within North Devon, there are two subtypes: Type 1, with relatively higher amounts of montmorillonite; and Type 2, with lower amounts of montmorillonite. The other major European plastic clay deposits are in the Westerwald region of Germany.

Asian ball clays have only recently been moving from small, local mines to the development and exploitation of larger deposits in southern China, Thailand, and Indonesia (Leonard, 2000).

Fireclay Fire and flint clays refer to light to dark gray clays that, when fired, can withstand temperatures of 1500°C (2730°F) or higher. The term *fireclay* may be used synonymously with refractory clay, and it may also include kaolin and plastic clays used in refractory applications. In general, fireclays consist primarily of poorly crystallized kaolinite with impurity minerals of illite or mica, aluminum hydroxides, and quartz. Minor impurities include organic matter, carbonates, goethite, and anatase (Worrall, 1982). The refractoriness of fireclays is enhanced by the amount of alumina present and is reduced by fluxing impurities such as calcium carbonate or iron minerals. Fireclays can range in plasticity, with the hard, nonplastic clays referred to as flint clays because they exhibit a conchoidal fracture habit similar to flint. Diaspore clay is a fireclay that is composed of a mix of diaspore (AlO[OH]) and kaolinite. Chamotte, also referred to as grog, is a synthesized flint clay made by calcining plastic clay to form a refractory aggregate to be used in the manufacture of firebrick. This is common where natural fireclay is scarce, such as Europe (Manning, 1995).

Flint clays are sedimentary and are found in basins or depressions under carbonate rocks or sandstones (Harben and Bates, 1984). The most common fireclays in the United States are underclays, which occur directly beneath coal seams of Pennsylvanian age. Younger (Cretaceous and Tertiary) fireclays are found in lenticular deposits of sedimentary basins (Pickering and Murray, 1994). Diaspore clay, which is normally found mixed with flint and plastic fireclays, is mined commercially in Missouri and Pennsylvania. In Europe, flint clays with Al_2O_3 greater than 42 percent are mined in Bavaria. A more complete review of refractory clay occurrences is found in Keegan (2000).

Bentonite Bentonites are composed mainly of the smectite series of minerals montmorillonite-beidellite, and the term *bentonite* is generally synonymous with a swelling clay. These clays are used only to a limited extent as a plasticizer in some ceramics and glazes and as a bonding agent for foundry sands. Smectite minerals may be regarded as undesirable impurities in kaolin and plastic clays, as even small amounts can drastically change the rheological characteristics of the clay when in suspension. In addition, smectites can cause excessive shrinkage and cracking during drying.

Bentonites are largely formed by the alteration of silicic volcanic ash and tuffs soon after they are deposited in water. The composition of the bentonite depends on the composition of the ash and whether the ash was deposited in fresh or marine water (Chang, 2002). Natural bentonite can contain minor amounts of illite, chlorite, kaolinite, and nonclay minerals such as quartz, calcite, and gypsum. Because bentonites are volcanic in origin, they can have appreciable amounts of cristobalite, the high-temperature polymorph of silica that is found naturally in volcanic eruptions.

The largest producing areas of bentonite are in the United States, specifically northeastern Wyoming, western South Dakota, central and eastern Texas, and northeastern Mississippi. The Wyoming and South Dakota bentonites are known as sodium bentonites, because the montmorillonite has a relatively high sodium content (Elzea and Murray, 1994). The Texas and Mississippi bentonites are calcium bentonites. Greece is the second largest global producer after the United States (Russell, 2000). Bentonite is mined in Mexico near Cuencame. Brazil and Argentina are the largest producers in South America.

Chloritic Clay, Common Clay, and Shale Chloritic clays from the Emilian Appennines in Italy are an important raw material for the tile industry (Ciullo, 1996). The gray-blue chlorites from these deposits are generally used for high porosity glazed wall tile, the gray clays of the Antognola series are used in bodies for both glazed and unglazed floor tiles, and the high-iron red chlorites are used for glazed floor tiles and roofing tiles.

Common clays and shales are found in many different areas and different ages and account for the majority of clay used in the United States (Virta, 2003a), and likely it is the same for the rest of the world. Nevertheless, its value is considerably less than that of kaolin, so this class of clay does not always have the attention of its higher-priced brethren. Common clay can be defined as "a clay or claylike mass that is plastic enough to be readily molded and that vitrifies below 2000°F (1100°C)" (Harben and Bates, 1984). Common clays consist of a mix of illite (or other nonplastic, mica-like minerals such as muscovite, sericite, and biotite), chlorite, kaolinite, and montmorillonite, plus a variety of nonclay minerals such as quartz, mica, and calcium carbonate. Common clay and shales consequently have wide variations in properties, depending largely on the composition of the clay and the particle size of the constituents. Fired color depends on the clay composition, specifically Fe_2O_3 and TiO_2 content. For example, common clays that fire to a buff color usually contain some TiO_2 and have 1 percent to 5 percent Fe_2O_3, while clays that fire red have greater than 5 percent Fe_2O_3 (Patterson and Murray, 1983).

Fuller's Earth Fuller's earth is a mix of calcium montmorillonite and the clays attapulgite and sepiolite. The term refers to a mix of clay that was originally used for *fulling,* or the cleaning of raw wool. Fuller's earth has limited use in ceramic manufacturing in some catalyst carriers, but finds uses in drilling mud, absorbents, and plastic and paint fillers.

Mining and Processing

Almost all the commercial clays are mined by open-pit methods. Before mining commences, drilling is first used to determine the dimensions of the deposit below the surface. Drilling is performed as a patterned grid and usually starts at 120 m spacing for initial exploration (Pickering and Murray, 1994). Subsequent examination of the drill cores will determine the quality and the degree of heterogeneity throughout the deposit. The drilling studies are then used to build a three dimensional model of the depth and lateral extent of the deposit.

After determining the specific location of areas to be mined, the overburden is removed mechanically, with overburden-to-clay ratios normally around 6:1, although it can be as high as 10:1 (Murray, 2000). The beds are individually removed using mechanical excavators (Figure 5.17). Depending on the site, the clay is mined using draglines, shovels, or bucket loaders. Explosives and high pressure water jets are often used to mine primary kaolin deposits. The water jets are directed at the mine face and the kaolin is entrained into the water, leaving the coarser particles behind (Murray, 2002). The slurry is collected and pumped to settling tanks. At the mine site, ozone may be used to brighten gray kaolins (Harben, 2002) by oxidizing organic material. The clay is transported from the mine to processing plants using several methods, including belt conveyors, pipelines, trucks, and rail.

Mined secondary kaolins and plastic clays are moved to bulk storage areas, with clays from different beds kept in separate bins. The clay in each bin is sampled and tested to determine its composition and physical properties. Individual bins are mixed to produce a blended product with consistent properties. Blending can be done simply by placing a mix of clays from different bins in a hopper and shredding with rotating

Figure 5.17 Excavation of an open-pit kaolin mine (courtesy of W. M. Carty).

disks similar to large cheese graters, which reduces the size to 5 cm (e.g., Echlin, 2002). For some applications, blending and shredding are the only processing steps (sometimes referred to as unprocessed).

Blended, unprocessed clays can be blunged (mixed with water and agitated to produce a slurry) and then screened to remove clots of clay, sand, and other coarse particles (Figure 5.18). Kaolins and some plastic clays can be further processed using either wet or dry methods (Murray, 2000). The dry method results in a less expensive product, but at the cost of lower and/or variable quality. Kaolins that contain more than 5 percent to 8 percent grit (i.e., particles retained in a 325 mesh screen) cannot be dry processed (Murray, 2002). The dry process (Figure 5.18) starts with the shredded clay that has been mined and dried in rotary driers. It is then pulverized and sent to an air cyclone to remove coarse sand and mica and classify the clay into specified mesh size. Air-classified clay is also referred to as *air-floated.* The material is shipped in bulk or in bags.

Wet methods are used to process primary kaolins and when a higher-value secondary kaolin or plastic clay is desired (Murray, 2000). Figure 5.19 gives a general outline of the processing steps that may be used, depending on the clay and the desired final product specifications. In the case of primary kaolins, the material is brought in as a slurry from the pit face and pumped into spiral or bucket-type classifiers to remove coarse sand (Bridgett, 2000). Secondary kaolins and plastic clays are blunged with a dispersant (such as sodium phosphate, sodium silicate, or sodium polyacrylate) at 25 to 60 weight percent solids loading.

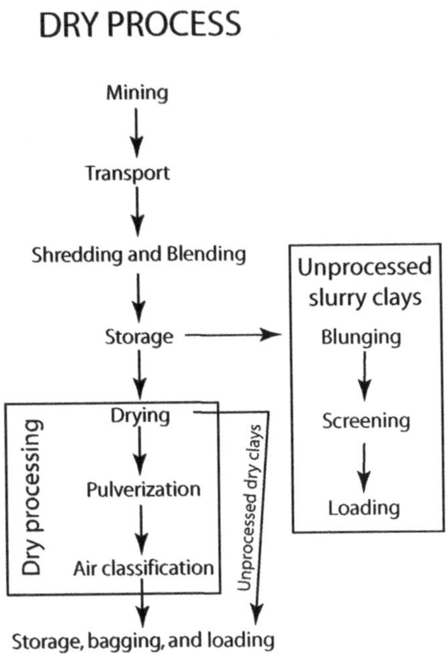

Figure 5.18 Flow diagram showing the route for unprocessed kaolins shipped in slurry form and dry processing for kaolins and ball clay (based on Murray, 2000; used with permission from the American Ceramics Society).

Degritting is then performed using classification boxes, which remove sand and mica particles coarser than 44 μm, followed by screening to remove coarse mica, lignite, and unblunged clay pieces. At this point, clay slurries may be blended to bring the product to the desired specifications.

Particle-size classification is done in a series of centrifuges or hydrocyclones to separate a coarse and fine clay fraction. To remove iron and titanium minerals, the slurry is passed through as series of high-intensity magnetic separators, either before size classification or after (Murray, 2000). The coarse fractions can either be dewatered for the final product or are processed to delaminate the coarse stacks. In either case, leaching can be performed to remove any iron oxide coatings and to brighten the material. For the fine fractions, the simplest processing is to add acid to selectively flocculate the suspensions for the final product. Further processing may include leaching to remove iron discoloration, treatment with ozone to oxidize organic material, and flotation or magnetic separation to remove iron and titanium minerals. At this point, the product can be shipped as a slurry, or be dewatered using filter presses and rotary driers, or by spray drying. Clays are shipped

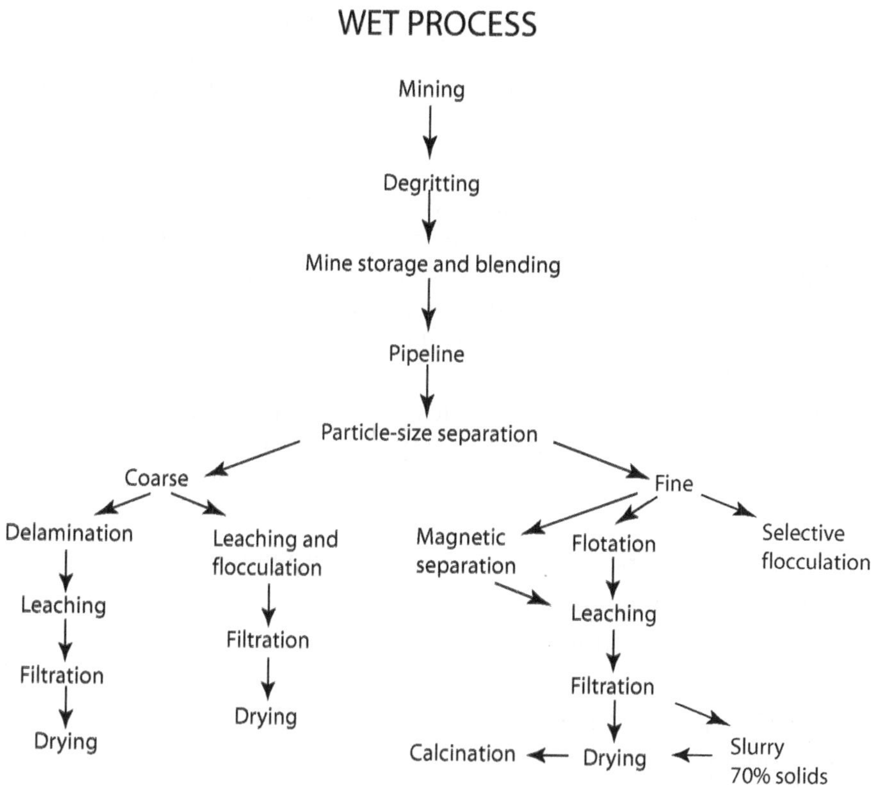

Figure 5.19 Flow diagram showing wet processing that is used for primary kaolin deposits and high-quality secondary kaolins (from Murray, 2000; used with permission from the American Ceramics Society).

either by rail in dry bulk or as a slurry at 70 percent solids in tank cars (Murray, 2002).

After fireclay is removed from the mine, it may be left in uncovered piles for up to a year to undergo additional weathering, which can assist in the comminution process. It is then crushed, dried, and often calcined. High-temperature calcining will convert the clay to a mixture of mullite and cristobalite. A glassy phase may also be present, depending on the calcine temperature and the presence of fluxing compounds, such as alkalis and iron.

Specifications and Uses

Throughout the entire mining and processing route, the quality and properties of the clays are continually checked to ensure proper selection and blending to produce the product desired by the consumer.

Kaolin is widely used as a filler material for glossy paper, plastics, and paint. In these applications, the kaolin is generally more highly processed than that used for ceramics and fiberglass. In this section, specifications and uses are restricted to ceramic and glass applications.

Whitewares Kaolin and plastic clays are the most important materials in whitewares. Manufacturers have a range of commercial clays available to them and several of the clays are blended to have properties that are desirable for a specific sector. For instance, a sanitaryware manufacturer would want coarser clays to promote rapid casting, and there would be differences between pressure casting and conventional plaster mold casting. The properties of a clay product that are of interest, and therefore should be in any specification sheet, include the following:

- Chemical composition (including loss on ignition)
- Mineralogical composition
- Particle-size distribution
- Slurry viscosity and casting behavior
- Surface area and methylene blue index (MBI)
- Plasticity
- Modulus of rupture (MOR)
- Cation exchange capacity
- Fired color
- Shrinkage upon drying and firing
- Pyrometric cone equivalent (PCE)

The details for the characterization methods for clay are discussed in Chapter 8.

The mineralogical composition refers to the specific minerals and the relative amounts present in a clay product. For example, the typical mineralogy of English kaolins is 84 percent kaolin, 14 percent mica, 1 percent feldspar, and 1 percent quartz (Bridgett, 2000). For most applications, kaolin is the desired clay mineral. X-ray diffraction (XRD) is the most precise method of obtaining the bulk mineralogy of a sample. For the most precise analysis, a sample may need to be divided into separate particle-size fractions and oriented slides produced. XRD may not be able to differentiate between halloysite and kaolin, so other methods such as electron microscopy and differential thermal analysis must be used (Murray, 2002). The mineralogical analysis will report the amounts of undesirable minerals, as well. Smectites such as montmorillonite can be found in secondary kaolins and are concentrated in the fine fraction (Elzea et al., 2000). The presence of montmorillonite can increase plasticity and green strength, but can also greatly affect slip rheology. Mica and illite are sources of potassium, which is undesirable in refractory clays because of its fluxing behavior. However, mica can improve the casting rate, which is beneficial if the clay is used in casting ware. Some transition metal oxide minerals such as goethite (FeO[OH]) and anatase (TiO_2) will discolor the clay upon firing and affect the transparency (Figure 5.20).

Mineralogy dictates the bulk composition of a clay. Table 5.6 gives examples of some commercial clays with their bulk compositions and mineralogy. Note that the clays with higher amounts of free quartz have higher bulk SiO_2 and those with higher mica have higher K_2O. Since kaolinite should ideally have 1:1 molar ratio of Al:Si, ratios that are lower can be attributed to quartz particles in the product. Iron may be present as discrete iron oxide and hydroxide minerals or as coatings on clay mineral surfaces.

Particle-size distribution is an important characteristic because changes in it can affect slurry viscosity, plasticity, shrinkage during drying and firing, and strength. This parameter is usually reported in terms of percent of the material that is less than a given screen size. For larger particle sizes, a grit test gives the percentage of particles that are retained by a 325-mesh screen (Murray, 2002). Particle-size distribution of clay particles can be done by the sedimentation methods that

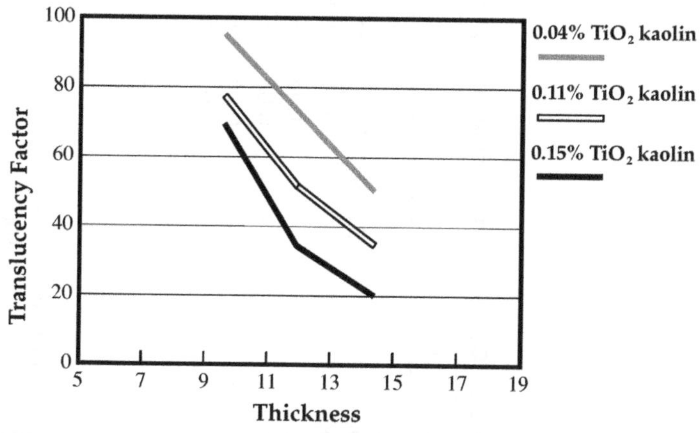

Figure 5.20 Graph showing the detrimental effect of TiO_2 on the translucency of three porcelain bodies (from Bridgett, 2000; used with permission from the American Ceramics Society).

Table 5.6 Compositions of some clay products

Oxide, wt. %	1	2	3	4	5	6	7	8	9	10	11	12	13
SiO_2	46.55	45.7	46.7	50.5	60.4	46.6	47.2	48.5	59.8	49.0	58.5	60.1	50.4
Al_2O_3	39.49	38.3	38.2	28.7	27.0	38.1	37.6	32.3	26.4	34.6	24.4	25.7	35.5
Fe_2O_3	—	0.41	0.60	0.91	0.93	0.69	0.50	0.98	1.00	0.71	1.26	1.7	0.25
TiO_2	—	1.55	1.42	1.48	1.62	0.07	0.05	1.16	1.39	0.02	0.92	1.2	0.05
CaO	—	0.08	0.12	0.40	0.28	0.19	0.20	0.18	0.20	0.35	0.05	0.4	—
MgO	—	0.06	0.20	0.30	0.26	0.20	0.08	0.21	0.51	0.34	1.05	0.5	—
K_2O	—	0.06	0.15	0.89	1.70	1.47	1.35	1.89	2.42	2.52	2.36	1.3	—
Na_2O	—	0.14	0.03	0.18	0.50	0.08	0.07	0.19	0.38	0.48	0.12	0.15	—
LOI	13.96	13.65	13.79	16.58	7.59	12.66	12.62	14.78	7.88	10.66	8.64	9.3	13.8
						Minerals, wt. %							
Montmorillonite	—	—	3	8	7	6	2	6	14	9	29	—	—
Kaolin group	100	96	93	58	44	80	82	60	34	60	23	52.6	—
Mica	—	2	2	10	21	12	11	18	25	27	22	12.6	—
Free quartz	—	trace	1	14	26	1	1	9	23	3	19	30	—
Organic	—	trace	trace	8	0.5	trace	trace	5	2	—	3	2.2	—

						Particle Size							
% < 20 μm	—	95	99	99	98	100	88	96	97	77	95	—	—
% < 5 μm	—	69	88	95	79	97	74	91	85	52	87	—	—
% < 2 μm	—	52	72	82	61	64	30	83	81	36	72	—	96
% < 1 μm	—	35	56	69	43	57	23	77	75	19	56	—	—
% < 0.5 μm	—	28	41	51	29	35	15	62	65	16	43	—	—
						Surface Area							
MBI, meq/100 g	—	1.6	10.5	12.1	5.6	5.4	2.4	8.7	12.8	3.4	16.5	—	10

1. Ideal kaolinite composition
2. Coarse kaolin, sedimentary, Washington County, Georgia, USA (Phelps and Wachtman, 2002)
3. Fine kaolin, sedimentary, Wilkinson County, Georgia, USA (Phelps and Wachtman, 2002)
4. Dark fine ball, Graves County, Kentucky, USA (Phelps and Wachtman, 2002)
5. Coarse light ball, Weakley County, Tennessee, USA (Phelps and Wachtman, 2002)
6. Fine china clay, Cornwall, England, UK (Phelps and Wachtman, 2002)
7. Coarse china clay, Cornwall, England, UK (Phelps and Wachtman, 2002)
8. Dark ball, Devonshire, England, UK (Phelps and Wachtman, 2002)
9. Light ball, Dorset, England, UK (Phelps and Wachtman, 2002)
10. China clay, primary, Thailand (Phelps and Wachtman, 2002)
11. Ball clay, Thailand (Phelps and Wachtman, 2002)
12. Fireclay, Yorkshire, UK (Worrall, 1982)
13. Halloysite china clay, New Zealand (Taylor, 2000)

are described in Chapter 8. Table 5.6 shows the particle-size distribution for some commercial clays. Note that, relative to kaolins, the ball clays have a larger fine fraction with 40 to 60 percent of particles being less than 0.5 μm and an overall wider particle-size distribution. The coarser kaolins are preferred in sanitaryware because the larger interstitial voids allow a higher rate of casting compared to a fine clay.

The viscosity of the clay in a slurry is important for casting and spray drying (Thompson, 2000) because the clays with their relatively high surface areas control the viscosity of a slip more than the nonplastic components. Viscosity is a function of several variables, including mineralogy, particle-size distribution, and the concentration of ions in the slurry. Viscosity at a given solids loading should be reported on a specifications sheet provided by the supplier. In kaolin slurries, the addition of illite and nonswelling divalent montmorillonites has little effect on viscosity, while even small additions (1 to 5 percent) of swelling sodium montmorillonite and magnesium smectites can greatly increase suspension viscosity (Schulz et al., 2003).

For casting clays, such as sanitaryware, the viscosity characteristics and filtration rate is important. ASTM test C866 (ASTM, 2002) can be used to determine the rate at which a clay slip forms a clay paste layer. The filtration rate will increase with increasing filter-press pressure and the number and size of interstitial voids in the casting. The rate will decrease with increasing surface area of the clay.

Modulus of rupture is used to determine the green (dry) strength of the clay. Tests are done on extruded round bars that are dry or under a determined relative humidity (ASTM, 2002; see C689-93). Dry strength will decrease with increasing proportions of nonclay minerals, decreasing particle size of the clay, and increasing organic matter. In general, ball clays have higher green strength than kaolins because of higher plasticity. The presence of minor amounts of montmorillonite generally increases the dry strength.

The methylene blue index (MBI) refers to the milliequivalents of methylene blue dye adsorbed per 100 grams of clay. MBI correlates with several clay properties including specific surface area, cation exchange capacity, and casting rate (ASTM, 2002; see C837-99). There is a direct correlation with specific surface area and MBI if kaolinite is the dominant colloidal element, but this correlation is not evident if there are significant amounts of illite or montmorillonite. Specific surface area is reported in m^2 per gram and is usually determined by nitrogen adsorption (Chapter 8). This determines the surface area, which is then divided by the mass of the powder sample.

Shrinkage during drying refers to the loss in volume when the clay dries and is dependent on the water content, the mineralogy of the clay product, and the particle-size distribution (Pickering and Murray, 1994). Tests are made on round bars or rectangular bars that are either cast or extruded (C326; ASTM, 2002). The shrinkage can be reported as percent linear shrinkage or percent volume shrinkage. Fired shrinkage should be accompanied by the firing temperature or pyrometric cone equivalent (see following paragraph). Plastic clays tend to have higher drying shrinkage rates, which can lead to warping and cracking. Swelling clays, such as sodium montmorillonite, can cause excessive shrinkage. Kaolins generally have lower shrinkage than ball clays because of the lower amounts of water required for workability (Worrall, 1982).

The behavior of clay under firing temperatures is measured by the firing shrinkage, glass formation, and color. Firing shrinkage depends on the particle packing (which is related to the particle-size distribution) and as well as the mineralogical and volatile content of the clay. Glass formation can be defined as the initiation of fusion of the low-temperature components in the clay and beginning of sintering. As the melt liquid forms, slumping of the material will begin, which is the basis of the pyrometric cone equivalent (PCE) system. Mineralogy plays the greatest role in firing range, and nonkaolinite minerals that are rich in volatiles, alkali, alkaline earth, and iron all act as fluxes and lower the firing temperature. Common minerals that reduce firing temperature include illite, montmorillonite, chlorite, feldspar, and calcite. The PCE range for kaolins is between cone 34 and 35 and for plastic clays, between cone 28 and 34 (Materials Handbook, 2005).

Fired color is also influenced by the presence of nonkaolinite minerals. Although it is commonly thought that iron-bearing minerals are the sole source of discoloration, Reynolds (2000) demonstrated the coloring increases linearly with titania content that is present as the TiO_2 minerals rutile and anatase or the mixed iron-titanium oxide ilmenite ($FeTiO_3$). This also affects the translucency of porcelain (Figure 5.20). Color is further controlled by the oxidation state of iron, the firing temperature and the atmosphere of the furnace.

Refractories Refractory clays can cover a wider range of definitions and compositions, making it difficult to accurately catalog. However, a general description is a clay with greater than 38 percent Al_2O_3 (averaging 42 percent to 47 percent) with low iron and total alkalis (Keegan, 2000). Calcined clays, which consist of mullite, cristobalite, and a glassy phase, have become common products as an alternative to as-mined fireclays. Patterson and Murray (1983) classified fireclays according to their service temperatures (or PCE) under these general groups:

- Low—PCE of 19 to 26
- Moderate—PCE of 26 to 31.5
- High—PCE of 31.5 to 33
- Superduty—PCE of 33 to 34

Fireclay may also need to adhere to specifications on fired shrinkage and plasticity.

Fireclays can be used as a cement or binder in refractory applications and for making refractory brick. Fireclay bricks are used in acidic applications such as linings of blast furnaces, coke ovens, and annealing ovens. Fireclay and bentonite are used to bind foundry sands used in metal casting molds. Fireclay-bonded molds result in a smoother metal surface than bentonite-bonded molds, but such a mold requires approximately three times as much fireclay than a bentonite-bonded mold.

Brick, Structural Ceramics, and Cement Common clays and shales are used in brick and structural clay products. Structural clay products include drainage tile, sewer pipe, and terra cotta roofing tiles. The specifications for bricks are not as strict as for other ceramics. In general, brick clays should have moderate plasticity, high dry strength, a long vitrification range, and a fired shrinkage of less than 10 percent (Harries-Rees, 1993). Free quartz should be present to give strength and durability, and some sand may be added to regulate the plasticity and increase the porosity of the brick.

Materials that are undesired in brick clays include (Harries-Rees, 1993): a high carbon content, which can lead to shattering during firing; calcium carbonate, which can cause defects after firing; pyrite, which will release sulfur dioxide (an air pollutant) leading to bloating and efflorescence during weathering of the brick; and fluorine, which is an environmental air pollutant and can corrode the furnace. Pipe made from common clay requires more exacting standards than bricks. The mineralogy should consist of equal parts kaolinite, mica, and quartz with less than 1 percent carbon and less than 0.2 percent SO_3 (Harries-Rees, 1993).

Lightweight aggregate can be made by rapidly heating common clays and shales, resulting in expansion of gases as water and organic materials volatilize. The result is a porous aggregate that can be used in concrete. Cement, the other ingredient in concrete in addition to aggregate, requires a source of silica and alumina that can be supplied by common clay and shale. The specifications are not stringent, and the essential requirement is that the quarry be located in close proximity to the cement plant.

Catalysts Kaolin and calcium bentonite can be used in the manufacture of some catalysts (Murray, 2002). For cracking catalysts, the clays should have very low iron content. Bentonite is reacted with acid and then washed, dried, and calcined to remove adsorbed alkali and alkaline-earth metals from the exterior of the crystal lattice and to remove iron and some of magnesium and aluminum from the lattice. The result is an increase in surface area.

Fiberglass In fiberglass, kaolin is used as a source of alumina and, to a lesser extent, silica. In general, a low-iron, air-floated product is used. Typical specifications are 37 percent Al_2O_3, 44 percent SiO_2, <1 percent Fe_2O_3, <2 percent Na_2O, and <1 percent H_2O (Harben, 2002).

Health and Environmental Concerns

Clays are not considered hazardous to human health. However, there are some environmental impacts associated with clays during mining and during firing in the kiln. Hydrogen fluoride (HF) and other fluoride compounds that are present in clay in the range of 0.01 percent to 0.06 percent will be released as HF in the kiln at temperatures of 500°C to 600°C (EPA, 1998). However, the presence of CaO in the ceramic body may react with the HF to form calcium fluoride (CaF_2) and reduce potential HF emissions.

Clay mining has its own set of environmental impacts. The removal of overburden and clay disturbs the local ecosystem through wildlife displacement and the potential for surface and groundwater contamination through erosion and leaching of metals. Clay processing produces wastewater and tailings that must be properly treated and impounded. In the United States, mines are required to level disturbed areas and to plant vegetation. The rules for disposal of coarse tailings are included within laws governing reclamation of the mined area (Virta, 2003a).

5.3 PYROPHYLLITE, TALC, AND WOLLASTONITE

Introduction

Talc, wollastonite, and pyrophyllite are mineralogically and compositionally distinct, but all three play a similar role as a primary or secondary flux in ceramic bodies and glazes. Pyrophyllite is a hydrous aluminosilicate with a nominal composition of $Al_2Si_4O_{10}(OH)_2$, talc is a hydrous magnesium silicate with a nominal composition of $Mg_3Si_4O_{10}(OH)_2$, and wollastonite is a calcium metasilicate with a nominal composition of $CaSiO_3$. Although they can be used in a similar fashion, care should be taken because the compositional and mineralogical differences lend advantages and disadvantages to each material. Because pyrophyllite contains alumina, it is superior in the manufacture of some refractory products, where a grade that is low in sericite (a fine-grained form of muscovite mica) is used. In porcelain, high-sericite pyrophyllite can reduce pyroplastic deformation and increase fired strength. Talc makes an excellent source of magnesium in ceramic and glaze formulations, particularly for slow-fired wall tile and cordierite products. Relatively pure or steatitic talc can been used to make thermal and electrical insulators. Wollastonite is valuable as a source of calcium for bodies and glazes that, unlike calcium carbonate, does not evolve carbon dioxide gas at high temperature. Wollastonite has been used commercially for only the past 50 years. Despite its relatively higher cost compared to talc or pyrophyllite, wollastonite is desirable in ceramic bodies because of its whiteness, its lack of compositional water, and its acicular or needlelike crystal habit that improves green strength, permeability, and dimensional stability. Pyrophyllite, talc, and wollastonite all find use as a filler and a reinforcing agent in other products, such as paints, paper, and plastics.

Composition and Mineral Structure

Despite appearing as white powders when delivered, pyrophyllite, talc, and wollastonite have differing compositions and, therefore, different crystal structures. Table 5.7 reports some of the properties of these minerals, along with some of the minerals commonly associated with talc. It should be noted here that commercial pyrophillite and talc products can contain several other minerals, and this is addressed later.

Pyrophyllite Pyrophyllite derives its name from the Greek words meaning fire and leaf, after its tendency to break into platy leaves when heated. It is a dioctahedral 2:1 sheet silicate with a structure similar

Table 5.7 Properties of pyrophyllite, talc, tremolite, chlorite, and wollastonite

	Formula	Specific Gravity	Mohs Hardness	H_2O (wt. %)	Heating Characteristics
Pyrophyllite	$Al_2(Si_4O_{10})(OH)_2$	2.65–2.90	1–2	5	450–850°C H_2O loss 1200°C conversion to mullite and cristobalite
Talc	$Mg_3\ Si_4O_{10}(OH)_2$	2.58–2.83	1	5–6	600–800°C H_2O loss >–800°C conversion to enstatite and cristobalite
Tremolite	$Ca_2(Mg,Fe)_5Si_80_{22}(OH)_2$	2.9–3.2	5–6	2	
Chlorite	$(Mg,Fe,Al)_6(Si,Al)_4O_{10}(OH)_8$	2.55–2.75	2	10–12	600–850°C H_2O loss
Wollastonite	$CaSiO_3$	2.8–2.9	5	0	1200°C conversion to pseudowollastonite

to muscovite mica, in which an octahedrally coordinated sheet of aluminum lies between two sheets of silica tetrahedra (Figure 5.21). The nominal composition is $Al_2Si_4O_{10}(OH)_2$, with very little ionic substitution in the lattice occurring in natural samples. The weak interlayer bonding leads to a mineral that is very soft with a greasy feel. It has a Mohs hardness of 1 to 2 and a specific gravity of 2.65 to 2.9. The color of pyrophyllite is pearly white to grayish, sometimes with a brownish tinge. It is regarded as chemically inert and resists attack from strong acid and basic solutions, although it will break down in the presence of sulfuric acid (Ciullo and Thompson, 1994). When heated, pyrophyllite begins to dehydrate between 450°C and 850°C. At 1200°C, it will convert to mullite and cristobalite, although the temperature of this phase change is affected by impurity elements such that the conversion in commercial products can occur between 1000°C and 1100°C (Singer and Singer, 1964).

Talc Talc is a sheet silicate mineral with a structure similar to pyrophyllite, but with magnesium inhabiting the octahedral sites rather than aluminum (Figure 5.21). It is a trioctahedral layered mineral, which means that all octahedral sites are occupied and the layers are electrically neutral. As a result, there are no interlayer cations and the layers are held together by weak electrostatic bonds. Similar to pyrophyllite, the weak interlayer bonding produces a very soft mineral with a greasy feel. In fact, talc is the reference starting point of the Mohs scale with a value of 1. However, talc products can be much harder due to the presence of associated minerals. For example, the hardness of talcose rocks from New York can be as high as 6 (Emrich and Hannon, 1964) as a result of the presence of tremolite. The mineral talc has a nominal composition of $Mg_3Si_40_{10}(OH)_2$ with little variation in natural specimens. There can be substitution of Si^{4+} by Al^{3+} and Ti^{4+} and substitution of Fe^{2+} and Fe^{3+} for Mg^{2+} (Deer et al., 1992).

Pure talc is foliated when viewed in a hand sample (Figure 5.22). Massive talcose rocks that are composed predominantly of the mineral talc are referred to as *steatite,* while an impure talcose rock is called *soapstone*. Both can be easily carved into shapes and have historically been used to make electrical insulators, wood stove walls, and acid-resistant laboratory counters and sinks. Individual talc platelets range in size from 1 μm to 100 μm. Natural talc deposits can be described based on the size of the crystal platelets. Large platelets make a *lamellar* or platy talc, whereas small platelets make a *microcrystalline* or compact talc rock. The size of the crystals is important because this can affect processing. Another physical characteristic of talc that can

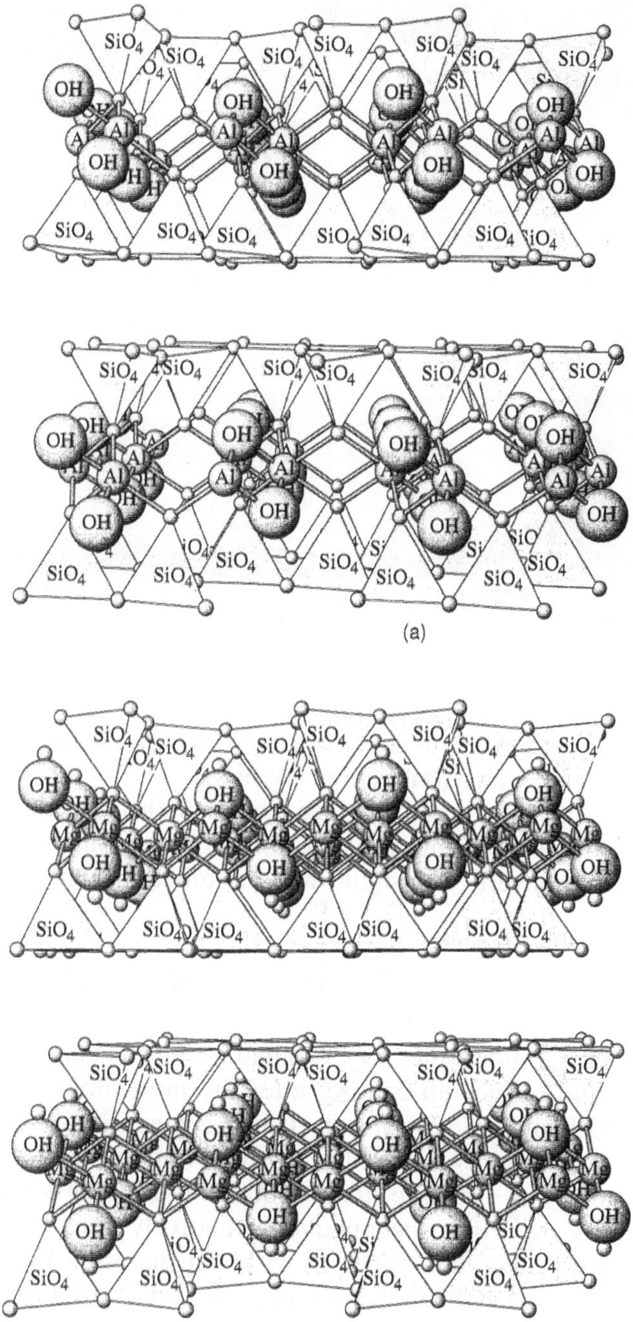

(a)

(b)

Figure 5.21 Idealized structure of pyrophyllite and talc (from Gaines et al., 1994; used with permission from John Wiley & Sons).

Figure 5.22 Photograph of a hand sample of talc (used with permission from R.T. Vanderbilt).

affect processing behavior is its natural hydrophobicity, so it is not readily wetted by water (but can be by oils).

Upon heating, talc begins to lose bound hydroxyl between 600°C and 800°C (Singer and Singer, 1964). On further heating beyond 800°C, talc begins to convert to enstatite and cristobalite represented by the reaction:

$$Mg_3Si_4O_{10}(OH)_2 \rightarrow 3MgSiO_3 + SiO_2 + H_2O$$

Wollastonite Wollastonite was originally called table spar, but was renamed in the nineteenth century after the mineralogist and crystallographer William Hyde Wollaston. It is a chain silicate similar to the pyroxenes (pyroxenoid) that, unlike pyrophyllite and talc, has no constituent water in its structure. Its nominal composition is $CaSiO_3$, and most naturally occurring specimens are relatively pure, although they are associated with other minerals within a deposit. A very limited amount of iron, manganese, strontium, and magnesium can replace calcium in the crystal lattice. Pure wollastonite is a brilliant white, but impurities will give it a grayish hue. Wollastonite occurs as acicular or needlelike crystals (Figure 5.23) with aspect ratios of processed materials ranging from 3:1 up to 20:1 and average diameter of 3.5 μm (Bauer et al., 1994). On grinding, wollastonite cleaves perpendicular to the crystal axis, which requires specialized processing to retain the higher aspect ratios. The mineral is harder than pyrophyllite or talc, with a hardness of 4.5 to 5.0 and a specific gravity of 2.9.

Upon heating, pure wollastonite will invert to pseudowollastonite at 1120°C to 1200°C and melt at 1540°C. The melting temperature of commercial wollastonite is somewhat lower (as low as around 1400°C) due to impurity minerals.

Geological Occurrence and Distribution

Pyrophyllite Pyrophyllite is a metamorphic mineral that forms during the alteration of aluminum-rich rocks such as granites, claystones, and bauxitic sediments. It can be found in zones where hydrothermal fluids were allowed to penetrate, such as faults and fractures, resulting in veins or lenses of ore within the host rock. Pyrophyllite occurs in three general forms (Deer et al., 1992):

Figure 5.23 Close-up photograph of wollastonite from New York State showing the acicular nature of the mineral (used with permission from R.T. Vanderbilt).

1. *Foliated,* which is fine grained and platy
2. *Radial* crystals and needles
3. *Massive,* compact spherulitic aggregates of smaller crystals

All three forms are used in ceramics, with the massive type preferred for refractory applications and the foliated variety is used in whiteware applications. Pyrophyllite deposits are generally impure and contain accessory minerals such as quartz, clays (kaolinite, illite, and sericite; sericite is a general term for a fine-grained, white mica), diaspore, kyanite, alunite (a potassium aluminum sulfate), feldspar, and hematite.

Pyrophyllite is found on virtually all the continents (Figure 5.24), but Asia is the greatest producer of pyrophyllite, with the Republic of Korea as the largest source, followed by Japan (Virta, 2002b). The deposits in both these countries appear to be geologically related and were formed by the hydrothermal alteration of silicic volcanic rocks and high-alumina shales (Ciullo and Thompson, 1994). The pyrophyllite is found in irregular lenses and veins and is associated with other minerals, including quartz, sericite, kaolinite, and diaspore.

Brazil is the largest producer of pyrophyllite in South America, and third in the world (Virta, 2002b). The majority of mined pyrophyllite comes from a 64 km-long belt of deposits in Minas Gerais (Ciullo and Thompson, 1994). Brazil produces a product called agalmatolite, which is a mix of pyrophyllite and either sericite, diaspore, quartz, or kyanite (O'Driscoll, 1993a). In North America, pyrophyllite is mined in North Carolina and Newfoundland. In North Carolina, it occurs as irregular lenses within a belt of metamorphosed volcanic rocks known as the Volcanic Slate Series, which lies northeast and southwest across the eastern Piedmont.

Talc Talc is the product of alteration of magnesium silicates in ultramafic rocks that contain olivine ($[Mg,Fe]_2SiO_4$) and enstatite ($MgSiO_3$) or of regional alteration of dolomitic carbonates by silica rich hydrothermal fluids. Where ultramafic rocks are the parent material, talc is often associated with serpentine ($Mg_3Si_2O_5[OH]_4$) and olivine. Where siliceous dolomites are the parent rock, talc may be associated with the amphibole minerals tremolite and anthophyllite (a hydrated magnesium silicate amphibole), serpentine (as antigorite), chlorite, carbonate minerals, and quartz. Tremolite (Table 5.7) has a composition of $Ca_2(Mg,Fe)_5Si_8O_{22}(OH)_2$. Tremolite and anthophyllite do not occur in the asbestiform in commercial talc products. Chlorite (Table 5.7) can have a widely variable composition approximated by the general formula $(Mg,Fe,Al)_6(Si,Al)_4O_{10}(OH)_8$ and has a layered structure simi-

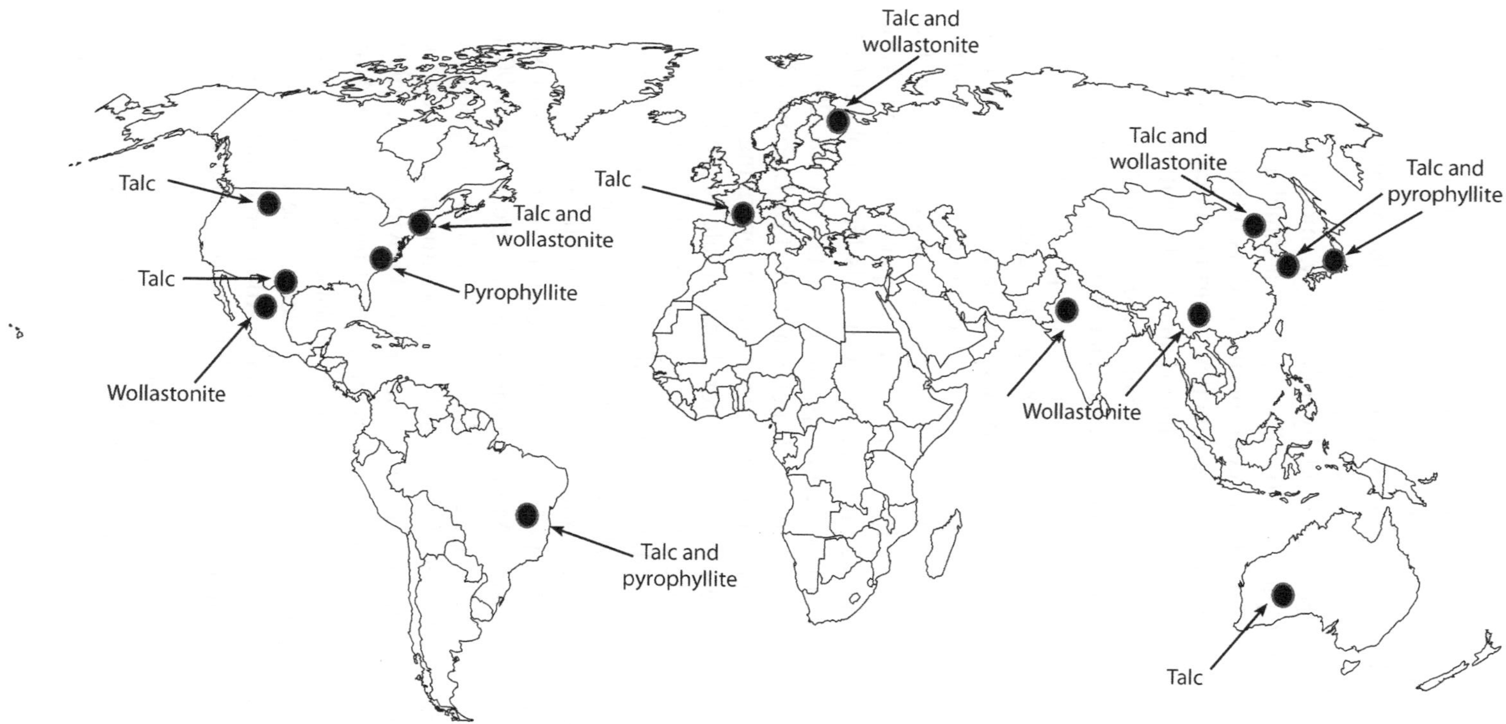

Figure 5.24 World map showing the major commercial deposits of pyrophyllite, talc, and wollastonite.

lar to talc, but is more aluminum rich (see section 5.2). Serpentine minerals, such as antigorite, have an approximate composition of $Mg_3Si_2O_5(OH)_4$. Antigorite is a platy or bladed variety of serpentine that is commonly found in New York talc.

Talc deposits are classified by mineralogy, morphology, and geographic source (Ciullo, 1996). *Platy talc* is a lamellar, soft talc, typically of more than 90 percent purity and is used in nonceramic, high-value products such as cosmetics, pharmaceuticals, and reinforcing fillers. *Tremolitic talc* is a fine-grained talc that usually contains less than 50 percent talc with the balance of tremolite and serpentine. Tremolitic and chloritic talcose rocks lack the degree of hydrophobicity of pure talc, which is beneficial in ceramic porcelain processing.

High-purity talc deposits (greater than 95 percent talc) are extremely rare. High-purity, massive, platy talc is mined in Montana (United States), Australia and China. Elsewhere, talc occurs in conjunction with other minerals, such as magnesite in the deposits of Vermont, Quebec, Ontario and Finland, with tremolite and serpentine in New York, and with chlorite in France and Austria (Materials Handbook, 2005).

China is the world's leading producer of talc (Virta, 2002b), with the majority coming from the Haicheng district of Liaoning Province. In Europe, fewer than 20 mines are currently operating (Figure 5.25), with Finland as the largest producer. Finnish talc is produced by froth flotation of soapstone and talcose schists found in the easter part of the country. In France, chlorite-bearing talc is mined in the Trimouns Mine in the Pyrenees (Piniazkiewicz et al., 1994). Italian talc is of very high purity and is generally not used in ceramic applications.

In North America, most talc used in ceramics comes from Montana, New York, and Texas. Montana talc contains greater than 90 percent talc, requires little beneficiation, and has a high dry brightness. It was derived from the alteration of Archean dolomitic marbles and is found as lenses and veins of steatitic talc (Piniazkiewicz et al., 1994). New York talc comes from the Governeur/Balmat region of the Adirondack Mountains. The commercial products consists of 35 to 60 percent talc, 30 to 55 percent tremolite, 20 percent serpentine (as antigorite), and minor anthophyllite. Texas talc from the Allamore District usually contains organic material (giving it a dark color), quartz, and dolomite (Ciullo, 1996).

Commercial talc deposits in South America are located in the Brazilian States of Bahia, Paraná, Rio Grande do Sul and São Paulo. Smaller commercial deposits are found in Minas Gerais and Goiás.

Figure 5.25 Map showing the major commercial talc deposits of Europe (based on Eurotalc, 2005).

Wollastonite Most commercial wollastonite deposits are formed by the metamorphism of siliceous limestones into skarn deposits, represented by the following reaction:

$$CaCO_3 + SiO_2 \rightarrow CaSiO_3 + CO_2$$

where SiO_2 is introduced through hydrothermal fluids. This reaction will occur only if the volatile CO_2 is allowed to escape from the system. Other minerals found with wollastonite are calcium-rich garnet, diopside, and calcite, depending on the original rock type that was altered. Wollastonite can also be found in volcanic carbonatites, but not as commercial deposits.

Commercial deposits of wollastonite are rare. In North America, the United States is the largest producer of wollastonite, with commercial production limited to two areas in northern New York State: the NYCO deposit in the eastern Adirondack region near Willsboro and the R.T. Vanderbilt deposit in Lewis County to the west. The NYCO deposit ore is made of about 60 percent wollastonite with 30 percent garnet and 10 percent diopside (Bauer et al., 1994). The R.T. Vanderbilt deposit was formed under similar conditions as the NYCO deposit, but instead of garnet as an accessory mineral, it is associated with calcite, prehnite (a hydrous calcium aluminosilicate mineral), and diopside. In Mexico, wollastonite is mined in Zacatecas State, and deposits in the Sonoran Desert are being developed.

In Asia, China is the largest global producer of wollastonite with three commercial mines operating, two in the northern Jilin Province and one in the southern Yunnan Province (Bauer et al., 1994). In India, a high-purity (96 percent) wollastonite is mined in the western state of Rajhastan.

In Europe, wollastonite is mined in southeastern Finland from a deposit that was formed by the alteration of an Archean limestone. Wollastonite mineral concentrate is a product of flotation separation of mined calcite. Most Finnish wollastonite is powder-grade with low aspect ratios (O'Driscoll, 1993b). Synthetic calcium metasilicates are of interest to some European countries due to the expense of importing natural wollastonite (Bauer et al., 1994). This can be made through hydrothermal means using limestone and a silica source, such as diatomite (Ibañez and Sandoval, 1998) or by heating a mixture of sand, chalk, and dolomite. However, these synthetic materials are hydrous, do not have the acicular crystal habit, and cost about twice that of natural wollastonite. Synthetic wollastonite contributes less than 2 percent of the world production capacity (Hawley, 2002).

Mining and Processing

Most pyrophyllite, talc, and wollastonite mines are open pit with a few exceptions where high-purity veins and lenses are worth the expense of underground mining (Figure 3.3). All three materials have different processing routes, depending on the nature of the deposit, the associated minerals, and the end use. Higher-quality products will require more processing steps and, therefore, carry a higher cost.

Pyrophyllite is usually processed using dry methods that include crushing, screening, and separation with air cyclones (Ciullo and Thompson, 1994). At the mine, ore is broken into manageable sizes using a bulldozer or small explosive charges. The material is sorted visually and placed into separate stockpiles, which are later blended and crushed to smaller than 16 mm using jaw crushers and gyratory crushers. Further size reduction is done in impact crushers, after which the ore is sent to ball-milling circuits. Air classifiers are used to obtain a 200 mesh particle size and magnets are used to remove tramp iron and undesirable ferromagnetic minerals.

Most talc ores undergo primary crushing at the mine or at the surface near an underground mine before being conveyed to the mill. In New York, up to three different grades of tremolitic talc ores are carried from the mine to the plant to be crushed and screened: hard, ceramic, and medium ore (Rieger, 1980). The hard and ceramic grades are destined for ceramic uses. At the mill, the material may be dried prior to secondary crushing in roller mills or ball mills. The material is then classified in air cyclones. Froth flotation may be used to obtain a high-purity concentrate from the less pure ores, but not those in the Gov-erneur, New York, deposit. Talcose rocks that contain calcite and organic material, such as those mined in Texas, may be calcined to burn off the organic material, release the carbonate from calcite, and reduce the platiness of the mineral.

For wollastonite, processing is dependent on the specific mine. In New York, the NYCO deposit contains garnet and diopside that is removed using high-intensity magnetic separators. The R.T. Vanderbilt deposit in Lewis County consists primarily of wollastonite with minor amounts of calcite, prehnite, and diopside. After primary crushing, the ore stream is handpicked to remove the accessory minerals. The remaining ore is milled using proprietary techniques to retain the acicularity of the crystals and to control the aspect ratio. Air classification is used to size the final products before packing and storing. However, care must be taken, as agglomeration of wollastonite can occur during storage.

Uses and Specifications

Pyrophyllite Pyrophyllite used in ceramics is available in two general grades: refractory and ceramic (Table 5.8). The grades are based on the presence of accessory minerals that, depending on the applications, may be beneficial or detrimental. In refractory applications, the presence of quartz and kaolin can increase performance while the presence of sericite mica, pyrite (FeS_2), hematite (Fe_2S_3), and feldspar can act as fluxes and decrease high-temperature performance. However, the fluxing minerals, with the exception of those that are iron-bearing, can be beneficial if pyrophyllite is used as a flux in porcelain ceramics.

For refractory grade, pyrophyllite must have less than 1 percent total alkali (and therefore a low sericite content) and less than 1 percent

Table 5.8 Sample analyses of commercial pyrophyllite, talc, and wollastonite products (wt. %)

	SiO_2	Al_2O_3	Fe_2O_3	CaO	MgO	MnO	Na_2O	K_2O	PCE	LOI
Talc										
Theoretical	63.4				31.9					4.7
Texas talc[1]	60.6	1.24	<0.69	9.59	27					11.5
Montana talc[2]	61	0.5	1.3	0.3	31					5.2
New York tremolitic talc[3]	55.2	0.31	0.16	8.42	30	0.17	0.34			5.4
Chloritic talc[4]	46.3	9.5	1.9	0.7	32					8.8
Pyrophyllite										
Theoretical	66.7	28.3								5.0
High-sericite pyrophyllite[5]	80.9	13.8	0.2				0.4	2.3	20–21	2.3
Refractory-grade pyrophyllite[6]	75	19.3	0.8	0.1	0.1		0.1	0.2	27–28	3.9
Wollastonite										
Theoretical	51.7			48.3						
New York wollastonite[7]	50.0	1.8	0.3	44	1.5		0.2		18	2.2
Finland wollastonite[8]	53.0	1.0	0.25	43.2	0.8					

[1] Zemex Pioneer 2882
[2] Luzenac Yellowstone as reported in IMIL (2000)
[3] R.T. Vanderbilt Nytal
[4] from Harben (2002)
[5] R.T Vanderbilt Pyrax HS from North Carolina—20–25% Sericite, 50–60% quartz, and 5–10% kaolin
[6] R.T Vanderbilt Pyrax RG—30–34% quartz and 5–15% kaolin
[7] R.T. Vanderbilt Vansil
[8] O'Driscoll (1993b)

other fluxing materials such as iron and titanium oxides (Ciullo, 1996; Harben, 2002). It is used in refractories due to its permanent expansion on firing, thermal stability, excellent heat cycling characteristics, creep resistance, low thermal conductivity, and high resistance to corrosion by molten metals and basic slags. In refractories, pyrophyllite is often mixed with kaolinitic clays. Specific applications include metal pouring ladles, kiln car furniture, insulating firebrick, ramming and gunning mixes, foundry mold coatings, and refractory mortars.

For *ceramic grade,* the preferred product is higher in alkali to facilitate faster firing, so higher sericite content is advantageous. In a triaxial porcelain body, pyrophyllite can replace some of the quartz and feldspar or nepheline syenite. Low FeO, Fe_2O_3, and TiO_2 is required for whiteware applications to retain fired whiteness. Pyrophyllite can improve thermal shock resistance due to a low coefficient of thermal expansion and reduced fired shrinkage and warping. It may decrease crazing due to thermal shock or moisture expansion. High-sericite pyrophyllite will form mullite at relatively low firing temperatures, resulting in an increase in fired strength (Rieger, 1996).

Talc Talc is used in ceramic bodies and glazes as a source of magnesium. Specific uses are wall tile, porcelain, and cordierite bodies. Desired characteristics of commercial talc products include the following:

- Low iron to retain fired whiteness
- Low carbonate content to reduce high-temperature gas evolution
- Uniform chemical and mineralogical composition
- Consistent fired shrinkage
- Good pressing characteristics, particularly if used in wall tile

Ceramic grade talc should have a minimum of 30 percent MgO and 60 percent SiO_2 and a maximum of 4 percent Al_2O_3, 1.5 percent Fe_2O_3, and 0.4 percent total alkali (Industrial Minerals Information, 2000; Harben, 2002).

Talc is often used in wall tile formulations because it is reported to reduce crazing in glazed semivitreous ware in two ways. First, the conversion to enstatite produces a high-thermal expansion body that places the glaze in compression. Second, the low-moisture expansion reduces delayed crazing (Rieger, 1996; Ciullo and Thompson, 1994). Tremolitic talc from New York and partially calcined talc from Texas are often preferred for wall tile because they can be dry pressed without lamination and because tremolite contributes further to high uniform thermal expansion and low moisture expansion that prevents crazing of glazed products (Ciullo, 1996). A typical fast-fired tile body (fired in less than 30 minutes) consists of 10 percent to 25 percent talc, 25 percent to 35 percent ball clay, 10 percent to 20 percent wollastonite, and 20 percent to 40 percent pyrophyllite. Wollastonite can be added to improve pressing and permeability of the body if platier talc is used in the formulation (Industrial Minerals Information, 2000).

In wet processing of ceramic bodies, talc is an excellent source of MgO because, unlike dolomite or magnesite, it is relatively insoluble in water. However, talc is naturally hydrophobic, and when used in a suspension, additives may be required to make it hydrophilic. The use of tremolitic talc will reduce the hydrophobicity of the talc product. In glazes, talc adds MgO without evolving carbon dioxide during firing, which can cause pinholing or other defects. However, high-temperature evolution of water from talc may contribute to these problems.

Historically, steatite was carved into shapes to make electrical insulators. Today, steatite is synthesized using a mix of 80 percent talc, 10 percent kaolin, and 10 percent barium carbonate that is fired at cone 12 to 13 (Materials Handbook, 2005). Purity of the talc is critical, as any addition of alkali or calcium can affect the performance of the product. The microcrystalline variety of talc rather than the platy type is preferred for this application. During firing, the transformation of talc to cristobalite and enstatite is desirable as both have excellent electrical insulation properties.

Talc is used in the manufacture of cordierite ($2MgO$-Al_2O_3-$5SiO_2$) materials, such as kiln furniture and catalysts supports. The important compositional parameters are high MgO and SiO_2 and low CaO and alkali elements. Iron in minor quantities (2 to 3 percent) is acceptable. It is incorporated into the cordierite crystal lattice without impairing the properties of the cordierite (Mondo Minerals, 2004). A typical body for cordierite used in automotive catalyst supports contains 44 percent talc, 41 percent plastic kaolin, and 15 percent alumina (Materials Handbook, 2005). Alternatively, a 50/50 mixture of kaolin and chloritic talc (containing 45 percent chlorite) can be used to make a cordierite body. The resulting product is preferred for kiln furniture because of its lower cost and better high temperature creep resistance (Materials Handbook, 2005).

Wollastonite Wollastonite is used for its chemical components, CaO and SiO_2, as well as for its acicular crystal morphology. Compared to other sources of CaO, such as calcium carbonate (whiting, calcite, limestone), wollastonite has low loss on ignition (LOI) due to its lack of constitutional water and carbonate. This reduces bubbling and associated defects during firing, such as pinholing. The low alkali content improves its dielectric characteristics, and is therefore useful in electrical insulators. Its low coefficient of expansion and almost linear thermal expansion curve allows rapid heating and cooling and reduces cracking and warping, which is desirable for fast-fired products. Wollastonite has virtually no moisture expansion, reducing delayed crazing when added to a ceramic body (Rieger, 1996).

The acicular wollastonite crystals are desirable in ceramic bodies for several reasons. The elongated crystals can improve permeability by providing a conduit for vapors that are released during drying and firing. The crystals can also serve as reinforcement for green and fired bodies to improve green strength, fired strength, and heat shock resistance. In tile and dry-pressed whitewares, wollastonite can improve production rates through improved die fill and pressing characteristics.

When used in glazes, wollastonite imparts whiteness and brilliance. It is preferred to other sources of CaO because there is essentially no high temperature degassing during firing. It is used in matte and satin finishes and can lower surface gloss without increasing opacity. The addition of wollastonite reduces glaze viscosity, which can enhance spraying.

Wollastonite products used in ceramic applications are available with high aspect ratios or as powders with an average aspect ratio of 1:5. The powdered form is used more where the composition is more important than the acicularity of the crystals. In glazes, the powdered form is normally used with a 325 mesh particle size.

Health and Environmental Concerns

Pyrophyllite, talc, and wollastonite should all be used with some care, and inhalation should be minimized as with all industrial mineral products. All materials should be handled in a manner that minimizes dust generation. Good ventilation and dust respirators should be used at all times. Prolonged exposure to excessive airborne concentrations of talc and pyrophyllite can result in scarring of the lungs (pneumoconiosis). Wollastonite may be a source of mechanical abrasion of the skin, and excessive exposure to airborne particles may aggravate preexisting respiratory conditions. There is no credible link of cancer to long-term exposure of these materials.

Talc products, particularly those of metasedimentary origin, may contain significant amounts of fluorine (Piniazkiewicz et al., 1994).

This is of concern during the firing of ceramics containing a large portion of talc in the body, such as glazed wall tile, because of the potential for releasing fluorine emissions into the air. Fluorine can react to form hydrogen fluoride (HF), which is classified as a hazardous air pollutant under the United States Clean Air Act because of its acute toxicity and severe corrosiveness. Fluorine compounds can corrode kiln and furnace surfaces.

5.4 FELDSPAR AND NEPHELINE SYENITE

Introduction

Feldspars are the most common minerals in the earth's crust and are a primary component of virtually all rocks. They are light-colored crystals that consist of aluminum, silicon, and a mix of potassium, sodium, and calcium. Feldspars are classified by composition. Nepheline syenite is a relatively rare, silica-undersaturated rock (contains no free quartz) composed of alkali feldspars and the mineral nepheline, a feldspathoid mineral that has a higher alkali to silica ratio relative to the feldspars.

Feldspar and nepheline syenite are widely used in glass as a source of alumina and in ceramics as a source of alkali. In soda-lime-silicate glassmaking alumina is desired in small amounts to increase the durability of the glass and reduce devitrification. The alkali content added by feldspar or nepheline syenite to the glass batch is minor relative to the alkali derived from soda ash. In porcelain ceramics and in glazes, feldspar or nepheline syenite is used as an alkali source while adding minor alumina.

Although feldspar is a common mineral, deposits that are suitable for economical extraction are limited and are found principally in pegmatites, intrusive igneous bodies associated with granites (see section 2.3). Feldspar may also be separated from coarse granitic rocks that are rich in alkali feldspar and are low in iron, such as aplite and alaskite. Weathered, low-iron granitic rocks that are mined for kaolin, such as Cornish stone in England, can be used as a feldspathic material. Feldspars can be separated from arkosic alluvial sands, although the grains may be partially weathered to clay. Commercial nepheline syenite is a relatively uncommon igneous intrusive rock with low-iron examples currently limited to deposits in Ontario and Norway. The majority of mined feldspars and nepheline syenite are used in glass and ceramics. These minerals can also be used as a mild abrasive and as a filler in paints and polymeric materials.

Composition and Mineral Structure

Feldspars are the most common minerals in the Earth's crust and are a primary component of virtually all rocks. They are light-colored crystals that consist of aluminum, silicon, and a mix of potassium, sodium, and calcium that are arranged in a silicate framework (tectosilicate) in which SiO_4 and AlO_4 tetrahedra are connected to make a continuous, three-dimensional crystal lattice (Figure 5.26). The general compositional formula for feldspar is XZ_4O_8 where X is an alkali or alkaline earth ion and Z is occupied by tetrahedrally coordinated aluminum and silicon in ratios of either 1:1 or 1:2.

Common feldspars are classified by two solid solution series, the alkali feldspars and the plagioclase feldspars, using three end-member compositions with a range of intermediate compositions (Figure 5.27). The alkali feldspars fall between albite ($NaAlSi_3O_8$) and orthoclase or microcline ($KAlSi_3O_8$). A wide range of alkali feldspar compositions occurs by the simple substitution of potassium and sodium, as well as some minor substitution of calcium on the albite end of the spectrum. Alkali feldspars with compositions that lie between albite and orthoclase may be called soda feldspars if they are more sodium-rich or potash feldspars if they are more potassium rich. One specific term is anorthoclase, which refers to a sodium-rich alkali feldspar with about 60 percent to 90 percent of the albite component. Plagioclase feldspars fall between the anorthite ($CaAl_2Si_2O_8$) and albite ($NaAlSi_3O_8$) end-members and are formed by the coupled substitution of Ca-Al for Na-

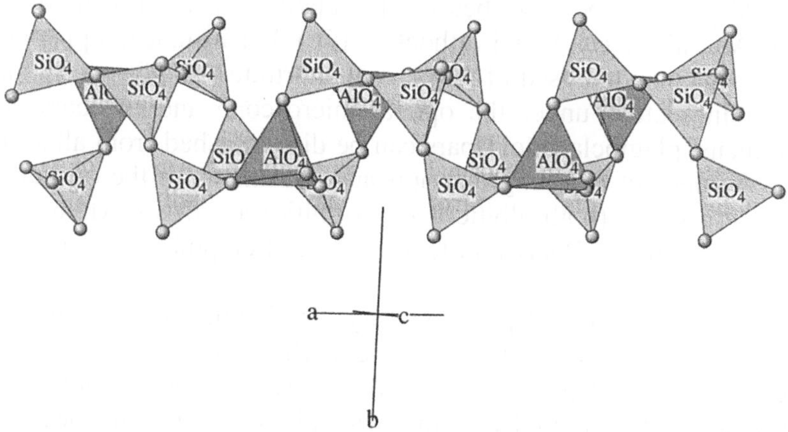

Figure 5.26 Partial structure of low albite showing "crankshaft" structure of silica and alumina tetrahedra (from Gaines et al., 1994; reprinted with permission from John Wiley & Sons).

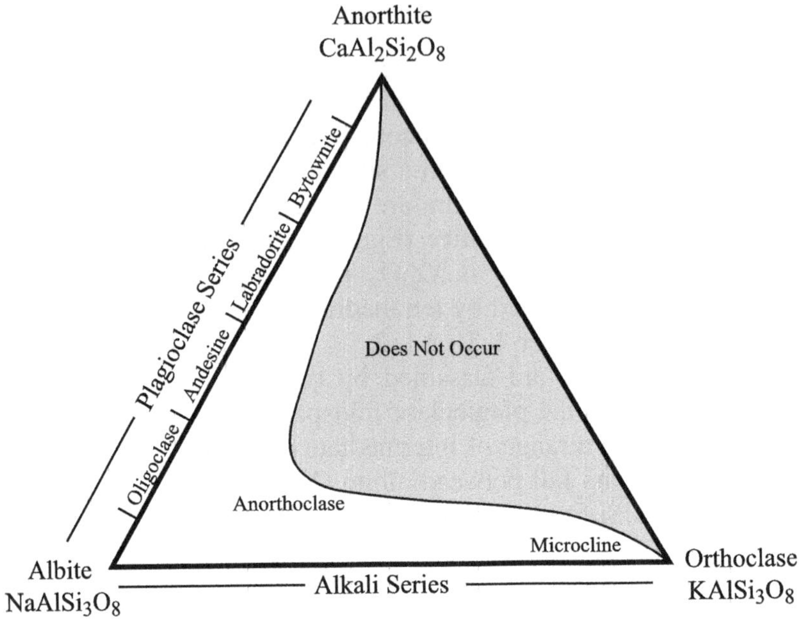

Figure 5.27 Ternary diagram showing the three end-member compositions of feldspars and names of some of the solid solution compositions.

Si. There are specific compositional names for the plagioclase series, which are, from calcium-rich to sodium-rich: anorthite, bytownite, labradorite, andesine, oligoclase, and albite. There are no feldspar compositions that fall between anorthite and orthoclase, although a small amount of the potassic orthoclase can be found in plagioclase feldspars. All feldspars can be described by the relative molar amounts of anorthite (An), albite (Ab), and orthoclase (Or). For instance, a plagioclase that is An_{50} is composed of 50 percent anorthite and 50 percent albite.

In thin section under the optical microscope and in coarse hand specimens, plagioclase feldspars can be distinguished from alkali feldspars by distinct twinning that appear as striations in the crystal. Plagioclase crystals (both albitic and anorthitic) are frosty white to gray in hand specimens. Potassium feldspars are often pink, but can be white as well.

The structure of feldspars can vary, depending on the temperature of crystallization and cooling rate under which they formed. The structure can be classified into high temperature, disordered minerals and low temperature ordered minerals with phase separation (Deer et al., 1992). The high-temperature potassium feldspar is called sanidine while the low-temperature feldspars of similar composition are microcline or orthoclase. Orthoclase and microcline have the same composition but differ in crystal symmetry, with the former being monoclinic and the latter triclinic. Orthoclase is stable at 500°C to 900°C and microcline is stable at 400°C or less. In general, alkali feldspars that form in a slowly cooling environment such as a granitic intrusion will become more ordered. In a hypersolvus granite or pegmatite, there may be one alkali feldspar that forms from the melt that will experience exsolution upon cooling, in which a sodium-rich and potassium-rich feldspar phase form intergrown lamellae. This is referred to as a *perthitic* texture and may be observed in hand sample or under the microscope. Many natural alkali feldspars and the majority of commercial feldspars come from slowly cooled plutonic rocks. Subsolvus rocks will form two separate alkali feldspars from the cooling melt, one potassium-rich (orthoclase) and one sodium-rich (albite). Crystals of both feldspars may have intimate intergrowths of quartz. Alkaline volcanic rocks that are cooled quickly, such as a volcanic rhyolite, will contain a disordered feldspar, such as sanidine, that will lack exsolution textures. Sanidines are not found in ore deposits, but can be an accessory mineral in bentonitic clays (bentonites are formed by weathing of rhyolite).

The plagioclase series do not display exsolution textures that are common in alkali feldspars, although they are commonly zoned such that concentric bands within a single crystal can have significant compositional differences. This is due to the exposure of the crystal to variations in temperature and composition during formation and the extremely slow kinetics that restricts compositional re-equilibration.

Substitutions within feldspar minerals can be of concern to the ceramist or glass scientist. In the potassium feldspars, TiO_2 content is low, but there may be up to 0.5 percent Fe_2O_3 held within the crystal lattice. Iron that is in excess of 0.5 percent may be attributed to sources outside of the crystal structure, such as inclusions, coatings, or material exsolved during cooling (Deer et al., 1992). In pegmatitic alkali feldspars, rubidium may be present in amounts up to 1 percent.

The alkali feldspars are relatively soft with a Mohs hardness of 1 to 2 compared to the hardness of 5 to 6 for anorthite (Table 5.9). Anorthite also has a higher specific gravity (2.9 to 3.2) compared to albite (2.58 to 2.83) and orthoclase (2.65 to 2.90).

Nepheline is a colorless to white sodium-rich feldspathoid that has a nominal composition of $Na_3(Na,K)(Al_4Si_4O_{16})$. It can be considered a silica-deficient feldspar. It has a hexagonal symmetry with a structure similar to tridymite where SiO_4 and AlO_4 tetrahedra form a three-

Table 5.9 Formula and some properties of feldspar and nepheline

	Formula	Specific Gravity	Mohs Hardness
Orthoclase	$KAl_2Si_3O_8$	2.65–2.90	1–2
Albite	$NaAl_2Si_3O_8$	2.58–2.83	1
Anorthite	$CaAl_2Si_3O_8$	2.9–3.2	5–6
Nepheline	$(Na,K)AlSiO_4$	2.63	2

dimensional framework (Figure 5.28) with charge-balancing cations occupying the interstices (Deer et al., 1992). Relative to the theoretical formula, almost all nepheline minerals contain more SiO_2 than Al_2O_3 on a molar basis. The structure can allow up to 1 percent Fe_2O_3 and CaO. Nepheline occurs in a wide range of alkalic igneous rocks, but its commercial value is its occurrence in nepheline syenite. Nepheline syenite is a light-colored, coarse- to medium-grained rock resembling granite but with a significantly different mineral composition. Syenite

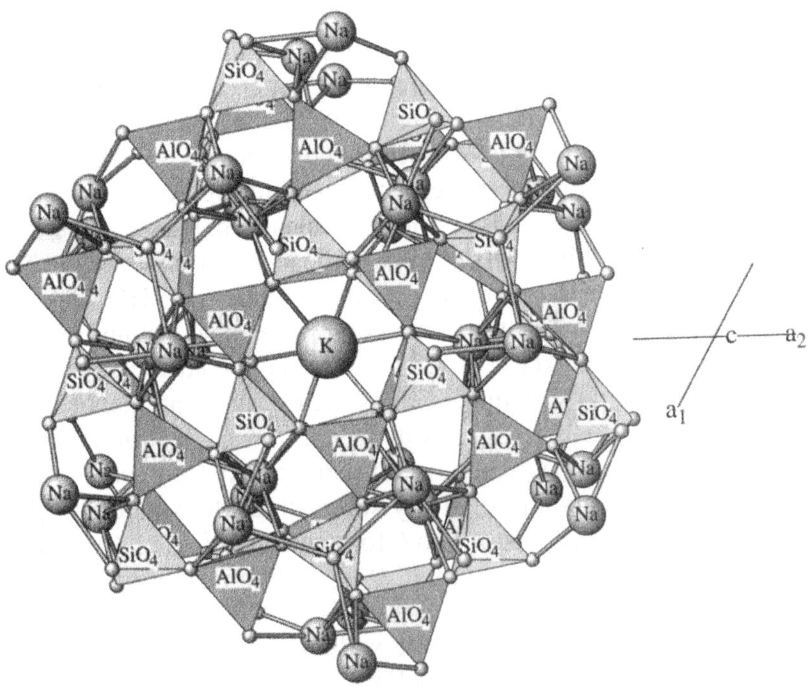

Figure 5.28 Idealized structure of nepheline (from Gaines et al., 1994; reprinted with permission from John Wiley & Sons).

is a general term describing an igneous plutonic rock of 55 percent albite, 20 percent alkali feldspar, 25 percent nepheline, and accessory minerals of mica, hornblende, corundum, and magnetite. There is no free quartz in the mineral assemblage because the magma from which syenite forms is deficient or undersaturated in silica. Compared to the bulk composition of granite, nepheline syenite has lower SiO_2, higher Al_2O_3, and higher total alkali.

The melting characteristics of soda feldspar, potash feldspar, and nepheline syenite differ due to the differences in composition. Pure orthoclase melts incongruently to leucite ($KAlSi_2O_6$) and liquid at 1150°C and albite melts congruently at about 1100°C. The minimum melting temperature of the alkali feldspar series is 1060°C at 38 percent potash feldspar (Deer et al., 1992). In the presence of silica (as tridymite), the binary eutectic between alkali feldspar and silica is 990°C. Anorthite, although not commonly used as a raw material, is the highest temperature feldspar, melting congruently at 1570°C. Nepheline syenite rock melts around 1020°C.

Geological Occurrence and Distribution

Feldspar Feldspars are an essential constituent of virtually all igneous and many metamorphic and sedimentary rocks. As a general rule, the alkali feldspars are more common in continental rocks, such as granite, graniodiorite, and syenite. Calcium rich plagioclase is usually associated with magnesium- and iron-rich igneous rocks such as basalt and andesite. The relatively rare rock anorthosite is composed predominantly of plagioclase feldspar and is found in New York's Adirondack Mountains, Quebec, and the moon.

Virtually all commercially mined feldspars are alkali feldspars associated with granitic and pegmatitic intrusions. Upon cooling of the intrusions, one or two feldspars crystallize along with quartz and a hydrous accessory mineral, such as a mica or hornblende. Any iron in the magma will usually partition into the accessory minerals. The total amount of alkali feldspar in granite is between 50 percent and 70 percent. However, granite is not normally considered as an ore for feldspar, although the weathering of granite can produce a mixed clay, feldspar, and quartz product, such as Cornish stone. By far the most sought-after feldspar deposits are pegmatites (see Chapter 2), as they host extremely large, high-purity crystals (Figure 5.29) that are relatively easy to process. Pegmatites may also be mined for high-purity quartz, lithium minerals, and rare earth minerals as well. Another type of rock that is

Figure 5.29 Photo of a single potassium feldspar crystal mined from a pegmatite in central Argentina. The scale bar is one inch (sample courtesy of Piedra Grande).

an ore for commercial feldspar is *alaskite*. Alaskite is a uniform pegmatitic granite that is composed of 45 percent plagioclase, 25 percent quartz, 20 percent potassium feldspar (microcline), 10 percent muscovite mica, and trace amounts of garnet (Parker, 1952; Marek and Sinton, 2000). Alaskite intrusions can be up to several kilometers in diameter and be closely associated with pegmatites. *Aplite* is a fine-grained, low-iron granitic rock with a sugary texture. It is used as a mixed feldspar and quartz product. Immature or *arkosic* sands contain quartz and 10 percent-35 percent partially weathered feldspar (Kauffman and Van Dyk, 1994). In areas where pegmatitic feldspar is not available economically, these sand deposits can be processed to produce commercial feldspathic products.

Feldspathic materials are commercially mined in most parts of the world (Figure 5.30). Italy is the leading feldspar and aplite producer in Europe and the rest of the world, supplying more than 25 percent of global consumption. Feldspars are mined in several other locations in Europe as well (Eurofel, 2005). In France, feldspar is produced in Saint-Paul de Fenouillet-Salvezines in the West of Perpignan, in Montebras in the Creuse, and in Etang sur Arroux near Autun. In Germany, feldspar is mined in Türkismühle, Saar. In Spain, feldspar sand deposits in Carrascal del Rio, Segovia were derived from weathering of granite

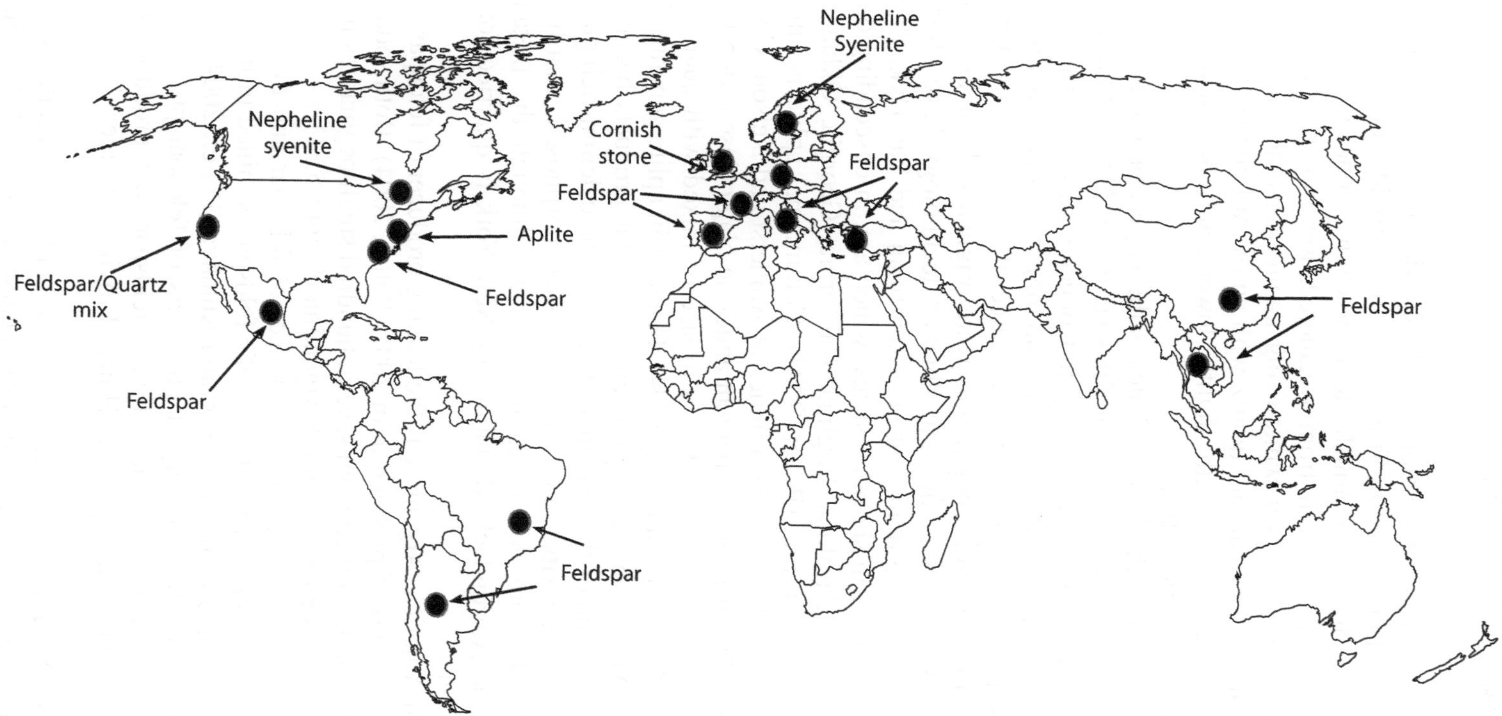

Figure 5.30 World map showing the major areas of feldspar and nepheline syenite mining.

and gneiss from the northern slopes of the Guadarrama Mountains. The ore is mined and processed by froth flotation to provide for the ceramics and glass industries. Cornish stone or China stone, is mined in Plymouth, United Kingdom from weathered, low-iron granitic rocks that are also the source for white-firing kaolin. Commercial Cornish stone is a mix of feldspar and quartz with an overall composition that is similar to feldspar, but with slightly lower total alkali and higher volatile content (Table 5.10).

In North America, the United States is the largest producer of feldspar and aplite. The production of feldspar in the United States first began in the early nineteenth century in Connecticut, where pegmatites were mined and hand sorted. These are no longer commercially mined, and now North Carolina is the major source of feldspar in the United States. As in Connecticut, pegmatites were originally the main source of feldspar, but advances in froth flotation technology have allowed alaskite to become the dominant ore (Kauffman and Van Dyk, 1994). Feldspar was mined in six other states, which are, in descending order of output, North Carolina, Virginia, California, Georgia, Oklahoma, Idaho, and South Dakota (Potter, 2004). Commercial aplite is limited to occurrences in the Virginia counties of Amherst and Nelson (Chang, 2002), consisting of about 90 percent andesine plagioclase (30–50 percent anorthite) crystals up to 35 cm in size of mixed with quartz and biotite (Kauffman and Van Dyk, 1994). Feldspar in aplite generally has the highest Al_2O_3 content of commercial feldspar products because it is a plagioclase rather than an alkali feldspar. The calcium-rich plagioclase feldspar deposits anorthosite and metamorphosed anorthosite (metanorthosite), are not mined for feldspar, although the metanorthosite in the Adirondack Mountains of New York is quarried for dimension stone.

In South America, Argentina is the leading producer of feldspar, with 60,000 metric tons of production in 2002 (Potter, 2004). Both soda and potash feldspar are mined from several hundred small pegmatites found in the San Luis region of central Argentina (Figure 5.31).

In Asia, Turkey is the largest producer of feldspar and the second largest global producer after Italy. Sodic feldspar is mined in the provinces of Manisa, Bilecik, Balikesir, and Kutahya (Potter, 2004). China and Thailand are the next largest producers of feldspar products followed by India and Australia. Chinese feldspars are produced from pegmatites, predominantly in the Hunan Province (Bolger, 1997).

Nepheline Syenite Commercial-quality nepheline syenite deposits are rare and should have a medium to coarse texture with less than 5 percent mafic (iron and magnesium silicate) minerals. The mafic min-

Table 5.10 Compositions of some feldspathic products (wt. %)

	SiO_2	Al_2O_3	Fe_2O_3	TiO_2	CaO	MgO	Na_2O	K_2O	LOI
				Potash Feldspar					
Theoretical	64.76	18.32						16.92	
United States[1]	66.3	18.5	0.08		0.81		3.04	10.75	0.16
Argentina[2]	66.2	18.5	0.06	0.1	0.2	0.1	3.2	11.3	
Norway[3]	65.9	18.6	0.07		0.4		2.9	11.8	0.2
				Soda Feldspar					
Theoretical (An_5)	67.39	20.35			1.07		11.19		
United States[4]	68.2	18.9	0.07		1.4		6.82	4.1	0.09
Argentina[5]	70.4	18.4	0.3	0.1	0.1	0.05	7.5	2.5	
Sweden[6]	75.7	14.1	0.15	0.02	0.3	0.1	5	3.8	0.5
				Anorthite					
Theoretical (An_{95})	44.40	35.84			19.20		0.56		
Canada[7]	47.9	33.1	0.32		16.1	<0.05	2.12	0.03	0.3
				Nepheline Syenite					
Norway[8]	56	24	0.1	0.1	1.6		7.8	8.8	
Canada[9]	60.2	23.5	0.08		0.6		10.2	4.9	0.45
China[10]	55.1	22.7	0.24	0.07	1.09	0.22	6.2	12.7	0.7
				Other Products					
Cornish stone[11]	72.8	15.5	0.16	0.05	1.5	0.1	3.1	4.3	1.7
Blast furnace slag[12]	36.5	12.0	0.25	0.5	40.0	9.0	0.5	0.5	1.1

Table 5.10 *(Continued)*

[1] Spruce Pine Zemex G200
[2] Piedra Grande Special potash grade
[3] Dalehall Mills NorfloatK (Sugden, 2001)
[4] Spruce Pine Zemex NC-4
[5] Piedra Grande APG #10
[6] Dalehall Mills Forshammer (Sugden, 2001)
[7] Avalon Ventures
[8] North Cape Minerals Ataflux 45 (IMIL, 2000)
[9] Sugden (2003)
[10] Geiju NS (Sugden, 2002)
[11] Cornish Stone 80:20 blend (Sugden, 2001)
[12] Kendall (1993)

Figure 5.31 Photograph of open-pit mining of a small pegmatite in central Argentina.

erals, such as hornblende and magnetite, can be separated after the rock is ground to 30 mesh (Harben, 2002). Currently, there are two major sources of nepheline syenite in the world. The 1.3 billion-year-old Blue Mountain deposit in Ontario, Canada, is mined by Unimin Canada, Ltd., which produced 710,000 metric tons in 2004 (Potter, 2004). The other deposit is on Stjernoya Island, Norway, and is mined by North Cape Minerals, producing 340,000 metric tons in 2004 (Potter, 2004). China also has nepheline syenite deposits, with a large reserve in Yunnan Province (Sugden, 2002), but these have yet to be fully exploited.

Mining and Processing

In feldspar processing, the source, the labor costs, and the desired final product specifications are going to dictate the processing route. Historically, pegmatite feldspars were hand sorted or cobbed, a method that continues today in areas where labor is inexpensive. Where labor is relatively expensive or the ore is too fine grained for hand sorting, crushing followed by froth flotation is more common.

The mining of feldspar and nepheline syenite is similar to most other hard rock mining operations. Once a potential deposit has been identified, it may take up to three years before a mine is brought to production (Marek and Sinton, 2000). Virtually all mining is done by open pit methods using explosives to initially break up the rock into a series of benches. The rock is collected by a front end loader then it is crushed and screened at the mine before being trucked to a central processing facility. Hand sorting may be done at pegmatite mines before being sent to the central facility.

At the central facility, hand-sorted pegmatitic feldspars are crushed and ground to the desired size. Feldspar ores, such as alaskite and feldspathic sands, must undergo processing to concentrate the feldspar. Froth flotation is the most common method of processing alaskite from the Spruce Pine deposit (Kauffman and Van Dyk, 1994; Marek and Sinton, 2000). Alaskite ore is initially crushed with jaw crushers, followed by secondary crushing in a cone or gyratory mill to a size of about 25 mm. Water may be sprayed on the ore to control dust. The material is then sent to a grinding circuit of steel rod mills or ceramic ball mills that can each process 70 tons of material per hour. After grinding, an air classifier is used to remove the 30 mesh (600 μm) particles and send the oversize back to the grinding mills. The finer material is at optimal size for the flotation process, although hydrocyclones are used to remove particles smaller than 38 μm (desliming) prior to flotation.

Mineral separation is done in three flotation steps. The first step is to remove the mica by adding an amine cationic collector and sulfuric acid to bring the pH to 3. The mica is removed and can be marketed for other products such as brake linings. The second step is the removal of iron-bearing minerals (garnet and biotite) by adding an anionic collector such as a petroleum sulfonate. At this point, the mix of quartz and feldspar can be dried and processed as a mixed product or the two minerals can be separated by adding more amine and hydrofluoric acid, which depresses the quartz from floating and allows the feldspar to be removed. Both the quartz and feldspar products are passed through high-intensity magnets to remove any tramp steel or iron that may have been introduced during processing. The quartz can be used in the semiconductor industry or even sold to golf courses for professional tournaments. The feldspar is further ground to the desired particle size.

The composition of feldspars within a group of mined pegmatites can differ, so careful testing and stockpiling needs to be done to control the final product. An example is the La Toma facility in Argentina. Feldspars are mined from up to 200 different, relatively small pegmatites (Figure 5.31). Feldspars from each pegmatite are tested for composition and stockpiled separately (Figure 5.32). To make a feldspar

Figure 5.32 A front-end loader moves feldspar from one of many stockpiles at the La Toma plant in Argentina. The material is sent to the mill in the background.

product of a desired composition, calculated amounts from specific stockpiles are delivered to the crusher, where they are mixed.

The processing of nepheline syenite is relatively simple. It is crushed first with a jaw crusher with secondary crushing in a cone crusher to 30 mesh. The iron-bearing accessory minerals are removed with magnetic and high intensity electrostatic separation techniques (Harben, 2002).

Uses and Specifications

Glass and ceramic manufacturers are the largest users of feldspar and nepheline syenite, accounting for 85 percent to 90 percent of global feldspar production and the majority of nepheline syenite production (Harben, 2002). Most commercial feldspars fall within the compositions of microcline/orthoclase, andesine, albite, and oligoclase. Commercial soda feldspars have between 5 percent to 7 percent Na_2O and 19 percent Al_2O_3 while potash feldspars have at least 10 percent K_2O and 17 percent to 18 percent Al_2O_3.

Glass In the glass industry, feldspar and nepheline syenite products are used as sources of alumina and secondary sources of alkali (soda ash is the primary source). Alumina is added to many common soda-lime-silicate glasses in amounts up to 1 percent to 3 percent of the final glass composition to improve hardness, durability, and resistance to chemical corrosion. Container glasses generally have higher alumina levels, up to 7 percent, to improve workability in high speed forming systems. However, even small additions of alumina can greatly increase the high-temperature viscosity of the melt, and therefore alumina-bearing minerals need to be added carefully. Feldspar and nepheline syenite are considered to be better sources of alumina than bauxite or Bayer alumina in soda-lime-silicate glass because the alkali in the rock promotes incorporation into the melt while bauxite or Bayer alumina can be more difficult to melt. Nevertheless, some specialty glasses use high purity alumina in order to strictly control undesirable elements such as iron. Fiberglass compositions that do not have sodium, such as E-glass, do not use feldspar for alumina, relying on kaolin instead.

The choice of which material to use (soda feldspar, potash feldspar, or nepheline syenite) depends mainly on the overall economics and some processing considerations, such as desired Na:K ratio of the final product. In terms of economics, soda feldspar is usually the least-expensive option, with potash feldspar almost twice the cost. In some glasses, a final composition that has a mixed alkali of both sodium and potassium may be beneficial in terms of durability and ion exchange strengthening (Sinton et al., 2000; Sinton and LaCourse, 2001). In this case, a potash feldspar can be used to add some K_2O to the composition while soda ash gives the majority of the alkali as Na_2O.

Nepheline syenite can be cost competitive with soda spar. However, the economic factors will be decided by the delivered cost, which is driven by the location of the end user relative to the potential suppliers. For instance, a manufacturer in the northeastern United States may want to use Canadian nepheline syenite because the source is closer than feldspar from North Carolina or aplite from Virginia. Nepheline syenite can deliver more Al_2O_3 per ton (~24 percent) compared to feldspar (<18 percent), which reduces overall costs because less material is used in the batch for the same amount of alumina, with the balance of silica added with less expensive sand. The higher proportion of total alkali in nepheline syenite (~16 percent) compared to soda feldspar (<11 percent) means a slight reduction in the need for soda ash, which is the most expensive component of a soda-lime-silicate glass batch.

Feldspar or nepheline syenite may be added to the batch in proportions of up to 0.5 percent for flat glass and 8 percent for containers (Harben, 2002). A common particle size range of the material is 20 to 40 mesh (600 to 425 μm) (Kendall, 1993). In general, a product with low fines is desirable because contaminants associated with grinding often concentrate in the fines. Table 5.11 shows compositional specifications for nepheline syenite and feldspathic materials used in container glass. In addition to the chemical composition and size (which must be consistent over time), refractory mineral particles, such as zircon and aluminum silicate minerals, must be as low as possible to reduce defects in the glass. Additional specifications for feldspar used in glass container manufacturing include (Harben, 2002):

- No quartz grains are larger than U.S. 16 mesh (1.19 mm).
- No clay particles are larger than U.S 20 mesh (0.84 mm), or more than 50 particles. larger than U.S. 30 mesh (0.59 mm) per 10 pounds of sample.
- No kyanite, mullite, sillimanite, or andalusite are larger than U.S. 30 mesh (0.59 mm), or more than 10 grains larger than U.S. 40 mesh (0.42 mm) per 10 pounds of sample.

Table 5.11 Standards for feldspathic materials used in container glass (from Harben, 2002)

	Low-iron Nepheline Syenite	High-iron Nepheline Syenite	Aplite	Feldspathic Sand[1]
SiO_2	±0.5%	±0.5%	±0.5%	±0.25%
Fe_2O_3	0.1% max.	0.35%	0.1% max.	0.05% max.
Al_2O_3	±0.5%	±0.5%	±0.5%	±0.25%
CaO+MgO	—	—	—	±0.05%
CaO	±0.5%	±0.5%	±0.5%	—
Na_2O+K_2O	±0.5%	±0.5%	±0.5%	±0.15%
ZrO_2	—	—	—	±0.01%
Cr_2O_3	0.001% max.	0.001% max.	0.001% max.	0.001% max.
Moisture	0.1% max.	0.1% max.	0.1% max.	0.1% max.

[1]Flint grade

- No refractory minerals like corundum are larger than U.S. 40 mesh (0.42 mm); and
- No zircon, cassiterite, or chrome particles are larger than U.S. 60 mesh (0.25 mm).

Substitutes for feldspar and nepheline syenite include clays, iron blast furnace slag, and glass cullet (cullet and slag are discussed in Chapter 7). Iron blast furnace slag contains about 12 percent Al_2O_3, but contains very little alkali and slightly higher Fe_2O_3 relative to feldspathic products (Table 5.10). Sulfur compounds in the slag can act as fining agents as well. Slag can be used in both flat and container glass batches in proportions of 4 percent to 8 percent, depending on the tolerance for iron (Kendall, 1993). Steel furnace slag is not suitable for glassmaking.

Ceramics In the ceramic industry, feldspar or nepheline syenite that is 200 mesh (75 μm) or smaller is the principal fluxing material in a triaxial porcelain body. Common proportions of either material are 15 percent to 30 percent for dinnerware, 25 percent to 35 percent for sanitaryware, 30 percent to 50 percent for electrical porcelain, and 10 percent to 50 percent for wall tile. The Fe_2O_3 content should not exceed 0.07 percent. Feldspar is used in ceramic glazes in amounts ranging from 30 percent to 50 percent to provide alumina and alkali (Kendall, 1993).

In the porcelain industry, soda feldspar has been the traditional flux because of the lower cost and more powerful fluxing ability (on a molar basis). However, soda feldspars are generally not as pure in the sense that there is always some calcium due to solid solution in the plagioclase series, whereas potash feldspars ideally have no calcium. In the production of electrical insulators, potash feldspar is preferred because of the lower conductivity of the final product. Because mole for mole, sodium reduces the viscosity of the glass phase more than potassium at a high temperature, potash feldspar can reduce pyroplastic deformation. Potash feldspar may also result in a more translucent, stronger and more heat resistant ceramic piece compared to soda feldspar. It is reported that the substitution of nepheline syenite for feldspar in ceramic bodies can result in lower firing temperatures, longer firing range, decreased warpage, lower variation in thermal expansion, and lower moisture expansion (Harben, 2002). In enamels, potash feldspars are preferred over soda spar, as the latter can produce fish scaling (Kendall, 1993). The addition of rubidium-rich feldspars in spark plugs and other electrical insulators promotes high dielectric capacity.

Health and Environmental Concerns

Feldspar and nepheline syenite are not associated with any environmental or health hazards. However, products that contain free crystalline silica of greater than 0.1 percent may be regulated.

5.5 LIMESTONE, DOLOSTONE, AND MAGNESIA

Introduction

Calcium and magnesium are two common alkaline earth elements that are widely used in both the ceramic and glass industry as a flux and, in the case of magnesia (MgO), as an important refractory material. These two elements are commonly found in nature in combination with carbon and oxygen as the calcium carbonate minerals calcite and aragonite, which make up the rock limestone, and the mixed calcium and magnesium carbonate, dolomite. Both limestone and the dolomite-bearing dolostone are very common sedimentary rocks that are found in virtually all regions of the earth. The pure magnesium carbonate mineral, magnesite, is somewhat less common. Other sources of magnesium carbonate are brines and seawater. Oxides of calcium and magnesium, which are rarely found in nature, are made by calcining the carbonate minerals to drive off carbon dioxide.

Limestone and dolomite are integral ingredients to commercial soda-lime-silicate glasses, serving to assist in fluxing the melt while enhancing the durability of the product. Ground calcium carbonate, or whiting, can be added to ceramic glazes or bodies as a source of calcia (CaO). Dead burned dolomite and magnesia (MgO or periclase) are important refractory materials, with magnesia having one of the highest melting points of the refractory oxides (up to 2800°C). Magnesia-based refractory products, sometimes mixed with chrome or alumina, are widely used in metal, glass, and cement manufacturing facilities.

Glass and ceramics make up only a very small portion of the market for limestone and dolostone, which are widely used in aggregate, dimension stone, cement, papermaking, flue gas desulfurization, and metallurgical fluxes.

Composition and Mineral Structure

Carbonates are common rock-forming minerals, all containing the carbonate (CO_3^{2-}) ion that is most often paired with an alkaline earth ion. Since calcium and magnesium are the most common alkaline earth elements, carbonates with these two are naturally the most common. These are calcite/aragonite ($CaCO_3$), dolomite ($CaMg[CO_3]_2$), and magnesite ($MgCO_3$). The carbonate ion will also combine with other metals, such as ferrous iron to form siderite ($FeCO_3$), which makes a solid solution with magnesite. Sodium carbonates are discussed in section 5.6.

Calcite and Aragonite Calcium carbonate ($CaCO_3$) minerals are some of the more abundant minerals on the Earth's crust and occur in at least five polymorphs, although the two most common at atmospheric pressure are calcite and aragonite (Deer et al., 1992). The structure of calcite is similar to halite (NaCl) and forms crystals with trigonal (hexagonal) symmetry. The carbonate group forms a triangle of a central carbon atom, with oxygen atoms at each apex and with the calcium atoms in octahedral coordination with the oxygen atoms (Figure 5.33). When pure, calcite is colorless and transparent to white with a specific gravity of 2.71 and a Mohs hardness of 3 (Table 5.12). Aragonite has the same composition as calcite, but has a more compact, orthorhombic structure, and the divalent cations are not strictly planar as they are in calcite. This structure can accommodate other cations such as strontium and lead. Pure aragonite has a specific gravity of

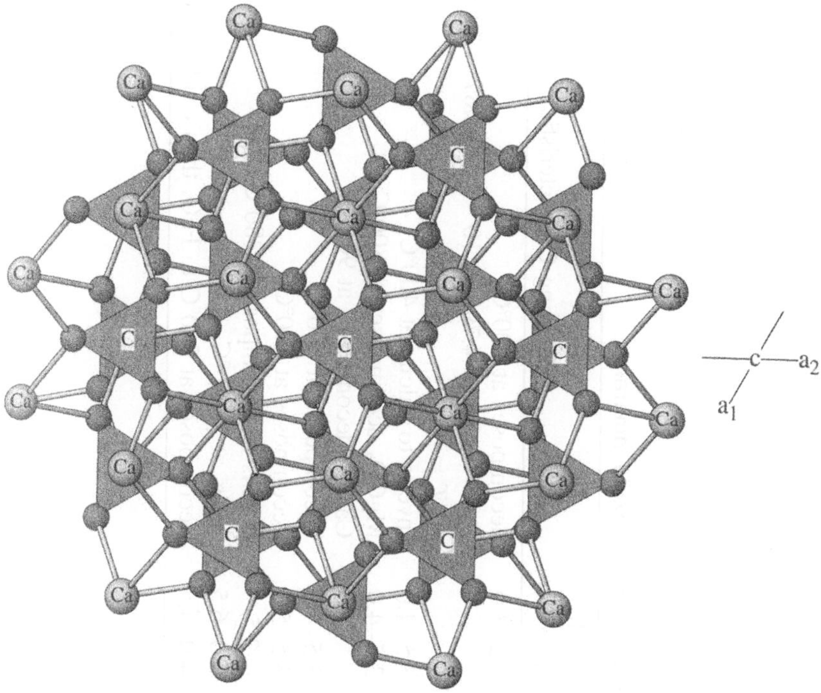

Figure 5.33 Idealized structure of calcite (from Gaines et al., 1994; used with permission from John Wiley & Sons).

2.91 and Mohs hardness of 3.5 to 4 (Table 5.12). Aragonite is metastable at atmospheric pressures and will invert to calcite over time.

In general, calcite will dissociate to CaO and CO_2 at just under 900°C:

$$CaCO_3 \rightarrow CaO + CO_2$$

However, the heating rate can affect the actual temperature of dissociation. At a heating rate of 10°C/min, the decomposition begins at 650°C and is complete at 758°C (Figure 5.34). At a heating rate of 200°C/min, calcite is present to 800°C, and is fully decomposed to CaO by 900°C (Misture, 2004).

Dolomite Dolomite is a mixed 1:1 molar ratio calcium and magnesium carbonate with a nominal composition of $CaMg(CO_3)_2$, which is 54 percent $CaCO_3$ and 46 percent $MgCO_3$ by weight. Its structure is similar to calcite, with alternating planar layers of carbonate and cations. It can be considered to have one layer of calcite alternating with

Table 5.12 Properties of some calcium, magnesium, and iron carbonate and oxide minerals

	Formula	Specific Gravity	Mohs Hardness	Heating Characteristics
Calcite	$CaCO_3$	2.71	2.5–3.0	decomposes at 900°C $Cp = 82.34 + 4.795 \times 10^{-2}T + 12.87 \times 10^5T^2$ (T = °K)
Aragonite	$CaCO_3$	2.95	3.5–4.0	inverts to calcite at 400°C
Dolomite	$CaMg(CO_3)_2$	2.85	3.5–4.0	$MgCO_3$ decomposes at 800°C $CaCO_3$ decomposes at 900°C
Magnesite (crystalline)	$MgCO_3$	3.0–3.2	3.5–4.5	decomposes at 800°C
Magnesite (cryptocrystalline)	$MgCO_3$	2.9–3.0	3.5–5.0	decomposes at 800°C
Magnesium hydroxide (brucite)	$Mg(OH)_2$	2.38	2.5	begins to decompose at 380°C
Periclase	MgO	3.65	6.0–6.5	melts at 2800 °C
Siderite	$FeCO_3$	3.9	4.0–4.5	decomposes at 580°C to FeO, then oxidizes to Fe_2O_3

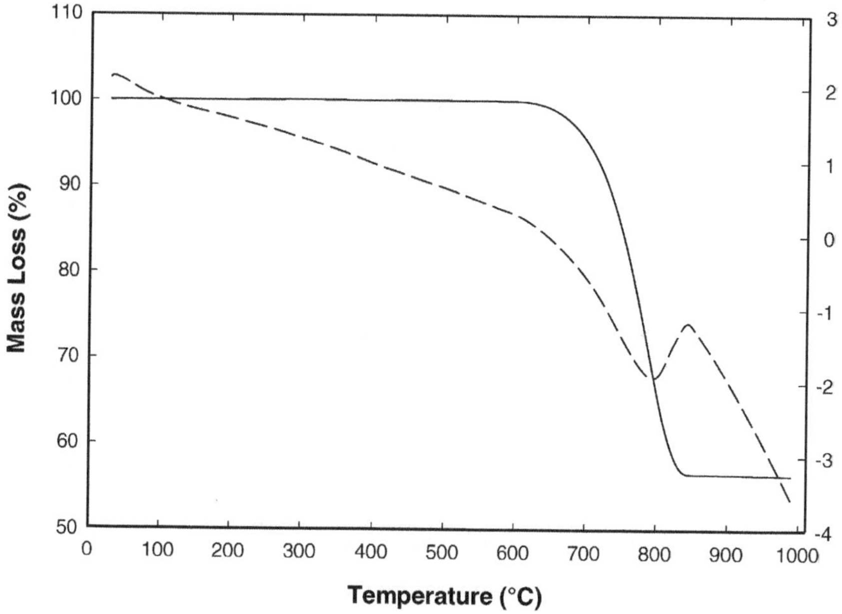

Figure 5.34 DTA-TGA curves of $CaCO_3$, with the solid line representing mass loss and the dashed line representing ΔT (10°C/minute heating rate). The endothermic peak at 758.3°C indicates the dissociation of $CaCO_3$. The melting temperature for CaO is above 1000°C; therefore, the endothermic peak associated with melting is not seen (courtesy of U. Kim).

one layer of magnesite (Deer et al., 1992). There can be some substitution of Fe^{2+} for Mg^{2+}, giving the normally colorless-to-white mineral a brownish hue. It has a specific gravity of 2.85 and is slightly harder than calcite with a Mohs hardness of 3.5 to 4 (Table 5.12).

Upon heating, dolomite follows a two-step decomposition (Figure 5.35). First, the magnesite component dissociates at 776°C:

$$CaMg(CO_3)_2 \rightarrow CaCO_3 + MgO \text{ (periclase)} + CO_2$$

followed by the breakdown of the calcite component at 883°C:

$$CaCO_3 \rightarrow CaO + CO_2$$

Periclase has a melting point of 2826°C, and CaO melts at 2927°C (Lide, 1995).

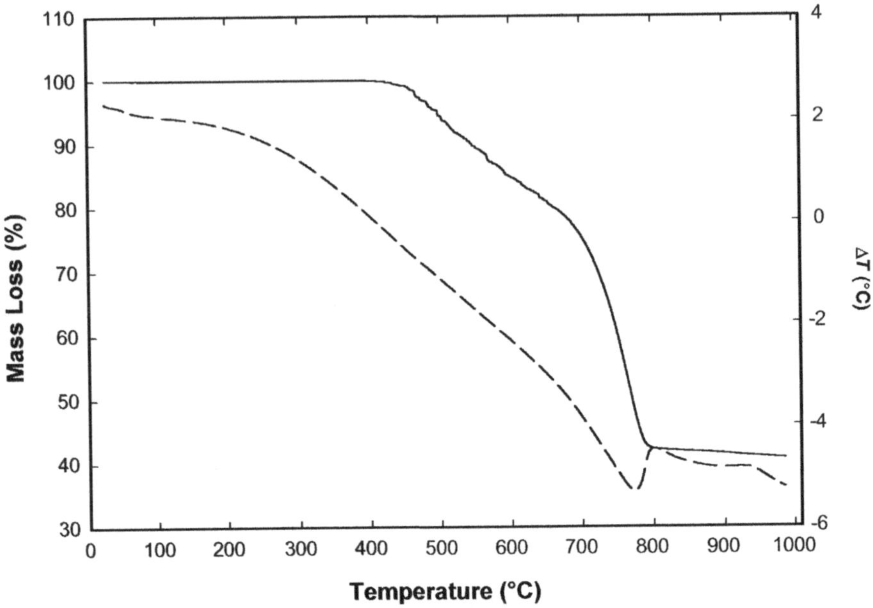

Figure 5.35 DTA-TGA curves of dolomite, with the solid line representing mass loss and the dashed line representing ΔT (10°C/minute heating rate). Two endothermic peaks are shown at 776°C and 883°C (courtesy of U. Kim).

Magnesite Magnesite ($MgCO_3$) has the same trigonal structure as calcite, but is slightly smaller due to the smaller size of the magnesium ion. There is a complete solid solution between magnesite and siderite ($FeCO_3$) and substitutions of iron will give the normally vitreous white to gray color a brownish tinge. The variation in cell dimensions using X-ray diffraction can be used to estimate the relative amounts of magnesite and siderite in a sample (Chang, 2002). Crystalline magnesite has a hardness value of 3.5 to 4.5 and specific gravity is 3.0 to 3.2, while the cryptocrystalline variety has a slightly lower a specific gravity of 2.9 to 3.0 and a similar hardness value (Table 5.12). Magnesite dissociates at between 650°C and 900°C to periclase and carbon dioxide:

$$MgCO_3 \rightarrow MgO + CO_2$$

Magnesite is slightly soluble in cold water (0.0106 percent by weight; Seeger and Werke, 2002) and soluble in acid. The thermal conductivity of the pure crystals is 22 W/m-°C at 300°C and 7 W/m-°C at 1000°C and for technical grade material 12 W/m-°C at 300°C and 5 W/m-°C at 1000°C (Seeger and Werke, 2002). A hydrated form of magnesite, referred to as hydromagnesite $3MgCO_3$-$Mg(OH)_2$-$3H_2O$, is found naturally or can be made by heating a mixed solution of magnesium sulfate and sodium carbonate.

Geological Occurrence and Distribution

Calcite is found in igneous, metamorphic, and sedimentary rocks, but it is the sedimentary sources that are of industrial importance. Sedimentary calcite and aragonite are formed in seawater and freshwater by chemical precipitation or from biogenic fixing by organisms such as plankton and shellfish. In the biogenic process, animals fix Ca^{2+} and carbon (as bicarbonate, carbonate, or carbon dioxide gas) dissolved in seawater to form either calcite or aragonite for shells and internal skeletons or tests. Whether or not calcite or aragonite is used depends on several factors, including the particular organism. When the animal dies, the shell or test (internal skeleton of plankton) will fall through the water column and, depending on the water depth and temperature, will either redissolve into the seawater or settle on the seafloor to form a sedimentary deposit. Present-day reefs are examples of current calcium carbonate deposits that may one day form limestone. Chemical precipitation of calcium carbonate occurs in warm seas to form oolites, spherical grains that appear to have crystallized around shell nuclei.

After sedimentation, the layers of calcium carbonate can be altered by post-depositional processes that will change the chemical and physical characteristics. The sediments will be compacted as more material collects on top and any aragonite will eventually invert to calcite. Water flowing through the deposit may dissolve some of the calcite and recrystallize it at the grain surfaces to cement the particles together in the process of diagenesis or lithification. Alternatively, silica-bearing fluids may deposit silica between adjoining grains. The resulting deposit is a *limestone,* a relatively broad term for a rock that consists of greater than 50 percent calcite or calcite and dolomite. As sedimentary layers, they are found in bedded units that may be bounded by other types of sediments, such as sands and clays. As with any other sedimentary deposit, they are often heterogeneous in chemical and physical characteristics.

The characteristics of the deposit can depend greatly on the initial composition of the sediments. If the deposit is made predominantly of the small tests of plankton, such as coccoliths and foraminifera, the deposit will be a soft, friable limestone referred to as *chalk.* Clay, sand (as quartz grains), detrital minerals (e.g., zircon, kyanite, ilmenite), and organic material may also be deposited intimately with the calcium carbonate to form a lower-purity limestone. Chert (amorphous SiO_2) can be particularly problematic as it can be disseminated throughout the rock as 1 to 10 μm particles. Chert can harbor impurities, and its high hardness can damage milling equipment (Carr et al., 1994).

Sediments deposited in high-energy environment (waves, currents) generally have fewer impurities of clay, sand, and organic material. Sediments that remain in low-oxygen conditions may preserve some of the organic material and sulfide minerals that pose significant problems in industrial applications. If a limestone is buried and subjected to elevated temperature and pressure, it will metamorphose and recrystallize to form marble. This can occur only if carbon dioxide is not permitted to escape. If carbon dioxide does escape, the carbonate is not stable and the limestone may be converted to wollastonite ($CaSiO_3$) if silica is present.

Dolostone is produced after deposition in a process referred to as *dolomitization,* in which magnesium-bearing fluids that penetrate the deposit exchange magnesium ions (and possibly iron) for calcium in the calcite structure. Virtually all dolomite is formed in this manner, and very little is actually formed as a primary mineral. This process will result in dolomite deposits, although varying amounts of calcite can still remain and a commercial dolomite product may be a mix of dolomite and calcite.

Since limestone and dolostone are intimately related, it is rare that a very pure deposit is encountered and most are a mix of limestone, dolostone, and noncarbonate minerals. Thus, these rocks can be classified by the varying amounts of these constituents (Carr et al., 1994) as follows:

- Ultra-high calcium carbonate limestone contains >97.5 percent $CaCO_3$
- High-calcium limestone contains >95 percent $CaCO_3$
- High-purity carbonate rock contains >95 percent combined $CaCO_3$ and $MgCO_3$
- High-magnesium dolostone contains >43 percent $MgCO_3$

Limestone and dolostone deposits vary widely in age from recent to hundreds of millions of years old. Many limestone deposits, such as the White Cliffs of Dover in the United Kingdom, were formed during high sea-level periods in the Cretaceous. Limestone and dolostone are found throughout the world and its relatively low value requires that sources be local to the manufacturer. Nevertheless, there are some high-purity limestone and dolostone deposits that have a high enough value for some products where significant transportation is feasible.

Examples in the United States include limestones of Missouri and Kentucky and dolostones found in Silurian-age reefs exposed in Ohio, northern Illinois, Indiana, and southern Michigan (Carr et al., 1994). Dolomite used in refractories is mined and processed in the United States in York, Pennsylvania. Off the coast of south Florida and the Bahama Bank are deposits of aragonite sand (oolites). Aragonite sands produce high-purity ores of 96 percent to 97 percent $CaCO_3$ with low organic and virtually no iron, and are sold to the glass industry for low-light attenuation glasses such as solar glass and clear flint containers.

In the United Kingdom, high-purity limestone deposits are found in Derbyshire, such as the Bee Low Limestone at Buxton, the Aldwarke Beds near Matlock (Antony, 2001), and an underground mine in Wirksworth (Manning, 1995). Higher-value dolostone is mined from the Permian Magnesian Limestone that is found from Nottingham to Yorkshire.

In other parts of Europe, high-purity dolostone is mined in Marone, Italy, the Cantabria region of Spain (Keegan, 1998) and offshore of Germany (Materials Handbook, 2005). In Belgium, there is a major dolostone mine near Neau, and a mine in Bodo, northern Norway, produces a dolostone of 99.5 percent purity (Chang, 2002).

Magnesite is mined as natural hard rock ores or is synthesized from seawater and brines. Hard rock magnesite is most often found as an alteration product of magnesium-rich silicate and carbonate minerals where CO_2 is present during metamorphism. The resulting deposits can be classified as: (1) cryptocrystalline, amorphous, or massive; and (2) crystalline or macrocrystalline (Duncan and McCracken, 1994). Cryptocrystalline magnesite is formed during the reaction of serpentine (formed by the alteration of ultramafic rocks) with CO_2-bearing fluids at shallow levels. It forms lenticular masses and veins and can be associated with other minerals in serpentine belts, such as talc and chlorite. Cryptocrystalline sedimentary deposits can form in semiarid basins that collect water that has weathered ultrabasic (MgO-rich) rocks (Seeger and Werke, 2002). Crystalline magnesite can form by the alteration of preexisting evaporitic sediments (Deer et al., 1992) or by the intrusion of magnesium-rich fluids into dolostone and limestone (Chang, 2002; Harben, 2002).

China is the largest global producer of magnesite, with more than a billion metric tons of hard rock, raw magnesite reserves, mainly in Liaoning province in northeast China (O'Driscoll, 2003). These low iron reserves are macrocrystalline with MgO content of 44 percent to 48 percent. Although the production of higher grades of magnesite in China have been limited in the past, froth flotation and other beneficiation techniques have been implemented to increase output (Keegan, 1998). In Australia, cryptocrystalline magnesite is mined in Queensland in the Kunwarara deposit located 70 km north of Rockhampton (O'Driscoll, 2003). India produces significant magnesite ores with mines in Madras State and Uttar Pradesh (Duncan and McCracken, 1994) and processing plants in Orissa, Tamil Nadu, and Uttar Pradesh (O'Driscoll, 2003). In North America, magnesite is found in Nevada, Washington, and British Columbia. In South America, Brazil is the largest producer with operations in Bahia. Other important magnesite deposits are found in Russia, Austria, Turkey, Slovakia, Spain, and Greece. The deposits in Austria and Slovakia are high iron and have a texture that does not lend itself to physical separation of the iron (Duncan and McCracken, 1994).

Synthetic magnesite and magnesia is derived from brines and seawater, the latter of which contains 1300 ppm magnesium. Magnesium chloride and magnesium sulfate dissolved in the brine can be precipitated to produce a magnesium hydroxide (see the following section). More than 50 percent of magnesium compounds produced in the United States are derived from these two sources (Kramer, 2002). Two-thirds of the production of refractory and caustic-calcined magnesia in the western countries comes from synthetic sources (Duncan and McCracken, 1994).

Mining and Processing

Most limestone and dolostone deposits are mined from open pits by blasting and mechanical shovels. However, there are many underground mines that are used to recover high purity limestone, chalk, and dolostone. Aragonite sands are retrieved by suction dredging. The hard stone is crushed and sized with screens. Some dolostone may have silicate impurities that can be removed by heavy media separation. Finer milling and classification can be subsequently used for the final product.

Magnesite is mined using both selective open pits and underground mines. Underground mines are used in mountainous terrain or where high-purity veins merit the cost. Cryptocrystalline magnesite can be sorted by hand or photometric means due to the pronounced color difference between the ore and gangue (Seeger and Werke, 2002). If the magnesite crystals contain intergrowths of undesirable minerals (such as talc, olivine, dolomite, and calcite) in the range of 0.03 to 1.0 mm, the ore must be ground to liberate the magnesite particles before being sent to a froth flotation system. Crystalline magnesite may be crushed and sorted by gravity concentration, photometric sorting, and/or magnetic separation (Harben, 2002).

Limestone, dolostone, and magnesite can be calcined to produce calcium and magnesium oxide materials. The temperature of calcination is determined by the end use, with lower temperature *burned* products having relatively high surface area and reactivity whereas the high temperature or *dead burned* variety have lower surface area, higher density, and are chemically inert. The lower temperature products may absorb water and therefore have variable mass, depending on storage conditions. Limestone or dolostone calcined at 1000°C will form quicklime (CaO) or dolomitic quicklime (CaO-MgO). These materials are very reactive in the presence of water and will become hydrated. Hydrated quicklime and dolomitic quicklime are referred to as *slaked lime* and *dolomitic hydrate,* respectively. Dolomitic hydrate is made with the addition of 20 percent water (Materials Handbook, 2005).

Dolomite calcined at 1800°C in coal-fired kilns produces dead-burned dolomite with a density of 3.25 g/cm^3. Calcining can be done in either a single or double burn process (Hopkins, 2002). In the double burn process, high-purity dolomite is heated to just above the decomposition temperature, then milled, briquetted, and calcined at 1800°C. Iron oxide greatly influences the rate and extent of sintering (Keegan, 1998). Dead-burned dolomite can be used in refractory applications.

Magnesite that is calcined at 600°C to 1000°C (light burned) forms caustic calcined magnesia with a loose bulk density of 0.3 to 0.5 g/cm^3 and a specific surface area of 10 to 65 m^2/g (Seeger and Werke, 2002). This form of magnesia is very reactive and readily absorbs water vapor and carbon dioxide (Duncan and McCracken, 1994). Before calcining, magnesite is crushed in mills, with the final particle size depending on the method of calcinations. For shaft kilns the desired range is 20 to 200 mm and for rotary kilns a particle size of 2 to 60 mm is desired (Seeger and Werke, 2002). When calcined to temperatures above 1450°C, dead burned (sintered) magnesia or periclase has a density of 3.35 to 3.40 g/cm^3 and specific surface area of <0.1 m^2/g (Harben, 2002). Fused magnesite is formed by heating to 2500°C to

3000°C in an electric arc furnace. Electrical grade fused magnesia, used in heating elements, may have 2 to 3 percent silica added to improve electrical properties.

The production of synthetic MgO can be done using several feedstocks. Seawater and natural brines (including the Dead Sea, natural deep well brines in the United States, and solution mining brines in the Netherlands) are common sources, as well as the end liquor from solar salt operations and residue from calcined dolomite (Duncan and McCracken, 1994). About 470 m^3 of seawater is required to produce one metric ton of MgO, although in practice 600 m^3 are needed (Seeger and Werke, 2002). The process of recovering the magnesium begins with the addition of calcium hydroxide (slaked lime) to the brine to produce an ion exchange reaction between the calcium and magnesium to produce $Mg(OH)_2$ (Evans, 1993):

$$Ca(OH)_2 + MgCl_2 \rightarrow CaCl_2 + Mg(OH)_2$$

The precipitated $Mg(OH)_2$ is then flocculated, filtered, and calcined to produce magnesia. However, there can be boron, calcium, silicon, alumina, and iron derived from the seawater and brines. Boron can be reduced from about 0.2 percent B_2O_3 to 0.03 percent B_2O_3 using several steps, including adding an excess of lime during processing and adding soda ash before prolonged calcining of the hydroxide (Seeger and Werke, 2002).

Uses and Specifications

Limestone, dolomite, magnesite, and their calcined derivatives are all important materials for glasses, glazes, and refractories. However, these manufacturing sectors make up only a small fraction of the total industrial demand for these materials, particularly limestone or calcium carbonate, which is used in a wide variety of industries including cement, fillers for paper and polymers, and flue gas desulfurization. Therefore the markets for these materials can be affected by factors other than the demand from glass and ceramics.

Glass Limestone and dolomite together are the main source of CaO and MgO in commercial soda-lime-silicate flat, container, fiber, and consumer glasses. Limestone and dolomite are used to assist melting, adjust melt viscosity, and impart desirable qualities to the final glass such as increased durability, increased strength, and reduced devitrification. The total amount of CaO and MgO depends on the application for the glass, but ranges from 10 percent to13 percent for most soda-lime-silicate glasses. In amounts higher than 13 percent, the properties of the glass begin to deteriorate.

A mix of CaO and MgO can improve the performance of the melt and the final glass product when compared to using only CaO. Substitution of MgO for some of the CaO in a soda-lime-silicate glass can increase high-temperature viscosity, add shine to the glass, and increase strength (Materials Handbook, 2005). The mixed CaO-MgO does not seem to have an affect on durability of commercial glasses (Sinton and LaCourse, 2001), but it has been reported that the replacement of MgO for CaO can enhance resistance to dissolution when exposed to water (Materials Handbook, 2005). The MgO/CaO ratio can affect chemical strengthening rate, with increasing MgO/CaO increasing the rate of ion-exchange when the other components are held constant (Sinton et al., 2000). The effect appears to be most dramatic between 0 and 0.5 MgO/CaO. The MgO/CaO for flat glass is about 0.4 by weight, although some low-light attenuation glasses have very little MgO, likely due to the generally higher FeO content of dolomitic sources. Some glasses use only dolomitic limestone to add CaO and MgO to the batch, while others, such as float glass, will have both limestone and dolostone in the batch.

The specifications for limestone and dolostone depend on the desired properties of the glass, particularly color. Table 5.13 gives the compositional and size specifications for limestone and dolostone used in some glasses. Iron is the contaminant of greatest concern because of its effect on color. Total iron should be below about 0.1 percent. Silica levels are not necessarily of concern, but the amount should not vary more than 0.5 percent between batches. The hardness of the limestone can affect performance. Softer limestones tend to be easier to melt and reduce abrasion of the refractory lining of the tank (Antony, 2001).

Burned lime (and burned dolomite) can be used in place of limestone in a glass batch. The major advantage is that the amount of material required per unit of CaO is reduced because the CO_2 has been removed. This reduces transportation costs and, to a certain degree, storage volume. In addition, the substitution of burned lime for limestone can cause a slight reduction in the reaction temperature between silica and soda ash in a glass batch (Kim and Matyáš, 2002). The disadvantages of substituting burned lime are the higher costs per ton and the problems of storage caused by hydration and dustiness. Hydration caused by exposure to humidity can affect weighing during the batching proc-

Table 5.13 Chemical specifications for limestone, dolostone, and magnesia in glass and ceramics (wt. %)

	CaO	MgO	Fe_2O_3	SiO_2	Al_2O_3	MnO	Cr_2O_3	B_2O_3	Sulfate	Moisture
					Calcite-Limestone					
Theoretical	56.03									
—flat glass[1]	≥54.85	≤0.8	≤0.075	—	≤0.35	—	—		≤0.05	≤0.05
—container glass[1]	±0.3	±0.3	≤0.1	±0.5	±0.5	—	≤0.001		—	≤0.1
—fiberglass[2]	≥55.4	—	≤0.05	—	≤0.2	≤0.1	—		≤0.1	≤0.4
					Dolomite					
Theoretical	30.41	21.86								
—flat glass[1]	≥29.50	≥21.40	≤0.25	—	≤0.40	—	—		≤0.20	≤0.1
—container glass[1]	±0.3	±0.3	≤0.1	±0.5	±0.5	—	≤0.001		—	—
Refractory dead-burned dolomite[3]	58–62	36–41	0.5–1.0	0.5–1.5	0.2–0.8	0.1–0.2	0.4		2.3	2.3
First-grade dead-burned magnesia[4]	0–2	>97	<0.2	<1				<0.01		

Size Specifications for Limestone and Dolostone—Cumulative % Retained

	2.4 mm 8 mesh	1.7 mm 12 mesh	1.2 mm 16 mesh	20 mesh 850 μm	100 mesh 150 μm	140 mesh 106 μm	170 mesh 90 μm	200 mesh 75 μm	325 mesh 50 μm
Limestone—flatglass[1]	0	≤0.1	—	—	—	≥88	—	≥95	—
Limestone—container glass[1]	—	0	—	≤20	—	—	≥95	—	—
Limestone—fiberglass[2]	0	0	0	0	≤2	—	—	≤30	100
Dolomite—flat glass[1]	0	—	≤0.5	≤20	≥88	—	—	≥95	—
Dolomite—container glass[1]	0	—	—	≤20	≤10	—	—	—	—

[1] Compiled from Harben (2002) and Skillen (1997)
[2] Skillen (1993)
[3] Harben (2002)—max. 2% combined SiO_2, Al_2O_3, and Fe_2O_3
[4] Keegan (1998)

ess and dusting is problematic in regenerative glass furnaces, which tend to blow fine particles into the exhaust stream that collect in the checkers. The high-temperature decomposition of burned lime does not result in the degassing of carbon dioxide that occurs when limestone is used. Many glass engineers believe that the degassing of limestone can contribute to the mixing of the glass melt.

Whitewares Calcium and magnesium carbonates are used as a minor ingredient in whiteware manufacturing. *Whiting,* or finely ground limestone, can be used in ceramic porcelain bodies and glazes as a source of calcia. In clay bodies, it is used in small amounts (<3 percent) as an additional flux to decrease porosity and increase densification, although evolution of CO_2 at higher temperatures is problematic as it can cause pinholing. For this reason, wollastonite ($CaSiO_3$) is often used as a source of calcia despite its higher cost. Hydromagnesite can be added as a flocculant in clay slurries in amounts of 0.125 percent to 0.250 percent. In glazes, whiting is used in leadless matte formulations in combination with other metal oxides. Magnesium carbonate can be used in glazes if a magnesium source is desired. Hydromagnesite is used as an electrolyte in both ground and cover coat enamels (Materials Handbook, 2005)

Refractories Pure periclase (MgO) is extremely useful in many refractory applications because of its high melting point of 2800°C, high temperature thermal conductivity and electrical resistivity, and a chemically inert behavior. It is particularly favored in basic refractories used in metal casting. Both magnesia and dead-burned dolomite have been used for many decades in steelmaking. Dead-burned magnesia resists attack from slags and is resistant to mechanical wear. The thermal properties of magnesia include high heat capacity (specific heat of 1.1. kJ/kg-°K at 600°C) and high thermal conductivity (9 W/m-°K at 600°C) (Seeger and Werke, 2002). These properties make magnesia-based refractories an excellent material for glass furnace checkers and other regenerator applications that are subjected to alkali attack.

Specifications for refractory magnesia include >97 percent MgO for first-grade material (Table 5.13). First-grade material should have a bulk density of 3.40 g/cm^3 and a crystal size greater than 120 μm (O'Driscoll, 2003). Boron, which is in brines and may find its way into synthetic MgO, needs to be kept to levels <0.01 percent. CaO and SiO_2 are undesirable because they will react to form a dicalcium silicate that will compromise the physical integrity of the grains. Dense, well burned grains that tend toward a larger grain size and are shown to be resistant to hydration are desirable (Duncan and McCracken, 1994). Chinese dead-burned magnesite grades are, from lower to higher grade (O'Driscoll, 2003):

- DBM9010 produced by direct sintering of magnesite ore in coal-fired shaft kilns at 1600° to 1700°C;
- DBM95 produced by calcining magnesite ore at 1000°C, followed by grinding, pelletizing, then sintering at 1700°C in a coke-fired shaft kiln; and
- DBM97 produced in the same manner as DBM95, but at temperatures of 1800°C to 1900°C and using an oil-fired kiln.

Magnesia and dead-burned dolomite can also be combined with other refractory materials, such as chrome, carbon, and alumina to produce a range of refractory products. Listed here are some of those products and their definitions (ASTM, 2002):

- *Chrome-magnesia brick.* A refractory brick, which may be burned or unburned, chrome-magnesia brick is manufactured predominantly of a mixture of refractory-grade chrome ore and refractory-grade magnesia in which the refractory-grade chrome ore predominates by weight. The refractory-grade magnesia may be seawater magnesia, dead-burned magnesite, fused magnesia, or a combination of these materials.
- *Refractory dolomite.* This is a refractory manufactured predominantly of dead-burned dolomite, synthetic dolomite, fused dolomite, or combinations of these materials.
- *Dolomite-carbon refractory.* This refractory brick is manufactured predominantly from a mixture of refractory-grade dolomite and 2 percent to 20 percent carbonaceous materials, with resin, tar, pitch, or a combination of these materials as the bonding agent. The refractory-grade dolomite may be either dead-burned dolomite, synthetic dolomite, fused dolomite, or combinations of these materials, and the carbonaceous material may be either graphite, carbon black, or a combination of these materials.
- *Magnesia-spinel brick.* The predominant component of this refractory brick is sintered or fused magnesia. Its lesser component is

sintered, fused, or in situ formed magnesium aluminate spinel ($MgO \cdot Al_2O_3$).

- *Magnesia-carbon brick.* This refractory brick is manufactured predominantly of a mixture of refractory-grade magnesia and 5 percent to 30 percent carbonaceous material, with resin, tar, pitch, or a combination of these materials as the bonding agent. The refractory-grade magnesia may be seawater magnesia, dead-burned magnesite, fused magnesia, or a combination of these materials. The carbonaceous material may be either flake or vein graphite, carbon black, petroleum or metallurgical coke, or a combination of these materials.
- *Magnesia-chrome brick.* This refractory brick, which may be burned or unburned, is manufactured predominantly of a mixture of refractory-grade magnesia and refractory-grade chrome ore in which the refractory-grade magnesia predominates by weight. The refractory-grade magnesia may be seawater magnesia, dead-burned magnesite, fused magnesia, or a combination of these materials.
- *Magnesia-dolomite refractory.* This refractory, which may be burned or unburned, is manufactured predominantly of a mixture of refractory-grade magnesia and refractory-grade dolomite in which the refractory-grade magnesia predominates by weight.

Dead-burned dolomite is used to make either fired bricks or bricks bonded with carbon or resin. These are used widely in the steel industry for ladles and linings for basic oxygen furnaces and in cement rotary kilns and lime shaft kilns (Keegan, 1998). Burned or fused dolomite is relatively stable in vacuum or reducing atmospheres and has a high resistance to basic slags, although fused magnesia is more resistant to iron-rich slags (Worrall, 1982). Bricks are made with sintered or fused dolomite, sometimes with added zirconia, by shaping and firing to 1500°C. Bonded bricks use sintered or fused dolomite, or fused dolomite-magnesia with coal pitch or a phenolic resin added as a binder. The carbon enhances its slag resistance, but once the carbon is burned off, the brick is susceptible to slag infiltration. Table 5.13 shows the general compositional specifications for refractory-grade burned dolomite. An excess of iron oxides can act as a flux and reduce the performance of the product. Chrome-bearing magnesia refractories are often used in the smelting of copper and other metals. However, the adverse health effects of chromium have forced the industry to seek chrome-free alternatives.

Health and Environmental Concerns

Calcium and magnesium carbonates and oxides are not regulated by any national policies. However, quicklime is caustic and can result in injury to exposed skin and eyes. Inhalation of finely ground materials should be avoided.

5.6 SODA ASH

Introduction

Sodium, one of the alkali metals and the sixth most abundant element in the earth's crust, is extremely reactive and will oxidize immediately when exposed to air. Because of its reactivity, it is never found naturally in its elemental metallic form, but rather as either a major constituent of many rock forming silicate and carbonate minerals or as the dominant cation in brines and seawater.

Sodium compounds are widely used as fluxes in glass and ceramic manufacturing, specifically to assist in the melting of refractory materials such as silica sand. The preferred source of sodium (or soda, referring to its oxide Na_2O) for soda-lime-silicate glasses is sodium carbonate (Na_2CO_3), or soda ash. Historically, a crude sodium carbonate called barilla was obtained from the burning of kelp and other plants that grow in a saline environment. In fact, the word *alkali* is derived from the Arabic term for the ashes of saltwort, a plant that grows along the edges of bays and marshes. Inorganically derived soda ash occurs naturally in least a dozen sodium carbonate minerals, with the two most important being the sodium sesquicarbonate mineral trona ($Na_3CO_3[HCO_3]\cdot 2H_2O$) and the sodium bicarbonate mineral nahcolite ($NaHCO_3$). Soda ash can be derived from natural, low-chloride brines that collect in alkaline lakes or that are produced during solution mining of underground deposits. Where natural sodium carbonate minerals are not available, soda ash can be synthesized through a reaction of sodium chloride, limestone, and carbon in the Solvay process.

Almost half of global soda ash production is used in glass making, with the balance used as a feedstock for other industrial processes such as detergent and paper manufacturing, metal ore processing, and water treatment. Soda ash products are sold as either a light or dense grade, which is determined by the bulk density of the crystals. The dense type ($\geq$925 kg/m^3) is most often used in the glass industry, making up 2 percent to 20 percent of a glass batching formulation. Feldspar or nepheline syenite (see section 5.4) can add some alkali to the glass batch, but there is too much alumina in these materials to provide the level of soda needed in the batch. Caustic soda (sodium hydroxide) is another source of soda for a glass batch if it is more economical than soda ash. Soda ash has a limited application in ceramics, although it may be used as a flux in enamels or as a deflocculant in clay-water slurries.

Composition and Mineral Structure

There are several minerals that are important in the production of soda (Table 5.14). Trona or sodium sesquicarbonate is the most important natural ore and has a nominal composition of $Na_3CO_3[HCO_3]\cdot 2H_2O$. On a weight basis, trona contains 70.4 percent Na_2CO_3. It is a monoclinic mineral that occurs as massive, fibrous, or bladed crystals that are white to yellowish gray with a vitreous luster. It has a specific gravity of 2.17 and a Mohs hardness of 2.5 to 3.0. Nahcolite, or natural baking soda, is of lesser importance, but still can be used as a source, particularly in solution mining. Its composition is $NaHCO_3$. It forms monoclinic minerals that are colorless to black with a vitreous luster. Other, less common sodium carbonate minerals include the hydrated sodium carbonate minerals natron ($Na_2CO_3\cdot 10H_2O$) and thermonatrite ($Na_2CO_3\cdot H_2O$) and mixed sodium and calcium carbonates such as shortite ($Na_2CO_3\cdot 2CaCO_3$).

When these ores are commercially refined, the resulting product is anhydrous sodium carbonate (Na_2CO_3), a white crystalline powder with a specific gravity of 2.53 and a melting point of 851°C to 854°C (Figure 5.36). Depending on the heating rate, soda ash can begin to dissociate at 400°C, releasing CO_2. The heat capacity of soda ash from 300° to 500°K can be expressed by the equation (Solvay Minerals, 2001):

Table 5.14 Properties of some sodium carbonate minerals

	Formula	Specific Gravity	Mohs Hardness	Heating Characteristics
Soda ash	Na_2CO_3	2.53		melts at 825–850°C
Trona	$Na_3CO_3(HCO_3)\cdot 2H_2O$	2.11	2.5–3.0	dissociates at 120 to 160°C
Nahcolite	$NaHCO_3$	2.21	2.5	

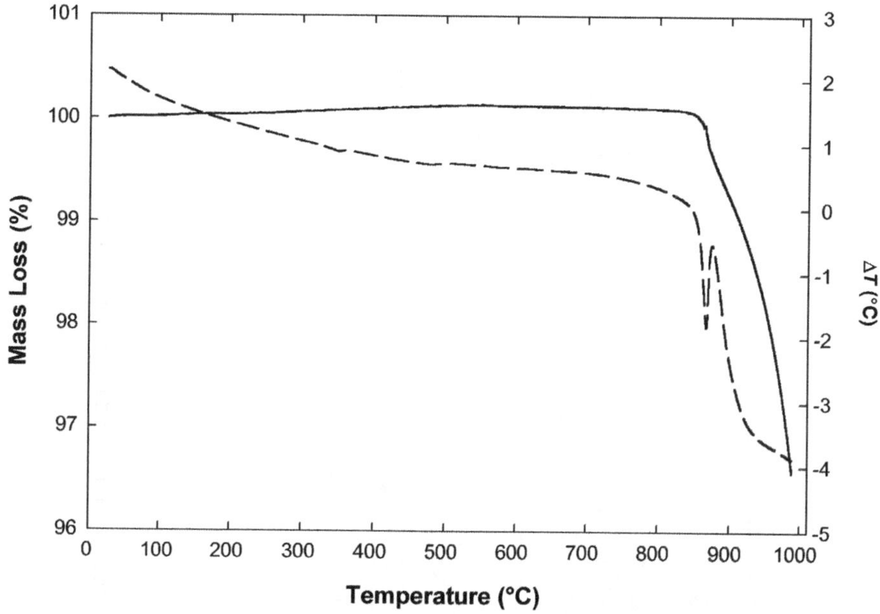

Figure 5.36 DTA-TGA curves for Na_2CO_3 (10°C/minute heating rate). The solid line represents mass loss and the dashed line is ΔT. The endotherm at 854°C is the melting point (courtesy of U. Kim).

$$Cp = 13.98 + 5.44 \times 10^{-2}T - 3.125 \times 10^{-5}T^2$$

The solubility of soda ash in water depends greatly on temperature. At 20°C, the solubility is 6.5 g per 100 g solution, and at 100°C it is 30.7 g per 100 g solution. The maximum solubility is at 35°C with 33.2 g per 100 g solution (Solvay Minerals, 2001).

Geological Occurrence and Distribution

Sodium is the most common of the alkali metals and is the sixth most abundant element in the Earth's crust at about 2.8 percent. It normally occurs as the highly soluble monovalent Na^+ cation. In magmatic environments, sodium is a constituent of all igneous melts, but is more abundant in continental alkalic rocks of granitic and syenitic composition, which have Na_2O content of 4 percent to 5 percent. The high solubility of sodium in water results in its concentration in seawater, brines, and evaporite rocks. Despite the ubiquity of sodium, economic deposits of natural sodium carbonate minerals and ores are rare (Figure 5.37).

Figure 5.37 World map showing the major resources of natural soda ash mineral ores and brines.

Natural Mineral Ores Sodium carbonate minerals, usually in association with other evaporitic chloride and sulfate salts, can precipitate from shallow inland alkaline lakes in warm, arid or semiarid regions. The lake basins are often formed in areas of tectonic extension, where block faulting of the crust forms depressions in which water collects and does not drain. Examples of alkaline lakes include the Basin and Range region of the western United States and the East African Rift valley. The tectonic subsidence of the depression needs to continue at the pace of sediment deposition, or else the depression will quickly fill with material. Not all alkaline lakes will form sodium carbonate minerals. To promote their formation, the water should have high sodium relative to other cations and high carbonate to (chloride + sulfate) ratios. Other important factors are the activity of algae and bacteria, which can produce carbon dioxide for the carbonate component (Eugster, 1971).

When evaporite minerals precipitate, they collect on the bottom of the lake or along the shoreline. Because alkaline lakes are dynamic with seasonal changes and varying influxes of dissolved material, large deposits that contain economic quantities of high-purity sodium carbonate minerals are somewhat rare (Kostick, 1994). However, there are several important occurrences of natural sodium carbonate mineral deposits around the world, which are fully described in Mannion (1983) and Garrett (1992).

The world's largest deposit of trona is in the Eocene age (more than 50 million years ago) Green River Basin of Wyoming, which covers over 3100 km^2 and contains approximately 47 billion tons of identified soda ash resources that could be recovered from the estimated 56 billion tons of bedded trona and the 47 billion tons of interbedded or intermixed trona and halite (Kostick, 2002). Because of the immense size of its deposits, Wyoming calls itself the "Saudi Arabia of sodium carbonate." The deposits reside in 42 beds that are 800 to 1,700 feet (240 to 520 m) below the surface. Wyoming trona resources are being depleted at the rate of about 15 million tons per year (Kostick, 2002), but reserves are expected to meet global demand for the next few centuries. An extension of the Green River Formation in the Piceance Creek Basin of northwestern Colorado is estimated to contain 30 billion tons of nahcolite.

In Africa, mining of trona occurs in rift valleys, salt pans, and volcanic crater lakes. Some of the larger operations include Lake Magadi in the Great Rift Valley of Kenya and Lake Natron in Tanzania. In Lake Magadi, the lake bed is underlain with 30 to 45 m of relatively pure trona interbedded with clays and other sodium salts (Kostick, 1994).

In Turkey, the Beypazari district in the interior of the Central Anatolia contains a buried, Miocene-age deposit of trona, minor nahcolite, lignite, and bituminous shale overlain by deposits of sodium sulfate and gypsum. The deposit, with an estimated 250 million tons of trona, is just beginning to be exploited, with an anticipated production of 500,000 tons per year by early 2006 (Kostick, 2002). A second trona deposit was recently discovered in 1998 at Kazan, with an estimated reserve of 600 million tons of trona.

In Asia, some trona and nahcolite deposits are worked in Henan Province in Anpeng and the Wucheng Basin (Mannion, 1983). However, the majority of China's demand for soda ash is met with synthetically produced material (see following section).

Sodium aluminosilicate minerals, such as nepheline, can be used as ores to convert to sodium carbonate. This is practiced in Russia (Polkan et al., 2003), where natural trona is scarce. The process requires a nepheline-bearing rock such as nepheline syenite and a source of limestone.

Brines Brines develop in lakes under conditions similar to those that form evaporite minerals. The lakes can be a mix of brines, evaporite minerals, and terrigenous materials such as mud and silt. The brines are processed to manufacture several salts, including soda ash. In the United States, Searles Lake and Owens Lake in California are the two main brine lakes that contain large soda ash reserves. Searles Lake is a complex body of interbedded evaporite minerals (such as borax, halite, thenardite, in addition to sodium carbonate) interstitial brines, and terrigenous muds (Kostick, 1994).

Mining and Processing

Trona In the Wyoming trona deposits, underground room-and-pillar mining is the primary method of obtaining the ore due to the considerable depth at which the deposits are located. After removal, the ore is sent to central underground crushers, at which point the particle size is reduced enough so that it can be conveyed to the surface for further processing. In lakes underlain with crystalline trona, such as Lake Magadi, the material is recovered using bucket dredgers mounted on pontoons, where the ore is crushed, slurried, and pumped to the shore. At this point, the ore often processed using froth flotation to remove unwanted chloride and sulfate minerals. The ore is crushed to a particle size small enough to allow rapid dissolution during refining.

To convert the trona to a clean soda ash, there are two processing routes that can be used (Figure 5.38) (Kostick, 1994; Harben, 2002; Chang, 2002). Both require approximately 1.8 tons of trona to produce 1 ton of soda ash. The most common method is the monohydrate process, which was developed by the FMC Corporation in 1960. In this method, trona is calcined at 200°C then dissolved in hot water to make a 30 percent solution. The solution may be further processed to remove organic carbon, followed by evaporation and precipitation of sodium carbonate monohydrate. The crystals are hydrocycloned, centrifuged, and dried at 150°C to produce sodium carbonate. The material is sized with screens and stored. In the less common trona or sesquicarbonate

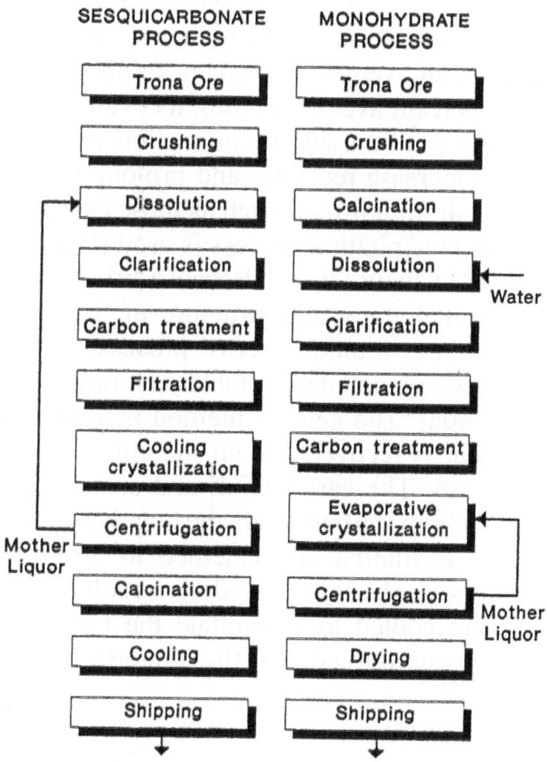

Figure 5.38 Flow diagram describing the two processing routes for converting trona ore to soda ash (from Kostick, 1994; courtesy of the Society for Mining, Metallurgy, and Exploration (SME), www.smenet.org).

process, trona ore is dissolved in water at 95°C to 98°C, clarified, filtered, and cooled to precipitate a cleaner trona (sesquicarbonate). The trona is then calcined to form soda ash.

Brines Sodium-carbonate-bearing brines can be naturally occurring liquids or can be the result of solution mining of trona or nahcolite. Solution mining of nahcolite had been done in Colorado, although operations have recently ceased. In solution mining, a pair of boreholes are drilled several hundred meters apart (Kostick, 1994). A pressurized stream of hot water is pumped down one well and the recovered brine is collected at the other well. A solvent, such as caustic soda, may be added to prevent the precipitation of sodium bicarbonate, which would coat the trona deposit and reduce the efficiency of the mining operation. There are some distinct advantages of solution mining, as it inherently has lower capital and labor costs, leaves insoluble tailings underground, and uses less energy (Aitala and Aitala, 1998). However, it cannot remove undesirable chloride and sulfate minerals. Room and pillar mining in Wyoming has an average 45 percent mining recovery, which is higher than the 30 percent average mining recovery from solution mining. Solution mining can be used in conjunction with mechanical mining in Wyoming to increase recovery and prolong project life.

To recover soda ash from brines, such as those extracted from Searles Lake, a carbonation precipitation process is used (Figure 5.39) (Garrett, 1992; Harben, 2002; Chang, 2002). Because brine compositions vary depending on the mineralogy and structure of the surrounding rock, each operation will have a unique set of processing steps. Additional material in the brines can include insoluble materials as well as chloride minerals (Figure 5.40a). The general method is to react the brine with process CO_2 gas that is recovered from the conversion of sodium bicarbonate to soda ash. The carbonation and subsequent cooling converts the soluble sodium carbonate to the relatively insoluble sodium bicarbonate, which is filtered and calcined to 260°C to produce an intermediate light soda ash. Nahcolite-derived solutions do not require carbonation, and are cooled to precipitate the bicarbonate. A monohydrate crystallizer is used to convert the light ash to sodium carbonate monohydrate.

Synthesized Soda Ash Just before the end of the eighteenth century, the French chemist Nicolas Leblanc developed a method of synthesizing soda ash from common sodium chloride, sulfuric acid, coal, and

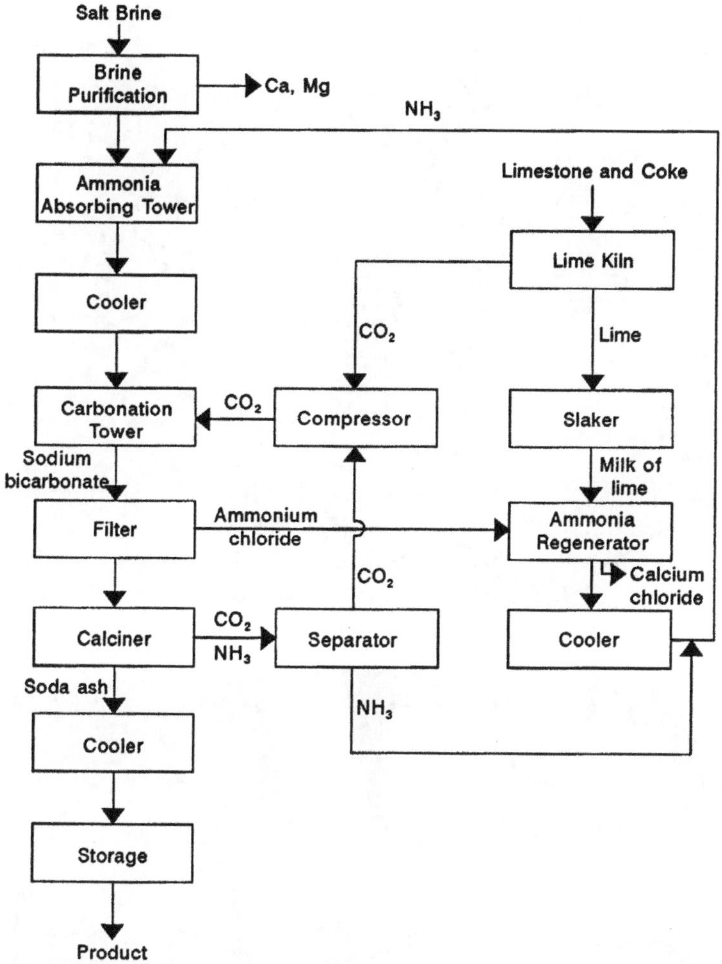

Figure 5.39 Flow diagram showing the general steps for the Solvay process (from Kostick, 1994; courtesy of the Society for Mining, Metallurgy, and Exploration (SME), www.smenet.org).

limestone. The salt and sulfuric acid reacted to form sodium sulfate (salt cake), which was then heated in water with limestone or chalk and coal to produce soda ash and calcium sulfate. The sodium carbonate dissolved in the water and was separated from the insoluble calcium sulfate. This process was used until the 1880s, at which point the Belgian chemist Ernest Solvay created a more efficient alternative that uses ammonia (a byproduct of the coke industry), sodium chloride, and limestone. This is commonly referred to as the *Solvay process* or am-

(a)

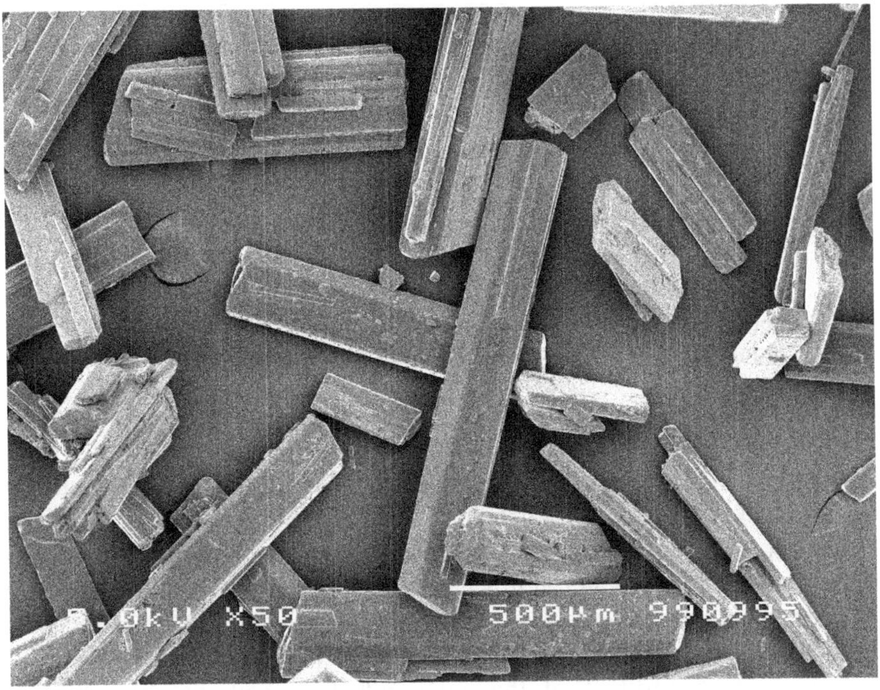

(b)

Figure 5.40 SEM photomicrographs of trona crystals. (a) Trona ore from Searles Lake, California. The ore contains 8% to 10% insoluble material and 3% to 5% of soluble impurities halite (NaCl) and thenardite (Na_2SO_4). The bar represents 100 μm; (b) Trona crystals produced from the Solvay process. The bar represents 500 μm (courtesy of R. Gärtner, Delft University of Technology).

monia process. The method uses many steps, but the overall reaction is

$$NaCl + NH_4HCO_3 \rightarrow NaHCO_3 + NH_4Cl$$

It requires a brine of sodium chloride that can be produced from solution mining of underground halite deposits. Ammonia used in the process is recycled, and limestone is mined from local sources and then calcined to produce CaO and CO_2. In order to produce 1 ton of soda ash, the method requires 1.7 tons of salt, 1.4 tons of limestone, 0.8 ton of coal for process heat, and an ammonia catalyst (Harben and Raleigh, 1998). The waste products consist of 1.7 tons of NaCl and CaCl. The resulting product can be very pure crystals of euhedral sodium carbonate (Figure 5.40b). A variation on the Solvay process was developed in Japan by Asahi Glass, which produces ammonium chloride as a byproduct that can be used in fertilizers. Relative to natural trona processing, the Solvay process is considerably more energy intensive and has a greater environmental impact. A more detailed explanation of the Solvay process and the environmental, energy, and material demands for the process can be found in ESAPA (2004). Due to the higher production costs, synthesized soda ash is generally produced and used in regions where natural soda ash deposits are not available and the transportation costs are too high, such as Europe and Asia.

Storage Soda ash is hygroscopic, so it will absorb water from the atmosphere during transport and storage. Over six months of storage, anhydrous soda ash with initially no moisture can have more than 11 percent water (Hou, 1942). Moisture absorption causes caking of the surface layer of a pile of soda ash. This reacted layer protects the material below, which retains its free-flowing characteristics (Hall, 2004). Excessive caking can cause problems during batching and conveying. Light ash is more susceptible to atmospheric moisture than dense soda ash. During hydration, the sodium carbonate transforms to sodium monohydrate or thermonatrite ($Na_2CO_3 \cdot H_2O$). As it also absorbs atmospheric CO_2, it will gradually transform to the sesquicarbonate ($Na_3CO_3[HCO_3] \cdot 2H_2O$), which is the stable form under ambient conditions.

Uses and Specifications

Commercial soda ash is marketed as either light or dense (granular) with different particle morphologies, depending on the processing route (see Figure 5.40). Light soda ash, which is used in soaps and detergents, has a particle-size distribution in which most grains are from 80 to 100 μm with a bulk density of approximately 500 kg/m^3, while dense soda ash has particle sizes ranging from 250 to 300 μm and a bulk density of 1,000 to 1,100 kg/m^3 (Harben, 2002). Dense soda ash is a major batch ingredient for all soda-lime-silicate (SLS) glasses.

Glass Almost 50 percent of global soda ash production is used for glassmaking, where it is added as the principal flux and glass modifier for SLS glass. Soda ash can make up as much as 20 percent to 22 percent of a typical SLS glass batch. Pressed and blown containers generally have a higher amount of soda in the batch compared to float glass, due to the need for reduced high-temperature viscosity in rapid production lines. However, too much soda can reduce the durability and strength of the final product and increases thermal expansion. Some glasses that do not use soda ash or use very little include textile fiberglass, borosilicate, and cathode ray tubes.

Soda ash is the highest cost component in a SLS glass batch, so any changes in the cost can significantly affect the cost of manufacturing. Dense soda ash is preferred to light soda ash because it mixes easier with the other batch components, reduces dusting, and requires less storage volume. As with all materials, soda ash used in glass must meet consistent physical and chemical specifications. The attributes of greatest concern in glassmaking include the following (Harben and Raleigh, 1997):

- Minimum Na_2O content of 58 percent
- Maximum Na_2SO_4 content of 0.2 percent
- Maximum NaCl content of 0.03 percent
- Maximum iron content of 0.001 percent or 100 ppm
- Water insolubles
- Particle-size distribution 1 percent on U.S. 20 (850 μm), 5 percent on U.S. 30 (600 μm), 20 percent through U.S. 100 (150 μm) and 1 percent through U.S. 200 (75 μm)

Some examples of soda ash specifications are given in Table 5.15. Natural soda ash derived from Wyoming is often considered the more desirable product, as it contains more than 99.5 percent sodium carbonate and little chloride. Solvay process soda ash tends to have a higher chloride content (0.4 percent to 0.6 percent), although some of

Table 5.15 Chemical specifications for dense soda ash (wt. %)

	Wyoming[1]	California[2]	European Solvay[3]	Brazilian Solvay[4]
Na_2CO_3	99.8	99.8	99.8	99.5
Na_2O	58.4	58.4	58.24	58.2
Na_2SO_4	0.05	0.08	0.016	0.07
NaCl	0.06	0.08	0.14	0.40
Insolubles (mg/kg)	500 max		8	300
Fe_2O_3 (ppm)	4	5	9	30
B_2O_3 (ppm)		150		

[1] FMC Grade 260
[2] Harben and Raleigh (1997)
[3] Brunner Mond Granular Standard Grade UK—typical
[4] Álcalis dense

the larger producers can produce low chloride grades with less than 0.15 percent chloride. Chloride is undesirable because it can form HCl during batch reactions in the furnace, leading to corrosion of checker refractories and exposed metal surfaces. Fluoride in soda ash can also pose similar problems. Solvay soda ash tends to be softer and easier to break during handling, which can produce dust that can be blown into the checkers.

The particle-size distribution of all glass batch materials (excluding cullet) should be similar to avoid segregation during mixing and conveying to the furnace, so soda ash granules should be in the range of 150 to 425 μm. Other physical characteristics specific to soda ash include bulk density and flowability. Bulk density is determined using the weight of sample divided by the volume. Flowability is defined as the angle of repose made by a pile of poured soda ash. A product with a high angle of repose may have become agglomerated, which then may pose problems in conveying. Soda ash processors may be able to tailor the crystal size and morphology of the particles. It can come in blocky to needlelike morphologies with the blocky, well-crystallized particles more desirable because of better flow characteristics and reduced dust generation.

The type of soda ash to use will largely be determined by the location of the plant and tolerance of chloride. Virtually all glass plants in the North America use Wyoming trona from at least five processors that distribute their product in a fleet of dedicated closed hopper, 100 ton capacity railcars and trucks. European, Asian, and South American manufacturing plants use almost exclusively Solvay process soda ash.

Whitewares In the preparation of clay-based ceramic bodies, sodium carbonate has been used in small quantities as a deflocculant and a dispersant. As a dispersant, it is dissolved in water before addition to the clay slurry. However, a comparison of different sodium-based dispersants found that soda ash was a relatively ineffective dispersant (Carty et al., 2000) and that dispersion is a function of the anionic polyelectrolytes and not the Na^{+} cation. Nevertheless, soda ash can react with massive lignite that may occur in plastic clays to form an effective dispersant (Materials Handbook, 2005).

Health and Environmental Concerns

Soda ash is not considered to be a very hazardous substance, but it should be handled with reasonable care. Direct contact with the product causes irritation of the eyes, and continuous contact may cause skin irritation (red, dry, cracked skin). Excessive levels of airborne dust may irritate the mucous membranes and upper respiratory tract. It may be harmful if swallowed in excessive amounts.

5.7 LITHIA

Introduction

Lithium is the lightest of the alkali metals, number three on the periodic table, and ranks 35th in order of abundance of the elements in the crust of the earth at an average of 20 ppm in the continental crust and 0.17 ppm in seawater. Like sodium and potassium, its preferred valence state of +1 makes it extremely reactive and therefore it is never found naturally in its elemental form. The lithium ion is the smallest of the alkalis and can act as a powerful flux. However, its small atomic and ionic radii (1.33 Å and 0.60 Å, respectively) and high ionic potential give it properties similar to the alkaline earth elements such as magnesium.

Although there are more than 150 lithium bearing minerals, only several are of commercial importance: two lithium aluminosilicates, spodumene ($LiAlSi_2O_6$) and petalite ($LiAlSi_4O_{10}$), the lithium phosphate amblygonite/montebrasite ($LiAlPO_4[F,OH]$), and the lithium mica lepidolite ($K[Li,Al]_3[Al,Si]_4O_{10}[F,OH]_2$). The minerals can be processed to make lithium carbonate or may be used directly as a raw material if the batch allows for additional silica, alumina, and, in the case of amblygonite/montebrasite, phosphorus. Another important source of lithium carbonate is geothermal brines in volcanically active regions and in alkaline lakes.

Lithia can be used in a wide variety of glass and ceramic products. In soda-lime-silicate glasses, it acts as a powerful flux that reduces the temperature of the points on the viscosity curve. It is an integral material for glass-ceramics that are based on the crystallization of beta spodumene. In traditional ceramics, lithia is not yet widely used, but it can be incorporated into low thermal expansion glazes and lead-free glazes. In metal casting, lithium minerals can be used in casting molds.

Composition and Mineral Structure

There are several lithium-bearing minerals that are of commercial importance and most of them occur together in pegmatites (see section 2.3). The details of these minerals are presented in this section, and properties are listed in Table 5.16.

Spodumene Spodumene is the most abundant commercial lithium ore mineral and occurs in at least two polymorphs, a low-temperature α form and a high-temperature β form. The α form of spodumene is a chain silicate belonging to the pyroxene group with a nominal formula of $LiAlSi_2O_6$. The structure consists of SiO_3^{2-} chains linked with Al^{3+} ions with the charge balanced by Li^+ ions. There can be some substitution of Li^+ by Na^+, but there is generally little variation from the nominal composition (Deer et al., 1992). Excess silica may be attributed to common inclusions of albite feldspar and mica. In pegmatites, spodumene often occurs as large crystals that can vary in color from translucent white to yellow, gray, green, and purple (Figure 5.41). It has a specific gravity of 3.03 to 3.23 and a hardness of 6.5 to 7.0 (Gaines et al., 1997). When heated to 1000°C to 1080°C, the α form will convert to the β form, increasing in volume by 30 percent and decreasing its specific gravity to 2.4. The β form has a structure similar to keatite (a synthetic, high-pressure silica polymorph) and has very low thermal expansion of 1×10^{-6}/°C for 25°C to 1000°C. Upon further heating, β-spodumene melts at 1420°C. Spodumene is only slightly soluble in water.

Eucryptite Eucryptite is a relatively rare lithium aluminosilicate mineral with a nominal composition of $LiAlSiO_4$. Although it has a relatively high Li_2O content (11 percent), its rarity does not allow it to be used as an ore mineral.

Table 5.16 Properties of some lithium-bearing minerals

	Formula	Specific Gravity	Mohs Hardness	wt. % Li_2O[1]	Heating Characteristics
Spodumene	$LiAlSi_2O_6$	3.2	6.5–7.0	8 (1.0–7.5)	irreversible transition to β form at 1000°C
Amblygonite	$LiAlPO_4(F,OH)$	2.98–3.11	5.5–6.0	10.1 (3.5–10)	
Lepidolite	$KLi_2AlSi_4O_{10}(F,OH)_2$	2.8	2.0–4.0	6.4 (3.0–4.5)	
Petalite	$LiAlSi_4O_{10}$	2.3–2.5	6.0–6.5	4.9 (3.6–4.7)	converts to β spodumene and SiO_2 at 1000°C
Lithium Carbonate	Li_2CO_3	2.11		18.8	decomposes at 600°C melts at 720°C

[1] Theoretical amount with actual range of natural minerals in parentheses; Chang (2002)

Figure 5.41 Hand sample of a natural spodumene crystal mined from a pegmatite (reprinted with permission from Amalgamet Canada).

Petalite Petalite is a monoclinic framework silicate of SiO_4 and AlO_4 tetrahedra, with the charge imbalance from the AlO_4 tetrahedra offset by the monovalent Li^+ cation. The nominal composition is represented by $LiAlSi_4O_{10}$, with minor substitution of Li^+ by Na^+ or K^+ and of Al^{3+} by Fe^{3+} (Gaines at al., 1997). It is whitish gray with a specific gravity of 2.4 and a hardness of 6.5. When heated to over 1000°C, it will convert to β spodumene and silica.

Amblygonite/Montebrasite Amblygonite and montebrasite are lithium phosphate minerals found as lenticular masses up to several meters in phosphate-rich pegmatites. Amblygonite is named after the Greek meaning for oblique or blunt angles and montebrasite is named after its type locality of Montebras, France. They make up the end members of a solid solution represented by the formula of $LiAlPO_4(F,OH)$, with complete solid solution between OH and F. Although there appear to be some discrepancies in the literature as to the naming of each end member, Gaines et al. (1997) place amblygonite as the fluorine-rich mineral and montebrasite as the hydroxyl-rich mineral. Amblygonite is white to gray, similar in appearance to albite, but with distinctive angled cleavages in at least four directions. Both have specific gravity values of around 3.0 and hardness of 6.0. Amblygonite is no longer used very often as a source of lithia in the production of lithium carbonate (Wietelmann and Bauer, 2002).

Lepidolite Lepidolite, which bears a name after the Greek words for scaly stone, is a lithium phyllosilicate (mica) with an theoretical composition approximated by $KLi_2AlSi_4O_{10}(F,OH)_2$. It forms a platy, monoclinic structure where potassium resides in the interlayer site. The lithium content in lepidolite can vary considerably. The relative amounts of Al^{3+} and Si^{4+} can vary within the octahedral sites, which may also host Fe^{3+} or Mg^{2+}. Substitutions for interlayer potassium include sodium, rubidium, and cesium (Deer et al., 1992), with the latter element in high enough abundance that it may be used as an ore for cesium. Substitution of fluorine for hydroxyl appears to correlate positively with lithium content. Lepidolite is readily identified by its typical violet to pink color and occurs as scaly masses within pegmatites. Lepidolite has a specific gravity of 2.8 to 2.9 and a hardness of 2.5 to 4.0.

Lithium Carbonate Lithium carbonate (Li_2CO_3) is produced through processing of lithium-bearing minerals and brines. It forms colorless monoclinic prisms with a specific gravity of 2.11 and bulk density of approximately 0.8 kg/l (Wietelmann and Bauer, 2002). Upon heating, it decomposes at 600°C and melts at 732°C. Lithium carbonate is slightly soluble in water, with solubility decreasing with increasing temperature (Table 5.17).

Geological Occurrence and Distribution

Lithium-bearing minerals are usually associated with granitic pegmatites, which are the residual magmatic fluids rich in incompatible elements and volatiles that represent the final crystallization of some granitic intrusions (see section 2.3). Pegmatites may be zoned, in which specific mineral assemblages may be restricted to concentric zones in the body, or unzoned. All or some of the minerals listed above except lithium carbonate, may occur within a pegmatite. Pegmatite deposits that yield lithium minerals in economically viable amounts are relatively rare and are distributed across the globe (Figure 5.42). The largest petalite deposit is the Bikita mine of Zimbabwe, which contains six lithium minerals with lepidolite and petalite as the major ores. In Canada, the Tanco mine at Bernic Lake in southeastern Manitoba, is a pegmatite containing close to 100 different minerals in eight discrete mineralogical zones. The mined pegmatite is one of a number of pegmatite sheets within a metagrabbro intrusive complex. It was initially mined for tin in 1929 and then later for tantalum in 1969. Although the lithium content of the Tanco pegmatite was first investigated in the last half of the 1950s, the production of ceramic-grade, low-iron spodumene concentrates did not commence until 1984 (Merivale, 2004). A similar pegmatite body in Ontario has been promoted as a source of mixed quartz-feldspar-petalite product for glass and ceramic applications. In the United States, high-iron spodumene had been mined in the Kings Mountain district of North Carolina, where a group of gen-

Table 5.17 Solubility of lithium carbonate in water

Temperature (°C)	g Li_2CO_3/100 g H_2O
0	1.54
20	1.33
40	1.17
60	1.01
80	0.85
100	0.72

From Wietelmann and Bauer (2002)

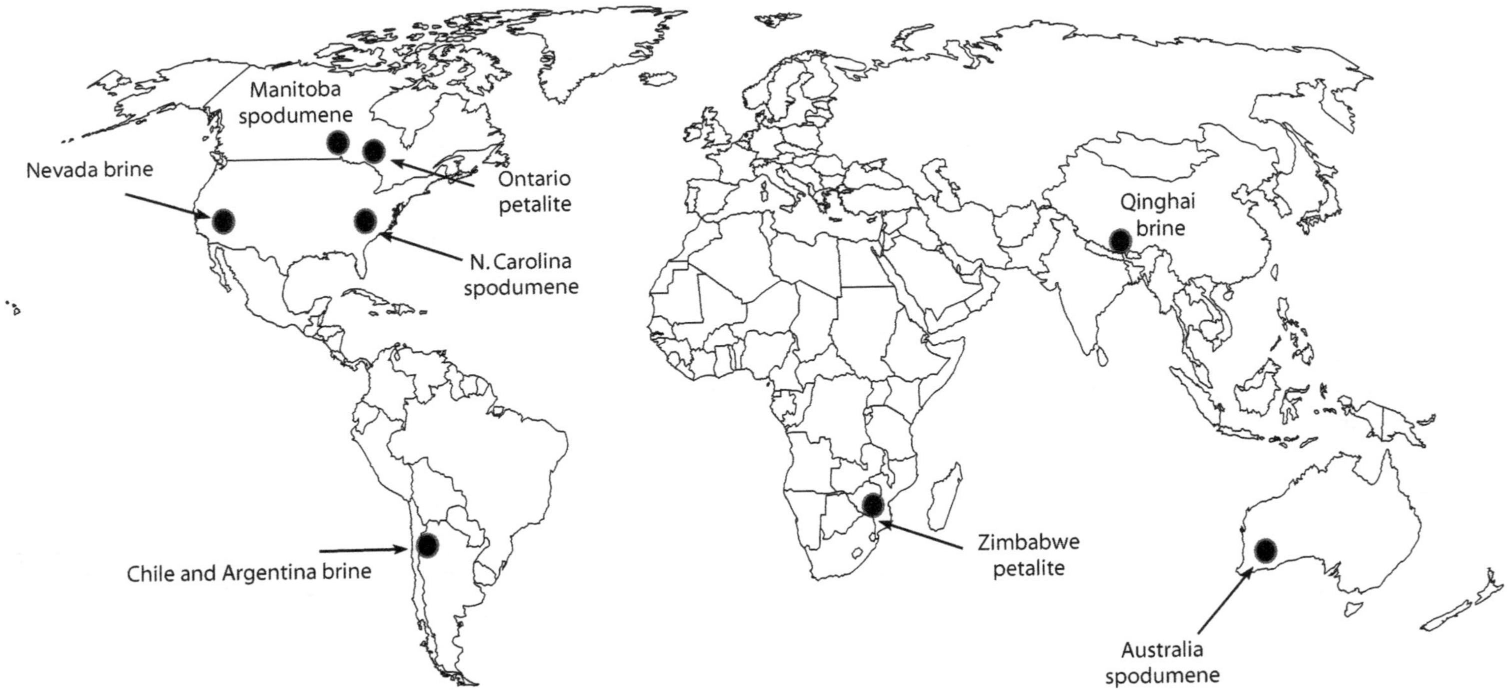

Figure 5.42 Locations of pegmatites and brines that yield lithium minerals and compounds.

erally unzoned pegmatites contain an average of 20 percent spodumene (Chang, 2002). Mining is currently on hold due to increased competition. In Brazil, petalite is mined from the Aracuai-Itinga district of Minas Gerais. In the Greenbushes region of western Australia 200 km south of Perth, high lithium, low iron spodumene is mined from a 100 km^2 open pit pegmatite mine (Meitz, 1989). Spodumene is mined in smaller quantities in Russia and China.

Brines have gradually become important sources of lithium because the extraction of lithium compounds from spodumene and petalite is very energy intensive compared to brine processing. Lithium can be present in nonmarine hydrothermal brines, particularly those in areas with both active volcanism and arid conditions. Groundwater and hydrothermal water leach the lithium from volcanic minerals or from sediments, such as the lithium-rich clay hectorite. The water collects in surface depressions and evaporates to form salars. In salars, the brines occur under a solid halite or sulfate crust. The largest lithium carbonate producer in the world is currently Chile, where the Salar de Atacama resides. It is estimated that the Salar de Atacama contains more than 1.1×10^6 tons of lithium (Wietelmann and Bauer, 2002). In the United States, the only active brine operation is in the Clayton Valley of Nevada, where water collects in an undrained depression. Other important brines occur in Quinghai Basin of China and the Salar de Hombre Muerto in Argentina (Kunasz, 1994). Brines usually range in lithium content (as dissolved lithium chloride) from 100 to 300 ppm, although the Chilean salars can have up to 5,000 ppm.

Mining and Processing

Lithium mineral extraction occurs in both underground mines (Canada) and open-pit mines (Zimbabwe, Australia, United States). The Canadian Tanco deposit is mined by the room and pillar method in an area 60 m under Bernic Lake (Figure 5.43). Both the underground and open pit operations use drilling and blasting to free the ore. The ore may be undergo primary crushing before being transported out of the mine to central processing plants for crushing and mineral separation (Figure 5.44).

Most pegmatite ores are beneficiated by flotation, although hand sorting and dense media separation may take place at some plants (Wietelmann and Bauer, 2002). The ores are run through several levels of crushing and grinding (Meitz, 1989), starting with a run-of-mine size of one meter boulders to less than 16 mm after primary and secondary

Figure 5.43 Underground room and pillar mining of a spodumene-bearing pegmatite (reprinted with permission from Amalgamet Canada).

Figure 5.44 The processing plant at Bernic Lake, Manitoba (reprinted with permission from Amalgamet Canada).

crushing. After crushing, the ore is ground in rod or ball mills. At this point, dry processed spodumene and petalite that is destined for glass making may be air classified and run through a magnetic separation process. The resulting material can be a mix of lithium minerals and quartz. For ores that are to be separated by flotation or other wet methods, the material may be sent to hydrosizers where a greater than 250 μm fraction is removed and processed for glass and the less than 250 μm fraction is deslimed and sent to the flotation cells to concentrate the spodumene or petalite (Meitz, 1989). Spirals may be used to remove heavy minerals such as tantalite ($[Fe, Mn][Ta, Nb]_2O_6$) and cassiterite (SnO_2).These two minerals as well as others may be mined concurrently from the pegmatite and the entire process would be designed to separate all desired minerals into separate concentrates and leave the undesired gangue in the tailings pile. After separation, some spodumene concentrates can be treated with a chlorination process to remove some of the iron.

There are several methods of producing lithium carbonate from lithium-bearing minerals (Fishwick, 1974; Kunasz, 1994; Wietelmann and Bauer, 2002). Spodumene and petalite are calcined at 1100°C to convert α spodumene to β spodumene and petalite to β spodumene and SiO_2, thus making the material easier to react in the subsequent steps. Lepidolite and amblygonite/montebrasite do not need calcining, but are ground and reacted with H_2SO_4 to form lithium sulfate (Li_2SO_4). The sulfate is then mixed with soda ash (Na_2CO_3) to force the following reaction:

$$Li_2SO_4 + Na_2CO_3 \rightarrow Na_2SO_4 + Li_2CO_3$$

resulting in the precipitation of relatively insoluble lithium carbonate.

To produce lithium carbonate from brines, the solutions are pumped to the surface into evaporation ponds. The brines are left to evaporate for up to 18 months to increase the amount of lithium to at least 6,000 ppm. Lime may be added to precipitate and remove magnesium compounds (Chang, 2002). The concentrated brine is treated with soda ash to convert lithium chloride to lithium carbonate, which precipitates from the solution. Chilean brines can be concentrated in a series of ponds to extract sodium and potassium chloride to the point that the brine contains 6 percent lithium (Kunasz, 1994). The brine is then transported to a central processing facility where lithium carbonate is extracted.

Uses and Specifications

Lithium is used in a variety of industrial processes, including batteries, concrete additives, and aluminum smelting. The glass and ceramic industries are some of the largest consumers as it is a major component of specialty glasses, glass ceramics, foundry molds, and specialty ceramics. Lithium is a minor additive in soda-lime-silicate glasses and porcelain bodies and glazes. As with other alkalis, its main use is as a flux. However, lithium also has characteristic similar to alkaline earth elements. For example, additions of lithia can raise the surface tension of glasses, while sodium and potassium reduce it (Fishwick, 1974). In a ternary system with alumina and silica, lithia forms low thermal expansion minerals such as beta spodumene that are indispensable for some glass-ceramics and low thermal expansion ware. To date, Fishwick (1974) gives the most comprehensive evaluation of the use of lithium in glass and ceramic production. A detailed history of the use of lithium minerals in the production of glass and glass-ceramics can be found in Edwards (1988). This gives an account of how changes in economics and politics can affect the procurement of high-value raw materials such as lithium minerals.

For any of the applications, lithia can be added as either lithium mineral concentrates or as synthesized lithium carbonate. Mineral concentrates appear to be the dominant source for ceramics and glass because of the general lower cost compared to lithium carbonate. Lithium carbonate is generally used in low-volume, high-value specialty glasses and ceramics where purity needs to be strictly controlled (Anon., 2001). Lithium minerals and lithium carbonate are available in several grades (Table 5.18).

Glass Lithium compounds are not a major component of soda-lime-silicate batch formulations. However, the addition of lithia as either a mineral concentrate or lithium carbonate can enhance melting and improve the properties of the final product. Because of the high ionic potential of lithium, it tends to be a more powerful flux than sodium. Added in small amounts (0.1 percent to 0.2 percent), lithia can lower melting times and furnace temperatures, resulting in a higher pull rate and lower energy consumption per unit produced. This is known as *chemical boosting*. One study of container glasses found that by adding lithium carbonate, a 100°C to 120°C reduction in furnace temperature and a 12 percent increase in pull rates could be achieved (Carroll and Angelo, 1983). Grahl (2002) summarized other studies that demonstrated that additions of 0.03 percent to 0.20 percent to the batch of

Table 5.18 Composition of lithium-bearing minerals (wt. %)

	SiO_2	Al_2O_3	Fe_2O_3	TiO_2	CaO	Li_2O	Na_2O	K_2O	P_2O_5	LOI
					Spodumene					
Theoretical	64.58	27.40				8.03				
Australia[1]	75.0	18.5	0.1			5.0	0.25	0.35		0.3
Australia[2]	68.0	22.0	0.4	0.2	0.4	6.1	1.0	1.0	0.3	
Manitoba[3]		24.0	0.06			7.25	0.35	0.30	0.27	
Zimbabwe[4]	65.00	27.00	0.04			7.30	0.10	0.08		
					Petalite					
Theoretical	78.48	16.65				4.88				
Zimbabwe[4]	77.00	16.00	0.03			4.30	0.50	0.30		
					Montebrasite					
Theoretical		34.82				10.21			48.48	4.61
Manitoba[5]		26.5	0.13			7.0	0.3	0.6	8.0	
					Mixed					
Ontario[6]	77.7	14.6	0.04	0.02	0.16	1.57	4.08	1.59	0.03	0.40

[1] Glass grade—Sons of Gwalia, Ltd.
[2] Universal grade—Sons of Gwalia, Ltd.—also contains 0.2% MnO
[3] 7.25% grade—TANCO Cabot Corporation
[4] Minerals Marketing Corporation of Zimbabwe
[5] Amalgamet Canada
[6] Glass Grade mixture of quartz, feldspar, petalite, and lithian muscovite (lepidolite)—Avalon Ventures; also contains 0.18% Rb_2O

lithium minerals provided energy savings of 4 percent to 10 percent. However, preliminary laboratory-scale experiments using a mix of feldspars and lithium minerals in a soda-lime-silicate glass demonstrated that a glass needs to have Li_2O_3 levels greater than 0.25 percent in order to see an improvement in melting compared to a glass batch without lithia (Shelby, 2004). In container glass, the addition of 0.025 percent to 0.050 percent lithia may lead to a stiffer gob with improved forming ability (Franklin and Klein, 1983). Lithia can increase the density of a glass by contracting the network, which can lead to a greater brilliance to glass and to increased hardness (Fishwick, 1974), although the amount needed to make a significant increase in these properties is not specified. On a molar basis, lithia can reduce the thermal expansion of a glass when substituted for sodium (Sun and Silverman, 1941; Fishwick, 1974). Because lithium minerals are more expensive than feldspars and nepheline syenite, the decision to use lithia as a flux in soda-lime-silicate glass comes down to economic considerations. If the combined benefits offset the increased raw material cost, then lithia may be used.

Lithia can be added to commercial glasses other than soda-lime-silicate glass. In fiberglass manufacturing, it can be added for the same reasons that it can be added to other glasses, with the addition that the reduced viscosity imparted by lithia can reduce temperatures and wear of refractories and bushings and that it can replace fluorite as a means of reducing fluorine emission (Skillen, 1993). Lithium carbonate is added as a flux for vitrification of high-level radioactive wastes (Crawford et al., 2002). Lepidolite is used in opal glasses as up to 28 percent of the batch (Materials Handbook, 2005). In high-barium cathode ray television tubes, lepidolite is added as a source of alumina and lithia to improve melting and to enhance the thermal stability and durabilityof the glass (Materials Handbook, 2005). Low thermal expansion borosilicate glasses used in electronic substrates contain potassium and lithium.

A lithia-lime-silicate flat glass can be made for the purposes of chemical strengthening. After forming, the glass is exposed to a molten sodium salt bath, resulting in the exchange of Li^{+1} cations in the glass for Na^{+1} cations in the molten salt. The larger Na^{+1} is "stuffed" into the smaller site previously occupied by the smaller Li^{+1}, resulting in compression that creates a higher-strength glass that is used in critical applications, such as aircraft canopies. In commercial float glass tanks, these glasses tend to be made at the end of a furnace campaign because the high lithia content is very corrosive to the refractory linings.

Glass-Ceramics Many of the commercial glass-ceramic (pyroceramic) materials used in mild refractory applications such as thermal shock-resistant cookware, range tops, and oven doors are made of lithium aluminosilicate materials. In the United States, the manufacturing of these glass-ceramics consumed the majority of lithium compounds used in the ceramics and glass industry (Ober, 2002). These materials are based on the conversion of spodumene to its β phase when heated to around 1000°C, creating a thermally stable material.

Whiteware Ceramics In a porcelain ceramic body, lithia can act as a flux or as a low expansion filler (Fishwick, 1974). When added as a flux—for instance 2 percent to 4 percent spodumene replacing feldspar or nepheline syenite—it can reduce the glass forming temperature and, therefore, reduce firing temperatures by up to 20°C to 30°C (Grahl, 2002). Since it dissolves into the glass phase completely, it does not form β-spodumene and will theoretically not greatly affect thermal expansion when fired. Higher amounts of added spodumene, if not melted during firing, will convert to β-spodumene and result in a low thermal expansion ceramic piece. Fishwick (1974) suggested that spodumene be used in a ceramic body, except where low iron is needed, in which lithium carbonate may be used.

Lithia can be added to both glazes and enamels as either a mineral concentrate or lithium carbonate. As with glasses, lithia lowers the viscosity of glass, permitting a thinner and more even glaze or enamel coating. Additions of up to 3 percent can reduce melting temperature by several cones. Lithia can increase surface tension, which improves reflectivity of enamels and glazes (Wietelmann and Bauer, 2002). When replacing sodium on a molecular basis, it will reduce the thermal expansion of the glaze. In ceramic pottery glazes, an addition of 1 percent lithium carbonate is reported to increase gloss, perhaps due to the increased fluidity of the glaze (Material Handbook, 2005). Lithium carbonate can be used instead of fluorspar (CaF_2) to reduce emissions of fluorine.

Health and Environmental Concerns

Lithium has been used for several decades to treat manic-depressive illnesses, but for individuals not previously exposed to lithium, a single large dose may result in vomiting and diarrhea. The more water-soluble lithium compounds are more dangerous. For example, if 12 to 60 g of lithium chloride are ingested, it can cause fatal poisoning (Opresko,

1995). However, exposure in an industrial setting to the inorganic lithium materials mentioned in this section does not appear to pose a hazard, and there are no regulations in North America or Europe concerning them.

5.8 ANHYDROUS ALUMINOSILICATES

Introduction

Natural compounds that contain only aluminum and silicon are the aluminosilicates. The hydrous aluminosilicates, clays and pyrophyllite, are discussed in previous sections (sections 5.2 and 5.3, respectively). Here, andalusite, kyanite, and sillimanite, which are three naturally occurring polymorphs of Al_2SiO_5, are examined. These three minerals, often referred to collectively as the sillimanite minerals, can be calcined to form mullite ($Al_3Si_2O_{10}$) and silica. Mullite is found only rarely in nature, but is commonly synthesized from the sillimanite minerals or by reaction of alumina with silica or bauxite with kaolin.

Composition and Mineral Structure

Andalusite, kyanite, and sillimanite are minerals that share the same nominal composition, Al_2SiO_5, but have different crystal structures and somewhat different properties. Sillimanite has a structure of chains of edge-sharing AlO_6 octahedra parallel to the c-axis that are linked laterally by a double chain of alternating silicon and aluminum atoms in tetrahedral coordination (Deer et al., 1992). It forms crystals with orthorhombic symmetry. Andalusite and kyanite contain chains of edge-sharing AlO_6 octahedral parallel to the c-axis as well, but these are linked by SiO_4 tetrahedra and AlO_5 in andalusite and AlO_6 in kyanite (Figure 5.45). Andalusite crystals are orthorhombic and kyanite crystals have triclinic symmetry. These minerals have little ionic substitution within the crystal lattice, with the exception that Fe^{3+} may substitute for Al^{3+} (Deer et al., 1992). *Chiastolite* is a term used for andalusite that contains inclusions of graphite, quartz, and ilmenite that are not expelled during crystal growth (Grobbelaar, 1994).

Sillimanite, andalusite, and kyanite have well-defined cleavage that allows them to splinter into prismatic pieces and laths. The color of andalusite can be pink, red, or green; kyanite is usually blue (Figure 5.46); and sillimanite is colorless. However, the colors can vary considerably depending on the presence of impurities. All aluminosilicate

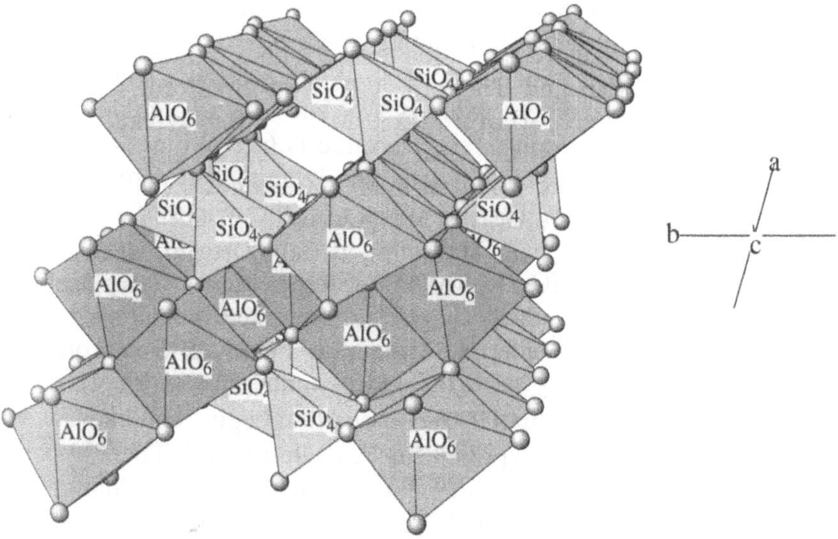

Figure 5.45 Idealized structure of kyanite (from Gaines et al., 1994; reprinted with permission from John Wiley & Sons).

Figure 5.46 Hand sample of elongated kyanite crystals intermixed with quartz (reprinted with permission from Kyanite Mining Corporation; www.kyanite.com).

minerals are relatively hard with Mohs values in the 6.5 to 7.5 range (Table 5.19). An exception is kyanite, which has varying hardness depending on the direction of the scratch. When scratched parallel to the long axis of the crystal, the value is 4.5, and when scratched perpendicular to or across the long axis the value is 6.5.

On heating, sillimanite, andalusite, and kyanite will convert to mullite (~88 percent by weight) and silica with the reaction:

$$3Al_2SiO_5 \rightarrow 3Al_2O_3 \cdot 2SiO_2 + SiO_2$$

The temperature at which this reaction begins and is completed ranges from 1100°C to 1550°C depending on the polymorph. Table 5.20 shows the temperature at which decomposition begins and ends, as well as the volume expansion experienced by each mineral. There appear to be several values for these temperatures in the literature (Chang, 2002; Harben, 2002; Grobbelaar, 1994; Sweet, 1994), so these values should be taken as a range of temperatures. The rate of the reaction is very much dependent on the temperature at which the material is exposed, with higher temperature requiring less time (Chang, 2002). Kyanite has the lowest transformation temperature, but also has the largest degree of expansion of 16 percent to 18 percent compared to the 4 percent and 6 percent expansion of sillimanite and andalusite, respectively. These properties are critical when using any of these materials in refractory applications.

Mullite has a formula of $Al_3Si_2O_{10}$ or $3Al_2O_3 \cdot 2SiO_2$ that forms an orthorhombic structure similar to sillimanite with partial replacement of silicon by aluminum atoms. The relative amounts of aluminum and silicon atoms can vary, so a more general formula of $Al_{4+2x}Si_{2-2x}O_{10-x}$ can be used, where x has values between 0.17 to 0.59 (Gaines et al., 1997). The common mullite with a molar ratio of 3:2 aluminum to silicon is referred to as 3:2 mullite. A variety of transition metal cations can substitute into the mullite structure, with minor amounts of Ti^{4+}

Table 5.19 Formula and some properties of aluminosilicate minerals

	Formula	Crystal Symmetry	Specific Gravity	Mohs Hardness
Sillimanite	Al_2SiO_5	orthorhombic	3.25	6.5–7.5
Kyanite	Al_2SiO_5	triclinic	3.58	4.5–6.5
Andalusite	Al_2SiO_5	orthorhombic	3.15	6.5–7.5
Mullite	$3Al_2O_3 \cdot 2SiO_2$	orthorhombic	3.11–3.26	6.0–7.0

Table 5.20 Heating characteristics of aluminosilicate minerals

	Decomposition Begins °C	Decomposition Ends °C	% Volume Expansion
Sillimanite	1300–1550	1625–1700	4–7
Kyanite	1100–1150	1300–1480	16–18
Andalusite	1250–1450	1500	6

From Harben (2002); Chang (2002) and Grobbelaar (1994)

and Fe^{3+} in natural specimens (Schneider et al., 1994). The color of mullite is transparent to white. The melting behavior of mullite is not clear (see Schneider et al., 1994, for a discussion). It can melt congruently at 1890°C or melt incongruently at 1828°C to α alumina and liquid. The melting behavior appears to depend on the Al_2O_3 content of the material and the condition of the starting materials used in the experiment. The coefficient of thermal expansion of mullite is about 5.3×10^{-6}/°C.

Geological Occurrence and Distribution

Sillimanite, andalusite, and kyanite occur in schists and gneisses that result from the metamorphism of aluminous sediments such as clays and shales. Which phase is stable depends on the final pressure and temperature regime under which metamorphism occurred. These minerals can form in both contact metamorphism, such as the area surrounding an igneous intrusion, and in regional metamorphism in which a larger terrane is buried and subjected to increased pressure and temperature. An example of the reaction that may occur is the combination of muscovite mica and quartz to form potassium feldspar and andalusite (Grobbelaar, 1994):

$$KAl_2(AlSi_3)O_{10}(OH)_2 + SiO_2 \rightarrow KAlSi_3O_8 + Al_2SiO_5 + H_2O$$

The sillimanite minerals are found with a variety of other minerals in addition to feldspars. Sillimanite can occur with quartz and mica, andalusite with cordierite and mica, and kyanite with quartz, rutile, corundum, pyrite, mica, and garnet. The sillimanite minerals are relatively resistant to weathering, so they are often found as detrital accessory minerals in sands. They are undesirable in quartz sands that are used in glass manufacturing due to the refractory nature of the minerals.

Commercial mining of anhydrous aluminosilicate minerals occurs throughout the world. Most of the exploited material is andalusite and kyanite. Sillimanite, while a common mineral in metamorphic rocks, often occurs in fine, fibrous aggregates intimately intergrown with other minerals, making it difficult to beneficiate (Sweet, 1994).

South Africa is the largest producer of andalusite, which produced approximately 220,000 metric tons in 2003 (Potter, 2003). Andalusite formed by contact metamorphism during the emplacement of the massive 1.95 billion-year-old Bushveld igneous complex, which itself is a major global source of strategic metals. The metamorphic halo extends for thousands of kilometers in the Transvaal in the northeastern region of the country, although only a few select areas have commercial-quality material (Grobbelaar, 1994). The result is that there are several mines that produce andalusite concentrates with less than 0.9 percent Fe_2O_3 and 58 percent to 60 percent Al_2O_3.

The United States is the largest producer of kyanite in the world. The most productive mines are in Buckingham County, Virginia, at two sites, Willis Mountain and the East Ridge. The material formed during the metamorphism of sandstones containing silt and clay, resulting in the highly resistant quartzite rocks that occur as mountains in the Appalachians (Sweet, 1994). Some andalusite is mined in a slate belt, along with pyrophyllite and sericite in North Carolina (Potter, 2003).

France produces andalusite (called kerphalite after the nearby town of Kerphales) from Ordovician age schists in the Glomel region in Britanny (Grobbelaar, 1994). The deposit contains up to 15 percent to 20 percent andalusite crystals that are several centimeters long, with the remaining mineralogy consisting of biotite, muscovite, hornblende, and feldspar (Sweet, 2004). The ore bodies can be as thick as 400 m with little overburden.

China has become a recent global supplier of sillimanite, kyanite, and andalusite (Zhou et al., 2005). There are 50 known deposits, 10 of which have reserves greater than 2 million tons. The majority of the material is andalusite. The largest andalusite deposit is mined in the Xinjiang Uygur Autonomous Region in the northwest part of the country. In India, kyanite is found in a metamorphic belt in Singhbhum (Klinefelter and Cooper, 1961). Sillimanite is mostly of the massive variety sawn in to shapes (Sweet, 1994).

Mining and Processing

Virtually all kyanite and andalusite are taken from open-pit mines (Figure 5.47) as the value of the minerals generally does not permit un-

Figure 5.47 Open-pit mining for kyanite (reprinted with permission from Kyanite Mining Corporation; www.kyanite.com).

derground mining. If the ore is well lithified, blasting is used to loosen the material. After collection of the ore, the particle size is reduced through primary crushing and screening. The subsequent beneficiation steps depend on the mineralogy and texture of the ore, and therefore each mine has a unique process (Grobbelaar, 1994; Chang, 2002). In South Africa, andalusite ore may go through two stages of heavy media separation followed by high intensity magnetic separation. Heavy media is used because the density difference between the andalusite and the shale gangue materials is not great enough for hydro-based spirals and jigs. The ore may be roasted before introduction into the magnetic process to increase the magnetic susceptibility of the weakly magnetic gangue minerals (Grobbelaar, 1994). Froth flotation is used to recover very fine andalusite. Kyanite in the United States can be processed by flotation to remove mica and pyrite. The concentrate is roasted and passed through high-intensity magnetics with final grinding in rod mills (Sweet, 1994).

Kyanite may be calcined to form mullite and silica. Mineral concentrates are heated to 1450°C in a rotary furnace to produce a granular material that is composed of 80 percent mullite, 11 percent fine amorphous silica, 7 percent quartz, and 2 percent cristobalite (Kyanite Mining Corporation, 2005).

Synthetic mullite can be made through three general routes: (1) sinter-mullite; (2) fused mullite; and (3) chemical or high-purity mullite (Schneider et al., 1994). The temperature of reaction depends greatly on the raw materials and method that is employed. Sintered mullite is made using silica and alumina-bearing minerals, such as kaolin, pyrophyllite, bauxite, and Bayer alumina, in the correct stoichiometric proportions to yield $3Al_2O_3 \cdot 2SiO_2$. For example, sintered mullite can be made by mixing low-iron kaolin and calcined alumina or bauxite (Sacks and Pask, 1982) and then shaping, drying, and firing at 1750°C. Pellets can be made in a rotary kiln then subsequently crushed and sized. Sintering aids or mineralizers such as alkalis, alkaline earths, TiO_2, and Fe_2O_3 may be used to decrease the temperature of mullitization (Schneider et al., 1994). Commercial-fused mullite is made by melting alumina and quartz sand or fused quartz in electric arc furnace at 2000°C then casting the material into shapes. The temperature of crystallization and the rate of cooling will define the composition of the mullite in terms of the molar Al:Si ratio (Schneider et al., 1994). Fused mullite with Al:Si ratios of 2.2 to 3.2 consist mainly of mullite and a glass phase whereas with ratios above 3.3, α alumina will crystallize as well. When mixed with zircon (21 percent to 36 percent ZrO_2) then fused, a zirconia mullite is produced. Chemical mullite is prepared using chemically refined precursor components in advanced processing methods such as sol-gels, precipitation, hydrolysis, and chemical vapor deposition (Schneider et al., 1994).

Uses and Specifications

Because the anhydrous aluminosilicates are, for the most part, used in refractory applications, the higher grades have higher Al_2O_3, lower SiO_2, and minimal fluxing compounds, such as Fe_2O_3 and the alkali and alkaline earth elements. Table 5.21 gives the compositions of some commercially available products, along with the theoretical compositions. There are a range of available grain sizes, with the coarser grain sizes of up to 8mm more desirable for refractories (Harben, 2002). Finer sizes are desired in foundry applications. Because of the nature of the ore body texture, U.S. kyanite is not offered in grain sizes larger than 0.5 mm (Materials Handbook, 2005).

Sillimanite, kyanite, and andalusite can be used directly in the manufacture of a variety of refractory pieces. The desired final product after

Table 5.21 Compositions of aluminosilicate minerals (wt. %)

	SiO_2	Al_2O_3	Fe_2O_3	TiO_2	CaO	MgO	Na_2O	K_2O	PCE
			Sillimanite						
Theoretical	37.08	62.92							
China—A[1]		54–55	1.2–1.5	0.25–0.3			0.25–0.3		
China—C[1]		58–59	0.8–1.0	0.35–0.45			0.35–0.45		
			Kyanite						
Theoretical	37.08	62.92							
China—Shuyang[1]		57–59	0.6–0.8	1.2–1.3			0.3–0.4		
United States[2]	39–42	54–60	0.4–1.0	0.5–1.6	<0.4	<0.4	<0.4	<0.7	36–38
			Andalusite						
Theoretical	37.08	62.92							
S. Africa-Andrafrax[3]	38.6	59.3	0.9	0.2	0.1	0.1	0.1	0.2	
France—KF[3]	38.4	60.0	0.5	0.2	0.1	0.1	0.1	0.2	
China—A[1]		54–55	1.2–1.5	0.3–0.35			0.6–0.7		
China—Xinjiang[1]	38.95	59.78	0.4	0.11			0.05	0.12	

Table 5.21 (Continued)

	SiO_2	Al_2O_3	Fe_2O_3	TiO_2	CaO	MgO	Na_2O	K_2O	PCE
			Mullite						
Theoretical	35.41	64.84							
Mullite clinker—China—M60-II[4]	38	58	0.5	0.8	0.35	0.1	0.1	0.1	
Sintered mullite—China—JLM75[4]	21–23	75–77	≤0.35	≤0.35			≤0.5		
Fused mullite—China—B[4]	20.09	78.10	trace	0.02	0.12	0.62	0.02	trace	
US—calcined[5]	26.8	68.8	1.22	2.82	0.6	0.8	0.07	0.05	39
Brazil—fused[6]	25.56	73.52	0.19	0.02	0.29	0.08	0.30	0.04	

[1] Zhou et al. (2005)—Na_2O here is reported as total R_2O in reference
[2] Kyanite Mining Corporation
[3] As reported in Harben (2002)
[4] Zhou et al. (2005)—made from kaolin and alumina or bauxite
[5] C-E Minerals—Mulcoa 70
[6] Elfusa

firing is a material consisting of mullite needles within a matrix of silica. The relatively low expansion of sillimanite and andalusite upon conversion to mullite and silica at high temperature allows both to be used directly to form refractory bricks. Depending on the application, kyanite may need to be calcined before using due to its 16 percent to 18 percent expansion, so it is available commercially as a raw ore concentrate or as a calcined product. In general, kyanite is used in North America with the exception of the andalusite material from North Carolina that consists of about 40 percent andalusite mixed with pyrophyllite and silica (Materials Handbook, 2005). Andalusite is more commonly used in Europe and Asia.

Refractories made from andalusite and kyanite can replace fireclay materials used in linings, pouring ladles, and furnace roofs for the iron and steel industry. Andalusite can be used in mixes with graphite, silicon carbide, zircon, and chrome oxide to make refractory castables and ramming and gunning mixes (Harben, 2002). When added to fireclays or other materials, kyanite can counteract fired shrinkage and can maintain a stable volume. Kyanite and andalusite can be used in foundry applications as molding sands or as mold coatings to allow casting removal. Because of its high thermal and electrical insulating qualities, mullite-based refractories from kyanite and andalusite were once used in spark plug insulators.

Because of its higher cost relative to the natural aluminosilicate minerals, synthetic mullite is used in critical refractory applications in steel and glass manufacturing. It is not easily wetted when in contact with metal and glass melts, retains strength at high temperature, is an excellent thermal and electrical insulator, and is resistant to corrosion. For example, mullite is used in fiberglass furnaces in the area just above the melt line that is subject to high temperature and chemical attack as well as regenerator brick (Keegan, 1998). In the steel industry, it is used in nozzles, sliding gates, and well blocks.

Health and Environmental Concerns

Sillimanite, kyanite, andalusite, and mullite are generally not considered hazardous to the environment. However, because these products can contain more than 0.1 percent crystalline silica, it may be labeled as a potential carcinogen. These materials can cause skin and eye abrasion and should not be inhaled. Adequate protection and ventilation should be used when the materials are handled.

REFERENCES

Aitala, R., and M. Aitala (1998). Process Selection Criteria for Refining Trona to Commercial Products. In *Proceedings of the First International Soda Ash Conference;* v. 1, ed. J. R. Dyni and R. W. Jones, R. W. Laramie: Geological Survey of Wyoming,

Anon. (2001). Lithium Minerals Market Is in Stable Mode. *Glass International.* July/August, pp. 28–29.

Antony, J. (2001). Calcium Carbonate: Not 'Just Limestone'. *Glass International,* v. 24, pp. 34–35.

ASTM (2002). *Glass; Ceramic Whitewares,* v. 15.02. West Conshohocken, PA: ASTM International.

Austin, R. S. (1998). Origin of Kaolin in the Southeastern U.S. *Mining Engineering,* v. 50, pp. 52–57.

Austin, C. R., H. Z. Schofield, and N. L. Haldy (1946). Alumina in Whiteware. *Journal of the American Ceramic Society,* v. 29(12), pp. 341–354.

Bailey, S. W. (1988). Chlorites: Structures and Crystal Chemistry. in *Hydrous Phyllosilicates,* Reviews in Mineralogy, v. 19, ed. S. W. Bailey. Washington, DC: Mineralogical Society of America, pp. 347–403.

Bauer, R. R., J. R. Copeland, and K. Santini (1994). Wollastonite. In *Industrial Minerals and Rocks,* 6th ed., ed. D. D. Carr. Littleton, CO: Society for Mining, Metallurgy, and Exploration, p. 1119.

Baumgart, W. (1984). Refractories. In *Process Mineralogy of Ceramic Materials,* ed. W. Baumgart, A. C. Dunham, and G. C. Amstutz, Amsterdam: Elsevier, pp. 80–103.

Beall, G. H. (1994). Industrial Applications of Silica. In. Reviews in Mineralogy, Mineralogical Society of America, v. 29, ed. P. Ribbe, pp. 469–505.

Bolger, R. (1997). Feldspar and Nepheline Syenite; A Choice of Alumina. In *Raw Materials for the Glass Industry,* ed. J. B. Griffiths. London: Industrial Minerals, pp. 44–48.

Bridgett. (2000). In *Science of Whitewares II,* ed. W. M. Carty and C. W. Sinton. Westerville, OH: American Ceramic Society, pp. 23–32.

Bruvel, F. (1999). *Silica Sand.* Queensland Department of Mines and Energy, Mineral Information Leaflet No. 3, 3 pp.

Burrows, S., and H. Borad (2000). Surrey Sand for the Glass Industry. *Glass Industry,* v. 81(12), pp. 24–25.

Carr, D. D., L. F. Rooney, and R. C. Freas (1994). Limestone and Dolomite. In *Industrial Minerals and Rocks,* 6th ed., ed. D. D. Carr. Littleton, CO: Society for Mining, Metallurgy, and Exploration, pp. 605–629.

Carroll, H. R., and J. Angelo (1983). Adding Lithium Can Improve Melting–Forming Performance. *Glass Industry,* pp. 14–19.

Carty, W. M., and B. Pinto (2002). Effect of Filler Size on Strength of Porcelain Bodies. *Ceramic Engineering and Science Proceedings,* v. 23(2), p. 95–104.

Carty, W. M., K. R. Rossington, and U. Senepati (2000). A Critical Review of Dispersants for Whiteware Applications. In *Science of Whitewares II*, ed. W. M. Carty and C. W. Sinton. Westerville, OH: American Ceramic Society, pp. 155–168.

Carty, W., and U. Senapati (1998). Porcelain: Raw Materials, Processing, Phase Evolution, and Mechanical Behavior. *Journal of the American Ceramic Society,* v. 81(1), pp. 3–20.

Chang, L. L. Y. (2002). *Industrial Mineralogy: Materials, Processes, and Uses,* Upper Saddle River, NJ: Prentice Hall.

Ciullo, P. A. (1996) *Industrial Minerals and Their Uses: A Handbook and Formulary*. Westwood, NJ: Noyes Publications.

Ciullo, P. A., and C. S. Thompson (1994). Pyrophyllite. in *Industrial Minerals and Rocks,* 6th edition, ed. D. D. Carr. Littleton, CO: Society for Mining, Metallurgy, and Exploration, pp. 815.

Crawford, C. L., D. M. Ferrara, R. F. Schumacher, and N. E. Bibler (2002). Vitrification of Low-Activity Radioactive Waste Streams and a High-Level Radioactive Waste Stream in Support of the Hanford River Protection Program, U.S. Department of Energy report WSRC-MS-2002-00449.

Deer, W. A., R. A. Howie, and J. Zussman (1992). *An Introduction to the Rock-Forming Minerals*. Essex: Longman Group.

Dickson, T., and P. W. Harben (2000). Global Kaolin. in *Industrial Clays,* ed. N. Keegan. Surrey: Industrial Mineral Information Ltd., pp. 39–47.

Dolley, T. P. (2002). Silica, Minerals Yearbook. Reston, VA: U.S. Geological Survey,

Dumont, M. (2001). Silica. *Canadian Minerals Yearbook,* pp. 46.1–46.8.

Duncan, L. R., and W. H. McCracken (1994). Magnesite and Magnesia. In *Industrial Minerals and Rocks,* 6th ed., ed. D. D. Carr. Littleton, CO: Society for Mining, Metallurgy, and Exploration, pp. 643–654.

Echlin, C. (2002). Evaluation of a Commercial Ball Clay Deposit. *American Ceramic Society Bulletin,* v. 81(8).

Edwards, G. H. (1988). The Application of Economic Geology in Glass Raw Materials—A Case History. In *Advances in the Fusion of Glass,* ed. D. F. Bickford et al. Westerville, OH: American Ceramic Society, pp. 44.1–44.9.

Elzea, J. M., and H. H. Murray (1994). Bentonite. In *Industrial Minerals and Rocks,* 6th edition, ed. D. D. Carr. Littleton, CO: Society for Mining, Metallurgy, and Exploration, pp. 233–246.

Elzea, J., J. Bruns, and S. Rice (2000). Mineralogy of Ceramic Grade Kaolins from the Southeastern United States. In *Science of Whitewares II,* ed. W. M. Cartyand C. W. Sinton. Westerville, OH: American Ceramic Society, pp. 77–89.

Emrich, E. W., and R. C. Hannon (1964). Talc as an Auxiliary Flux in Plastically Formed Ceramics. *American Ceramic Society Bulletin,* v. 43, pp. 186–190.

Encyclopedia Britannica Online (2004). Clay Mineral. *Encyclopedia Britannica,* Table 28: Cation-Exchange Capacities and Specific Surface Areas of Clay Minerals.

EPA (1998). Ceramic Products Manufacturing, Emission Factor Documentation for AP-42, Section 11.7. Washington, DC: U.S. Environmental Protection Agency.

ESAPA (2004). *Process BREF for Soda Ash.* European Soda Ash Producers Association, European Chemical Industry Council, Brussels.

Eurofel (2005). The European Association of Feldspar Producers, Brussels.

Eurotalc (2005). The European Association of Talc Producers, Brussels, http://www.ima-eu.org/en/talcindex.html

Eugster, H. P. (1971). Origin and Deposition of Trona. In *Contributions to Geology,* ed. R. B. Parker. Laramie: University of Wyoming, Laramie, v. 10, p. 15–23.

Evans, A. M. (1993). *Ore Geology and Industrial Minerals,* 3rd ed. Oxford: Blackwell Science.

Fishwick, J. H. (1974). *Applications of Lithium in Ceramics.* Boston: Cahners Publishing.

Flörke, O. W., and H. Graetsch (2002). Silica. In *Ullmann's Encyclopedia of Industrial Chemistry,* 6th ed. Weinheim, Germany: Wiley-VCH.

Fournier, R. O. (1985). The Behavior of Silica in Hydrothermal Solutions. *Reviews in Economic Geology,* v. 2, pp. 45–61.

Franklin, B. R., and L. C. Klein (1983). Lithia in Container Glass, *American Ceramic Society Bulletin,* v. 62, pp. 209–211.

Gaines, R. V., H. C. W. Skinner, E. E. Foord, B. Mason, and A. Rosenzwieg (1997). Dana's New Mineralogy, 8th ed. New York: John Wiley and Sons, Inc.

Garrett, D. E. (1991). *Natural Soda Ash: Occurrences, Processing, and Use.* New York: Van Nostrand Reinhold.

Giese, R. F., Jr. (1988). Kaolin Minerals: Structures and Stabilities. In *Hydrous Phyllosilicates,* ed. S. W. Bailey. Reviews in Mineralogy, v. 19, Mineralogical Society of America, Washington, DC, pp. 29–66.

Grahl, C. L. (2003). Raw and Manufactured Materials: 2003 Review and Forecast. *Ceramic Industry,* v. 153(1), pp. 19–22.

Grahl, C. L. (2002). Saving Energy with Raw Materials. *Ceramic Industry,* v. 152(7), pp. 13–16.

Grobbelaar, A. P. (1994). Andalusite. In *Industrial Minerals and Rocks,* 6th ed., ed. D. D. Carr. Littleton, CO: Society for Mining, Metallurgy, and Exploration, pp. 913–920.

Güven, N. (1988). Smectites. In *Reviews in Mineralogy,* v. 19, ed. S. W. Bailey. Washington, DC: Mineralogical Society of America, pp. 497–559.

Hall, D. (2004). Personal communication.

Harben, P. W. (2002). *The Industrial Minerals Handybook,* 4th ed., Industrial Minerals Information, Surrey.

Harben, P. W., and R. L. Bates (1984). *Geology of the Nonmetallics*. New York: Metal Bulletin, Inc.

Harben, P. W., and C. W. Raleigh (1998). Soda Ash—Wyoming Sets the Standard. In *Raw Materials for the Glass Industry,* ed. J. B. Griffiths. London: Industrial Minerals, pp. 28–34.

Harries-Rees, K. (1993). Structural Ceramic Clays. In *Raw Materials for the Glass and Ceramic Industries,* 2nd ed., ed. A. D. Skillenand J. B. Griffiths. London: Industrial Minerals Information, pp. 12–14.

Hawley, G. C. (2002). Wollastonite. *Mining Engineering,* v. 54, pp. 55–56.

Heaney, P. J. (1994). Structure and chemistry of the low-pressure silica polymorphs. In *Silica, Physical Behavior, Geochemistry and Materials Application,* ed. P. J. Heaney, C. T. Prewitt, and G. V. Gibbs, Washington, DC: Mineralogical Society of America, pp. 41–49.

Hopkins, D. A. (2002). Value Added Refractory Dolomite: A Case Study. In *Industrial Minerals and Extractive Industry Geology,* ed. P. W. Scott and C. M. Bristow. Bath, UK: The Geological Society, pp. 167–168.

Hou, T. (1942). *Manufacture of Soda with Special Reference to the Ammonia Process*. American Chemical Society, New York.

Ibañez, A., and F. Sandoval (1998). Producing Synthetic Wollastonite by Autoclaving. *American Ceramic Society Bulletin.,* v. 77(9), www.ceramicbulletin.org.

Industrial Minerals Information (2000). *Glass and Ceramics Sourcebook,* Surrey, UK: Industrial Minerals Information Ltd.

Kauffman, R. A., and D. Van Dyk (1994). Feldspars. In *Industrial Minerals and Rocks,* 6th ed., ed. D. D. Carr. Littleton, CO: Society for Mining, Metallurgy, and Exploration, pp. 473–481.

Keegan, N., ed. (1998). *Raw Materials for the Refractories Industry,* 4th ed. London: Industrial Minerals Information.

Keegan, N., ed. (2000). *Industrial Clays,* 3rd ed. London: Industrial Minerals Information.

Kendall, T. (1993). Feldspar and Nepheline Syenite. In *Raw Materials for the Glass and Ceramics Industry,* 2nd ed., ed. A. D. Skillen and J. B. Griffiths. London: Industrial Minerals Information, pp. 37–43.

Kesler, S. E. (1994). *Mineral Resources, Economics, and the Environment.* New York: Macmillan Publishing.

Kim, D., and J. Matyáš (2002). *Batch Reactions of a Soda-Lime Silicate Glass*. Richland, WA: Pacific Northwest National Laboratory.

Klein, C., and C. S. Hurlbut Jr. (1993). *Manual of Mineralogy,* 21st ed. New York: John Wiley and Sons.

Kostick, D. S. (1994). Soda Ash. In *Industrial Minerals and Rocks,* 6th ed., ed. D. D. Carr. Littleton, CO: Society for Mining, Metallurgy, and Exploration, Inc., pp. 929–958.

Kramer, D. A. (2002). Magnesium Compounds, Minerals Yearbook, Reston, VA: U.S. Geological Survey.

Kunasz, A. (1994). Lithium Resources. In *Industrial Minerals and Rocks,* 6th ed., ed. D. D. Carr. Littleton, CO: Society of Mining, Metallurgy, and Exploration, pp. 631–642.

Kyanite Mining Corporation (2005). www.kyanite.com.

Leonard, R. S. (2000). Plastic Clays. In *Industrial Clays,* ed. N. Keegan. Surrey, UK: Industrial Mineral Information Ltd., pp. 77–84.

Lide, D. R., ed. (1995). *CRC Handbook of Chemistry and Physics,* 76th ed. Boca Raton, FL: CRC Press.

Loughbrough, R. (1993). Kaolin. In *Raw Materials for the Glass and Ceramic Industries,* 2nd ed., ed. A. D. Skillen and J. B. Griffiths. London: Industrial Minerals Information, pp. 20–25.

Manning, D. A. C. (1994). *Introduction to Industrial Minerals.* London: Chapman and Hall.

Mannion, L. E. (1983). Sodium Carbonate Deposits. In *Industrial Minerals and Rocks,* 5th ed., ed. S. J. Lefond. New York: AIME, pp. 1187–1206.

Marek, D., and C. W. Sinton (2000). Feldspar Extraction and Processing. In *Science of Whitewares II,* ed. W. M. Carty and C. W. Sinton. Westerville, OH: American Ceramic Society, pp. 33–39.

Maslennikova, G. N., N. F. Solodkii, M. N. Solodkaya, and A. S. Shamrikov (2004). Use of Kaolins from Different Deposits in the Production of Fine Ceramics. *Glass and Ceramics,* v. 61(7–8), pp. 257–266.

Materials Handbook (2005). *Ceramic Industry,* v. 155(1), pp. 23–125.

Meitz, H. (1989). The Use of Spodumene Ore as a Flux for Glass Melting, *Glass Technology,* v. 30(1), pp. 6–9.

Merivale, C. (2004). Personal communication.

Misture, S. (2004). Personal communication.

Mondo Minerals (2004). The Suitability of Various Talc Grades for the Manufacture of Highly Porous Ceramics with High Cordierite Content, Technical Bulletin 1401, Helsinki, www.mondominerals.com

Mostafa, A. (2003). Influence of Alumina Substitution on Phase Composition, Microstructure and Properties of Some Porcelain Bodies, *Interceram,* v. 52(1), p. 36–41.

Murray, H. H. (2000). Processing Kaolins and Ball Clays for Ceramic Markets, in Carty, W. M. and Sinton, C. W., eds., *Science of Whitewares II,* American Ceramic Society, Westerville, Ohio, pp. 13–22.

Murray, H. H. (2002). Clays, in *Ullmann's Encyclopedia of Industrial Chemistry,* 6th ed., Wiley-VCH, Weinheim.

Murray, H. H., C. Harvey, and J. M. Smith (1977). Mineralogy and Geology of the Maungaparerua Halloysite Deposit in New Zealand. *Clays and Clay Minerals,* v. 25, pp. 1–5.

Newman, H. R. (2002). The Mineral Industry of Germany. Reston, VA: U.S. Geological Survey.

Ober, J. A. (2002). Lithium Minerals Yearbook. Reston, VA: U.S. Geological Survey.

O'Driscoll, M. (2003). In the Hall of the Magnesite King: Asian DBM Reviewed. *Industrial Minerals,* July, pp. 26–37.

O'Driscoll, M. (1993a). Pyrophyllite: Far East Dominates Market. In *Raw Materials for the Glass and Ceramics Industry,* 2nd ed., ed. A. D. Skillen and J. B. Griffiths. London: Industrial Minerals Information, pp. 26–28.

O'Driscoll, M. (1993b). Wollastonite: Ceramics Remains Top Consumer. In *Raw Materials for the Glass and Ceramics Industry,* 2nd ed., ed. A. D. Skillen and J. B. Griffiths. London: Industrial Minerals Information, pp. 77–80.

Opresko, D. M. (1995). *Toxicity Summary for Lithium.* Risk Assessment Information System report. Washington, DC: U.S. Department of Energy.

OSHA (2004). Crystalline Silica. U.S. Department of Labor, Occupational Safety and Health Administration, www.osha.gov/SLTC/silicacrystalline/index.html.

Ottonello, G. (1997). *Principles of Geochemistry.* New York: Columbia University Press.

Paige, M. L. (2004). XRF Applications in the Silica Industry, presented at the 53rd Annual Denver X-Ray Conference, The International Centre for Diffraction Data, Newtown, Pennsylvania.

Parker, J. M. (1952). *Geology and structure of part of the Spruce Pine District, North Carolina,* North Carolina Division Mineral Resources Bulletin, v. 65., North Carolina Geological Survey, Raleigh, North Carolina.

Patterson, S. H., and H. H. Murray (1983) Clays. In *Industrial Minerals and Rocks,* ed. S. J. Lefond New York: American Institute of Mining, pp. 585–651.

Patterson, S. H., and H. H. Murray (1984). *Kaolin, Refractory Clay, Ball Clay, and Halloysite in North America, Hawaii, and the Caribbean Region.* Reston, VA: U.S. Geological Survey Professional Paper 1306.

Pereira da Silva, S. (2002). Kaolin, Mineral Summary, National Department of Mineral Production of Brazil, www.dnpm.gov.br.

Phelps, G. W., and J. B. Wachtman (2002). Ceramics. In *Ullmann's Encyclopedia of Industrial Chemistry*, 6th ed. Weinheim: Wiley-VCH.

Pickering, S. M., and H. H. Murray (1994). Kaolin. In *Industrial Minerals and Rocks,* 6th edition, ed. D. D. Carr. Littleton, CO: Society of Mining, Metallurgy, and Exploration, pp. 255–277.

Piniazkiewicz, R. J., E. F. McCarthy, and N. A. Genco, (1994). Talc. In *Industrial Minerals and Rocks,* 6th ed., ed. D. D. Carr. Littleton, CO: Society for Mining, Metallurgy, and Exploration, Inc., pp. 1049–1069.

Polkan, G. A., S. V. Zavarina V. A. Gorokhovskii, V. L. Pentko, L. D. Bulanova, L. A. Orlova, and V. A. Gorshkov (2003). Soda Ash Based on Nepheline Material Used in the Glass Industry. *Glass and Ceramics,* v. 60(7–8). pp. 208–212.

Potter, M. J. (2004). Feldspar and Nepheline Syenite, Minerals Yearbook. Reston, VA: U.S. Geological Survey.

Potter, M. J. (2003). Kyanite and Related Minerals, Minerals Yearbook. Reston, VA: U.S. Geological Survey.

Prasad, M. S., K. J. Reid, and H. H. Murray (1991). Kaolin: Processing, Properties, and Applications. *Applied Clay Science,* v. 6, pp. 87–119.

Reynolds, M. (2000). Effect of Auxiliary Fluxes on Whiteware Bodies, unpublished M.S. thesis, Alfred University, Alfred, New York.

Rieger, K. C. (1980). Quality Control of Talc for the Ceramic Industry. *American Ceramic Society Bulletin,* v. 59, pp. 519–520.

Rieger, K. C. (1996). Wollastonite, Pyrophyllite, and Talc for Rapid-Fire Wall Tile Bodies. *Ceramic Engineering Science Proceedings,* v. 17, pp. 180–182.

Routschka, G. (2002). Refractory Ceramics. In *Ullmann's Encyclopedia of Industrial Chemistry,* 6th ed. Weinheim, Germany: Wiley-VCH.

Sacks, M. D., and J. A. Pask (1982). Sintering of Mullite-Containing Materials. *Journal of the American Ceramic Society,* v. 73, pp. 65–70.

Sadowski, J. (2003). Processing of Silica Sand, 5th Chinese Industrial Minerals Conference, Shanghai, China, Outokumpu Minerals, www.outokumpu.com.

Sawyer Technical Materials (1999). *Hydrothermal Growth of Quartz.* Eastlake, Ohio: Technical Brief.

Sawyer Technical Materials, (1997). *General Materials Specifications for Cultured Quartz Crystals,* Technical Brief, Eastlake, Ohio.

Schneider, H., K. Okada, and J. Pask (1994). *Mullite and Mullite Ceramics.* New York: John Wiley and Sons.

Schüller, K. H. (1984). Silicate ceramics. In *Process Mineralogy of Ceramic Materials,* ed. W. Baumgart, A. C. Dunham, and G. C. Amstutz. Elsevier: Amsterdam, pp. 1–27.

Schulz, B. M., M. Brumbach, C. Caughel, and W. Carty (2003). Adsorption of Poly(acrylic acid) on Clay Surfaces, the Effects of Washing to Remove Impurity Species Present in the Clay. *Ceramic Engineering Science Proceedings,* v. 24(2), pp. 149–175.

Seeger, M., and D. Werke (2002). Magnesium Compounds. In *Ullmann's Encyclopedia of Industrial Chemistry,* 6th ed. Weinheim, Germany: Wiley-VCH.

SEGEMAR, (2002). Servicio Geológico Minera Argentina, www.segemar.gov.ar.

Shelby, J. (2004). Personal communication.

Singer, F., and S. Singer (1964). *Industrial Ceramics.* London: Chapman and Hall.

Sinton, C. W., and W. C. LaCourse (2001). Experimental Survey of the Chemical Durability of Commercial Soda-Lime-Silicate Glasses. *Materials Research Bulletin,* v. 36, pp. 2471–2479.

Sinton, C. W., W. C. LaCourse, and M. J. O'Connell (2000). Variations in K^+ - Na^+ Ion Exchange Depth in Commercial and Experimental Float Glass Compositions. *Materials Research Bulletin,* v. 34, pp. 2351–2359.

Skillen, A., (1993). Lithium Minerals: Fluidity Is Key. In *Raw Materials for the Glass and Ceramics Industry,* 2nd ed., ed. A. D. Skillen and J. B. Griffiths. London: Industrial Minerals Information, pp. 87–92.

Solvay Minerals (2001). *Heat Effects of the Trona System.* Solvay Minerals Technical Publication.

Sugden, A. (2002). Changes and Developments of Non-Plastic Raw Materials for Whiteware Part 2: The Identification and Evaluation of New Sources of Feldspar, *International Ceramics* v. 1, pp. 9–16.

Sugden, A. (2001). Changes and Developments of Non-Plastic Raw Materials for Whiteware Part 1: the Development of Better Value Products, *International Ceramics,* v. 2, pp. 10–14.

Sun, K. H., A. Silverman (1941). Additive Factors for Calculating the Thermal Expansion of Glass from its Composition. *Glass Industry,* v. 22, pp. 114–115.

Sweet, P. C. (1994). Kyanite and Related Minerals. In *Industrial Minerals and Rocks,* 6th ed., ed. D. D. Carr, Littleton, CO: Society for Mining, Metallurgy, and Exploration, Inc., pp. 921–927.

Taylor, C. (2000). New Zealand Halloysite. In *Industrial Clays,* ed. N. Keegan. Surrey, UK: Industrial Mineral Information Ltd., pp. 62–64.

Thompson, T. (2000). The Rheology of Ceramic Kaolins. *International Ceramics,* v. 1, pp. 12–13.

Virta, R. L. (2002a). Clay and Shale, Minerals Yearbook. Reston, VA: U.S. Geological Survey.

Virta, R. L. (2002b). Talc and Pyrophyllite, Minerals Yearbook. Reston, VA: U.S. Geological Survey.

Wallis, D. S. (1999). *Kaolin.* Mineral Information Leaflet No 10, Queensland Department of Mines and Energy.

Warshaw, S. I., and R. Seider (1967). Comparison of Strength of Triaxial Porcelains Containing Alumina and Silica. *Journal of the American Ceramic Society,* v. 50(7), pp. 337–342.

Wenk, H., and A. Bulakh (2004). *Minerals: Their Constitution and Origin.* Cambridge University Press, Cambridge.

Wietelmann, U., and R. J. Bauer (2002). Lithium and Lithium Compounds. In *Ullmann's Encyclopedia of Industrial Chemistry,* 6th ed., Weinheim, Germany: Wiley-VCH.

Worrall, W. E. (1982). *Ceramic Raw Materials*. Oxford: Pergamon Press.

Zdunczyk, M. J., and M. A. Linkous (1994). Industrial Sand and Sandstone. In *Industrial Minerals and Rocks,* 6th ed., ed. D. D. Carr. Littleton, CO: Society for Mining, Metallurgy, and Exploration, Inc., pp. 879–891.

Zheng Z., L. Zhentang, Y. Weishun, X. Jiquan, Z. Renjun, and F. Yesen (1996). Clay Deposits of China. *Mineral Deposits of China,* v. 4, pp. 235–287.

Zhou, N., J. Liu, and B. Lin (2005). Refractory Raw Materials in China. *American Ceramic Society Bulletin,* v. 84(2). pp. 20–24.

6

PROCESSED AND SYNTHETIC MATERIALS

6.1 ALUMINA

Introduction

Aluminum is the most abundant metal and third most abundant element in the Earth's crust, after silicon and oxygen. Like many metals, aluminum is never found naturally in its elemental form, as it readily combines with oxygen and other elements to form a variety of minerals. Its oxide, alumina or Al_2O_3, was first used as a dye fixative for textiles by Egyptians and Babylonians 4,000 years ago and was named *alumen* by Pliny around 80 A.D. (Hudson et al., 2002). In the late eighteenth century, alumina and aluminum hydroxides were further described and named. The naturally occurring alumina mineral is termed *corundum,* which can occur as the gemstones ruby (colored by chromium) and sapphire (colored by iron and titanium). More common are the aluminum hydroxide minerals, such as the commercially exploited gibbsite, boehmite, and diaspore, which make up the ore rock bauxite. Bauxite is formed by intensive weathering of rock and soil and is found on every continent.

Beneficiated bauxite can be used as a raw material for some alumina products, such as refractories and abrasives. For applications that require a more refined product, bauxite is used in the Bayer process, in which the ore is ground and digested in a caustic solution under pressure that yields a higher purity alumina. Bayer alumina can be made into a wide array of products with a range of physical properties.

The largest market for bauxite and Bayer alumina is the production of aluminum metal, which uses about 85 percent of bauxite used in the United States. In ceramics, alumina is used for its high hardness, resistance to heat and chemical attack, and high dielectric constant. Tabular and fused alumina products are therefore widely applied to abrasives, insulators, seals, nozzles, cutting and grinding tools, and refractories. As a filler in whiteware bodies, it replaces some of the silica to increase refractoriness and strength and to increase the firing range. Activated alumina is important in catalyst manufacturing. Colloidal alumina hydroxides or gels are important precursors for activated alumina and high-purity alumina materials. Single crystal corundum (sapphire) is used for high-temperature optics. Although Bayer alumina can be used in specialty glass where low iron is required, most commodity soda-lime-silicate glasses, fiberglasses, and ceramic glazes have alumina added by aluminosilicate minerals such as kaolin, feldspar, and nepheline syenite.

Composition and Mineral Structure

The most abundant natural alumina minerals are the aluminum hydroxides, with natural corundum relatively rare. The common hydroxides that are used in ceramics or processed to make alumina are gibbsite, boehmite, and diaspore. Properties of these minerals are given in Table 6.1.

Gibbsite Gibbsite, named after the American mineralogist, George Gibbs, is a trihydroxide mineral with a composition of $Al(OH)_3$ or $Al_2O_3 \cdot 3H_2O$. Other terms for gibbsite include γ $Al(OH)_3$ and hydrargillite. Gibbsite is made of layers of octahedra in which a layer of Al^{3+} ions are sandwiched between two hydroxyl layers forming a monoclinic symmetry (Gaines et al., 1997), though some crystals can be triclinic (Hudson et al., 2002). Only two-thirds of the octahedrally coordinated sites are occupied (Figure 6.1). The layers of hydroxyl octahedra in gibbsite are linked through hydrogen bonding. Gibbsite has a specific gravity of 2.42 and a Mohs hardness of 2.5 to 3.0. It is white when pure but will become reddish or yellow when it contains impurities such as iron oxides. Gibbsite is very soluble in hot alkali solutions such as NaOH. Bayerite, or α $Al(OH)_3$, and nordstrandite, or β $Al(OH)_3$, are related minerals with identical composition but with different crystal symmetries resulting from different stacking arrangements of the layers.

Table 6.1 Properties of alumina minerals

	Formula	Specific Gravity	Mohs Hardness	Crystal System	Heating Characteristics
Gibbsite	$Al(OH)_3$	2.42	2.5–3.5	monoclinic	dehydrates at 150°C
Boehmite	AlO(OH)	3.01	3.5–4.0	orthorhombic	dehydrates at 350°C
Diaspore	AlO(OH)	3.44	6.5–7.0	orthorhombic	dehydrates at 450°C
Corundum	Al_2O_3	3.98	9.0	hexagonal	Melts at 2000 to 2050°C

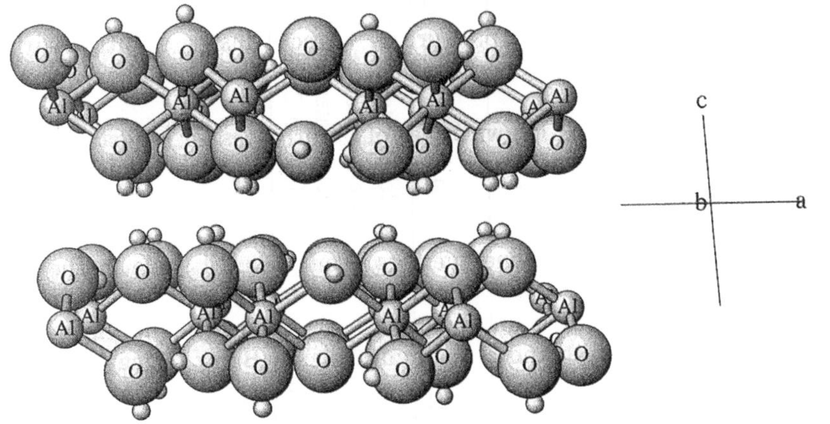

Figure 6.1 Structure of gibbsite (from Gaines et al., 1994; reprinted with permission from John Wiley & Sons).

Boehmite and Diaspore The other common aluminum hydroxide minerals are boehmite and diaspore, which both have the nominal composition of AlO(OH) and may be referred to as γ AlO(OH) and α AlO(OH), respectively. Both have Al^{3+} in octahedral coordination within the crystal lattice. In diaspore, every oxygen atom is hydrogen bonded to another oxygen atom in hexagonal close packing with hydrogen atoms occupying the interstitial sites not occupied by aluminum. In boehmite, half of the oxygen atoms are hydrogen bonded to another oxygen atom in cubic close packing (Chang, 2002). Both form orthorhombic crystals and are white, although they can appear yellowish to grayish. Boehmite and diaspore have respective specific gravity values of 3.01 and 3.44 and Mohs hardness of 3.5 to 4.0 and 6.5 to 7.0.

Corundum Aluminum oxide (Al_2O_3) forms the mineral corundum, which consists of octahedrally coordinated Al^{3+} that occupy two-thirds of the available sites. In the α form, oxygen ions are bonded to the aluminum ions to form alternating layers that make a crystal of hexagonal symmetry. α Al_2O_3 is the naturally stable form, although a hexagonal β form (which contains some sodium) and a cubic γ form can be synthesized and remain stable in ambient conditions. However, these two will convert to α Al_2O_3 when heated (Deer et al., 1992). In fused α alumina products, undesired sodium will react with alumina to form inclusions of softer β alumina (Gitzen, 1970). Corundum crystals are white and can be prismatic to tabular with no cleavage (Figure 6.2).

Figure 6.2 SEM photomicrograph of hydrothermally synthesized corundum (α alumina) crystals. The median size of the particles is 40 μm (printed with permission from Sawyer Technical Materials, LLC).

Small degrees of ionic substitution of Al^{3+} by Cr^{3+}, Fe^{3+}, and Ti^{4+} substitution can make bright-colored gemstones; Cr^{3+} makes blue sapphire while Fe^{3+} and Ti^{4+} make red ruby. Corundum is extremely hard with a Mohs hardness of 9.0 and a specific gravity of 4.0. It is essentially insoluble in acids.

Heating Characteristics When the aluminum hydroxide minerals are heated under ambient conditions, they undergo a series of compositional and structural changes before ultimately being converted to corundum (Gitzen, 1970; Hudson et al., 2002). Impurities, heating rates, pressure, and atmospheric composition all affect the rates and paths of the decomposition. The trihydroxides, like gibbsite, contain 34 percent H_2O and begin to decompose at 200°C. Boehmite and diaspore contain 15 percent H_2O and begin to decompose at about 450°C. All the hydroxide minerals go through at least two transitional structures upon heating until α Al_2O_3 is formed. When gibbsite is heated, it first transforms to boehmite and then to γ-Al_2O_3 at around 450°C (Deer et al., 1992). Diaspore and α Al_2O_3 have similar crystal structures, which allow a relatively simple conversion to occur by 600°C. α

Al_2O_3 formed in this manner grows epitaxially on the decomposing diaspore, with the (0001) plane of Al_2O_3 parallel to the (010) plane of AlO(OH) (Hudson et al., 2002). The transformation of gibbsite and boehmite to α Al_2O_3 requires more structural reorganization than diaspore. An increase in porosity occurs with dehydration as the open spaces caused by the removal of water molecules are stable until phase transformation. At temperatures greater than 1100°C, the transition minerals are transformed into corundum and porosity decreases. Above 1600°C, the material will sinter into a tabular alumina, and at 2050°C, the alumina begins to fuse.

Geological Occurrence and Distribution

Because aluminum is so common in the Earth's crust, it is a major component of many minerals, particularly aluminosilicates such as feldspars and clays. However, it is difficult to separate silica from alumina in these minerals, and therefore the aluminum hydroxides previously listed are currently the only commercially exploited sources of alumina. Natural corundum is not found in concentrated ore deposits; rather, it is a common accessory mineral in igneous and metamorphic rocks, particularly in pegmatites and nepheline syenites (Deer et al., 1992).

The formation of aluminum hydroxides begins with the development of laterites, which are soil horizons that are intensely weathered and have had most insoluble elements removed. Aluminum is one of the most insoluble elements, so it is a major component of laterites if the original rock contained aluminum. Unfortunately, other metals have very low solubilities in water, so laterites may contain appreciable amounts of iron, chromium and nickel. *Bauxites* are defined as high-aluminum laterites. They form by the weathering of virtually any rock (Evans, 1994), particularly in tropical areas that are flat and experience little erosion. The term is named for Les Baux in southern France, where it was first mined for alumina. In the process of weathering, aluminum will accumulate in the clay minerals, but with further weathering, clays such as kaolinite will lose silica and form aluminum hydroxides. An idealized weathering sequence leading to bauxite would be:

$$\text{aluminosilicate minerals} \rightarrow \text{kaolinization} \rightarrow \text{laterization} \rightarrow \text{bauxitization}$$

The sequence begins with removal of alkali and alkaline earths to form kaolinite. With further weathering, laterization occurs, leaving only hydroxides and oxides of silicon, aluminum, titanium, and iron. Bauxitization follows with the removal of most or all of the silica.

The mineralogy of bauxite deposits is dependent on several variables, including the composition of the original bedrock, temperature, and age. In general, young bauxites are gibbsitic and older deposits consist of boehmite and diaspore. Bauxite deposits vary greatly in terms of mineralogy and bulk composition, both within and between deposits (Table 6.2). Silica, as shown in the bulk composition, is most often sequestered in clay minerals such as kaolinite and halloysite but also can occur as free quartz. Other accessory minerals include hematite (Fe_2O_3), goethite (FeOOH), and refractory minerals such as rutile and anatase (TiO_2).

Bauxite deposits have been classified under several systems (e.g., Maynard, 1983). The simplest classifications divide them into two broad groups: lateritic and karst (Mediterranean). The lateritic bauxites, which make up more than 75 percent of global resources, can be subdivided into high-level varieties formed by weathering of volcanic plateaus and producing blanket deposits with gibbsitic mineralogy, and low-level varieties formed in tropical lowlands and producing boehmitic deposits. Lateritic bauxites may be associated with kaolinite deposits. Karst bauxites are formed by the weathering of limestone, which seems counterintuitive because limestone normally does not contain appreciable amounts of aluminum. It is thought that the aluminum comes from volcanic tuff (ash) or clays deposited with the calcium carbonate (Comer, 1974).

Bauxite can have a variety of colors and textures (Patterson et al., 1986). Many have what is termed a pisolitic structure, which consists of spherical aggregates occurring in concentric bands. Karst-type deposits are often massive, earthy, and dense ores that are colored brown and red, although deposits with lower iron are gray to white. The lithification of deposits is variable, with some ores occurring as soft, friable material while others are dense and hard.

Although bauxites are relatively common, particularly along the equatorial belt, about two-thirds of mined bauxite comes from four countries, in descending order: Australia, Guinea, Brazil, and Jamaica (Figure 6.3). In all, bauxite is mined in twenty-two countries. Sources of nonmetallurgical grade bauxite, which has more stringent purity specifications than metallurgical grade, is somewhat more limited with abrasive grade produced in Australia, China, Greece, Guinea, Guyana,

Table 6.2 Composition of alumina minerals and bauxite (wt. %)

	SiO_2	Al_2O_3	Fe_2O_3	TiO_2	CaO	MgO	Na_2O	LOI	H_2O
			Minerals (Theoretical)						
Corundum		100							
Gibbsite		65.4							34.6
Diaspore/boehmite		85.0							15.0
			Bauxite						
Australia—Darling Range[1]	26.5	37	16.4	1.1				19.3	
Australia—Weipa[1]	4.5	58	6.9	2.5				26.8	
Brazil—Trombetas, Pará[1]	5.1	52	13.9	1.2				28.1	
Guyana[1]	4.9	59	2.9	2.4				30.4	
Suriname—Onverdacht[1]	4.3	59	3.1	2.5				30.9	
Jamaica[1]	2.6	47.8	17.6	2.3				27.3	
Guinea—Boke[1]	1.5	56	7.9	3.7				30.1	
India—Orissa[1]	2.7	46	22.4	1.1				24.2	
Arkansas[2]	4.6	62	0.9	0.4				31.6	

			Calcined Bauxite					
Guyana—RASC[3]	5.34	90.50	1.07	3.00				
Refractory spec.[4]	<7.5	>84.5	<2.5	<4.0				
Abrasive spec.[4]	<5	80–88	2–5	2–4				<1
			Alumina					
White fused[5]	0.02	99.5	0.08	0.01	0.02	0.01	0.30	
White fused[6]	0.09	99.4	0.06	0.01	0.04	0.04	0.38	

[1] From Hudson et al. (2002)
[2] Tan granular textured (Patterson et al., 1986)
[3] From Minelco
[4] Harben (2002)
[5] Cichy (1990)
[6] Elfusa—Brazil

Figure 6.3 Global map showing the locations of the major bauxite deposits.

and Italy and refractory grade in Brazil, China, and Guyana (Plunkert, 2002). A compilation of world bauxite deposits can be found in Patterson et al. (1986).

Australia is the largest producer of bauxite, accounting for 39 percent of global output in 2001, and is largest exporter of alumina in the world. Major bauxite resources are found in Weipa in northern Queensland, the Darling Range in the southwest part of the country, and Gove on the northeastern tip of Arnhem Land in the Northern Territory (Loughnan and Sadleir, 1984). In Weipa, the predominantly gibbsitic bauxite occurs as cemented nodules in a 1 to 9 m thick horizon overlying kaolinite beds (Queensland Government, 2002; Harben and Bates, 1990). The parent material of most of the deposit is a kaolinitic sandstone. Weipa was first mined in 1955 and now produces up to 15 million metric tons per year. Bauxite from the Darling Range is usually a yellow-brown earthy material that is in beds of 2 to 10 m thick overlain by a one-meter crust.

In Asia, there are a series of bauxite deposits in the Central Urals and Kazakhstan. Much of the material has high silica and requires additional processing to produce alumina. A belt in southern China and Vietnam contains predominantly boehmitic bauxites that are important for nonmetallurgical applications (Hill, 1994). China is an important source for refractory grade and abrasive grade bauxite. Most Chinese bauxite comes from the provinces of Shanxi, Henan, Guanxi, and Guizhou, which produce a predominantly diaspore material with an average Al:Si ratio of 7:1, compared to the 10:1 average Al:Si ratio of global commercial bauxites. The majority of refractory-grade bauxite comes from Guizhou and Shanxi, which are similar in composition with the exception that the Guizhou material has higher TiO_2 (4.2 percent) and alkalis (0.6 percent) (McCracken, 1999). In India, most of the bauxite is found in gibbsitic blanket deposits in the western Deccan Plateau region, formed by the weathering of the flat lying basaltic Deccan traps. Because basalts are iron rich and silica poor, the bauxites are low in silica (Patterson et al., 1986). The East Coast bauxite province lies in Andra Pradesh and southern Orissa. It is composed of 10 to 15 m thick lateritic deposits of gibbsitic composition. The parent rock is a schist.

In South America, Brazil, Guyana, and Suriname are the largest producers of refractory grade bauxite. These deposits, and those in Guinea and Cameroon in Africa, are predominantly associated with laterization of Precambrian shield rocks, such as granitic gneisses, that are common to both continents. The shield rocks were related but split during the opening of the South Atlantic Ocean. The mainly gibbsitic Guyana deposits were discovered in 1869 and commercially exported by 1917 (McCracken, 1999). The resources are found in a 200-mile-long belt within 100 miles of the coastline. The deposit is underlain by kaolinitic clays, which is likely the parent material that was weathered to produce the bauxite. Impurities include siderite with minor goethite, pyrite, hematite, and heavy minerals (Harben and Bates, 1984). The high purity of the Guyana and Suriname deposits is such that a particular grade, Refractory Aggregate Super Calcined (RASC), is considered to be the standard for the industry. Guyana bauxite is found in two areas (McCracken, 1999): the Linmines mine from which RASC is produced, and the Berbice/Bermine mine, which is primarily a source of abrasive and chemical-grade bauxite. Brazil has large bauxite mines in the lower Amazon states of Pará and Amazonas and the southeastern state of Minas Gerais (Patterson et al., 1986). The material in lower Amazon are pisolitic, mainly gibbsitic bauxites. The Poços de Caldas plateau of Minas Gerais contains bauxite from low iron silicic rocks and is nodular, gibbsitic, and may contain quartz and kaolinite.

Guinea is the most important bauxite source in Africa and has the largest bauxite reserves in the world. Thick lateritic deposits occur on the Sangaredi Plateau (Harben and Bates, 1984) with a mineralogical composition of gibbsite and boehmite with <7 percent goethite, <1 percent quartz, and 2 percent to 4 percent titania minerals.

Much of central Jamaica contains karst-type bauxite deposits that are predominantly gibbsitic in mineralogy. The relatively high percentage of iron of the ore (20 percent Fe_2O_3 in high grade material) is due to very fine-grained hematite and goethite (Patterson et al., 1986), making it unsuitable for refractory applications. However, the low SiO_2 content makes it desirable as a feedstock for refined alumina. There are few commercial bauxite deposits found in North America, due in part to the lack of a tropical climate. Gibbsitic bauxite is mined in Arkansas from rocks derived from weathered nepheline syenite, although virtually all bauxite used in the United States is imported. Arkansas mines have supplied less than 1 percent of the demand, and all the domestic bauxite production is used in nonmetallurgical products (Plunkert, 2002).

There are numerous small bauxite mines in southern and central Europe, with most consisting of diaspore and boehmite of Upper Cretaceous age (Bardossy, 1984). France, where the bauxite mining originally began, has mines in the Pyrenees. Greece and Italy mine material that is used in the refractory and abrasive industries.

Mining and Processing

Bauxite is mined from open pits with initial overburden removal followed by digging or scraping to remove the ore. Blasting may be necessary for harder deposits. Subsequent processing routes depend on the ore composition and the desired use for the final product. For instance, refractory-grade bauxite from Guyana is first crushed to less than 9 cm, washed, crushed to less than 4 cm, washed again, then calcined in a rotary kiln (McCracken, 1999). Steps for other ores are similar with additional magnetic separation and froth flotation used for lower grade ores (Harben, 2002).

Converting bauxite to alumina is done using the *Bayer process* (Figure 6.4), which was first used in the late nineteenth century and is named for its inventor, the Austrian chemist Karl Josef Bayer. The general process involves crushing, grinding, and digesting the ore in a caustic solution. The process is summarized here, but a more complete description can be found in Hudson et al. (2002). Bauxite ore is crushed in cone crushers and hammermills to less than 2 cm then ground to particles less than 1.5 mm in rod or ball mills after the addition of some of the caustic solution. The fine bauxite is then digested in hot, concentrated caustic soda (NaOH) and Na_2CO_3 (caustic liquor) under pressure. The alumina hydroxide minerals dissolve in the liquor, forming sodium aluminates and leaving the insoluble iron, silicon, and titanium compounds as solids:

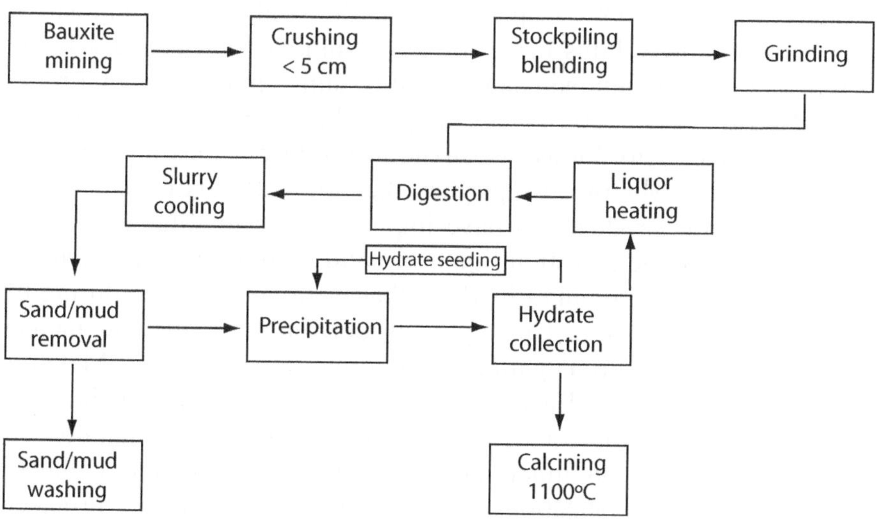

Figure 6.4 Flow diagram of the Bayer process.

$$Al_2O_3 \cdot xH_2O + 2NaOH \rightarrow 2NaAlO_2 + (x + 1)H_2O$$

The concentration, temperature, and pressure of the process depend on the properties of the ore. Gibbsitic bauxites are the easiest to digest, requiring lower temperatures and a lower concentration caustic solution than boehmitic and diasporic ores. Most Bayer plants operate at 200°C to 240°C at pressures of 30 atm (440 psi) with 100 to 120 grams NaOH per liter of solution. For boehmitic bauxites, temperatures up to 280°C at pressures up to 47 atm (700 psi) and 300 grams NaOH per liter of solution are needed (Chang, 2002). CaO may be added to assist in the digestion of diaspore. If the feed contains a mixed mineralogy, the digestion conditions will be based on the least-soluble phase. After digestion, the liquor is separated from the insoluble residue or red mud by filtering. The caustic liquor is removed from the residue and recycled into the digestion process. Silica as clay or free quartz is undesirable because it reacts with sodium aluminate solutions to form insoluble sodium aluminum silicates during digestion, causing loss of sodium hydroxide and extractable alumina (Hudson et al., 2002).

After the removal of insoluble solids, the liquor is cooled in a tank where crystalline alumina trihydrate is precipitated from the digestion liquor by seeding with fine crystals of $Al(OH)_3$. This is essentially a controlled reversal of the digestion process. The alumina trihydrate crystals agglomerate and are washed to remove the caustic solution. The separated crystals are size classified and calcined in rotary or stationary kilns at 1100°C to produce anhydrous alumina. Alumina calcined in stationary kilns has a lower α alumina content and a higher surface area than alumina calcined at the same temperature in a rotary kiln (Hudson et al., 2002). Depending on the composition of the bauxite ore, it takes 4 to 7 tons of bauxite to produce 2 tons of alumina.

While the Bayer process purifies the original bauxite ore, the resulting alumina product can have impurities due to soluble materials remaining from the digestion liquor or solid particles incorporated into the crystallites of alumina trihydrate that were not removed during filtering and clarification. Contaminants derived from the ore include silica, phosphorus, iron, gallium, and organic compounds. Materials introduced through the Bayer process include sodium and calcium. Sodium is the most problematic contaminant because of its high concentration in the digestion liquor and its solubility. Any sodium in alumina used in refractory applications can significantly reduce its performance.

Uses and Specifications

Alumina has been used in commercial ceramic applications for almost a century (Dörre and Hübner, 1984), with the first practical application in spark plug insulators. Since then, bauxite and synthesized alumina have found uses in a wide range of ceramic materials. Calcined bauxite is used predominantly in abrasives and high alumina refractories. Bayer alumina is used to make extra-high-alumina refractories, tabular and fused alumina, high-pressure sodium vapor lamp tubes, catalyst supports, and whiteware body and glaze fillers. Bauxite and Bayer alumina are not widely used in commercial glass batches where alumina comes from feldspathic materials or clays.

Bauxite is classified according to its intended commercial application (Plunkert, 2002). At least 90 percent of bauxite is metallurgical grade, which is used to make Bayer process alumina that is destined for the smelter to make aluminum metal. Because the Bayer process removes undesirable elements such as iron, titanium, and silicon, the specifications for metallurgical grade bauxite are considerably less stringent than the nonmetallurgical grades that are used for refractories and abrasives.

Calcined Bauxite Refractories Bauxite used in nonmetallurgical applications is usually calcined at temperatures from 1100°C to 1700°C (for refractory grade) to remove water and transform the aluminum hydroxide minerals to α alumina. The calcining conditions depend on the mineralogical makeup of the bauxite. Calcining can be done in shaft, round, and rotary kilns. Shaft kilns may have no temperature control that can adversely affect the quality of the material (Hill, 1994).

Deposits of bauxite that are pure enough for refractory applications are relatively rare. Most refractory grade bauxite comes from Surinam and China (Hill, 1994). Table 6.2 shows the composition of a typical refractory bauxite. Some general requirements for refractory bauxite include the following (Andrews, 1984):

- Alumina content is high.
- There is a maximum 2.5 percent Fe_2O_3 after calcination.
- There is a maximum 4 percent TiO_2.
- Only traces of alkali or alkaline earths are present.
- Silica may exceed 10 percent, but should be present as clay and not free quartz.

Any free silica present in the raw bauxite will be converted to mullite when calcined, which may be desirable depending on the final product. Physical properties of interest include pyrometric cone equivalent and other thermal properties, porosity, bulk density, and modulus of rupture (Chapter 8).

Refractories based on calcined bauxite are widely used in iron and steel production, which uses 60 percent or more of the bauxite-based refractory products manufactured globally (McCracken, 1999). Refractory bricks can be fashioned with calcined bauxite bonded with plastic clay or a mix of alumina and hydrogen phosphate. The phosphate ion is very reactive with alumina and will make a layer of $AlPO_4$ between the grains (Baumgart, 1984). Andalusite may be added to bauxite-based refractories to enhance the creep-resistance (McCracken, 1999).

Fused Bauxite and Alumina Abrasives With a Mohs hardness of 9.0, corundum is the second hardest natural mineral after diamond; it therefore makes an excellent abrasive. Natural corundum had been used historically, but it has given way to corundum produced from calcined bauxite. Abrasive alumina produced from 94 percent to 97 percent pure bauxite is commonly called brown fused alumina (BFA). White fused alumina is made from 99 percent pure Bayer alumina. BFA is produced by melting and reducing calcined bauxite (>80–82 percent Al_2O_3; <6 percent SiO_2) in an electric arc furnace. The production of fused alumina requires that the impurities, such as iron oxides, silica, and titania be removed. Carbon (as coke) is added as a reducing agent for the impurities so that they will react to form a ferrosilicon material that is immiscible in the alumina. This precipitates out to pool at the bottom of the furnace or the cooled mixture is crushed and the ferrosilicon is removed magnetically (Andrews, 1984). The alkalis and alkaline earth elements are undesirable because they can promote the formation of a glassy phase in the alumina and react with alumina to form β alumina, which reduces the toughness of the BFA (Lunghofer and Wolfe, 2000). Sulfur may also be an undesirable contaminant, introduced where coal is used as a reducing agent, which is common in China. The amount of reduction needs to be closely controlled to create the desired ferrosilicon slag. If over reduction occurs, aluminum carbide and aluminum oxycarbide will form (Lunghofer and Wolfe, 2000). This poses a problem if the BFA product is placed in service in an oxidizing atmosphere because the carbides will oxidize and lead to loss of structural integrity. The cooling rate of the fused bauxite determines the grain size, which is available in 300 μm to 1000 μm sizes. A rapidly cooled material that is poured into molds will form a microcrystalline product.

Calcined Alumina Bayer alumina trihydrate that is calcined at 1400°C to the point that 75 percent to 100 percent of the material is transformed to α alumina is referred to as *calcined alumina.* It is used in a wide variety of products, such as refractories, porcelain, glazes, electrical insulators, catalyst supports, and feedstocks for tabular alumina and synthetic mullite. Additives, such as fluorine and boron, may be used to control the crystallite size. These calcined aluminas can be categorized broadly according to their sodium content (Harben, 2002). The chemical and physical characteristics, such as the size and morphology of the crystallites (usually 1 to 10 μm), depend on the variables from the Bayer process (soda content, crystal size) and the calcining conditions. There are two general types: normal to low-soda grades and high purity. Normal soda content (<0.45 percent Na_2O) is used in applications where soda is not a problem or is desired, such as porcelain bodies, glazes, glass, and some refractories. Medium to intermediate soda content (<0.15 percent Na_2O) are used in products where soda is not desirable but small amounts are tolerated, such as electrical insulators (e.g., spark plugs), catalyst supports, and metal filters. Low soda (<0.10 percent Na_2O) is made by adding boric acid or silica before calcinations (Naito et al., 2003). This grade of alumina is used in electronic substrates and other components. For example, tape cast alumina that is widely used in electronic substrates requires low sodium levels.

Calcined alumina (and aluminum hydroxide) can be used as a filler in triaxial whiteware bodies as a substitute for quartz and flint. Early studies showed an increase in fired strength and other mechanical properties for a wide range of porcelain applications (Austin et al., 1946). Subsequent studies have supported these findings (e.g., Warshaw and Seider, 1967; Mostafa, 2003). It is possible that the increase in strength is due to the reduced effects of the β to α inversion at 573°C, although it has been suggested that the increased strength is due to the fact that alumina particles are stronger than quartz particles (Pinto, 2005).

High-Purity Alumina For products where very-high-purity alumina is required (>99.9 percent Al_2O_3), such as high-purity chemicals and advanced electronic ceramics, Bayer alumina undergoes several washing steps before it is calcined. Ultra-high-purity alumina of greater than 99.99 percent Al_2O_3 can be made through several distillation processes.

One method is to react high-purity Bayer alumina trihydrate with sulfuric acid to form aluminum sulfate. This is then mixed with ammonia to form a crystalline compound that is calcined at 1000°C. A second method uses high-purity alumina gels that are derived from aluminum metal. A third process involves dissolving alumina in hydrochloric acid and precipitating a hydrated aluminum chloride, which is subsequently calcined. Ultra-high-purity alumina is a raw material for synthetic sapphires, electronic substrates, and in polishing compounds used in the manufacture of electronic equipment (Harben, 2002).

Synthetic Single Crystal Alumina Single crystal corundum (white sapphire), ruby, and sapphire occur in nature, but most industrial applications use a synthesized crystal. These crystals are transparent and can be either clear or colored with added transition metals. Single crystals can be made from several processes, all using high-purity alumina as a raw material and a corundum or sapphire seed crystal (Altech, 2004). The most common and least expensive method is the Verneuil process, in which the crystal grows slowly in an oxyhydrogen flame on a rotating seed crystal. There is no crucible that would be a potential source of impurities, and crystals grow at a rate of 5 to 20 mm/h. In the Czochralski process, a single crystal is pulled from a gob of molten alumina that is held close to the liquidus temperature. The temperature and the pulling speed determine the diameter of the boule. The Bridgman process directs the crystallization by passing a crucible of molten alumina through a horizontal or vertical temperature gradient, thus moving the solidification front emanating from the seed crystal to the liquid. In the Stepanov or Edge-Defined Film Fed Growth (EFG) process, monocrystalline plates, rods, or tubes are drawn from molten alumina using a molybdenum template that is dipped in the liquid.

Single crystal corundum or sapphire can be used for laser optics and transparent windows for high-temperature and corrosive environments as well as precision, wear-resistant tools. Synthetic corundum can be purchased in a variety of shapes, including boules, wafers, ribbons, and rods. Synthetic sapphire should have no twinning or marks. Several grades are available (Maintech, 2004):

- *Optical Grade 1.* There are no optical dispersal centers such as bubbles.
- *Optical Grade 2.* There are some micro-bubbles up to 10 μm.
- *Optical Grade 3.* Bubbles not larger than 50 μm and not closer than 500 μm from each other are uniformly distributed.

- *Optical Grade 4.* Bubbles are not larger than 200 μm and not closer than 20 mm from each other.
- *Technical Grade.* Uniformly distributed bubbles are less than 50 μm and not closer than 500 μm, and bubble concentrations are not larger than 500 μm and not closer than 20 mm.
- *Industrial Grade.* Bubbles are not larger than 50 μm, with a total diameter less than 200 μm, and not closer than 20 mm.

Tabular Alumina Tabular alumina is made by forming spheres or pellets of finely ground, calcined alumina and an organic binder. The spheres are heated to just below the fusion point (around 1900°C) in a vertical shaft kiln for approximately 15 hours. The result is a tabular crystal of sintered α alumina. There are three overlapping stages of sintering: (1) neck growth between adjacent particles; (2) densification; and (3) grain growth that eliminates the final voids. Other metal oxides, particularly those with ionic radii similar to Al^{3+}, may be added to influence the grain growth and to control the physical properties (Baumgart, 1984). For instance, up to 2 percent Cr_2O_3 may be added to increase hardness, thus making pink or red alumina. To inhibit grain growth, MgO is added in amounts ranging from 0.05 percent to 0.5 percent (Schüller, 1984). This forms a spinel phase at the grain boundary and inhibits diffusion of Al^{3+}. The result is a material with very low porosity and a density of about 3.55 g/cm^3. Tabular alumina can be crushed to particle sizes of 50 to 300 μm to be used in refractories for steel making, molten metal filters, and catalyst supports.

Activated Alumina Activated alumina is a high-porosity, high-surface-area material that finds widespread use as a catalyst carrier, catalyst, and adsorbent. It is made by thermal decomposition of aluminum hydroxide, usually derived from the Bayer process, at temperatures between 250°C and 800°C in rotary or fluidized bed kilns. It can also be made by pyrolisis of hydrolyzed aluminum salts or alkoxides (Naito et al., 2003). The temperature and size of the particles are tailored to produce the desired surface area, which ranges from about 100 to 400 m^2/g (Hudson et al., 2002). Activated alumina is commonly used for dechlorination and defluorination in natural gas and petroleum processing. Spherical activated alumina is used as a Claus catalyst. In the Claus process, hydrogen sulfide that is present in natural gas and petroleum is converted to elemental sulfur over a catalyst bed that has a surface area of more than 300 m^2/g.

Alumina Gels Colloidal alumina hydroxides or gels are important as precursors for activated alumina and high-purity alumina materials. Alumina gels are produced by neutralization of aluminum sulfate or ammonium alum by NH_4OH, or from sodium aluminate by neutralization with acids, CO_2, $NaHCO_3$, and $Al_2(SO_4)_3$ (Hudson et al., 2002). The resulting gel is filtered, washed, and dried. The dried gel can then be formed into a variety of shapes and particle sizes.

Polycrystalline corundum for grinding can be made from boehmite sols (Krell, 2000). Using different doping compounds, a fine-grained, stable, and low-porosity material can be made that outperforms fused corundum in grinding applications.

Health and Environmental Concerns

Alumina powders can be irritating to skin, eyes, and, if inhaled, the lungs. However, it is not carcinogenic and overall poses little health risk to humans. Environmental concerns arise from the mining of bauxite and the problem of disposing of the residue produced during the Bayer process. The Bayer process results in a highly alkaline, iron-rich, claylike material.

6.2 BORATES

Introduction

Boron is a semimetallic element that is not particularly abundant on Earth, with an average concentration of 8.7 ppm in continental crust and 4.4 ppm in seawater (Table 2.1). Although boron is well dispersed in the Earth's crust, it can be relatively concentrated in some volcanic rocks. Boron is highly soluble, particularly at higher temperatures, and will readily dissolve in hydrothermal fluids. If boron-bearing fluids collect in enclosed lake basins that are exposed to arid conditions, the evaporation of water will eventually result in the precipitation of a variety of borate minerals. There are at least 230 naturally occurring borate minerals (Garrett, 1998), but the most common commercial minerals are colemanite, kernite, tincal (natural borax), ulexite, and datolite. Economic concentrations of these are very rare, and currently borate minerals and brines are mined from less than 10 global locations, with the United States, Russia, and Turkey being the largest producers (Lyday, 2003). Natural borate minerals and brines can be processed to make refined borate chemicals.

Borate minerals and chemicals are used as a source of boric oxide (B_2O_3) for glass and ceramic products. The largest user of borates is the fiberglass industry, where they are added for the glass-forming properties of boric oxide and, in the case of sodium and calcium borates, for the fluxing properties of the alkali and alkaline earth elements. Boric oxide can help prevent devitrification of the glass and increase durability. For similar reasons, it is used in the vitrification of radioactive wastes for long-term sequestration. Boric oxide reduces thermal expansion and is therefore widely used in heat-resistant borosilicate glass and in some porcelain glazes and enamels. Refined borate compounds are used as feedstocks for boron nitride, boron carbide, and metal borides, which are important high-performance materials (section 6.5). In addition to glass and ceramics, boron is used in other industries such as flame-retardants plant fertilizers, detergents, and corrosion inhibitors.

Composition and Mineral Structure

Boron does not occur in its elemental form in nature because it readily combines with oxygen, water, and metallic elements, particularly the alkali and alkaline earth elements. Boron bonds strongly to oxygen in either threefold coordination as trigonal planar BO_3 or fourfold coordination as BO_4 tetrahedra. In the BO_3 structure, each oxygen is bonded to boron with one valence site left for bonding with other species, leading to a polymerization sequence of triangles (Strunz, 1997). The crystal structure of many of the nonsilicate borate minerals is represented by chains of interlocking triangles and tetrahedra with the metal cations and water molecules located between the chains. As a result of this behavior and the ability of boron to form double or multiple salts with many compounds (Garrett, 1998), there are more than 230 different varieties of borate minerals. Most of these are rare, and only several of them are found in great enough abundance to be commercially important. In this section, the more frequently used minerals are described.

Sodium Borate Hydrate Minerals When sodium combines with boron, oxygen, and water, several sodium borate hydrate minerals may form. One of the more common is borax, which comes from *baurach* or *buraq,* the Arabic term for "white." Borax is used as a general expression for sodium tetraborate decahydrate, which has a nominal formula of $Na_2B_4O_7 \cdot 10H_2O$. Tincal, which comes from *tincana,* the

Sanskrit word for "borax," refers to the naturally occurring borax mineral. Tincalconite is borax with five moles of water ($Na_2B_4O_7 \cdot 5H_2O$), and commercial products may be called sodium tetraborate pentahydrate or 5 mol borax. Tincalconite can form through the dehydration of tincal because borax decahydrate will gradually dehydrate to a water content between 7 mol and 5 mol. Kernite ($Na_2B_4O_7 \cdot 4H_2O$), named after Kern County in California, is a sodium borate hydrate with four moles of H_2O. Under ambient conditions, kernite will hydrate to tincalconite.

Borax occurs as white, monoclinic prisms with an average specific gravity of 1.71 and a Mohs hardness of 2.0 to 2.5. Kernite also is found as white, monoclinic prisms with a pronounced cleavage that can lead to the formation of needles and fibers when crushed. Kernite has an average specific gravity of 1.95 and a Mohs hardness of 2.5. Tincalconite crystals have trigonal symmetry with a specific gravity of 1.88 and a Mohs hardness of 2.0.

The behavior of sodium borate hydrate minerals when heated can be complex and may follow several pathways (Smith, 2002a). If borax is heated in a closed system, it will dissolve in its own water of hydration at about 60°C. In an open system that allows the water to escape, borax loses 5 mol of water between 50°C and 100°C, an additional 3 mol up to 160°C, and the remaining 2 mol of water up to 400°C. Upon further heating, anhydrous borax melts at 742°C.

Colemanite Colemanite is the most common calcium borate mineral with a nominal composition of $Ca_2B_6O_{11} \cdot 5\ H_2O$. It is named after William T. Coleman, one of the first developers of borate minerals in California and owner of the mine in which the mineral was first found. Colemanite crystallizes in the monoclinic system to make transparent to translucent prismatic crystals. It has a specific gravity of 2.42 and Mohs hardness of 4.5. If one of the calcium cations is replaced with magnesium, the mineral is hydroboracite (Table 6.3). Upon heating, colemanite decrepitates violently at 350°C as it loses its water of hydration.

Ulexite Ulexite is a mixed sodium and calcium borate mineral with a nominal composition of $NaCaB_5O_9 \cdot 8H_2O$ that forms triclinic, needlelike crystals that can occur as masses that resemble a cotton ball. This mineral appears to be formed by the alteration of borax in the presence of calcium. Ulexite has a density of 1.96 and a Mohs hardness of 2.0. On heating, ulexite dehydrates by initially losing 4 mol of water

Table 6.3 Properties of some borate minerals

	Formula	Specific Gravity	Mohs Hardness	wt.% B_2O_3	wt.% H_2O
Kernite	$Na_2B_4O_7 \cdot 4H_2O$	1.95	2.5	51.0	26.4
Tincalconite	$Na_2B_4O_7 \cdot 5H_2O$	1.88	2.0	48.8	29.5
Tincal (natural borax)	$Na_2B_4O_7 \cdot 10H_2O$	1.71	2–2.5	36.5	47.2
Ulexite	$NaCaB_5O_9 \cdot 8H_2O$	1.96	2.0	43.0	35.6
Colemanite	$Ca_2B_6O_{11} \cdot 5H_2O$	2.42	4.5	50.8	21.9
Hydroboracite	$CaMgB_6O_{11} \cdot 6H_2O$	2.2	2.0	50.5	26.2
Datolite	$CaBSiO_4(OH)$	2.9	5.0–5.5	21.8	5.6
Anhydrous borax	$Na_2B_4O_7$	2.37	—	69.2	0
Boric acid	H_3BO_3	1.51	—	56.3	43.7
Boric oxide	B_2O_3	1.84	—	100	0

at 70°C to 90°C, and then the remaining water is gradually lost up to about 400°C (Smith, 2002a).

Datolite Datolite is a calcium borosilicate mineral with a nominal composition of $CaBSiO_4(OH)$. It forms monoclinic crystal with an average density of 2.9 and Mohs hardness of 5.0 to 5.5. It is transparent to translucent and is often found in the pores of basalt flows (vesicles).

Boric Acid Boric acid is found naturally as sassolite, a relatively rare mineral that precipitates near hydrothermal springs. Commercial boric acid is a synthetic material made from the more common natural borate minerals. Boric acid can be considered as a hydrated boric oxide that is either a trihydrate ($B_2O_3 \cdot 3H_2O$ or $B[OH]_3$) or a monohydrate ($B_2O_3 \cdot H_2O$ or HBO_2) (Smith, 2002a). The trihydrate is more stable, and therefore the more commercially available product. Boric acid occurs as monoclinic, white, platy crystals in which sheets of hydrogen bonded $B(OH)_3$ are held together by weak Van der Waal's bonds.

When heated, boric acid loses water between 100°C to 130°C and first forms an orthorhombic metaboric acid (HBO_2) and then forms a coarsely crystalline monoclinic HBO_2 at 160°C (Smith, 2002a). At 200°C to 250°C, cubic HBO_2 slowly crystallizes from a high-viscosity melt. With continued heating, any remaining water is removed and the material is then anhydrous boric oxide (B_2O_3).

Boric Oxide Boric oxide (B_2O_3), also referred to as anhydrous boric acid, is a synthetic material made from natural borate minerals or brines. It is available commercially as a vitreous glass that is made by the dehydration of boric acid. It is white and has a specific gravity of 1.84. Boric oxide is hygroscopic and will absorb moisture when exposed to humidity. When the crystalline material is heated, it will soften at about 325°C and melt at 450°C while the vitreous form has no definite melting point, softening at about 450°C and becoming fluid at about 500°C. However, it is still viscous at temperatures above 1000°C (Smith, 2002a).

Solubility of Borates The borate minerals and compounds have varying solubility behavior in water (Table 6.4). With the exception of ammonium borates and boric acid, only the alkali borates are very soluble in water (Kistler and Helvaci, 1994). Kernite hydrates more slowly than tincal or tincalconite and it dissolves incongruently in water, leaving a solid pentaborate salt after dissolution of the kernite crys-

Table 6.4 Solubility of some borate materials in water

	Solubility in % by Weight[1]					
	0°C	20°C	40°C	60°C	80°C	100°C
Borax pentahydrate	1.52	3.59	8.57	23.16	33.85	50.13
Borax decahydrate	1.99	4.71	11.22	30.33	44.31	65.64
Boric acid	2.52	4.72	8.08	12.97	19.10	27.53

[1]Compiled from U.S. Borax product profiles

tal (Noirot, 2005). The solubility of colemanite in water at 25°C is about 0.18 percent B_2O_3 and 0.38 percent B_2O_3 at 100°C.

Geological Occurrence and Distribution

The largest concentration of commercially valuable borate minerals (colemanite, ulexite, and borax) occur in evaporite deposits that formed in arid, nonmarine settings such as playas and isolated alkaline lakes. Most of these formed during the Cenozoic era, or between 65 and 1.5 million years ago (Smith, 2002b). The deposits form when groundwater and hydrothermal fluid permeate boron-enriched volcanic rocks, leaching the boron (and alkali elements). If the fluids reach the surface and drain into enclosed, arid basins, the boron and other soluble elements will become concentrated as the host water evaporates. When all of the water has evaporated, borate minerals such as borax and ulexite will crystallize. Depending on the composition of the brine, other evaporite minerals may be deposited along with the borates. For example, some marine evaporate deposits may contain appreciable amounts of the magnesium borate szaibelyite ($Mg_2B_2O_5 \cdot H_2O$) due to the high magnesium content of seawater. Chlorine and sulfate are undesirable contaminants that can be associated with borate minerals formed in evaporitic settings (Table 6.5).

After deposition, successive influxes of sediments, particularly clays, cover the minerals and protect them from dissolution. As a result, borate deposits are often associated with other terrigenous sediments such as clays and ash tuffs. After deposition, the buried borates minerals may be altered by either reaction diagenesis or thermal diagenesis (Smith, 2002b). Reaction diagenesis occurs when borate minerals react with fluids enriched in calcium and/or magnesium, resulting in the removal of some of the sodium and the formation of secondary calcium and/or magnesium borate minerals. Examples include the transforma-

Table 6.5 Composition and specifications for borate materials

	B_2O_3 wt. %	SiO_2 wt. %	Al_2O_3 wt. %	CaO wt. %	MgO wt. %	Na_2O wt. %	H_2O wt. %	Fe ppm	Cl ppm	SO_4 ppm
				Borax Decahydrate						
Theoretical	36.51					16.25	47.24			
Turkey[1]	>36.47					>16.48		<10	<50	<70
California[2]	36.5–38.3					16.2–17.1		≤30	≤700	≤600
				Borax Pentahydrate						
Turkey[1]	>47.74					>21.41		<10	<50	<70
California[2]	48.6–49.3					21.6–21.9		≤10	≤120	≤100
				Anhydrous Borax						
California[2]	68.5–69.4					30.5–30.9	≤1.0	≤40		≤150
				Colemanite						
Theoretical	50.9			27.2			21.9			
Turkey[1]	42–43	2–6		24–29						
Argentina[3]	38	7.0	1.0	22.5	1.25	0.2		500		

Boric Acid					
Theoretical	56.30	43.7			
Turkey[1]	56.25		20	20	500
California[2]	56.3–56.8		≤6	≤18	≤350
B_2O_3					
Theoretical	100.0				
California[2]	97.5–100.0	≤2.5	≤10		≤250

[1] Ab Etiproducts Oy; www.etiproducts.com
[2] U.S. Borax; borax.com
[3] Ceramic Materials Info, ceramic-materials.com

tion of borax to ulexite and ulexite to colemanite. Thermal diagenesis occurs when the primary minerals are heated and dehydrated, which will reduce the amount of water in borax minerals.

There is a wide array of borate occurrences around the world (see Garrett, 1998, for a complete description), but global commercial mining of borates is limited to only several sources (Figure 6.5). In the United States, California is the largest producer of borates, with the majority of mining occurring near the town of Boron. Boron, located in the northwestern Mojave Desert, is one of the world's largest producers of sodium borate minerals, including tincal, kernite, and ulexite. The deposit is located at the edge of a Tertiary basin and consists of a lenticular mass of the borate minerals interbedded with clays and covered by sands (Figure 6.6: Kistler and Helvaci, 1994). There are more than 23 borate deposits near Death Valley, California (Garrett, 1998), with commercial mining of colemanite and ulexite presently taking place at one of the underground deposits. Searles Lake, which contains thick deposits of evaporite minerals and brines, is a commercial source of both soda ash (see section 5.6) and borate minerals. There are two general structural horizons within the deposit, the Upper Salt and Lower Salt, with a B_2O_3 content of 1.0 percent and 1.2 percent, respectively (Smith, 1979), although recirculated brine pumped into the upper unit is used to increase the concentration to a theoretical 1.45 percent (Lyday, 2002). Borax, boric acid, and boric oxide are produced from the brines.

Turkey is another major borate producer, with large sodium borate deposits located within five main districts in Western Anatolia: Bigadic, Kestelek, Sultancayiri, Emet, and Kirka (Kistler and Helvaci, 1994). The geologic setting of the Turkey deposits is similar to those of California. Colemanite, ulexite, tincal and kernite are formed in enclosed, arid basins and are interbedded with clays and other sediments. Colemanite and ulexite are the principal minerals mined (Smith, 2002a). The Kirka reserve is the largest known borate deposit in the world, with a principal mineral of tincal. Another producer in Asia is China, where borate deposits are found in magnesium-rich, iron-poor carbonates within metamorphosed evaporite deposits in Liaoning and Jilin Provinces (Lyday, 2003). In Qinghai province, ulexite, borax, and hydroboracite are found in playa lakes. There are potential deposits in the Tibetan Plateau, but these have not yet been developed (Kistler and Helvaci, 1994).

Numerous borate deposits are found in South America along the volcanically active Andes Mountains and the arid, enclosed basins near

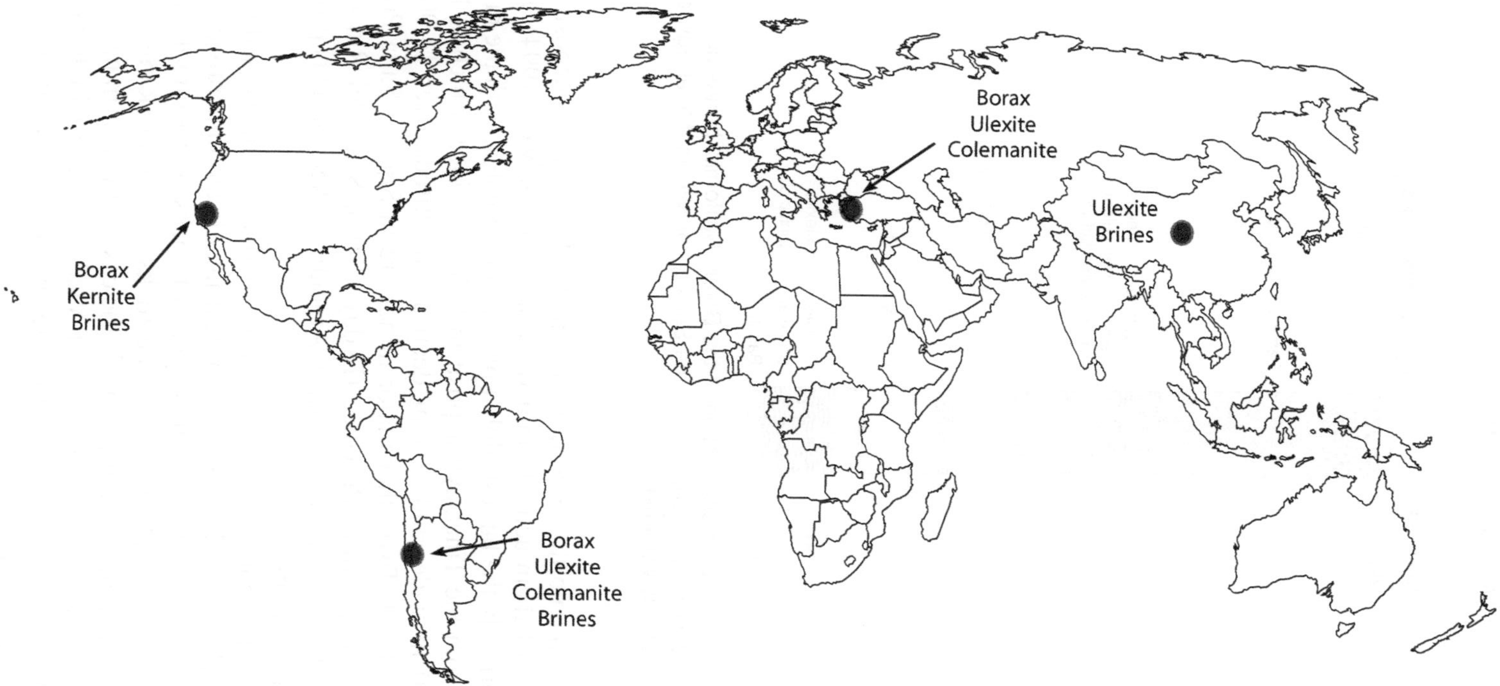

Figure 6.5 Global map showing the major sources of borates.

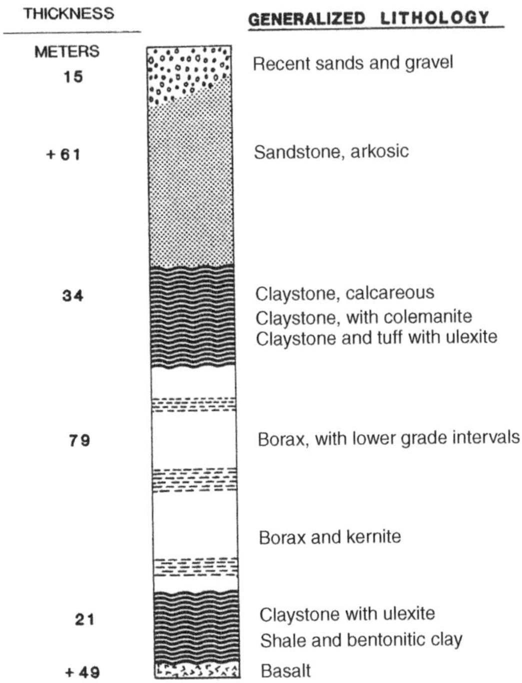

Figure 6.6 Schematic cross-section of the Boron, California, borate deposit (from Kistler and Helvaci, 1994; courtesy of the Society for Mining, Metallurgy, and Exploration (SME), www.smenet.org).

the base of the north–south trending range. There are 40 deposits, with at least 14 of them actively mined (Kistler and Helvaci, 1994). Argentina is currently the largest global producer of borates, with the majority coming from the Tincalayu deposit in Salta Province, which is 1.5 km long, 500 m wide, and 100 m deep at an altitude over 14,000 feet above sea level. Two other major mines are in Sijes in Salta Province and Porvenir in Jujuy Province (Lyday, 2003). Salar del Rincon is a 500 km^2 playa lake in northwestern Argentina where brine is processed to produce borates.

Most of the commercial borates are derived from the sedimentary deposits. However, there are boron-bearing silicate minerals that occur in commercial amounts in other types of rocks. For example, contact metamorphism of calcium carbonate rocks may form skarns (see section 2.3) that contain datolite. This is the case at the Bor mine in eastern Russia near Dalnegorsk, northeast of Vladivostok, in which datolite is mined. Szaibelyite and hydroboracite are also found in Kazakhstan (Kistler and Helvaci, 1994).

Mining and Processing

A summary of the general mining and processing routes is presented here, because detailed descriptions at each particular operation are presented by Garrett (1998). Most natural borate minerals are mined at the surface, except in some cases where it has been worth the expense of underground mining, such as the Billie mine in Death Valley and small mines in Bigadic, Turkey. The Boron, California, mine was originally an underground mine but was converted to a surface mine in the 1957. At the surface mines of Boron and Kirka, ore and overburden are loosened with explosives and removed with bucket loaders. Some of the thin salars of Argentina are mined by hand. Ore is crushed to ~2 cm pieces and then transported to refineries by truck or conveyer (Kistler and Helvaci, 1994).

After mining, the subsequent processing route depends on the ore mineral and the desired final product. Some minerals, such as ulexite and colemanite, may be beneficiated and marketed directly, while the remainder of the ore is used to make refined boron chemicals. To make refined borax, crushed tincal and kernite ores are dissolved with soda ash in 100°C water (Smith, 2002a). The borates dissolve in the water and the coarse insoluble material is removed with screens. The saturated borate solution is transferred to thickening tanks to allow the fine clay particles to settle out. The liquor is then cooled to the point at which borax pentahydrate crystal precipitates. The slurry of crystals and water is vacuum filtered or centrifuged to remove the water and then the material is dried. To reduce caking, the dried material is washed with a weak boric acid solution that leaves a thin surface coat of borax pentahydrate.

The calcium-bearing borate minerals, colemanite and ulexite, can be used as a feedstock for synthetic borax where natural tincal is not available. Ulexite is preferred over colemanite because less alkali is required and fewer insoluble calcium salts are produced (Smith, 2002a). Both minerals are crushed to finer than 2.5 cm and calcined at 450°C to 500°C (Garrett, 1998). After cooling, the material is mixed with soda ash and then dissolved in water. The insoluble calcium carbonate is filtered, and the liquor containing the dissolved sodium borate is sent to crystallizers to precipitate the borax pentahydrate.

Borax can be synthesized from brines by passing the solution through a series of controlled crystallization steps to progressively remove NaCl, KCl, Na_2CO_3, and Na_2SO_4. The remaining solution, which is supersaturated with borax, is reacted with $NaHCO_3$ to force the crystallization of borax pentahydrate.

Anhydrous borax is produced by heating borax to its melting point (742°C) to form a borate melt (Smith, 2002a). This is quenched to form a borate glass or cooled slowly to make a crystalline product.

Boric acid can be manufactured from either purified borax or directly from natural borate minerals. The feedstock depends greatly on which is the most economical option. In the United States, natural sodium borate minerals and brines are the feedstocks. In Europe, boric acid is commonly made from imported Turkish colemanite (Smith, 2002a). In China and Russia, magnesium borates are used. A general processing route for boric acid production is the initial digestion of borate minerals with sulfuric acid. After the undesirable minerals, such as clay, are removed from the liquor, controlled cooling allows the crystallization of boric acid crystals, which are then removed by filtration. Where colemanite is used, calcium sulfate ($CaSO_4$) is produced during the reaction with the sulfuric acid. This is removed with the gangue. Where magnesium borates such as ascharite, szaibelyite, and hydroboracite are used, magnesium sulfate needs to be precipitated and removed before boric acid is crystallized. Where the calcium borosilicate mineral datolite is used as the feedstock, silica needs to be removed after acid digestion. Boric acid is synthesized from brines through an extraction process using a solvent, such as kerosene. Through several steps, a solution of boric acid and sodium sulfate is concentrated and cooled to crystallize boric acid.

When handling borate materials, one must keep in mind any reactions that can occur under ambient conditions. Borax pentahydrate can lose water while in storage, so careful determination of its water content should be performed prior to batching. Anhydrous borax is readily hydrated under humid conditions, which can lead to caking.

Uses and Specifications

Borate materials are used in both the glass and ceramic industries to add boric oxide and, in the case of borax, colemanite, and ulexite, to add alkali and alkaline earth fluxes. There are several products that require boric oxide as a major component, while others use minor additions to enhance some of the properties. In the latter case, it is usually economic considerations that determine if the benefits outweigh the added cost of borates. Boron can be added using any of the materials listed in the previous sections, depending on the desired composition and purity of the product. The price of boron minerals and refined chemicals is based on B_2O_3 content and the purity of the material. Of the refined compounds, borax pentahydrate or 5 mol borax is produced in the highest volumes, followed by boric acid. Borax decahydrate, anhydrous borax, and anhydrous boric acid are produced in smaller volumes (Smith, 2002b). All products must be low in iron to reduce coloring and low in chlorine and sulfate (Table 6.5), which can corrode refractories and contribute to emissions from glass-producing furnaces.

Glass In glasses containing silicon and boron, both will bond with oxygen and act as network formers. In the presence of alkali ions, the SiO_4 tetrahedra will combine with the alkali ion to form a nonbridging oxygen (NBO), reducing the connectivity of the melt structure. With the addition of boron, the BO_3 triangles combine with the alkali ion but form BO_4 tetrahedra and no increase in NBO (Shelby, 1997; Varshneya, 1994). This behavior is an example of the complex interactions of boron in glass melts. Some of these interactions result in glasses with desirable traits, such as low thermal expansion, reduced devitrification, and increased durability. Therefore, boron is an important material for the commercial glass industry, which accounts for 78 percent of boron consumption in the United States (Lyday, 2003). It is used primarily in fiberglass and borosilicate glasses, as well as some specialty glasses and radioactive material vitrification (Table 6.6).

Borosilicate glasses contain 5 percent to 30 percent B_2O_3, with typical content of 12 percent to 15 percent (Bühler, 1997). Borosilicate glass is important for heat-resistant service, such as cookware, laboratory glass, range tops and oven windows, displays, and lighting applications. Borosilicate glasses have low thermal expansion and resistance to thermal shock compared to soda-lime-silicate glass.

In fiberglass manufacturing, borates are used as a flux to lower melting temperatures, to control viscosity, to lower surface tension, to reduce devitrification, and to increase chemical durability. The lower melting temperatures that borates allow decrease wear on the expensive spinners and bushings. The lower surface tension improves fiberization

Table 6.6 Boric oxide content of common glass formulations (wt. %)

E-glass (textile fibers)[1]	5–10
Glass wool (insulation fibers)[1]	5–15
Borosilicate[2]	12.5
Lead crystal[2]	2.5

[1] Gonterman and Wolf, 1988
[2] Harries-Rees, 1993

and reduces beading. In the 1940s, the sources of boric oxide used in fiberglass were borax and boric acid. Since then, the products have diversified (Garrett, 1998). Borates, in the form of ulexite or borax pentahydrate, are used in the formulation of discontinuous insulation fiberglass melts, which have 5 percent to 15 percent B_2O_3 (Table 6.6). Borax pentahydrate is the most common borax material that is used, making up at least 80 percent of the borates used in insulation glass (Bühler, 1997). Continuous glass fibers used in reinforcement applications, such as E-glass, are based on the CaO-Al_2O_3-SiO_2 eutectic system, so sodium is not part of the batch formulation. As a result, boric acid or colemanite are used to make E-glass formulations with 5 percent to 10 percent B_2O_3. Colemanite can replace limestone and boric acid and reportedly allow faster melting (Garrett, 1998). Anhydrous borax may be used because of the reduced energy needs to drive off the water relative to hydrated borax, and the reduced mass will reduce transportation costs. Anhydrous borax may also be a desirable component because the lack of water decreases swelling of the batch during initial melting (Materials Handbook, 2005).

Similar to lithium compounds, small additions to soda-lime-silicate glass compositions can reduce energy costs and prolong the life of the furnace. It is reported that glasses containing 1 percent to 1.5 percent B_2O_3 can significantly increase production rates, brilliance, strength, durability, and shock resistance (Harries-Rees, 1993).

The encapsulation of radioactive waste into a durable glassy material is the preferred long-term storage option for these dangerous materials. The glass must have a high chemical durability and have favorable low-temperature viscosity characteristics to reduce volatilization of hazardous materials during the vitrification process. Processed radioactive materials are mixed with a preformed frit that contains between 7 percent and 8 percent B_2O_3 (Plodinec, 1988).

Ceramics Borate minerals are commonly used as fluxing compounds in leadless porcelain glazes and enamels. B_2O_3 can reduce glaze viscosity, reduce thermal expansion, and create a higher gloss (Harries-Rees, 1993). Because borax is soluble in glaze suspensions, boric oxide can be added as a frit (powdered glass) or as the minerals colemanite and ulexite to reduce dissolution. The use of a frit with 5 percent to 15 percent B_2O_3 avoids flocculating the glaze by entrapping the borax in the glassy matrix (Noirot, 1999). Borax is best used in relatively small amounts, as excessive use can lead to crazing and blistering.

Borax is used in porcelain enamel frits as a flux and glass former (Table 6.7). It can make up from 15 percent to 45 percent of the frit batch, depending on the application. It imparts important physical properties, such as strength, toughness, and durability while adding brilliance to colors. Increasing the amount of B_2O_3 can reduce scumming of enamels caused by low-temperature firing (Materials Handbook, 2005).

The addition of borates into porcelain ceramics is not as well known as their use in glazes, perhaps due to the higher cost of borates relative to traditional raw materials. The addition of borate minerals should be viewed as an addition of both a flux (from the sodium and/or calcium) and a glass former (from the boric oxide) (Noirot, 1999). There is some evidence that the addition of boric oxide to aluminosilicate gels can decrease the temperature of mullite formation (Hong and Messing, 1997). The addition of borates to a whiteware body can increase the solubility of silica and alumina in the glass phase, leading to increased chemical and mechanical durability (Noirot, 1999). However, the high solubility of borates in clay slurries can be problematic. Noirot (2000) found that addition of borates of 20 to 200 ppm B_2O_3 in kaolin and illite slurries can affect viscosity characteristics, although this is dependent on the concentration of Na^+ dissolved in the suspension. Additional challenges include the rapid decomposition or puffing of borax during firing. This and the problems associated with borate solubility can be avoided by adding fritted or synthetic borates, although this adds to the cost of the material.

Refined borates are components in the manufacturing of some advanced ceramic materials such as boron nitride, boron carbide, and metal borides. High-purity boric acid is the source of boron in the synthesis of these materials. In the production of rare earth magnets, metallic boron is combined with iron and a rare earth element such as neodymium or samarium. Boron metal can be made either through vapor phase reduction of BCl_3 or BBr_3 or by the reduction of boric acid and carbon (Harben, 2002).

Table 6.7 Borax in porcelain enamel frit formulations (wt. %)

Sheet steel ground coat	20–45
Sheet steel cover coat	15–40
Dry-process ground coat	20–45

Harries-Rees (1993)

Health and Environmental Concerns

In mammals, sodium borates and boric acid are quickly absorbed into the body when ingested, but also excreted rapidly. Incidental exposure to small amounts of borates through inhalation or ingestion appears to pose no significant threat to human health. Borates have low acute toxicity. Boron is an essential micronutrient for plants, and may be valuable for higher organisms. Large doses can be detrimental to plants, and certain aquatic life forms are particularly sensitive to elevated boron levels. As a result, borates should not be released into the environment.

6.3 ZIRCONIA

Introduction

Zirconium is the ninth most abundant metal in the earth's crust with an average abundance of 130 ppm (Table 2.1). It is never found in its elemental state and combines with oxygen and other elements to form minerals. The zirconium cation Zr^{4+} is relatively incompatible in most rock-forming minerals and is more likely to be found in one of up to 40 zirconium-bearing minerals (Nielsen, 2002), of which zircon ($ZrSiO_4$) and baddeleyite (ZrO_2) are the most widely used commercially. Zircon is the most common and is also the ore for hafnium. While it is a common accessory mineral in igneous and metamorphic rocks, it is found in economic concentrations only in sedimentary heavy mineral sand deposits. Zircon can be used directly in some applications, or zirconia (ZrO_2) is extracted from it by several different methods that remove the silica.

Zirconia is an important material that is used for a wide variety of ceramic and glass products. Because of its ability to withstand high temperature, zircon and zirconia are used in refractories, metal casting, and applications where high strength, hardness, and thermal stability are required. Its hardness allows it to be used for cutting and grinding tools. In whitewares, zircon is used in ceramic glazes as an opacifier and pigment. Zirconia that is stabilized with calcia, magnesia, or yttria is used in some refractory applications as well as solid electrolytes that are used in oxygen sensors and fuel cell applications. Highly purified zirconia compounds are used in catalyst coatings and supports.

Composition and Mineral Structure

Because of its high valence state and ionic radius of 730 Å, the Zr^{+4} cation does not partition readily into any of the major rock-forming igneous minerals, but prefers to form its own minerals. While there are several zirconium-bearing minerals, the most common and those commercially exploited are zircon ($ZrSiO_4$) and, to a lesser extent, the oxide mineral baddeleyite (ZrO_2).

Zircon is a nesosilicate with a tetragonal structure consisting of chains of alternating SiO_4 tetrahedra and ZrO_8 triangular dodecahedra (Deer et al., 1993) (Figure 6.7). There is always some substitution of hafnium for zirconium with all zircon having a Hf/Zr ratio of 0.02 to 0.04 and an average 1 percent Hf. Thorium (up to 10 percent) and uranium (up to 7 percent) are both present in zircon as well (Gaines et al., 1997). The color of zircon is variable, usually brown to reddish

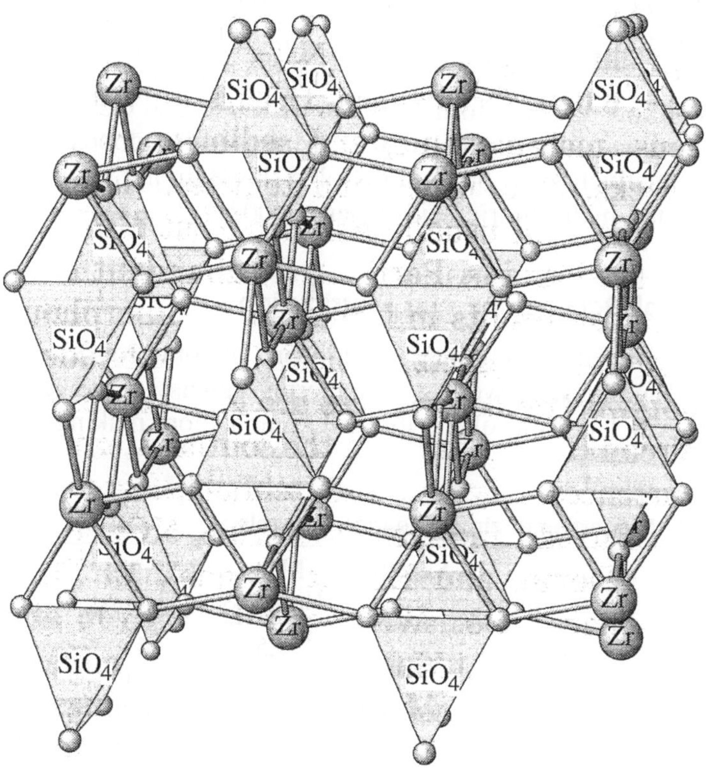

Figure 6.7 Structure of zircon (from Gaines et al., 1994, p. 1055; reprinted with permission from John Wiley & Sons).

brown, but gem quality specimens are colorless to yellow and green. It has a hardness of 7.5 and specific gravity of 4.6 to 4.7 (Table 6.8). When heated above 1450°C, it decomposes to ZrO_2 and SiO_2. Zircon has a coefficient of thermal expansion of 1.4×10^{-6}/°C.

Although baddeleyite is composed of zirconia (ZrO_2), most commercial zirconia is produced from zircon. Zirconia is stable under ambient pressure in three different crystal structures, depending on temperature (e.g., Subbarao, 1981). Up to about 1170°C, it occurs as a monoclinic crystal that has a hardness of 6.5 and a specific gravity of 5.5 to 6.0 (Table 6.8). When heated to between 1150°C and 1170°C, the structure converts to tetragonal symmetry with an abrupt change in the lattice structure. This transformation occurs with a 3 percent to 5 percent decrease in volume. The tetragonal to monoclinic transformation occurs on cooling, but occurs gradually between about 1000°C and 800°C, although rapid cooling can keep the tetragonal phase metastable (Nielsen, 2002). The tetragonal phase is stable up to about 2370°C. From 2370°C to a melting point near 2680°C, cubic zirconia is stable. However, the cubic form cannot be retained through quenching and must be stabilized by the addition of dopants. The coefficient of thermal expansion for the cubic form is 10.5×10^{-6}/°C and 6.5×10^{-6}/°C for the monoclinic form up to 1000°C (Materials Handbook, 2005). Zirconia is relatively inert but will dissolve slowly in hot sulfuric acid and concentrated hydrofluoric acid (Chang, 2002).

Geological Occurrence and Distribution

Zircon is an accessory mineral in igneous plutonic rocks such as syenites and granites, as well as metamorphic gneisses. It usually makes up less than 1 percent of the mineral composition, making these rocks

Table 6.8 Properties of zircon-bearing minerals

Name	Formula	Specific Gravity	Mohs Hardness	Heating Characteristics
Zircon	$ZrSiO_4$	4.6–4.7	7.5	>1450°C decomposes to ZrO_2 and SiO_2
Baddeleyite—Zirconia	ZrO_2	5.5–6.0	6.5	1170°C converts from monoclinic to tetragonal; 2370°C converts to cubic; melts at ~2700°C

poor sources for commercial zircon mining. However, the processes of weathering, erosion, and hydraulic sorting provide a natural method of concentrating zircon. Because of zircon's resistance to weathering, it remains intact during the physical and chemical breakdown of granite. The crystals that are freed by weathering are eroded and transported along with quartz grains and other resistant minerals that eventually settle and accumulate in residual sedimentary deposits on beaches and placers. Such deposits may also contain other heavy minerals of economic interest, including the titania ore minerals rutile and ilmenite and the rare earth mineral monazite.

Zircon is mined commercially throughout the world (Figure 6.8), with world reserves estimated to be 38 million tons of ZrO_2 and identified world resources of approximately 72 million tons of ZrO_2 (Hedrick, 2003a). Australia is the largest producer of heavy minerals, including zircon, from Pleistocene to Recent sands on both its east and west coasts. Total Australian production of zircon in 2003 was approximately 452,000 tons (Hedrick, 2003a). The richest deposits on the east coast are found in an area between Yamba and North Stradbroke Island (Garnar, 1994).

South Africa is the second largest producer of zircon after Australia, with deposits mined from a large dune north of Durban in Richard Bay. Baddeleyite has been recovered as a byproduct of copper and phosphate mining in South Africa as well (Garnar, 1994). Smaller mining operations are found in other African nations such as Mozambique.

In North America, zircon mining occurs in the United States in Florida, Georgia, and Virginia (Hedrick, 2003). In Florida and Georgia, heavy minerals including zircon are mined from a linear sand ridge that crops out near the border between the two states. In Florida, zircon is found near Jacksonville in the Trail Ridge and Green Cove Springs deposits (Force and Rich, 1986). In Georgia, mining is done in Lulaton, Brantley County. In Virginia, ilmenite and zircon mining takes place in Stony Creek south of Richmond. In South America, a mixture of baddeleyite and zircon called caldasite that contains 65 percent to 80 percent ZrO_2, is mined in Brazil (Nielsen, 2002). The largest Brazilian mining operations are in the state of Amazonas and with lesser production in several other states including Paraíba and Rio de Janeiro.

In India, zircon is mined with other heavy minerals, such as ilmenite and rutile, in the coastal states of Kerala, Orissa, Andhra Pradesh, Tamil Nadu (Hedrick, 2003a). Both dry, open-pit mines and dredging are used. In Russia, baddeleyite is mined as a co-product of iron ore mining in the Kola Peninsula.

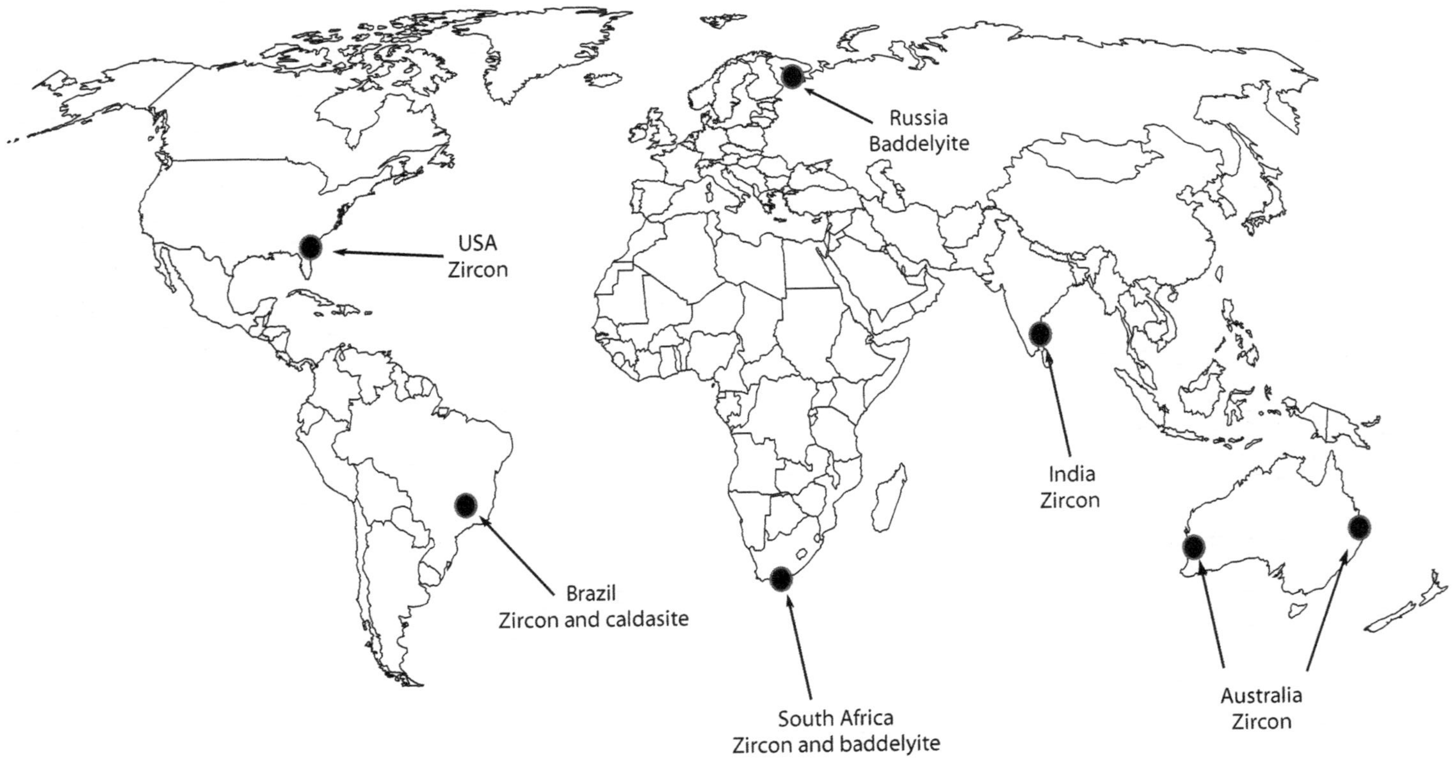

Figure 6.8 Map showing the major global producers of zirconium-bearing minerals.

Mining and Processing

Zircon is normally produced during the mining and processing of unconsolidated heavy mineral sands that contain a mixture of mineral ores used to produce titanium, phosphate, zirconium, and rare earth elements. In dry mining operations, scrapers are used to collect and transport ore to concentrator plants. Dry mining is suitable where deposits are shallow or are partially cemented or lithified. Dredging, or wet mining, is best suited for recovering loose sand deposits that are below the water table. Dredges mine in artificial ponds, pumping ore in slurry form to a floating concentrator (Figure 3.2). A dredge can consist of a rotating cutter that digs into the deposit, loosening the material and mixing it with water into a slurry that is pumped to the concentrator. The desired minerals may make up only 1 percent of the sand, so the concentrator serves to upgrade the material to 90 percent heavy minerals by gravity separation in spiral separators (Figure 6.9).

After gravity separation, the heavy mineral concentrate may be scrubbed to remove any surface coatings (Garnar, 1994). The ilmenite is removed by high-intensity magnets and rutile is removed by electrostatic methods. The remaining mineral assemblage can consist predominantly of zircon and monazite, a rare earth ore mineral (see section 6.5). The zircon and monazite can be separated using induced magnetic roll separators (Chang, 2002).

At this point in the process, some of the zircon is graded and crushed to the desired size to be used directly in refractories, ceramic glazes, and foundry sands. Finely ground zircon flour can be spray-dried to make a granulate that is used as an opacifier in glazes. Because of its high hardness, grinding zircon is difficult such that finer grinding can add substantially to the cost of the material.

For other ceramic applications that require zirconia, baddeleyite can be used, but more commonly zircon is processed to remove its compositional silica. There are several techniques that can be used to synthesize zirconia from zircon, including a caustic extraction, a lime fusion method, a carbo-chlorination technique, and a plasma arc process (Chang, 2002; Harben, 2002). In the caustic extraction process, sodium hydroxide or sodium carbonate reacts with zircon at 600°C (with sodium hydroxide) to 1000°C (with sodium carbonate) to produce a sodium zirconate and a sodium silicate (Nielsen, 2002):

$$ZrSiO_4 + 4NaOH \rightarrow Na_2ZrO_3 + Na_2SiO_3 + 2H_2O$$

Water is added to dissolve the sodium silicate and hydrolyze the sodium zirconate, resulting in the precipitation of an insoluble zirco-

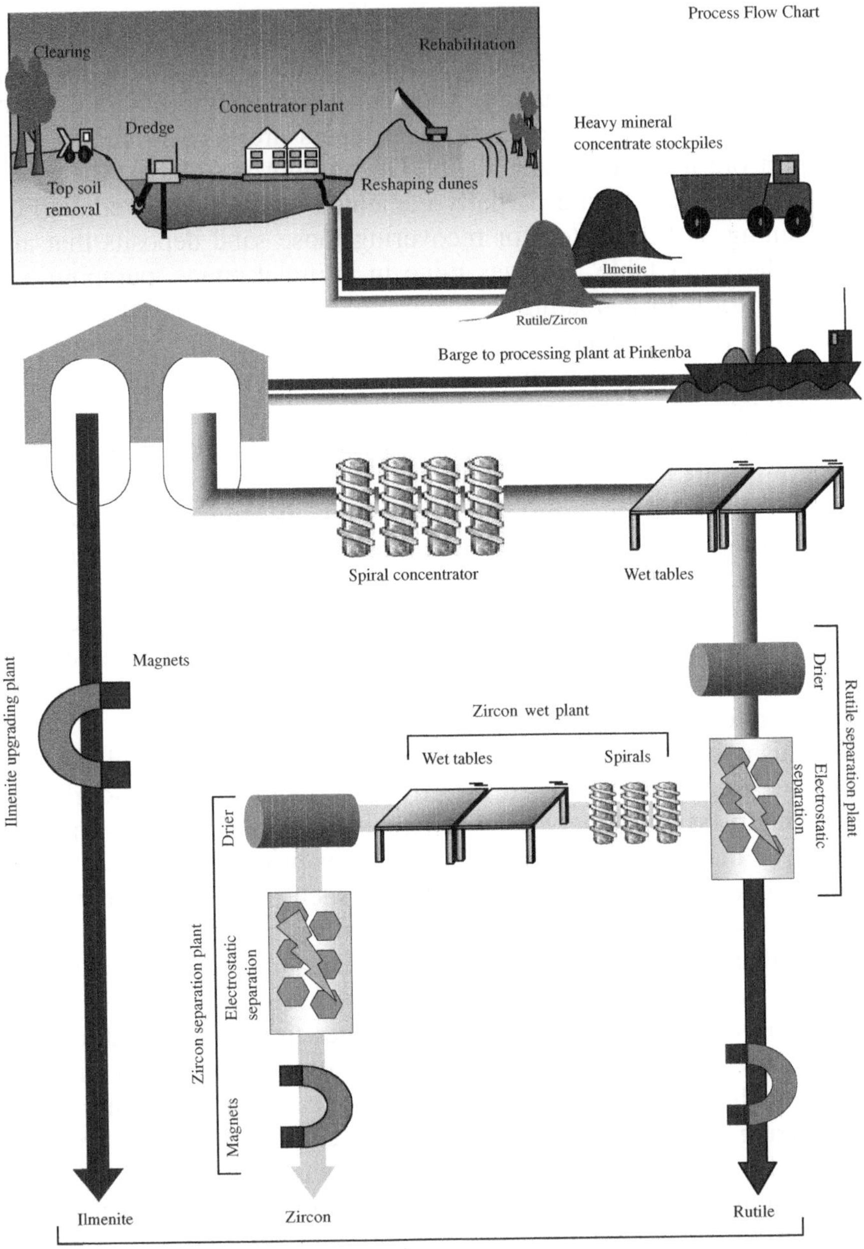

Figure 6.9 Schematic of a floating dredge and concentrator used to mine heavy minerals, including zircon, from unconsolidated sands (photographs courtesy of Consolidated Rutile Limited).

nium hydroxide. This material can be sold directly to be used in some applications, such as catalysts, or it can be dissolved in HCl to form zirconium oxychloride ($ZrOCl \cdot 8H_2O$), also called zirconyl chloride or zirconium oxide dichloride. This is calcined to produce zirconia.

In the lime fusion method, zircon is mixed with limestone or dolomite and heated to fusion to produce calcium zirconate and calcium or magnesium silicate (Nielsen, 2002):

$$2ZrSiO_4 + 5CaCO_3 \rightarrow CaZrO_3 + (CaO)_3(SiO_2) + CO_2$$

The material is cooled to produce a mixed calcium silicate powder and coarse calcium zirconate crystals. The calcium zirconate is separated and converted into zirconium oxide through an acid digestion process.

In the carbo-chlorination method, zircon is reacted with chlorine and carbon to form zirconium tetrachloride ($ZrCl_4$).

$$ZrSiO_4 + 2C + 4Cl_2 \rightarrow ZrCl_4 + SiCl_4 + 2CO_2$$

The tetrachloride gases are cooled to 200°C, at which point the zirconium tetrachloride powder is collected. The silicon tetrachloride is collected at much lower temperatures as a liquid, which is then purified and used in high-purity silica applications (Nielsen, 2002).

The zirconium tetrachloride can be hydrolyzed to make a zirconium oxychloride, which is used as an intermediate feedstock for many zirconium chemicals, including zirconia and zirconium hydroxide.

The plasma arc process is one of several thermal processes that can be used to dissociate zircon into zirconia and silica. Finely ground zircon is injected into an arc plasma to form zirconia particles in a matrix of liquid silica. Rapid quenching produces a mixture of zirconium oxide crystallites and silica fume (amorphous silica). A similar process is performed to make fused zirconia, in which zircon sand is melted to 2700°C. At this temperature, the silica is volatilized, removed, and precipitated as silica fume, leaving a fused zirconia with a small percentage of hafnia. The material is usually quenched quickly to form a granulate that can be crushed and sized. Dopants used to make stabilized zirconia may be added before quenching.

Fully stabilized or partly stabilized zirconia refers to zirconia that has been mixed with up to 20 percent calcia (CaO), magnesia (MgO), or yttria (Y_2O_3). Zirconia forms a solid solution with these dopant metal oxides and results in the reduction of the transition temperatures of the monoclinic-tetragonal and tetragonal-cubic transformations and a reduction of the volume change associated with the first transformation (Subbarao, 1981). The result is the stabilization of the cubic form of the mineral and a material that does not undergo phase inversions during heating and cooling. Partially stabilized zirconia (PSZ) contains less than the amount of dopant needed to promote the complete transformation to the cubic form. PSZ contains 8 mol percent CaO or MgO or 3 to 4 mol percent Y_2O_3 (Harben, 2002), which results in a mix of cubic and up to 50 percent tetragonal and monoclinic structures. Fully stabilized zirconia contains 16 mol percent CaO or MgO or 8 mol percent Y_2O_3 (Harben, 2002). The partially stabilized material is preferred to the fully stabilized material in refractory applications because of its relatively lower thermal expansion and improved thermal shock resistance (Baumgart, 1984). Commercial PSZ may consist of 70 percent to 80 percent cubic zirconia with a coefficient of thermal expansion of 5.5×10^{-6}/°C. Stabilized tetragonal zirconia is made by adding less than 4 percent yttria and creating a fine grain structure that results in a material made completely of the tetragonal phase (Cannon et al., 2002). This material can be stronger than other PSZ materials.

Fused alumina-zirconia-silica (AZS) is made by melting a mix of zircon sand and Bayer process alumina (see section 6.1) in an electric arc furnace to more than 1780°C. The melt is cast into silica molds and cooled to make refractory bricks. The silica in the zircon and alumina form a mixture of elongated mullite crystals intermixed with monoclinic zirconia and approximately 5 percent interstitial glass (Materials Handbook, 2005). Magnesia-zirconia bricks are made by sintering or fusing magnesia (see section 5.5) and zirconia.

High-purity, fine zirconia powders can be made using chemical precipitation techniques (Naito et al., 2003). Zirconium oxychloride produced by the carbo-chlorination method is reacted with ammonium hydroxide to cause the precipitation of zirconium hydroxide. Yttrium salts are added to the solution before mixing in order to produce PSZ. The precipitating phase is filtered and calcined to produce zirconia. The conditions of calcining and precipitation will determine the physical characteristics of the powder.

Uses and Specifications

Zircon and zirconia are used in a wide array of ceramic and glass applications. The classification of zircon grades is not always coherent, and has changed over time as different applications and their specifications are developed (Skillen, 1993). Traditionally, zircon was graded as premium, intermediate, and standard based on the relative amounts of iron and titanium compounds (Table 6.9). However, this has given way to a variety of specifications due in part to a lack of continuity between the grades of different suppliers, the wide range of applications, and the increased concern of the amount of radioactive uranium and thorium present in the product. Zircon is available in two standard sizes, sand (75 to 200 μm) and flour (45 to 75 μm) (Harben, 2002).

Refractories Zircon and zirconia-bearing materials are used in a variety of refractory applications in traditional metal and glass manufacturing processes, as well as some advanced materials production equipment such as setter plates used in the production of ferrite and titanate (Materials Handbook, 2005). With its combination of low thermal expansion, high thermal conductivity, and resistance to molten metals, zircon sand is used in the facings of foundry molds to reduce the penetration of molten metal and to give a uniform finish to castings (Hedrick, 2003). Zircon bricks are used in casting ladles, casting nozzles, and hearths that are in contact with molten metals. Refractories had been one of the largest markets for zircon, but this market decreased due to competition from less expensive refractories such as alumina spinels (Nielsen, 2002).

Zirconia is resistant to attack by most common glass melts, and therefore it is used in glass furnace refractories. For this application, there are three groups of zirconia-based refractories: sintered zircon, sintered zircon-mullite, and fused-cast AZS (Keegan, 1998). Sintered zircon materials are based on ground zircon and additives such as clay, silica, or titania. Bricks are pressed or slip cast and fired up to 1600°C. The porosity of the bricks can range from less than 1 percent to 20 percent, with the higher-quality bricks having lower porosity. Sintered zircon-mullite refractory materials are made from mixtures of raw zircon sand, ground zircon, and alumina that are cast or pressed into shapes. Upon firing up to 1600°C, the grain boundaries between the alumina and zircon sinter and form mullite and zirconia. Fused-cast alumina-zirconia-silica (AZS) is used in the critical areas in the glass furnace. There are two grades: a standard grade with 32.5 percent ZrO_2 and a higher grade of 41 percent ZrO_2 (Keegan, 1998).

Partially stabilized zirconia is used in high-temperature environments where good insulating qualities are required. Yttria-stabilized zirconia (YSZ) can be used in crucibles for melting quartz boules for optical fiber manufacturing (Nielsen, 2002). YSZ fibers are made by a process in which a continuous cellulose fiber is infused with zirconium salts

Table 6.9 Composition of ideal and some commercial zircon and zirconia products (wt. %)

	ZrO_2 ($+HfO_2$)	SiO_2	Al_2O_3	Fe_2O_3	TiO_2	SO_3	LOI	Th + U
Theoretical zircon	67.22	32.78						
USA standard[1]	66.4	32.4	0.46	0.04	0.12			350 ppm
USA ceramic[1]	66.4	32.5	0.35	0.04	0.13			350 ppm
E. Australia premium[1]	66.4		0.20	0.055	0.11			540 ppm
E. Australia ceramic[1]	66.35	32.65	0.40	0.29	0.08			450 ppm
South Africa standard[1]	65			0.25	0.3			
India[2]	65.8	32.5	1.20	0.10	0.25			
Zircon flour[3]	64–65	32–33		0.08–0.13	0.3–0.4			
Theoretical zirconia	100							
Electronic grade[4]	99	0.10		0.01	0.15	0.10	0.3	
Glass grade[5]	99	0.10		0.006	0.15	0.25	0.3	
Pigment grade[6]	99	0.10		0.01	0.15	0.25	0.3	
Oxygen sensors[7]	99.6	0.02	0.03	0.002	0.11	0.05	0.5	

[1] USA as reported in Harben (2002)
[2] Indian Rare Earths Limited—Grade MK typical
[3] Indian Rare Earths Limited—Typical
[4] MEL Chemicals E30
[5] MEL Chemicals G60
[6] MEL Chemicals S grade
[7] Zirconia Sales (America) Ltd. DK-2

and then calcined in an oxidized environment to burn away the cellulose and oxidize the zirconium (Fitzer, 2002). These can be made into fabric for a variety of high-temperature applications.

Whitewares Zircon flour, finely ground fused zirconia, or stabilized zirconia are used in whiteware glazes and pigments. Low iron is critical, so higher grades of material with <0.05 percent Fe_2O_3 and low TiO_2 are used. Zircon has been one of the chief components of low lead or lead-free glazes because of its relatively low toxicity. It serves as an opacifier due to the precipitation of tetragonal crystallites during firing. For high-temperature glazes, ultra-fine zircon with high surface area is used. However, zircon with a specific surface area higher than 26,000 cm^2/gram tends to agglomerate such that even vigorous mixing in water can result in only partial dispersion (Skillen, 1993). In low-temperature glazes, zircon can be added as a frit or fused zirconia (Krause, 1984). Grades for zircon used in glazes are based on size. Zircon flour (45 to 75 μm) is used for frits whereas opacifier grades are based on four sizes: 9 μm ceramic glaze, 6 μm ceramic glaze, 5 μm benchmark grade unglazed tile and sanitaryware, and 3 to 1 μm sanitaryware and special applications (Harben, 2002).

Zircon-based pigments are derived from the substitution of coloring metal cations in the zircon structure. The zircon lends high heat resistance, low thermal expansion and increased durability to the pigment while reducing the amount of expensive or toxic colorants. For instance, using zircon in cadmium-based pigments can allow higher temperature service while reducing the dissolution of cadmium (Krause, 1984). Other colorants include praseodymium (yellow), iron (pink), and vanadium (blue and yellow) (Skillen, 1993).

Catalysts Zirconia is used in a variety of catalyst applications because of its adjustable physical properties, such as specific surface area and porosity, and adjustable chemical properties, such as surface acidity and basicity (Moles, 1999). Zirconia may be combined with silica, rare earth metals, tungsten, and sulfate in refined chemicals used in environmental and industrial catalyst materials. Calcined mixtures of zirconia, silica, and lanthanum oxide will form stabilized, tetragonal, or monoclinic zirconia with a surface area of 70 to 85 m^2/gram (Franklin et al., 1999). Because of its thermal stability and oxygen storage capability, ceria-doped zirconia is used in the part of the wash coat of three-way catalysts that are used in gasoline engines.

Advanced Structural Ceramics Because of its high melting temperature, zirconia can be used for structural applications that are subjected to temperatures higher than those in which alumina can be used (Cannon et al., 2002). For instance, YSZ may find use in combustion engine components such as cylinder and valve linings, although vanadium, sodium, magnesium, and sulfur impurities in fuel oil can cause zirconia or yttria to become unstable (Materials Handbook, 2005). YSZ may be plasma sprayed as a thermal barrier for critical turbine engine parts.

Biomedical prosthesis, such as hip joints, can be made from zirconia materials (Kokubo et al., 2003). Sintered YSZ or a magnesium partially stabilized zirconia hip joints are used commercially with polyethylene cups. Initial concerns of the radioactivity that is caused by the decay of uranium and thorium that occurs naturally with zirconia have proven to be negligible.

Abrasives Alumina-zirconia abrasives used in steel finishing are made by fusing alumina with zirconia in electric arc furnaces. The improvement in properties relative to alumina is due to the reduction in brittleness brought about by the addition of zirconia, which toughens the alumina (Nielsen, 2002). YSZ that is typically 94.9 percent ZrO_2 and 5.1 percent Y_2O_3 is used in high-density grinding media (Harben, 2002). Zircon can be used as a sand-blasting media, although it tends to be more expensive than other alternative materials.

Electronic Ceramics At high temperatures, YSZ is a solid electrolyte that is used in the manufacture of oxygen sensors that control combustion in furnaces and automobile engines. This is due to the ability of the material to conduct oxygen anions through the vacancies in the lattice that are caused by the substitution of Y^{3+} for Zr^{4+}. Zirconia is used to make the piezoelectric material lead zirconate titanate (PZT) used in sound devices, ultrasonic transducers, and igniters. Electronics grade zirconia is on the order of 99 percent to 99.5 percent $Zr(Hf)O_2$ (Table 6.9).

Glass Zirconia can be used in a variety of glass applications as a glass former that increases refractive index without compromising durability. It is used in optical glass, specialty glass fibers, cathode ray tubes (to absorb X-rays), and as a nucleating agent in some glass-ceramics.

Health and Environmental Concerns

Zircon and zirconia are not considered to pose a hazard to human health. Some of the halogenated zirconium compounds and zirconium salts such as zirconium tetrachloride may be corrosive if exposed to the skin or are inhaled. Research reports on the chronic inhalation of zirconia are variable, ranging from no effect to a link to fibrosis (Nielsen, 2002).

6.4 RARE EARTH ELEMENTS

Introduction

The 15 lanthanide elements beginning with lanthanum (atomic number 57) and ending with lutetium (atomic number 71) are grouped together as the rare earth elements (REE) (Figure 6.10). Discovery and isolation of the individual REE began at the end of the eighteenth century and throughout the subsequent century. The term *rare earth* was suggested by Johann Gadolin in 1794; "rare" because when the first of the REE were discovered they were thought to be present in the Earth's crust

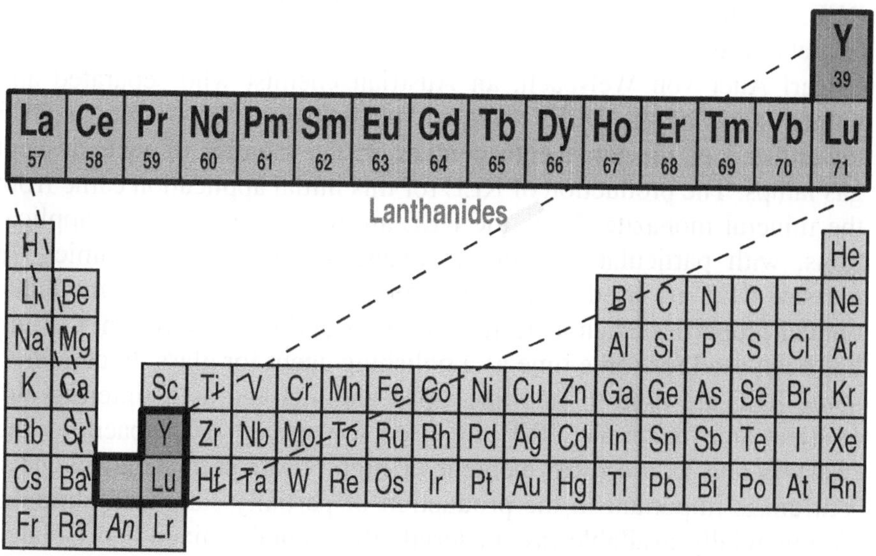

Figure 6.10 Periodic table showing the location of the REE (from Haxel et al., 2002; reprinted with permission from the U.S. Geological Survey).

only in small amounts, and "earths" because that was the term used for a homogenous element before the periodic table was established (Baumgartner, 1996). Thorium, yttrium, and scandium are often grouped with the REE, as these elements occur together naturally and possess similar chemical properties. Because of the chemical similarities between the REE, researchers required more than a century from the first discovery to complete isolation and classification of this group. It was also realized that rare earth is somewhat of a misnomer because some of them, particularly cerium, are relatively abundant in the continental crust.

The REE can be subdivided into the cerium or light group (lanthanum to gadolinium and sometimes including thorium) and the yttrium or heavy group (terbium to lutetium including yttrium and sometimes including scandium). The most common valence state for REE cations is +3 with the exception of europium, which occurs as +3 and +2, and cerium, which occurs as +3 and +4. The relatively large ionic radii of the REE preclude their accumulation in common rock-forming minerals and as a result, they often partition into phosphate and carbonate minerals associated with pegmatites and carbonatites. REE-bearing minerals can be mined from a variety of rock types, with the more common ore minerals being bastnäsite and monazite, which contain the light REE, and xenotime, which is more abundant in the heavy REE. Weathering of primary rocks result in REE deposits in laterites and clay ores.

Carl Auer von Welsbach, an Austrian chemist who separated and named some of the REE, invented the first commercial use of REE toward end of the nineteenth century in the mantles of incandescent gas lamps. The production of REE for this initial application came from the mineral monazite. Now, the REE are used in a variety of applications, with particular growth in advanced glasses and ceramics. In glasses, REE are used to add or control color as well as to modulate optical properties, such as refractive index and UV absorption. Cerium has been used for some time as a polishing agent for glass. In ceramics, some REE are used to stabilize alpha alumina and as a sintering aid. In electronic materials, REE are used for specialized capacitors and resistors and as a critical element of high-power permanent magnets. Yttrium is important in the production of partially stabilized zirconia. Commercially available raw materials that contain mixed and isolated REE include mineral concentrates of bastnäsite and monazite, as well as refined oxides, hydroxides, nitrates, chlorides, and fluorides.

Composition and Mineral Structure

Despite the name, some of the REE are relatively abundant elements. For example, cerium is about the twenty-fifth most abundant element in the earth's crust and has an average abundance of 60 ppm, compared to copper, which has an average abundance of 68 ppm. The REE follow the Oddo-Harkins effect, which is the observation that the elements with even atomic numbers are more abundant than those with odd atomic numbers (Figure 6.11). The REE are subdivided into the light REE, which include those from lanthanum to europium, and the heavy REE, which include those from gadolinium to lutetium as well as yttrium, with the light REE generally more abundant than the heavy REE.

The REE are grouped together in part because of their similar chemical behavior (Henderson, 1996). The ground state of all the neutral REE atoms can be approximated beginning with $[Xe]4f^1 5d^1 6s^2$. As atomic number increases, electrons are added to the 4f orbitals, leaving the three outer electrons in a position to be removed to produce a valence state of +3, which is the most common for the REE. The relative increased stability of empty $4f^0$ (Ce^{4+}), half-full $4f^7$ (Eu^{2+}), and completely full $4f^{14}$ (Yb^{2+}) shells can allow stable oxidation states

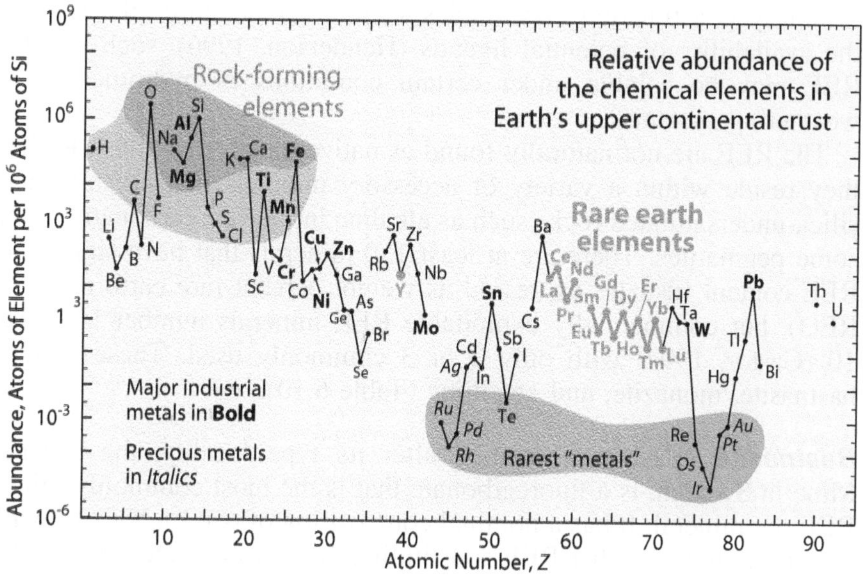

Figure 6.11 Plot of atomic number versus crustal abundance for the REE (from Haxel et al., 2002; reprinted with permission from the U.S. Geological Survey).

other than 3+. Therefore, europium occurs as both 3+ and 2+, and cerium occurs as both +3 (cerous) and +4 (ceric).

The ionic radii of the REE are relatively large, between 0.115 and 0.100 nm depending on the valence state. The ionic radii decrease as atomic number increases due to increasing attraction of electrons in the 4f orbitals as the positive charge of the nucleus increases. This is known as the *lanthanide contraction.* Compared to the heavy REE, the light REE are slightly more incompatible in most common rock-forming silicate minerals because their ionic radii are larger. For instance, during partial melting of the mantle or lower crust, residual garnet or clinopyroxene can preferentially retain the heavy REE, leading to light REE enrichment in the resulting magma. This has led to a higher concentration of the light REE in the continental crust than the heavy REE.

An exception to the general incompatibility of the REE in minerals is the substitution of Eu^{2+} for the larger alkaline earth cations, such as Ca^{2+} and Sr^{2+}. For instance, magmas that have undergone significant precipitation and removal of plagioclase, a calcium aluminosilicate mineral, may show a relative depletion of europium relative to the other REE in the residual magma.

In aqueous systems, the REE are generally insoluble, evident from their almost insignificant abundance in seawater (Table 2.1). However, solubility is a function of variables such as temperature, pH, Eh, and the availability of potential ligands (Henderson, 1996), such that the REE may be soluble under certain conditions in hydrothermal or weathering systems.

The REE are not naturally found as native metals or oxides. Rather, they reside within a variety of accessory minerals that crystallize in silica-undersaturated rocks such as alkaline intrusives, carbonatites, and some pegmatites. There are at least 150 minerals that have significant REE content (usually expressed as weight percent rare earth oxide or REO), but commercially exploitable REE minerals number less than 10 (Castor, 1994) with only 2 or 3 commonly used. These include bastnäsite, monazite, and xenotime (Table 6.10).

Bastnäsite Bastnäsite, named after its type locality, the Bastnäs Mine in Sweden, is a fluorcarbonate that is the most commonly mined REE ore mineral. It has a nominal composition of $(Ce,La,Y)CO_3F$ with minor amounts of other REE substituting in the cation site. The crystal structure consists of layers of CO_3 groups and layers of REE and fluorine (Figure 6.12). The REE are in ninefold coordination with six oxygen atoms and three fluorine atoms (Gaines et al., 1997). Bastnäsite can have about 50 percent of its REO content as cerium, with the

Table 6.10 Properties of some rare earth ore minerals and oxides

	Formula	Specific Gravity	Mohs Hardness	wt. % REO[1]
Bastnäsite	(Ce,La,Y)CO_3F	4.9–5.2	4.0–4.5	75
Monozite	(Ce,La,Nd,Th)PO_4	4.8–5.5	5.0–5.5	70
Xenotime	YPO_4	4.4–5.1	4.0–5.0	60–70
Ceric oxide	CeO_2	7.3	6.5	100

[1]The total wt. % REO will depend on the proportions of the different REE.

remainder as 33 percent lanthanum, 13 percent neodymium, 4 percent praseodymium, 0.5 percent samarium, 0.2 percent gadolinium, 0.1 percent europium, and 0.2 percent others (Molycorp, 1992). It has a hardness of 4.0 to 4.5 and a specific gravity of 4.9 to 5.2. It is soluble in strong, hot acids.

Monazite Monazite is an important REE and yttrium ore mineral, with its name derived from the Greek *monazein,* meaning to be solitary, in reference to its presence as an isolated accessory mineral in some rocks.The monazite group of minerals has the general formula of ATO_4, where the A site can be occupied by Ce^{3+}, La^{3+}, Nd^{3+}, Th^{4+}, Ca^{2+},

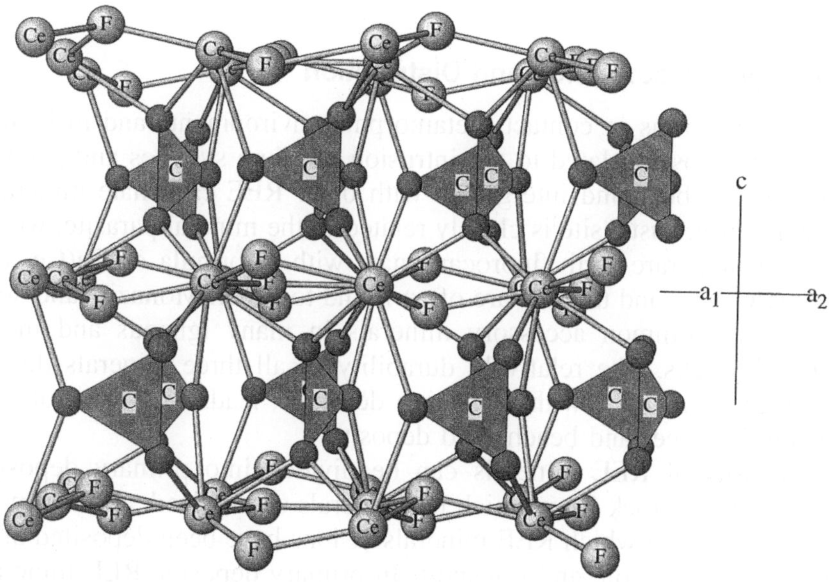

Figure 6.12 Idealized structure of bastnäsite (from Gaines et al., 1997 used with permission from John Wiley & Sons).

and other REE cations and the *T* site is occupied by phosphorus with minor substitution of silicon or arsenic (Gaines et al., 1997). The atoms in monazite form a three-dimensional monoclinic framework where cerium or the other *A*-site cations are in ninefold coordination with oxygen. Monazite usually contains thorium, which may be present in amounts up to 30 percent, leading to potential problems posed by radioactivity. In general, all monazite minerals are richer in the light REE relative to the heavy REE. Three general subdivisions of monazite are used—cerium, lanthanum, and neodymium—to denote the relative proportion of the dominant REE within the mineral. For example, in monazite-Ce cerium is the principal REE. In terms of the overall natural abundance of types of monazite, the cerium-rich variety is the most common. It can occur in varying colors, usually yellow to reddish brown, with a hardness of 5.0 and a specific gravity of 4.6. It is only slightly soluble in acid.

Xenotime The xenotime group of minerals follows the general formula of YTO_4, where the *Y* site can be yttrium, a REE, or bismuth and the *T* site is usually occupied by phosphorus but can also host arsenic, vanadium, or silicon (Gaines et al., 1997). It forms crystals with a framework similar to monazite but crystallizes in the tetragonal system. It has a hardness of 4.0 to 5.0 and a specific gravity of 4.4 to 5.1. It is only lightly attacked by acids and is very resistant to weathering.

Geological Occurrence and Distribution

Bastnäsite forms in contact metamorphic environments and in hydrothermal deposits related to the intrusion of alkali syenites and pegmatites. It can be found intergrown with other REE carbonate minerals. For instance, bastnäsite is closely related to the mineral parasite, which is a calcium rare-earth fluorocarbonate with a formula of $Ca(Ce, La, Nd)_2(CO_3)_3F_2$, and the two are often found together. Monazite and xenotime are common accessory minerals in many igneous and metamorphic rocks. The relatively durability of all three minerals during weathering, coupled with their high densities, leads to a habit of collecting in placer and beach sand deposits.

Deposits of REE minerals can be divided into primary deposits, which contain rock from which the minerals crystallized, and sedimentary deposits, in which REE minerals or ions have been deposited after weathering, erosion, and transport. In primary deposits, REE minerals crystallize in magmatic and metamorphic systems, particularly silica undersaturated, alkalic rocks such as syenites, pegmatites, carbonatites, and hydrothermal veins. Most of the primary deposits are very old and formed during the Precambrian or prior. Primary deposits that are rich enough in REE for mining on their own are rare and more often than not, REE minerals are recovered during mining for other ores. The most important primary deposits are carbonatites and iron-REE deposits (Castor, 1994). Not all carbonatites have appreciable amounts of REE and there is essentially only one location, Mountain Pass in California, that has recently been mined only for these minerals. In the iron-REE deposits, a complex banded mineralogy hosts both REE minerals as well as iron ore minerals such as magnetite and hematite.

Secondary deposits consist of bastnäsite, monazite, and xenotime that have collected in detrital sediments such as placers and beach sands. Many of the placer and sand deposits that are mined for zircon and titanium minerals also have appreciable amounts of monazite and xenotime.

Intense weathering of carbonatites, syenites, and granites can produce laterites or residual rock that has had the more soluble material removed, resulting in an enrichment of insoluble elements such as yttrium and the REE. These deposits are referred to as ion adsorption lateritic crusts, which have the advantage of being easily mined and having relatively high proportions of the heavy REE (Figure 6.13) compared to bastnäsite and monazite ores (Haxel et al., 2002).

A general characteristic of most rare earth deposits is that they are light REE enriched such that the first four REE (lanthanum, cerium, praseodymium, and neodymium) constitute 80 percent to 99 percent of the total REE content (Haxel et al., 2002). Therefore, deposits relatively enriched in the scarcer and therefore more valuable heavy REE, such as lateritic crust and xenotime ores, are particularly desirable. Figure 6.13 shows the relative proportions of REE from a bastnäsite ore, dominated by the light REE, and a lateritic ion adsorption ore, which is dominated by yttrium and the heavy REE.

A catalog of REE mines is found in Orris and Grauch (2002). Despite the abundance and diversity of REE deposits, there are only a few commercial producers in the world (Figure 6.14), with China currently producing the majority of global output with lesser amounts from Brazil, India, and Russia (Hedrick, 2003b).

China has the largest single deposit of REE minerals at the Bayan Obo iron mines of Inner Mongolia (Wu et al., 1996). It is located within Proterozoic schists and gneisses of the northern Sino-Korean craton. The ore occurs in dolomitic marble and consists of the iron ore mineral

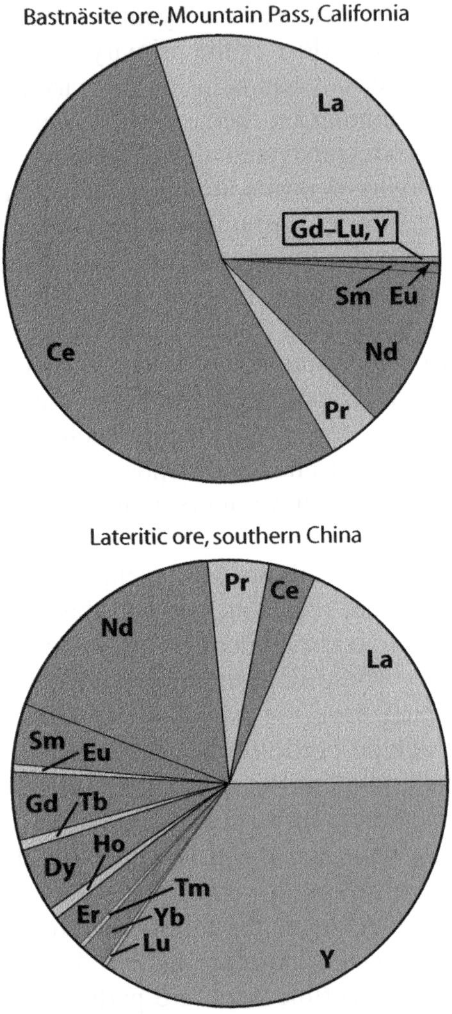

Figure 6.13 Pie charts showing the percentage of the different REE found in California and China mines (from Haxel et al., 2002; reprinted with permission from the U.S. Geological Survey).

magnetite mixed with bastnäsite and monazite (Figure 3.4). Fluorite and the niobium ore, columbite ($[Fe,Mn][Ta,Nb]_2O_6$), can also be present. The average ore grades are 6 percent REO, 0.13 percent Nb, and 35 percent iron. The Jiangxi province in southern China hosts important reserves of lateritic ore deposits rich in yttrium and the heavy REE (Figure 6.13). The ore bodies consist of 3 m to 10 m thick horizons of completely weathered granitic rocks that have been altered to mostly

Figure 6.14 Global map showing the locations of major rare-earth mineral producers.

halloysite and kaolinite with minor quartz and feldspar (Wu et al., 1996). The REE are adsorbed to these clays.

In the United States, there is currently one important commercial mine at the Mountain Pass deposits in the upper Mojave Desert of California. First mined in 1949, this is an 8 km by 30 km carbonatite deposit with a thickness of about 80 m. It is located in a block of Precambrian metamorphic rocks (Chang, 2002), and is the only deposit that is mined exclusively for REE minerals (Castor, 1994). The deposit has an average REO content of 8.9 percent in a mix of up to nine REE-bearing minerals, although the only recovered mineral is bastnäsite. The ore consists of 12 percent bastnäsite, 10 percent strontianite, 40 percent calcite, 8 percent silica, and 25 percent barite (Molycorp, 1992). Compared to the Mountain Pass ore minerals, the Bayan Obo bastnäsite has higher samarium and europium (Minggin and Xuehong, 1997). Bastnäsite mining in the United States at the Mountain Pass deposit is presently on hold, although material has been stockpiled and is currently processed and sold.

Australia is a major potential source of REE minerals with reserves in carbonatite, heavy mineral, and lateritic ore deposits (Hedrick, 2003b). Monazite and xenotime are recovered as a byproduct of heavy mineral sands processing on east and west coasts, although during low demand it is either not recovered or stockpiled. Concerns of the radioactivity in thorium within the monazite may make it a less-favorable feedstock for the production of refined REE compounds. The Mount Weld carbonatite complex located 35 km south of Laverton in Western Australia is being exploited for REE minerals. Australia's lateritic deposits are not currently mined but provide alternative potential reserves.

In Russia, much of the REE are derived predominantly from loparite, a mixed metal oxide with a composition of $(Ce,Na,Ca)_2(Ti,Nb)_2O_6$ that is found in silica undersaturated rocks. Loparite is mined from the Kola Peninsula, which is located east of the border with Finland. REE can also be derived from the significant phosphate rock (apatite) mining in this region as well (Evans, 1993).

Over the past 50 years, the global production of REE has risen dramatically from essentially nothing to more than 80,000 tons per year (Figure 6.15). Throughout the 1960s, production of REE came predominantly from monazite mined from placers and small pegmatites and carbonatites (Haxel et al., 2002). From 1965 to the 1980s, Mountain Pass bastnäsite was the dominant source of REE. Since then, China has gradually come to dominate the global market.

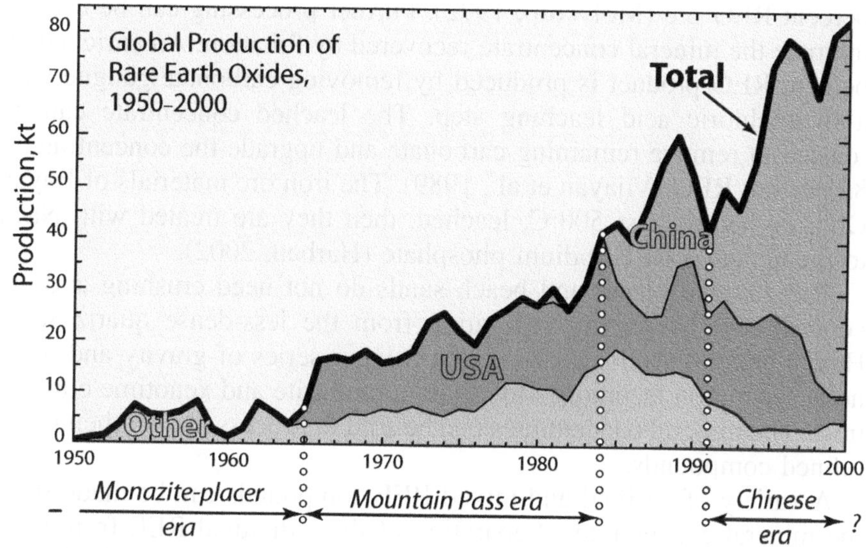

Figure 6.15 Graph of global production of REE versus time. There are four periods of production: the monazite-placer era, starting in the late 1800s and ending abruptly in 1964; the Mountain Pass, California, era, starting in 1965 and ending about 1984; a transitional period from about 1984 to 1991; and the Chinese era, beginning about 1991 (from Haxel et al., 2002; reprinted with permission from the U.S. Geological Survey).

Mining and Processing

Deposits are evaluated and mapped in three dimensions using core drilling techniques to determine the size as well as the REO content over the extent of the deposit. The Mountain Pass and Bayan Obo deposits are open-pit operations that use blasting and power shovels to remove the ore. The Mountain Pass ore is collected with a ratio of approximately eight tons of waste per ton of ore (Molycorp, 1992). The ore is crushed and blended at the site to be used at the nearby mill. The Bayan Obo ore is shipped by rail 150 km to a mill due to a lack of water at the mine (Argall, 1980; Castor, 1994). Placer and sand deposits are mined using bucket and suction dredges in water (Figure 3.2) and excavating equipment in dry sands.

For the hard rock ores such as those from Mountain Pass and Bayan Obo, primary crushing to >1 cm is followed by secondary crushing, classification, and ball milling. The material from Mountain Pass is slurried and heated to prepare it for froth flotation, where bastnäsite is collected to produce a 60 percent REO concentrate from an initial 8.5

percent REO ore (Molycorp, 1992). Further processing can be used to upgrade the mineral concentrate recovered by flotation. An enriched 70 percent REO product is produced by removing carbonate gangue with a hydrochloric acid leaching step. The leached concentrate can be roasted to remove remaining carbonate and upgrade the concentrate to 85 percent REO (Vijayan et al., 1989). The iron ore materials of Bayan Obo are heated past 500°C, leached, then they are treated with NaCl to precipitate a REE sodium phosphate (Harben, 2002).

The dredged placer and beach sands do not need crushing and are concentrated by gravity separation from the less-dense quartz sand. Heavy mineral sands can be sorted using a series of gravity and magnetic separation techniques to obtain a monazite and xenotime concentrate. The mineral concentrates can be sold to processors that then make refined compounds.

A variety of purified and mixed REE compounds can be made from the mineral concentrates. Separation of the individual REE from each other is relatively difficult because of the similarities in chemical properties. Bastnäsite processing begins with calcination to remove carbonate and to oxidize Ce^{3+} to Ce^{4+}. After cooling, HCl or H_2SO_4 is added to leach the trivalent REE. The insoluble portion is removed, which is about 62 percent CeO_2, 10 percent other REE oxides, 6 percent CaO, 4 percent other oxides, 10 percent F, and 8 percent LOI (Harben, 2002). The dissolved portion is precipitated to produce a material that is 46 percent La_2O_3, 22 percent other REE oxides, 12 percent CeO_2, 12 percent chlorine, and 8 percent LOI.

Monazite and xenotime have a slightly different processing route compared to bastnäsite. The mineral concentrates are finely ground and reacted with HCl to produce a mixed rare-earth chloride. This can be reacted with sulfuric acid to produce a REE sulfate. The sulfate is dissolved in water and separated from insoluble material. An alternative is the alkaline process in which monazite and xenotime are dissolved in 150°C sodium hydroxide for several hours (Molycorp, 1992; Vijayan et al., 1989). The REE and thorium are concentrated in an insoluble hydroxide after the soluble phosphate is removed according to the following reaction:

$$\text{(REE,Th)PO}_4 + \text{NaOH} \rightarrow \text{Na}_3\text{PO}_4 + \text{REE(OH)}_3 + \text{Th(OH)}_4$$

The hydroxides are reacted with hydrochloric acid, in which the $Th(OH)_4$ remains insoluble and can be removed, leaving a mixed REE chloride.

The mixed REE compounds resulting from the digestion of mineral concentrates can be used as is or can be further processed to separate the different REE. A mixed-light REE product can be marketed under the name didymium. Separation of the individual REE can be done using solvent extraction (SX) and, to a lesser extent, ion exchange resins. SX is an ion exchange process that is based on the differing behavior of the REE between an aqueous phase and an organic liquid phase. The process uses three major components: an extracting agent, a modifier, and a diluent (Vijayan et al., 1989). The extracting agent, usually an organic phosphate such as tributyl phosphate, selectively complexes with the individual REE. Modifiers reduce emulsion formation and improve the exchange rate while the diluent is used to improve mixing of the aqueous phase. Countercurrents of the organic solvent are placed in contact with the aqueous stream containing mixed rare earths. Rare earth complexes that are soluble in the organic phase develop until an equilibrium is established between the REE ions in the aqueous solution and the complex ions in the organic phase. As the two streams flow past each other, the desired group of elements is separated (Harben, 2002). Some elements, such as yttrium, lanthanum, cerium, and europium, have equilibrium constants that are more amenable to separation than others. SX can be used in one stage to remove a mixed REE product or in several stages to produce individual light REE. The heavy REE can be separated using ion exchange columns that use resins specific to each element. Once separated, the REE are further processed to make compounds such as hydroxides, chlorides, fluorides, or carbonates, which can be calcined to produce an oxide.

Uses and Specifications

The REE group of elements has unique properties that make them indispensable for many modern glass and ceramic applications. Some examples include the strong paramagnetic behavior of most rare earth compounds, various optical properties, and high coordination numbers in compounds. Increased use of REE can also be attributed to their ability to substitute for more toxic heavy metals in some applications. In the United States, the major users include: automotive catalytic converters, 14 percent; glass polishing and ceramics, 30 percent; metallurgical additives and alloys, 19 percent; permanent magnets, 3 percent; petroleum refining catalysts, 28 percent; rare-earth phosphors for lighting, televisions, computer monitors, radar, and X-ray intensifying film, 3 percent (Hedrick, 2003b). The specifications for REE compounds are driven by the requirements from the consumer. Commercial REE products can be purchased from virtually any degree of processing, from a mineral concentrate, mixed element compound, to highly purified oxides, chlorides, fluorides, oxalates, and other compounds of up to 99.99 percent purity. Of course each refining steps adds to the cost of the material.

Catalysts Automotive catalysts that use REE include gasoline engine catalytic converters that reduce NO_x and oxidize CO and hydrocarbons, as well as the relatively new diesel catalytic converters that are designed to reduce particulate emissions. Gasoline engine catalytic converters use about 5 percent cerium and lanthanum compounds to protect the platinum group metals, such as rhenium, that are embedded on the catalyst substrate from oxidation. In addition, cerium oxide acts as promoter of the water-gas shift reactions and an enhancer of the NO_x reduction capacity of rhenium (Harben, 2002). The diesel catalytic converters can use injection of cerium compounds to assist in the oxidation of soot particulates.

A variety of catalysts in the petrochemical and chemical processing industries use REE (Materials Handbook, 2005). Stabilized zeolite fluid-cracking catalysts (FCC), used in petroleum refining, are composed of faujustite Y-type zeolites with embedded cerium and lanthanum at levels of 1 percent to 5 percent. The REE can be added by immersing the zeolitic material into a solution of REE salts, such as chlorides and nitrates. The REE stabilize the zeolite structure by making it more resistant to degradation by contaminants in crude oil, such as sulfur, vanadium, and nickel. The REE allow the zeolite surface to retain acidic properties and increase the surface area. The surface area increases from 100 m^2/g at 5 percent REO to 700 m^2/g at 20 percent REO (Chang, 2002). Desulfurization of fuels such as diesel is becoming more prevalent, and catalysts containing cerium embedded in an aluminosilicate spinel structure can be used to remove sulfur from crude oils (Rhodia, 2004).

Glass The REE have many applications in glass manufacturing. Cerium is widely used, primarily as a polishing agent and decolorizer. Glass had long been polished using iron oxide compounds, and later zirconium and titanium compounds. However, cerium compounds have proven to be superior for polishing critical glass components, such as optical lenses, plate glass, mirrors, data disk drives, analytical samples, and video display glass. Bastnäsite powder, CeO powder, or other cerium compounds in proportions of 50 percent to 90 percent of the polishing agent are slurried to make pastes (Chang, 2002). It is not the hardness of CeO that is responsible for this ability, but rather, it is thought to be a mix of mechanical and chemical activity or of a catalytic nature (Silvernail and Goetzinger, 1971). In the presence of water, the CeO reacts with the silicon to break the O-Si-O bonds using the following reaction (Harben, 2002):

$$SiO_2 + Ce(OH)_4 \rightarrow CeO_2 . + Si(OH)_4$$

The REO concentrate derived from the digestion of bastnäsite is predominantly cerium oxide and works very well as a polish and is less costly than a higher-purity cerium oxide product (Molycorp, 1992). Polishing of precision optical materials may require higher purity CeO.

Ce^{4+} is used in some glass batch formulations to counteract the color caused by iron. In glass, the ferric Fe^{3+} adds a brownish-yellow tinge although ferrous Fe^{2+} produces a blue-green tinge that is 10 times more powerful discoloring ability than the ferric. Cerium is a powerful oxidizer that helps to keep the iron in the ferric state, thus reducing the amount of coloring. Cerium hydroxide or oxide can be used in amounts of 1500 ppm for decolorization (Varshneya, 1994), although this will depend completely on the level of iron in the batch materials. Cerium cations are stable in molten glass, and since neither valence state absorbs light in the visible region, it adds no color to the glass.

Some of the other REE can be used to tailor the color absorption of glass. For instance, Nd_2O_3 adds a light pink-violet color, Er_2O_3 adds pink, Eu_2O_3 adds red, and Pr_6O_{11} adds green. Neodymium and praseodymium are used to filter certain wavelengths of light. Neodymium is used as a filter for yellow light in welder's goggles. In cathode ray tube (CRT) glasses, Nd_2O_3 can be added to enhance picture brightness by absorbing yellow light waves. Neodymium has a strong absorption band centered at 580 nm wavelength, which helps clarify the contrast between reds and greens (Molycorp, 1992).

Cerium-doped glass can be used in a variety of glassware to store or protect materials that degrade under ultraviolet exposure. Ce^{+4} in particular provides strong absorption below 400 nm and makes glass opaque to near ultraviolet radiation (Molycorp, 1992). An example is automotive glass, where reduced ultraviolet exposure can protect interior plastic components (IMI, 2000).

Lanthanum can be added to optical glass and borosilicate glass to improve resistance to alkali attack, increase the refractive index, decrease dispersion, and increase visible light transmission. Applications include photography, microscopes, video, and optical fiber. Gd_2O_3 and Y_2O_3 can be added as optical glass as stabilizers to prevent glass containing lanthanum from recrystallizing (Rhodia, 2004).

Laser and Photonic Materials Erbium is used in optical fiber amplifiers as a dopant. This enables amplification of the optical signal without any conversion between optical and electronic signals. Many lasers use materials containing yttrium and the REE. Yttrium aluminum garnet (YAG) is the most widely used laser crystal for solid-state crystal material. YAG can be doped with neodymium, holmium, erbium, thulium, and ytterbium to tailor the optical properties such as laser output wavelength. Dopant levels can be around 1 percent by mole. Neodymium-doped YAG lasers can be used in semiconductor etching. YAG are used in a variety of other uses, including microwave generators, selective emitters and filters, and high-temperature UV-IR optical windows. Yttria and lanthanum oxide together can be used to make optical windows that are transparent to light wavelengths from 280 nm to near 8,000 nm (Harben, 2002).

Ceramics The REE can be used in traditional ceramics as coloring agents. Cerium is used as an opacifier and replacement for tin in enamels (Materials Handbook, 2005). A variety of pigments for glazes can be made with REE, such as praseodymium yellow, praseodymium green, and neodymium violet (Minggin and Xuehong, 1997). In dental ceramics, cerium can be added in amounts of 0.1 percent CeO_4 to increase fluorescence to match natural teeth (Harben, 2002).

In advanced structural ceramics, lanthanum and neodymium are used as sintering aids in nitrides and carbides, resulting in increased flexural strength and fracture toughness (Zhou et al., 2002). Yttria in partially stabilized zirconia (YSZ) serves to convert the tetragonal zirconia to the more stable cubic form.

Electronic Materials The REE are used in production of phosphors for CRTs and lamps that are used in television screens, computer monitors, and X-ray imaging devices. The interiors of the CRTs are coated with phosphors that generate the colors red, blue, and green. A Eu-Y compound is the only available red phosphor, while other REE, including europium, cerium, terbium, and gadolinium, can be used for blue and green phosphors. A typical 48 cm television screen can contain 0.5 g of Eu_2O_3 (Harben, 2002). Terbium is used in some phosphors that are excited by UV light, finding applications in mercury arc fluorescent lighting.

Rare earths are used in several dielectric materials. Multilayer ceramic capacitors for printed circuit board and other electronic devices use high-purity lanthanum and neodymium. Lanthanum aluminate is used as a substrate for microwave filters used in global positioning systems and cellular phones. Neodymium (or lanthanum and praseodymium) can be used in barium titanate, where the Nd^{3+} ion can substitute for the Ti^{3+} ion and can flatten the temperature/capacitance curve (Rhodia, 2004).

Synthetic rare earth garnets are essential for some electronic applications. Yttrium iron garnets (YIG) with nominal compositions of $Y_3Fe_5O_{12}$ can be used in many magneto-optical and microwave applications, such as frequency meters, cellular equipment, magnetic field measurement, and tunable transistors. YIG crystals have very high Q resonance with low loss at microwave frequencies. Gadolinium is used in Gd-Ga garnets (GGG) with a nominal composition of $Gd_3Ga_5O_{12}$. GGG crystals are used in substrates for infrared optical isolators, microwave isolators, and thin film magnetic bubble memory systems.

Energy storage devices incorporate REE, particularly lanthanum. Alloys of nickel and lanthanum are used in nickel metal hydride (NiMH) rechargeable batteries. The alloy forms a cubic structure with lanthanum atoms on the corners and nickel occupying the interior. Hydrogen atoms occupy the interstices within the cubic structure.

Magnetic Materials The REE have been critical to the development of high-resolution magnetic resonance imaging (MRI). Initial rare earth magnets were high-temperature Sm-Co magnets ($SmCo_5$ or Sm_2Co_{17}) with Curie temperatures greater than 700°C. Advances in the latter part of the twentieth century led to the development of Nd-Fe-B ($Nd_2Fe_{14}B$) magnets (Rhodia, 2004). Dysprosium and holmium, which have the highest magnetic moment of all the elements, can be added to increase the coercivity. The magnetic wave generation systems of MRI instruments use neodymium-based magnets. The advances in REE magnet technologies have allowed for the miniaturization of electric motors and electronic devices used in virtually all products, including automotive, medical, and military.

Health and Environmental Concerns

The REE generally are considered to be only slightly toxic (Harben, 2002). Human respiratory exposure to lanthanides can cause itching, insensitivity to heat, and a sharpening of taste and smell. Inhalation of REE has been proven to cause acute pneumonitis in the lungs of animals. Long-term inhalation to REE dust seems to cause pneumoconiosis in humans (Hirano and Suzuki, 1996). Oral ingestion of REE appears to have little effect because of poor absorption from the gastrointestinal tract (Molycorp, 1992).

6.5 CARBIDES AND NITRIDES

Introduction

Carbides and nitrides are some of the best-performing ceramic materials, both of which are often placed in the general group of nonoxide synthetic hard materials. The borides are also included within this group, but these materials are not discussed here. The structures of the metal carbides and nitrides can be considered interstitial materials because the carbon or nitrogen atoms occupy interstitial sites within a metallic structure (Baumgart, 1984). The isolation of the carbon and nitrogen means that there are no carbon-to-carbon or nitrogen-to-nitrogen bonds. Rather, the carbon or nitrogen bonds strongly with the metal atoms and serves to increase the metal-to-metal bonds, resulting in an increase in hardness relative to the pure metal.

The properties of carbide and nitride materials are useful for a wide variety of applications, including refractories, high-temperature structural applications, electronics, cutting, and grinding. Most of these materials have been discovered over the past century or so, but it was not until the availability of high-quality, cost-effective powdered feedstocks that these advanced ceramics were made in commercial amounts (Srinivasan and Rafaniello, 1997). Because carbides and nitrides are considerably more expensive than many other ceramic raw materials, they are more commonly used in critical applications or in those that require low volumes of material. For instance, using aluminum nitride for heat sinks in electronic chip modules may be worth the extra cost over a less expensive and less effective material such as alumina.

For the most part, carbides and nitrides do not occur in nature except for the rare silicon carbide mineral, moissanite. Commercial carbide and nitride synthesis is done under completely oxygen-free conditions where the metal, metal oxide, or other metallic compound reacts with carbon or nitrogen. The atmosphere in which the reaction occurs can either be a vacuum, inert gas (nitrogen), hydrogen, or a hydrocarbon gas. This chapter provides an overview of the synthesis and uses of some of the more common carbides and nitrides. More detailed discussion of these materials can be found in other sources such as Weimer (1997) and Riedel (2000).

Carbides

Carbides are made by combining carbon with both nonmetallic and metallic elements. The metal carbides combine carbon with transition metals from Group 4 through Group 6 on the periodic table. These form either 1:1 metal-to-carbon compounds that make cubic crystals or 2:1 metal-to-carbon compounds that make hexagonal crystals. Some can have nonstoichiometric compositions, such as boron carbide. Most metal carbides are produced through a carbothermal reduction process where the metal and a carbon source react at high temperature in a variety of reactor designs (Weimer, 1997). Nonmetal carbides include those made with boron and silicon. The carbon source for carbides can be derived from graphite, petroleum coke, or organic material. Carbides have properties of both ceramics and metals, such as high melting points and high hardness, which are useful in grinding and cutting applications.

Boron Carbide Boron carbide is one of the hardest commercially available materials, second only to diamond and boron nitride. It is also one of the lightest ceramic materials. Boron carbides can come in a wide range of stoichiometric boron to carbon ratios, depending on the application, with commercial materials usually having a B:C ratio of 4:1. The crystal structure of boron carbide is complex (Clark and Hoard, 1943) but can be described as a unit cell of $B_{12}C_3$ with 12-atom icosahedra that occupies the points of a rhombohedron linked by a chain of atoms that runs along the c-axis of the rhombohedron (Thevenot, 1990) (Figure 6.16). Because of the substitution between boron and carbon, a single phase can be stable with carbon content ranging from 8.8 to 20 molar percent.

Pure crystals of boron carbide are black with a specific gravity of 2.52 and a Knoop hardness of 2750 to 3000 and a Mohs hardness greater than 9.5. The highest hardness values are found in stoichiometric ratios of B:C ratio of 4:1 (Srinivasan and Rafaniello, 1997). The average coefficient of thermal expansion is 4.6×10^{-6}°C^{-1} at temperatures between 25°C and 800°C (Schwetz, 2002). Boron carbide retains a high degree of hardness at high temperatures. For example, above 1300°C, it is harder than diamond and cubic boron nitride

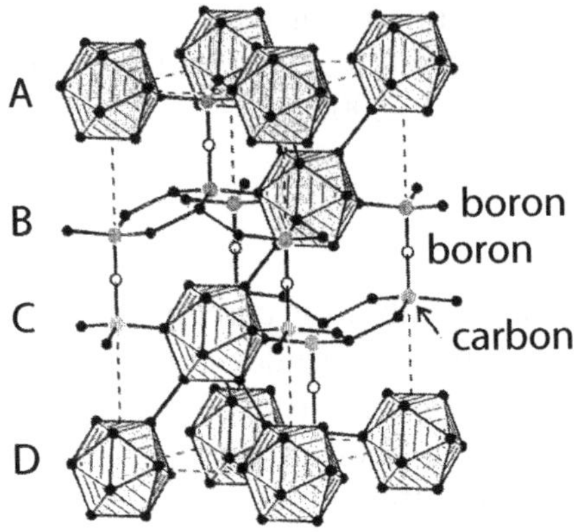

Figure 6.16 Idealized crystal structure of hexagonal $B_{13}C_2$ (B_4C). Filled circles are boron atoms that form empty icosahedra. Open circles are boron atoms at the center of linear C-B-C units. Gray shaded circles are carbon atoms (from Jeitschko et al., 2000, used with permission from Wiley-VCH).

(Materials Handbook, 2005). Although it melts congruently at about 2450°C, it will start oxidizing at 600°C in an oxidizing environment, forming a thin B_2O_3 film that serves to protect the underlying material to a certain extent (Thenevot, 1990).

The production of boron carbide is done in a furnace by carbothermal reduction of anhydrous boric oxide using this general reaction:

$$2B_2O_3 + 7C \rightarrow B_4C + 6CO$$

The boric oxide should be at least 99 percent pure, have LOI no greater than 1.5 percent, and a particle size around 4 mesh to reduce losses associated with dusting (Wilson and Guichelaar, 1997). The carbon source can be graphite or petroleum coke, although chips recovered from the machining of graphite electrodes are desired for their higher purity (99.9 percent). The particle size of the carbon should be 8 mesh (2.36 mm). The boric oxide is reacted with the carbon in an electric arc furnace at 2763°C (Thenevot, 1990). The process is similar to the Acheson process used to manufacture silicon carbide (see following section). The heating rate within the reactor controls the crystallite nucleation rate and the final particle size (Weimer, 1997). The resulting ingot is removed and broken into chunks by hand. The higher purity material is located in the center of the ingot, so the chunks are graded based on location in the ingot. The chunks are crushed and milled to the desired particle size. Because of the high hardness of boron carbide, the final milled product is washed with sulfuric acid to remove tramp iron and any free boron (Wilson and Guichelaar, 1997).

Boron carbide powders can be synthesized below the B_4C melting point (Schwetz, 2002) using methods such as a modified carbothermic reduction at 1600°C to 1800°C using H_3BO_3, acetylene black or sugar, and ethylene glycol. A magnesiothermic reduction is done at 1000°C to 1800°C using boric oxide, magnesium metal, and carbon and yielding B_4C and magnesia:

$$2B_2O_3 + 6Mg + C \rightarrow B_4C + 6MgO$$

The magnesia acts as a grain growth inhibitor, resulting in particles of 0.1 to 1.5 μm.

Boron carbide is relatively stable when in contact with acids, although it is attacked slowly in mixtures of hydrofluoric and sulfuric acid or hydrofluoric and nitric acid. However, it will react with molten metals and will dissociate when in high-temperature contact with molten alkalis and acidic salts. Its maximum operating temperature in air is about 1000°C for coarse particle-size materials, with decreasing temperature range for finer particle sizes (Schwetz, 2002).

Boron carbide can be made into a variety of shaped parts by sintering, hot pressing, and hot isostatic pressing at 2200°C to 2250°C. Unlike most other carbides and nitrides, boron carbide can be densified without sintering additives (Srinivasan and Rafaniello, 1997). However, pressureless sintering is easier when done with additives, such as carbon and silicon carbide. Cermets can be made by sintering boron carbide with metals such as aluminum. Thin films can be fabricated using sputtering techniques.

Boron carbide is used extensively in abrasive materials, cutting tools, and spray and blasting nozzles. However, because it oxidizes at higher temperatures, it is not used in bonded abrasive wheels (Materials Handbook, 2005). Boron carbide coatings can be used to reduce wear in automotive engine components such as valve train parts and cylinder liners. Because B_4C is a good absorber of neutrons, it is used for radiation shielding components and nuclear reactor control rods. Boron carbide is commercially available in a range of particle sizes, from 50 mm to submicron powder grades. Several grades are available based on B:C ratios ranging from 3.9 to 4.2 (Materials Handbook, 2005; Reade, 2005). Technical or general industrial grades have 65 percent to 78 percent boron, standard grade has at least 74 percent boron, nuclear grade has at least 76.5 percent boron, and high-purity grades have 77 percent to 80 percent boron.

Silicon Carbide Silicon carbide, sometimes referred to as *carborundum,* is made of 1 : 1 stoichiometric ratio of carbon and silicon in which each carbon is tetrahedrally coordinated with four silicon atoms, and vice versa. This combination can form several structural polymorphs and polytypes (Liethschmidt, 2002) as a result of the unit cells being arranged in planes perpendicular to the *c*-axis in different stacking sequences (Figure 6.17). The nomenclature for the different structural arrangements was first set by Thiebault (1944), in which there are two general polymorphs, a α and a β phase. The β phase is cubic while the α phase can have several hexagonal and rhombohedral structural modifications or types. Ramsdell (1947) offered a naming system in which C, H, and R are used to denote the cubic, hexagonal, and rhombohedral symmetry, respectively, and a digit in front of the number refers to the number of SiC layers per unit cell. For instance, 6H is used to denote the hexagonal symmetry with six layers in the unit cell. The most common types of silicon carbide are 3C, 4H, 6H, 15R, and 27R (Rafaniello, 1997). The pure α phase is colorless and the β phase is yellow, although the presence of boron and aluminum gives a blue-black color while nitrogen will give a green color to the 3C and 6H types and a yellow color to 4H and 15R (Schwetz, 2002) (Figure 6.18).

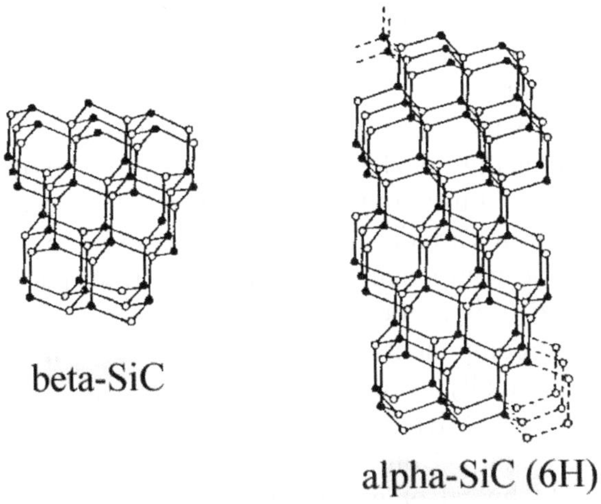

Figure 6.17 Idealized crystal structure of the SiC polytypes 3C and 6H (from Schwetz, 2000; reprinted with permission from Wiley-VCH).

Figure 6.18 Sample of black silicon carbide formed by the Acheson process (used courtesy of ESK-SIC GmbH, www.esk-sic.com).

Silicon carbide is one of the hardest materials (Mohs hardness of 9.5 to 9.75). Despite its hardness, silicon carbide is very brittle and is relatively easily crushed.

The heating characteristics of silicon carbide can depend on whether it is heated in an open or closed system. In all systems, the β phase is stable at low temperatures and will convert to the α form when heated above 1900°C. When heated in a closed system at 1 atm, silicon carbide will decompose into graphite and a silica melt at 2830 $\pm$ 40°C. In an open system, it will decompose at about 2300°C (Schwetz, 2000).

Silicon carbide occurs very rarely in nature as the mineral moissanite, named after the French chemist, Henri Moissan. Moissanite can be found in some iron meteorites, such as the Canyon Diablo meteorite, and ultramafic, diamond-bearing igneous rocks called kimberlites. Synthetic silicon carbide was first produced in 1892 by E.G. Acheson as he passed an electrical current through a mixture of coke and clay (Acheson, 1893). He assumed that the crystals that formed were made of carbon and corundum (Al_2O_3), and named the material carborundum. Further study showed that the material was a compound of silicon and carbon made through the general reaction:

$$SiO_2 + 3C \rightarrow SiC + 2CO$$

There are some intermediate steps in this reaction where a gaseous silicon carbide forms and reacts with the carbon (Weimer, 1997). The Acheson process is the basis for modern synthesis of silicon carbide. The starting materials that are used need to be very pure in order produce a high-quality green silicon carbide. The silica comes from sand or ground lump quartz of >99.7 percent purity and low Fe_2O_3 (0.01 percent to 0.05 percent). For black silicon carbide, Fe_2O_3 can be present in amounts up to 0.30 percent, Al_2O_3 can be present up to 0.50 percent, and total CaO and MgO is limited to 0.25 percent (Guichelaar, 1997). The carbon is usually sourced from petroleum coke with more than 90 percent carbon available for the reaction. Alternatively, rice hulls, which are predominantly silica and carbon, can be used as a raw material (Guichelaar, 1997). Sawdust is mixed with the batch to promote the release of carbon monoxide and to loosen the material (Lieths-chmidt, 2002). The size of the initial carbon particles can greatly control the reaction rate and morphology of the final product (Weimer, 1997).

The heating of the raw material is done in a large electric resistance furnace with removable walls and graphite electrodes to feed the electric current (Guichelaar, 1997). A variant of the Acheson furnace, the ESK process, uses electrodes in the floor (Schwetz, 2000). This is a batch process, not a continuous process, in which the raw materials are loaded from the top in layers to allow the placement of a carbon resistance element across the center of the mass. After the batch is loaded, a current is applied from the electrodes to the resistance element that runs through the middle of the batch. Silicon carbide forms in the core and is insulated by the surrounding batch. Carbon monoxide is released or, in the ESK process, collected or burned. When the process is complete, the mass cools for several days and the side walls of the furnace are removed to expose a roll of zoned material with a graphite core (Figure 6.19), and inner zone of the highest quality α silicon carbide (purest and largest crystals) and an outer zone of fine, β phase crystals.

After removal from the furnace, the cylinder of silicon carbide is placed on the floor and cut by hand to sort the chunks by grade. The recovered lumps of material are sent through a series of crushers and mills for size reduction. The crusher design affects the resulting particle shape, with roller mills producing high-aspect ratio particles and impact mills producing blocky particles (Guichelaar, 1997). Because of its high

Figure 6.19 Photograph of a U-shaped silicon carbide roll (used courtesy of ESK-SIC GmbH, www.esk-sic.com).

hardness, silicon carbide can result in a significant amount of abrasion of the milling and sizing equipment surfaces, leading to the introduction of tramp iron and steel. Therefore, the material is passed through a series of magnets to remove the ferrous contaminants.

In addition to the Acheson process, silicon carbide can be synthesized using chemical vapor deposition (CVD) methods. The particle size and morphology can depend on the gas compositions and the reaction temperature (Schwetz, 2000). Volatile carbon and silicon compounds react at high temperature in a hydrogen or other nonoxidizing atmosphere to form either the α or β structure. This process can be used to produce coatings and high-purity silicon carbide crystals that can be sawn into wafers. Defect-free, large single crystal silicon carbide is made by depositing the volatile reactants onto the face of a seed crystal. This has been problematic, and many methods end up with linear defects that render the material unfit for use in electronic components such as transistors. However, a recent method (Nakamura et al., 2004) uses steps of crystal growth and crystal face cutting to reduce the number of defects.

Most silicon carbide whiskers are made by vapor-solid reactions that result in 1 μm diameter and up to 30 μm particles. However, a more expensive method can produce whiskers with 4 to 6 μm diameter and up to tens of millimeters long (Schwetz, 2002). This is done with a vapor-liquid-solid (VLS) method where methane (CH_4) and a silicon source (SiO or $SiCl_4$) react at 1400°C. The growth is initiated on a liquid iron drop placed on a graphite substrate.

Reaction bonded silicon carbide is made by impregnating shaped mixtures of silicon carbide and carbon with liquid silicon. The silicon fills the pores of the shape and reacts with the carbon to form silicon carbide that cements or bonds the preexisting silicon carbide particles. Pressureless sintered silicon carbide can be produced by mixing spray-dried silicon carbide granulate with sintering aids, such as boron and aluminum, and then firing up to 2000°C in a nonoxidizing atmosphere (Srinivasan and Rafaniello, 1997).

Grading of commercial silicon carbide can be done by application or by purity. There are at least four general grades based on the amounts of free silicon, free carbon, oxygen, transition metals, aluminum, and nitrogen. These are, with increasing purity, metallurgical, black, green, and semiconductor (Table 6.11). The higher performance materials have higher proportions of SiC and lower amounts of free carbon (that is, carbon not bound within the silicon carbide crystals) and non-SiC materials. Oxygen is present as an oxide layer up to 10 nm thick at the particle surface, so the amount of oxygen tends to increase with the specific surface area of the powder (Rafaniello, 1997).

Metallurgical grade is typically less than 94 percent SiC and is used as an additive in iron and steel manufacturing. The black and green grades are named because of their color, and these can be further subdivided by regular and premium grades. Black silicon carbide contains 97.5 percent to 98.5 percent SiC while green silicon carbide contains about 98.5 percent to 99.8 percent SiC. At 99.5 percent purity, the material is dark green and at 99.8 percent purity, it becomes light green (Materials Handbook, 2005).

Semiconductor grade silicon carbide is the purest, with some CVD silicon carbide up to 99.9995 percent SiC. All products should be designated as either alpha or beta structure on the specification sheet. Alpha silicon carbide may allow a broader processing window (Rafaniello, 1997). A variety of particle sizes (and therefore surface area) can be found. In general, a surface area of 5 to 15 m^2/g is needed for well densified products (Prochazka, 1974). Very high purity siliconized silicon carbide is used in semiconductor applications. This material contains extremely low levels of impurities (Table 6.11) but has a larger proportion of free silicon (15 to 20 percent) compared to the other grades of silicon carbide. The addition of free silicon lowers the service temperature of the material, but there are several advantages including (Vayda, 2005):

- Increases the modulus of rupture;
- Increases the modulus of elasticity;

Table 6.11 Composition of commercial silicon carbide products (wt. %)

Grade	Semiconductor[1]	Green[2]	Premium Green[3]	Black[2]	Premium Black[2]	Metallurgical
SiC	80–85	98.6	99.5	97.5	98.5	<94
Free Si	15–20	0.15	0.03	0.5	0.3	
Free C		0.36	0.1	0.6	0.3	3–6
SiO_2		0.63	0.2	0.8	0.5	
Fe	3 ppm	0.08	0.04	0.2	0.08	
Al	50 ppm	0.08		0.2	0.1	
Ca	1 ppm	0.05		0.05		
Mg	0.04 ppm	0.03		0.05		

[1] Siliconized silicon carbide, Saint Gobain Crystar, www.crystar.com and personal communication (2005)
[2] Materials Handbook (2005)
[3] Electro Abrasives, www.electroabrasives.com

- Increases thermal conductivity;
- Increases electrical conductivity;
- Makes the body impermeable;
- Improved oxidation resistance; and
- Reduces particle generation

There are a variety of structural, refractory, abrasive, electric, and electronic applications for silicon carbide. It is used in super-duty refractories such as kiln furniture, combustion chambers, heat exchangers, and high-temperature nozzles. Its corrosion resistance and hardness make it useful in casting applications, armor, pump seals and, when applied as sputtered thin films, as a scratch-resistant surface for glass and other materials. Black silicon carbide is used in abrasives and stone-cutting saws, although green grade may be used in some specialty grinding wheels. Green silicon carbide is used in electrical resistance heating elements. Electronic applications include heat sinks, ultrafast Schottky diodes, field-effect transistors, thermistors, varistors and blue-light-emitting diodes.

Silicon carbide can be added to other materials to enhance their strength and heat resistance. For instance, silicon carbide whiskers and platelets can be used with ceramic oxides and nitrides, as well as with metals and polymers to create high strength composites with a variety of physical and chemical properties. Future applications for silicon carbide include gas sensors and a variety of semiconductor functions using single crystals. This is due to its wide energy band gap of 3.26 eV for 4H SiC and 3.03 eV for 6H SiC (Schwetz, 2000).

Tungsten Carbide Transition metals such as titanium, zirconium, and tantalum can be used to make carbides, but perhaps the most common is tungsten carbide. Tungsten carbide can be made in two stoichiometries: WC and W_2C. Only the WC structure is stable at room temperature (Baumgart, 1984), and W_2C may be undesirable as its presence in a sintered piece reduces toughness (Rafaniello, 1997). WC occurs as gray hexagonal crystals with a density of 15.6 and a Knoop hardness of 1880 (100g). Upon heating, it is stable to about 1000°C in oxidizing atmospheres and up to 1500°C in nonoxidizing atmospheres.

Tungsten carbide is not found in nature, but is synthesized from a rare metal, tungsten, and a common nonmetal, carbon. The average abundance of tungsten is only about 1.1 ppm in the Earth's crust (Table 2.1) but it is concentrated in the ore minerals scheelite ($CaWO_4$) and wolframite $(Fe_5Mn)WO_4$, which are associated with pegmatites and hydrothermal-magmatic deposits. These ores are subjected to hydrometallurgical processes to produce tungsten compounds. Wolframite concentrates are dissolved in concentrated sodium hydroxide, while scheelite concentrates are dissolved under pressure with sodium carbonate (Lassner, 2002). The resulting sodium tungstate solution is purified and converted to ammonium paratungstate (APT) with a composition of $(NH_4)_{10}(H_2W_{12}O_{42}) \cdot 4H_2O$. For the synthesis of tungsten metal, tungsten oxide (W_2O_5), APT, or tungstic acid (H_2WO_4) is reduced in hydrogen at 800°C to 1100°C to produce tungsten powder (Baumgart, 1984; Eschnauer 2002). The particle size of the powder will greatly affect the final texture of the tungsten carbide. Tungsten carbide is made by carburizing the tungsten metal powder with high-purity carbon at up to 2200°C in an electric arc furnace.

Tungsten carbide is used in a variety of abrasive and cutting applications. The most common use is to make cemented carbides, or hardmetals, by mixing and heating powdered tungsten carbide with powdered cobalt metal. The higher grades of tungsten carbide have lower free carbon, contaminant metals and oxygen. Some products are available doped or premixed with metal powders. A range of particle sizes are available, including fine (1 μm), medium (1.5 to 2 μm), and coarse (3 to 4 μm) (Rafaniello, 1997). Particle size and particle-size distribution are important because they control many of the physical properties of the final manufactured pieces. Tungsten carbide coatings are made using thermal spray, plasma spray, and CVD methods.

Nitrides

The metallic nitrides and the nonmetallic nitrides, boron nitride, and silicon nitride are important hard materials. Metallic nitrides can be made with transition metals from the fourth and fifth group of the periodic table (Baumgart, 1984). All are characterized by predominantly covalent bonds.

Aluminum Nitride Aluminum nitride (AlN) has been a known compound over the past century, but it has only been in the last two decades or so that it has become a commercially produced ceramic material. It has a hexagonal crystal structure with a greater ionic bonding character compared to other nitrides (Srinivasan and Rafaniello, 1997). Its characteristics include a specific gravity of 3.26, a Knoop hardness of 1225, and a coefficient of thermal expansion of 4.6×10^{-6} to 5.6×10^{-6}/°C. It has a very high thermal conductivity of 200 W/m-°K for single crystal AlN and 66 W/m-°K for 97 percent densified AlN (Borom et al., 1972), which is two to six times that of alumina. When heated, it will remain stable at atmospheric pressures up to 2500°C (Naito et el., 2003). Aluminum nitride will not react with acid, but will react with bases. A drawback of the material is that AlN powders will hydrolyze in the presence of water, a process that occurs more rapidly at higher temperature and at a higher surface area (Fukumoto et al., 2000).

Aluminum nitride is made by several methods (Naito et al., 2003; Materials Handbook, 2005), including the direct nitridation of aluminum, carbothermal reduction of alumina in the presence of nitrogen or ammonia, and CVD. With direct nitridation, aluminum metal powder is heated to 1200°C to 1500°C in a nitrogen gas atmosphere. The carbothermal reduction uses aluminum metal, nitrogen gas, and elemental carbon as represented by:

$$Al_2O_3 + 3C + N_2 \rightarrow 2AlN + 3CO$$

Resulting powders may be heated in air to remove free carbon and to produce an oxidize layer on the particles that serves as protection from hydrolysis (Naito et al., 2003).

Although it is relatively easily sintered without additives, greater densification is achieved with sintering aids such as yttria and calcia. The additives react with the powder to form an aluminate phase (Rafaniello, 1997). Sintering is done at 1600°C to 1900°C, depending on the particle size and the composition and proportion of additives (Naito et al., 2003). Aluminum nitride whiskers with aspect ratios of 5:1 to 20:1 and one μm diameter can be made using vapor deposition methods.

The relatively higher proportion of ionic bonds relative to the other nitrides results in aluminum nitride having relatively lower strength and hardness. Thus, it is not used in high wear applications but is very useful in microelectronic packaging and refractories. Both applications take advantage of the material's high thermal conductivity and resistance to chemical attack (except for water). The ability to be metallized, high resistivity and dielectric strength, and the low toxicity of AlN makes it a good replacement for toxic beryllia in microelectronic packaging substrates used in power modules and multichip modules. However, it may require nonaqueous cleaners during processing due to its tendency to hydrolyze. AlN is resistant to most molten metals, such as aluminum, and molten salts, making it useful in some metal refractories. AlN whiskers can be used in composites where high thermal conductivity is needed.

Commercial aluminum nitride can be sold as powders or as shapes and sheets. Powders can be in a variety of sizes. Since much of the material sold is for its ability to conduct heat, grades may be based not on composition but on thermal conductivity, ranging from about 80 to 190 W/m-°K. The lower grades have lower conductivity, which is directly related to the presence of oxygen, calcium, silicon, and iron (Rafaniello, 1997). Compositional data may include yttria for powders used for sintered materials.

Boron Nitride Boron nitride is an insulating and extremely hard material that is used in a wide variety of ceramic applications. Boron nitride has many properties that are similar to carbon. Forming a 1:1 B:N stoichiometry, there are three known polymorphs (e.g., Schwetz, 2002):

1. Hexagonal α-BN or h-BN, with a layered structure similar to graphite, which is why it is often called white graphite
2. Cubic β-BN or c-BN, a high-pressure diamond-like form with a cubic zinc blende structure
3. Dense hexagonal γ-BN or w-BN, with a wurtzite (close packed hexagonal) structure

h-BN is the phase that is stable under ambient conditions. Transformation of h-BN to w-BN occurs at pressures above 12 GPa at around 230°C and the transformation to the cubic BN occurs above 1300°C and pressures above 5.5 GPa.

The hexagonal h-BN has a density of 2.27 g/cm^3. It is insoluble in water and acids and is resistant to attack from molten metals. When heated, it is stable in air up to 1000°C, in a carbon monoxide atmosphere up to 1800°C, and in inert atmospheres up to 2700°C (Schwetz, 2002). It has high thermal conductivity and extremely low electrical conductivity, making it a good electrical insulator. The dielectric constant of h-BN is 3 and its electrical resistance varies from 1.7 × 1013 ohm-cm at 25°C to 3 × 104 ohm-cm at 1000°C (Materials Handbook, 2005).

h-BN is commonly produced by reacting boric acid or boron oxide (see section 6.2) and nitrogen or nitrogen compounds (Schwetz, 2002). One method uses boron oxide, ammonia and a tricalcium phosphate carrier at 900°C:

$$B_2O_3 + NH_3 \rightarrow 2BN + 3H_2O$$

followed by treatment at >1500°C in a nitrogen atmosphere. A second method reacts boron oxide with organic nitrogen compounds:

$$B_2O_3 + CO(NH_2)_2 \rightarrow 2BN + 2H_2O + CO_2$$

Some of the graphite-like properties of h-BN include excellent lubricating ability and machinability. Although it is a good electric insulator, it has high thermal conductivity (15 W/m-°K) and shock resistance. Pressed forms of h-BN can be machined into complex shapes. h-BN is chemically resistant to most high temperature metal and glass environments, though it will oxidize at high temperatures. It will not react with iron and iron alloys up to 1000°C to 1300°C (Srinivasan and Rafaniello, 1997). As a result, it is used in several refractory applications, including molds, mold coatings and release agents, crucibles (in vacuum settings), and thermocouple protection.

c-BN is manufactured using h-BN feedstocks that are subjected to high pressures of 50 to 90 kilobars and temperatures of 1500°C to 2200°C in the presence of a catalyst (Jeitschko et al., 2000). Cubic boron nitride is second in hardness only to diamond, so it has applications in cutting and grinding applications, particularly ferrous metals where diamond may react with the metal. In addition, diamond can be used to 600°C whereas c-BN can be used to 1100°C (Jeitschko et al., 2000). The thermal conductivity of c-BN is higher than that of copper (Srinivasan and Rafaniello, 1997), making it useful as a substrate for electronic components where heat dissipation is needed. Its coefficient of thermal expansion is on the order of 3.8×10^{-6}/°C. c-BN can be deposited as thin films on silicon substrates using a variety of physical vapor deposition, chemical vapor deposition, and sputtering methods (Bewilogua and Richter, 2000).

h-BN is available in powders of various sizes and grades, depending on the application, or as machinable shapes, such as rods and plates. Some products may have a sintering aid added, such as boric oxide. Grading is done by the weight percent BN, which can range from 86 percent to >99.5 percent. Specification sheets will show the amounts of free boron oxide and oxygen, with the higher grades generally have lower amounts of these. There is a cosmetic grade that is high purity and milled to submicron particle sizes.

Silicon Nitride Silicon nitride is a synthesized hard material with a composition of Si_3N_4. There are two structures of silicon nitride, an alpha and a beta phase (Figure 6.20). The structure of both polymorphs is based on SiN_4 tetrahedra connected at the corners that make up a three-dimensional network with hexagonal symmetry (Herrmann et al., 2000), differing only in the stacking sequence. The alpha phase has larger spacing within the structure that allows for alloying much more readily than the beta phase (Srinivasan and Rafaniello, 1997). Silicon nitride is a dark gray to black material with a specific gravity of about 3.2 and a Knoop hardness of 2200.

When subjected to heat in air, the alpha phase is stable up to 1400°C and then converts to the beta phase, which is stable up to 1500°C to 1800°C at which point it dissociates in air. It is stable up 1850°C at 1 atm in a nitrogen environment (Materials Handbook, 2005).

The most common method for manufacturing silicon nitride powders is by the direct nitridation of silicon or silicon di-imide ($Si[NH]_2$) at 1200°C to 1500°C (Naito et al., 2002). Another method is through the carbothermal reduction of silica:

$$3SiO_2 + 3C + 2N_2\ (g) \rightarrow Si_3N_4 + 3CO_2$$

and vapor deposition methods represented by (Materials Handbook, 2005):

$$3SiH_4 + 4NH_3 \rightarrow Si_3N_4 + 12H_2$$

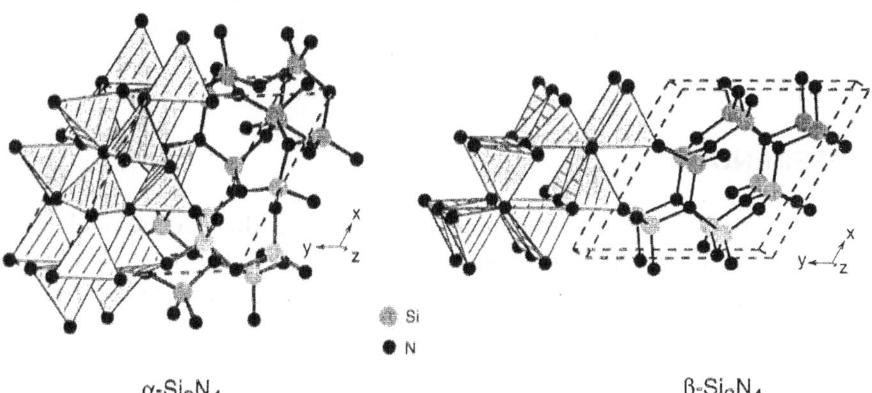

Figure 6.20 Crystal structure of alpha-silicon nitride and beta-silicon nitride (from Jeitschko et al., 2000; used with permission from Wiley-VCH).

Low diffusion rates in silicon nitride powders make them difficult to sinter below the temperature of dissociation. Therefore, sintering aids such as magnesia and yttria are added in order to make hot pressed silicon nitride (HPSN), sintered silicon nitrides (SSN), or hot isostatically pressed silicon nitride (HIPSN). The alpha phase is the main starting material (Herrmann et al., 2000). The sintering aids promote the formation of a liquid phase that bonds the silicon nitride particles together. In general, the higher the amount of the alpha phase in the starting material, the better the densification of the final product (Baumgart, 1984). Additionally, the amount of silicon dioxide on the particle surface affects sintering. An oxygen content of 1 percent to 2 percent of the powder is preferred to optimize densifications (Rafaniello, 1997). The most common sintering method is gas-pressure sintering, whereas HIPSN is used only for specialty applications such as ball bearings due to its expense (Herrmann et al., 2000). An alternative to sintered powders is reaction bonded silicon nitride (RBSN) in which a compacted form is made using silicon powder and then nitrided directly. The advantage of RBSN is that it can be made into near net shapes prior to nitridation. The disadvantage is that full densification is difficult to achieve, and therefore RBSN may not be as tolerant of use as the materials made directly from silicon nitride powders.

The high wear resistance, high strength at high temperature, and good thermal shock resistance have made silicon nitride an excellent material for turbine and reciprocating engine parts, particularly those prone to wear such as turbochargers, valves, and valvetrain parts. Other applications include nonlubricated bearings, metal cutting, and molten, nonferrous metal handling. There are many potential applications, but the relatively high cost of silicon nitride can be a deterrent to its use.

REFERENCES

Acheson, E. G. (1893). Carborundum: Its History, Manufacture, and Uses. *Journal of the Franklin Institute,* v. 136, pp. 194–214.

Andrews, W. H. (1984). Uses and Specifications of Bauxite. In *Bauxite,* ed. L. Jacob, Jr. Littleton, CO: Society of Mining Engineers AIME, pp. 49–66.

Argall, G. O. (1980). Three Iron Ore Bodies of Bayan Obo. *World Mining,* January, pp. 38–41.

Austin, C. R., H. Z. Schofield, and N. L. Haldy, (1946). Alumina in Whiteware. *Journal of the American Ceramic Society,* v. 29(12), pp. 341–354.

Bardossy, G. (1984). European Bauxite Deposits. In *Bauxite,* ed. L. Jacob, Jr. Littleton, CO: Society of Mining Engineers AIME, pp. 411–435.

Baumgart, W. (1984). Refractories. In *Process Mineralogy of Ceramic Materials,* ed. W. Baumgart, A. C. Dunham, and G. C. Amstutz. Amsterdam: Elsevier, pp. 80–103.

Baumgartner, E. (1996). Carl Auer von Welsbach: A Pioneer in the Industrial Application of Rare Earths. In *Episodes from the History of the Rare Earth Elements,* ed. C. H. Evans. Dordrecht, Netherlands: Kluwer Academic, pp. 113–129.

Bewilogua, K., and F. Richter (2000). Vapor Phase Deposition of Cubic Boron Nitride Films. In *Handbook of Ceramic Hard Materials,* ed. R. Riedel. Weinheim, Germany: Wiley-VCH, pp. 420–445.

Borom, M. P., G. A. Slack, and J. W. Szymaszek (1972). Thermal Conductivity of Commercial Aluminum Nitride. *American Ceramic Society Bulletin,* v. 51(11), pp. 852–856.

Bühler, W. (1997). Borates Still Going Strong. In *Raw Materials for the Glass Industry,* ed. J. B. Griffiths. London: Industrial Minerals, pp. 53–56.

Cannon, W. R., E. Gugel, G. Leimer, and G. Woetting (2002). Ceramics, Advanced Structural Products. In Ullmann's Encyclopedia of Industrial Chemistry, 6th ed. Weinheim, Germany: Wiley-VCH.

Castor, S. B. (1994). Rare Earth Minerals. In *Industrial Minerals and Rocks,* 6th ed., ed. D. D. Carr. Littleton, CO: Society for Mining, Metallurgy, and Exploration, Inc., pp. 827–839.

Chang, L. L. Y. (2002). *Industrial Mineralogy: Materials, Processes, and Uses*. Upper Saddle River, NJ: Prentice Hall.

Clark, H. K., and J. L. Hoard (1943). The Crystal Structure of Boron Carbide. *Journal of the American Chemical Society,* v. 65, pp. 2115–2119.

Comer, J. B. (1974). Genesis of Jamaican Bauxite. *Economic Geology,* v. 69, pp. 1251–1264.

Dörre, E., and H. Hübner (1984). *Alumina: Processing Properties, and Applications*. Berlin: Springer-Verlag.

Eschnauer, H. (2002). Hard Materials. In *Ullmann's Encyclopedia of Industrial Chemistry,* 6th ed. Weinheim, Germany: Wiley-VCH.

Evans, A. M. (1993). *Ore Geology and Industrial Minerals.* Oxford: Blackwell Science.

Fitzer, E. (2002). Synthetic Inorganic Fibers. In Ullmann's Encyclopedia of Industrial Chemistry, 6th ed. Weinheim, Germany: Wiley-VCH.

Force, E. R., and F. J. Rich, (1986). Geologic Evolution of Trail Ridge Eolian Heavy Mineral Sand and Underlying Peat, Northern Florida. Reston, Virginia: U.S. Geological Survey Professional Paper 1499.

Franklin, R., P. Goulding, J. Haviland, R. W. Joyner, I. McAlpine, P. Moles, C. Norman, and T. Nowell (1999). Stabilization and Catalytic Properties of High Surface Area Zirconia. *Catalysis Today,* v. 10, pp. 405–407.

Fukumoto, S., T. Hookabe, and H. Tsubankino (2000). Hydrolysis Behavior of Aluminum Nitride in Various Solutions. *Journal of Materials Science,* v. 35(11), pp. 2743–2748.

Gaines, R. V., H. C. W. Skinner, E. E. Ford, B. Mason, and A. Rosenzwieg (1997). *Dana's New Mineralogy,* 8th ed. New York: John Wiley and Sons, Inc.

Garnar, T. E. (1994). Zirconium and Hafnium Minerals. In *Industrial Minerals and Rocks,* 6th ed., ed. D. D. Carr. Littleton, CO: Society for Mining, Metallurgy, and Exploration, Inc., pp. 1159–1165.

Garrett, D. E. (1998). *Borates: Handbook of Deposits, Processing, Properties, and Use.* San Diego, CA: Academic Press.

Gitzen, W. H. (1970). *Alumina as a Ceramic Material.* Westerville, OH: American Ceramic Society,

Gonterman, J. R., and W. W. Wolf (1988). The Technology of Glass Fibers, In *Advances in the Fusion of Glass,* ed. F. Bickford et al. Westerville, OH: American Ceramic Society, pp. 7.1–7.15.

Guichelaar, P. J. (1997). Acheson Process. In *Carbide, Nitride, and Boride Materials Synthesis and Processing,* ed. A. W. Weimer. London: Chapman and Hall, pp. 115–129.

Harben, P. W. (2002). *The Industrial Minerals Handybook,* 4th ed. Industrial Minerals Information, Surrey.

Harben, P. W., and R. L. Bates (1984). *Geology of the Nonmetallics.* New York: Metal Bulletin, Inc.

Harries-Rees, K. (1993). Borates: Withstanding the Heat. In *Raw Materials for the Glass and Ceramics Industry,* ed. A. D. Skillen and J. B. Griffiths. London: Industrial Minerals, p. 37–43.

Haxel, G. B., J. B. Hedrick, and G. J. Orris (2002). Rare Earth Elements—Critical Resources for High Technology. Reston, VA: U.S. Geological Survey Fact Sheet 087–02.

Hedrick, J. B. (2003a). Zirconium and Hafnium. Minerals Yearbook. Reston, VA: U.S. Geological Survey.

Hedrick, J. B. (2003b). *Rare Earths.* Minerals Yearbook. Reston, VA: U.S. Geological Survey.

Henderson, P. (1996). The Rare Earth Elements: Introduction and Review. In A. P. Jones, F. Wall, and C. T. Williams. *Rare Earth Minerals: Chemistry, Origin, and Ore Deposits.* London: Chapman and Hall, pp. 1–19.

Hermann, M., H. Klemm, and C. Schubert (2000). Silicon Nitride Based Hard Materials. In *Handbook of Ceramic Hard Materials,* ed. R. Riedel. Weinheim, Germany: Wiley-VCH, pp. 749–801.

Hirano, S., and K. T. Suzuki (1996). Exposure, Metabolism, and Toxicity of Rare Earths and Related Compounds. *Environmental Health Perspectives,* v. 104, supplement 1.

Hong, S. H., and G. L. Messing (1997). Mullite Transformation Kinetics in P_2O_5-, TiO_2-, and B_2O_3-Doped Aluminosilicate Gels. *Journal of the American Ceramic Society,* v. 80, pp. 1551–1559.

Hudson, L. K., C. Misra, A. J. Perrotta, K. Wefers, and F. S. Williams (2002). Aluminum Oxide. In *Ullmann's Encyclopedia of Industrial Chemistry,* 6th ed. Weinheim, Germany: Wiley-VCH.

Hill, V. G. (1994). Bauxite. In *Industrial Minerals and Rocks,* 6th ed., ed. D. D. Carr. Littleton, CO: Society for Mining, Metallurgy, and Exploration, Inc., pp. 135–147.

International Mineral Information (IMI) (2000). *Glass and Ceramics Sourcebook.* Surrey, UK: International Mineral Information, Ltd., 131 pp.

Jeitschko, W., R. Pöttgen, and R. D. Hoffmann (2000). Structural Chemistry of Hard Materials. In *Handbook of Ceramic Hard Materials,* ed. R. Riedel. Weinheim, Germany: Wiley-VCH, pp. 3–40.

Keegan, N., ed. (1998). *Raw Materials for the Refractories Industry,* 4th ed. London: Industrial Minerals Information.

Kistler R. B., and C. Helvaci (1994). Boron and Borates. In *Industrial Minerals and Rocks,* 6th ed., ed. D. D. Carr. Littleton, CO: Society for Mining, Metallurgy, and Exploration, Inc., pp. 171–186.

Kokubo, T., H-M. Kim, and M. Kawashita (2003). Ceramics for Biomedical Applications. In *Handbook of Advanced Ceramics,* v. II, ed. S. Somiya. Oxford: Elsevier, pp. 385–416

Krause, H. J. (1984). Ceramic Pigments. In *Process Mineralogy of Ceramic Materials,* ed. W. Baumgart, A. C. Dunham, and G. C. Amstutz. Amsterdam: Elsevier, pp. 196–207.

Krell, A. (2000). Ceramics Based on Alumina: Increasing the Hardness for Tool Applications. In *Handbook of Ceramic Hard Materials,* ed. R. Riedel. Weinheim, Germany: Wiley-VCH, pp. 648–682.

Lassner, E. (2002). Tungsten, Tungsten Alloys, and Tungsten Compounds, in *Ullmann's Encyclopedia of Industrial Chemistry,* 6th ed. Weinheim, Germany: Wiley-VCH.

Liethschmidt, K. (2002). Silicon Carbide. In *Ullmann's Encyclopedia of Industrial Chemistry,* 6th ed. Weinheim, Germany: Wiley-VCH.

Loughnan, F. C., and S. B. Sadleir (1984). Geology of Established Bauxite-Producing Areas in Australia. In *Bauxite,* ed. L. Jacob, Jr. Society of Mining Engineers AIME, pp. 436–450.

Lunghofer, E. P., and L. A. Wolfe (2000). Fused Brown Alumina Production. In China. *Ceramic Industry,* July, www.ceramicindustry.com.

Lyday, P. A. (2003). Boron. Minerals Yearbook. Reston, VA: U.S. Geological Survey.

Maintech Sapphires (2004). www.maintechsapphires.com.

Materials Handbook (2005). *Ceramic Industry,* v. 155 (1), p. 23–125.

Maynard, J. B. (1983). *Geochemistry of Sedimentary Ore Deposits.* New York: Springer-Verlag.

McCracken, W. (1998). Bauxite—A Refractory Workhorse. *American Ceramic Society Bulletin,* v. 77(11), pp. 57–61.

Minggin, W., and Xuehong, D. (1997). The History of China's Rare Earth Industry. In *Episodes from the History of the Rare Earth Elements,* ed. C. H. Evans. Dordrecht, Netherlands: Kluwer Academic, pp. 131–147.

Moles. P. (1999). Zirconium Based Catalysts. *Chemical Week,* August, chemweek.com.

Molycorp (1992). *Cerium: A Guide to Its Role in Chemical Technology.* Molycorp, Inc. Mountain Pass, California.

Nakamura, D., I. Gunjishima, S. Yamaguchi, T. Ito, A. Okamoto, H. Kondo, S. Onda, and K. Takatori (2004). Ultrahigh-Quality Silicon Carbide Single Crystals. *Nature,* v. 430, pp. 1009–1012.

Naito, M., N. Shinohara, and K. Uematsu (2003). Raw Materials. In *Handbook of Advanced Ceramics,* v. I, ed. S. Somiya. Oxford: Elsevier, pp. 81–129.

Nielsen, R. (2002). Zirconium and Zirconium Compounds In *Ullmann's Encyclopedia of Industrial Chemistry,* 6th ed. Weinheim, Germany: Wiley-VCH.

Noirot, M. (2005). Personal communication.

Noirot, M. D. (2000). Interaction Between Borates and Clay Suspensions. In *Science of Whitewares II,* ed. W. M. Carty, and C. W. Sinton. Westerville, OH: American Ceramic Society, pp.189–198.

Noirot, M. (1999). Borates: Advancing Traditional Ceramics. *American Ceramic Society Bulletin,* v. 78(8), pp. 79–84.

Orris, G. J., and R. I. Grauch (2002). Rare Earth Element Mines, Deposits, and Occurrences. Reston, VA: U.S. Geological Survey Open-File Report 02-189.

Patterson, S. H., Kurtz, H. F., Olson, J. C., and Neeley, C. L., (1986). World Bauxite Resources, U.S. Geological Survey Professional Paper 1076B, Reston, Virginia.

Pinto, B. (2005). Personal communication.

Plodinec, M. J. (1988). Advances in Processing Nuclear Waste Glasses. In *Advances in the Fusion of Glass,* ed. F. Bickford et al. Westerville, OH: American Ceramic Society, pp. 9.1–9.11.

Plunkert, P. A. (2002). Bauxite and Alumina. Minerals Yearbook. Reston, VA: U.S. Geological Survey.

Prochazka, S. (1974). Sintering of Silicon Carbide. In *Ceramics for High-Performance Applications,* ed. J. J. Burcke, A. E. Gorum, and R. N. Katz. Chestnut Hill, MA: Brook Hill Publishing, pp. 239–252.

Queensland Government (2002). Bauxite. Natural Resources and Mines, Mineral Information Leaflet No. 6.

Rafaniello, W. (1997). Critical Powder Characterization. In *Carbide, Nitride, and Boride Materials Synthesis and Processing,* ed. A. W. Weimer. London: Chapman and Hall, pp. 43–73.

Ramdsdell, L. S. (1947). Studies on Silicon Carbide. *American Mineralogist,* v. 32(1–2) 64–82.

Reade, (2005). Reade Advanced Materials. Providence, RI, www.reade.com.

Riedel, R., ed. (2000). *Handbook of Ceramic Hard Materials.* Weinheim, Germany: Wiley-VCH,

Rhodia (2004). www.rhodia.com.

Schüller, K. (1984). Oxide Ceramics. In *Process Mineralogy of Ceramic Materials,* ed. W. Baumgart, A. C. Dunham, and G. C. Amstutz. Amsterdam: Elsevier, pp. 104–124.

Schwetz, K. A. (2002). Boron Carbide, Boron Nitride, and Metal Borides. In *Ullmann's Encyclopedia of Industrial Chemistry,* 6th ed. Weinheim, Germany: Wiley-VCH.

Schwetz, K. A. (2000). Silicon Carbide Based Hard Materials. In *Handbook of Ceramic Hard Materials,* ed. R. Riedel.Weinheim, Germany: Wiley-VCH, pp. 683–748.

Shelby, J., (1997). *Introduction to Glass Science and Technology.* Cambridge: The Royal Society of Chemistry.

Silvernail, W. L., and N. J. Goetzinger, (1971). The Mechanism of Glass Polishing. *Glass Industry,* v. 52, pp. 130–152.

Skillen, A. (1993). Zirconium Minerals. In *Raw Materials for the Glass and Ceramics Industry,* ed. A. D. Skillen and J. B. Griffiths. London: Industrial Minerals, pp. 81–85.

Smith, G. I. (1979). Surface Stratigraphy and Geochemistry of Later Quaternary Evaporites. Searles Lake, California, Professional Paper 1047, Reston, VA: U.S. Geological Survey,

Smith, R. A. (2002a). Boric Oxide, Boric Acid, and Borates. In *Ullmann's Encyclopedia of Industrial Chemistry,* 6th ed. Weinheim, Germany: Wiley-VCH.

Smith, R. A. (2002b). Basic Geology and Chemistry of Borate. *American Ceramic Society Bulletin,* v. 81(8), pp. 61–64.

Srinivasan, M., and W. Rafaniello (1997). Nonoxide Materials: Applications and Engineering. In *Carbide, Nitride, and Boride Materials Synthesis and Processing,* ed. A. W. Weimer. London: Chapman and Hall, pp. 1–42.

Strunz, H. (1997). Classification of Borate Minerals, *European Journal of Mineralogy,* v. 9, p. 225–235.

Subbrarao, E. C. (1981). Zirconia—An Overview. In *Science and Technology of Zirconia,* ed. A. H. Heuer and L. W. Hobbs. Westerville, OH: American Ceramic Society, pp. 1–13.

Thenevot, F. (1990). Boron Carbide—A Comprehensive Review. *Journal of the European Ceramic Society,* v. 646, pp. 205–225.

Thiebault, N. W. (1944). Morphological and Structural Crystallography and Optical Properties of Silicon Carbide: Part I: Morphological Crystallography and Etching Figures. *American Mineralogist,* v. 29(7–8), pp. 249–278.

Upadhyaya, G. S. (1998). Cemented Tungsten Carbides: Production, Properties, and Testing, Noyes Publications. Westwood, New Jersey.

Varshneya, A. K. (1994). *Fundamentals of Inorganic Glasses.* Boston: Academic Press.

Vayda, J., (2005). Personal communication.

Vijayan, S., A. J. Melnyk, R. D. Singh, and K. Nuttall (1989). Rare Earths: Their Mining, Processing, and Growing Industrial Usage. *Mining Engineering,* January, pp. 13–18.

Warshaw, S. I., and R. Seider (1967). Comparison of Strength of Triaxial Porcelains Containing Alumina and Silica. *Journal of the American Ceramic Society,* v. 50(7), pp. 337–342.

Weimer, A. W. (1997). Thermochemistry and Kinetics. In *Carbide, Nitride, and Boride Materials Synthesis and Processing,* ed. A. W. in Weimer. London: Chapman and Hall, pp. 79–113.

Wilson, W. S., and P. J. Guichelaar (1997). Electric Arc Furnace Processes. In *Carbide, Nitride, and Boride Materials Synthesis and Processing,* ed. A. W. Weimer. London: Chapman and Hall, pp. 131–136.

Wu, C., Z. Yuan, and G. Bai, (1996). Rare Earth Deposits in China. In *Rare Earth Minerals: Chemistry, Origin, and Ore Deposits,* ed. A. P. Jones, F. Wall, and C. T. Williams. London: Chapman and Hall, pp. 281–310.

Zhou, Y., K. Hirao, Y. Yamauchi, and S. Kanzaki (2002). Tailoring the Mechanical Properties of Silicon Carbide Ceramics by Modification of the Intergranular Phase Chemistry and Microstructure. *Journal of the European Ceramic Society,* v. 22, pp. 2689–2696.

7

RECYCLED AND WASTE MATERIALS

The previous sections presented information about virgin raw materials—that is, those that have not been through a cycle of manufacturing and consumer use. There is, however, a place for some materials that have been through one or more cycles of use. These can be wastes, which are for the most part materials normally rendered to the landfill, or recycled materials that have been collected for the purpose of using again. The line between what is waste material and what is recyclable material changes with time, technology, economics, and cultural/social expectations, so both are considered here simultaneously. The waste and recycled materials that are directly relevant to glass and ceramic manufacturing include consumer and industrial glass cullet, slags, and ashes. There are low-grade minerals that are not used, such as clays that are a byproduct of mining. However, as these materials are generally not landfilled or transported far from the mine site, they can be considered low value raw materials that become usable when economic conditions allow.

7.1 GLASS CULLET

Pre-Consumer Cullet

Before discussing the use of recycled glass, some terms should be defined. *Cullet* is the general term used for broken scrap glass that comes from either pre-consumer or post-consumer sources. *Pre-*

consumer cullet comes from in-house scrap generated in the glass manufacturing plant or from scrap produced by the secondary users of glass commodities such as bottlers and producers of architectural and automotive glass. The in-house cullet comes from breakage, trimmed edges, flawed product, or from the production of transition material between changes in batch formulation, specifications, or forming processes. All glass-manufacturing plants have a return system along the forming lines to transport in-house cullet to a storage area adjacent to the facilities (Rosenthal, 2002). In-house cullet naturally has a composition similar to that which is being produced, and this is the same case for cullet that is returned by secondary users if it originated at the plant. For secondary cullet, there are costs associated with transportation that can add up quickly if the source is far from the plant. In addition, there is the possibility of contamination at the site of the secondary user and during transport back to the melter. For instance, a bottling plant has the potential of including metal caps, and an automotive windshield manufacturer may return cullet that contains plastics or polymeric materials. From a purity standpoint, pre-consumer cullet, particularly in-house material, is preferred to post-consumer material because its composition is known and the potential for contamination is lower. However, too much in-house cullet is a reflection of inefficiency in the whole process, so there is a balance that needs to be maintained. Virtually all glass manufacturers have a stockpile of cullet as a component of the batch and to keep the furnace charged in case there is a problem with the raw material supply chain.

Post-Consumer Cullet

Post-consumer cullet refers to glass that is recovered after its use in a commercial product, such as a bottle, jar, or window. Commercially available cullet consists almost exclusively of containers from beverages and food that are divided into three general colors (flint, amber, and green) and a mixed-color product. There is currently no economical method of recovering post-consumer flat glass and fiberglass at the end of its lifecycle.

Not long ago, *glass recycling* referred to the reuse of an unbroken glass container to be refilled food or beverages. This was particularly the case for beverage containers that, when emptied, were returned to the grocery store or picked up by the milkman, washed, and refilled by the bottler. In fact, in many countries this still holds true, particularly for mineral water. However, the change from refillable bottles to throw-away bottles changed between the 1950s and 1970s in the United States. For example, in 1958, 98 percent of soft drinks and 58 percent of beer were packaged in returnable glass bottles. By the early 1970s, these proportions had been reduced to about 30 percent and 20 percent, respectively (Hannon, 1972). The reasons for this include competition from other packages, a desire by the container companies to produce more product, and changes in consumer habits. This meant that bottles and jars were sent to the trash rather than back to the bottler.

Very little of this nonreturnable container glass was incorporated into recycling in the United States until the legislation of state-mandated regulations, beginning with Oregon and Vermont, which require a deposit on some beverage bottles. The goals of the legislation were to reduce litter, promote the use of refillable containers, and promote the recycling of nonrefillable glass. A similar story follows for other countries. Currently, nine of the United States have "bottle bills" that require deposits and drop off of used containers at the retailer for refund. These bottles are not refilled but are sorted and crushed to cullet. Another driver for glass recovery has been overall changes in solid waste disposal practices and regulations. The cost of placing refuse in landfills has become more expensive, due in part to tighter environmental restrictions as well as a reduction in available landfill space, requiring waste to be hauled further and further from the source. As a response, local municipalities may mandate the collection of recyclable materials, including glass containers.

The recovery of post-consumer glass containers can be accomplished in several ways, and the chosen method can greatly affect the quality of the cullet. These methods include the following (Hayes et al., 2001):

- Deposit and refund upon return ("bottle bills")
- Decentralized drop-off and collection stations for separated materials
- Curbside sorting and collection of recyclable material
- Collection from commercial sources such as bars and restaurants
- Recovery from mixed recyclable material
- Recovery from mixed solid waste

Glass recovery ideally should keep colors separate and remove any contaminants that have been introduced. Post-consumer glass contaminants of concern include organic matter, ferrous metals, nonferrous metals, and ceramic materials. Organic materials include paper labels, plastics, and residual food. Although these will burn off when the cullet is placed into the furnace, their oxidation will reduce the furnace atmosphere and cause potentially harmful redox reactions. Ferrous metals include iron and steel that may be introduced during collection and sorting processes. Steel bottle caps and jar lids are a particularly common contaminant. Nonferrous materials include the nonferrous metals such as aluminum that make up neck rings of some containers. Trace amount of colorant metals, such as cobalt, may be in labels painted onto the container, causing streaking in flint glass. Ceramics, glass-ceramics, porcelain, and stones (sometimes referred to collectively as CPS) are undesirable because their high melting point means that they can remain unmelted in the glass, leading to defects and potential fouling of melting and forming equipment.

The best way to minimize the presence of contaminants and reduce color mixing is through careful collection. In general, containers collected from deposit and refund programs are the highest quality and demand the highest prices (Gitlitz, 2001) because they need to be cleaned, whole, and carefully sorted in order to receive the deposit. Drop off centers (bring banks) and curbside sorted glasses can result in clean glass, but suffers from having contaminants of different colored glass or nonglass materials. This drawback can be somewhat alleviated by transporting the material to a central dedicated glass-sorting facility. In the United States, residential curbside collection of mixed recyclable material (glass, metal, paper, plastic, and cardboard) is moving from the more expensive hand sorting to single-stream collection of mixed materials that are sent to a *material recovery facility* (MRF). At a modern, automated MRF, the mixed paper, metal, cardboard, and glass are sorted by a system of human manual labor, screens, air classifiers, magnets, and eddy currents (Knox and Brockway, 2001). Glass color sorting is often done by hand, although optical sorters have been developed recently. However, this kind of equipment is expensive to install and is unable to detect transparent glass-ceramics that pose a considerable hazard if introduced into a commercial container or fiberglass melter because they melt at higher temperatures than soda-lime-silicate and borosilicate glass.

One potential method of reducing the effect of color mixing is to control or decolorize the glass (e.g., Dalbey and Purser, 1996; Rapp et al., 2003). The color of amber glass is produced in a reduced state with a chromophore of sulfur and iron. If zinc is added, the S^{-} ligand will be attracted preferentially to the Zn^{2+} cation and will reduce the amber color and promote a blue/green color of the free Fe^{2+} cation (Dalbey and Purser, 1996). A combination of 2.0 percent ZnO and 2.0 percent Er_2O_3 added to amber glass can produce a nearly colorless glass.

Users of Cullet

The container and fiberglass insulation industries are the only commercial users of post-consumer glass cullet that make new glass products. Post-consumer cullet does find additional use in nonglass materials, such as filler in low-end ceramics, as aggregate in concrete and asphalt, and as filtration media. The applications that are not relevant to glass and ceramic manufacturing are considered elsewhere (e.g., Dhir et al., 2001).

The use of cullet in making new glass products has direct economic benefits because it can reduce the energy requirements and, depending on the cullet price, reduce the cost of raw materials. Since the heat of fusion of the crystalline minerals has already been added, it is not required when the cullet is remelted. As a result, the energy to melt cullet requires only about half the energy needed for virgin raw materials (USDOE, 2002). Theoretically, there should be a 0.36 percent reduction in energy needed to melt glass for every 1 percent use of cullet in the batch, but in industrial experience the energy reduction is about 0.25 percent to 0.30 percent for every 1 percent use of cullet (Glüsing and Conradt, 2001). In addition to the lower energy needs, there are fewer emissions of dust and criteria air pollutants per unit of glass produced compared to the melting of virgin minerals. Cullet can decrease the operating temperature of the furnace, thereby increasing the life of the refractories in the tank and checkers and increasing the time between campaigns.

With all of the benefits of using cullet, there is still a wide range in the amount of cullet used in new glass products by individual manufacturing plants and different countries. The differences can be attributed to geography, proximity, and relative cost of virgin raw materials and government regulations. Most manufacturers will use as much cullet as possible as long as it meets strict specifications of consistency and purity and is available a reasonable price. In addition, the source of cullet needs to be consistent, as is the case with any raw material supplier. In the European Union, cullet use in container glass production varies from less than 20 percent to more than 90 percent, with a regional average of 48 percent (IPPC, 2000). In the United States, containers are recycled at a rate of about 35 percent (GPI, 2004). Geography plays a considerable role in large countries, such as Canada and the United States, because the bottle manufacturing plant may be far from the source of post-consumer cullet (such as large urban centers). At larger distances, the overall energy savings from using cullet are negated by the energy required to transport the material (Gaines and Mintz, 1994).

Although post-consumer insulation fiberglass is not recycled, fiberglass manufacturing has been one of the largest uses of post-consumer container cullet (Guter, 1993). Cullet can make up from 10 percent to over 90 percent of the batch, with the latter consisting of container cullet and borates. The use of cullet has the energy and environmental benefits listed previously. In addition, cullet forms a melt quickly, which is beneficial for electric cold top melters often used in fiberglass manufacturing. However, there is increased concern of contaminants to the point that some manufacturers are reducing or eliminating the use of cullet. Also, because the industry is slowly moving from electric melters to hot-top, oxy-fuel furnaces, the practical advantage of cullet is decreased.

Ceramic products can incorporate some cullet as a nonplastic filler and glass former. Its general composition is similar to a high-silica, low-alumina flux (Table 7.1). Finely ground cullet from container glass and cathode ray tube glass can be used as a flux in bricks and for porcelain tiles.

One set of experiments using several clay-based bodies showed that the replacement of talc and nepheline syenite with ground glass did not affect fired shrinkage, demonstrated reduced absorption with increasing glass content, and reduced the maturation temperature (CWC, 1999). However, with substitutions of more than 20 percent, the reduction in body plasticity was detrimental. In bricks, 5 percent to 10 percent ground glass can be added to reduce the firing temperature by up to 50°C while maintaining the desired durability and absorption characteristics (Jackson and Smith, 2001).

Cullet use is not always solely based on economic considerations. In addition to the laws regarding the recovery of post-consumer glass, some laws also require that glass products contain a certain amount of post-consumer cullet. For instance, the state of California requires fiberglass producers to use at least 30 percent post-consumer cullet in insulation fiberglass that is made or sold in California and all glass food and beverage containers manufactured for sale in California must use at least 35 percent post consumer cullet. In the European Union, by the end of 2008, containers should contain 60 percent by weight of recycled material (EU, 2004).

Table 7.1 Examples of the composition of cullet (wt. %)

	SiO_2	Al_2O_3	Fe_2O_3	CaO	MgO	Na_2O	K_2O	TiO_2	Cr_2O_3
Flint container[1]	72.4	1.6	0.05	10.0	0.49	14.1	0.26	0.03	
Flint container[2]	70.9	1.59	0.7	9.78	1.84	12.7	0.47		
Green container[1]	71.2	1.93	0.34	10.6	1.67	13.0	0.66	0.06	0.26
Amber container[1]	71.2	1.9	0.25	11.4	0.18	14.6	0.50	0.03	0.004
Amber container[2]	70.6	1.73	0.31	10.0	1.70	12.8	0.59		
Mixed container[1]	72.8	2.77	0.35	9.6	1.01	11.6	0.63	0.10	0.021
Mixed container[2]	70.2	1.84	0.29	10.1	1.51	12.7	0.72		
Flat[1]	71.3	0.37	0.31	9.2	3.9	14.0	0.06	0.02	0.005
Mixed flat[2]	70.5	0.73	0.27	8.20	3.97	12.7	0.23		

[1] Guter (1993)
[2] Suomen Uusioaines Ltd http://www.uusioaines.com/

Supply and Specifications

The quality, availability, and price of cullet can differ from country and region, depending on methods of collection and sorting, governmental regulations and incentives, and overall demand. Because of the high transportation costs, post-consumer cullet cannot be shipped over long distances. Thus, for cullet that is destined to be used in new containers, the demand will be affected by the balance of regional container usage versus regional container production. For instance, the consumption of glass in the United Kingdom results in a large proportion of green glass, much from imported wine and oil, while production is dominated by flint glass for spirits, of which almost 90 percent is exported (Barlow, 2001). Because of the imbalance of consumption versus production and the fact that green cullet is unusable in making flint glass, there is a surplus of green cullet in the United Kingdom that cannot be consumed by the demand. There is a similar situation in the eastern United States, where there is essentially no market for green cullet. About 85 percent of recovered green cullet is used for landfill cover and the rest is used in other, nonglass applications (Gitlitz, 2001)

The value of cullet depends on the color. In general, clean flint glass has a strong demand, while green and amber have lower demand. Mixed or commingled colors have almost no value and, in fact, have a negative value because of costs associated with sending the material to the landfill. In some areas, green cullet can have a negative value as well.

There are specifications for post-consumer cullet used in new containers (Table 7.2) and in insulation fiberglass (Table 7.3). *Unprocessed* cullet (Figure 7.1) is either whole or broken and has not been through cleaning, crushing, and contaminant removal. *Processed* or furnace-ready cullet has been cleaned and beneficiated to the appropriate specifications. Specifications are set by each manufacturer, but published specifications are issued in the United States by the Institute of Scrap Recycling Industries (ISRI, 2003).

Fiberglass manufacturers have different cullet specifications than container manufacturers (Table 7.3). Cullet used in fiberglass needs to have a known and consistent composition, because even small changes composition can cause changes in properties important to fiberization, such as melt viscosity and emissivity (Guter, 1993). Changes in viscosity will affect fiber diameter. Specifically, increases in silica and alumina will increase viscosity, while increases in alkalis will decrease viscosity at a given temperature point. The amount of transition metals, such as those present in colored glass cullet, will affect the infrared

Table 7.2 Standards and measured impurities in processed cullet

	CPS[1] (g/ton)	Aluminum (g/ton)	Lead (g/ton)	Iron (g/ton)	Organics (g/ton)	Other Colors (%)
U.S. processed flint[2]	prohibited	prohibited	prohibited	prohibited	per agreement	0–5 amber 0–1 green Total nonflint <5
U.S. processed amber[2]	prohibited	prohibited	prohibited	prohibited		0–10 flint 0–10 green Total nonamber <10
U.S. processed green[2]	prohibited	prohibited	prohibited	prohibited		0–15 flint 0–15 amber Total nongreen <30
German standard[3]	< 25	< 5	< 1	< 5	< 500	0.5% off-color in flint 15% off-color in green 10% off-color in amber
German flint[4]	27	2	0.1	1.4		
Finnish flint[5]	< 10	< 10	< 5	< 10	< 500	< 0.5
German green[4]	22	7	1.1	2.6		
German amber[4]	74	4.4	0.4	3.2		
Finnish amber[5]	< 10	< 10	< 5	< 10	< 500	< 10

[1] CPS = ceramics, porcelain, and stones. Can also include opal glass and glass-ceramics
[2] ISRI (2003); processed cullet also referred to as furnace-ready
[3] Drescher (2003)
[4] Schaeffer (1999); average measured values for German cullet for 1996
[5] Finnish container cullet as reported in specifications for Suomen Uusioaines Ltd

Table 7.3 Specifications for cullet used in insulation fiberglass

Material	Minimum (wt. %)	Maximum (wt. %)
SiO_2	70.0	none
Fe_2O_3	none	0.5
CaO/MgO	11.0	none
Na_2O	13.0	none
PbO	none	0.2
H_2O	none	2.0
Carbon	none	0.1
FeO	none	0.1
Ag_2O	none	0.05
Cr_2O_3	none	0.1
CoO	none	0.05

From CWC (1996)

Figure 7.1 Example of unprocessed mixed container glass at a glass recovery facility.

emissivity, which relates to the rate of cooling of the fibers. If an electric furnace is used in melting, compositional changes can affect the power settings and thermal flow within the tank. Cullet color can affect the redox conditions of the tank due to the different oxidation states of flint (highly oxidized), green (slightly oxidized), and amber (highly reduced) cullet. Any change in the redox conditions in the tank can result in the release of SO_2 gas bubbles, which can upset cold-top melters because a foam layer will form between the cold batch on top and the melt below (Guter, 1993).

One of the largest problems associated with use of cullet in fiberglass manufacturing is contamination of metals and refractory particles. In general, ceramic particles larger than 1.7 mm will not melt in the glass tank (CWC, 1996). These can plug the bushings or fiber spinners, leading to considerable down time on the production line. Transparent glass-ceramic articles, such as those used in oven doors and range tops, are particularly problematic because they are nearly impossible to differentiate from soda-lime-silicate glass during cullet processing. Metal contaminants will not dissolve into the glass but, rather, will melt and collect on the bottom of the glass tank. These can lead to holes in the refractory lining, creating dangerous conditions (e.g., shorting of electric melters), and an expensive furnace rebuild. Particularly difficult sources of metals are silvered or mirrored glass.

7.2 SLAGS AND ASHES

Slag

Slag is the glassy material that is produced during the smelting of metals such as iron and steel. To make iron, iron oxide ores are reduced to elemental iron while the silicate mineral impurities are separated and removed. This is done in a blast furnace, in which a mixture of iron ore, coke, and a flux, such as limestone or dolomite, are heated to temperatures up to 1650°C under limited oxygen conditions. When the coke is burned, it produces carbon monoxide, which, in turn, reduces the iron oxide ore into molten iron metal. The lime and magnesia combine with the silicate minerals in the ore to form an immiscible liquid, or slag, that floats on top of the molten iron. The slag is removed and cooled to form a mixture of glass and microcrystalline silicates, aluminosilicates, calcium aluminosilicates, and sulfur compounds. The composition of the slag will depend on the initial composition of the ore and, to a certain degree, the composition of the flux. Table 7.4

Table 7.4 Examples of the composition of blast furnace slag used in glass (wt. %)

	SiO_2	Al_2O_3	Fe_2O_3	CaO	MgO	Na_2O	K_2O	TiO_2	MnO	S_2
United Kingdom[1]	36.0	14.0	0.23	38.5	8.5	0.3	0.3	0.6	0.5	0.7
Czech Reublic[2]	40.5	7.3	0.22	37.2	13.3	0.2	0.5	0.5	0.6	0.64

[1] Calumite Limited—Scunthorpe, UK
[2] Calumite Limited—Ostrava, Czech Republic

gives the compositions of two blast-furnace slags that are used in the glass industry. Note the twofold difference in Al_2O_3 between the two examples.

Blast-furnace slag can be cooled by several methods to produce commercial products (van Oss, 2003). The dominant kind is *air cooled,* in which the molten slag is poured and slowly cooled in air or with the assistance of a water spray. The slag crystallizes to form hard, vesicular lumps which can then be crushed and sized. *Granulated* slag is formed by quickly quenching molten slag in water, resulting in sand-sized glassy particles. The third type of blast-furnace slag is *pelletized* or expanded slag. This type is cooled in such a way that the resulting particles have a high concentration of vesicles.

Most commercial slag products are used in the concrete and cement industries as either aggregate or as a cementitious additive (pozzolan) to Portland cement. However, because it is mainly composed of oxides that are components of soda-lime-silicate glasses, it is possible to use slag in part of the batch if the slag is low in iron (Jones, 2003). Iron can be reduced by milling the slag to less than 40 mesh (425 μm) and passing the material through a magnetic separation system (Wang et al., 1989). In container glass, 6 percent to 7 percent of the sand can be replaced by slag for flint glass, 10 percent to 13 percent for green glass, and up to 18 percent in amber glass. In float glass, 4 percent to 8 percent of sand can be replaced by slag; for insulation fiberglass up to 18 percent can be replaced. Used alone, air-cooled blast furnace slag can be made into mineral wool. The slag is melted and spun or blown to produce glassy fibers (van Oss, 2003). A glass purported to be useful for containers was made with 40 percent to 50 percent beneficiated blast furnace slag mixed with silica, lithia, zirconia, and sodium sulfate (Wang et al., 1989)

Combustion Ashes

The combustion of organic materials such as wood, coal, and municipal waste generates ashes composed of inorganic oxides that may be used in some lower-value ceramic products. In the kraft process for manufacturing paper, wood chips are digested to release the desired cellulose fibers, and the resulting liquid from this process (black liquor) is concentrated and then burned in a boiler to recover inorganic process chemicals and to produce steam and electrical power. Wastes from the recovery process include ashes from the boiler consisting of predominantly crystalline and amorphous SiO_2, Al_2O_3, and CaO (Table 7.5). A

Table 7.5 Examples of the composition of ashes (wt. %)

	SiO_2	Al_2O_3	Fe_2O_3	CaO	MgO	Na_2O	K_2O	TiO_2	MnO	P_2O_5
Coal Fly Ash										
Spain[1]	51.56	29.13	4.78	4.21	1.62	0.26	0.82	1.66		
Italy[2]	15.17	7.14	3.76	23.71	1.35	0.9	0.44	0.31		
U.S. bituminous[3]	20–60	5–35	10–40	1–12	0–5	0–4	0–3			
U.S. sub-bituminous[3]	40–60	20–30	4–10	5–30	1–6	0–2	0–4			
Coal Bottom Ash										
Bitminous—Ohio[4]	47.1	28.3	10.7	0.4	5.2	0.8	0.2			
Sub-bituminous—Texas[4]	45.4	19.3	9.7	15.3	3.1	1	—			
Paper Boiler Ash										
Bottom ash[5]	49.79	6.78	3.23	29.72	3.49	1.05	3.61	0.90	0.58	0.84
Fly ash[5]	25.89	4.83	2.20	54.32	3.24	0.94	4.46	0.68	1.46	1.97

[1] Tereul, Barbieri et al. (1997)
[2] Rete 2, Barbieri et al. (1997)
[3] North Carolina Division of Pollution Prevention and Environmental Assistance
[4] Moulton (1973)
[5] From the author, from a U.S. plant, on a volatile-free basis; bottom ash LOI is 3% to 10% and LOI for fly ash is 40% to 50%

secondary waste stream called *recaust grit* consists primarily of CaO and Na_2O compounds.

Like wood, coal is composed of mostly combustible hydrocarbon material but contains some noncombustible matter. When coal is burned to generate steam for electricity generation, the noncombustible material will be retained as slag or ashes. The heavier ashes and slag are retained at the bottom of the boiler (bottom ash), and fly ash is the smaller particulate matter recovered from the exhaust stream. The composition of the coal fly ash is broadly dominated by SiO_2, Al_2O_3, Fe_2O_3, CaO, and carbon, but the proportions can be highly variable depending on the type of coal that is used (Table 7.5). In general, ash derived from sub-bituminous coals has higher CaO compared to ash from bituminous coal. Ashes derived from paper plant boilers have lower iron relative to coal ashes. Coal fly ash can have widely varying amounts of carbon (Andreola et al., 2004).

Like blast furnace slag, coal fly ash and bottom ash are widely used in cement and concrete as either aggregate or as a cementitious additive. Fly ash can also be used on roadways for snow and ice control. Although coal and wood ashes are not widely incorporated in commercial products, several studies have shown their potential for use in ceramic products. Coal fly ash mixed with plastic clays in proportions of 40:60, respectively, results in the production of a ceramic with 50 percent mullite (Queralt et al., 1997). Other studies have shown the use of coal fly ash in the production of glass ceramics by the addition of up to 50 percent fly ash mixed with glass cullet and dolomite (Barbieri et al., 1999). Municipal incinerator bottom and fly ashes can make glass-ceramics that crystallize wollastonite or anorthite (Andreola, 2004). Paper mill boiler ash can be used to make bricks (Liaw et al., 1998). Paper mill ash, which has a higher CaO content, may be more desirable material for ceramics because of much lower iron and heavy metal concentrations compared to coal fly ash. It should be noted that while it is physically possible to make ceramic and glass articles from waste ashes, a major hurdle is the lack of a consistent product due to heterogeneities in the composition between sources and the change in composition over time due to fluctuations in the coal or, to a lesser extent, tree composition.

REFERENCES

Andreola, F., L. Barbieri, A. Corradi, and I. Lancellotti (2004). New Marketable Products from Inorganic Residues. *American Ceramic Society Bulletin,* v. 83(3), p. 26.

Barbieri, L., I. Lancellotti, T. Manfredini, I. Queralt, J. Rincon, and M. Romero (1999). Design, Obtainment and Properties of Glasses and Glass–ceramics from Coal Fly Ash. *Fuel,* v. 78, pp. 271–276.

Barlow, D. K. (2001). Glass Recycling—The UK Scene. In *Recycling and Reuse of Glass Cullet,* ed. R. K. Dhir, M. C. Limbachiya, and T. D. Dyer. London: Thomas Telford, pp. 11–14.

CWC, (1996). Cullet Specifications for Fiberglass Insulation Manufacturing, Best Practices in Glass Recycling BP-GL3-01-03. Seattle, WA: Clean Washington Center, www.cwc.org.

CWC, (1999). Glass Clay Body Flux, GL99-01. Seattle, WA: Clean Washington Center, www.cwc.org.

Dalbey, R. and R. Purser, (1996). Color Modification of Post-Consumer Glass Cullet. Report GL-96-4. Seattle, WA: Clean Washington Center, www.cwc.org.

Dhir, R. K., M. C. Limbachiya, and T. D. Dyer, eds. (2001). *Recycling and Reuse of Glass Cullet.* London: Thomas Telford.

Drescher, H. (2003). Modern Recycling Technologies in Glass: A Survey of the State of the Art. In 63rd Conference on Glass Problems, ed. C. H. Drummond. *Ceramic Engineering and Science Proceedings,* v. 24(1), p. 115–128.

EU (2004). Directive 2004/12/EC of the European Parliament and of the Council of 11 February 2004 amending Directive 94/62/EC on packaging and packaging waste, *Official Journal of the European Union,* v. L47, February 18 2004, p. 26–31.

Gaines, L. L., and M. M. Mintz (1994). *Energy Implications of Glass-Container Recycling,* Argonne National Laboratory report ANL/ESD-18 or NREL/TP-430-5703, Argonne, Illinois.

Geiger, G. (1994). Environmental and Energy Issues in the Glass Industry. *American Ceramic Society Bulletin,* v. 73(2), pp. 32–37.

Gitlitz, J. (2001). Glass Recycling Market Trends, Contamination Problems Discussed. Northeast Recycling Council, September Email Bulletin, Brattleboro, Vermont.

Glüsing, A. K., and R. Conradt (2001). Dissolution Kinetics of Impurities in Recycled Cullet. In *Recycling and Reuse of Glass Cullet,* ed. R. K. Dhir, M. C. Limbachiya, and T. D. Dyer. London: Thomas Telford, pp. 29–41.

GPI, (2004). Glass Packaging Institute, http://www.gpi.org.

Guter, E. (1993). Quality Cullet Is Required for Fiberglass, Too. *Glass Industry,* January, pp. 13–25.

Hannon, B. M. (1972). Bottles Cans Energy. *Environment,* v. 14 (2), p. 11–21.

Hayes, T. L., A. Lame, and R. Lehman, (2001). Glass Beverage Bottles. In *The McGraw-Hill Recycling Handbook,* ed. H. F. Lund, New York: McGraw-Hill, pp. 13.1–13.21.

Integrated Pollution Prevention Control (IPPC). (2000). *Reference Document on Best Available Practices in the Glass Manufacturing Industry.* Seville: European Commission, Institute for Prospective Technological Studies.

ISRI, (2003). Scrap Specifications Circular 2003. Washington, DC: Institute of Scrap Recycling Industries, www.isri.org.

Jackson, P. R., and A. Smith (2001). The Waste Swap. *International Ceramics,* v. 1, pp. 16–19.

Jones, S. (2003). A Must for Any Batch, *Glass International,* July/August, p. 44.

Knox, T. D., and R. C. Brockway (2001). Materials Recovery Facilities. In *The McGraw-Hill Recycling Handbook,* ed. H. F. Lund. New York: McGraw-Hill, pp. 24.1–24.28.

Liaw, C., H. Chang, W. Hsu, and C. Huang (1998). A Novel Method to Reuse Paper Sludge and Co-generation Ashes from Paper Mill. *J. of Hazardous Materials,* v. 58, p. 93–102.

Moulton, L. K., (1973). Bottom Ash and Boiler Slag. *Proceedings of the Third International Ash Utilization Symposium.* U.S. Bureau of Mines, Information Circular No. 8640, Washington, DC.

Queralt, I., X. Querol, A. López-Soler, and F. Plana (1997). Use of Coal Fly Ash for Ceramics: A Case Study for a Large Spanish Powerstation. *Fuel,* v. 76 (8), p. 787–791.

Rapp, D. B., M. A. Dorsey, M. M. Ashton-Patton, and J. E. Shelby, (2004). Decolorization of Amber Glass. In *Advances in Fusion and Processing of Glass III, Ceramic Transactions,* v. 141, p. 323–330.

Rosenthal, J., (2002). Cullet as a Raw Material: Treatment of Factory and Foreign Cullet. *International Glass Journal,* v. 118, p. 22–29.

Schaeffer, H. A. (1999). The Glass Industry in Germany: Environmentally Sound Melting and Recycling of Glass. *Ceramic Engineering Science Proceedings,* v. 20 (1), p. 207–214.

Tyrrell, M. E., and A. H. Goode (1972). Waste Glass as a Flux for Brick Clays, U.S. Bureau of Mines, Report of Investigation 7701.

USDOE (2002). *Energy and Environmental Profile of the U.S. Glass Industry.* Washington, DC: U.S. Department of Energy.

van Oss, H. G. (2003). Slag—Iron and Steel. *Minerals Yearbook.* Reston, VA: U.S. Geological Survey.

Wang, M. C., J. H. Liaw, N. C. Wu, and M. H. Hon (1989). Blast Furnace Slag as a Raw Material for Glass Melting. *Glass Technology,* v. 30(1), pp. 29–32.

8

CHARACTERIZATION AND QUALITY CONTROL

In glass and ceramic manufacturing, precise knowledge of the characteristics of the raw material is critical to ensuring consistency of the process that, in turn, results in higher quality and fewer rejected products. Each processing step, regardless of the product, broadly entails the physical or chemical change of the material(s), and the flow between steps is expected to be smooth. Deviations from the expected characteristics will result directly in processing problems that, at best will cause a brief production of defective material or, at worst, cause damage to expensive equipment or refractory. Constant awareness and control of the quality of raw material is therefore one of the most important aspects of manufacturing. The further up the supply chain that this information is known, the better one will be able to exert control on the process.

The initial point of control is with the vendor from which the material is purchased. A manufacturer will normally have negotiated the desired specifications prior to shipment. When a material is shipped, the vendor should send a certificate of analysis with each shipment, but it should be made clear on what basis the analysis was made. For instance, does the information come from that particular shipment, or from a large volume at the processing plant? What was sampled, and how many analyses were conducted? In some cases, a preshipment sample of the material may be sent to give the manufacturer a chance to test it prior to delivery. The exact procedure will vary with each material and application, but the intent is the same: to know precisely the characteristics of each shipment.

The characteristics of raw materials can be divided into compositional and physical. The *compositional* characteristics include mineralogy and bulk chemistry (including moisture). The *physical* properties of interest can include particle-size distribution, surface area, particle morphology, color, thermal behavior, and bulk density. In this chapter, some of the more general methods used to sample and characterize raw materials are described. There are standard tests, such as those from the American Society for Testing and Materials (ASTM) and the International Organization for Standardization (ISO). These are referenced in this section, but the reader is referred to the specific standard to learn how to conduct the test. Tests that are specific to the whiteware industry are explained in Ryan and Radford (1987). Any standard test should be viewed as a general best practice, and common sense will dictate its applicability to each specific situation.

8.1 INSPECTION AND SAMPLING

When a material arrives at the manufacturing plant, each truckload or rail car should be inspected before releasing the shipment. If the material is in sacks or supersacks, it may be difficult or impossible to inspect each one. Inspection of a delivery needs to be done as quickly as possible, because keeping the hauler waiting to offload the shipment can incur demurrage fees (see Chapter 4). At the same time, for large shipments that are transferred to silos, it is very difficult to make changes after unloading. At minimum, a visual inspection should be done, although it is better to take a sample for a quick screening to get a general idea of particle-size distribution and to inspect for off-color particles. It is good practice to retain a sample from each truck or rail car in case there are problems discovered at a later time (Ninos, 2005).

Once a delivery has been transferred, the manufacturer should know about the characteristics of the shipment. Because it is impossible to know the characteristics of each particle in an entire railcar or truckload, one must rely on the analysis of a sample that is representative of the shipment. Thus, proper sampling is a critical part of any quality-control system. A sample taken from a shipment of material can be either *random,* which is selected in an arbitrary manner, or *representative,* which is taken as a result of a planned strategy. A representative sample may include the selection of a random sample, followed by processing that is designed to increase the chances of the sample accurately representing the full load. A detailed discussion of the theory of sampling is found in Pitard (1993). Sampling procedures specific to the glass industry, but useful as well to ceramic applications, are found in Edwards and Harben (2002) and ASTM standard C429-01 (ASTM, 2002). Guidelines for sampling bulk shipments of lump, shredded, and air-floated clays are found in ASTM standard C322-82 (ASTM, 2002).

The overall goal in obtaining a representative sample is that all portions that make up a raw material shipment must be given equal probability of being collected in the sample. A basic requirement for a representative sample is to take it from a falling stream of material, where all of the stream is taken several times. A general rule of thumb is to collect about 500 to 1,000 grams of sample per ton of material. The success of obtaining a representative samples depends both on the degree of homogeneity of the shipment and on the methods used to collect the sample or samples. The less homogenous the shipment is, the more difficult it is to obtain a representative sample.

Once a sample is collected, it needs to be methodically split to obtain a subsample that can be readily handled for analysis. These processes include *coning* and *quartering,* in which a powdered or granular material is poured into a cone shape, inverted, and poured onto a flat surface several times. The cone is flattened, cut into quarters, and one quarter is retained and coned and quartered further until the desired sample size is attained. Another method is to use a splitting device such as a riffler, which divides a stream of material into two halves by passing it through a battery of chutes with alternating exit directions.

A random sample from a sack, truck bed, or railcar can be taken to make a quick visual inspection and screening test before offloading. However, one must realize that considerable settling and particle size segregation can occur during transportation. If a sample is taken from the margins of a shipment container, such as from the top of a railcar or from the bottom hopper gate, it is likely that such a sample is not representative of the bulk. The user may take core samples from within the load using a cylindrical sampler. However, Edwards and Harben (2002) warn that this type of sampling almost never works satisfactorily. A caveat to this is if the purchaser wants to determine if there is contamination from a prior load in a railcar or truck, a sample near the bottom or in contact with the inside surface of the container may be taken. There is no correct way to sample such loads and from a theoretical point of view it is an unsolvable sampling problem (Pitard, 1993). In the end, the sampling method will be determined by the plant manager depending on experience, time constraints, economics, and the particular material that is sampled.

8.2 MINERAL AND PHASE IDENTIFICATION

Most glass and ceramic raw materials consist of crystalline mineral phases. Therefore, it is important to know the mineralogy and particle morphology of the sample in order to determine the presence of unwanted minerals and to assess any changes in the processing parameters, such as the level of dispersants and other materials used to control properties. The mineralogy can be quickly assessed qualitatively using an optical petrographic microscope, but detailed and quantitative observations are made using X-ray diffraction and scanning electron microscopy. If the raw material is poorly crystallized, amorphous, or vitreous, then mineralogical analysis is not conducted and, rather, the bulk composition is determined (see following section)

Optical Microscopy

Mineralogists routinely use optical microscopes to determine the mineralogy and texture of rock samples. This is done with thin sections of rock mounted on a glass slide that is viewed with transmitted light (i.e., the light is passed through the slide from below). The behavior of the light as it passes through is diagnostic for each mineral. For powdered materials—which raw materials often are—a sample can be viewed with transmitted light by attaching it to a glass slide with epoxy then grinding and polishing to a thin section. Reflected light (i.e., the light is directed from above the sample and reflected into the optics) can be used with thin sections to determine mineral phases and textures (Figure 3.4) and it can also be used to quickly assess granulated or powdered minerals, with the exception of clays and very fine powders. This kind of analysis is useful in scanning for undesirable mineral grains and coatings and to determine particle morphology.

In general, the microscope is most effective for particles that are less than 50 mesh (300 μm) tp greater than100 mesh (150 μm) (Hurlbut and Klein, 1993) but can be used down to 0.2 μm to examine surface topography and microstructure of polished and etched samples (Reed, 1995). The advantage of optical microscopy is that it is relatively inexpensive and does not require specialized facilities or training, although knowledge of mineral phase identification is needed.

X-Ray Diffraction

For very fine particles and for quantitative determination of mineralogy, a sample is analyzed using powder X-ray diffraction (XRD). XRD is based on the interaction between X-rays of a specific wavelength and the atoms in a crystal structure. The X-rays are diffracted at angles and intensities that are characteristic of each individual type of crystalline material. The angle depends on the lattice spacing and the wavelength of the incident X-ray according to Bragg's Law:

$$n\lambda = 2d \sin\theta$$

where n is an integer, d is the lattice spacing, and θ is the diffraction angle.

A powdered sample is mounted in a holder and rotated within the instrument while a fixed beam of X-rays is directed at it. The X-rays are diffracted by the crystal structure and collected by the analyzer. A pattern of the relative intensity of the diffracted beam and the diffraction angle are plotted. The diffraction pattern is compared to reference patterns of known crystalline solids in order to determine the mineralogy of the unknown sample. For a thorough description of the techniques, the reader is referred to Jenkins and Snyder (1996) and Cullity and Stock (2001). XRD is the best method for accurately obtaining the mineralogical composition of a sample, but it requires expensive equipment and a dedicated lab. For manufacturers that do not have XRD equipment, it is common to send samples to a dedicated analytical lab.

Scanning Electron Microscopy (SEM)

For higher resolution imaging of particles and for analysis of single grains, scanning electron microscopy (SEM) is the preferred instrument. SEM uses a narrow beam of accelerated electrons to scan the surface of a sample that has been attached to a mount and coated with a thin layer of conductive material, usually carbon. The coating allows the electrons to be conducted away from the sample; otherwise, a negative charge will build up on the sample. Some of the electrons from the beam are bounced back from the material, or backscattered, while some are absorbed. Absorbed electrons may knock electrons in the material out of their orbitals. These secondary electrons are collected and give information on the topography of the sample, from which a very-high-resolution image is made with magnification up to 20000X. The number of backscattered electrons is dependent on the atomic number of the material, such that higher atomic number elements appear brighter than lower atomic number elements. Thus, a backscattered image shows differences in composition in the exposed surface.

The SEM can be coupled with an energy dispersive X-ray spectrometer (EDS), which is a system that detects the energy of X-rays that are emitted when electrons fill orbital vacancies left by ejected secondary electrons. These X-ray photons are collected and parsed by energy to produce a spectrum to show peaks or intensities that are characteristic of elements present in the material. The height of the peaks is related to the concentration of the element. A wave dispersive spectrometer (WDS) uses crystals and detectors to tune to desired wavelengths of emitted energy. WDS provides accurate quantitative analysis of the material when it is calibrated to known standards. WDS is the basis for electron microprobes, which are routinely used for single mineral analysis.

8.3 BULK CHEMICAL ANALYSIS

There are several methods that can be used to determine bulk chemistry of raw materials. For most oxide-based materials, the common methods include X-ray fluorescence (XRF), inductively couple plasma optical emission spectroscopy (ICP-OES), and inductively couple plasma mass spectroscopy (ICP-MS). These work well for materials that are composed predominantly of metal oxides. For nonoxide materials, such as nitrides, these instruments will not work and these are analyzed using one of several combustion and gas detection methods (Pei et al., 1999).

Historically, major element oxide determinations were done using wet chemical methods that involve fusing the powdered sample with a flux, digesting with acids, and determining the relative proportions of oxides by gravimetry and titration. These methods are currently used where XRF and ICP are not readily available and some standards, such as ASTM C323-56 (ASTM, 2002), still refer to them. A materials supplier may have small XRF systems, which requires a trained technician. The equipment needed for the ICP systems can be expensive to own, operate, and maintain. It is often the case that a company does not need ICP on-site and rather sends out samples to an analytical laboratory on a regular basis. It is a good policy to also send a sample with a known composition with each shipment of unknown materials in order to monitor the reproducibility of the laboratory.

For raw materials that consist predominantly of metal oxides, the bulk composition of a sample is usually reported in terms of the weight percent of the oxide. For iron, it is normally assumed that it is fully oxidized to Fe_2O_3. For nonoxide ceramic materials, such as the nitrides, carbides, and borides, the composition is reported as the weight percent of the element.

X-Ray Fluorescence

X-ray fluorescence (XRF) is one of the most popular techniques for determining the elemental composition of minerals and inorganic materials because it is relatively rapid, accurate, and inexpensive. This method uses a beam of X-rays directed on to the face of a powdered sample that has been compressed into a disk. Some of the incident X-rays have enough energy to eject electrons from inner orbitals of the target material, creating vacancies that are filled by an electron dropping from a higher energy orbit. When this happens, X-rays are emitted with a wavelength that is characteristic of the difference between the energies of the corresponding electron orbitals. The measurement of the characteristic emitted radiation allows the instrument to quantitatively determine the composition of the unknown if it has been calibrated to known material standards of similar composition. For instance, if feldspar is being analyzed, the machine should be calibrated with several feldspar standards of precisely known compositions.

Inductively Coupled Plasma Analysis

Inductively coupled plasma optical emission spectrometry (ICP-OES) and mass spectrometry (ICP-MS) are two very powerful methods for obtaining accurate and precise determination of up to 60 elements simultaneously in rock and mineral materials (e.g., Broekaert, 2002). ICP-OES is often used for major elements, while ICP-MS is better suited for trace and ultra-trace elements. Sample dissolution for both methods is done by three general methods: a fluxed fusion with acid digestion, a microwave bomb, or acid digestion. In the fusion method, the powdered sample is mixed with a lithium borate or carbonate compound and fused in a graphite crucible. The fused sampled is poured into a mixture of acids, usually hydrochloric and nitric. However, if lithium and boron are elements of interest, a different flux needs to be used or the microwave bomb method is used. In the microwave bomb method, the sample powder is mixed with acids without a flux in an encapsulated PTFE container or bomb that is sealed in a high strength vessel. The apparatus is subjected to microwaves, raising the temperature and pressure in the bomb to allow dissolution of the material. For trace elements in silicate materials, hydrofluoric acid digestion is used.

In this method, the sample is first dissolved in hydrofluoric acid and then heated to drive off the silica. The removal of silica can allow for more precise measurements of elements that are present in small amounts.

The sample solutions are injected into the ionization chamber of the machine, which houses a 7000°C argon gas plasma contained in a magnetic field formed by a water-cooled, radio frequency (RF) coil. The solvent and dissolved load are completely atomized and ionized by the plasma. For ICP-OES, the radiation emitted by the excited atoms is passed through a slit to a series of spectrometers or to a single adjustable spectrometer where the wavelength of the photons and their intensity are measured. Each element has a set of characteristic wavelengths, which are selected for measurement of the desired elements in such a way as to reduce the potential for interference between elements. For ICP-MS, the mass of the ions rather than the wavelength of the photons is measured. The ions produced in the ionization chamber are passed into a high-vacuum mass analyzer that uses the charge-to-mass ratio to determine the isotopes that are contained in the sample. The detectors measure the signal intensity for each charge-to-mass ratio value. Like XRF and WDS, quantitative information for both ICP-OES and ICP-MS is based on calibration of the machines to known standard materials that have similar compositions to those that are analyzed.

Volatile Analysis

Raw materials will almost always contain some type of volatile material, either as an integral component or as an undesirable contaminant. The most common volatile component is water. Others are gases included in the material, or compounds that volatilize during heating, such as organic carbon that will burn to carbon dioxide when heated.

There are several terms used to describe water. *Absorbed* water is that which resides in the interstices of particles. This can be removed by squeezing, such as in a filter press. *Adsorbed* water is that which adheres to the grain surface as a thin layer through surface tension. *Constitutional* or intrinsic water resides within the structure of the material, such as the water molecules within a clay or hydroxide mineral. The *water of crystallization* refers to the water molecules surrounding stable hydrated minerals, such as borax pentahydrate ($Na_2B_4O_7 \cdot 5H_2O$). Free moisture is water that is absorbed and adsorbed and can be measured by heating a sample to 110°C for at least 24 hours in a drying oven (e.g., ASTM C324-01; ASTM, 2002). The mass loss that occurs can be attributed to the evaporation of this water.

Discriminating between the different "types" of water is important for both economic and technical reasons. In terms of economics, undesired moisture increases the cost of transportation and may require additional energy for dewatering or drying. In addition, if purchasing a material by weight, one does not want to pay for water. On the technical side, knowing how much and where water is incorporated into the material is needed to predict the behavior of the material during processing, forming, and heating. As an example, materials that are too wet can cause fouling during conveying into silos or mixers.

The loss of all volatiles after the removal of free moisture is referred to as loss on ignition (LOI). This is determined by weighing a dried powder contained in a crucible and then placing the crucible into a furnace heated up to at least 1000°C for several hours. Afterward, the crucible is weighed and the difference between the original mass and the mass after heating is calculated. The mass loss can be attributed to the intrinsic water, water of crystallization, and the combined other materials that can be volatilized at high temperature such as organic carbon, sulfur, fluorine, chlorine, and carbon dioxide. The specific composition of each volatile constituent can be measured using an instrument that used solid-state infrared spectrometers and ion selective electrodes.

Nonoxide ceramics, such as nitrides and carbides have their carbon and nitrogen analyzed by heating the sample to combustion and then the proportions of the volatilized constituents are determined. Carbides are heated in an oxidized atmosphere so that the carbon combines with the oxygen to form carbon dioxide, which can be measured using infrared absorption (ASTM, C1494-01; ASTM, 2002). Alternatively, the material is heated and the carbon dioxide is absorbed by $Ba(ClO_4)_2$ and determined using coulometry (ISO 9286; ISO, 1997) The nitrogen and oxygen content of silicon nitride is determined by heating the sample in an inert, helium atmosphere at greater than 1900°C, which is sufficient to release oxygen, nitrogen and hydrogen from the sample. The nitrogen is released as elemental nitrogen gas, and the oxygen combines with carbon of the graphite crucible to form carbon monoxide. The gases are measured by thermal conductivity (ASTM, C1494-01; ASTM, 2002). For metal carbides and nitrides, after the volatile components have been removed, the residue can be analyzed by ICP methods to determine the metal content, though it must be calculated in its elemental form rather than the oxide form.

8.4 THERMAL ANALYSIS

Thermogravimetric and Differential Thermal Analysis

Thermogravimetric analysis (TGA) and differential thermal analysis (DTA) techniques are usually done in tandem to determine the mass and phase changes, respectively, that occur as a material or mix of materials is heated or cooled at a desired rate (usually 10°C per minute). This is done by placing the unknown sample and a known, inert reference material (usually alumina powder) in contact with a set of thermocouples in either an oxidizing, inert, or reducing atmosphere. For TGA, changes in the sample mass are recorded as temperature is changed. For DTA, the change in temperature of the unknown is measured relative to the inert reference. Differences between the two will indicate when endothermic or exothermic reactions occur at specific temperatures. That is, the temperature change should be approximately the same for both the unknown and the reference until a reaction occurs in the unknown sample that either requires energy in excess of heat capacity (endothermic) or evolves energy (exothermic). Figure 8.1 shows an example of kaolinite that exhibits an endotherm around 600°C due to dehydroxylation or the removal of constitutional water. Alumina is used as a reference because there are no reactions during heating and its heat capacity is similar to most materials used in glass and ceramics. Care should be taken not to have an alumina reference with high surface area, because any adsorbed water will dehydrate on heating and result in an erroneous apparent exothermic reaction to the unknown (Carty, 2005).

The DTA-TGA method is often used to determine the behavior of a batched mixture of material during heating and cooling. Specifically, one can identify the temperature and respective magnitudes of critical reactions, such as decomposition, phase transitions, and binder burnout. However, DTA-TGA is not commonly used as a regular tool for characterizing raw materials. This is unfortunate because it is a rather straightforward and uncomplicated procedure and has the potential to be used to identify the mineralogy of a sample. For instance, the characteristic endotherm of kaolinite at 600°C can be used to test for the presence of other clays if other endotherms are displayed.

Fired Color and Refractoriness

For whiteware ceramics, the fired color, the refractoriness (for silica minerals, alumina, and fireclay), and the fluxing ability (for feldspar

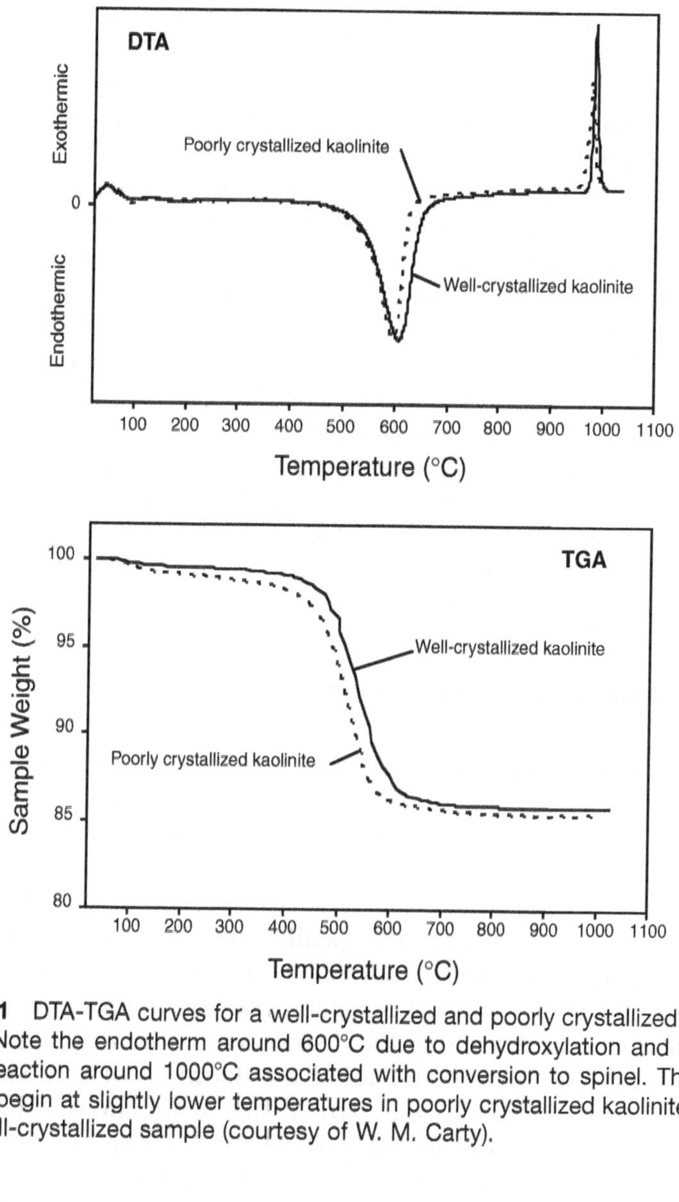

Figure 8.1 DTA-TGA curves for a well-crystallized and poorly crystallized kaolinite sample. Note the endotherm around 600°C due to dehydroxylation and the exothermic reaction around 1000°C associated with conversion to spinel. The phase changes begin at slightly lower temperatures in poorly crystallized kaolinite relative to the well-crystallized sample (courtesy of W. M. Carty).

and nepheline syenite) are important characteristics. Although these can be monitored by bulk chemical analysis, the firing of test samples is a rather quick method (Ryan and Radford, 1987). The refractoriness of a material has been characterized using pyrometric cone equivalents (PCE) for more than a century. In this method, a cone or slim pyramid of test material is fabricated and heated with a set of standard cones

(ASTM C24-01; ASTM, 2002). A binder might need to be added to nonplastic materials to keep the cone cohesive. The use of standard cones allows precise metering of the temperature by comparing the highest number (i.e., highest temperature) cone with a tip that has slumped to the base.

As heat is applied, the tip of the cone begins to bend due to gravity-induced slumping, first slowly and than relatively rapidly as bending continues. It normally takes 15 to 25 minutes for a cone to bend once it starts. When the cone tip slumps to the base, it is considered properly fired, and this is the point for which temperature equivalents are determined (Orton, 2005). Standard charts for several heating rates are provided by the cone manufacturer. Popular cones include Orton cones (United States), Seger cones (Germany), and Staffordshire cones (United Kingdom). As an alternative to cones, bars of material can be extruded and placed in a furnace with both ends supported. The temperature at which deformation and slumping occurs is then observed to obtain an idea of refractoriness of the material.

8.5 PARTICLE-SIZE DISTRIBUTION

Because most glass and ceramic raw materials consist of powders and grains, the determination of the particle-size distribution is extremely important. For instance, sand used to make a float glass must be within a specified particle-size range because material that is too fine may cause dusting, and particles too large will remain unmelted and form stones in the glass. Particle size usually refers to diameter, suggesting that all particles are spherical. However, because few particles are spherical, diameter is subject to interpretation and this should be kept in mind whenever discussing the distribution of sizes (Orr, 1978).

The size of particles and the relative proportions of each size within a sample are referred to as the particle-size distribution, which is usually reported graphically as the particle size versus cumulative mass percent (denoted as larger than or smaller than) on a log or semilog plot. The relationship can be expressed as either a curve of the cumulative distribution or as a histogram (Figure 8.2). The percentage of the total mass is calculated and plotted on a semilog plot to yield a particle size distribution curve. Another method of displaying the particle-size distribution is to plot the cumulative weight percent versus percent greater than the opening size.

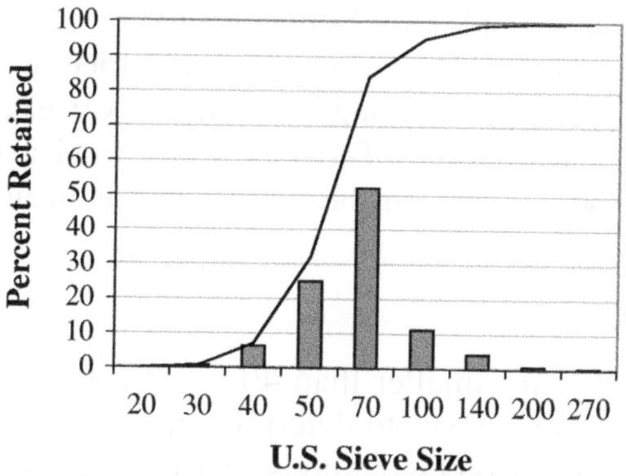

Figure 8.2 A graph showing the particle-size distribution of silica sand as reported on a mesh-size versus percent-retained plot. The bars show the percent retained for each sieve, and the line shows the cumulative.

Following are a variety of instruments and techniques that are used to determine the particle size distribution of granulated materials and powders (Allen, 1993; Reed, 1995). However, the results obtained can depend on the instrument and technique used (Jillavenkatesa et al., 1999), so it is best to ensure that any comparisons are consistent with the methodology.

Sieving

The simplest and most common particle size distribution analysis is done with sieves. This method is useful for particles between 2.5 cm and 40 μm in diameter as agglomeration of particles is a problem for sizes below 40 μm (Reed, 1995). The analysis is conducted using a series of cylindrical sieves that are stacked so that those with larger openings are successively set above those with smaller openings. A pan at the bottom catches all of the undersize material. The material retained on a given sieve is considered to be larger than the size of that sieve opening and smaller than the sieve opening directly above. The sieves can be designated by the number of holes per inch or by the size of the sieve openings. See Table 3.1 for standard sieve sizes.

For a dry analysis, the stack of sieves is placed on an agitator that rotates and taps the column until all of the particles have been properly sorted, a process that takes up to 15 minutes. The mass of material retained on each sieve is then measured and recorded. Clays (ASTM C325-81; ASTM, 2002) and nonclay materials that are not free flowing such as alumina, talc, and pyrophillite (ASTM C371-89; ASTM, 2002) use a wet sieve process. The material is first dried and then dispersed in a slurry by mechanical mixing. The slurry is washed through a set of stacked sieves with water. The sieves are dried and the weight of residue on each screen and on the bottom pan is measured.

Sedimentation

When the particles are smaller than 40 to 30 μm, it is essentially impossible to use testing sieves to determine the particle size of the powder. There are several other methods that can be used to analyze the smaller particles. One of these is *sedimentation,* a process that is seen widely in nature where sand, silt, and clay particle entrained in moving water will settle out at rates directly proportional to particle size. This relationship was first expressed by the British scientist Sir George Stokes, who came up with a mathematical expression for the settling velocity of spherical particles in a fluid. The expression is known as the Stokes equation, which can be arranged to solve for velocity:

$$v = \frac{d^2(\rho_p - \rho_f)g}{18\eta}$$

where d is the particle diameter; η is the viscosity of the suspending fluid; v is the velocity of the particle; ρ_p and ρ_f are the densities of the powder and the suspending fluid, respectively; and g is the gravitational constant or the force from a centrifuge. The Stokes equation assumes that the particle is spherical and no turbulence occurs during descent in the fluid. Since most particles are not spherical, the diameter of a particle determined using Stokes settling is referred to as an equivalent spherical diameter. It is apparent from the Stokes equation that the settling velocity is directly proportional to the square of the particle diameter. From the velocity, one can determine the time required for a particle of a certain size to settle a given distance and a particle-size distribution can be expressed.

The simplest method of using sedimentation to determine particle-size distribution of clays between 44 μm and 0.1 μm is with hydrometers (e.g., ASTM C775-79; ASTM, 2002). However, it is common to use one of the available commercial particle size analyzers that use X-rays to determine the concentration of particles within the fluid. These instruments can measure particles as small as 0.1 μm. The analysis is done by projecting a beam through the mixed suspension and measuring the amount of attenuation, or loss of energy, as particles absorb or reflect the radiation. It is necessary first to break up and eliminate agglomerates by dispersing the powder in the fluid (usually water) to make a suspension of about 2.5 volume percent. Materials that chemically react with water should be dispersed in an alternative medium. The particles in the suspension will absorb some of the X-rays so the higher the concentration of particles, the higher the attenuation of the beam and the lower the measured intensity. The suspension is allowed to settle and the intensity of the transmitted beam is measured over time. During the sedimentation process, the largest particles will first settle out and gradually the finer particles fall until the fluid is clear and the intensity of the beam matches that of the fluid prior to the addition of the powder.

Electric Sensing Zone Method

A technique originally developed for blood cell counting can also be used to determine the number and size of nonconductive mineral particles (Orr, 1978). This method, referred to as the *electrical sensing zone* or *Coulter method,* is based on the changes in electrical resistance produced by passing a suspension of particles in an electrolyte between two electrodes. As a particle passes through the opening or the sensing zone, each particle displaces a volume of electrolyte that is proportional to its size. The change in resistance results in a voltage pulse that can be related to the particle volume. The technique is independent of the density and shape of the particle. The size of the opening between the electrodes will affect the accuracy of the method, so powders with large particle-size distributions may require more than one orifice size. With this method it difficult to measure particles that are less than 2 μm (Rawle, 2005).

Light Scattering Analysis Technique

The particle size distribution for some nonplastic materials such as quartz and alumina can be determined using a light-scattering technique, often referred to as low-angle laser light scattering (LALS or LALLS) (ASTM C1070-01; ASTM, 2002: ISO 13320-1; ISO, 2005). In this method, particles with diameters ranging from 500 μm to 0.1 μm are dispersed in a liquid and passed in front of a laser beam. The light is scattered in a manner that is inversely proportional to particle size and with an intensity directly proportional to particle size. Detectors collect the scattered light and convert the optical signals to electrical pulses. The data are converted to a particle-size distribution using Fraunhofer diffraction or Mie scattering, or a combination of both. Fraunhofer diffraction is the simplest method of determining particle size using LALS and can be used for particles $>1\,\mu m$ (Kippax, 2004). Mie theory is used for particles that are smaller than the wavelength of the incident light. The information is processed with software models (supplied by the instrument manufacturer) assuming spherical particles and the results are reported as equivalent spherical diameters.

8.6 POWDER DENSITY

The apparent density of a material is often given in the specifications of dry, granular, or powdered materials. The volume of a powder can be determined using a helium pycnometer. By dividing the mass of the powder by the calculated volume, the apparent powder density can be determined. The helium pycnometer method determines the volume of a sample by measuring the pressure change associated with the volume of gas excluded by the solid. The following equation is used to calculate the excluded volume V_p:

$$V_P = V_C - V_R\left(\frac{P_1}{P_2} - 1\right)$$

where P_1 and P_2 are the measured pressures and V_C and V_E are the cell volume and expansion volumes, respectively. The manufacturer provides values for V_C and V_E. It is assumed that there are no closed pores and that the helium gas can access all pore space.

8.7 SPECIFIC SURFACE AREA

The specific surface area of a material refers to the total exposed surface of a particle and is defined as the area divided by the powder mass in m^2/g. This is an important parameter because as the specific surface area increase, the reactivity of the material increases. The traditional technique for determining specific surface was developed by Brunauer, Emmett, and Teller, known as B.E.T., using the adsorption behavior of nitrogen on a substrate. The analysis is done at the temperature of liquid nitrogen. A more complete description of the method can be found in Lowell and Shields (1991). For samples with low specific surface area, krypton and argon are used instead of nitrogen to improve accuracy (Giesche, 1999).

8.8 QUALITY MANAGEMENT SYSTEMS

Quality management systems refer to an organized set of methods instituted by the manufacturer to address quality control of the entire production process. The most widely recognized is the ISO quality assurance system *ISO 9000,* which is a set of standards that is accepted around the world. Vendors that are certified by the ISO have implemented a quality system that incorporates a set of specified requirements in order to comply with the standards. The implemented quality system is accredited by an external auditor to maintain the objectivity of the program. The goal of the system is to assure the buyer of the raw materials that their needs and expectation on the specifications are being met by methodical and internationally accepted guidelines. In addition, the vendor has a means of continually improving its performance through audits and reviews. Many ceramic manufacturers not only demand materials from companies that are ISO 9000 certified, but also are certified themselves (e.g., Cargill, 2001). In addition to ISO 9000, some companies comply with the ISO 14000, which is a set of standards concerned with environmental management such that the supplier is minimizing environmental impacts and working to continually improve its environmental performance.

REFERENCES

ASTM (2002). Glass; Ceramic Whitewares, v. 15.02. West Conshohocken, PA: The American Society for Testing and Materials.

Allen, T. (1993). *Particle Size Measurements,* 4th ed. London: Chapman and Hall.

Broekaert, J. A. C. (2002). *Analytical Atomic Spectrometry with Flames and Plasmas.* Weinheim, Germany: Wiley-VCH.

Cargill, G. (2001). ISO Certification: ISO 9001:2000—Beginning the Transition. Ceramic Industry, www.ceramicindustry.com.

Carty, W. M., (2005). Personal communication.

Cullity, B. D., and S. R. Stock (2001). *Elements of X-Ray Diffraction,* 3rd ed. Upper Saddle River, NJ: Prentice-Hall.

Edwards, G. H., and P. W. Harben (2002). Sampling Glass Raw Materials. In 63rd Conference on Glass Problems, ed. C. H. Drummond. *Ceramic Engineering and Science Proceedings,* v. 24(1), pp. 175–193.

Edwards, G. H., and P. W. Harben (1999). *Sampling Bulk Industrial Minerals and Ores; A Practical Approach.* Surrey, UK: Industrial Minerals Information Ltd..

Giesche, H. (1999). Surface Area and Porosity Characterization by Gas Adsorption: An Overview. In *Advances in Process Measurements for the Ceramic Industry,* ed. A. Jillavenkatesa and G. Y. Onoda. Westerville, OH: American Ceramic Society, pp. 69–86.

ISO (2005). International Organization for Standardization, http://www.iso.org.

Jenkins R., and R. L. Snyder, (1996). *X-Ray Powder Diffractometry.* New York: Wiley-Interscience.

Jillavenkatesa, A., V. A. Hackley, and G. Y. Onoda, (1999). Identification and Characterization for a Model Powder System for Assessment of Particle Size Measurement Techniques and Procedure. In *Advances in Process Measurements for the Ceramic Industry,* ed. A. Jillavenkatesa and G. Y. Onoda. Westerville, OH: American Ceramic Society, pp. 29–56.

Kippax, P. (2004). Measuring Particles with Laser Diffraction. Ceramic Industry, www.ceramicindustry.com.

Klein, C., and C. S. Hurlbut Jr., (1993). *Manual of Mineralogy,* 21st ed. New York: John Wiley and Sons.

Lowell, S., and E. Shields (1991). *Powder Surface Area and Porosity.* 3rd ed. London: Chapman Hall.

Ninos, N. (2005). Personal communication.

Pei, P., L. Lin, and G. Y. Onoda (1999). Chemical Analysis of Carbon, Nitrogen, and Oxygen in Silicon Nitride. In *Advances in Process Measurements for the Ceramic Industry,* ed. A. Jillavenkatesa, A. and G. Y. Onoda. Westerville, OH: American Ceramic Society, pp. 127–139.

Pitard, F. F. (1993). *Pierre Gy's Sampling Theory and Sampling Practice: Heterogeneity, Sampling Correctness, and Sampling Practice,* 2nd ed. Boca Raton, FL: CRC Press.

Orr, C., Jr., (1978). Physical Characterization Techniques for Particles. In *Ceramic Processing Before Firing,* ed. G. Y. Onoda Jr. and L. L. Hench. New York: John Wiley and Sons.

Rawle, A. (2005). Basic Principles of Particle Size Analysis. Malvern Instruments, www.malverninstruments.com.

Reed, J. S. (1995). *Principles of Ceramics Processing,* 2nd ed. New York: Wiley-Interscience.

Ryan, W., and C. Radford (1987). *Whitewares Production, Testing and Quality Control.* Oxford, UK: Pergamon Press.

GLOSSARY

Alaskite a pegmatitic granite that is composed of 45 percent plagioclase, 25 percent quartz, 20 percent potassium feldspar (microcline), and 10 percent muscovite mica; used an ore rock for feldspar

Acheson process used to make silicon carbide in which an electrical current is passed through a mixture of high purity quartz and coke (carbon)

Albite the sodium-rich endmember of the plagioclase and alkali feldspars with a nominal composition of $NaAlSi_3O_8$

Amblygonite a lithium phosphate ore mineral with a nominal composition of $LiAlPO_4(F,OH)$; montebrasite has a similar composition

Anatase crystalline TiO_2, polymorphous with rutile and brookite

Andalusite the orthorhombic polymorph of Al_2SiO_5; the other polymorphs of this composition are kyanite and sillimanite

Anorthite the calcium-rich endmember of the plagioclase feldspars with a nominal composition of $CaAl_2Si_2O_8$

Anorthosite a rare rock consisting predominantly of plagioclase feldspar; a metamorphosed variant is called metanorthosite

Aplite a fine-grained, low-iron granitic rock with a sugary texture; used as a mixed source of alkali feldspar and quartz

APT ammonium paratungstate

Aragonite *see* Calcite

Arkosic a general term to describe a sand or sandstone with at least 25% feldspar

Asbestiform refers to the habit of some minerals, such as chrysotile, that easily separate into strong, thin fibers

Attrition scrubbing the process of washing particles by intensively mixing at slurry densities of 70% to 80% solids, forcing individual particle abrasion resulting in surface cleaning and disintegration of agglomerates. Often used in the removal of iron mineral stains from sands

AZS Fused alumina-zirconia-silica made by melting a mix of zircon sand and alumina

B.E.T. refers to a technique for determining specific surface developed by Brunauer, Emmett, and Teller

Baddeleyite a zirconia mineral with a nominal composition of ZrO_2

Ball Clay *see* Plastic clay

Bastnäsite a fluorocarbonate rare earth mineral with a general composition of $(Ce,La,Y)CO_3F$

Bauxite a high-aluminum laterite formed by the intense weathering of rock, particularly in flat, tropical areas

Bayer process used to manufacture alumina by digesting bauxite in a caustic solution with subsequent crystallization and calcination of purified aluminum hydroxide

Bentonite primary deposit, where the weathering of silicic volcanic ash deposits alters the amorphous silica and alkali rich ash particles into a mix of clays that is predominantly montmorillonite

BFA refers to brown fused alumina produced by melting and reducing calcined bauxite in an electric arc furnace

Biotite a common mica mineral with a nominal formula of $K(Fe,Mg)_3AlSi_3O_{10}(F,OH)_2$

Borax a sodium tetraborate decahydrate mineral with a nominal formula of $Na_2B_4O_7 \cdot 10H_2O$; the naturally occurring form is called tincal

Boric acid a trihydrate boric oxide with a nominal composition of $B_2O_3 \cdot 3H_2O$ or $B(OH)_3$

Calcite the common calcium carbonate mineral with a composition of $CaCO_3$; aragonite has the same composition but will eventually invert to calcite

Caldasite a mixture of zircon and baddeleyite mined in Brazil

Carbonatite a carbonate rock of magmatic origin; an important source of rare earth minerals

Cation exchange capacity (CEC) refers to the total amount of cations that can be retained by a given amount of material and is reported either as moles per unit mass, or more often in milli-equivalents per 100 grams (meq/100g)

Caustic soda sodium hydroxide

CBN cubic boron nitride

Chalcedony a cryptocrystalline, microscopically fibrous variety of quartz; chert and flint are often composed of chalcedony

Chalk a soft, fine-textured limestone of marine origin consisting primarily of calcium carbonate tests (internal skeletons) of planktonic animals such as foraminifer and coccoliths

Chamotte calcined clay or shale used as a refractory aggregate; also called grog

Chemical boosting in glass manufacturing, refers to the addition of materials, such as lithia, that can lower melting times and furnace temperatures, resulting in a higher pull rate and lower specific energy consumption

Chert a hard, dense microcrystalline to cryptocrystalline siliceous sedimentary rock composed of chalcedony and may contain opal (amorphous hydrated silica)

Chlorite a 2:1 clay group with a wide range of compositions given as $(Mg,Fe,Mn,Al)_{12}(Si,Al)_8O_{20}(OH)_{16}$

Chromite $FeCr_2O_4$

Clay hydrous aluminosilicate sheet minerals that characteristically have fine particle size and plasticity; also refers to a sediment particle size of less than 2 microns

Colemanite the most common calcium borate mineral with a nominal composition of $Ca_2B_6O_{11} \cdot 5\ H_2O$

Comminution the reduction in particle size of a material; pulverization

Constitutional water water molecules that reside within the structure of a material

Cornish stone a mixture of kaolin, quartz, and feldspar formed by the weathering of granite; also known as Cornwall stone

Corundum crystalline alumina mineral Al_2O_3; sapphire and ruby are gem varieties

Cristobalite a high temperature polymorph of SiO_2

Cullet the general term used for broken scrap glass that comes from either pre-consumer or post-consumer sources

Datolite a calcium borosilicate mineral with a nominal composition of $CaBSiO_4(OH)$

Dead-burned dolomite dolomite that has been calcined at 1800°C

Dead-burned magnesia magnesite that has been calcined to 1450°C; periclase

Demurrage a fee levied by the shipping company if the buyer does not unload the shipment by a specified date

Desliming removal of the very fine particles (slimes) from a slurry

Diagenesis a general term to describe the post-depositional compaction and cementing of sediments

Diaspore an aluminum hydroxide mineral, diamorphous with boehmite and a major constituent of some bauxites

Diaspore clay a fireclay that is composed of a mix of diaspore and kaolinite

Diatomite a soft, siliceous sedimentary deposit consisting of the silica tests (internal skeletons) of diatoms, a phytoplankton; also referred to as kieselguhr

Dickite a kaolinite mineral

Dioctahedral refers to clay minerals that within the unit cell have two of three available octahedral sites filled with cations

Dolomite a mixed calcium-magnesium carbonate mineral with a nominal composition of $CaMg(CO_3)_2$

Dolostone a sedimentary rock composed of dolomite

DTA differential thermal analysis; used to determine phase changes during heating and cooling

Electrical sensing zone a method of particle size analysis based on the changes in electrical resistance produced by passing a suspension of particles in an electrolyte between two electrodes; also known as the Coulter method

Evaporite minerals or mineral assemblages (rocks) that are deposited after the evaporation and subsequent precipitation of dissolved ions

Ex works refers to the cost of the raw material at the vendor's site

Feldspar a group of framework silicate minerals that can be described by three compositional endmembers: anorthite, albite, and orthoclase

Ferromagnetic extremely high magnetic permeability

Fire clay a refractory clay capable of withstanding high heat without deforming; used synonymously with refractory clay

Flint a dark type of chert; more often used in ceramics as a general term for finely ground quartz; flint glass refers to clear glass

Flint clay a hard, compact kaolinitic clay with a pronounced conchoidal fracture surface when broken; it is used in refractories and the term is often used synonymously with fireclay

Fly ash the smaller ash particles recovered from the exhaust stream of a combustion process, such as a coal-fired boiler

FOB acronym for Free On Board, which refers to the price for which the seller undertakes the risk and expense to load the goods onto a carrier at a specified point and title passes to the buyer after loading

Free moisture water that is absorbed and adsorbed and can be measured by heating a sample to 110°C for at least 24 hours

Froth flotation a mineral separation method, based on surface chemistry properties, in which slurried minerals are selectively attached to bubbles after addition of chemicals to the slurry

Fuller's earth a mix of calcium montmorillonite, attapulgite and sepiolite; used in some catalyst carriers

Fused quartz refers to glass that is produced by melting naturally occurring crystalline quartz; formerly used to clear silica glass

Fused silica refers to glass that is produced by melting purified silica; formerly used to describe opaque silica glass

Gangue the minerals or rock that have no economic value

Gannister term used for a type of fine-grained sandstone that is associated with coal deposits; also spelled ganister

Garnet a group of silicate minerals with the general formula $A_3B_2(SiO_4)_3$ where A is commonly Ca, Mg, Fe^{2+} and B is commonly Al, Fe^{3+}

Gibbsite an aluminum trihydroxide mineral with a composition of $Al(OH)_3$ or $Al_2O_3 \cdot 3H_2O$; an important mineral in some bauxites

Granite a silica rich, intrusive (plutonic) rock made of quartz, feldspars, and mica

Grizzly screen a set of parallel steel bars used to screen out very course material

Grog *see* Chamotte; can also refer to ground rejected or recycled ceramic material that is used as a raw material

Halloysite a kaolinite mineral that can incorporate interlayer water molecules; occurs as coils, tubes, and fibers within kaolinite plates

Hematite an iron oxide mineral with the composition Fe_2O_3

Hornblende the most common group of amphibole minerals with a general composition of $Ca_2(Mg, Fe, Al)_5(Al, Si)_8O_{22}(OH)_2$

Hydrosizer a particle size classifier that uses a current of water directed up from the bottom of a vessel to separate fine and coarse particle size fractions

ICP inductively coupled plasma spectrometry, a method of determining the bulk composition of a sample

Illite a clay group related to mica with 2:1 structure and variable amounts of potassium in the interlayer site

Ilmenite $FeTiO_3$

Industrial minerals a general term for economically valuable natural minerals, excluding metallic minerals and fossil fuels

Intermodal transportation the combination of two or more methods of moving freight in order to reach distant market and distribution points

Ion adsorption lateritic crusts laterites rich in adsorbed rare earth elements

Kaolin refers to a high-grade, white-firing clay that is made predominantly (85% to 95%) of kaolinite minerals, including halloysite; used synonymously with the term China clay

Kaolinite a 1:1 clay mineral with a nominal composition of $Al_2Si_2O_5(OH)_4$

Kyanite the orthorhombic polymorph of Al_2SiO_5; the other polymorphs of this composition are sillimanite and andalusite

Lascas Portuguese for "chips", refers to natural quartz crystal pieces used in the production of synthetic quartz crystal manufacturing

Laterite a highly weathered subsoil rich in oxides and hydroxides of aluminum and iron

Lepidolite a lithium phyllosilicate (mica) with a nominal composition approximated by $KLi_2AlSi_4O_{10}(F,OH)_2$

Leucoxene a general term for a fine-grained alteration product of ilmenite consisting of rutile and partly sphene ($CaTiSiO_5$)

Lignite a soft, brownish-black coal that is intermediate between peat and sub-bituminous coal

Limestone a relatively broad term for a rock that consists of greater than 50 percent calcite or calcite and dolomite

Limonite a mixture of hydrated iron oxide minerals, mostly goethite (FeO[OH])

Lithification the conversion of sediment into solid rock through compacting and cementing; similar to diagenesis

Loparite a mixed metal oxide with a composition of $(Ce,Na,Ca)_2(Ti,Nb)_2O_6$ that is found in silica undersaturated rocks

Loss on ignition (LOI) loss of all volatiles after the removal of free moisture, i.e. after a sample is heated to 110°C for 24 hours

Magnesite a magnesium carbonate mineral with a nominal composition of $MgCO_3$

Magnetite a strongly magnetic spinel mineral with a nominal composition of $(Fe,Mg)Fe_2O_4$

Methylene blue index refers to the milliequivalents of methylene blue dye adsorbed per 100 grams of clay

Microcline *see* Orthoclase

Middlings rock particles that contain both ore and gangue

Mineral a naturally occurring, solid material that has characteristic crystal structures and defined compositions

Monazite a rare earth phosphate mineral with a nominal composition of $(Ce,La,Th)PO_4$

Montebrasite *see* Amblygonite

Montmorillonite a dioctahedral smectite mineral in which some of the Al^{3+} of the octahedral sites in the pyrophyllite structure is occupied by a divalent cation such as Mg^{2+} ; sodium montmorillonites are swelling and calcium montmorillonites are non-swelling

Mullite an aluminosilicate mineral with a general formula of $Al_3Si_2O_{10}$; it is not found in economical concentrations and is commercially synthesized by reacting alumina with silica

Muscovite a common mica mineral with a nominal formula of $KAl_3Si_3O_{10}(OH)_2$

Nacrite a kaolinite mineral

Nahcolite the natural sodium bicarbonate mineral with a nominal composition of $NaHCO_3$

Nepheline a sodium-rich feldspathoid mineral with a nominal composition of $Na_3(Na,K)[Al_4Si_4O_{16}]$.

Nepheline syenite a relatively rare, silica-undersaturated rock (contains no free quartz) composed of alkali feldspars and the mineral nepheline

Opal a hydrated silica mineral with a nominal composition of SiO_2-nH_2O

Ore naturally occurring material, usually mineral, that can be economically extracted

Orthoclase the potassium-rich endmember of the alkali feldspars with a nominal composition of $KAlSi_3O_8$; microcline has the same composition; sanidine is the high temperature structure of the orthoclase composition

Orthoquartzite a sedimentary rock composed of greater than 95% quartz grains cemented with silica

OSHA the U.S. federal government's Occupational Safety and Health Administration

Palygorskite a chain lattice clay mineral with a nominal composition of $(Mg,Al)_2Si_4O_{10}(OH)\cdot 4(H_2O)$; used synonymously with the term attapulgite

Paramagnetic having a small positive magnetic susceptibility; paramagnetic materials are attracted to a magnet

Pegmatite an exceptionally coarse-grained igneous intrusive rock that represents the last portion of a granite to crystallize.

Periclase a magnesium oxide mineral with a composition MgO

Petalite a lithium aluminosilicate mineral with a nominal composition of $LiAlSi_4O_{10}$

Pisolitic describes a bauxite deposit consisting of spherical aggregates occurring in concentric bands

Place value a qualitative assessment that takes into account the monetary value for the commodity and the costs of transporting it to market

Placer a sedimentary deposit that concentrates heavy minerals

Plastic clay a clay than can be easily molded and composed primarily of 50% to 70% kaolinite with the remainder consisting of illite, mica, smectite, and quartz

Playa Spanish for "beach", but here refers to a the lowest part of an arid, undrained basin

Polymorph minerals that have identical compositions but different crystal structures

Pyrite an iron sulfide mineral with a composition FeS_2; fool's gold

Pyrometric cone a pyramid made of a material for use in determining its refractoriness relative to a set of reference cones

Pyrometric cone equivalent (PCE) the number of the pyrometric reference cone that touches the support base simultaneously with a cone of the tested material

Pyrophyllite dioctahedral 2:1 sheet silicate with a structure similar to muscovite mica with a nominal formula of $Al_2Si_4O_{10}(OH)_2$

PZT lead zirconate titanate

Quartz the crystalline silicon dioxide (SiO_2) polymorph that is stable at ambient conditions; the most common mineral in the earth's crust

Quartz inversion the sudden, displacive transformation from α-quartz to high or β-quartz at 573°C

Quartzite a hard, metamorphosed quartz sandstone; an orthoquartzite is a quartz sandstone cemented with silica

Quicklime calcium oxide (CaO) made by calcining limestone to 1000°C

RASC an acronym for Refractory Aggregate Super Calcined, a high quality calcined bauxite from Guyana and Suriname

REE rare earth element

Refractory clay *see* Fire clay

Rhyolite a fine-grained volcanic rock with the composition of granite; usually consists of quartz (or cristobalite) and alkali feldspar (sanidine) in a glassy to cryptocrystalline matrix

Rutile a titanium oxide mineral with a composition TiO_2; a major ore of titanium and an undesirable detrital mineral in quartz sands

Salar term used in South America and the southwestern United States to describe a salt flat or salt basin

Sanidine the high temperature potassium feldspar with the composition of orthoclase

SEM scanning electron microscopy

Sepiolite a fibrous, hydrated magnesium silicate clay with a nominal composition of $Mg_4Si_6O_{15}(OH)_2 \cdot 6(H_2O)$

Sericite a white, fine-grained potassium mica similar to muscovite

Serpentine magnesium and iron silicate mineral group with a nominal composition of $Mg_3Si_2O_5[OH]_4$

Shale a fine-grained sedimentary rock formed by the compaction and clay, silt, or mud

Shortite a mixed sodium and calcium carbonate with a nominal composition of $Na_2CO_3 \cdot 2CaCO_3$

Siderite an iron carbonate mineral with an endmember composition of $FeCO_3$; it can form a solid solution with magnesite ($MgCO_3$)

Silica silicon dioxide

Silicosis a noncancerous but nonreversible and sometimes fatal lung disease caused by the chronic inhalation of crystalline silica

Sillimanite the orthorhombic polymorph of Al_2SiO_5; the other polymorphs of this composition are kyanite and andalusite

Skarn an ore deposit formed by the metamorphism of limestone or dolostone; source for wollastonite

Slag the glassy material that is produced during the smelting of metals such as iron and steel

Slaked lime calcium hydroxide or hydrated quicklime

Slimes refers to the finest particles suspended in a slurry

Smectite clay minerals that are 2:1 dioctahedral or trioctahedral, which can be described by isomorphous substitution of cations between the end members of pyrophyllite $Al_2Si_4O_{10}(OH)_2$ and talc $Mg_3Si_4O_{10}(OH)_2$

Soda ash sodium carbonate, the primary flux in soda-lime-silicate glasses; derived from trona, natural brines, and the Solvay process

Solid solution the complete free substitution of similar elements between two compositional endmembers of a mineral

Solvay process used to synthesize soda ash (sodium carbonate) from ammonia, sodium chloride, and limestone

Spodumene a lithium aluminosilicate mineral with a nominal composition of $LiAlSi_2O_6$; a significant ore for lithium compounds

Steatite a compact, massive rock consisting primarily of talc and traditionally used to make insulators; synthetic steatite is made from talc, kaolin, and barium carbonate

Super sack a tightly woven mylar or cloth bag with a volume of one cubic yard

Syenite a general term to describe an intrusive (plutonic) igneous rock that is similar to granite but contains little or no free quartz

Sylvite a potassium chloride (KCl) mineral that is a major ore for potassium compounds; occurs with halite in evaporite deposits

Szaibelyite a magnesium borate with a nominal composition of $MgBO_2OH$

Talc an extremely soft, platy mineral with a nominal composition of $Mg_3Si_4O_{10}(OH)_2$

TGA Thermogravimetric analysis; used to determine mass changes during heating

Tincal *see* Borax

Tourmaline a group of closely related, complex silicate minerals with the general composition of $Na(Al,Fe,Li,Mg,Mn)_3Al(Si_6O_{18})(BO_3)_3(OH,F)_4$

Tremolite a monoclinic amphibole mineral with a nominal composition of $Ca_2(Mg,Fe)_5Si_8O_{22}(OH)_2$; common in some commercial talc products

Tridymite a silica (SiO_2) polymorph that is stable between quartz and cristobalite

Trioctahedral refers to clay minerals that within the unit cell have all three available octahedral sites filled with cations

Tripoli a silica material formed by leaching calcareous cherts or siliceous limestones, which removes the calcia and leaves the silica

Trona the sodium sesquicarbonate mineral with a composition of $Na_3CO_3HCO_3 \cdot 2H_2O$

Ulexite a mixed sodium and calcium borate mineral with a nominal composition of $NaCaB_5O_9 \cdot 8H_2O$

Vermiculite a hydrated magnesium-aluminum-iron silicate with a general composition of $(Mg,Fe^{+2},Al)_3(Al,Si)_4O_{10}(OH)_2 \cdot 4H_2O$; often heated to make an expanded material

Water of crystallization water molecules surrounding stable hydrated minerals, such as borax pentahydrate ($Na_2B_4O_7 \cdot 5H_2O$)

Whiting finely ground limestone (calcium carbonate)

Wolframite $[Fe,Mn]WO_4$

Wollastonite a chain silicate with a nominal composition is $CaSiO_3$

XRD X-ray diffraction; the method for determining mineralogy

XRF X-ray fluorescence, a method of determining the bulk composition of a sample

Zircon a zirconium silicate mineral witha nominal composition of $ZrSiO_4$; used on its own or as the predominant ore for zirconia

INDEX

Printed and bound by CPI Group (UK) Ltd, Croydon, CR0 4YY
16/09/2026
14963457-0002